Multicomponent Reactions towards Heterocycles

Multicomponent Reactions towards Heterocycles

Concepts and Applications

Edited by

Erik V. Van der Eycken
Upendra K. Sharma

WILEY-VCH

Editors

Prof. Dr. Erik V. Van der Eycken
Katholieke Universiteit Leuven
Department of Chemistry
Celestijnenlaan 200F
3001 Heverlee
Belgium

Dr. Upendra K. Sharma
Katholieke Universiteit Leuven
Department of Chemistry
Celestijnenlaan 200F
3001 Leuven
Belgium

Cover Image: DenisKot (globe)

Library of Congress Card No.: applied for

British Library Cataloguing-in-Publication Data
A catalogue record for this book is available from the British Library.

Bibliographic information published by the Deutsche Nationalbibliothek
The Deutsche Nationalbibliothek lists this publication in the Deutsche Nationalbibliografie; detailed bibliographic data are available on the Internet at <http://dnb.d-nb.de>.

Print ISBN: 978-3-527-34908-1
ePDF ISBN: 978-3-527-83242-2
ePub ISBN: 978-3-527-83244-6
oBook ISBN: 978-3-527-83243-9

Cover Design: Adam-Design, Weinheim, Germany
Typesetting Straive, Chennai, India
Printing and Binding CPI Group (UK) Ltd, Croydon, CR0 4YY

Printed on acid-free paper

C113177_171121

Contents

Preface

Owing to the countless plausible combinations of carbon, hydrogen, and various heteroatoms, heterocyclic chemistry has remained as the foundation of novel chemical compounds in the sphere of natural product chemistry, pharmaceuticals, agrochemicals, and material sciences. They can serve as useful tools to facilitate tunable interactions with biological targets, thereby providing improved pharmacological and physicochemical properties of biomolecules as well as drug candidates. Recently, heightened cognizance of environmental issues is directing our society toward more sustainable solutions. Ever since the 12 principles of Green Chemistry were articulated, chemists answered the call to play their part in generating more sustainable syntheses. Multicomponent reactions (MCRs) appear as an obvious solution since most of the reactants' atoms are often incorporated in the final product. Moreover, a vast literature has been produced showing the power of MCRs as well as post-MCRs, in simplifying the synthetic design and yet obtaining high complexity and diversity in the construction of privileged structures. This is crucial in the development of novel bioactive molecules, wherein the production of libraries of compounds is necessary for the search of optimal drug candidates. Given the broad applications of heterocycles in the plethora of scientific fields, the current book title "*Multicomponent Reactions towards Heterocycles. Concepts and Applications*" is well warranted. In addition, recent advances in the field of MCR chemistry along with the plausible scope toward the synthesis and functionalization of biologically relevant heterocycles has encouraged us to compile this volume.

As the vast majority of small molecule drugs are of heterocyclic nature, the interplay of heterocycles with MCRs becomes therefore significant. The first chapter focuses on the recent progress made in the area according to the main reactivity mode involved in the transformation: concerted, radical, metal-catalyzed, carbonyl/imine, and isocyanide-based processes. The chapter itself provides an overview of heterocycles as input in MCRs. The next chapter "Heterocycles and Multi-Component Polymerizations" highlights some of the latest examples in this emerging field. The third chapter highlights examples from the viewpoint of target-oriented synthesis, the use of MCR in medicinal chemistry, from drug discovery, synthesis of drugs, to screening libraries, and biopharmaceutical applications. Further, heterocyclic chemistry has traditionally relied on solution-phase synthesis as technological platform to discover and produce bioactive scaffolds. The next chapter "*Solid-phase Heterocycle Synthesis using Multicomponent Reactions*" highlights methodological aspects of the implementation of on-resin MCRs to produce heterocycle compounds. Different name reactions, synthetic strategies, and solid-supports are analyzed critically in this chapter.

In the synthesis of heterocyclic compounds, MCRs have inherent advantage on pot, atom, and step economy (PASE) and are simple in operation, consume less energy, and release a reduced amount of waste. The fifth chapter discusses MCR-based green synthetic methods, including high-order MCRs, consecutive MCRs, MCRs followed by cyclization, and cycloaddition reactions, for the synthesis of heterocycles. Further, the use of enabling methods viz. continuous flow approaches has been beneficial in terms of yield, selectivity, reaction time, real-time monitoring. The next chapter is focused on different methodologies that can be used to perform heterocycle multicomponent syntheses in a continuous flow, to highlight the advantages over batch synthesis. Similarly, the next chapter analyzes a merging of C–H functionalization and MCRs approaches toward synthesis and modification of heterocyclic compounds.

MCRs have demonstrated their reliability and effectiveness as a synthetic approach that provides rapid access to chemical complexity. Among the many factors that bring this about, the MCR effect stands out, based on the fact that a changed number of reagents becomes the main differentiating factor of the reaction direction, that enables *multicomponent-switched heterocyclizations*, the topic of the next chapter in the book. Alkynoyl functionalities are densely substituted bifunctional electrophiles and prerequisite in many heterocycle syntheses via cyclocondenzations or cycloadditions. The 10th chapter of the book summarizes, explains, and highlights recent endeavors in the catalytic alkynoyl generation and their application to diversity-oriented multicomponent syntheses of heterocycles. The next chapter of the book is focused on the MCR synthesis of saturated heterocycles, encompassing the synthesis of small ring size (from 3 to 7) heterocycles, macroheterocycles, fused rings, and spirocyclic compounds. This is followed by the topic *asymmetric MCRs* where the development of enantioselective versions of these reactions have led to optimized reaction conditions, broader scopes, and increased chemo- and enantioselectivities. At last, the final chapter focuses on the recent advances made in MCRs built upon transition metal-catalysis directed toward the synthesis of heterocycles. In conclusion, this volume offers a versatile overview of the topic alongside discussing the recent progress in the flourishing area of MCR chemistry. This would, in turn, provide a platform for future innovations toward the designing of novel green transformations for heterocyclic synthesis.

Finally, we are extremely grateful to all authors for their excellent contributions to this volume. We are also thankful to the Wiley editors in particular Dr. Elke Maase and Ms. Katherine Wong for their professional support and assistance during this endeavor.

University of Leuven (KU Leuven)
Belgium
21 June 2021

Dr. Upendra K. Sharma
Prof. Dr. Erik V. Van der Eycken

1

Heterocycles as Inputs in MCRs: An Update

Ouldouz Ghashghaei, Marina Pedrola, Carmen Escolano, and Rodolfo Lavilla

University of Barcelona, Laboratory of Medicinal Chemistry, Faculty of Pharmacy and Food Sciences, and Institute of Biomedicine IBUB, Av. de Joan XXIII, 27-31, Barcelona 08028, Spain

1.1 Introduction

Multicomponent reactions (MCRs) hold a privileged position in organic synthesis and are currently gaining momentum in the fields where a fast access to high levels of structural diversity is needed. This is especially important in medicinal chemistry and key to drug discovery. In this endeavor, as the vast majority of small-molecule drugs are of heterocyclic nature, the interplay of heterocycles with MCRs becomes significant [1]. Although the majority of work has been devoted to the synthesis of heterocyclic adducts from non-heterocyclic reactants [2, 3], we will focus, however, on the intrinsic reactivity of basic heterocycles as a source of synthetically useful MCRs (Scheme 1.1). This approach, still quite unexplored in the MCR context, is arguably a rich source of novel, complex scaffolds. There is a wide choice of commercially available heterocyclic inputs, which together with their often-exclusive reactivity make this perspective simple, conceptually attractive, and synthetically productive. In this chapter, we describe a representative selection of relevant results in the last six years, as the field has experienced impressive growth since our last revision [4], and an exhaustive account is out of scope. This update groups the highlighted processes according to the main reactivity modes defining the MCRs: concerted, radical, metal-catalyzed, carbonyl/imine, and isocyanide-based processes. Finally, a miscellany section is included to cluster those MCRs that do not clearly fit in the classification. Occasionally, some significant post-transformations and applications have been detailed.

1.2 Concerted MCRs

The impact of heterocycle-based concerted MCRs in organic synthesis is quite relevant, with recent contributions arising from Povarov reactions, hetero Diels–Alder processes, and dipolar cycloadditions. The Povarov MCR, the interaction of an

Multicomponent Reactions towards Heterocycles: Concepts and Applications, First Edition.
Edited by Erik V. Van der Eycken and Upendra K. Sharma.

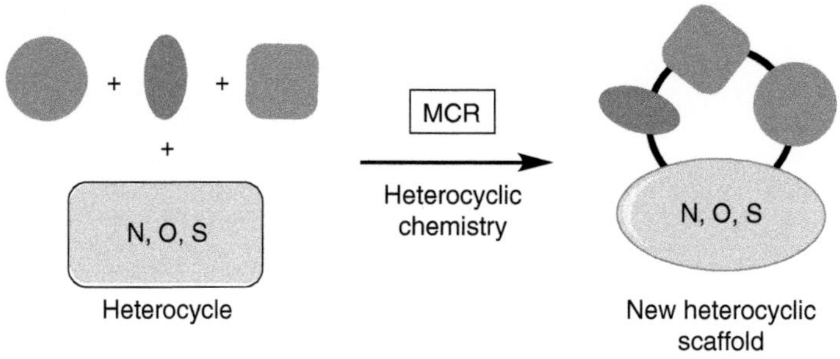

Scheme 1.1 Heterocycles as inputs in MCRs.

aromatic amine, an aldehyde, and an activated alkene, remains one of the best synthetic approaches to access tetrahydroquinolines (THQs) [5] and is especially productive in medicinal chemistry [6]. Although the concerted cycloaddition is a well-founded hypothesis for the reaction mechanism, there is evidence on polar stepwise processes in some cases, and both pathways are considered here.

For instance, a double Povarov process led to julolidine derivatives: the first MCR generates a secondary amine, which under calixarene-based polysulfonic acid catalysis spontaneously triggers a second MCR, leading to the final five-component adducts with good yields and modest stereoselectivity (Scheme 1.2) [7].

Indole derivatives participate in Povarov MCRs not only as aldehyde or olefin inputs, but also as aniline surrogates. Their specific structural arrangement, and the catalytic conditions used, determines the outcome. In this way, while indole-3-carbaldehyde gives the expected Povarov adduct [8], indole-7-carbaldehyde reacts in a different way, leading to fused adduct where the indole nitrogen closes a six-membered ring [9]. Interestingly, indole-2-carbaldehyde, depending on the catalysts used, may lead to the *normal* Povarov adduct or to a different scaffold, with a distinct connectivity through an alternative [3 + 2] cycloaddition mode (Scheme 1.3) [10].

As olefin inputs, indoles unsubstituted at C2 and C3 yield the THQ adduct, losing the aromaticity at the pyrrole ring [11]. In this respect, 2-vinylindoles react exclusively at the olefin moiety to yield the expected THQ adduct [12]. However, the isomeric 3-vinyl derivatives react quite differently, leading to bisindole-piperidines in a stereo- and enantio-controlled fashion, using chiral catalysts (Scheme 1.3) [13].

Regarding heterocyclic inputs, the interaction of aldehydes, 1,4-dhydropyridines as activated olefins, and aminocoumarin, as aniline surrogate, leads to complex

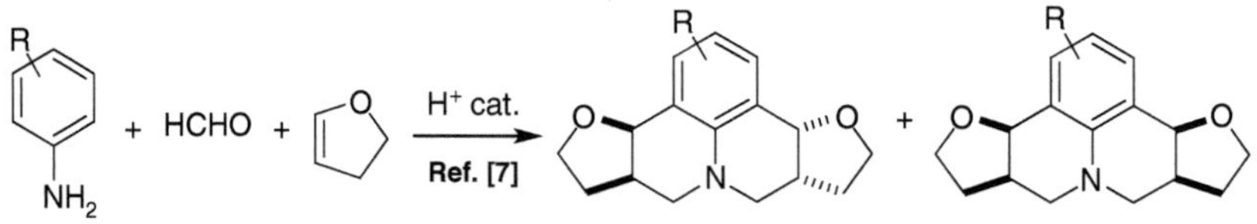

Scheme 1.2 Access to julolidines via double Povarov MCRs.

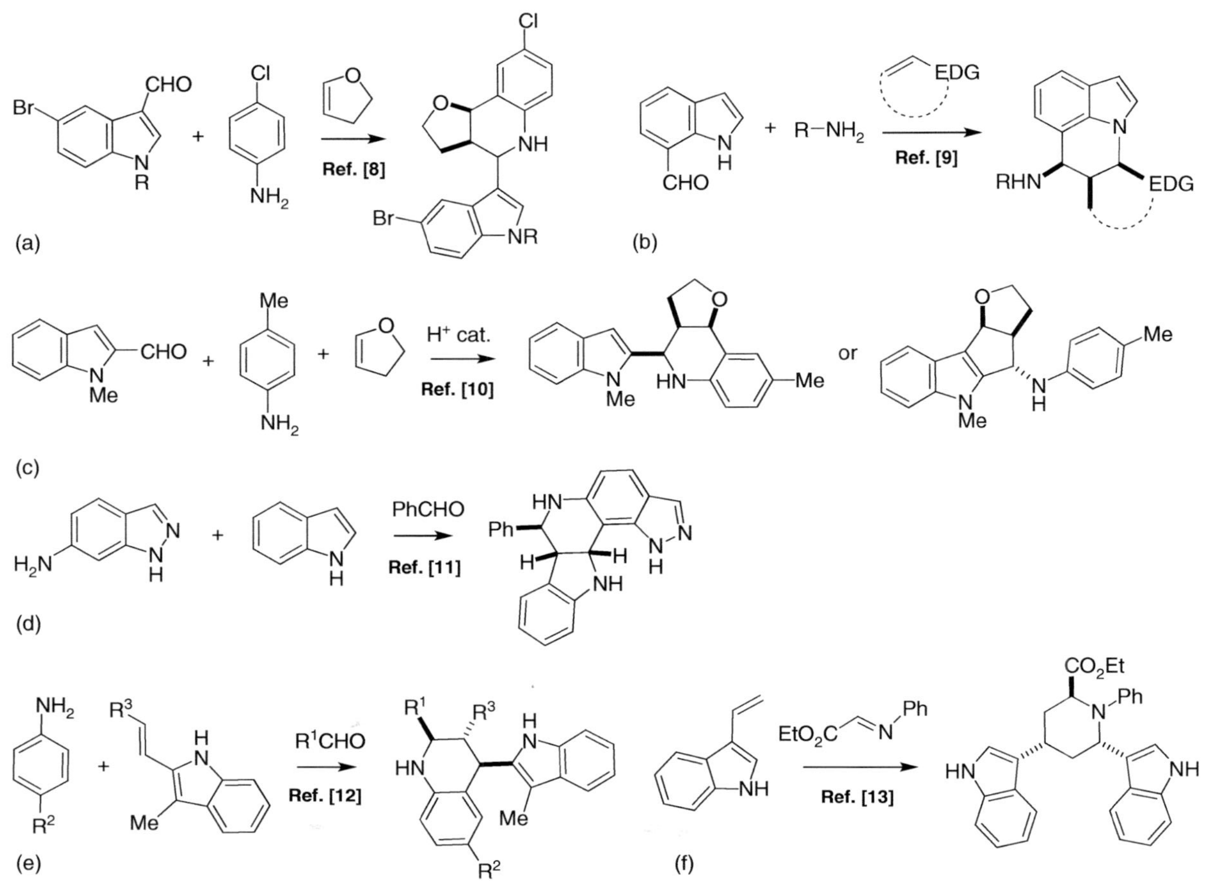

Scheme 1.3 Indoles as inputs in Povarov MCRs.

functionalized chromenonaphthyridines [14]. Relevantly, 3-aminopyridine imines react with alkynes (terminal or internal) to regioselectively afford the naphthyridine scaffold [15]. Similarly, 3-aminopyridones also lead to oxidized Povarov adducts (Scheme 1.4) [16].

There are mechanistic variations that dramatically modify the connectivity pattern of standard Povarov MCRs. For instance, a Ferrier rearrangement was promoted during a Povarov process involving glycals [17]. An interesting example of interrupted Povarov process with salicylaldehydes, anilines, and dihydrofurans, instead of yielding the expected THQ adduct, follows a Mannich-type process with the enol ether, and the resulting intermediate is trapped by the phenolic hydroxyl, yielding the MCR adduct in a stereoselective fashion (Scheme 1.5) [18].

In a remarkable photoredox-catalyzed process, aldimines, dihydrofurans and trimethylsilyl azide, afforded azidotetrahydrofurans. The observed polarity reversal can be explained through a mechanism involving an azido radical, which adds on the β-position of the enol ether to promote the imine addition (Scheme 1.5) [19].

Finally, the Povarov MCR has enabled the selective tagging of benzaldehyde-functionalized DNA chains through the reaction with anilines and an *N*-protected dihydropyrrole [20].

Isochromenylium ions react with dienophiles in a [4 + 2] cycloaddition to yield adducts, which go through a Ritter-type domino process with acetonitrile to afford complex tetracyclic compounds [21]. Also, a formal concerted MCR connects *in situ* generated isoquinolinium salts with unsaturated aldehydes and alcohols in a process promoted by *N*-heterocyclic carbenes to give bridged azaheterocycles [22]. A [4 + 3] cycloaddition process is triggered by the condensation of an iminoindole with aldehydes to give an azadiene that reacts *in situ* with a sulfur ylide to yield azepinoindoles (Scheme 1.6) [23].

MCRs involving [3 + 2] cycloadditions have produced a substantial number of new transformations. The processes involving azinium ions have been reviewed [24]. The interaction of heterocyclic secondary amines with carbonyl inputs to generate dipoles is a common motif in the field. For instance, THQs, aldehydes, and ketomalonate afford the corresponding oxazolidine adducts [25].

Azomethine ylides, mostly generated by condensation or decarboxylation of α-amino acids, have been thoroughly used in MCRs in the presence of suitable dipolarophiles, often with applications in drug discovery [26]. The synthesis of pyrrolizidines and indolizidines through this MCR methodology has been reviewed [27]. A remarkable five-component interaction based on a double [3 + 2] cycloaddition of azomethine ylides has led to tetracyclic adducts in high yields in a stereoselective manner (Scheme 1.7) [28].

Azines are also present in this reactivity. α-Methylquinolines, aldehydes and alkynoates yield a fused adduct in a domino process starting with the formation of the dehydrated aldol-like intermediate [29]. Moreover, quinoline and pyridine dipoles react with azomethine ylides in an unprecedented fashion to yield complex fused pyrrolidine cycloadducts [30]. Finally, isatin undergoes a series of complex transformations triggered by the initial [3 + 2] cycloadduct generated through its interaction with proline and alkynoates (Scheme 1.8) [31].

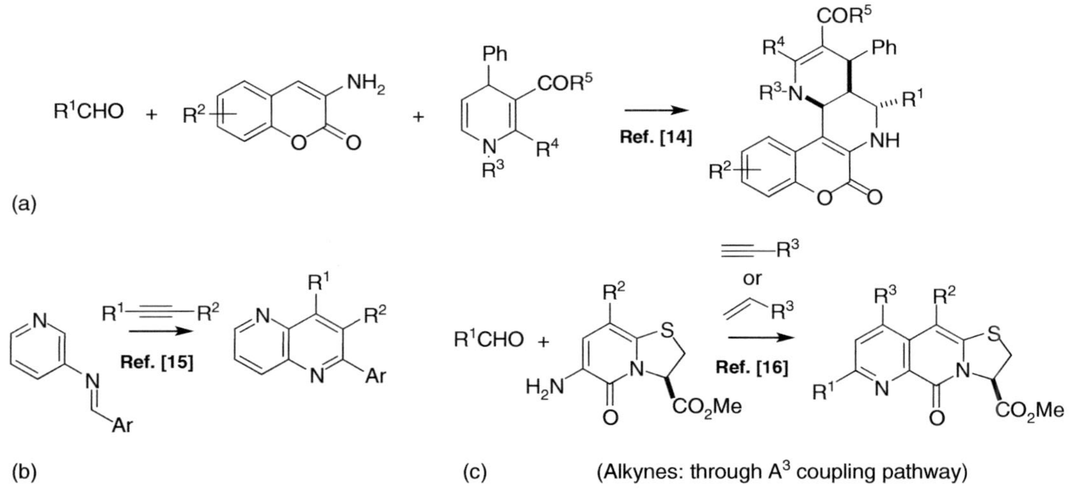

Scheme 1.4 Aminoheterocycles in Povarov MCRs.

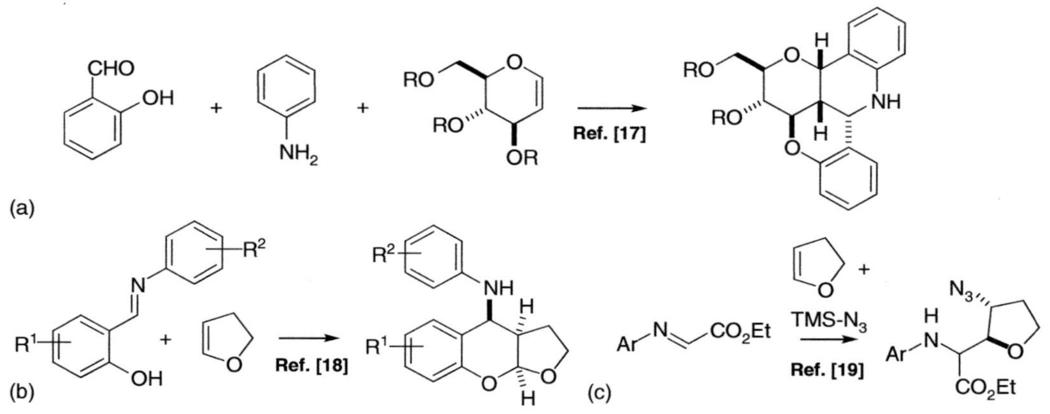

Scheme 1.5 Mechanistic variations of the Povarov-type processes.

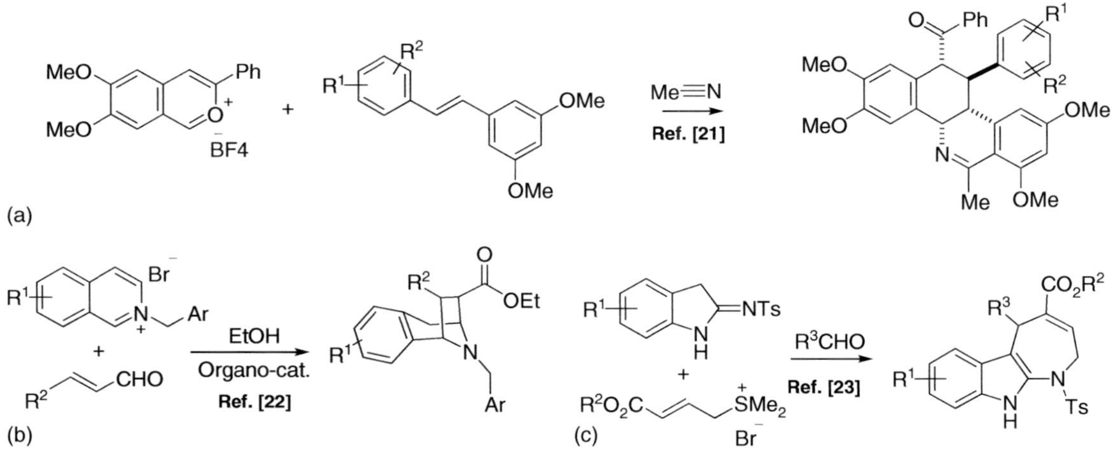

Scheme 1.6 Cycloaddition-type MCRs.

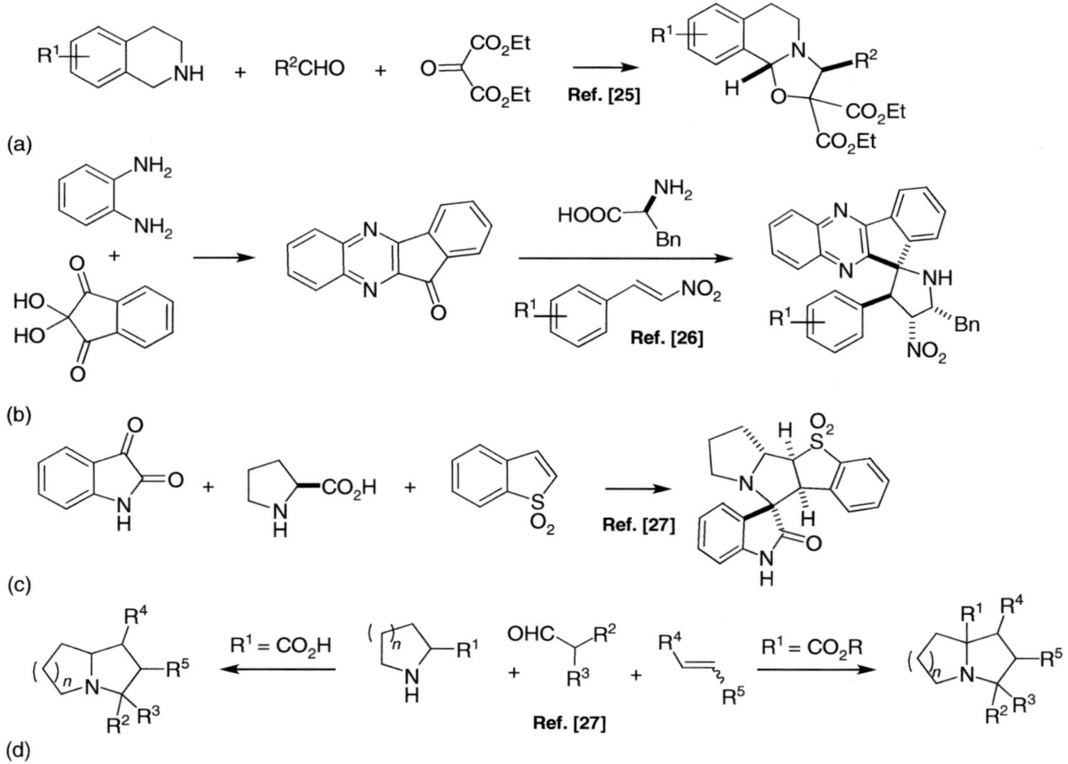

Scheme 1.7 [3 + 2] Dipolar cycloaddition MCRs.

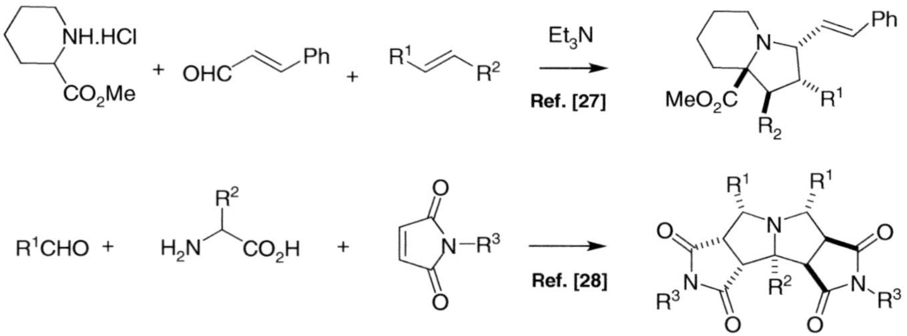

(e)

(f)

Scheme 1.7 (*Continued*)

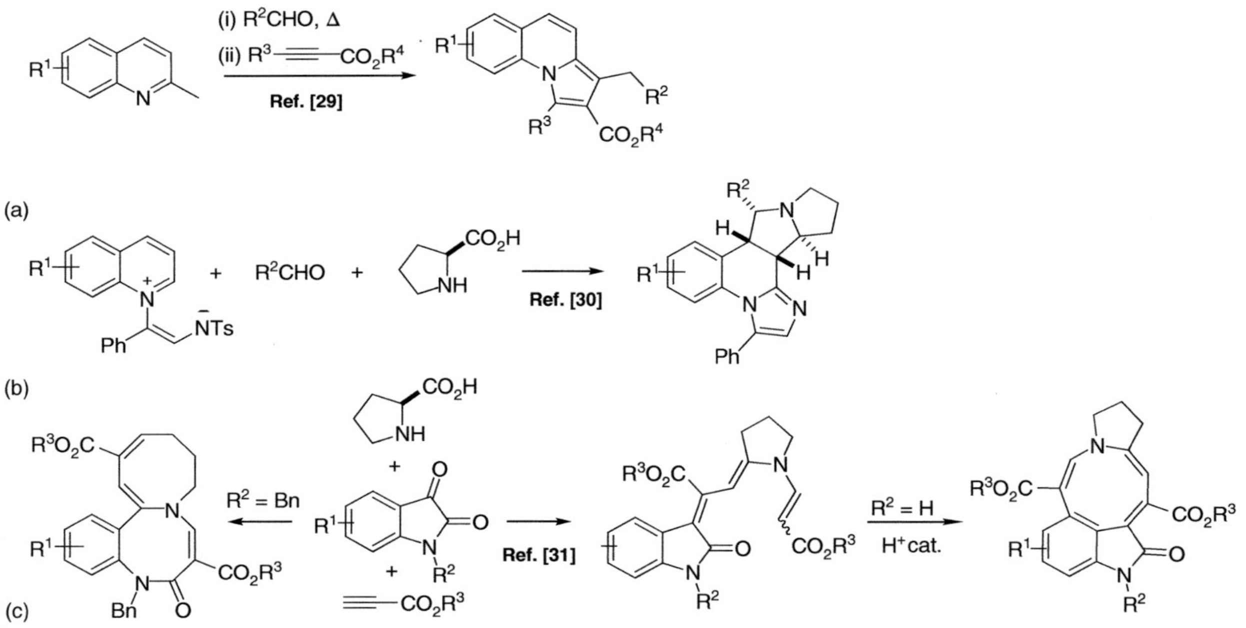

Scheme 1.8 Azines and isatins in dipolar MCRs.

Arynes yield dipoles through interaction with nucleophilic species. Their participation in MCRs has been recently reviewed [32]. Azines are *N*-arylated, and the resulting dipole interacts with carbonyl groups in an addition/cyclization mode or through proton transfer to generate second nucleophiles that trap the azinium intermediate. Also, the azine dipoles react with the aryne in [3 + 2] dipolar cycloaddition MCRs (Scheme 1.9).

In a series of related processes, epoxides, aziridines, and also four-membered cyclic amines and (thio)ethers react with arynes and protonucleophiles leading to the corresponding adduct featuring a substituted chain originated in the heterocycle (Scheme 1.10) [32].

1.3 Radical MCRs

The incorporation of radical chemistry into MCRs has unlocked access to new synthetic pathways unavailable through conventional polar reactions. Radical MCRs generally consist of a proradical, a relay reagent, and a trapping component [33]. Novel radical MCRs exploiting photochemical approaches have experienced rapid growth in recent years [34]. However, their pairing with heterocyclic inputs has been mainly restricted to the functionalization of the heterocyclic component. In this regard, the multicomponent versions of Minisci reaction stand out [35]. In these processes pyridine-type heterocycles get alkylated in the presence of a suitable alkene and an initiator amenable to produce the radical species [36]. β-Dicarbonyl radicals [37] as well as heteroatomic radicals including azido [38], sulfonyl, and phosphonyl [39] species have been reported to yield Minisci adducts in a similar fashion. As for the alkene components, *N*-vinylacetamide has been coupled with suitable azines and the proradical, to enantioselecitvely afford γ-aminoesters in the presence of a chiral phosphoric acid (Scheme 1.11) [40].

The scope of the heterocycic inputs in Minisci MCRs is mainly restricted to pyridine-type systems, usually substituted at some reactive positions (C2/C4) to block undesired regioisomer formation. In an alternative approach, the use of 4-cyanopyridine allows the γ-selective functionalization under a variety of conditions, involving the favored generation of pyridyl radicals [41, 42]. Interestingly, the use of Tf_2O as the azine activator and a CF_3 radical source results in the regioselective *p*-trifluoromethyl-alkylation of pyridines and quinolines [43]. In a related process, the use of pyridyl halides directs the functionalization upon the C4 position in a Ni-catalized radical process. It also features an interesting [1,5]-H shift that enables the heteroatom addition upon the β position of the initiating carbon radical (Scheme 1.12) [44].

Other heterocyclic systems have also been functionalized through radical MCRs. For instance, the C-sulfonylation of imidazoles has been reported in an Eosin-catalyzed photoredox transformation (Scheme 1.13) [45].

Dearomatization of indoles and related heterocycles has also been achieved through radical MCRs. In a remarkable approach, C3-spiro trifluoromethylindolines have been assembled in a copper-catalyzed radical MCR with β-aminomethylindoles,

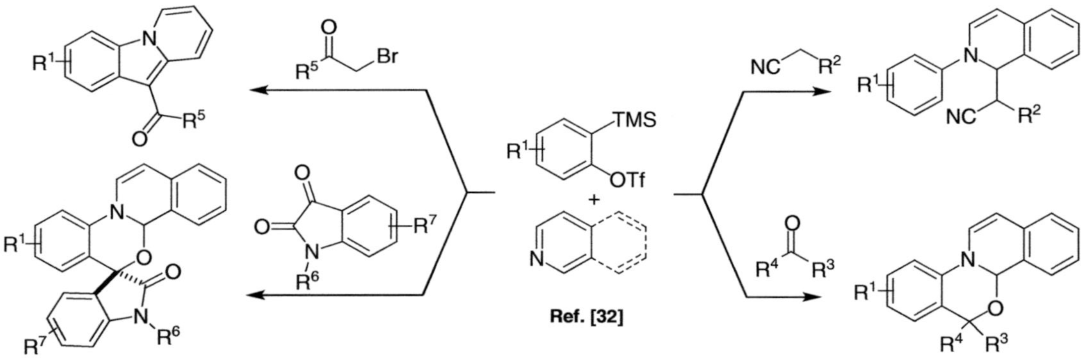

Scheme 1.9 Azine-aryne MCRs.

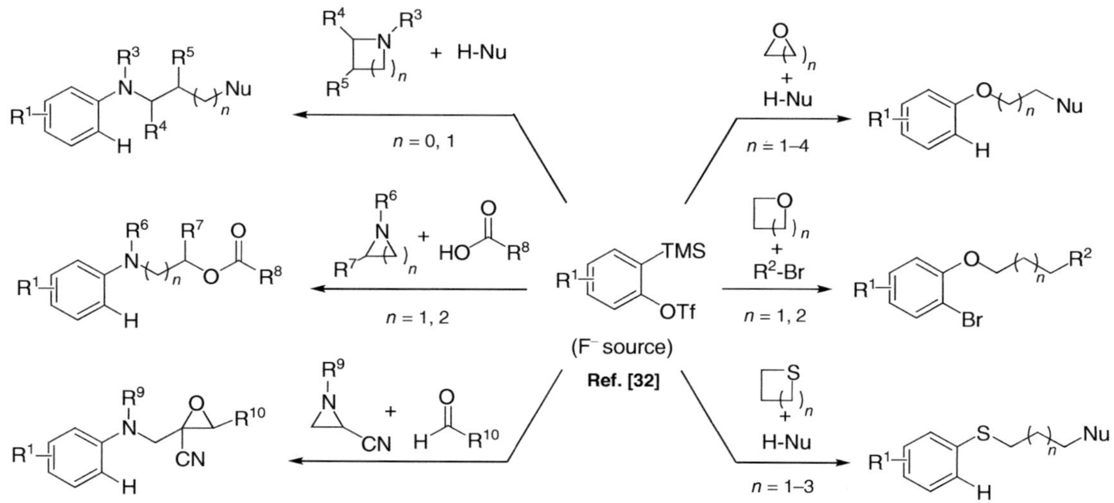

Scheme 1.10 3/4-Membered heterocycles in aryne MCRs.

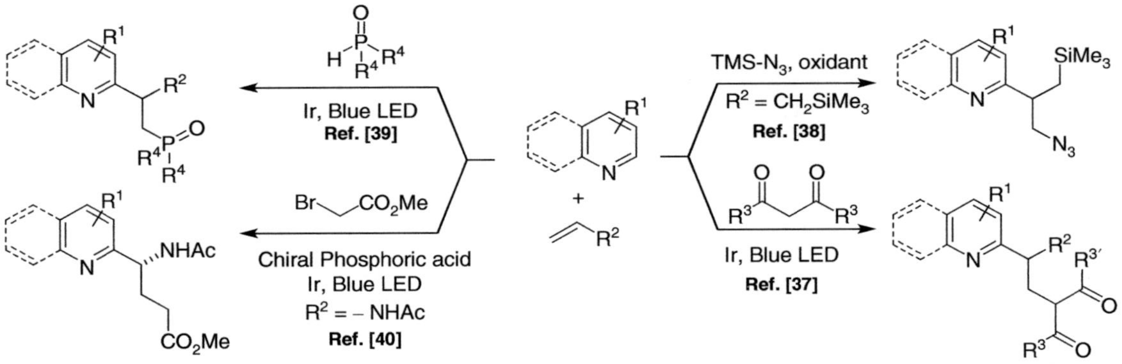

Scheme 1.11 Minisci-type radical MCRs.

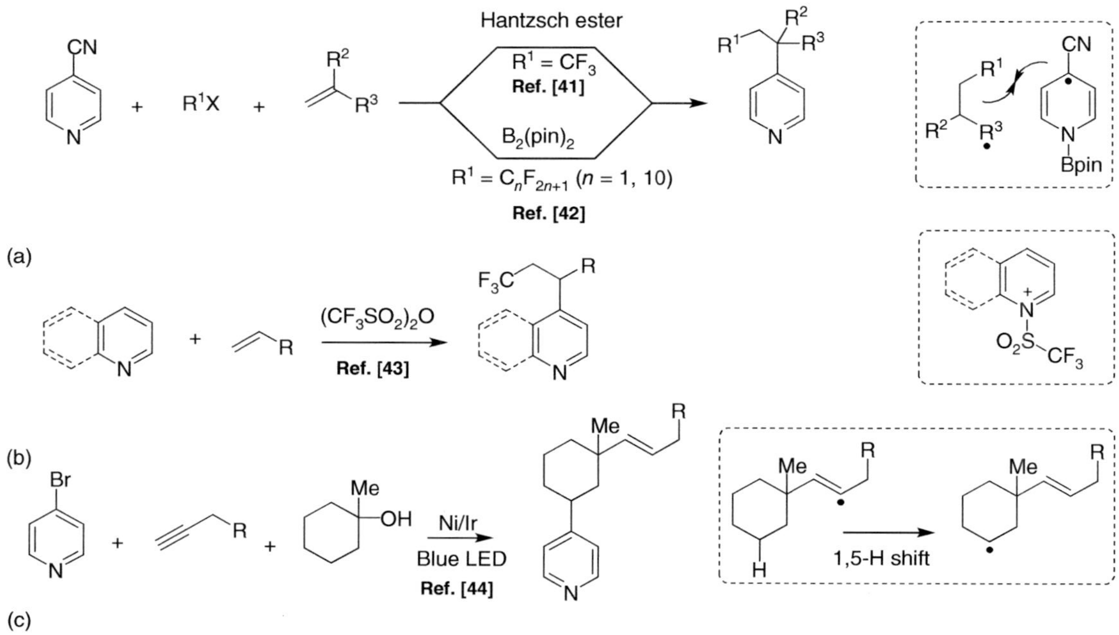

Scheme 1.12 Site-selective azine-based radical MCRs .

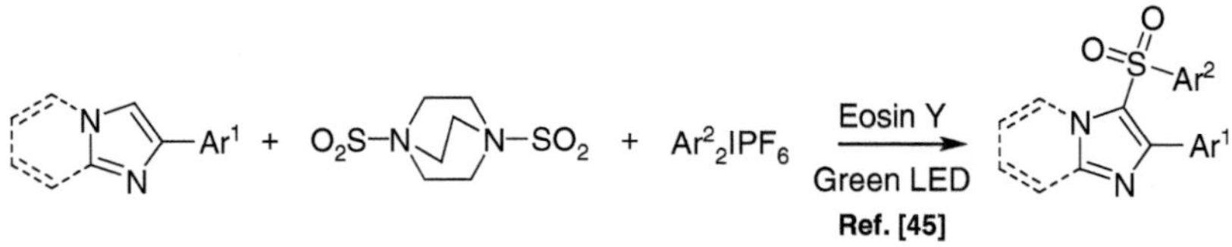

Scheme 1.13 SO_2 photoredox MCR.

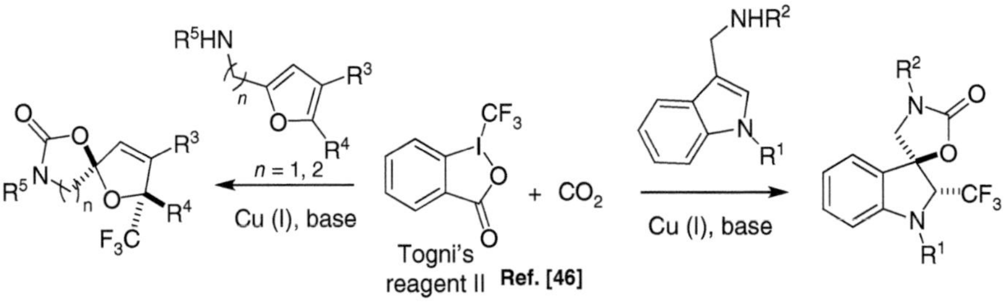

Scheme 1.14 Heterocycle dearomatization in radical MCRs.

carbon dioxide, and a trifluoromethyl radical source. The CF_3-indole radical is intramolecularly trapped by the copper carbamate, which is formed *in situ*, through the condensation of amine and CO_2. Furans with similar side chains have successfully afforded the corresponding spiro adducts (Scheme 1.14) [46].

Finally, maleimides have been involved in a remarkable Minisci-type MCR, in which the initiating alkyl radical was generated through a novel mild process [47]. Moreover, the assembly of fused quinolines through the condensation of 3-arylaminoacrylates, maleimides, and an electrophilic radical source has been achieved, matching the radical affinities in a domino process (Scheme 1.15) [48].

1.4 Metal-catalyzed MCRs

Transition metal-catalyzed MCRs featuring heterocyclic inputs have also experienced immense progress in recent years. Regarding the C–H activation processes, the direct functionalization of azoles through the insertion of an isocyanide, followed by the attack of a heterocycle, has been reported for the synthesis of di(hetero)aryl-ketones and-alkylamines [49]. The methodology involves the reaction of azoles, haloarenes, and isocyanides resulting in the formation of an imine, which can be hydrolyzed or reduced to yield the final adducts. Other examples of C—H bond functionalization include the preparation of fused imidazo-heterocycles starting from methyl ketones, *o*-tosylhydroxylamine and 2-pyridinone or thiazo/benzo[d]thiazol-2(3H)-ones [50]. This MCR consists of the copper catalyst coordination, the formation of the C–H functionalized intermediate, followed by a tandem addition-cyclization process. A relevant C–H glycosylation via a Catellani-type arylation allows the synthesis of C-aryl glycosides, which can undergo further transformations, such as Heck, Suzuki, and Sonogashira cross-couplings (Scheme 1.16) [51].

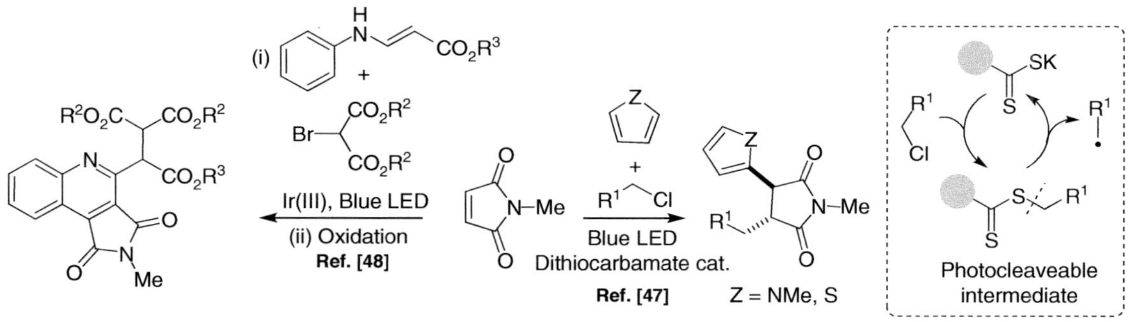

Scheme 1.15 Maleamides as inputs in radical MCRs.

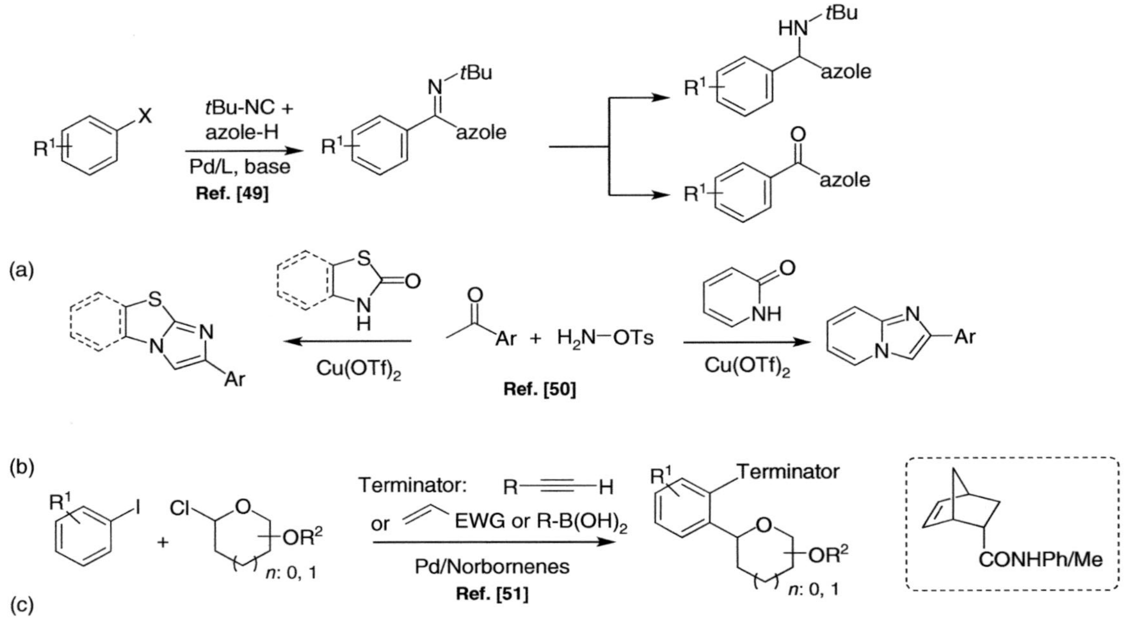

Scheme 1.16 C–H activation MCRs.

Progress in the A3-related MCRs, the interaction of aldehydes, amines, and alkynes, includes the use of isoquinolines, suitably activated by a chloroformate as amine inputs through an enantioselective copper-catalyzed protocol [52]. Remarkably, the interaction of azine-2-carbaldehydes with secondary amines and terminal alkynes starts via the A3 MCR, and the adduct undergoes a formal Cu-catalyzed hydroamination to yield indolizines [53, 54]. Terminal alkynes are also useful inputs in the MCR coupling of *N*-heteroaromatics (quinolines) with alkyl halides. The tandem process is catalyzed by CuI and allows the formation of 1,2-difunctionalized quinoline-type derivatives (Scheme 1.17) [55].

Some carbonylative MCR processes dealing with heterocyclic inputs have also been disclosed: a Pd-catalyzed four-component coupling involving tryptamine leads to alkaloid-like compounds featuring the quinazolinone core [56]. Divergent PdI_2/KI-catalyzed aminocarbonylation-cyclization pathways starting from alkynylthioimidazoles yield functionalized imidazo-thiazinones and -thiazoles (Scheme 1.18) [57].

Although not strictly belonging to this section, metal-catalyzed post-modifications of MCR adducts constitute a powerful and versatile synthetic tool. For instance, Au/Ag-catalysis on phenol-alkynyl Ugi adducts efficiently promotes an intramolecular dearomative cyclization, followed by an aza-Michael addition yielding the tetracyclic scaffold [58]. Similarly, a concerted [4+2] cyclization on an indole substrate terminates the assembly of complex bridged polycyclic alkaloid arrangements [59]. In this approach, the key step involves the alkyne hydroarylation. A related process involving a furane-alkyne Ugi adduct, undergoes the Au-domino process ending with a ring fragmentation to yield unsaturated 2-pyridones (Scheme 1.19) [60].

1.5 Carbonyl/Imine Polar MCRs

In this section a variety of multicomponent transformations involving carbonyl and/or imine substrates (not specifically related to the rest of the sections) is analyzed. For instance, a Mannich-type MCR of indoles, amines, and substituted aldehydes followed by a lactamization leads to a bicyclic a adduct [61]. A cascade three-component reaction of 6-aminouracil, aldehydes and tetrahydroisoquinolines allows the formation of a new pyrimidine ring through the functionalization of the C–H adjacent to the nitrogen by a 1,5-hydride transfer and concomitant oxidation [62]. A domino process involving heterocyclic *N*-acylenamines, formaldehyde, and primary amines builds conjugated pyrimidine rings in a stepwise Mannich-aminal MCR [63]. A mechanistically related process connects tryptamines, alkyl propiolates, and nitroalkenes, yielding indolizino-indoles or chromeno-indolizinoindoles (Scheme 1.20) [64].

The Petasis MCR, the interaction of *in situ* generated imines and boronic acids has been reviewed [65]. A highly diastereoselective three-component Petasis/intramolecular Diels–Alder tandem reaction involving allyl amines,

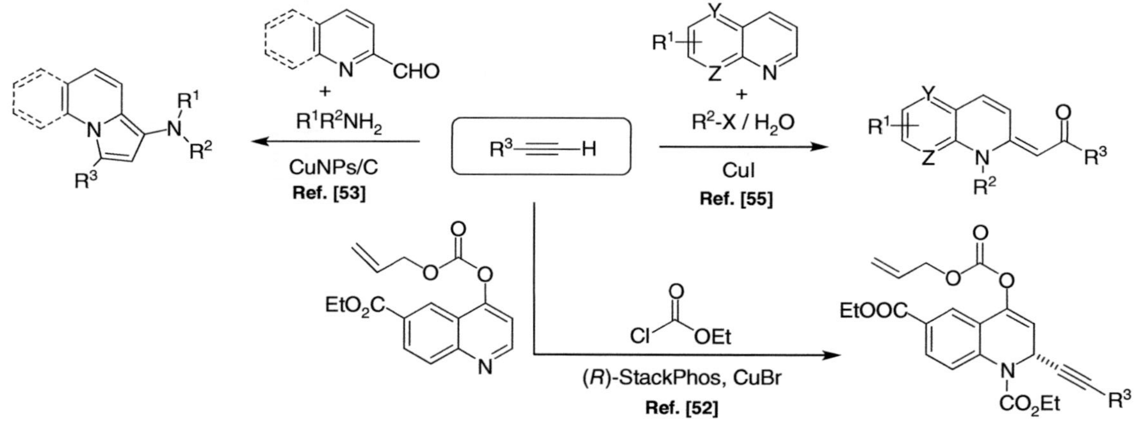

Scheme 1.17 A3-type MCRs.

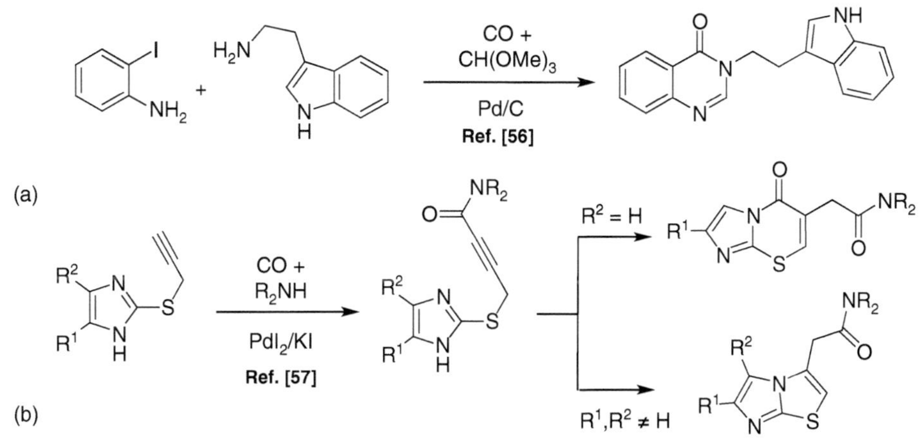

Scheme 1.18 Carbonylative transition-metal catalyzed MCRs.

furylboronic acid, and α-hydroxylated aldehydes led to a compact functionalized tricyclic system (Scheme 1.21) [66].

The Biginelli-type MCRs stand for the interaction of urea or urea-like compounds, aldehydes, and dicarbonyl derivatives. In this way, thiazolo-quinazolinones are generated from aminothiazoles [67]. Diversely substituted aminotriazoles [68] and aminopyrazoles [69] are active in Bigineli MCRs leading to the corresponding pyrimidine adducts (Scheme 1.22).

In a Hantzsch-type MCR, fused-tricyclic pyrans, and dihydropyridines were prepared by an indium (III)-catalyzed protocol involving a Knoevenagel adduct that cyclized to the final *N*- or *O*-tricyclic core (Scheme 1.22) [70]. Similarly, coumarin-fused pyrimidines were prepared by Biginelli-type MCRs (Scheme 1.23) [71, 72].

The Yonemitsu MCR has been employed for the synthesis of indole-based triarylmethanes. As an example, coumarines, diversely substituted indoles and quinoline-aldehydes led to highly crowded adducts [73]. Cyclic thio-substituted β-enaminoesters reacted in a diastereoselective manner with isatins cyclic β-diketones or 4-hydroxychromen-2-one to furnish complex polycyclic spiroindolines (Scheme 1.23) [74].

In a variation of this reactivity pattern, a nitro-Michael acceptor is introduced, and azolopyrimidines were synthetized by a BF_3-catalyzed MCR involving aminoazoles, aldehydes, and morpholinonitroalkenes, through the *in situ* generation of the reactive nitroalkynes (Scheme 1.24) [75].

The Reissert-type reactions involve the addition of nucleophiles to *in situ* *N*-activated azines to yield covalent adducts, usually at the α-position. Progress in the area deals with the regioselective phosphonylation of quinolines upon activation with chloroformates. Thus, reaction of the intermediate with differently substituted *N*-heterocyclic phosphines, where the substituent at the oxygen atom determines the α- or γ-attack (Scheme 1.25) [76].

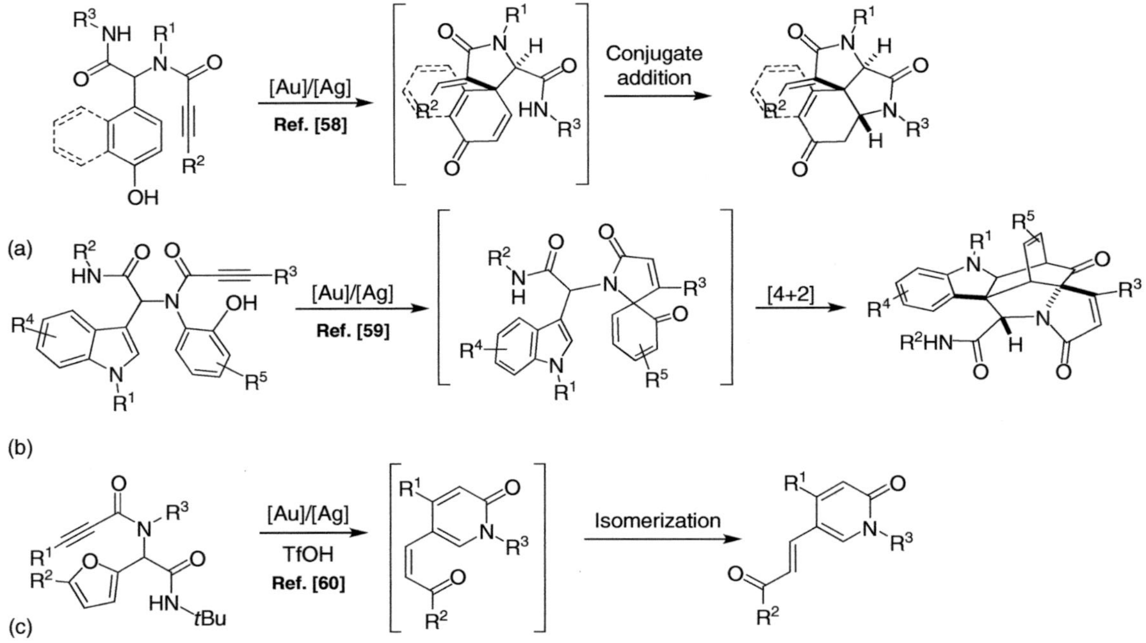

Scheme 1.19 Au(Ag)-catalyzed MCR post-transformations.

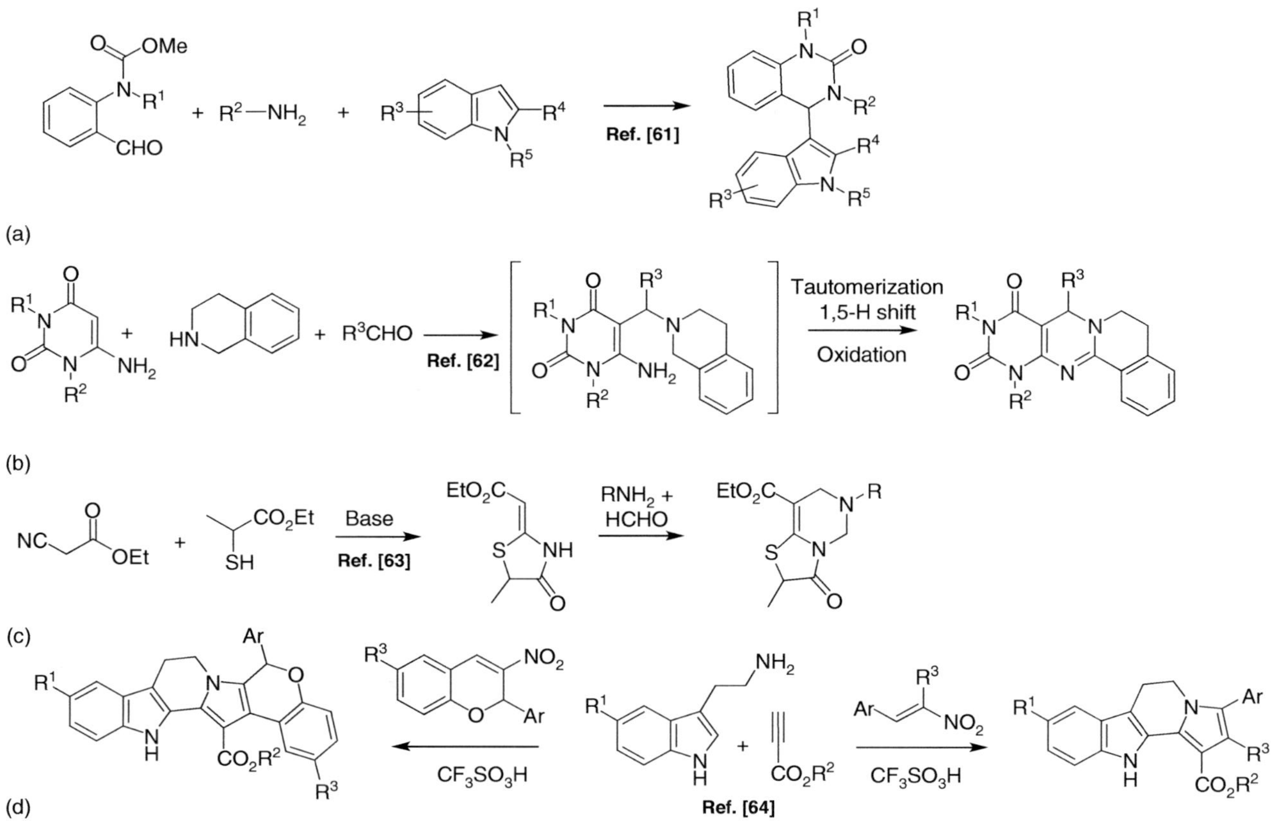

Scheme 1.20 Mannich-type MCRs.

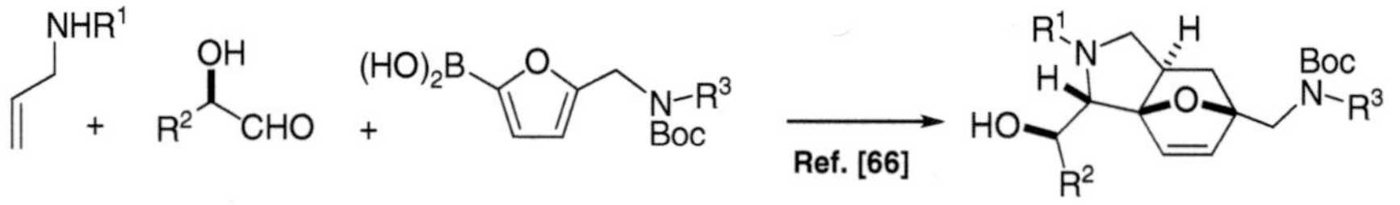

Scheme 1.21 Petasis MCR.

New enantioselective catalytic methods have been disclosed for this transformation. Chiral phosphoric acid promotes the *N*-addition of indoles upon *in situ* generated *N*-Boc-isoquinolinium ions [77]. Also, chloroformate promoted silyl ketene acetal additions to isoquinolines and other azines catalyzed by chiral anion-binding triazoles (Scheme 1.25) [78].

1.6 Isocyanide-based MCRs

Isocyanides stay as the most fruitful functional group in the MCR field. The classic Ugi and Passerini processes are still matter of active research, mainly dealing with mechanistic modifications and novel substrates. With respect to the Ugi-type MCRs, relevant results have appeared in this period. Regarding novelties in the reaction modes, interrupted processes have gained much importance. In these transformations, the usual trapping of the intermediate nitrilium ion by carboxylates and the subsequent Mumm rearrangement are replaced by a variety of nucleophilic additions affording structural diversity [79]. For instance, indole can efficiently trap the nitrilium, leading to spiroindolines, which continue a domino process to complex alkaloid-like compounds (Scheme 1.26) [80].

Heterocyclic amines have provided a series of oxidative [81] and redox neutral [82, 83] processes, involving the *in situ* formation of the active imine/iminium species, which subsequently react in an Ugi fashion. The Ugi–Smiles approach has been extended to functionalize thiouracil derivatives [84]. The interaction of an *in situ* generated Knoevenagel adduct with isocyanides and maleimides leads to a convenient preparation of a polycyclic adduct, which spontaneously evolves into a variety of isoindolocarbazoles (Scheme 1.26) [85].

Oxochromones react in a similar manner, upon the interaction with isocyanides and alkenes, through a series of [4 + 1]/[4 + 2] cycloadditions [86]. Interestingly, the closely related formylchromones can react with amines and isocyanides in a different way, leading to adducts arising from the initial conjugate addition of the isocyanide or the amine input to the conjugate carbonyl (Scheme 1.27) [87, 88].

Activated aziridines have produced polysubstituted tetrahydropyridines upon the interaction with cyanomalonates and isocyanides [89]. Finally, 2-bromo-6-isocyanopyridine is a general and affordable convertible isocyanide, since it suitably participates in Ugi processes, and the resulting aminopyridine unit can be replaced by a variety of nucleophiles in the adduct (Scheme 1.27) [90].

The Joullié MCR involves the interaction of imines with carboxylic acids and isocyanides [91]. From a synthetic point of view, this process is not just a simplification of the Ugi MCR, but rather a way to promote novel reactivity pathways. For

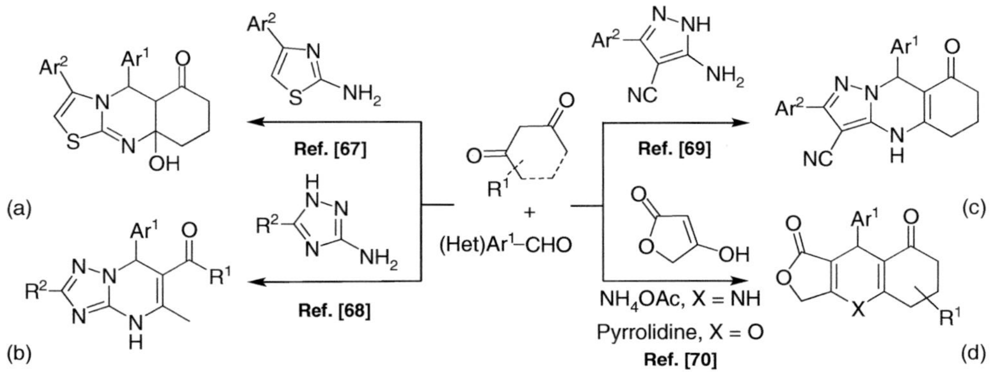

Scheme 1.22 Dicarbonyl derivatives in Biginelli- and Hantzsch-type MCRs.

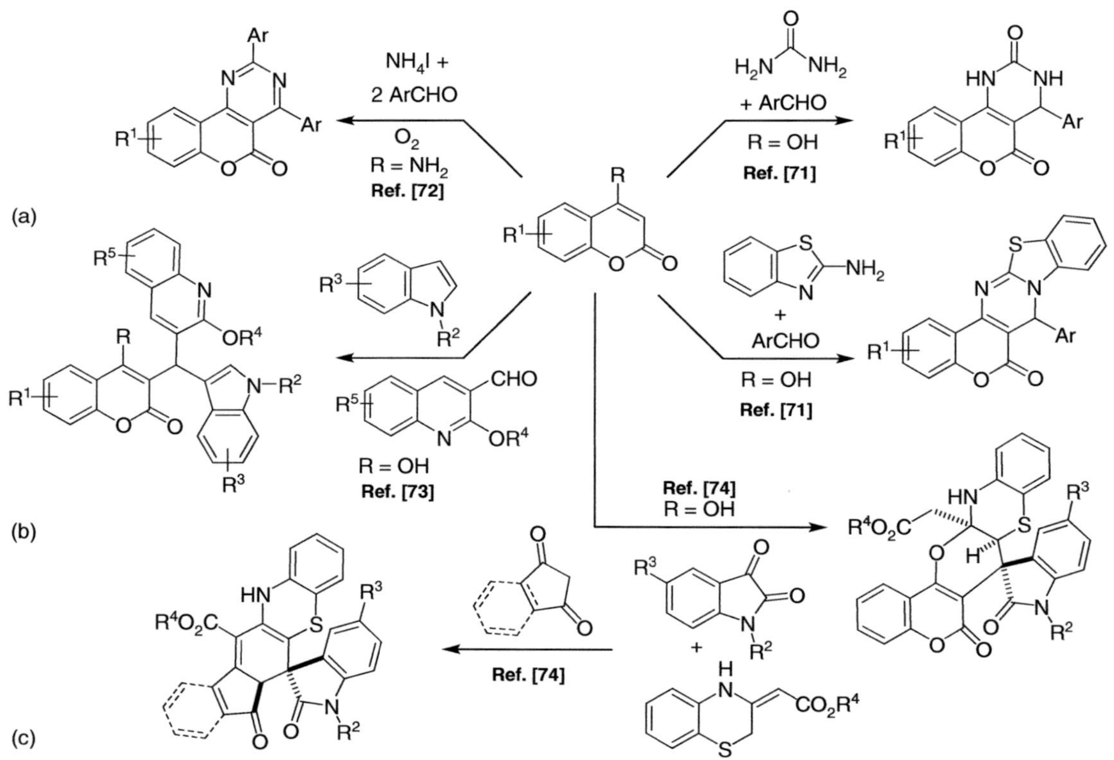

Scheme 1.23 Coumarins and isatins in MCRs.

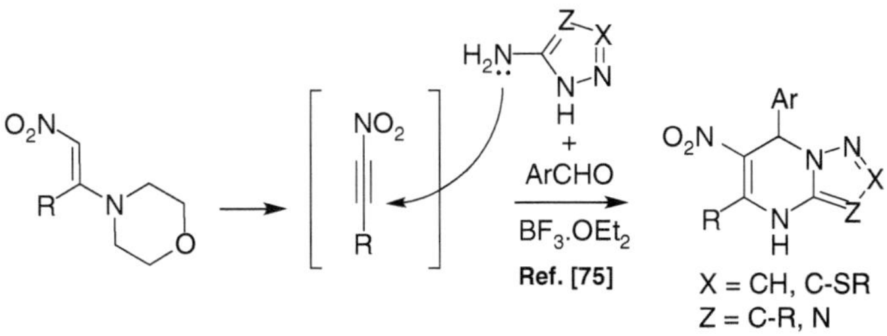

Scheme 1.24 Aminoazoles in conjugated addition-type MCRs.

instance, the *in situ* generation of cyclic imines from the (electro) chemical oxidation of secondary amines and their subsequent transformation has been described [92].

With respect to the different substrates engaged in the Joullié MCR, it is worth mentioning the spiroindolenines, which have suitably reacted to yield fused diketopiperazines, linked diamides, and tetrazoles [93, 94]. A remarkable approach to bicyclic hydantoines from *in situ* generated iminium salts resulting from the trifluoroacetic acid (TFA) treatment of *N*-Boc protected dihydropyrazines and β-aminoketones has been disclosed [95]. Azirines have recently been brought into this chemistry through Lewis acid activation, allowing a stereoselective access to functionalized *N*-acyl-aziridinecarboxamides (Scheme 1.28) [96].

The Groebke–Bienaymé–Blackburn (GBB) MCR, the interaction of α-aminoazines, aldehydes, and isocyanides, yield fused aminoimidazoles, a highly privileged scaffold in medicinal chemistry [97]. Mainly nonconventional examples of these transformations are mentioned here. Regarding aminoazines, GBBs with *N*-Boc-3-aminoindole followed by an oxidative cascade have resulted in the one-pot access to pyridodindoles [98]. The reaction mechanism features an interesting azirine intermediate, which evolves into the final adduct through a radical Neber-type rearrangement followed by a [1,2]-hydrogen shift/cyclization (Scheme 1.29).

Tetrahydroquinolin-8-amine, aldehydes, and isocyanides react to afford fused tricyclic quinoxaline adducts in a homoGBB transformation, in the presence of DMAP. The mechanism relies on the condensation of the aldehyde with the aniline to form the initial iminium ion, followed by the attack of the isocyanide [99].

Polyaminopolyazines have also been reported to afford multiple GBB adducts. The innate selectivity [100] observed in the case of 2,4-diaminopyrimindine results in the exclusive formation of a single monoadduct, enabling a selective second GBB upon the former product [101]. Incidentally, the monoadduct with the alternative regioselectivity was prepared through a protection strategy [102]. In a similar manner, triple-GBB transformations upon melamine yield unprecedented tripodal scaffolds, amenable to further diversification via suitable post-modifications (Scheme 1.29) [101].

Carbonyl components further diversify the reach of these transformations. GBBs with isatine and aminopyridines result in the one-pot formation of fused imidazopyrimidone salts through the expected spiroimidazole GBB adduct that

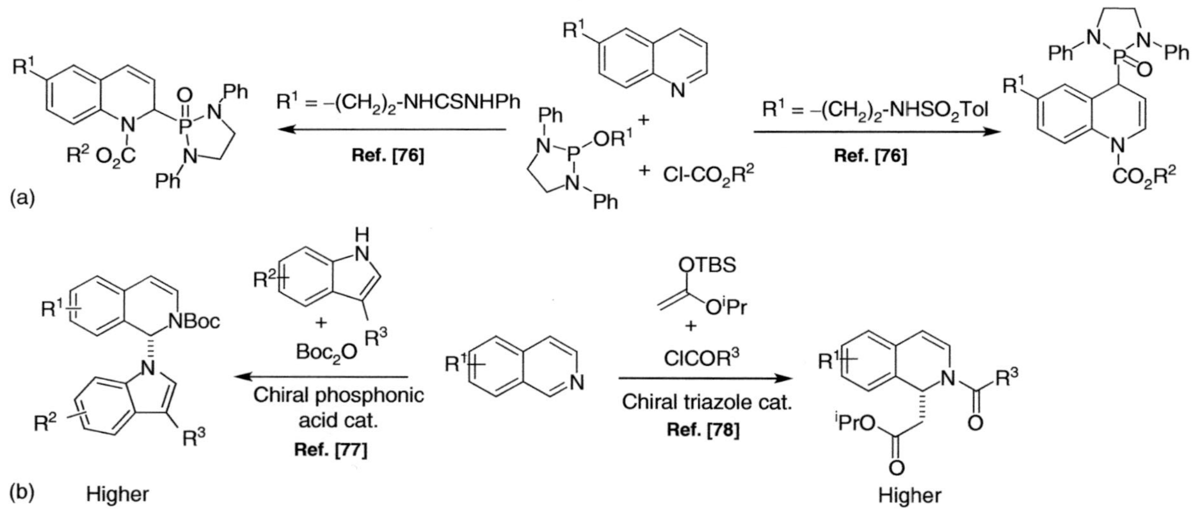

Scheme 1.25 Reissert-type reactions.

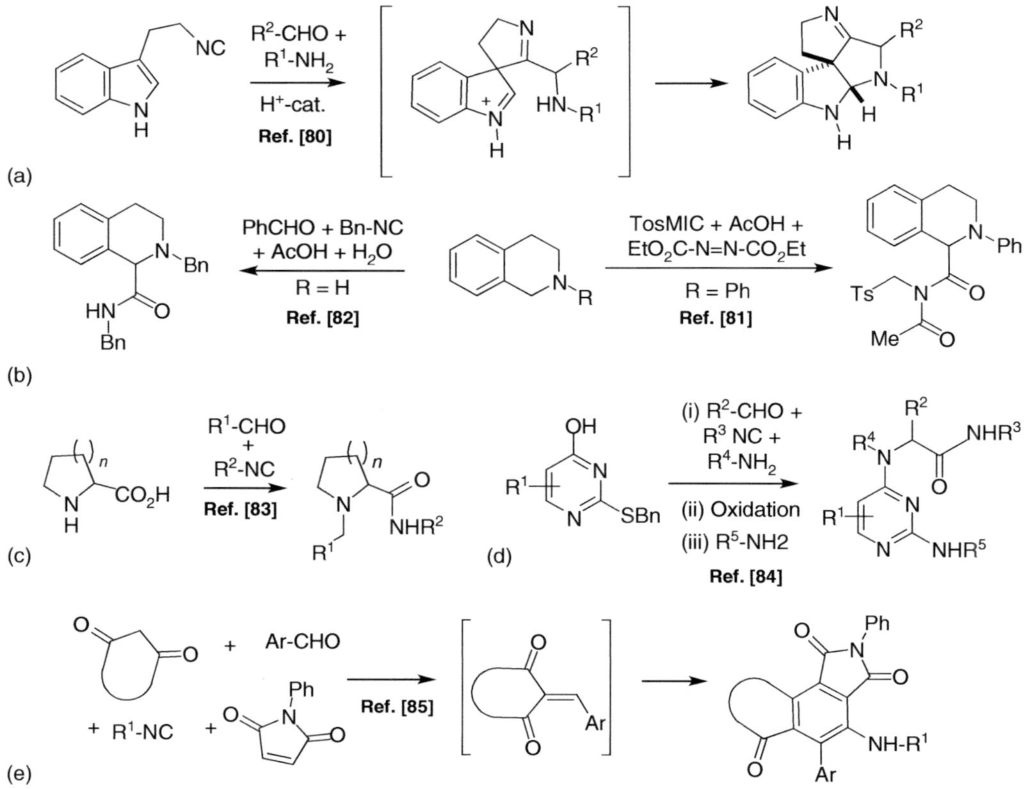

Scheme 1.26 Ugi-type MCRs. Mechanistic variations.

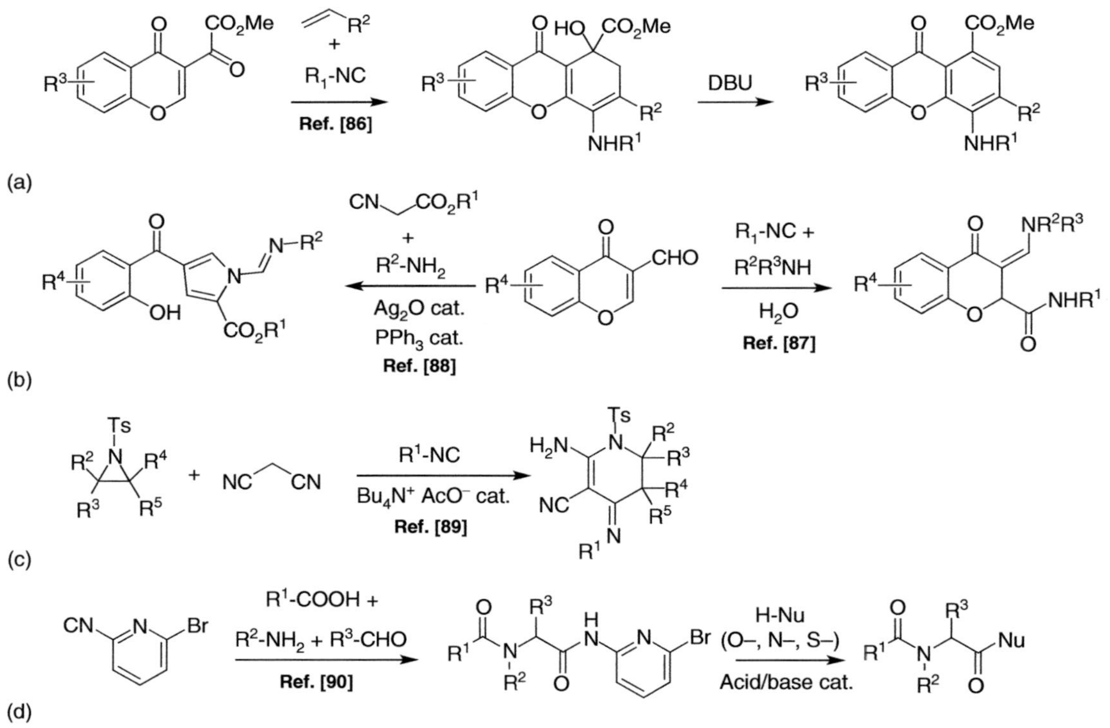

Scheme 1.27 Ugi-type MCRs. Substrate variations.

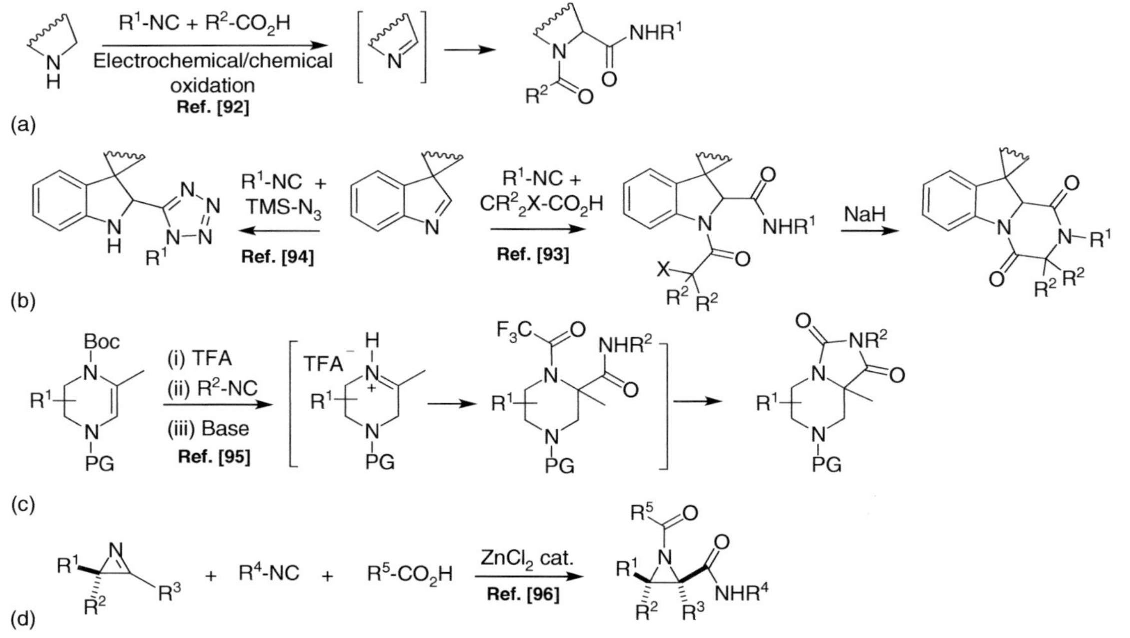

Scheme 1.28 Joullié-type MCRs. Substrate variations.

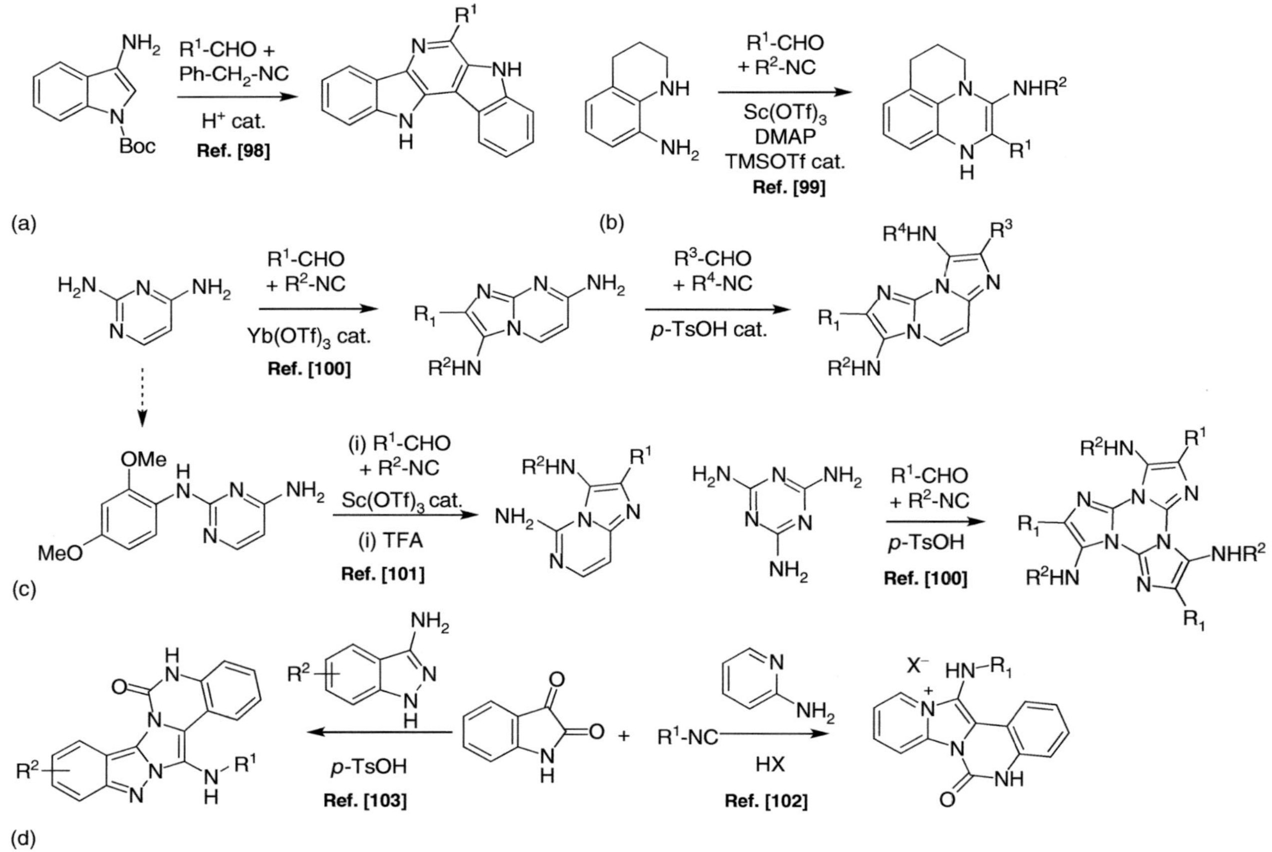

Scheme 1.29 GBB-type MCRs. Mechanistic variations.

suffers a [1,5]-H shift, followed by the intramolecular trapping of the isocyanate intermediate [103]. Aminoindazoles give the corresponding adduct in a similar manner (Scheme 1.29) [104].

The combination of 2-(2-bromoethyl)benzaldehyde and aminopyridines yields a dihydroquinolinium ion that continues the GBB reaction upon the attack of the isocyanide and the cyclization/aromatization step to afford the corresponding adduct [105]. Similar transformations have been reported with aminoindazole [106].

GBB reactions with propynals afford interesting post-condensation modifications exploiting the triple bond transformations, to yield iodo-substituted fused imidazopyrroles upon the cyclization with iodine [107]. In a recent report, however, the participation of benzyl isocyanide in GBB MCRs with propynals activates an alternative route: upon the *in situ* oxidation of the benzyl residue to give an imine, and the TBAB-mediated activation of the triple bond, the cyclization affords imidazo-dipyridines instead (Scheme 1.30) [108].

The incorporation of indole carbaldehydes enables GBB adducts to undergo a variety of domino processes, due to a key polarity inversion of the indole residue after the MCR. In this way, indole 3-carbaldehyde GBB adducts suffer a spontaneous oxidative Pictet–Spengler transformation to afford a variety of fused polyheterocyclic systems. However, indole 2-carbaldehyde GBB adducts yield indolocabazoles in an AB_3C fashion, and a striking bicyclic scaffold at higher temperature (Scheme 1.30) [109].

A variety of formal [3 + 2] cycloadditions involve novel interactions of isocyanides with heterocyclic substrates, forming part of the either dipole or the dipolarophile. For instance, the interaction of dipoles, *in situ* generated through the addition of isocyanides to unsaturated carbonyls, with isoquinolines leads to single adducts with high stereoselectivity, under chiral organocatalysis [110]. Similarly, the participation of acetylenedicarboxylate and sulfamate-derived cyclic imines leads to fused pyrroles in a regioselective manner via the corresponding isocyanide-dipole (Scheme 1.31) [111].

Insertion processes, not included in previous sections, account for meaningful transformations. For instance, indole, selenium, isocyanides, and secondary amines lead to a four-component adduct through a selenourea radical intermediate, under oxidative conditions [112]. Another mechanistic variation involves the insertion of isocyanides upon N—Si bonds, in a TMSCl-promoted interaction. In this way, azines and two equivalents of isocyanides, which can be differentiated because of having different roles in the process, lead to fused isoquinoline-imidazolium salts (Scheme 1.31) [113].

1.7 Miscellany Processes

Among the processes that do not fit in the precedent sections, we may mention the productive research in the chemistry of the BODIPY derivatives as a source of smart fluorophores. Formyl-BODIPYs have been prepared and reacted in Passerini MCR [114]. Interesting contributions to the synthesis of BODIPYs through MCRs

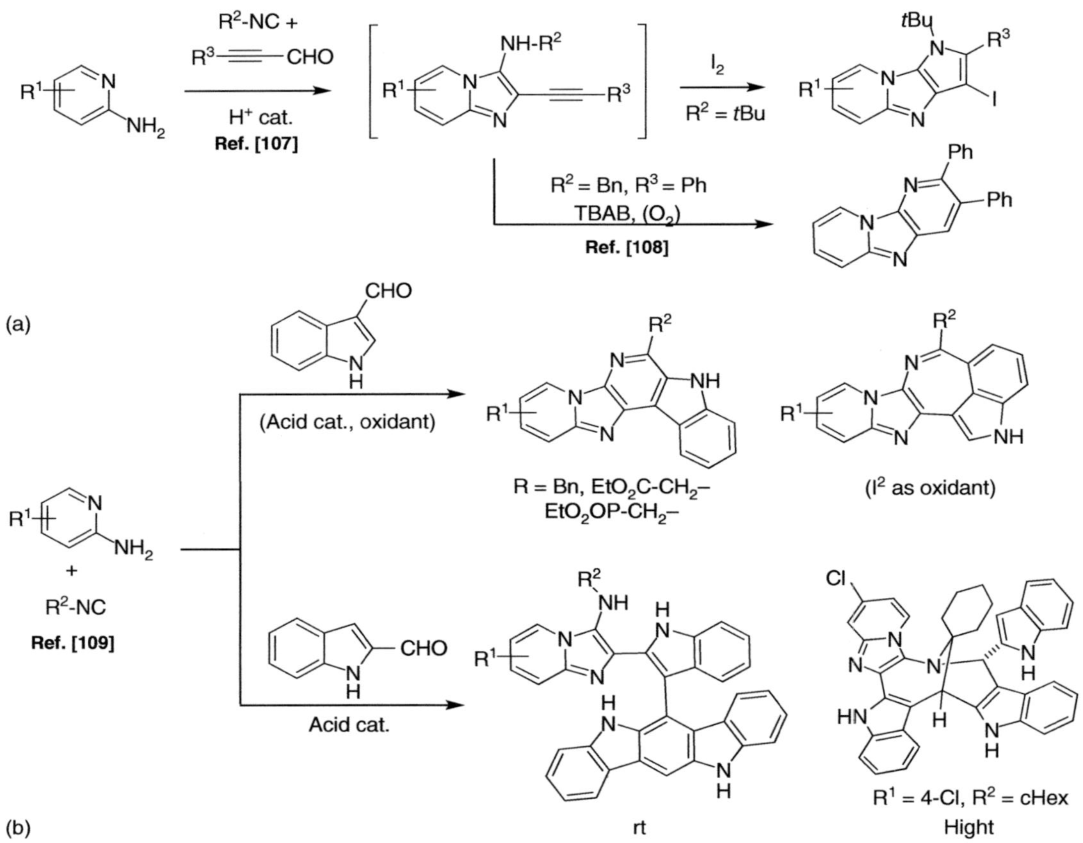

Scheme 1.30 GBB-type MCRs: post-transformations.

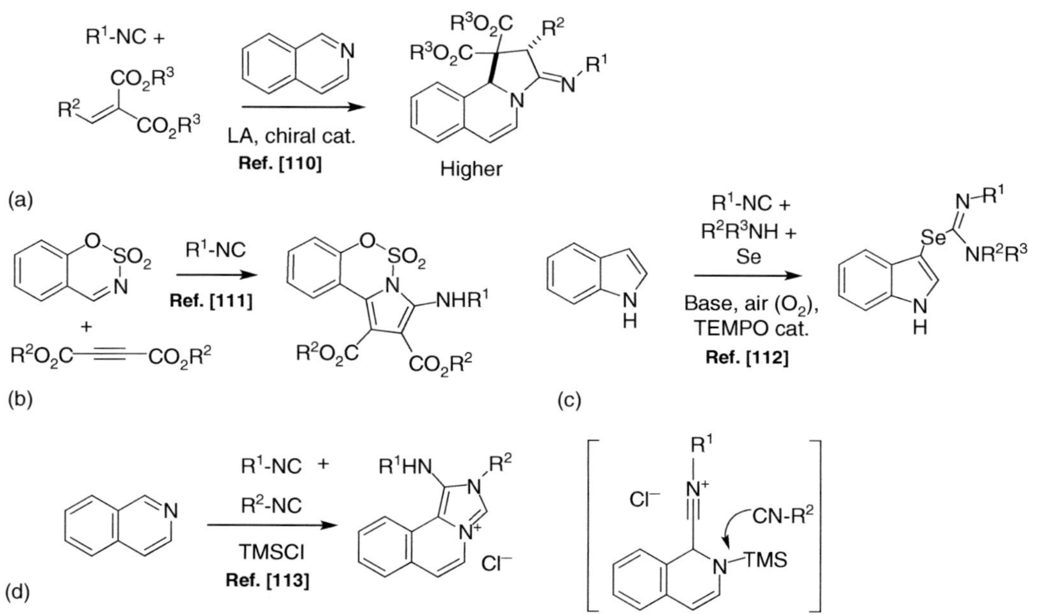

Scheme 1.31 Isocyanide MCRs based on cycloadditions and insertions.

have also appeared: the Lewis acid–catalyzed condensation of lactones with pyrrole [115] and the interaction of a phenol-substituted pyrrole with boronic acids and another pyrrole unit to yield the dye in one step [116]. The known interaction of isocyanides, azines, and trifluoroacetic anhydride leading to dipolar acid fluorides has been performed on a fluorophore-linked isoquinoline to afford a probe capable of selectively labeling amines, allowing its direct visualization in cell environments (Scheme 1.32) [117].

Fused imidazoazines have been prepared through aldol-type condensation followed by a SNAr (or Ullmann coupling) and a click reaction to yield complex MCR adducts [118]. Also, the interaction of aminopyridines with ketones and thiols yields thio-substituted imidazopyridines via a remarkable coupled I_2-Flavin catalysis (Scheme 1.33) [119]. Some MCR processes are based on a series of nucleophilic displacements: the interaction of azines with indoles and dichloroethane triggers a domino process, which is oxidatively terminated to yield cationic azahelicenes [120]. Furthermore, the participation of DABCO, α-chloroazines, and sulfide anions promotes an orchestrated sequence of SNAr-SN2-SNAr processes leading to a four-component scaffold (Scheme 1.33) [121].

Unsaturated nitro derivatives display a rich chemistry, enabling new MCRs based on conjugated additions and subsequent Henry or NO_2 elimination processes. For instance, their interaction with cyclohexanone and aminopyrazole leads to heterocyclic spiroadducts [122]. Furthermore, 3-nitro-indole or -benzothiazole reacts with *in situ* generated cyclic ketimines leading to indole-fused heteroacenes (Scheme 1.34) [123].

To finish this section, we mention an impressive contribution to the field of reaction discovery where, among other transformations, novel MCRs were described using a high-throughput autonomous organic synthesis robot fitted with analytical tools and controlled by a machine learning algorithm (Scheme 1.34) [124].

1.8 Conclusion

Merging the synthetic power of MCRs with the particular reactivity of heterocycles leads to an impressive array of new transformations and unprecedented connectivity patterns. These new scaffolds are produced in a straightforward manner, often by simply mixing the reactants and, due to their combinatorial nature, are amenable to parallelization. Furthermore, the highlighted processes show high levels of structural variability. The reaction discovery based on these heterocycle-based MCRs is already very fruitful and, although in its infancy, the description of this uncharted reactivity is paving the way to a systematic use of these processes in synthetic chemistry.

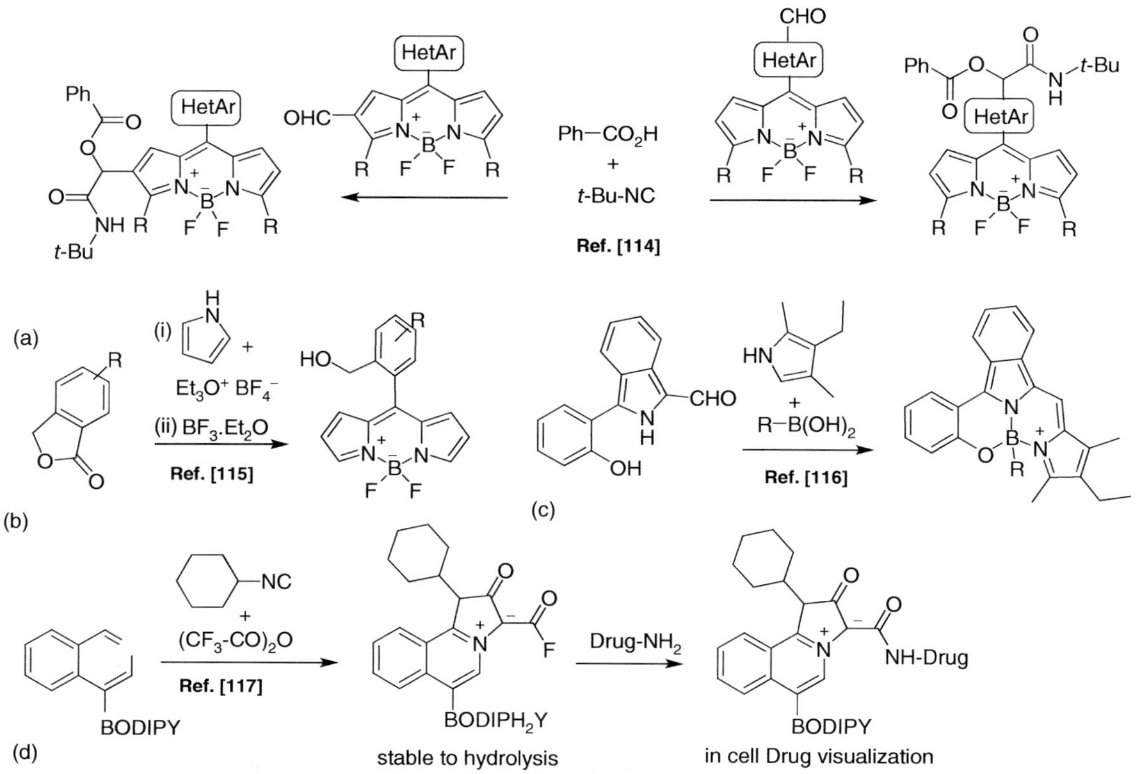

Scheme 1.32 MCRs featuring BODIPYs.

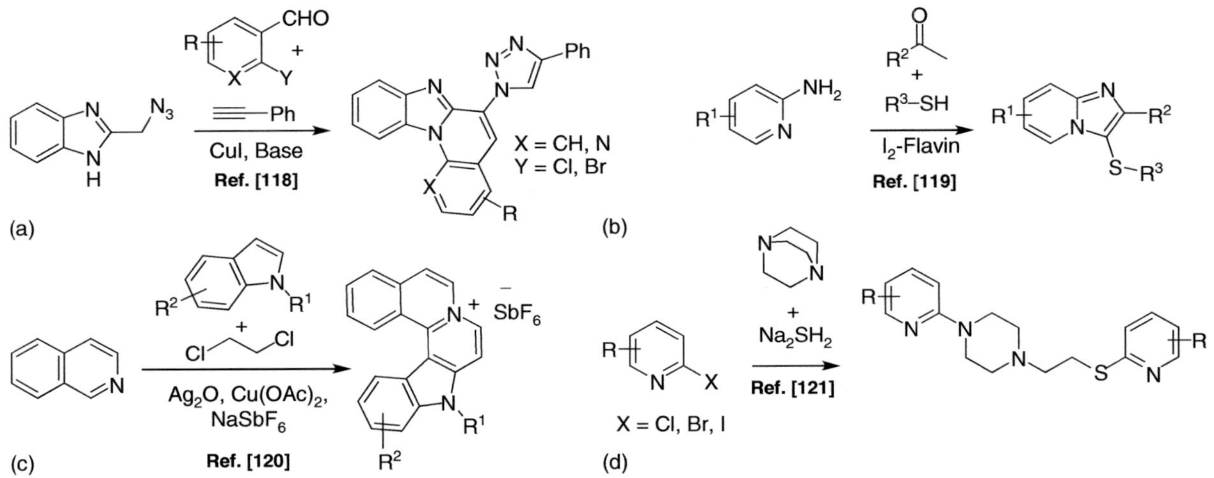

Scheme 1.33 Azole and azine nucleophilic MCRs.

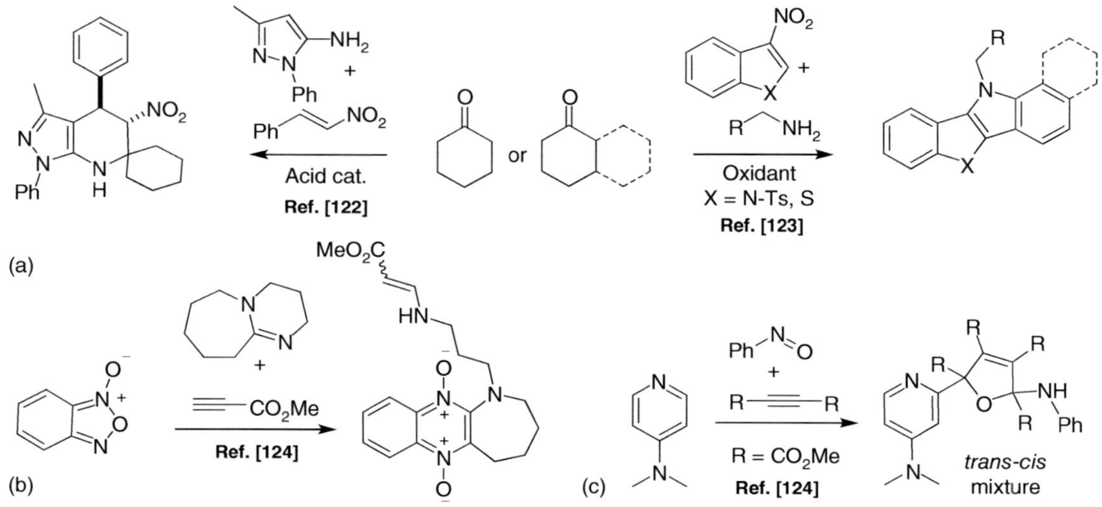

Scheme 1.34 MCRs based on nitro derivatives and AI reaction discovery.

Acknowledgment

Support from the Ministerio de Ciencia e Innovación (Spain, PID2019-107991RB-I00) is acknowledged. Profs. Gemma Fabriàs and Ignacio Alfonso (IQAC, CSIC Barcelona, Spain) are thanked for support.

References

1 Isambert, N. and Lavilla, R. (2008). *Chem. A Eur. J.* 14: 8444.
2 Sunderhaus, J.D. and Martin, S.F. (2009). *Chem. A Eur. J.* 15: 1300.
3 Rotstein, B.H., Zaretsky, S., Rai, V., and Yudin, A.K. (2014). *Chem. Rev.* 114: 8323.
4 Vicente-García, E., Kielland, N., and Lavilla, R. (2015). Functionalization of Heterocycles by MCRs. In: *Multicomponent Reactions in Organic Synthesis* (eds. J. Zhu, Q. Wang and M.-X. Wang), 159. Wiley.
5 Muthukrishnan, I., Sridharan, V., and Menéndez, J.C. (2019). *Chem. Rev.* 119: 5057.
6 Ghashghaei, O., Masdeu, C., Alonso, C. et al. (2018). *Drug Discov. Today Technol.* 29: 71.
7 Abranches, P.A.D.S., De Paiva, W.F., De Fátima, Â. et al. (2018). *J. Org. Chem.* 83: 1761.
8 Hoemann, M.Z., Xie, R.L., Rossi, R.F. et al. (2002). *Bioorg. Med. Chem. Lett.* 12: 129.
9 Galván, A., Fontaneda, R., Fañanás, F.J., and Rodríguez, F. (2016). *Adv. Synth. Catal.* 358: 1741.
10 Galván, A., Calleja, J., González-Pérez, A.B. et al. (2015). *Chem. A Eur. J.* 21: 16769.
11 Pasha, J., Kandagatla, B., Sen, S. et al. (2015). *Tetrahedron Lett.* 56: 2289.
12 Dai, W., Jiang, X.L., Tao, J.Y., and Shi, F. (2016). *J. Org. Chem.* 81: 185.
13 Retich, C. and Bräse, S. (2018). *Eur. J. Org. Chem.* 2018: 60.
14 Islam, K., Das, D.K., Akram, E., and Khan, A.T. (2015). *Synthesis* 47: 2745.
15 Tejería, A., Pérez-Pertejo, Y., Reguera, R.M. et al. (2018). *Eur. J. Med. Chem.* 152: 137.
16 Singh, P., Adolfsson, D.E., Ådén, J. et al. (2019). *J. Org. Chem.* 84: 3887.
17 Moshapo, P.T., Sokamisa, M., Mmutlane, E.M. et al. (2016). *Org. Biomol. Chem.* 14: 5627.
18 Jiang, H.J., Liu, K., Wang, J. et al. (2017). *Org. Biomol. Chem.* 15: 9077.
19 Yang, S., Zhu, S., Lu, D., and Gong, Y. (2019). *Org. Lett.* 21: 8464.
20 Bobers, J., Škopić, M.K., Dinter, R. et al. (2020). *ACS Comb. Sci.* 22: 101.
21 Chen, G.G., Wei, J.Q., Yang, X., and Yao, Z.J. (2016). *Org. Lett.* 18: 1502.
22 Xu, J.-H., Zheng, S.-C., Zhang, J.-W. et al. (2016). *Angew. Chem. Int. Ed.* 55: 11834.
23 Li, J.L., Dai, Q.S., Yang, K.C. et al. (2018). *Org. Lett.* 20: 7628.

24 Mirzaei, A., Turczel, G., Nagyházi, M. et al. (2021). *Eur. J. Org. Chem.* 2021: 326.

25 Wu, X., Zhu, Z.H., He, H. et al. (2020). *J. Org. Chem.* 85: 6216.

26 Arumugam, N., Almansour, A.I., Suresh Kumar, R. et al. (2020). *Bioorg. Chem.* 99: 103799.

27 Nájera, C. and Sansano, J.M. (2019). *Pure Appl. Chem.* 91: 575.

28 Zhang, X., Qiu, W., Evans, J. et al. (2019). *Org. Lett.* 21: 2176.

29 Wu, F.S., Zhao, H.Y., Xu, Y.L. et al. (2017). *J. Org. Chem.* 82: 4289.

30 Samala, S., Ryu, D.H., Song, C.E., and Yoo, E.J. (2019). *Org. Biomol. Chem.* 17: 1773.

31 Cao, J., Yang, F., Sun, J. et al. (2019). *J. Org. Chem.* 84: 622.

32 Ghorai, S. and Lee, D. (2020). *Synlett* 31: 750.

33 Liautard, V. and Landais, Y. (2015). Free-radical multicomponent processes. In: *Multicomponent Reactions in Organic Synthesis* (eds. J. Zhu, Q. Wang and M.-X. Wang), 401. Wiley.

34 Garbarino, S., Ravelli, D., Protti, S., and Basso, A. (2016). *Angew. Chem. Int. Ed.* 55: 15476.

35 Proctor, R.S.J. and Phipps, R.J. (2019). *Angew. Chem. Int. Ed.* 58: 13666.

36 Lear, J.M., Buquoi, J.Q., Gu, X. et al. (2019). *Chem. Commun.* 55: 8820.

37 Li, T., Liang, K., Zhang, Y. et al. (2020). *Org. Lett.* 22: 2386.

38 Liu, Z. and Liu, Z.Q. (2017). *Org. Lett.* 19: 5649.

39 Buquoi, J.Q., Lear, J.M., Gu, X., and Nagib, D.A. (2019). *ACS Catal.* 9: 5330.

40 Zheng, D. and Studer, A. (2019). *Angew. Chem. Int. Ed.* 58: 15803.

41 Chen, D., Xu, L., Long, T. et al. (2018). *Chem. Sci.* 9: 9012.

42 Cao, J., Wang, G., Gao, L. et al. (2019). *Chem. Sci.* 10: 2767.

43 Lee, K., Lee, S., Kim, N. et al. (2020). *Angew. Chem. Int. Ed.* 59: 13379.

44 Li, H., Guo, L., Feng, X. et al. (2020). *Chem. Sci.* 11: 4904.

45 Breton-Patient, C., Naud-Martin, D., Mahuteau-Betzer, F., and Piguel, S. (2020). *Eur. J. Org. Chem.* 2020: 6653.

46 Ye, J.H., Zhu, L., Yan, S.S. et al. (2017). *ACS Catal.* 7: 8324.

47 Cuadros, S., Horwitz, M.A., Schweitzer-Chaput, B., and Melchiorre, P. (2019). *Chem. Sci.* 10: 5484.

48 Choi, J.H. and Park, C.M. (2018). *Adv. Synth. Catal.* 360: 3553.

49 Sharma, U.K., Sharma, N., Xu, J. et al. (2015). *Chem. A Eur. J.* 21: 4908.

50 Kumar, G.S., Ragini, S.P., Kumar, A.S., and Meshram, H.M. (2015). *RSC Adv.* 5: 51576.

51 Lv, W., Chen, Y., Wen, S. et al. (2020). *J. Am. Chem. Soc.* 142: 14864.

52 Pappoppula, M. and Aponick, A. (2015). *Angew. Chem. Int. Ed.* 54: 15827.

53 Chinna Rajesh, U., Purohit, G., and Rawat, D.S. (2015). *ACS Sustain. Chem. Eng.* 3: 2397.

54 Albaladejo, M.J., Alonso, F., and González-Soria, M.J. (2015). *ACS Catal.* 5: 3446.

55 He, Q., Xie, F., Xia, C. et al. (2020). *Org. Lett.* 22: 7976.

56 Natte, K., Neumann, H., and Wu, X.F. (2015). *Catal. Sci. Technol.* 5: 4474.

57 Veltri, L., Mancuso, R., Altomare, A., and Gabriele, B. (2015). *ChemCatChem* 7: 2206.
58 He, Y., Li, Z., Tian, G. et al. (2017). *Chem. Commun.* 53: 6413.
59 He, Y., Liu, Z., Wu, D. et al. (2019). *Org. Lett.* 21: 4469.
60 Du, X., Yu, J., Gong, J. et al. (2019). *Eur. J. Org. Chem.* 2019: 2502.
61 Ramana, D.V., Vinayak, B., Dileepkumar, V. et al. (2016). *RSC Adv.* 6: 21789.
62 Deb, M.L., Borpatra, P.J., and Baruah, P.K. (2019). *Green Chem.* 21: 69.
63 Yıldırım, M. and Çelikel, D. (2015). *Mol. Divers.* 19: 1.
64 Sun, J., Jiang, W., and Yan, C.G. (2018). *RSC Adv.* 8: 28736.
65 Wu, P., Givskov, M., and Nielsen, T.E. (2019). *Chem. Rev.* 119: 11245.
66 Flagstad, T., Azevedo, C.M.G., Troelsen, N.S. et al. (2019). *Eur. J. Org. Chem.* 2019: 1061.
67 Mohanta, P.P., Pati, H.N., and Behera, A.K. (2020). *RSC Adv.* 10: 15354.
68 Wang, H., Lee, M., Peng, Z. et al. (2015). *J. Med. Chem.* 58: 4194.
69 Kolosov, M.A., Beloborodov, D.A., Orlov, V.D., and Dotsenko, V.V. (2016). *N. J. Chem.* 40: 7573.
70 Shearer, C., Desaunay, O., Zorc, S. et al. (2019). *Tetrahedron* 75: 130606.
71 Sahu, P.K., Sahu, P.K., Kaurav, M.S. et al. (2018). *ACS Omega* 3: 15035.
72 Chen, J., Ouyang, C., Xiao, T. et al. (2019). *ChemistrySelect* 4: 7327.
73 Abonia, R., Gutiérrez, L.F., Insuasty, B. et al. (2019). *Beilstein J. Org. Chem.* 15: 642.
74 Sun, Q.S., Sun, J., Pan, L.N., and Yan, C.G. (2020). *J. Org. Chem.* 85: 12117.
75 Lyapustin, D.N., Ulomsky, E.N., Zanakhov, T.O., and Rusinov, V.L. (2019). *J. Org. Chem.* 84: 15267.
76 Shetty, M., Huang, H., and Kang, J.Y. (2018). *Org. Lett.* 20: 700.
77 Cai, Y., Gu, Q., and You, S.L. (2018). *Org. Biomol. Chem.* 16: 6146.
78 Zurro, M., Asmus, S., Bamberger, J. et al. (2016). *Chem. A Eur. J.* 22: 3785.
79 Giustiniano, M., Moni, L., Sangaletti, L. et al. (2018). *Synthesis* 50: 3549.
80 Saya, J.M., Oppelaar, B., Cioc, R.C. et al. (2016). *Chem. Commun.* 52: 12482.
81 Wang, J., Sun, Y., Wang, G., and Zhen, L. (2017). *Eur. J. Org. Chem.* 2017: 6338.
82 Zhu, Z. and Seidel, D. (2016). *Org. Lett.* 18: 631.
83 Dighe, S.U., Kumar, A.K.S., Srivastava, S. et al. (2015). *J. Org. Chem.* 80: 99.
84 Sidhoum, M.A., El Kaïm, L., and Grimaud, L. (2018). *Tetrahedron* 74: 5222.
85 Bornadiego, A., Diáz, J., and Marcos, C.F. (2019). *J. Org. Chem.* 84: 7426.
86 Bornadiego, A., Díaz, J., and Marcos, C.F. (2015). *J. Org. Chem.* 80: 6165.
87 Lei, J., Li, Y., Li, Y. et al. (2020). *Green Chem.* 22: 3716.
88 Liao, J.Y., Yap, W.J., Wu, J.'.E. et al. (2017). *Chem. Commun.* 53: 9067.
89 Samzadeh-Kermani, A. (2019). *Monatsh. Chem.* 150: 1495.
90 van der Heijden, G., Jong, J.A.W., Ruijter, E., and Orru, R.V.A. (2016). *Org. Lett.* 18: 984.
91 Nazeri, M.T., Farhid, H., Mohammadian, R., and Shaabani, A. (2020). *ACS Comb. Sci.* 22: 361.
92 Pan, N., Ling, J., Zapata, R. et al. (2019). *Green Chem.* 21: 6194.
93 Golubev, P. and Krasavin, M. (2017). *Eur. J. Org. Chem.* 2017: 1740.
94 Estévez, V., Kloeters, L., Kwietniewska, N. et al. (2017). *Synlett* 28: 376.

95 Firth, J.D., Zhang, R., Morgentin, R. et al. (2015). *Synthesis* 47: 2391.
96 Angyal, A., Demjén, A., Wéber, E. et al. (2018). *J. Org. Chem.* 83: 3570.
97 Boltjes, A. and Dömling, A. (2019). *Eur. J. Org. Chem.* 2019: 7007.
98 Chen, Z.Z., Li, S.Q., Zhang, Y.J. et al. (2018). *Org. Lett.* 20: 7811.
99 Azad, C.S. and Narula, A.K. (2017). *Eur. J. Org. Chem.* 2017: 6413.
100 Ghashghaei, O., Seghetti, F., and Lavilla, R. (2019). *Beilstein J. Org. Chem.* 15: 521.
101 Ghashghaei, O., Caputo, S., Sintes, M. et al. (2018). *Chem. A Eur. J.* 24: 14513.
102 Konstantinidou, M., Boiarska, Z., Butera, R. et al. (2020). *Eur. J. Org. Chem.* 2020: 5601.
103 Yang, B., Tao, C., Shao, T. et al. (2016). *Beilstein J. Org. Chem.* 12: 1487.
104 Balwe, S.G. and Jeong, Y.T. (2018). *Org. Chem. Front.* 5: 1628.
105 Sagar, A., Nagarjuna Babu, V., Shinde, A.H., and Sharada, D.S. (2016). *Org. Biomol. Chem.* 14: 10366.
106 Balwe, S.G., Vagh, S.S., and Jeong, Y.T. (2020). *Tetrahedron Lett.* 61: 152101.
107 Tber, Z., Hiebel, M.A., El Hakmaoui, A. et al. (2015). *J. Org. Chem.* 80: 6564.
108 Li, Y., Huang, J.H., Wang, J.L. et al. (2019). *J. Org. Chem.* 84: 12632.
109 Ghashghaei, O., Pedrola, M., Seghetti, F. et al. (2021, 2603). *Angew. Chem. Int. Ed.*: 60: 2603.
110 Xiong, Q., Dong, S., Chen, Y. et al. (2019). *Nat. Commun.* 10: 2116.
111 Wang, C., Ma, Z., Qu, Y. et al. (2020). *Chem. An Asian J.* 15: 560.
112 Liu, H., Fang, Y., Wang, S.Y., and Ji, S.J. (2018). *Org. Lett.* 20: 930.
113 Kishore, K.G., Ghashghaei, O., Estarellas, C. et al. (2016). *Angew. Chem. Int. Ed.* 55: 8994.
114 Ramírez-Ornelas, D.E., Alvarado-Martínez, E., Bañuelos, J. et al. (2016). *J. Org. Chem.* 81: 2888.
115 del Río, M., Lobo, F., Lopez, J.C. et al. (2017). *J. Org. Chem.* 82: 1240.
116 Chen, N., Zhang, W., Chen, S. et al. (2017). *Org. Lett.* 19: 2026.
117 Sintes, M., De Moliner, F., Caballero-Lima, D. et al. (2016). *Bioconjug. Chem.* 27: 1430.
118 Nagesh, H.N., Suresh, A., Reddy, M.N. et al. (2016). *RSC Adv.* 6: 15884.
119 Okai, H., Tanimoto, K., Ohkado, R., and Iida, H. (2020). *Org. Lett.* 22: 8002.
120 Wang, Z., Jiang, L., Ji, J. et al. (2020). *Angew. Chem. Int. Ed.* 59: 23532.
121 Zhu, Q., Yuan, Q., Chen, M. et al. (2017). *Angew. Chem. Int. Ed.* 56: 5101.
122 Zhang, F., Li, C., and Qi, C. (2020). *Org. Chem. Front.* 7: 2456.
123 Santhini, P.V., Akhil Krishnan, R., Babu, S.A. et al. (2017). *J. Org. Chem.* 82: 10537.
124 Granda, J.M., Donina, L., Dragone, V. et al. (2018). *Nature* 559: 377.

2

Heterocycles and Multicomponent Polymerizations

Susan Sieben, Jordy M. Saya, Dean Johnson, and Romano V.A. Orru

Maastricht University, Aachen-Maastricht Institute for Biobased Materials (AMIBM), Urmonderbaan 22, Geleen 6167 RD, The Netherlands

2.1 Introduction

Multicomponent reactions (MCRs) are well-recognized synthetic tools to make sets of diversely functionalized complex small-molecule scaffolds. As it has become clear from the other chapters in this book, MCRs have been widely applied for the synthesis of heterocycles. Moreover, many MCR products can undergo so-called post-reaction modifications. These can proceed either stepwise or *in situ* generating advanced second-generation one-pot processes. The latter deliver higher-order complexity generating reactions ideal for diversity-oriented synthesis (DOS) strategies. This offers many opportunities to introduce additional desired (often predesigned) functionality in the targeted heterocyclic compounds, which accounts for the popularity of MCRs and their post-modifications in medicinal chemistry and catalysis.

The utilization of MCR chemistry to access functionalized polymers for applications in advanced materials with clever properties is much less common practice. However, the use of MCRs in polymer chemistry, also known as multicomponent polymerization (MCP), has gained quite some interest in recent years. In this chapter, we highlight some of the latest examples in this area. As the main topic of this book deals with MCR applications in heterocyclic chemistry, we focused our contribution on MCPs that result in polymers that include heterocyclic fragments.

Synthetic polymers are often designed with particular characteristics in mind to accommodate specific properties. This may be an up-front designed and calculated feature but can also be the result of serendipitous discovery. This means that the specific unique selling points of MCRs and their post-modifications as the methodology of choice for DOS-based strategies also make them efficient tools to accommodate function-directed MCPs.

Typical characteristics of modern polymeric materials include strength, toughness, resistance to corrosion, lack of conductivity, color, and transparency [1]. Such polymers or polymer-based advanced materials have indeed been synthesized via MCRs. The MCRs themselves are the familiar set of reactions that

Multicomponent Reactions towards Heterocycles: Concepts and Applications, First Edition.
Edited by Erik V. Van der Eycken and Upendra K. Sharma.

we know, e.g. from the heterocyclic chemistry arena and can be divided into three groups: (i) isocyanide-based MCRs; (ii) non-isocyanide-based MCRs, and (iii) metal-catalyzed MCRs [2]. Isocyanide-based MCRs include the Passerini and the Ugi reaction, non-isocyanide-based MCRs include the Biginelli and the Hantzsch reaction, and metal-catalyzed MCRs include the Cu-catalyzed reaction between alkyne, sulfonyl azide, and a nucleophile. The fundamental aspects and scope and limitation details of these MCRs have been discussed in the other chapters, but for convenience, the most widely employed MCRs including appropriate references are listed in Table 2.1.

A large proportion of literature available on MCPs are from the first group: the isocyanide-based MCRs. Especially the well-known Passerini and Ugi chemistries have been widely employed. However, many of these examples produce linear polymers that do not necessarily include heterocycles in the backbone. Most of these examples are therefore outside the scope of this chapter and book. Thus, only limited Ugi-type MCPs examples are discussed in this chapter.

Generally, three main MCP strategies can be identified to include heterocycles in polymers (Figure 2.1). Strategies I and II generate the heterocycle during the polymerization reaction and strategy III includes a heterocycle in one of the initial starting monomers. The examples discussed in this chapter classify under one of these three categories.

There are some challenges in the synthesis of polymers via MCPs, such as limited availability of suitably functionalized monomers, low molecular weight of the resulting polymeric products, low polymer solubility, and defects in the structure caused by side reactions due to the complexity of the reaction [14]. Despite this, several polymers have been reported via different MCPs. In order to discuss the advantages and disadvantages of certain MCRs and their post-reaction modifications in MCP processes, we have decided to analyze the MCP efficiency and utility to produce a specific polymer using a set of well-established parameters. For the reactions discussed in the subsequent sections of this chapter, we report (and wherever appropriate compare) the following physical properties:

(1) The number and weight average molecular weight (M_n and M_w, respectively) are used to determine the molecular weight of a polymer calculated from the molar mass distribution [15].
(2) The polydispersity ($Đ$) is used to predict the homogeneity of polymers. Monodisperse or uniform polymers contain chains with the same lengths and molecular weights, while nonuniform polymers contain chains of different length and molecular weights. The closer the $Đ$ is to 1, the more the uniform it is [16].
(3) The glass transition temperature (T_g) is the temperature at which the polymer transits from the solid state into a rubbery state, before becoming a liquid state [15, 17].
(4) The thermal degradation temperature (T_d) is the temperature at which the polymer starts to degrade [17].
(5) The degree of polymerization (DP) is the number of repeating units present in the polymer [16], which is an indicator for the efficiency of a polymerization reaction.

Table 2.1 Different MCRs utilized for MCPs.

Passerini[3a,4]	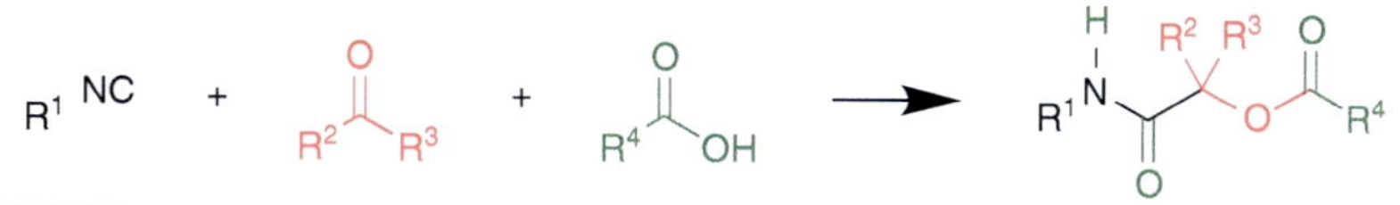
Ugi[3a,5]	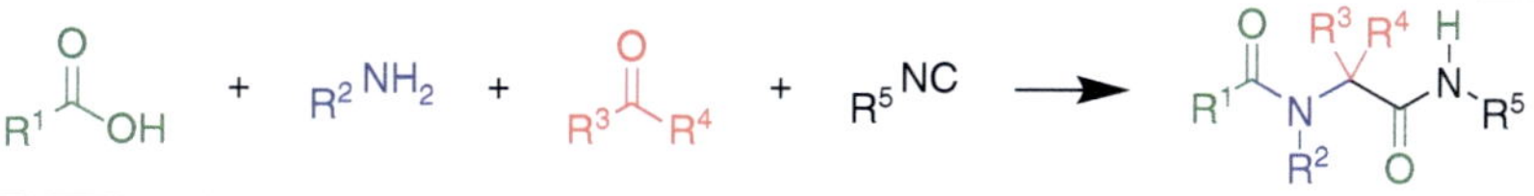
Mannich[3a,6]	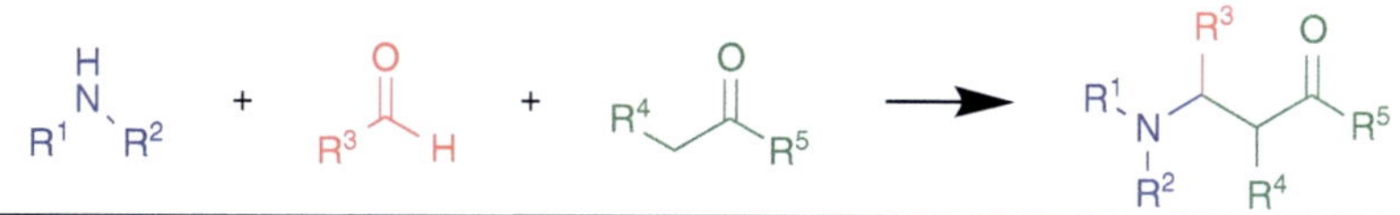
Biginelli[3a,7]	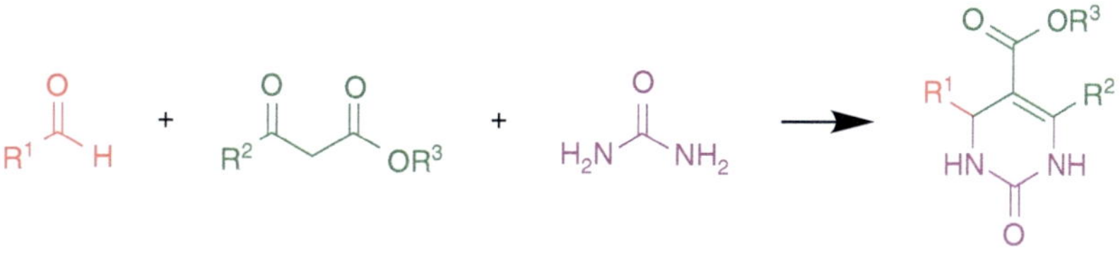
A^3-coupling[3a,8]	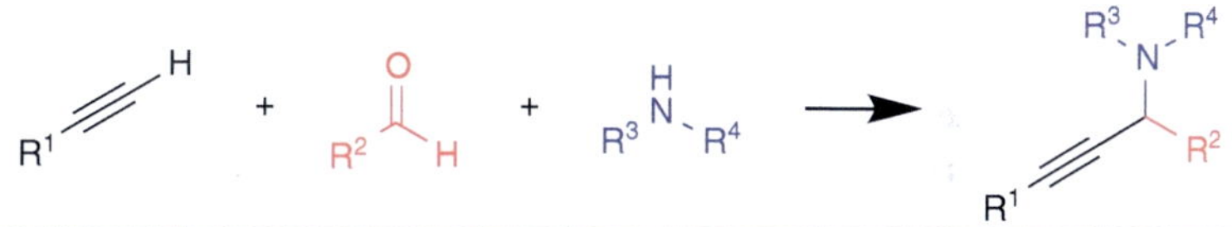
Kabachnik – Fields[9]	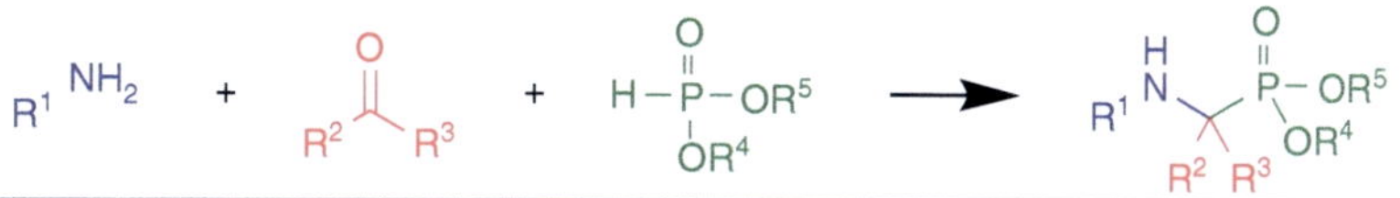
Hantzsch[3a,10]	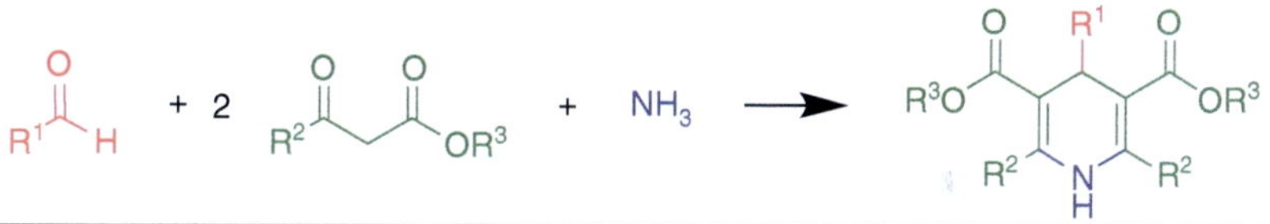
Debus–Radziszewski[3b,3c]	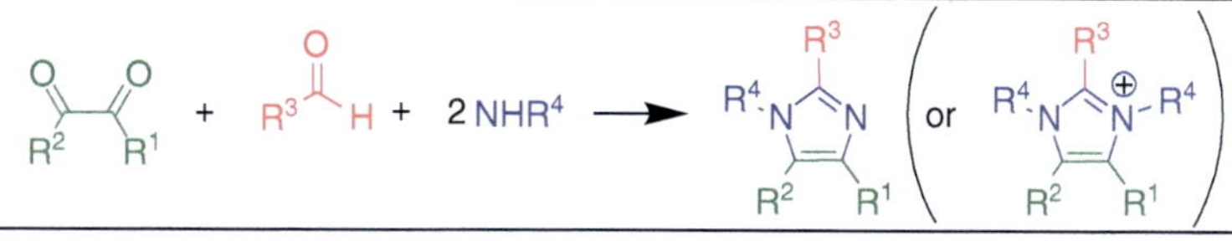
Amine-thiol-ene[11,12]	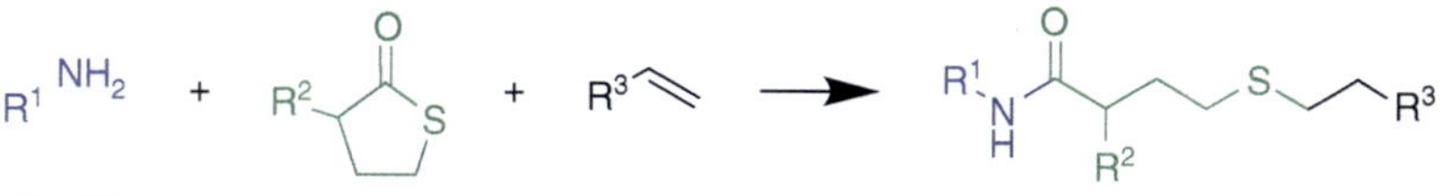
Cu-catalyzed reaction between alkyne, sulfonyl azide and a nucleophile[13]	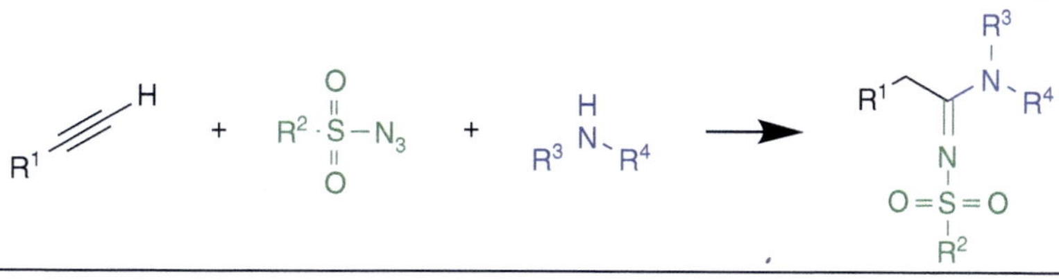

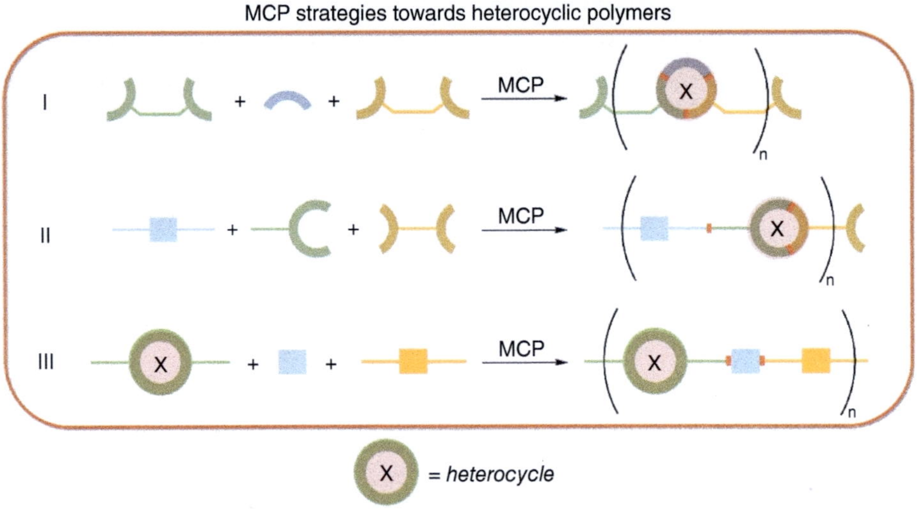

Figure 2.1 MCP strategies toward polymers with heterocyclic fragments.

In general, the efficiency of a polymerization reaction and thus of a MCP is determined by both the M_n and M_w. The higher these values are, the more efficient the polymerization reaction. Nevertheless, a polymer with high molecular weight side groups has intrinsically a larger M_n and M_w. In addition, the polymer chains of every polymer fold in a unique manner. Some of the polymer chains fold to a more compact structure and are thus smaller, while other polymer chains do not fold at all and are therefore larger. Due to this, the M_n and M_w cannot be used to directly compare the efficiency of different MCPs. We would like to stress that M_n values determined by NMR better reflect the actual size of a polymer compared to other analytical techniques (e.g. GPC) that are often used. This is because the signals in the NMR spectrum can be quantified and are independent of the polymer shape. To compensate for the molecular weight of the monomer, and really look at the MCP, the DP is a better parameter for the efficiency of a polymerization. The higher the DP, the more the repeating units are present and therefore the more efficient the MCP. The DP of every polymer is calculated in the same way, and the side groups with high molecular weights are included in the calculation, which makes it possible to compare the DPs of different polymers.

We will use the above-mentioned parameters to analyze the value of MCRs for a certain MCP and compare characteristics of different MCPs for the synthesis of polymers containing heterocyclic cores.

2.2 Ugi-type Multicomponent Polymerizations

The Ugi reaction is a well-established MCR of carboxylic acids, aldehydes/ketones, isocyanides, and amines to form α-aminoacyl amides (Table 2.1). It received recent attention in polymer science [18–20].

Table 2.2 Synthesis of polyamides **4**.

Polymer	Monomer[a)]	Yield (%)	M_n (g/mol)[b)]	$Đ$[b)]	DP[c)]	T_g (°C)[d)]	$T_{d,max}$ (°C)[e)]
4a2	**1 + 2a2 + 3a**	>99	7530	1.42	19	120	449
4a6	**1 + 2a6 + 3a**	93	12 320	1.66	27	86	462
4a12	**1 + 2a12 + 3a**	95	11 760	1.63	22	63	462
4b	**1 + 2b + 3a**	87	5230	1.36	10	52	377
4c	**1 + 2c + 3a**	96	—	—	—	—	—
4d	**1 + 2d + 3a**	89	6540 and 1840	1.09 and 1.15	15 and 4	—	—
4e	**1 + 2e + 3a**	89	10 300	1.78	—	—	—

a) [**1**] : [**2**] : [**3**] = 2 : 1 : 1.
b) Measured by GPC with linear narrow poly(methyl methacrylate) standards.
c) Calculated using the M_n.
d) Measured by DSC (10 °C/min).
e) Measured by TGA (10 °C/min).
Source: Adapted from Hartweg and Becer [21]; Hartweg and Becer [22].

For example, Becer and coworker reported the synthesis of polyamides **4.** They used an Ugi reaction under microwave heating of levulinic acid **1**, diamines **2**, and 1,6-diisocyanohexane **3a** (Figure 2.2 and Table 2.2) [21, 22]. Levulinic acid is a renewable biobased building block available in multiton quantities from sugar-based cellulose by rehydration of 5-(hydroxymethyl)furfural (HMF) at a cost-competitive level [23, 24]. This makes the MCP toward **4**, where levulinic acid is used directly as a bifunctional monomer, appealing from a green chemistry perspective [21, 22].

An increasing chain length (*R*) of the diamine **2** results in more flexibility of the polymer backbone and less steric hindrance. Consequently, the polyamides **4a6** and **4a12** have higher M_ns and DPs than polyamide **4a2**. Besides primary amine functionality, the spermine **2b** contains secondary amines, which may also participate in the Ugi reaction. However, NMR analysis shows that the MCP occurs selectively at the primary amine sites, giving polyamide **4b**. The use of triamine tris(2-aminoethyl)amine **2c** gave an insoluble crosslinked polyamide **4c** that hampered further analysis. The use of *p*-phenylenediamine **2d** did not gave a polyamide but oligomeric **4d** (DP = 4). This can be ascribed to the poor nucleophilicity of anilines. Hence, these amine inputs generally proceed less efficiently in Ugi reactions. Shorter chained polyamide **4a2** has a higher T_g than polyamides **4a6** and **4a12**, while the polyamide **4b** has a relatively low T_g due to the flexible polymer backbone caused by the incorporation of several heteroatoms [21, 22].

Meier and coworkers isolated polyamides **7** after an Ugi reaction of 2,5-furandicarboxylic acid (FDCA, **5**), isobutyraldehyde **6a**, decane-1,10-diamine **2a10**, and *t*-butyl isocyanide **3b** (Figure 2.3 and Table 2.3) [25].

Next to **7**, size exclusion chromatography-electrospray ionization mass spectrometry (SEC-ESI MS) showed that different macrocyclic side products were formed as well [25, 26]. The FDCA (**5**) can be synthesized from biomass, making it an attractive

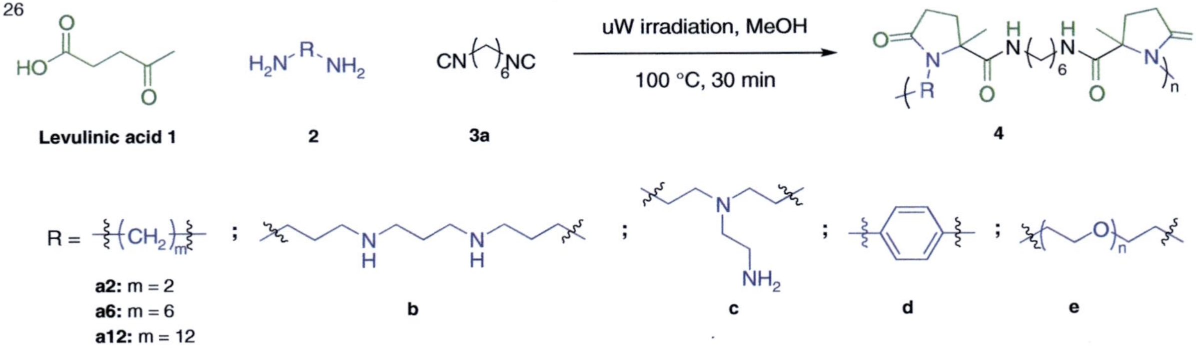

Figure 2.2 Synthesis of polyamide **4** via the Ugi reaction of levulinic acid **1**, diamines **2**, and **3a**. Source: Adapted from Llevot et al. [18]; Hartweg and Becer [21]; Hartweg and Becer [22].

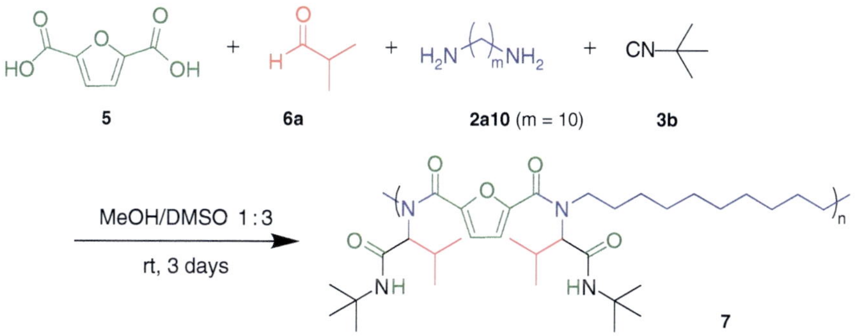

Figure 2.3 Synthesis of polyamide **7** via the Ugi reaction of **5**, **6a**, **2a10**, and **3b**. Source: Adapted from Schade et al. [25].

Table 2.3 Synthesis of polyamide **7**.

Polymer	Monomer[a)]	Yield (%)	M_n (g/mol)[b)]	Đ[b)]	DP[c)]	T_g (°C)[d)]
7	**5** + **6a** + **2a10** + **3b**	75	11 100	2.73	18	82.2

a) [**5**] : [**6a**] : [**2a10**] : [**3b**] = 1 : 3 : 1 : 3.
b) Measured by GPC with poly(methyl methacrylate) standards.
c) Calculated using the M_n.
d) Measured by DSC (10 °C/min).
Source: Data from Schade et al. [25].

renewable monomeric starting material [25]. Meier and coworkers showed that **5** could be prepared from HMF (see above) via hydrolysis and dehydration of sucrose, followed by oxidation with an Au/ZrO_2 catalyst [25]. During the synthesis of HMF via this approach, no additional purification steps were required [25].

Deng and coworkers prepared biocompatible polyamides **9** via the Ugi reaction in a MCP of succinic acid-end-capped polyethylene glycols (S-PEGs) **8**, aldehydes **6**, 1,4-bis (3-aminopropyl)piperazine **2f**, and *t*-butyl isocyanide **3b** (Figure 2.4 and Table 2.4). The polyamides **9** showed thermosensitive and antibacterial properties, which could be used as a nanodrug carrier or as a smart bactericidal surface. This avoids lengthy and complicated reaction steps employed in traditional synthesis of comparable thermosensitive polymers. However, big variations in efficiency when varying the inputs are observed. For example, the yield of **9a** is very low with a DP of 1, meaning that it cannot be considered a polymer. There is no clear reason why some polymerizations give high DPs and some much lower. Different combinations of **8** and **6** tune the hydrophobicity and hydrophilicity, while the PEG fragments render **9** biocompatible. Modification of the polyamides **9** by reaction with HBr introduces quaternary ammonium groups increasing the antibacterial activity even against multidrug-resistant *Escherichia coli* and *Staphylococcus aureus* [27].

Table 2.4 Synthesis of polyamides **9**.

Polymer	Monomer[a)]	Yield (%)	M_n (g/mol)[b)]	Đ[b)]	DP[c)]
9a	**8a2** + **6a** + **2f** + **3b**	16.4	1400	1.55	1
9b	**8a3** + **6a** + **2f** + **3b**	85.1	18 600	1.66	22
9c	**8a9** + **6a** + **2f** + **3b**	46.2	13 200	1.58	12
9d	**8a9** + **6d** + **2f** + **3b**	75.9	14 900	1.53	12
9e	**8a18** + **6b** + **2f** + **3b**	55.1	24 400	1.75	17
9f	**8a18** + **6c** + **2f** + **3b**	47.0	13 000	1.57	8
9g	**8a18** + **6a** + **2f** + **3b**	31.8 and 63.0	20 300 and 21 900	1.75 and 1.80	13 and 14
9h	**8a18** + **6d** + **2f** + **3b**	74.2 and 59.9	24 600 and 26 000	1.59 and 1.65	15 and 16
9i	**8a18** + **6e** + **2f** + **3b**	75.2	46 200	1.78	28
9j	**8a22** + **6a** + **2f** + **3b**	20.2	13 400	1.68	7
9k	**8a22** + **6d** + **2f** + **3b**	60.5	32 300	1.61	18
9l	**8a45** + **6a** + **2f** + **3b**	29.3	25 200	1.72	9
9m	**8a45** + **6d** + **2f** + **3b**	79.7	24 400	1.68	8

a) [**8**] : [**6**] : [**2f**] : [**3b**] = 1 : 2 : 1 : 2.
b) Measured by GPC.
c) Calculated using the M_n.
Source: Liu et al. [27].

2.3 Mannich-type Multicomponent Polymerizations

The Mannich reaction employs amines, aldehydes, and enolizable carbonyl compounds to form β-amino carbonyls (Table 2.1) and is frequently used in MCPs to synthesize polymers. Since its discovery in 1912, β-amino carbonyls receive much interest due to their biological activity, their use as synthetic building blocks, and their use as precursors of pharmaceutical products [28].

Phenolic resins are applied as a matrix in fiber-reinforced plastics, structural materials, adhesives, and paints due to their high heat resistance, high flame retardance, excellent electrical resistance, good chemical resistance, and low costs. However, they are brittle and require a catalyst for polymerization, and during the curing process, volumetric shrinkage occurs. These issues can be accommodated by using polybenzoxazines (PBAs) (Figure 2.5, structure **13**). These PBAs possess not only the positive characteristics of traditional phenolic resins but also unprecedented characteristics such as, excellent dimensional stability, low water absorption, and low dielectric constant. PBAs represent a new class of high-performance thermoset resins that resemble conventional phenolics but do not generate volatiles and undergo less shrinkage during cure, which results in higher-quality laminates and fiber composites [30].

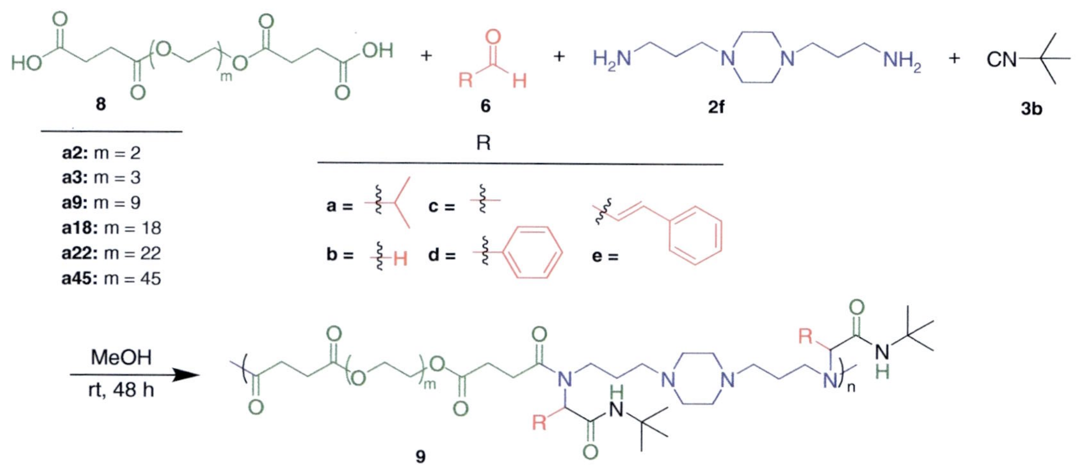

Figure 2.4 Synthesis of polyamides **9** via the Ugi reaction of S-PEGs **8**, aldehydes **6**, piperazine **2f**, and **3b**. Source: Liu et al. [27].

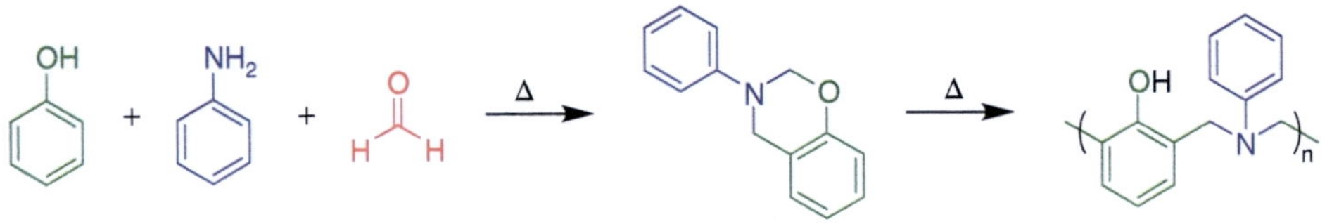

10a 11 6b 12 13

Figure 2.5 Synthesis of polybenzoxazine **13** from **12**, using a MCP of **10a**, **11**, and **6b**, followed by thermal ring-opening polymerization. Source: Agag and Takeichi [29].

Table 2.5 Synthesis of benzoxazine **12** and subsequent ring-opening polymerizations to PBA **13**.

Polymer	Monomer	T_g (E″) (°C)[a)]	T_g (tan δ) (°C)[a)]	$T_{d,5\%}$ (°C)[b)]	$T_{d,10\%}$ (°C)[b)]
13	**12**	146	160	342	369

a) Measured by dynamic viscoelastic measurements (4 °C/min).
b) Measured by TGA (5 °C/min).
Source: Agag and Takeichi [29].

The PBAs can be easily synthesized via the Mannich MCR. For example, the reaction of phenol **10a**, aniline **11**, and formaldehyde **6b** (Figure 2.5) provides **12** efficiently. Thermal ring-opening polymerization of benzoxazine **12** gives then PBA **13** (Figure 2.5 and Table 2.5) [29].

This process is formally a four-component MCP (see below), while the traditional Mannich reaction is a three-component reaction. However, the process is discussed in detail in this section as an enol and an iminium (Figure 2.6) species react, which are typical Mannich intermediates. The formation of PBA occurs via subsequent thermal ring-opening polymerization. First phenol **10a** reacts with iminium **I** (and **III**) losing aromaticity. Then, rearomatization regenerates the phenolic group that reacts with another iminium species. Nevertheless, benzoxazine is not required for polymerization. In both the formation of **12** and PBA, the same iminium intermediate **III** is involved, which could react with another enol directly. However, the formation of benzoxazine is thermodynamically favored and therefore the polymerization reaction will likely proceed via this route.

Under mild reaction conditions, the formation of hydrogen bonds terminates chain growth and the PBAs produced in this way have low molecular weights [34]. Crosslinked PBAs from difunctional benzoxazines results in much higher molecular weight polymeric material [34]. Difunctional benzoxazine **15** can be synthesized from bisphenol-A **14**, aniline **11**, and aldehyde **6b** (Figure 2.7 and Table 2.6) [30]. The use of bisphenol-A has serious health concerns but is nevertheless commonly employed in this MCP. Later in this section, we outline suitable alternatives. Thermal ring-opening polymerization of **15** gives brittle PBA **16** [35, 36], due to the rather high temperatures required [30].

Takeichi et al. addressed this issue in their synthesis of high molecular weight PBA precursors with benzoxazine in the polymer backbone. Thus, synthesis of PBA

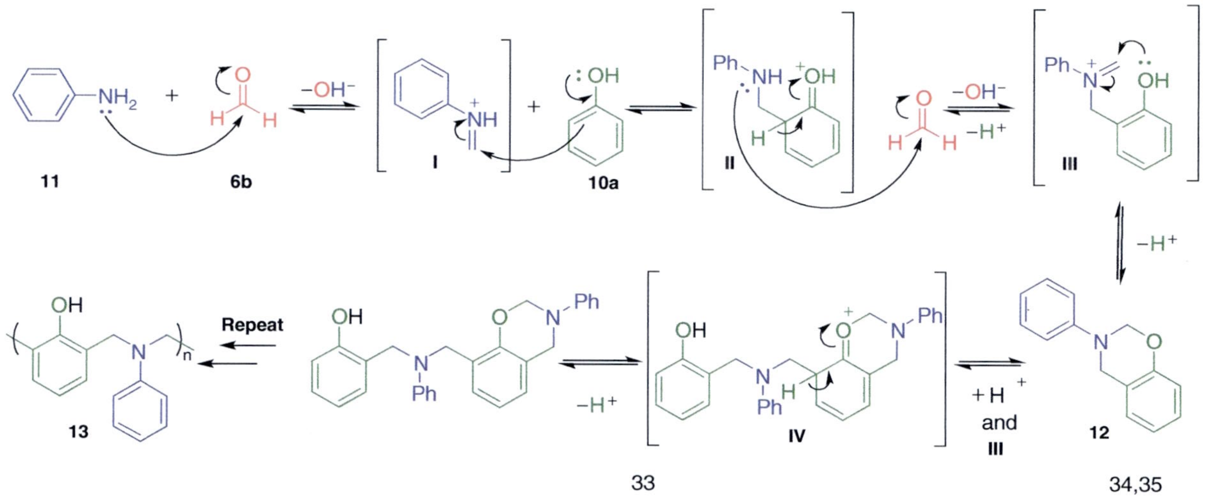

Figure 2.6 Reaction mechanism of the synthesis of benzoxazine **12** [31]. Reaction mechanism of the synthesis of PBA **13**. Source: Adapted from Chirachanchai et al. [31]; Sun et al. [32]; Han et al. [33].

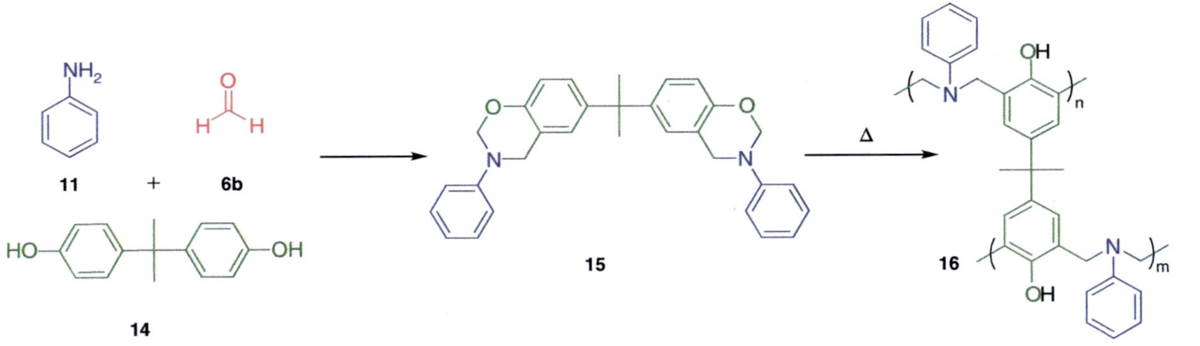

Figure 2.7 Synthesis of benzoxazine **15** from bisphenol-A **14**, aniline **11**, and formaldehyde **6b**, followed by thermal ring-opening polymerization to PBA **16**. Source: Takeichi et al. [30].

Table 2.6 Synthesis of benzoxazine **15** and subsequent ring-opening polymerizations to PBA **16**.

Polymer	Monomer	T_g (E″) (°C)[a)]	T_g (tan δ) (°C)[a)]	$T_{d,5\%}$ (°C)[b)]	$T_{d,10\%}$ (°C)[b)]
16	**15**	153	171	308	338

a) Measured by dynamic viscoelastic measurements (4 °C/min).
b) Measured by TGA (5 °C/min).
Source: Takeichi et al. [30].

precursors **18a–c** proceeds smoothly via the Mannich reaction of **15**, diamines **2a2**, **2a6**, or **2h1**, and paraformaldehyde **17** (Figure 2.8a and Table 2.7) [30]. Thermal ring-opening polymerization of **18a–c** gave the corresponding PBAs **19a–c** [30]. The syntheses of precursors **18a** and **18c** via bisphenol-A **14**, 1,6-diaminohexane **2a6** or 4,4′-diaminodiphenylmethane **2h1**, and paraformaldehyde **17** were also reported by others [34, 43].

Standard PBAs often lack flexibility, which limits their applications. Alloying with flexible polymers, using molecular predesign of benzoxazine monomers or by the synthesis of PBAs from a high molecular weight precursor may overcome this property. Takeichi et al. improved the toughness of the product by using a precursor with a polysiloxane unit. They reported the synthesis of PBA precursors **18d–h** via the Mannich reaction of bisphenol-A **14**, diamines **2h1** and/or **2j1-2j3**, and paraformaldehyde **17** (Figure 2.8a and Table 2.7). Thermal ring-opening polymerization of **18d–h** gave the corresponding PBAs **19d–h** with improved toughness, high T_gs, and high degradation temperatures. These were used as matrix resin of advanced composite materials and high-temperature adhesives [40].

The toughness of PBAs also significantly improves by introducing a polyether unit as was shown by Agag et al. [41]. They reported the synthesis of PBA precursors **18i–k** via the Mannich reaction of **15**, **2k1–3**, and **17** (Figure 2.8a and Table 2.7). After the thermal ring-opening polymerization of **18i–k** (physical properties were not all reported here), the corresponding PBAs **19i–k** with unique characteristics were obtained. These proved useful as toughening agents, coating materials, and membranes in various applications. In addition, they were employed as protective colloids for the stabilization of heterogeneous droplet mixtures with a diameter between 50 and 500 nm in o/w mini emulsions based on benzoxazine resins [44].

Materials with a low dielectric constant (*K*) are suitable for insulating applications [45]. The *K*-value depends on polarizability of functional groups and is lower for materials with high degrees of saturated hydrocarbons compared to materials containing unsaturated, conjugated, or phenyl groups [45, 46]. By breaking the conjugated system or decreasing the number of phenyl groups in the monomer, *K* can be decreased [45, 46]. However, this results in lower thermal stability. The C—H, C—C, and C—N bonds in aliphatic polymers are unstable at a temperature above 400 °C, while these bonds in non-aliphatic polymers can resist higher temperatures [46]. By incorporation of fluorinated substituents into the polymer backbone, PBAs with high thermal stability and a low *K* can be accessed [45]. Furthermore, Ishida

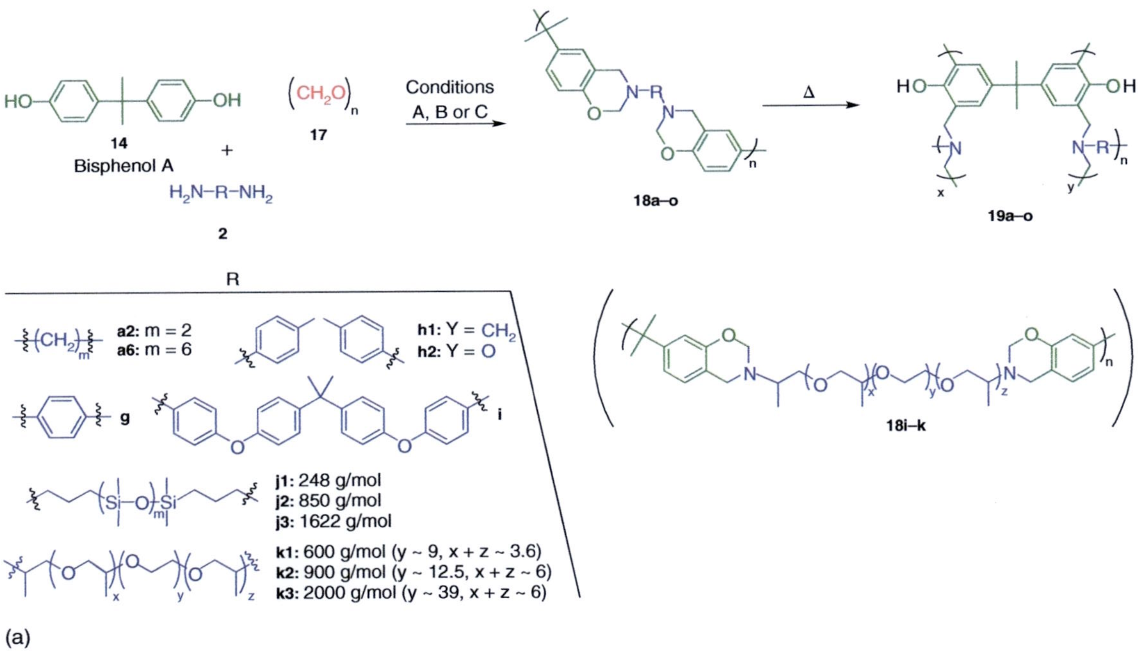

Figure 2.8 (a) Synthesis of PBA precursors **18** via the Mannich reaction of bisphenol A **14**, diamines **2**, and **17**, followed by thermal ring-opening polymerization to PBAs **19**. Conditions: [A]: $CHCl_3$, reflux, five hours; [B]: toluene/ethanol (2/1, v/v), 80 °C, eight hours; [C]: toluene, 90 °C, eight hours. Source: Adapted from Takeichi et al. [30]; Alhassan et al. [37]; Demir et al. [38]; Lin et al. [39]; Takeichi et al. [40]; Agag et al. [41]. (b) Synthesis of benzoxazine precursors and the corresponding PBAs via the Mannich reaction of more sustainable bisphenol-F isomers and selected other inputs. Source: Adapted from Demir et al. [38]; Liu et al. [42].

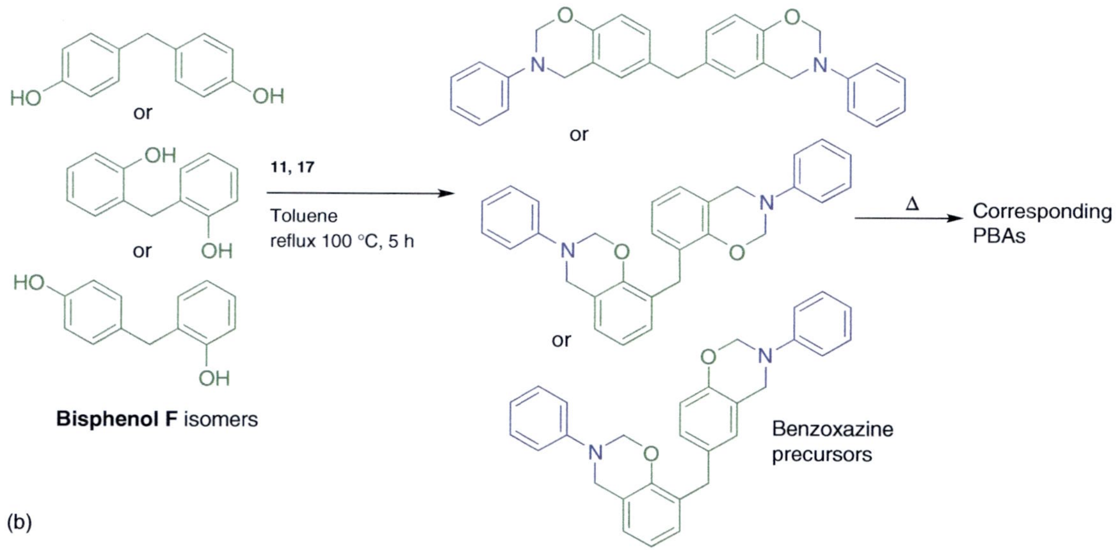

Figure 2.8 (*Continued*)

Table 2.7 Synthesis of PBA precursors **18** and subsequent ring-opening polymerizations toward **19** [30, 39–41].

Polymer	Monomer/PBA precursor	Con-ditions	Yield (%)	M_n (g/mol)	M_w (g/mol)	DP[d)]	T_g (E″) (°C)	T_g (tan δ) (°C)	$T_{d,5\%}$ (°C)	$T_{d,10\%}$ (°C)
18a	**15 + 2h1 + 17**[a)]	**A**	83	2232[b)]	6733[b)]	3	—	—	—	—
19a	**18a**	**Δ**	—	—	—	—	228[e)]	238[e)]	341[h)]	375[h)]
18b	**15 + 2a2 + 17**[a)]	**A**	77	2571[b)]	6092[b)]	5	—	—	—	—
19b	**18b**	**Δ**	—	—	—	—	249[e)]	259[e)]	282[h)]	301[h)]
18c	**15 + 2a6 + 17**[a)]	**A**	87	2358[b)]	8858[b)]	4	—	—	—	—
19c	**18c**	**Δ**	—	—	—	—	247[e)]	260[e)]	301[h)]	338[h)]
18d	**15 + 2j1 + 17**[a)]	**A**	75	3000[b)]	5700[b)]	4	—	—	—	—
19d	**18d**	**Δ**	—	—	—	—	225[e)]	238[e)]	324[h)]	364[h)]
18e	**15 + 2j2 + 17**[a)]	**A**	75	4200[b)]	9800[b)]	2	—	—	—	—
19e	**18e**	**Δ**	—	—	—	—	—	140[e)]	348[h)]	378[h)]
18f	**15 + 2j3 + 17**[a)]	**A**	79	4200[b)]	8900[b)]	1	—	—	—	—
19f	**18f**	**Δ**	—	—	—	—	—	—	384[h)]	426[h)]
18g	**15 + 2h1/2j2 + 17**[a)]	**A**	78	3500[b)]	10 000[b)]	—	—	—	—	—
19g	**18g**	**Δ**	—	—	—	—	279[e)]	283[e)]	333[h)]	370[h)]
18h	**15 + 2h1/2j3 + 17**[a)]	**A**	83	3600[b)]	11 000[b)]	—	—	—	—	—
19h	**18h**	**Δ**	—	—	—	—	272[e)]	276[e)]	344[h)]	380[h)]
18i	**15 + 2k1 + 17**[a)]	**C**	—	—	—	—	—	—	—	—
19i	**18i**		—	—	—	—	-19[f)]	—	285[i)]	303[i)]
18j	**15 + 2k2 + 17**[a)]	**C**	—	—	—	—	—	—	—	—
19j	**18j**		—	—	—	—	-40[f)]	—	324[i)]	344[i)]
18k	**15 + 2k3 + 17**[a)]	**C**	—	—	—	—	—	—	—	—
19k	**18k**						-45/0[f)]	—	350[i)]	368[i)]
18l	**15 + 2g + 17**[a)]	**B**	77	4200[c)]	25 000[c)]	10	—	—	—	
19l	**18l**	**Δ**	—	—	—	—		354[g)]	382[j)]	—
18m	**15 + 2h1 + 17**[a)]	**B**	80	3800[c)]	23 000[c)]	8	—	—	—	
19m	**18m**	**Δ**	—	—	—	—		303[g)]	374[j)]	—
18n	**15 + 2h2 + 17**[a)]	**B**	79	3400[c)]	32 000[c)]	7	—	—	—	
19n	**18n**	**Δ**	—	—	—	—		303[g)]	369[j)]	—
18o	**15 + 2i + 17**[a)]	**B**	78	3300[c)]	10 000[c)]	4	—	—	—	
19o	**18o**	**Δ**	—	—	—	—		272[g)]	407[j)]	—

a) [**15**] : [**2**] : [**17**] = 1 : 1 : 4.
b) Measured by SEC with polystyrene standards.
c) Measured by GPC.
d) Calculated using the M_n.
e) Measured by dynamic viscoelastic measurements (4 °C/min).
f) Measured by dynamic mechanical analysis (DMA) (3 °C/min).
g) Measured by DMA (5 °C/min).
h) Measured by TGA (5 °C/min).
i) Measured by TGA (10 °C/min).
j) Measured by TGA (20 °C/min).
Conditions: [A]: $CHCl_3$, reflux, five hours; [B]: toluene/ethanol (2/1, v/v), 80 °C, eight hours; [C]: toluene, 90 °C, eight hours.

and coworkers reported that in fluorinated PBAs (examples not shown), the T_g and the thermal stability significantly increase [46, 47].

The rigidity of the polymer chains is an important shortcoming of PBAs and results often in poor solubility of the materials. When, however, monomers are used that create more flexible polymer chains, this results in better solubility characteristics. In addition, the health and environmental concerns of bisphenol A (**14**) must be considered. Bisphenol F isomers are commercially available and inexpensive without severe health and environmental drawbacks. As a result, they are attractive alternatives. Ishida and coworkers reported the synthesis of benzoxazines based on bisphenol F isomers via a Mannich process (Figure 2.8b). Thermal ring-opening polymerization of these precursors gave the corresponding PBAs. These PBAs show excellent mechanical and physical properties and are easily processed and therefore can be used in resin transfer molding for high-performance applications [42].

Lin et al. investigated the synthesis of related but slightly different PBA precursors with an ether unit under different conditions. They prepared **18l–o** via the Mannich reaction of bisphenol-A **14**, diamines **2**, and paraformaldehyde **17** (Figure 2.8a and Table 2.7), respectively [39]. Subsequent thermal ring-opening polymerization gave the corresponding PBAs **19l–o** [39]. Next to these reports, Ishida and coworkers also studied these MCPs [48]. Looking at the characteristics, PBA **19l** has a relatively high T_g due to the rather low molecular weight of the crosslinks. On the other hand, PBA **19o** has a low T_g due to the high molecular weight of the crosslinks. In general, PBAs from precursors with a high molecular weight have optimal T_g and thermal stability. The PBAs **19l–o** are thus quite flexible and their high T_gs makes them suitable for electronic applications and high T_g coatings.

An alternative approach that produces PBAs with a high crosslink density employs high molecular weight precursors prepared from amine-containing phenols. In continuation of their research, Agag and Takeichi reported the synthesis of PBA precursor **21a–c** via the Mannich reaction of *p*-aminophenol **20a** and/or tyramine **20b** and **17** (Figure 2.9 and Table 2.8). Thermal ring-opening polymerization of **21a–c** gave PBAs with excellent thermomechanical characteristics and high thermal stability [29].

Besides the use of aliphatic and aromatic amines, Raj et al. investigated the use of a cyclic amine in the synthesis of PBA. They prepared **23a–d** (Figure 2.10 and Table 2.9) via the MCR of phenols **10a–c** or 1,5-dihydroxy naphthalene **10d**, piperazine **22**, and formaldehyde **6b** [49]. Epoxies are commonly used polymers for reinforced composites in load-bearing and aerospace materials in various applications [49]. Blending and curing of epoxy resins achieves high quality and performance materials. For example, Mannich bases efficiently cure epoxy resins and afford polymers with increased flexibility and improved mechanical characteristics [49, 50]. Epoxy resin modified by blending with PBAs **23** thus gave a polymer with better mechanical properties and higher thermal stability than epoxy resins cured with diamines. These polymers are used as matrix glass fiber laminates.

It is noteworthy that the M_ns of PBAs **23c** and **23d** are higher than the other PBAs due to the higher reactivity of resorcinol **10c** and 1,5-dihydroxy naphthalene **10d**.

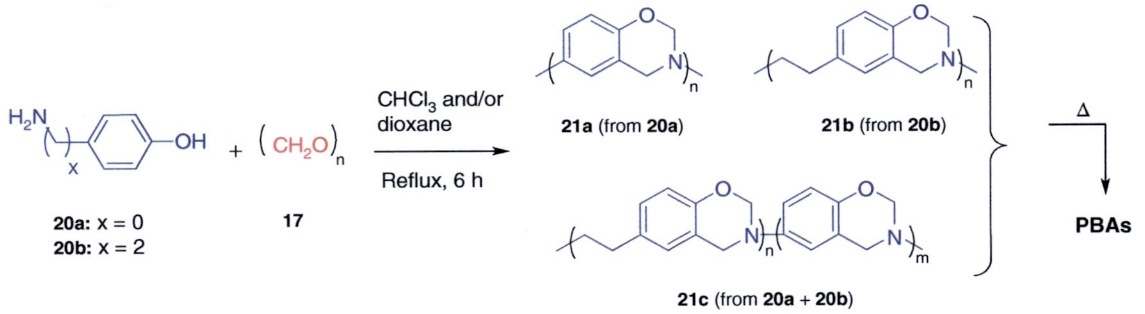

Figure 2.9 Synthesis of PBA precursor **21a-c** via the Mannich reaction of **20a** and/or **20b** and **17** and subsequent thermal ring-opening polymerization to PBAs. Source: Adapted from Agag and Takeichi [29]; Alhassan et al. [37].

Table 2.8 Synthesis of PBA precursor **21a–c** and ring-opening polymerization to the corresponding PBA.

Polymer	Monomer/PBA precursor	Yield (%)	M_n (g/mol)[c]	M_w (g/mol)[c]	DP[d]	T_g (E'') (°C)[e]	T_g (tan δ) (°C)[e]	$T_{d,5\%}$ (°C)[f]	$T_{d,10\%}$ (°C)[f]
21a	**20a** + **17**[a]	72	1953	3296	14	—	—	—	—
PBA	**21a**	—	—	—	—	301	312	355	401
21b	**20b** + **17**[a]	75	1957	4479	11	—	—	—	—
PBA	**21b**	—	—	—	—	261	291	315	363
21c	**20a** + **20b** + **17**[b]	68	1033	1374	3	—	—	—	—
PBA	**21c**	—	—	—	—	—	—	353	387

a) [**20a or 20b**] : [**17**] = 1:2.
b) [**20a**] : [**20b**] : [**17**] = 1 : 1 : 2.
c) Measured by SEC.
d) Calculated using the M_n.
e) Measured by dynamic viscoelastic measurements (4 °C/min).
f) Measured by TGA (5 °C/min).
Source: Agag and Takeichi [29].

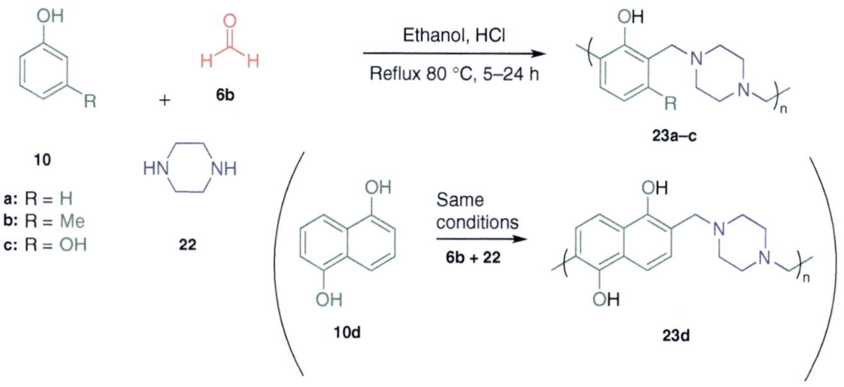

Figure 2.10 Synthesis of PBA **23** via the Mannich reaction of phenol **10**, piperazine **22**, and formaldehyde **6b**. Source: Raj et al. [49].

The numerous examples of the MCP involving a Mannich three-component reaction followed by thermal ring-opening polymerization show that this is a very interesting approach to produce quite diversely functionalized PBAs. It must be pointed out that in general, the DPs for these benzoxazine polymers are actually very low; however, for the material that is produced, it is not important because afterward when the ring-opening of the benzoxazine takes place, a crosslinking network is formed.

Table 2.9 Synthesis of PBAs **23**.

Polymer	Monomer[a)]	M_n (g/mol)[b)]	DP[c)]	$T_{d,0\%}$ (°C)[d)]	$T_{d,10\%}$ (°C)[d)]
23a	**10a** + **22** + **6b**	1021	4	23	275
23b	**10b** + **22** + **6b**	1528	6	28	238
23c	**10c** + **22** + **6b**	1540	6	28	190
23d	**10d** + **22** + **6b**	1620	5	28	55

a) [**10**] : [**22**] : [**6b**] = 1 : 1 : 2.
b) Measured by nonaqueous conductometric titration.
c) Calculated using the M_n.
d) Measured by TGA (10 °C/min).
Source: Raj et al. [49].

2.4 Biginelli-type Multicomponent Polymerizations

The Biginelli reaction is one of the most well-known MCRs and combines aldehydes, β-dicarbonyl compounds, and urea to form 3,4-dihydropyrimidin-2(1*H*)-ones (DHPMs) (Table 2.1). The interest in this reaction has developed due to the therapeutic and pharmacological properties of the products. Furthermore, several DHPMs have biological activities, such as antiviral, antitumor, antibacterial, anti-inflammatory, antimalaria, antitubercular, antidiabetic, and antiepileptic [51].

However, relevant to this review is that DHPMs have been used in the development of functional materials, such as polymers, adhesives, and dyes [51]. Sometimes re-examining an old reaction for new purposes brings benefits, so research groups have reinvestigated successful MCRs from the past and have found that the Biginelli reaction has almost all the "clickable" features reported in the well-known "click chemistry" [52]. Through the modification of polymer chains (side or chain end groups), and the combination with living radical polymerization in a one-pot strategy, functional homo- and copolymers can be prepared. The synthesis of polymers via the Biginelli reaction was introduced in selected fields of polymer chemistry and chemical biology by Tao and coworkers [52].

In addition to these Biginelli MCPs, direct polymerizations can also be performed. Although these processes can formally not be regarded as an MCP, we report them here, as they are quite relevant for the discussions in this section. The synthesis of polydihydropyrimidin-2(1*H*)-one (poly(DHPM)) **26a** via a direct Biginelli-type reaction of monomer **24a** and urea **25a** (Figure 2.11 and Table 2.10) is a nice example of this [53].

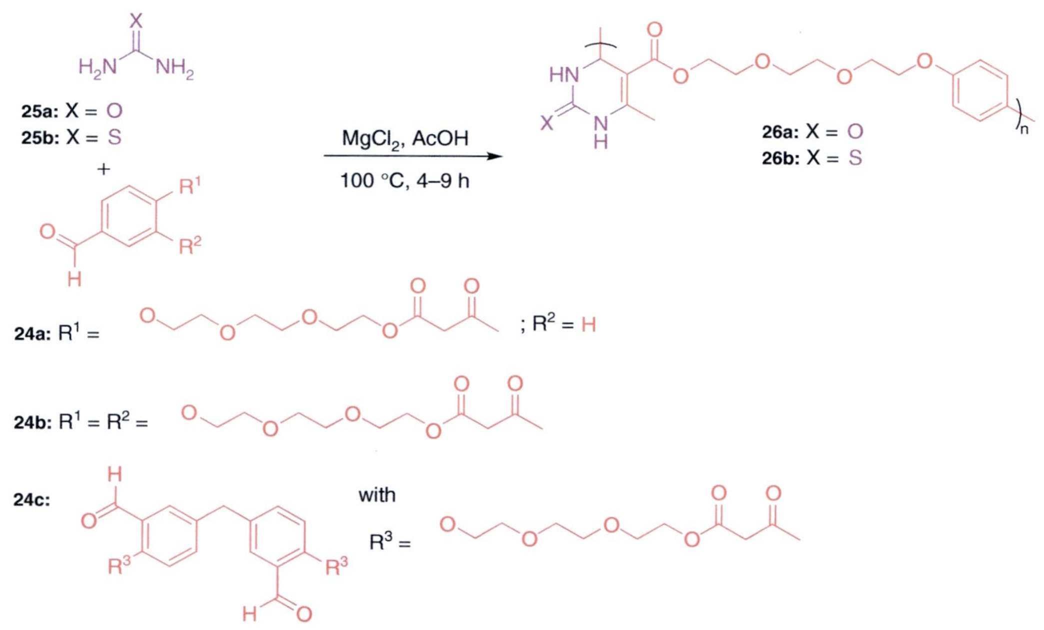

Figure 2.11 Synthesis of poly(DHPM(T)) **26** via the Biginelli reaction of **24** and **25**. Source: Adapted from Nagarajaiah et al. [51]; Zhao et al. [53]; Tunca et al. [56].

Table 2.10 Synthesis of poly(DHPM(T)) **26**.

Polymer	Monomer[a)]	Yield (%)	M_n (g/mol)	$Đ$[b)]	DP	T_g (°C)[e)]	T_d (°C)[f)]	$T_{d,2}$ (°C)[f)]
26a	**24a + 25a**	81	22 300[b)] and 30 400[c)]	1.59	61[d)]and 84[c)]	96.5	300	—
26b	**24a + 25b**	—	23 100[b)] and 36 700[c)]	1.45	60[d)]and 97[c)]	98.5	243	516

a) [**24**] : [**25**] = 1 : 2.
b) Measured by GPC with narrow molecular weight distribution (200–10^6 g/mol) polystyrene standards.
c) Measured by NMR.
d) Calculated using the M_n measured by GPC.
e) Measured by DSC (10 °C/min).
f) Measured by TGA.
Source: Adapted from Zhao et al. [53]; Zhao et al. [55].

Poly(DHPM) **26a** has a specific interaction with metals and can therefore be used as a metal adhesive [51, 53, 54]. The best bonding effects were observed with brass. Besides monomer **24a**, Tao and coworkers also used monomers **24b** and **24c** in the polymerization between two brass sheets [53]. These monomers (not shown) contain more functional groups and generate a higher molecular weight and stronger polymer structure due to crosslinked networks. The poly(DHPMs) **26a** derived from monomers **24b** and **24c** show stronger adhesive effects than those from monomer **24a**. This confirms that an increase in molecular weight and crosslink degree has a positive effect on the bonding strength [53]. Due to the ease of polymerization and excellent metal bonding abilities of poly(DHPM) **26a** and its analogs, these materials have great potential for future industrial applications.

New multifunctional polymers can be prepared via postpolymerization modification (PPM) by modifying the reactive chain ends or side chains of the polymer precursor. However, PPM of poly(DHPMs) is not possible due to the unreactive urea group in the Biginelli ring. By using the thiourea group, a reactive polymer precursor for PPM can be produced. In continuation of their research, Tao and coworkers investigated the synthesis of poly(dihydropyrimidin)-2(1*H*)-thione) (poly(DHPMT)) **26b** via the Biginelli reaction of monomer **24a** and thiourea **25b** (Figure 2.11 and Table 2.10) [55]. Due to its nucleophilicity, the thiourea group in the Biginelli ring can react with different haloalkene or haloalkynes to incorporate alkene or alkyne groups into the polymer [51, 54–56]. These alkene or alkyne groups can undergo a thiol–ene reaction or Cu(I)-catalyzed azide-alkyne cycloaddition (CuAAC) click reaction, respectively. Moreover, by reacting 3-phenylpopargyl chloride directly with poly(DHPMT) **26b**, a fluorescent polymer with a new conjugated heterocycle can be produced [54, 55].

In recent years, organic chemists try to develop more sustainable synthetic methodology by minimizing the production of waste, promoting the use of renewable resources, and achieving the highest energy efficiency as possible. By using renewable resources, a sustainable and eco-friendly Biginelli reaction can

Table 2.11 Synthesis of poly(DHMPs) **30**.

Polymer	Monomer	Yield (%)	M_n (g/mol)	M_w (g/mol)[c]	$Đ$[c]	DP	T_g (°C)[g]	$T_{d,5\%}$ (°C)[h]
29a	**27a** + **28a** + **25a**[a]	37.7	11 500[c] and 10 000[d]	27 600	2.39	27[e] and 23[f]	—	250
29b	**27a** + **28b** + **25a**[a]	30.6	14 600[c] and 20 500[d]	26 000	1.78	31[e] and 43[f]	193	280
29c	**27a** + **28c** + **25a**[a]	86.2	14 900[c] and 14 800[d]	38 500	2.57	30[e] and 30[f]	165	250
29d	**27a** + **28d** + **25a**[b]	89.3	34 600[d]	—	—	52[f]	141	255
29e	**27b** + **28a** + **25a**[a]	25.3	3400[c] and 14 600[d]	7000	2.04	5[e] and 24[f]	—	260
29f	**27b** + **28b** + **25a**[a]	27.4	7700[c] and 14 500[d]	14 500	1.89	12[e] and 22[f]	203	265

a) [**27**] : [**28**] : [**25a**] = 1 : 1 : 3.5.
b) [**27a**] : [**28d**] : [**25a**] = 1 : 1 : 2.2.
c) Measured by GPC with linear poly(methyl methacrylate) standards.
d) Measured by NMR.
e) Calculated using the M_n measured by GPC.
f) Calculated using the M_n measured by NMR.
g) Measured by DSC (15 °C/min).
h) Measured by TGA–DTA (5 K/min).
Source: Boukis et al. [57].

be performed. Meier and coworkers reported the synthesis of several poly(3,4-dihydropyrimidin-2(1*H*)-ones (poly(DHMPs)) **29** via a MCP Biginelli reaction of renewable dialdehydes **27**, diacetoacetates **28**, and urea **25a** (Figure 2.12 and Table 2.11) [57].

In this respect, it is noteworthy that diacetoacetates **28** are synthesized from a renewable diol produced from glucose [59] and castor oil [60]. In addition, terephthalaldehyde **27a** can be synthesized via the oxidation of *p*-cymene (from citrus peels [61]) or via the oxidation of *p*-xylene [62, 63]. Divanillin **27b** can be synthesized via the enzymatic dimerization of vanillin (from lignin [64]) [57].

Poly(DHMPs) **29e** and **29f** from divanillin **27b** have lower M_ns and DPs than poly (DHMP)s **29a–d** derived from terephthalaldehyde **27a** and therefore, divanillin **27b** is less reactive toward the MCP Biginelli reaction than terephthalaldehyde **27a**. The influence of the structure of dialdehydes **27** and the chain length of diacetoacetates **28** on the T_gs of poly(DHMPs) **29** was investigated in more detail. The authors found a decrease in the T_g values with an increasing aliphatic chain length introduced via diacetoacetate [57]. However, the T_g values increase if divanillin **27b** is used instead of terephthalaldehyde **27a** [57]. Furthermore, the T_gs of poly(DHMPs) are higher than the T_gs of common commercial thermoplastic polymers [15], which is rare for renewable polymers [57].

Combinatorial synthesis is a technique that rapidly prepares a large number of compounds by a combination of various reactants. Although originally developed as a tool for rapid production of compound libraries for high throughput screening

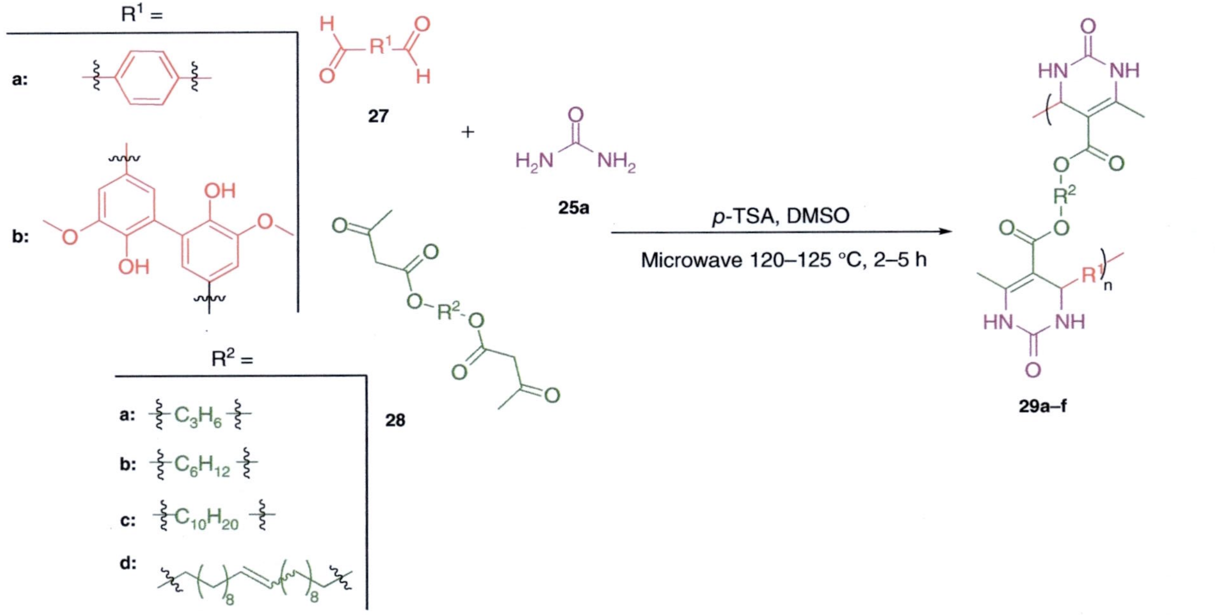

Figure 2.12 Synthesis of poly(DHMPs) **29** via the MCP Biginelli reaction of **27**, **28**, and **25a**. Source: Adapted from Nagarajaiah et al. [51]; Zhao et al. [54]. Boukis et al. [57]; Krieg and Manecke [58].

Table 2.12 Synthesis of poly(DHPM(T)) **30**.

Polymer	Monomer[a)]	M_n (g/mol)[b)]	$Đ$[b)]	DP[c)]	T_g (°C)[d)]
30a	**31a** + **32** + **25a**	7400–11 700	1.50–1.91	11–15	123.4–159.3
30b	**31b** + **32** + **25a**	9800–13 300	1.52–1.68	14–17	95.6–138.4
30c	**31c** + **32** + **25a**	13 700–24 800	1.62–1.99	18–31	76.5–109.3
30d	**31d** + **32** + **25a**	12 100–16 800	1.51–1.81	13–20	64.3–98.3
30e	**31e** + **32** + **25a**	17 500–29 800	1.73–2.03	20–32	46.2–66.6
30f	**31a** + **32** + **25b**	14 200–21 200	1.45–1.77	20–28	105.4–158.3
30g	**31b** + **32** + **25b**	15 200–21 500	1.42–1.54	20–27	89.7–117.8
30h	**31c** + **32** + **25b**	17 900–25 400	1.49–1.72	22–30	74.5–101.6
30i	**31d** + **32** + **25b**	15 200–18 200	1.47–1.57	16–21	67.1–95.1
30j	**31e** + **32** + **25b**	17 500–25 000	1.61–1.95	20–26	56.5–83.4

a) [**31**] : [**32**] : [**25**] = 1 : 1 : 3.
b) Measured by GPC with narrow molecular weight distribution (200–10^6 g/mol) polystyrene standards.
c) Calculated using the M_n.
d) Measured by DSC (10 °C/min).
Source: Xue et al. [65].

(HTP) in the pharmaceutical industry, it has been introduced in polymer chemistry to develop HTP polymerization systems. This prepares polymers in a time- and labor-saving manner. By combining HTP polymerization with MCP (HTP–MCP), a new combinatorial system that improves the efficiency of the overall synthesis can be achieved. Tao and coworkers reported the synthesis of a vast range of 80 different structurally distinct poly(DHPM(T)s) **30** via the Biginelli HTP–MCP of dialdehydes **31**, diacetoacetates **32**, and (thio)urea **25** (Figure 2.13 and Table 2.12) [65].

The Biginelli reaction is so efficient that in company with a click reaction, it can yield diverse organic scaffolds. We have summarized some of the current applications of the Biginelli reaction in polymer chemistry, including polymer coupling, post polymer modification, and new functional polymer synthesis. It is most likely that this ancient reaction will continue to draw attention from polymer chemists and play new roles in the polymer science. We anticipate that the current research paves a straightforward way to synthesize new libraries of polymers via MCRs and might prompt the broader study of these MCRs in MCPs for applications in a range of interdisciplinary fields.

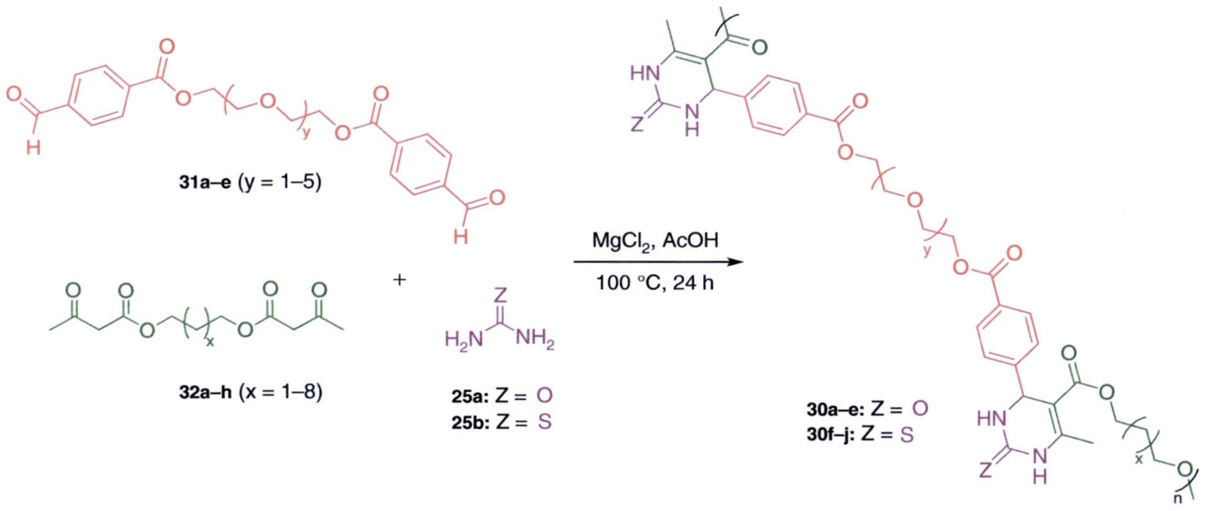

Figure 2.13 Synthesis poly(DHPM(T)) **30** via the Biginelli HTP–MCP of **31**, **32**, and **25**. Source: Xue et al. [65].

2.5 Hantzsch-type Multicomponent Polymerizations

The Hantzsch reaction, discovered in 1881, involves the reaction between aldehydes, β-ketoesters, and ammonia to produce 1,4-dihydropyridines (1,4-DHPs) (Table 2.1). Some 1,4-DHPs have been used as medicines for the treatment of cardiovascular diseases, and some have promising bioactivities, such as anti-inflammatory, antimicrobial, antioxidant, and antiulcer activities. However, until recently, the Hantzsch reaction has not been applied in polymer chemistry. Examples of this reaction in polymer chemistry are therefore scarce [66].

The Hantzsch reaction is suitable for direct polymerization due to its high efficiency and easy availability of the monomers [66]. Recently, Tao and coworkers reported the synthesis of poly(1,4-dihydropyridine) (poly(1,4-DHP)) **35a** via an MCP Hantzsch reaction of monomer **24a**, dimedone **34**, and ammonium acetate (NH_4OAc, **33**) (Figure 2.14 and Table 2.13) [67].

The Biginelli and the Hantzsch reaction use similar reactants (aldehydes and β-ketoester) and can take place under the same reaction conditions. This makes them compatible synthetic sequences. This was nicely demonstrated via a coupled Biginelli and Hantzsch MCP of monomer **24a**, (thio)urea **25a/b**, dimedone **34**, and NH_4OAc (Figure 2.14 and Table 2.13) [67]. This process delivers poly(1,4-dihydropyridine)-*co*-poly(3,4-dihydropyrimidin-2(1*H*)-(thi)one)-(poly(1,4-DHP)-*co*-poly(3,4-DHPM(T))) **35b.** Most MCRs are concentration-dependent and therefore the slower Biginelli reaction cannot proceed as efficiently as the faster Hantzsch reaction. As a result, less 3,4-dihydropyrimidin-2(1*H*)-one(3,4-DHPM) was introduced in the polymer, which gave a 1,4-DHP and 3,4-DHPM ratio of 1 : 0.83 (theoretical value 1 : 1), respectively. Tao and coworkers extended the scope by using thiourea **25b** instead of urea **25a**. Thiourea is less reactive than urea and therefore the Biginelli reaction is slowed even further. The ratio of 1,4-DHP and 3,4-DHPM in the polymer is 1 : 0.54, respectively [67].

The thermal properties of the polymers **35b** could be tuned by changing the monomer ratios giving materials with different ratios of 1,4-DHP and 3,4-DHPM(T) [51, 66, 67]. Furthermore, when urea **25a** is used as monomer, the T_g and T_d of **26** and **35b(O)** are higher than **35a** due to the rigid Biginelli ring [51, 67]. When thiourea **25b** is used as monomer, the T_gs of **27** and **35b(S)** are higher than **35a** [67]. However, the T_ds of **27** and **35b(S)** are lower than **35a** due to the less stable thiourea group in the Biginelli ring [67].

Copolymerization permits the synthesis of an almost unlimited range of polymers and is therefore often used to obtain a better balance of properties for the commercial application of polymeric materials. Copolymerization offers flexibility with respect to possible monomers and a great diversity of polymeric products, but also composition drift, which may need to be minimized or exploited in a controlled way, for specific structure–property relationships can also be addressed.

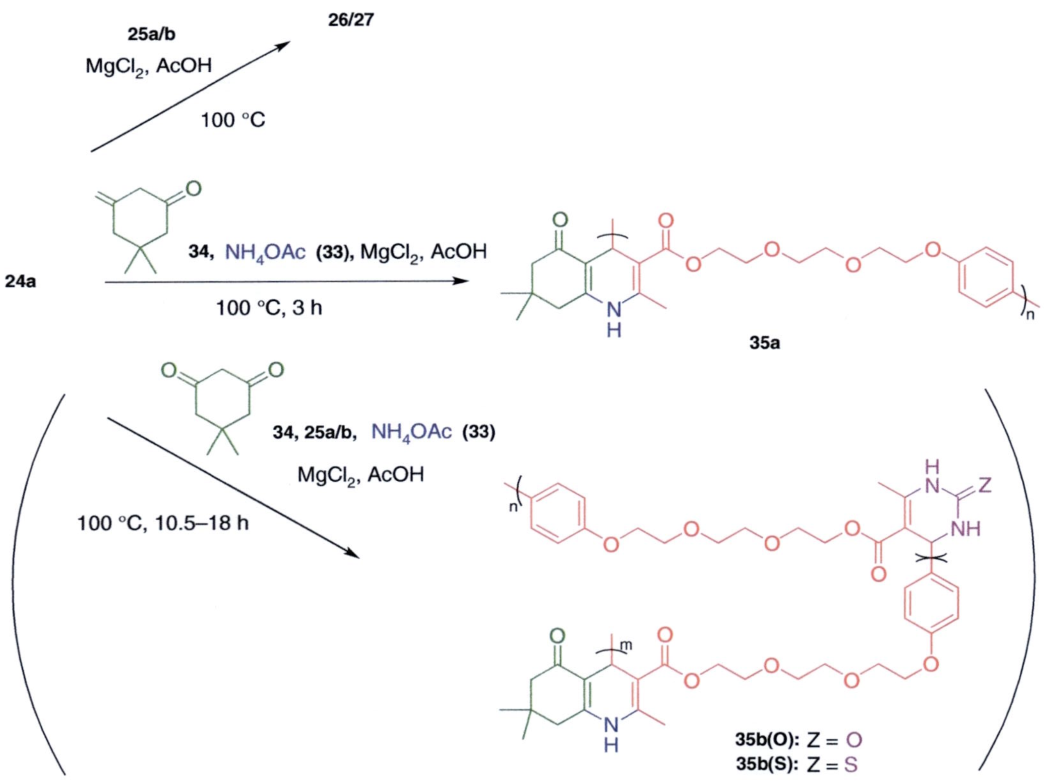

Figure 2.14 Synthesis of poly(3,4-DHMP(T)) **26/27** via the Biginelli reaction of **24a** and **25a/b** [67], poly(1,4-DHP) **35a** via the Hantzsch reaction of **24a**, **34**, and NH_4OAc (**33**) [58, 66, 67], or a coupled Hantzsch-Biginelli MCP of **24a**, **34**, **25a/b**, and NH_4OAc (**33**). Source: Adapted from Nagarajaiah et al. [51]; Wu et al. [66]; Wu et al. [67];Krieg and Manecke [58].

Table 2.13 Synthesis of poly(3,4-DHMP(T)) **26/27**, poly(1,4-DHP) **35a**, and poly(1,4-DHP)-*co*-poly(3,4-DHMP(T)) **35b**.

Polymer	Monomer	Yield (%)	M_n (g/mol)	$Đ$[d]	DP	T_g (°C)[g]	T_d (°C)[h]
26	**24a** + **25a**[a]	—	21 400[d]	1.36	58[f] and 102[e]	102	320
27	**24a** + **25b**[a]	—	28 800[d]	1.49	75[f] and 124[e]	101	260
35a	**24a** + **34** + **33**[b]	91	17 300[d] and 30 200[e]	1.52	38[f] and 64[e]	78	280
35b(O)	**24a** + **25a** + **34** + **33**[c]	87	13 900[d] and 25 000[e]	1.39	17[f] and 63[e]	103	320
35b(S)	**24a** + **25b** + **34** + **33**[c]	—	16 800[d] and 50 400[e]	1.46	20[f] and 132[e]	102	270

a) [**24a**] : [**25a** or **25b**] = 1 : 1.2.
b) [**24a**] : [**34**] : [**33**] = 1 : 1 : 1.
c) [**24a**] : [**34**] : [**25a** or **25b**] : [**33**] = 2 : 1 : 1 : 1.
d) Measured by GPC with narrow molecular weight distribution (200–10^6 g/mol) polystyrene standards.
e) Measured by NMR.
f) Calculated using the M_n measured by GPC.
g) Measured by DSC (10 °C/min).
h) Measured by TGA (20 °C/min).
Source: Wu et al. [67].

2.6 Debus–Radziszewski-type Multicomponent Polymerizations

The Debus–Radziszewski is another important MCR discovered in the nineteenth century that produces a hetereocyclic scaffold. In this reaction, an aldehyde, two amines, and an α-dicarbonyl compound (i.e. where carbonyl is either ketone or aldehyde) react to generate imidazole/imidazolium rings (Table 2.1). The reaction conditions depend on the nature of the reagents. When employing glyoxal as the α-dicarbonyl compound reactions proceed under mild conditions; however, when using 1,2-diketones, higher temperatures are needed to complete the reaction. One of the first examples of the Debus–Radziszewski reaction in polymer chemistry was reported by Krieg and Manecke in 1967 [58]. Interested by the properties of imidazole heterocycles in polymers, the authors performed this polycondensation with tetraketones **36**, dialdehydes **27**, and ammonium acetate **33** (Figure 2.15). Low molecular weights were obtained when mixing the reagents at room temperature (Table 2.14). Marestin and coworkers found that this reaction is much more efficient upon MW irradiation at 500 W for 15 minutes [68]. The authors later also showed that anilines could be added to generate N1-aryl imidazoles [69].

Instead of extending the polymer via the monomers with two α-dicarbonyls and two aldehydes, another option is to use a combination of diamines and dialdehydes. This, however, will not provide the imidazole ring, but rather gives a charged

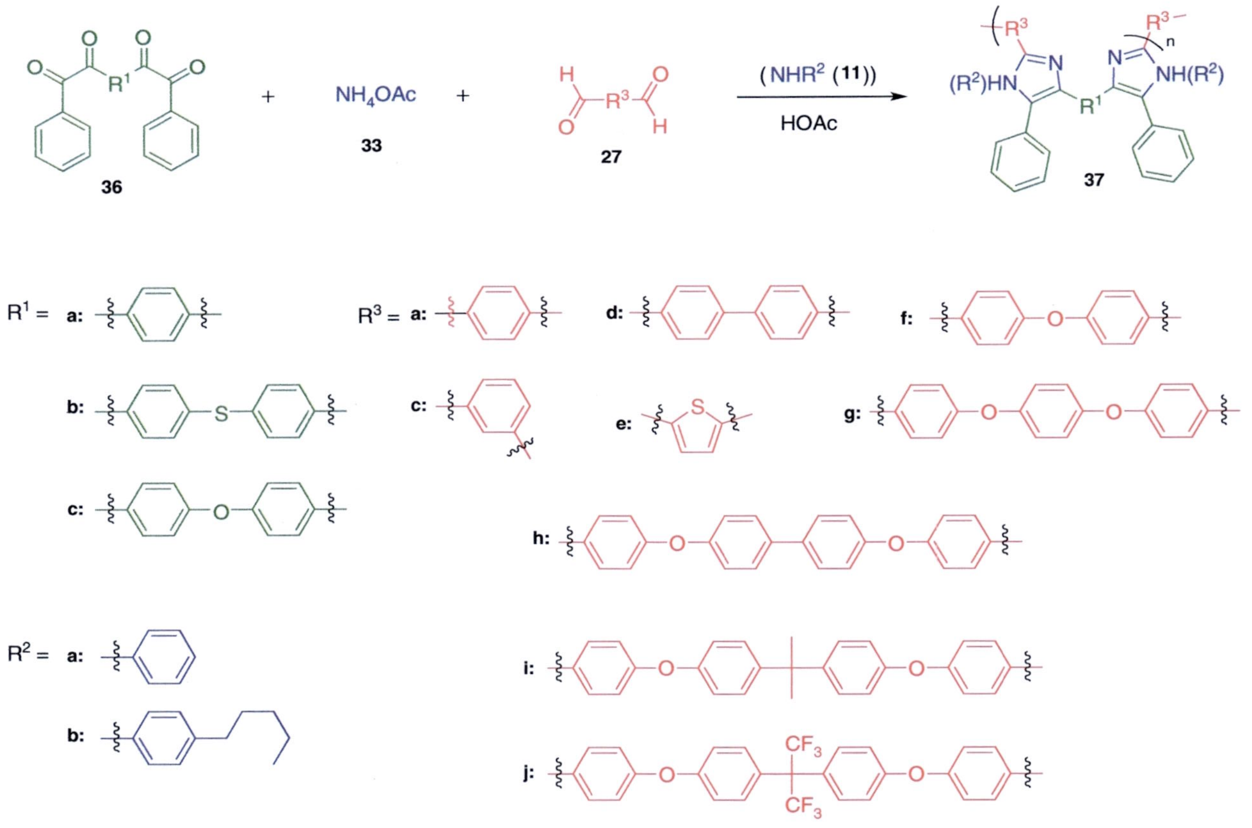

Figure 2.15 Synthesis of polyimidazoles **37** via the Debus–Radziszewski reaction of **36**, **27**, and NH_4OAc (**33**). Also four-component MCPs were demonstrated by adding an additional amine component (**11**). Source: Adapted from Chauveau et al. [68]; Chauveau et al. [69].

Table 2.14 Synthesis of polyimidzaoles **37**.

Polymer	Monomer	Conditions	M_n (g/mol)	Đ	DP[d]	T_g (°C)[e]
37a	**36a + 27a + 33**	rt	~2500[a]	—	5	—
37b	**36a + 27c + 33**	rt	~2500[a]	—	5	—
37c	**36a + 27d + 33**	rt	<3000[a]	—	—	—
37d	**36a + 27e + 33**	rt	<3000[a]	—	—	—
37e	**36b + 27a + 33**	MW (500 W)	—[b]	—	—	295
37f	**36b + 27f + 33**	MW (500 W)	—[b]	—	—	278
37g	**36b + 27g + 33**	MW (500 W)	—[b]	—	—	258
37h	**36b + 27h + 33**	MW (500 W)	—[b]	—	—	228
37i	**36b + 27i + 33**	MW (500 W)	16 700[c]	2.4	18	237
37j	**36b + 27j + 33**	MW (500 W)	9600[c]	1.8	9	239
37k	**36c + 27g + 33 + 11a**	MW (500 W)	31 000[c]	2.9	35	176
37l	**36c + 27f + 33 + 11a**	MW (500 W)	19 300[c]	3.4	25	189
37m	**36a + 27f + 33 + 11a**	MW (500 W)	25 500[c]	3.2	37	199
37n	**36c + 27g + 33 + 11b**	MW (500 W)	21 600[c]	2.6	23	217
37o	**36a + 27h + 33 + 11b**	MW (500 W)	—[b]	—	—	263
37p	**36c + 27f + 33 + 11b**	MW (500 W)	17 900[c]	2.7	20	225
37q	**36b + 27g + 33 + 11a**	MW (500 W)	19 000[c]	2.9	21	156
37r	**36b + 27f + 33 + 11a**	MW (500 W)	25 500[c]	2.6	32	181

a) Estimated by IR CO stretch.
b) Polymers were insoluble for in solvents used for SEC.
c) Measured by SEC with polystyrene standards.
d) Calculated using the M_n.
e) Measured by DSC (5 °C/min).
Source: Chauveau et al. [68]; Chauveau et al. [69].

imidazolium heterocycle. Antonietti and coworkers were the first to apply this approach as a strategy to crosslink amine side groups in polymers (Figure 2.16 and Table 2.15) [70].

Lindner noticed the potential of the Debus–Radziszewski in the synthesis of imidazolium-based polymers as polymeric ionic liquids [71]. In his report, several bisamines react with formaldehyde and glyoxal. The reaction tolerates several bisamines ranging from hydrophobic to zwitterionic backbones. Yuan and coworkers later showed that this polymerization was also compatible with methylglyoxal (**38b**) [72]. Polymeric materials were synthesized from several alkyl (**2a**) variations and **2d**; however, the authors did not report M_n values. By introducing magnetic nanoparticles coated with amine groups, Lucena and Cárdenas and coworkers were able to synthesize a nanocomposite which showed interesting sorptive properties toward salicylic acid, ketoprofen, and naproxen [73]. Also dendrimeric structures have been made by employing tetraamines [74]. These materials show great potential in capturing CO_2.

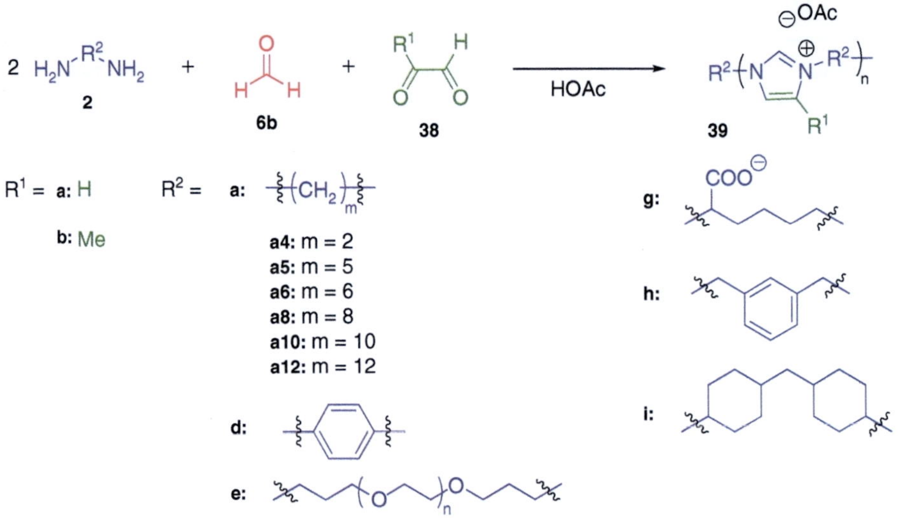

Figure 2.16 Synthesis of polyimidazoliniums **39** via the Debus–Radziszewski reaction of **2**, **6b**, and **38**. Source: Adopted from Krannig et al. [70]; Lindner et al. [71]; Grygiel et al. [72].

Table 2.15 Synthesis of polyimidzaoles **39**.

Polymer	Monomer	M_w (g/mol)[a)]	M_n (g/mol)[a)]	$Đ$	DP[b)]	T_g (°C)[d)]
39a	**2a4 + 38a + 6b**	2700–53 900	1500–9600	1.7–5.8	9–61	—
39b	**2a6 + 38a + 6b**	24 600	7200	3.4	47	−45.3
39c	**2e + 38a + 6b**	20 400	7500	2.7	—	−51.7
39d	**2g + 38a + 6b**	11 000	5400	1.9	29	48.3
39e	**2h + 38a + 6b**	26 000	8125	3.2	33	−45.3
39f	**2a12 + 38a + 6b**	70 000	28 000	2.5	94	−51.7
39g	**2i + 38a + 6b**	19 000	10 000	1.9	31	45.3
39h	**2a5 + 38b + 6b**	24 000	—	—	115[c)]	—

a) Measured by SEC with polystyrene standards.
b) Calculated using the M_n.
c) Calculated by using the M_w.
d) Measured by DSC (5 °C/min).

2.7 Other Multicomponent Polymerizations

In this section, a number of additional reactions will be considered introducing the use of metal-catalyzed MCRs for polymerization. Such reactions accelerate the development of efficient and economical one-pot MCPs toward functional polymeric materials.

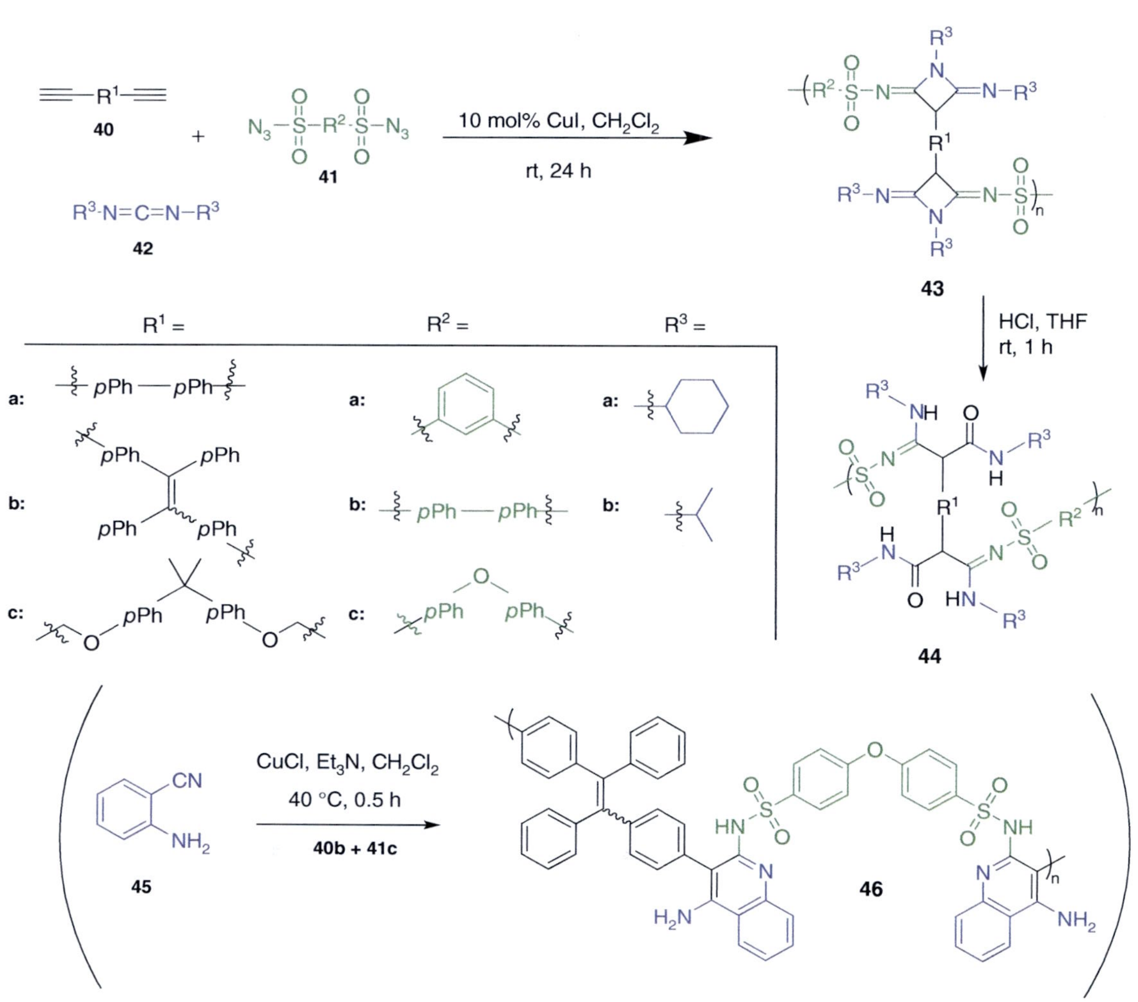

Figure 2.17 Cu-catalyzed MCP-synthesis of polymers **43** (or **46**) with **40**, **41**, and **42** (or **45**) followed by ring-opening to **44**. Source: Adapted from Zhang et al. [77]; Han et al. [76a]; Xu et al. [75].

2.7.1 The Cu(I)-catalyzed MCP of Diynes, Azides, and Carbodiimides/Nitriles

Tang and coworkers [76a] (based on the approach of Xu et al. [76b]) reported the Cu-catalyzed MCP synthesis of polymers **43** by reacting diynes**40**, disulfonyl azides **41**, and carbodiimides **42** (Figure 2.17 and Table 2.16). An additional base is not required due to the weak basicity of carbodiimides **42** [76a].

The 2,4-diiminoazetidine rings in **43** can undergo ring-opening with hydrogen chloride (HCl) to deliver polymers **44** with amide and amidine groups [76a]. Tang and coworkers [75] (based on the approach of Zhang and coworkers [78]) extended the scope of this versatile MCP by the Cu-catalyzed synthesis of quinoline-containing poly(*N*-sulfonylimine) **46** by reacting 1,2-bis(4-ethynylphenyl)-1,2-diphenylethene **40b**, 4,4′-disulfonylazidophenyl ether **41c**, and 2-aminobenzonitrile **45** [75].

2.7.2 The Pd-catalyzed MCP of Imines, Acyl Chlorides, and *N*-Sulfonyl Imines

Palladium can be used as catalyst in MCP reactions as was shown nicely by Arndtsen and coworkers who reported the Pd-catalyzed synthesis of polymer **50** by reacting imines **47**, di(acyl chlorides) **48**, and di(*N*-sulfonyl imines) **49** (Figure 2.18 and Table 2.17) [79].

Polymers **50** show different fluorescence emissions caused by π conjugation. Polymer **50a** has a blue emission, while polymers **50c**, **50e**, and **50h** have a red-shift in the

Table 2.16 Synthesis of the selected polymers **39**, **40**, and **42**.

Polymer	Monomer	Yield (%)	M_n (g/mol)	M_w (g/mol)	$Đ$	DP[e)]	T_g (°C)[f)]	$T_{d,5\%}$ (°C)[g)]
43a	**40b** + **41c** + **42a**[a)]	81.3	8100[c)]	21 600[c)]	2.7[c)]	7	—	272
43b	**40b** + **41c** + **42b**[a)]	83.5	11 000[c)]	74 500[c)]	6.8[c)]	11	—	254
43c	**40b** + **41b** + **42a**[a)]	47.9	6100[c)]	9000[c)]	1.5[c)]	5	—	279
43d	**40b** + **41a** + **42a**[a)]	72.4	7900[c)]	13 800[c)]	1.7[c)]	7	182	280
43e	**40a** + **41c** + **42a**[a)]	76.4	5000[c)]	10 200[c)]	2.0[c)]	5	—	277
43f	**40c** + **41c** + **42a**[a)]	78.9	9400[c)]	29 200[c)]	3.1[c)]	9	111	256
44a	**43a**	—	7600[c)]	12 100[c)]	1.6[c)]	6	192	272
44b	**43e**	—	5000[c)]	8100[c)]	1.6[c)]	5	—	230
44c	**43f**	—	6000[c)]	9700[c)]	1.6[c)]	5	96	199
46	**40b** + **41c** + **45**[b)]	82	15 399[d)]	32 800[d)]	2.13[d)]	16	—	314

a) [**40**] : [**41**] : [**42**] = 1 : 1 : 2.4.
b) [**40b**] : [**41c**] : [**45**] = 1 : 1 : 2.5.
c) Measured by GPC with monodisperse polystyrene standards.
d) measured by GPC with monodisperse poly(methyl methacrylate) standards.
e) Calculated using the M_n.
f) Measured by DSC (10 °C/min).
g) Measured by TGA (10 °C/min).
Source: Adapted from Han et al. [76a]; Xu et al. [75].

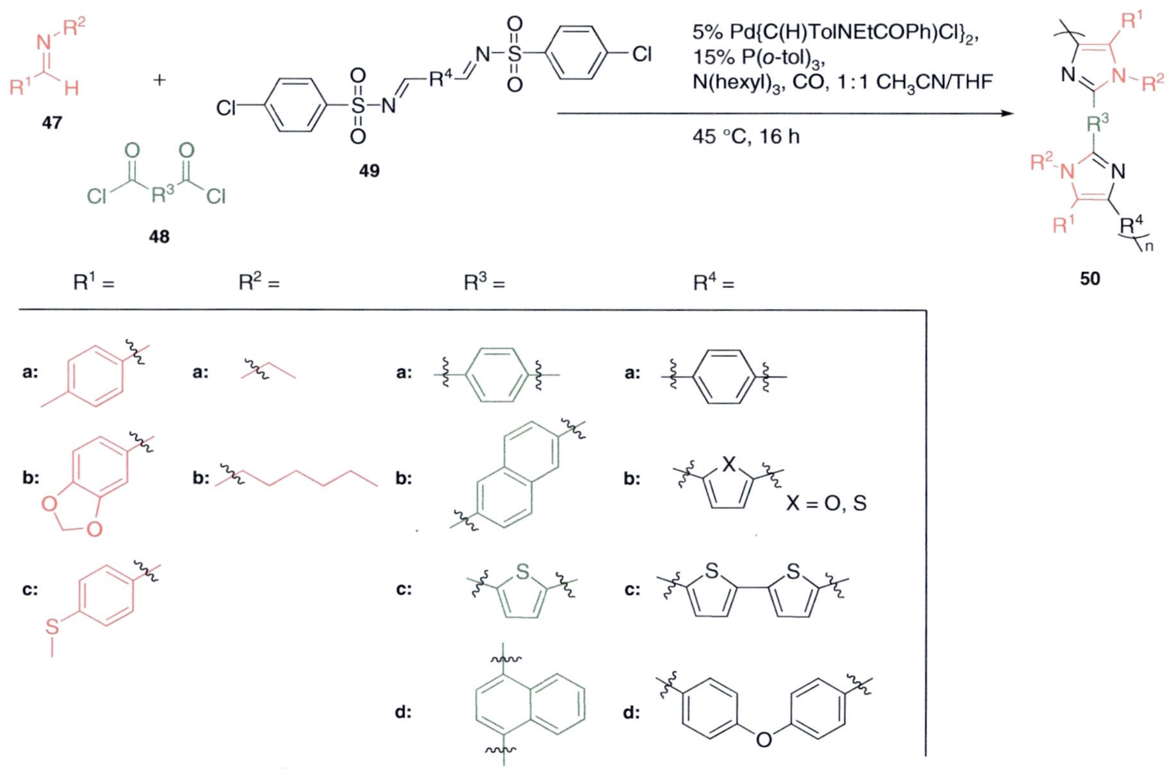

Figure 2.18 Pd-catalyzed synthesis of **50** reacting **47**, **48**, and **49**. Source: Based on Siamaki et al. [79].

Table 2.17 Synthesis of selected polymers **50**.

Polymer	Monomer[a]	Yield (%)	M_n (g/mol)[b]	M_w (g/mol)[b]	Đ[b]	DP[c]
50a	**47aa + 48a + 49a**	72	2680	3110	1.16	4
50b	**47ab + 48a + 49a**	70	4370	5410	1.24	6
50c	**47ab + 48a + 49b(O)**	61	3930	4450	1.13	6
50d	**47ab + 48b + 49d**	65	4150	4830	1.16	5
50e	**47ab + 48c + 49b(S)**	54	4630	5470	1.18	6
50f	**47ab + 48d + 49a**	46	3950	4670	1.18	5
50g	**47bb + 48b + 49a**	68	4430	5150	1.16	5
50h	**47cb + 48a + 49c**	74	3790	4350	1.15	4

a) [**47**] : [**48**] : [**49**] = 2 : 1 : 1.
b) Measured by GPC with narrow molecular weight distribution polystyrene standards.
c) Calculated using the M_n.
Source: Siamaki et al. [79].

Table 2.18 Synthesis of polymer **53**.

Polymer	Monomer	Yield (%)	M_n (g/mol)[a]	M_w (g/mol)[a]	Đ[a]	DP[b]
53a	**51a + 52b + 49a**	64	3840	4450	1.16	7
53b	**51b + 52a + 49a**	72	4120	4840	1.17	6

a) Measured by GPC with narrow molecular weight distribution polystyrene standards.
b) Calculated using the M_n.
Source: Siamaki et al. [79].

fluorescence emission due to the presence of different heterocycles. This represents an alternative approach to access imidazole-containing π-conjugated materials and a family of new conjugated materials [79].

Arndtsen and coworkers extended the scope of this MCP by the synthesis of polymers with different backbones via the combination of different monofunctional and bifunctional monomers. They reported the Pd-catalyzed synthesis of polymer **53** by reacting diimines **51**, acyl chlorides **52**, and 1,4-*N*-benzylidene-bis(4-chlorobenzenesulfonamide) **49a** (Figure 2.19 and Table 2.18) [79]. It should be noted that both polymers **50** and **53** have low DPs. Therefore, they are not polymers but oligomers.

2.7.3 The Mercaptoacetic Acid Locking Imine Reaction

Copolymer synthesis via traditional chemical reactions to link two different polymer chains is often slow and inefficient, usually leading to partially modified products because of the intrinsic steric hindrance of the polymer chain. Click reactions are the key solution to this traditional problem and two different polymer chains can be efficiently stitched together to generate the target copolymers.

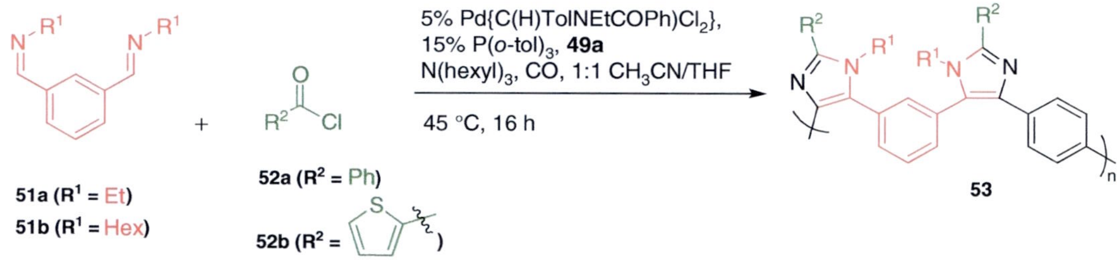

Figure 2.19 Pd-catalyzed synthesis of **53** reacting **51**, **52**, and **49a**. Source: Based on Siamaki et al. [79].

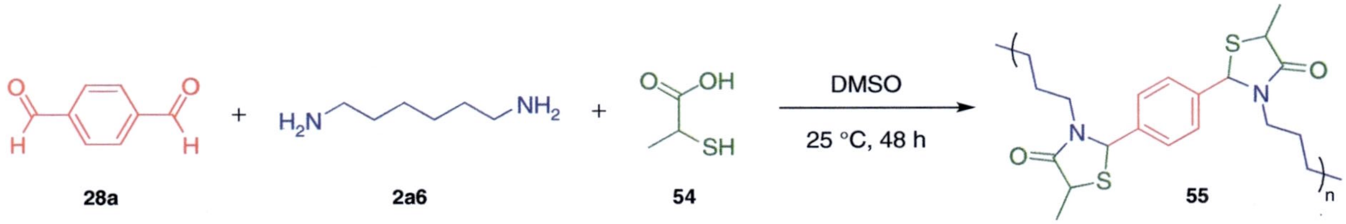

Figure 2.20 Synthesis of polymer **55** via the MALI reaction of **28a**, **2a6**, and **54**. Source: Zhao et al. [80].

Table 2.19 Synthesis of polymer **55**.

Polymer	Monomer[a)]	M_n (g/mol)[b)]	$Đ$[b)]	DP[c)]
55	**28a** + **2a6** + **54**	18 800	1.49	48

a) [**28a**] : [**2a6**] : [**54**] = 1 : 1 : 4.
b) Measured by GPC with narrow molecular weight distribution (200–10^6 g/mol) polystyrene standards.
c) Calculated using the M_n.
Source: Zhao et al. [80].

Some MCRs behave like two-component click reactions: they are modular, efficient, with high atom economy, generate water as only byproduct and can be performed under environmentally friendly reaction conditions, particularly at room temperature and without a catalyst. In this respect, the mercaptoacetic acid locking imine (MALI) reaction is interesting. This MCR combines a (di)aldehyde, a (di)amine, and mercaptoacetic acid to prepare 4-thiazolidiones (Figure 2.20 and Table 2.19), which have anti-human immunodeficiency (HIV), antituberculotic, and anticonvulsant features [80].

This MALI reaction is a modular, efficient, and a high atom economic MCR which behaves like other well-known two-component click reactions such as thiol–ene and CuAAC reactions. Of particular interest, the MALI reaction needs only common starting compounds (aldehyde, amine, and mercaptoacetic acid) and can be carried out under benign reaction conditions. Hence, the MALI reaction can therefore be considered as a green click reaction. The applications of this new click reaction in other fields such as material surface modification, protein conjugation, and synthesis of new polymers can now be accessed [80].

2.8 Conclusions and Outlook

In this chapter, an overview of different MCRs used in polymerization processes was discussed. We focused on those MCPs that involved heterocyclic fragments or produced polymeric materials that contained heterocyclic moieties. Main physical properties like the number and weight average molecular weight (M_n and M_w, respectively), the polydispersity ($Đ$), the glass transition temperature (T_g), the thermal degradation temperature (T_d), and the DP were calculated from $M_n$values that were available in the reports and served to make a comparative analysis. Especially, the Ugi-, Mannich-, Biginelli-, and Hantzsch MCRs have proved valuable in MCP synthesis of a wide range of different polymers with heterocycles. In addition, some more specialized reactions were discussed. The MCPs covered all deliver products with high levels of structural complexity and they can be used satisfactorily to access diversely substituted sets of small focused libraries of heterocyclic polymers. Key to a successful MCP is to monitor the DP. The reported DPs are often based on the M_n data that from the literature. Depending on how the M_n was measured (GPC or NMR) we have, therefore, taken caution with definite conclusions. Sometimes, these values are remarkably different as in the case of polymer **35b(S)** (20 vs. 132). We believe that especially in these cases M_n values determined by NMR end groups are more reliable. Analyzing the DPs of the polymers prepared via MCPs, it is remarkable that these cascades of elementary reaction steps are able to proceed such efficiencies (up to DPs of 115). Still, some MCPs like the Mannich-type PBA polymers generally display low efficiencies and it would be interesting to see if these reaction types can be improved. In general, this field can be considered to be still in its infancy and we envision that in the (near) future, many more interesting and useful MCPs will be discovered to produce (also in a scalable fashion) heterocyclic polymers that can serve as functional building blocks for advanced materials of the future.

References

1. Brinson, H.F. and Brinson, L.C. (2008). Characteristics, applications and properties of polymers. In: *Polymer Engineering Science and Viscoelasticity*, 55–97. Springer US https://doi.org/10.1007/978-0-387-73861-1_3.
2. Kakuchi, R. (2015). Metal-catalyzed multicomponent reactions for the synthesis of polymers. *Adv. Polym. Sci.* 269: 1–16. https://doi.org/10.1007/12_2014_300.
3. (a) Li, J.J. (2014). *Name Reactions*. Springer International Publishing https://doi.org/10.1007/978-3-319-03979-4; (b) For the original papers by Debus and Radzisewski see:Debus, H. (1858). Ueber die Einwirkung des Ammoniaks auf Glyoxal. *Justus Liebigs Ann. Chem.* 107 (2): 199–208. https://doi.org/10.1002/jlac.18581070209. (c) Radzisewski, B. (1882). Ueber Glyoxalin und seine Homologe. *Ber. Dtsch. Chem. Ges.* 15 (2): 2706–2708. https://doi.org/10.1002/cber.188201502245.
4. Passerini, M.S.G. (1921). Isonitrili (I). Composto Del p-Isonitrilazobenzolo Con Acetone Ed Acido Acetico. *Gazz. Chim. Ital.* 51 (2): 126–129.
5. Ugi, I. (1962). The α-addition of immonium ions and anions to isonitriles accompanied by secondary reactions. *Angew. Chem. Int. Ed. Eng.* 1 (1): 8–21. https://doi.org/10.1002/anie.196200081.
6. Mannich, C. and Krösche, W. (1912). Ueber Ein Kondensationsprodukt Aus Formaldehyd, Ammoniak Und Antipyrin. *Arch. Pharm.* 250 (1): 647–667. https://doi.org/10.1002/ardp.19122500151.
7. Biginelli, P. (1891). Ueber Aldehyduramide Des Acetessigäthers. *Ber. Dtsch. Chem. Ges.* 24 (1): 1317–1319. https://doi.org/10.1002/cber.189102401228.
8. Wei, C., Li, Z., and Li, C.J. (2004). The development of A3-coupling (aldehyde-alkyne-amine) and AA3-coupling (asymmetric aldehyde-alkyne-amine). *Synlett* 2004 (9): 1472–1483. https://doi.org/10.1055/s-2004-829531.
9. Fields, E.K. (1952). The synthesis of esters of substituted amino phosphonic acids. *J. Am. Chem. Soc.* 74 (6): 1528–1531. https://doi.org/10.1021/ja01126a054.
10. Hantzsch, A. (1881). Condensationsprodukte Aus Aldehydammoniak Und Ketonartigen Verbindungen. *Ber. Dtsch. Chem. Ges.* 14 (2): 1637–1638. https://doi.org/10.1002/cber.18810140214.

11 Espeel, P., Goethals, F., and Du Prez, F.E. (2011). One-pot multistep reactions based on thiolactones: extending the realm of thiol-ene chemistry in polymer synthesis. *J. Am. Chem. Soc.* 133 (6): 1678–1681. https://doi.org/10.1021/ja1098098.

12 Espeel, P., Goethals, F., Driessen, F. et al. (2013). One-pot, additive-free preparation of functionalized polyurethanes via amine-thiol-ene conjugation. *Polym. Chem.* 4 (8): 2449–2456. https://doi.org/10.1039/c3py00004d.

13 Bae, I., Han, H., and Chang, S. (2005). Highly efficient one-pot synthesis of N-sulfonylamidines by Cu-catalyzed three-component coupling of sulfonyl azide, alkyne, and amine. *J. Am. Chem. Soc.* 127 (7): 2038–2039. https://doi.org/10.1021/ja0432968.

14 Hu, R., Li, W., and Tang, B.Z. (2016). Recent advances in alkyne-based multicomponent polymerizations. *Macromol. Chem. Phys.* 217 (2): 213–224. https://doi.org/10.1002/macp.201500291.

15 van der Vegt, A. K.; Govaert, L. E. (2005) *Polymeren: Van Keten Tot Kunststof*; VSSD.

16 Shrivastava, A. (2018). Polymerization. In: *Introduction to Plastics Engineering*, 17–48. Elsevier https://doi.org/10.1016/B978-0-323-39500-7.00002-2.

17 Tomoda, B.T., Yassue-Cordeiro, P.H., Ernesto, J.V. et al. (2020). Characterization of biopolymer membranes and films: physicochemical, mechanical, barrier, and biological properties. In: *Biopolymer Membranes and Films*, 67–95. Elsevier https://doi.org/10.1016/b978-0-12-818134-8.00003-1.

18 Llevot, A., Boukis, A.C., Oelmann, S. et al. (2017). An update on isocyanide-based multicomponent reactions in polymer science. *Top. Curr. Chem.* 1: 1–29. https://doi.org/10.1007/s41061-017-0153-4.

19 Sehlinger, A. and Meier, M.A.R. (2015). Passerini and Ugi multicomponent reactions in polymer science. In: *Multi-Component and Sequential Reactions in Polymer Synthesis*, vol. 269 (ed. P. Theato), 61–86. Springer International Publishing https://doi.org/10.1007/12_2014_298.

20 Jiang, X., Feng, C., Lu, G., and Huang, X. (2015). Application of named reactions in polymer synthesis. *Sci. China Chem.* 58 (11): 1695–1709. https://doi.org/10.1007/s11426-015-5447-.

21 Hartweg, M. and Becer, C.R. (2016). Direct polymerization of levulinic acid via Ugi multicomponent reaction. *Green Chem.* 18 (11): 3272–3277. https://doi.org/10.1039/c6gc00372a.

22 Hartweg, M. and Becer, C.R. (2018). Levulinic acid as sustainable feedstock in polymer chemistry. *Polym. Chem.* 1310: 331–338. https://doi.org/10.1021/bk-2018-1310.ch020.

23 Rose, M. and Palkovits, R. (2011). Cellulose-based sustainable polymers: state of the art and future trends. *Macromol. Rapid Commun.* 32 (17): 1299–1311. https://doi.org/10.1002/marc.201100230.

24 Palkovits, R. (2010). Pentenoic acid pathways for cellulosic biofuels. *Angew. Chem. Int. Ed.* 49 (26): 4336–4338. https://doi.org/10.1002/anie.201002061.

25 Schade, O.R., Dannecker, P.K., Kalz, K.F. et al. (2019). Direct catalytic route to biomass-derived 2,5-furandicarboxylic acid and its use as monomer in a multicomponent polymerization. *ACS Omega* 4 (16): 16972–16979. https://doi.org/10.1021/acsomega.9b02373.

26 Wessjohann, L.A., Rivera, D.G., and Vercillo, O.E. (2009). Multiple multicomponent macrocyclizations (MiBs): a strategie development toward macrocycle diversity. *Chem. Rev.* 11: 796–814. https://doi.org/10.1021/cr8003407.

27 Liu, N., Fu, C., Zhang, Q. et al. (2020). Multifunctionalized polyamides prepared by facile Ugi reaction as thermosensitive, biocompatible and antibacterial biomaterials. *Chemistry Select* 5 (9): 2725–2734. https://doi.org/10.1002/slct.201904329.

28 Filho, J.F.A., Lemos, B.C., de Souza, A.S. et al. (2017). Multicomponent mannich reactions: general aspects, methodologies and applications. *Tetrahedron* 14: 6977–7004. https://doi.org/10.1016/j.tet.2017.10.063.

29 Agag, T. and Takeichi, T. (2007). High-molecular-weight AB-type benzoxazines as new precursors for high-performance thermosets. *J. Polym. Sci. Part A Polym. Chem.* 45 (10): 1878–1888. https://doi.org/10.1002/pola.21953.

30 Takeichi, T., Kano, T., and Agag, T. (2005). Synthesis and thermal cure of high molecular weight polybenzoxazine precursors and the properties of the thermosets. *Polymer (Guildf).* 46 (26): 12172–12180. https://doi.org/10.1016/j.polymer.2005.10.088.

31 Chirachanchai, S., Phongtamrug, S., Laobuthee, A., and Tashiro, K. (2011). Mono-substituted phenol-based benzoxazines: inevitable dimerization via self-termination and its metal complexation. In: *Handbook of Benzoxazine Resins*, 111–126. Elsevier https://doi.org/10.1016/B978-0-444-53790-4.00049-7.

32 Sun, J., Wei, W., Xu, Y. et al. (2015). A curing system of benzoxazine with amine: reactivity, reaction mechanism and material properties. *RSC Adv.* 5: 19048–19057. https://doi.org/10.1039/c4ra16582a.

33 Han, L., Salum, M.L., Zhang, K. et al. (2017). Intrinsic self-initiating thermal ring-opening polymerization of 1,3-benzoxazines without the influence of impurities using very high purity crystals. *J. Polym. Sci. Part A Polym. Chem.* 55 (20): 3434–3445. https://doi.org/10.1002/pola.28723.

34 Chernykh, A., Liu, J., and Ishida, H. (2006). Synthesis and properties of a new crosslinkable polymer containing benzoxazine moiety in the main chain. *Polymer* 47 (22): 7664–7669. https://doi.org/10.1016/j.polymer.2006.08.041.

35 Ning, X. and Ishida, H. (1994). Phenolic materials via ring-opening polymerization: synthesis and characterization of bisphenol-A based benzoxazines and their polymers. *J. Polym. Sci. Part A Polym. Chem.* 32 (6): 1121–1129. https://doi.org/10.1002/pola.1994.080320614.

36 Ishida, H. and Allen, D.J. (1996). Physical and mechanical characterization of near-zero shrinkage polybenzoxazines. *J. Polym. Sci. Part B Polym. Phys.* 34 (6): 1019–1030. https://doi.org/10.1002/(SICI)1099-0488(19960430)34:6<1019::AID-POLB1>3.0.CO;2-T.

37 Alhassan, S., Schiraldi, D., Qutubuddin, S. et al. (2011). Various approaches for main-chain type benzoxazine polymers. In: *Handbook of Benzoxazine Resins*, 309–318. Elsevier https://doi.org/10.1016/B978-0-444-53790-4.00060-6.

38 Demir, K.D., Kiskan, B., Aydogan, B., and Yagci, Y. (2013). Thermally curable main-chain benzoxazine prepolymers via polycondensation route. In: *Reactive* *and Functional Polymers*, vol. 73, 346–359. Elsevier https://doi.org/10.1016/j.reactfunctpolym.2012.04.016.

39 Lin, C.H., Chang, S.L., Shen, T.Y. et al. (2012). Flexible polybenzoxazine thermosets with high glass transition temperatures and low surface free energies. *Polym. Chem.* 3 (4): 935–945. https://doi.org/10.1039/c2py00449f.

40 Takeichi, T., Kano, T., Agag, T. et al. (2010). Preparation of high molecular weight polybenzoxazine prepolymers containing siloxane unites and properties of their thermosets. *J. Polym. Sci. Part A Polym. Chem.* 48 (24): 5945–5952. https://doi.org/10.1002/pola.24408.

41 Agag, T., Geiger, S., Alhassan, S.M. et al. (2010). Low-viscosity polyether-based main-chain benzoxazine polymers: precursors for flexible thermosetting polymers. *Macromolecules* 43 (17): 7122–7127. https://doi.org/10.1021/ma1014337.

42 Liu, J., Agag, T., and Ishida, H. (2010). Main-chain benzoxazine oligomers: a new approach for resin transfer moldable neat benzoxazines for high performance applications. *Polymer (Guildf).* 51 (24): 5688–5694. https://doi.org/10.1016/j.polymer.2010.08.059.

43 Wang, L. and Zheng, S. (2010). Morphology and thermomechanical properties of main-chain polybenzoxazine-block-polydimethylsiloxane multiblock copolymers. *Polymer (Guildf).* 51 (5): 1124–1132. https://doi.org/10.1016/j.polymer.2010.01.008.

44 Sawaryn, C., Landfester, K., and Taden, A. (2011). Benzoxazine miniemulsions stabilized with multifunctional main-chain benzoxazine protective colloids. *Macromolecules* 44 (14): 5650–5658. https://doi.org/10.1021/ma200973g.

45 Su, Y.C. and Chang, F.C. (2003). Synthesis and characterization of fluorinated polybenzoxazine material with low dielectric constant. *Polymer* 44 (26): 7989–7996. https://doi.org/10.1016/j.polymer.2003.10.026.

46 Velez-Herrera, P., Doyama, K., Abe, H., and Ishida, H. (2008). Synthesis and characterization of highly fluorinated polymer with the benzoxazine moiety in the main chain. *Macromolecules* 41 (24): 9704–9714. https://doi.org/10.1021/ma801253a.

47 Kanchanasopa, M., Yanumet, N., Hemvichian, K., and Ishida, H. (2001). The effect of polymerization conditions on the density and Tg of bisphenol-A and hexafluoroisopropylidene-containing polybenzoxazines. *Polym. Polym. Compos.* 9 (6): 367–376.

48 Liu, J., Agag, T., and Ishida, H. (2011). Main-chain type benzoxazine oligomers: a new concept for easy processable high performance polybenzoxazines. In: *Handbook of Benzoxazine Resins*, 355–362. Elsevier https://doi.org/10.1016/B978-0-444-53790-4.00063-1.

49 Raj, M.M., Raj, L.M., Shah, T.B., and Patel, P.M. (2010). Synthesis, characterization of mannich base oligomers used with epoxy resin for glass fibre-reinforced laminates. *J. Therm. Anal. Calorim.* 101 (3): 1003–1009. https://doi.org/10.1007/s10973-009-0591-5.

50 Lin, J., Lin, S., Kuo, T. et al. (2000). Synthesis and epoxy curing of mannich bases derived from bisphenol a and poly(oxyalkylene)diamine. *J. Appl. Polym.*

Sci. 78 (3): 615–623. https://doi.org/10.1002/1097-4628(20001017)78:3<615::AID-APP170>3.0.CO;2-N.

51 Nagarajaiah, H., Mukhopadhyay, A., and Moorthy, J.N. (2016). Biginelli reaction: an overview. *Tetrahedron Lett.* 23: 5135–5149. https://doi.org/10.1016/j.tetlet.2016.09.047.

52 Zhu, C., Yang, B., Zhao, Y. et al. (2013). A new insight into the biginelli reaction: the dawn of multicomponent click chemistry? *Polym. Chem.* 4: 5395–5400. https://doi.org/10.1039/c3py00553d.

53 Zhao, Y., Yu, Y., Zhang, Y. et al. (2015). From drug to adhesive: a new application of poly(dihydropyrimidin-2(1H)-one)s via the biginelli polycondensation. *Polym. Chem.* 6 (27): 4940–4945. https://doi.org/10.1039/c5py00684h.

54 Zhao, Y., Wu, H., Wang, Z. et al. (2016). Training the old dog new tricks: the applications of the biginelli reaction in polymer chemistry. *Sci. China Chem.* 59: 1541–1547. https://doi.org/10.1007/s11426-016-0219-4.

55 Zhao, Y., Wu, H., Zhang, Y. et al. (2015). Postpolymerization modification of poly(dihydropyrimidin-2(1H)-thione)s via the Thiourea–Haloalkane reaction to prepare functional polymers. *ACS Macro Lett.* 4 (8): 843–847. https://doi.org/10.1021/acsmacrolett.5b00428.

56 Tunca, U. (2018). Click and multicomponent reactions work together for polymer chemistry. *Macromol. Chem. Phys.* 219 (16): 1800163. https://doi.org/10.1002/macp.201800163.

57 Boukis, A.C., Llevot, A., and Meier, M.A.R. (2016). High glass transition temperature renewable polymers via biginelli multicomponent polymerization. *Macromol. Rapid Commun.* 37 (7): 643–649. https://doi.org/10.1002/marc.201500717.

58 Krieg, V.B. and Manecke, G. (1967). Synthese und Eigensehaften einiger Polyimidazole. *Makromol. Chem.* 108: 210–217. https://doi.org/10.1002/macp.1967.021080118.

59 Mendes, F.S., González-Pajuelo, M., Cordier, H. et al. (2011). 1,3-propanediol production in a two-step process fermentation from renewable feedstock. *Appl. Microbiol. Biotechnol.* 92 (3): 519–527. https://doi.org/10.1007/s00253-011-3369-1.

60 Mutlu, H. and Meier, M.A.R. (2010). Castor oil as a renewable resource for the chemical industry. *Eur. J. Lipid Sci. Technol.* 1: 10–30. https://doi.org/10.1002/ejlt.200900138.

61 Martín-Luengo, M.A., Yates, M., Martínez Domingo, M.J. et al. (2008). Synthesis of P-cymene from limonene, a renewable feedstock. *Appl. Catal. B Environ.* 81 (3–4): 218–224. https://doi.org/10.1016/j.apcatb.2007.12.003.

62 van Leeuwen, P.W.N.M. (2004). Oxidation with dioxygen. In: *Homogeneous Catalysis*, 319–336. Springer Netherlands https://doi.org/10.1007/1-4020-2000-7_15.

63 Simmons, K. E. (1977). Oxidation of P-Xylene to Terephthalaldehyde. US4017547A.

64 Krings, U., Esparan, V., and Berger, R.G. (2015). The taste enhancer divanillin: a review on sources and enzymatic generation. *Flavour Fragr. J.* 30 (5): 362–365. https://doi.org/10.1002/ffj.3251.

65 Xue, H., Zhao, Y., Wu, H. et al. (2016). Multicomponent combinatorial polymerization via the biginelli reaction. *J. Am. Chem. Soc.* 138 (28): 8690–8693. https://doi.org/10.1021/jacs.6b04425.

66 Wu, H., Wang, Z., and Tao, L. (2017). The hantzsch reaction in polymer chemistry: synthesis and tentative application. *Polym. Chem.* 21: 7290–7296. https://doi.org/10.1039/c7py01718a.

67 Wu, H., Fu, C., Zhao, Y. et al. (2015). Multicomponent copolycondensates via the simultaneous Hantzsch and Biginelli reactions. *ACS Macro Lett.* 4 (11): 1189–1193. https://doi.org/10.1021/acsmacrolett.5b00637.

68 Chauveau, E., Marestin, C., Martin, V., and Mercier, R. (2008). Microwave-assisted polymerization process: a way to design new, high molecular weight poly(arylimidazole)s. *Polymer* 49: 5209–5214. https://doi.org/10.1016/j.polymer.2008.09.042.

69 Chauveau, E., Marestin, C., and Mercier, R. (2014). Microwave-assisted synthesis of tetrasubstituted aryl imidazole based polymers via cascade polycondensation process. *Polymer* 55: 6435–6438. https://doi.org/10.1016/j.polymer.2014.10.042.

70 Krannig, K.-S., Esposito, D., and Antonietti, M. (2014). Highly efficient transfer of amino groups to imidazolium entities for polymer coupling and cross-linking. *Macromolecules* 47: 2350–2353. https://doi.org/10.1021/ma500269k.

71 Lindner, J.-P. (2016). Imidazolium-based polymers via the poly-Radziszewski reaction. *Macromolecules* 49: 2046–2053. https://doi.org/10.1021/acs.macromol.5b02417.

72 Grygiel, K., Kirchhecker, S., Gong, J. et al. (2017). Main-chain polyimidazolium polymers by one-pot synthesis and application as nitrogen-doped carbon precursors. *Macromol. Chem. Phys.* 218: 1600586. https://doi.org/10.1002/macp.201600586.

73 Castro-Grijalba, A., Reyes-Gallardo, E.M., Wuilloud, R.G. et al. (2017). Synthesis of magnetic polymeric ionic liquid nanocomposites by the Radziszewski reaction. *RSC Adv.* 7: 42979–42985. https://doi.org/10.1039/C7RA07947H.

74 Talapaneni, S.N., Buyukcakir, O., Je, S.H. et al. (2015). Nanoporous polymers incorporating sterically confined N-heterocyclic carbenes for simultaneous CO_2 capture and conversion at ambient pressure. *Chem. Mater.* 27: 6818–6826. https://doi.org/10.1021/acs.chemmater.5b03104.

75 Xu, L., Zhou, T., Liao, M. et al. (2019). Multicomponent polymerizations of alkynes, sulfonyl azides, and 2-hydroxybenzonitrile/2-aminobenzonitrile toward multifunctional iminocoumarin/quinoline-containing poly(N-sulfonylimine)s. *ACS Macro Lett.* 8 (2): 101–106. https://doi.org/10.1021/acsmacrolett.8b00884.

76 (a) Han, T., Deng, H., Qiu, Z. et al. (2018). Facile multicomponent polymerizations toward unconventional luminescent polymers with readily openable small heterocycles. *J. Am. Chem. Soc.* 140 (16): 5588–5598. https://doi.org/10.1021/jacs.8b01991. (b) Xu, X., Cheng, D., Li, J. et al. (2007). Copper-catalyzed highly efficient multicomponent reactions: synthesis of 2-(sulfonylimino)-4-(alkylimino)azetidine derivatives. *Org. Lett.* 9 (8): 1585–1587. https://doi.org/10.1021/ol070485x.

77 Zhang, Z., You, Y., and Hong, C. (2018). Multicomponent reactions and multicomponent cascade reactions for the synthesis of sequence-controlled polymers. *Macromol. Rapid Commun.* 39 (23): 1800362. https://doi.org/10.1002/marc.201800362.

78 Yi, F., Zhang, S., Huang, Y. et al. (2017). An efficient one-pot protocol for the synthesis of polysubstituted 4-amino-iminocoumarins and 4-aminoquinolines by a copper-catalyzed three-component reaction. *Eur. J. Org. Chem.* 2017 (1): 102–110. https://doi.org/10.1002/ejoc.201601217.

79 Siamaki, A.R., Sakalauskas, M., and Arndtsen, B.A. (2011). A palladium-catalyzed multicomponent coupling approach to π-conjugated oligomers: assembling imidazole-based materials from imines and acyl chlorides. *Angew. Chem. Int. Ed.* 50 (29): 6552–6556. https://doi.org/10.1002/anie.201100558.

80 Zhao, Y., Yang, B., Zhu, C. et al. (2014). Introducing mercaptoacetic acid locking imine reaction into polymer chemistry as a green click reaction. *Polym. Chem.* 5 (8): 2695–2699. https://doi.org/10.1039/c4py00058g.

3

Multicomponent Reactions in Medicinal Chemistry

Zefeng Wang and Alexander Domling

University of Groningen, Groningen Research Institute Pharmacy, Department of Drug Design, Antonius Deusinglaan 1, Groningen, The Netherlands

3.1 Introduction

Multicomponent reactions (MCRs) are defined as reactions where more than two starting materials react to give a product where most of the atoms appear in the product. Many MCRs are name reactions of their inventors (Figure 3.1). Many general reviews were written on MCR chemistry [1–15]. Comprehensive and authoritative reviews are available for the name reactions: Ugi [4–6, 9, 16, 17], Passerini [18], Betti [19], Mannich, Barginelli [20], Biginelli [21], Strecker, Groebcke–Blackburn–Bienayme [10], Pauson–Khand [22–26], Petasis [27–29], Kabachnik–Fields [30, 31], Povarov–Grieco [32–35], Castagnolli–Cushmann [36, 37], Hantzsch [26, 38], Asinger [39], Gewald [13], and Doebner [40].

Moreover, many MCR reviews were written for specific areas, including photochemistry [41], ionic liquids [42], oxidation-based MCR [43], dearomatization-based MCR [44], copper-catalyzed [45], medicinal chemistry [46, 47], stereoselectivity in MCR [48], green aspects [8], polymers [49], ligation and bioconjugation [50], specific compound classes such as crop-protecting products [51], piperazine [52], benzodiazepine [53], Meldrum acid [54], tetrazole [9], γ-lactame [55], quinoxaline [56], peptides [57], and macrocycles [58], just to name a few.

MCRs feature several advantages over sequential multistep syntheses. The most important descriptor of MCR is its multicomponent character. It means that several different building blocks are reacted to a complex product in "one-pot" by simply mixing. MCRs typically use simple, mild, and robust conditions with no need of very low or high temperature, with no need of protecting atmosphere and often not affected by moisture. The tolerated electronic, steric features, and functional group compatibility is surprisingly high.

Complex target molecules such as drugs are assembled by a sequential multistep synthesis. In average, a drug molecule is synthesized by a 7–8-step synthesis. Due to the increasing complexity of recent drugs, the tendency is toward increasing number of steps. During the discovery phase, the so-called design-make-test-analyze cycle

Multicomponent Reactions towards Heterocycles: Concepts and Applications, First Edition.
Edited by Erik V. Van der Eycken and Upendra K. Sharma.

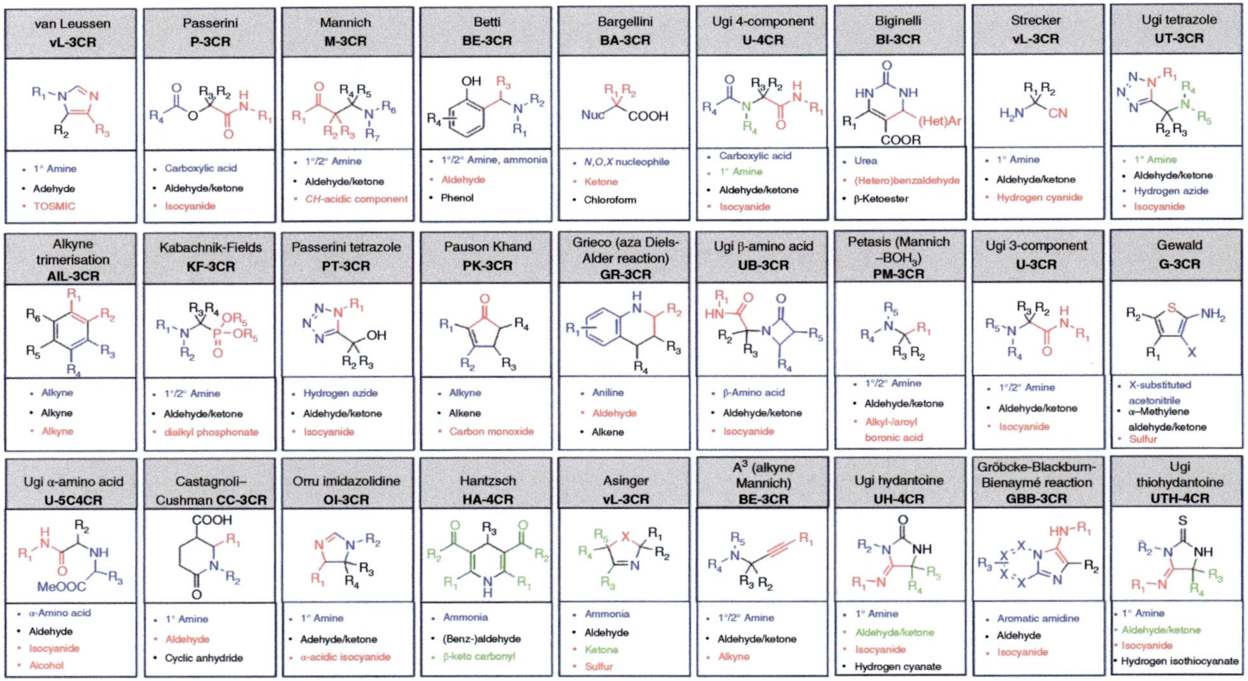

Figure 3.1 Major MCRs associated with their inventors (name reactions). The name, the MCR abbreviation, the general structure, and the atom origin from the starting materials are indicated.

(DMTA) is followed and hundreds or even thousands of derivatives have to be synthesized and evaluated [59]. Generally speaking, the more steps are used to reach the target; the more effort, time, and money have to be invested. MCRs are different from classical sequential syntheses, as they allow for the one-pot assembly of complex molecules. MCR can be compared to the assembly of complex structures via combinations of building blocks like Lego. Many different building blocks can be used to easily assemble all different figures, forms, and buildings. Thus, the quality of a MCR can be measured at first instance, by the number of available building blocks. Indeed, many MCRs use building blocks which are abundantly available. For example, the archetypical Ugi reaction combines carboxylic acids, primary amines, and aldehydes/ketones with isocyanides. All building blocks are commercially available as may different derivatives. Even so isocyanides are also commercially available in large number, they can also be easily synthesized from the primary amine or oxo-precursors [60–62]. Even mono-, di-, or trisaccharide isocyanides have been synthesized in high yields and stereochemically purity [61]. MCRs typically use simple, mild, and robust conditions with no need of very low or high temperature, with no need of protecting atmosphere and often not affected by moisture. The tolerated electronic, steric features, and functional group compatibility is surprisingly high.

Based on the underlaying combinatorial mathematical principles, very high numbers of products can be synthesized in MCRs of high order and general availability of building blocks. This was formulated by Ivar Ugi already in 1961: "reaction of 1.000 educts of each of the four groups of starting materials in the classical α-acylamino carbonamide scaffold yields theoretically 1×10^{12} products." Hence, MCR provides a very large chemical space to find compounds with customized properties.

Another important aspect of MCR is their sustainability. As reviewed recently, MCR obeys the principles of green chemistry [63]. For example, the principle of prevention is followed as target compounds can be often synthesized using much less steps. According to the definition of MCR, it can be hardly imagined that there are more atom economical reactions. Generally, MCR uses safe solvents and auxiliaries. MCR often can be performed in alcohols or even water [64]. Often MCR can be performed under protecting group free conditions. Renewable feedstock has been introduced in MCR.

While one could argue, the scaffold diversity of MCRs is not very high due to the low number of MCRs (Figure 3.1), the fact is that hundreds of highly variable scaffolds have been described based on the combination of an initial primary MCR, combined with a secondary reaction. MCR scaffolds are versatile synthetic hubs. Due to the excellent functional group compatibility, combined with the very large number of available building blocks, MCR chemistry can be easily combined with other reactions leading to a shear infinite scaffold diversity accessible in just a few steps (two to three). Hence, the combination of a few basic MCRs with other reactions provides a treasure trove for displaying substituents in 3D for optimal receptor interactions. A few examples are discussed to underscore this idea.

Additionally, MCR chemistry allows for many ways of late-stage functionalization [65].

Figure 3.2 Word cloud with the major attributes of MCR chemistry.

Due to the large number of publications, a rather personal, non-comprehensive essay discussing chances and opportunities for medicinal chemistry and related areas is given.

The building block principle of MCR allows for the fast and efficient assembly of molecules of almost all shape, substitution pattern, and physicochemical properties. Moreover, a particular scaffold can be presented in very high numbers of derivatives. High scaffold diversity and numbers of derivatives are ideally suited to optimize for properties. Scaffold diversity allows for rational scaffold hopping. Often, the 3D special and pharmacophoric occupancy of substituents responsible for binding to the receptor pocket can be represented by chemically very different scaffolds. The generally accepted advantages of MCR chemistry are summarized in the word cloud in Figure 3.2.

3.1.1 Example: Protein–Protein Interaction p53-MDM2

The protein–protein interaction (PPI) of the transcription factor p53 and its negative regulator MDM2 was an important target in oncology drug discovery over the past 15 years, and almost every pharma company with a cancer franchise developed antagonists for clinical trials [66–68]. In general, PPIs are difficult to target by small molecules based on their large and often featureless interfaces. Even so there are a few approved small-molecule drugs for PPI, they are largely the domain of biologics, such as monoclonal antibodies. Examples of such "undruggable" targets include receptor (Her, EGFR, PD1/PDL1, IL17) [69]. P53/MDM2 is such an exception which is indeed druggable by small molecules and is particularly suitable for MCR chemistry (Figure 3.3). In Figure 3.4, the cocrystal structure of the interface of the PPI p53/MDM2 epitope is shown [70]. The Phe19Trp23Lys26 triad of an amphiphatic α-helix of p53 is binding into the MDM2 receptor. The amino acid triad undergoes mostly hydrophobic interaction with the receptor, and the indole-H of Trp23 forms a hydrogen bond with p53-Lys54. Amongst the three amino acids, particularly Trp23 binds deeply into the MDM2 receptor and constitutes an anchor point. All small

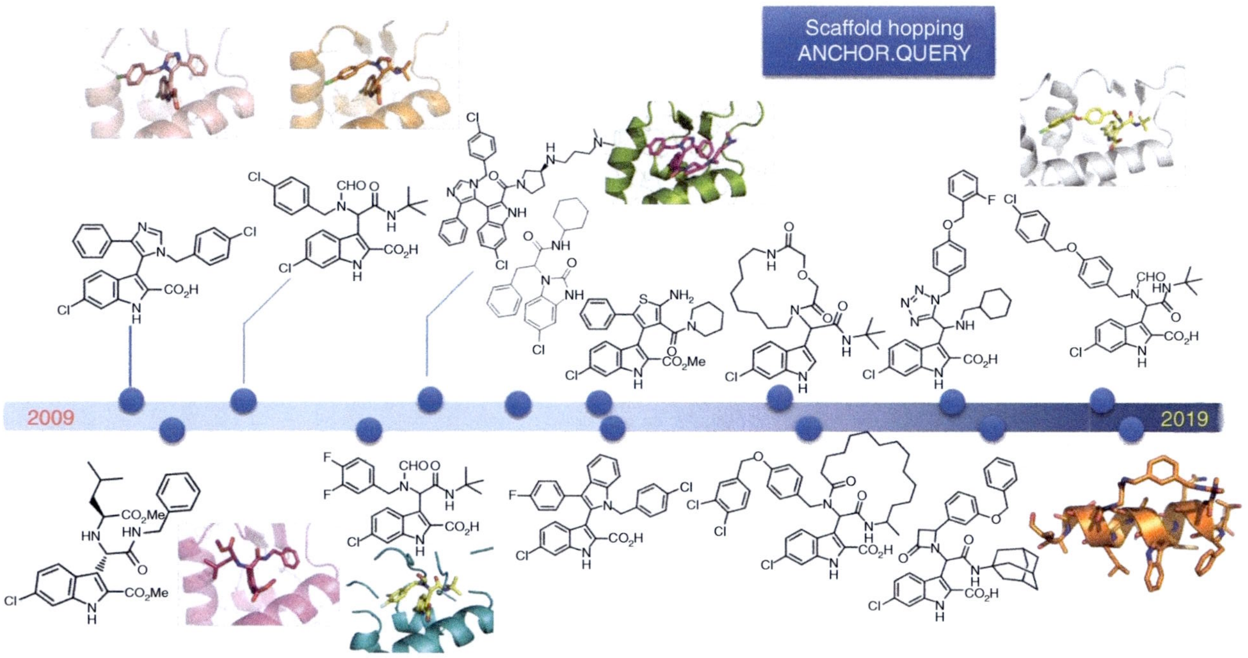

Figure 3.3 Timeline of p53/MDM2 antagonist discoveries in the Dömling laboratory.

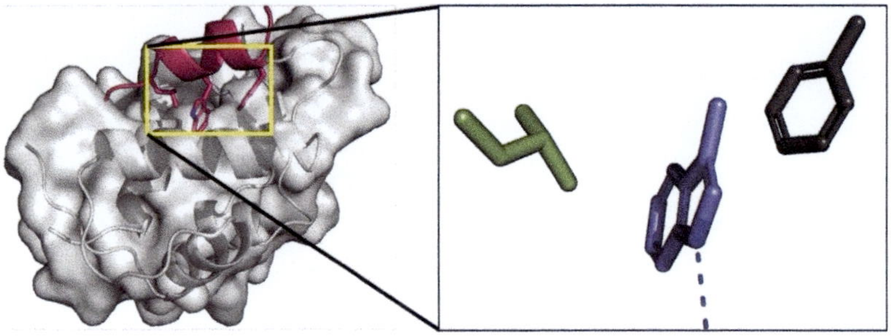

Figure 3.4 The protein–protein interaction between the negative regulator MDM2 and transcription factor p53 (PDB ID 1YCR) is a popular drug target. Antagonizing the protein–protein interaction can be accomplished by closely mimicking the hotspot triade $Y^{19}W^{23}L^{26}$ by small molecules.

molecules antagonizing the PPI p53/MDM2 indeed mimic the Trp23. Several scaffolds mimicking the Phe19Trp23Lys26 triade in the PPI p53/MDM2 were discovered by different MCR chemistries and can be used to antagonize the PPI in the same interface.

The van Leusen 1,4,5-trisubstituted imidazole is suitable to antagonize p53-MDM2. It can be synthesized in one step from an aldehyde, primary amine, and an α-substituted tosmic derivative at mild basic conditions [71]. The use of indole-3-carbaldehyde in the vL-3CR provides an anchor mimicking Trp23, and the other two components can be optimized to mimic the other two amino acids of the triad Phe19 and Lys26. Multiple potent molecules were described, and several cocrystal structures proof the binding mode and revealed new approaches for p53-MDMX/MDM2 antagonist drug discovery (Figure 3.5) [72–75].

Benzodiazepinediones were described as potent antagonists of p53/MDM2 [76–78]. The *p*-chlorophenyl side chain directly appending the diazepine ring protrudes deeply into the Y23 pocket in MDM2, while the second *p*-chlorophenyl group mimics the L26 as can be seen in a cocrystal structure. The iodo substituent of the benzodiazepinedione scaffold forms a halogen bond with the backbone carbonyl oxygen atom of Gln72. Compound **17** exhibits a potent $IC_{50} = 0.22\,\mu M$ and $K_d = 80\,nM$. The benzodiazepinedione series was synthesized by a Ugi of suitably protected anthranilic acids **12** and convertible cyclohexyl isocyanide **11**, followed by cyclization. The further development of this compound series was halted due to low cellular potency likely due to cell penetration issues (Scheme 3.1).

The benzimidazole-2-one has been introduced as a novel anchor principle to indole mimic W23. Such benzimidazole-2-ones can be synthesized by a U-3CR followed by a microwave-assisted cyclization (Scheme 3.2) [79]. From a small library of 15 analogues, enantiomer **22** revealed a Ki of ~2 μM, whereas the antipode was inactive by protein NMR titration. The separation of the enantiomers was accomplished by chiral supercritical fluid chromatography on a semipreparative scale with a run time of less than three minutes.

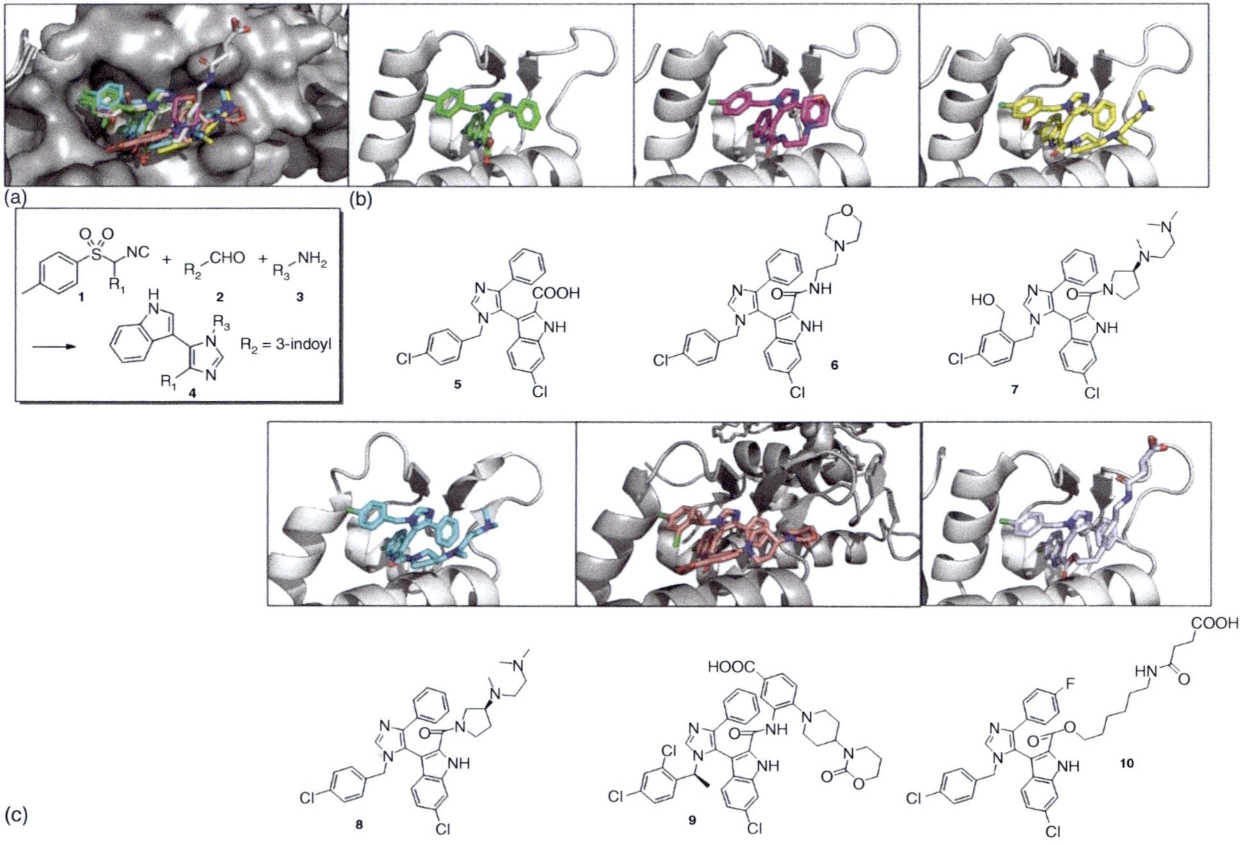

Figure 3.5 The van Leusen trisubstituted imidazole scaffold to antagonize p53/MDM2. (a) Alignment of six cocrystal structures of vL-3CR molecules in MDM2/x. The protein is shown as gray surface, the molecules in stick presentation. (b) One-by-one structural presentation showing the protein as gray cartoon: basic type **5** (PDB ID 3LBK), morpholinoethyl amide **6** with improved water solubility (PDB ID 3DIJ), pyrrolidine amide **7** in MDMX (PDB ID 6Q9L), pyrrolidine amide II **8** in MDMX (PDB ID 3LBJ), benzoic acid derivative **9** (PDB ID 6Q9S), and ester **10** (PDB ID 5J7G). (c) The van Leusen reaction (vL-3CR) of α-substituted tosmics, aldehydes, and primary amines.

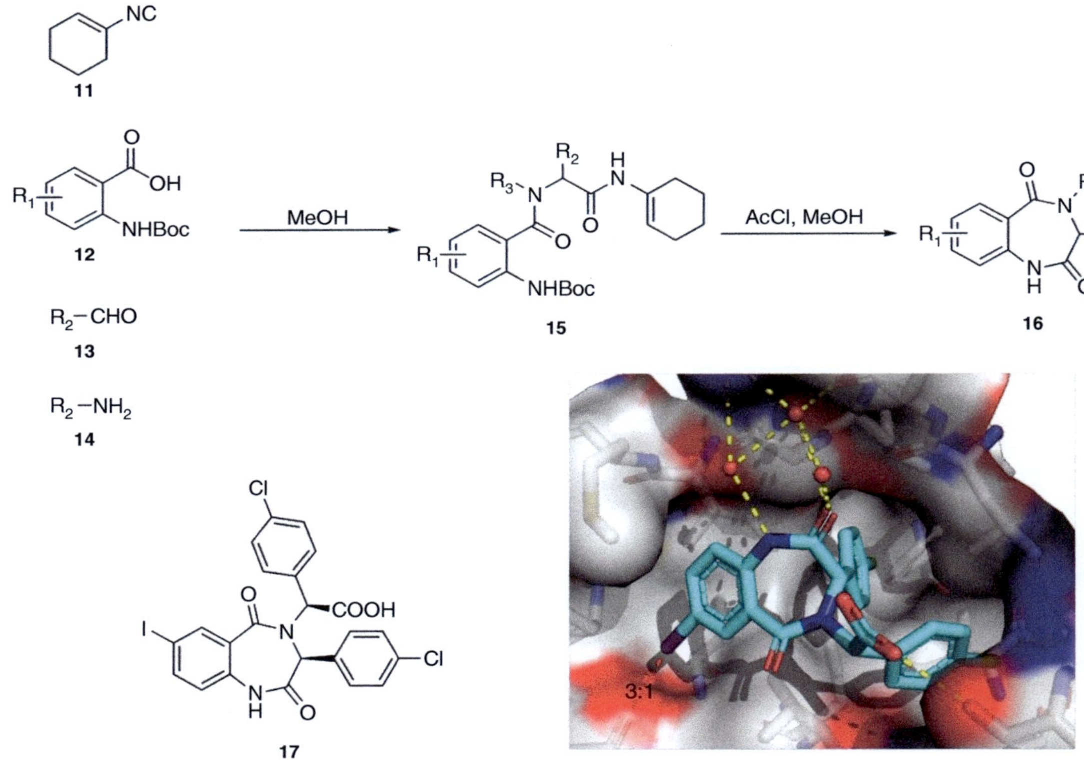

Scheme 3.1 Benzodiazepinediones as potent p53/MDM2 antagonists: 2-Step MCR synthesis and cocrystal structure (PDB ID 1T4E).

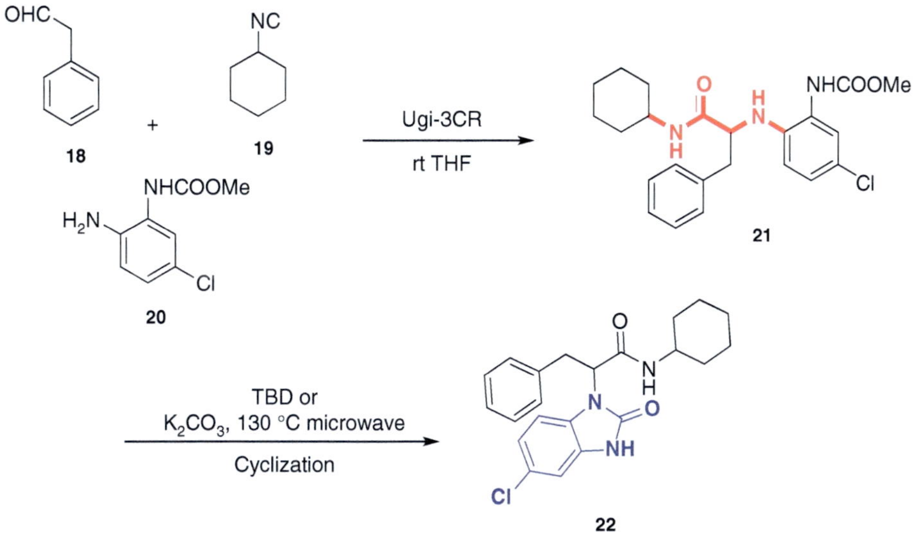

Scheme 3.2 Synthesis of benzimidazolones as p53 MDM2 antagonists. Source: Wang et al. [79].

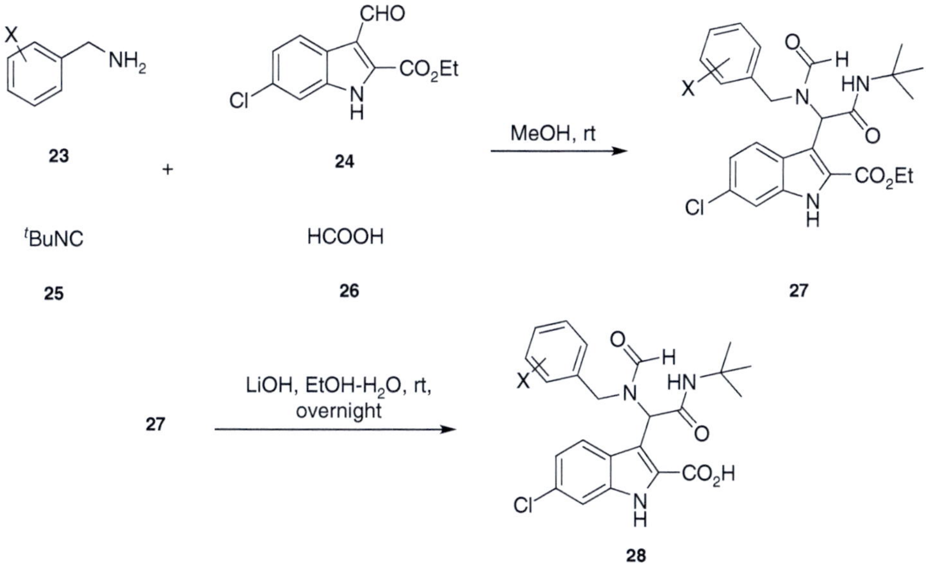

Scheme 3.3 Ugi-4CR for synthesis of potent p53/MDM2 antagonists.

α-Aminoacyl amides **27** generated by the Ugi-4CR were first discovered using a computational docking search technology named ANCHOR.QUERY that brings interactive virtual screening of novel protein–protein inhibitors to the desktop [75, 80, 81]. It can be synthesized by Ugi-4CR of benzylamine derivatives **23**, 1*H*-indole aldehyde **24**, isocyanide **25,** and formic acid **26** (Scheme 3.3).

A fluorinated indole-based MDM2 antagonist selectively inhibits the growth of p53(wt) osteosarcoma cells [82].

Fluorine scanning around the benzylamine component was used to probe the dipolar interaction with His96. All possible 20 fluorinated benzyl derivatives could be consciously synthesized using U-4CR in one-pot (Scheme 3.4). A cocrystal structure of (***S***)-**34** with MDM2 reveals the structural basis of the potent interaction. An antiparallel dipole momentum interaction between the His96 and the 3,4-difluorobenzyl group of (***S***)-**34** is leading to an attractive interaction. This findings underscore the principally known but often surprising effects that fluorine can exert on the biological activity of small molecules [80].

Dihydroimidazole can be synthesized by a three-component reaction of α-acidic isocyanides, aldehydes, and primary amines, which has been made popular by Orru (O-3CR). The use of suitable components, 4-chlorobenzaldehyde **35**, amine **36,** and isocyanide **37**, yields p53/MDM2 antagonists (Scheme 3.5) [75].

α-Aminoacyl amides were found as potent p53/MDM2 antagonists. In a scaffold hopping exercise and supported by the pharmacophore-based virtual screening platform ANCHOR.QUERY, the Ugi-4CR scaffold was morphed toward a β-lactam scaffold with potent p53-MDM2 antagonizing activities. 2D-HSQC and FP measurements confirm potent MDM2 binding. Molecular modeling studies were used to understand the observed structure-activity relationship (SAR) in the β-lactam series [83]. The synthesis involves the Ugi reaction of β-amino acids **41**, aldehydes **42,** and isocyanides **43** yielding **44** (Scheme 3.6).

Occupying the MDM2 hotspot FWL is key for obtaining active p53/MDM2 antagonists. However, the hotspot surrounding comprises a shallow extended binding site which can be occupied by a macrocyclic scaffold. These macrocycles, alternatively to stapled peptides, target for the first time the large hydrophobic surface area formed by Tyr67, Gln72, His73, Val93, and Lys94 yielding derivatives with affinity to MDM2 in the nanomolar range. Their binding affinity with MDM2 was evaluated using fluorescence polarization (FP) assay and ^{1}H–^{15}N two-dimensional HSQC nuclear magnetic resonance experiments [84]. Their synthesis involves a U-4CR of benzylamine **45**, indolecarbaldehyde **46,** alkene-substituted isocyanide **47,** and carboxylic acid **48** building blocks, followed by a ring-closing metathesis yielding 12- 24-membered macrocycles (Scheme 3.7).

Suitable Gewald thiophenes were also predicted to yield p53/MDM2 antagonists by using the ANCHOR.QUERY software [75]. Indeed, **54** showed a $K_d = 5$ μM and can be synthesized by reaction of the cyanoacetamide **51**, ketone **52,** and sulfur **53** (Scheme 3.8). The cyanoacetamides can be conveniently synthesized in great diversity by solventless simply mixing cyanoacetic acid ester and the amine component and filtering and washing the solid products [85].

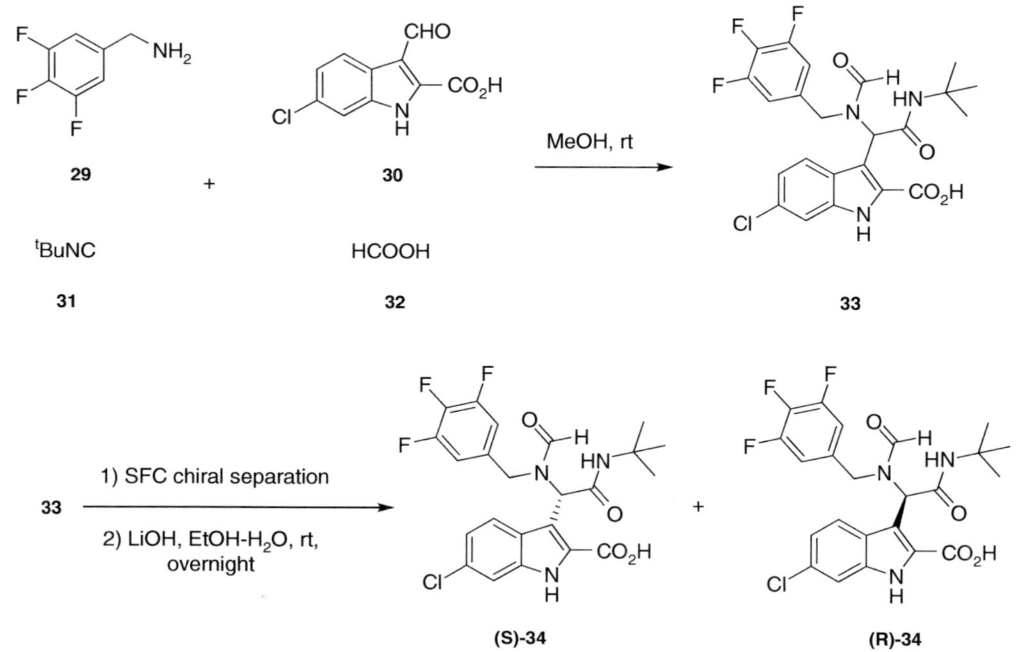

Scheme 3.4 U-4CR synthesis of fluorinated indole-based MDM2 antagonist.

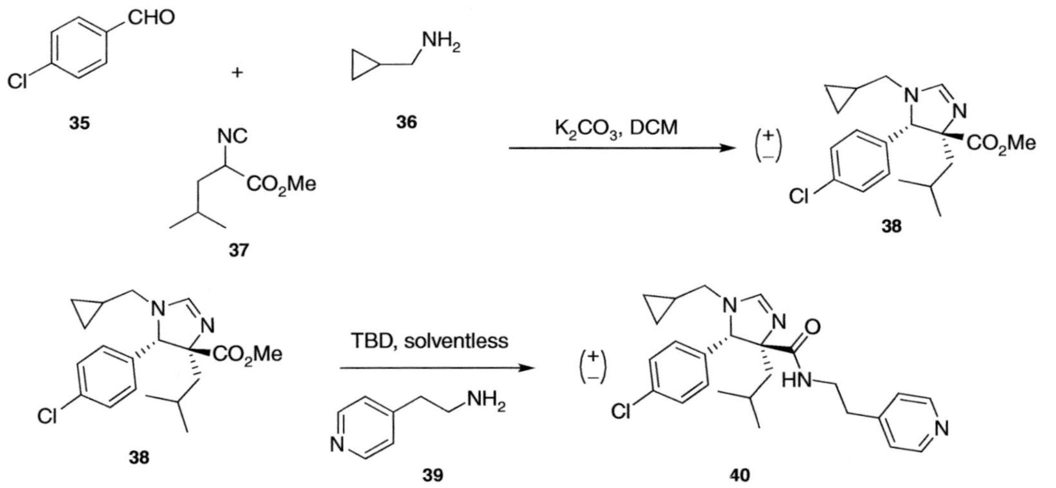

Scheme 3.5 One-pot Orru-3CR reaction followed by amidation for the synthesis of p53/MDM2 antagonists (TBD = 1,5,7-triazabicyclo(4.4.0)dec-5-en). Source: Adapted from Popowicz et al. [72]; Czarna et al. [73]; Furet et al. [74]; Czarna et al. [75].

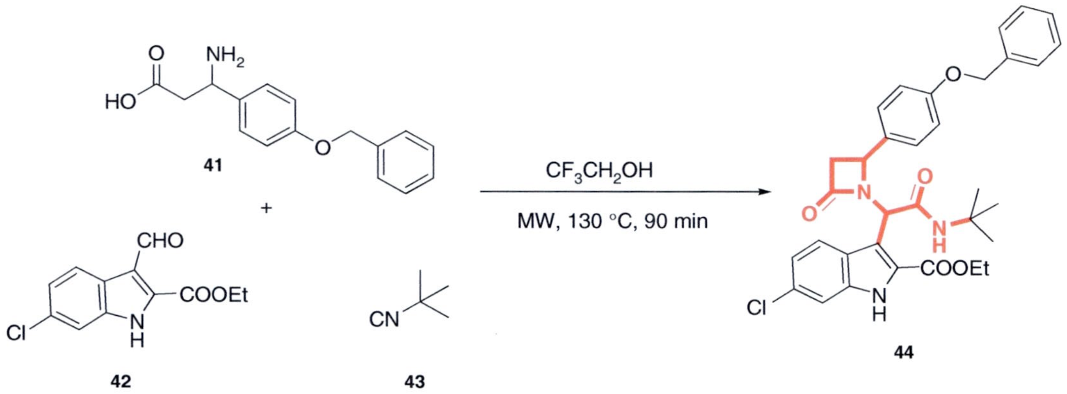

Scheme 3.6 β-Lactams as p53/MDM2 antagonists.

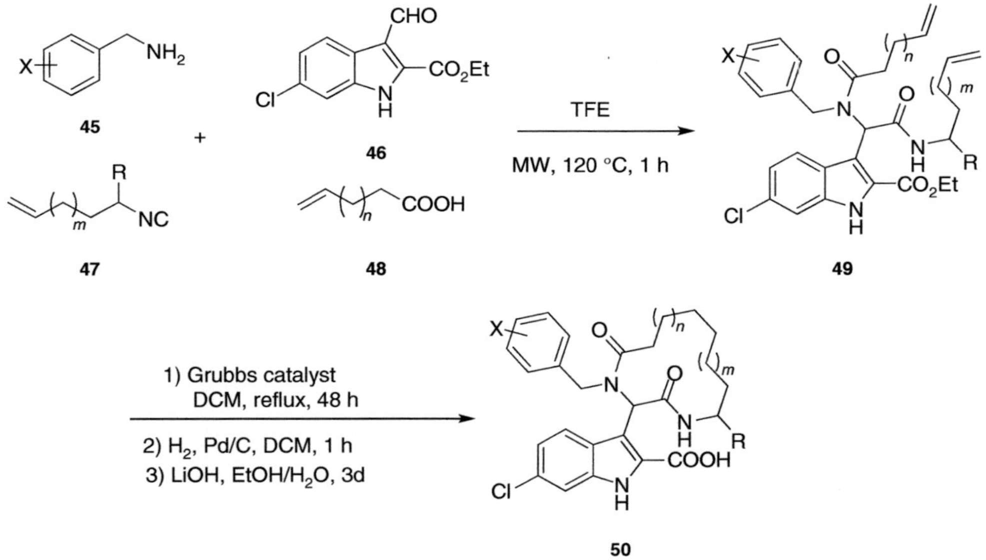

Scheme 3.7 Macrocyclic p53/MDM2 antagonists synthesized by U-4CR followed by RCM.

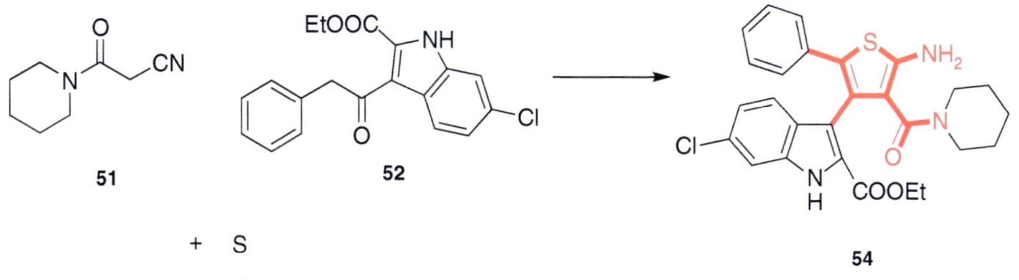

51 52 53 54

Scheme 3.8 G-3CR product as p53/MDM2 antagonist.

In another work, Gewald products **55** were used in the U-4CR with isocyanides **56**, glyoxalethylester **57**, and primary amines **58** to yield after deprotection of the intermediate and base-induced cyclization thienodiazepine-2,5-diones as p53/MDM2 antagonists **59** (Scheme 3.9) [86].

Stapled peptides are especially suitable to mimic α-helices and have been adopted to become "drug-like," including increased stability, solubility, and cell penetration. A stapled peptide potently able to antagonize p53-MDM2 and its congener MDMX was entering clinical trials [70]. Stapling of peptides is often done by metathesis of suitable alkene side chains leading to hydrophobic macro rings. Stapling based on MCR chemistry was first introduced by Rivera and Domling [87]. It was demonstrated that the MCR-based stapling is an effective strategy for the development of α-helical peptides with highly potent dual antagonistic action of MDM2 and MDMX binding p53 (Figure 3.6a). The satisfactory macrocyclization yields were obtained for stapled peptides 8 (**63**) after 72 hours of reaction via Ugi-4CR of isocyanide **60,** amine **61,** and formaldehyde **62**. The inhibitory activity of p53-MDM2/X interactions was assessed by FP, microscale thermophoresis, and 2D NMR, while several cocrystal structures with MDM2 were obtained (Figure 3.6b). This MCR stapling protocol proved efficient and versatile in terms of diversity generation at the staple, as evidenced by the incorporation of both exo- and endo-cyclic hydrophobic moieties at the side chain cross-linkers. Potential advantage of MCR stapling is the convergent introduction of beneficial properties into the staple.

DNA-encoded library chemistry (DEL) is a technology platform which allows to create very large chemical space on solid support and displayed on a very low volume [88]. Multiple applications based on sequential split and pool syntheses have been described [89]. Brunschweiger and coworkers introduced MCR chemistry to DEL [90]. An application of MCR DEL was the discovery of PPI antagonists for p53/MDM2 and YIAP [91].

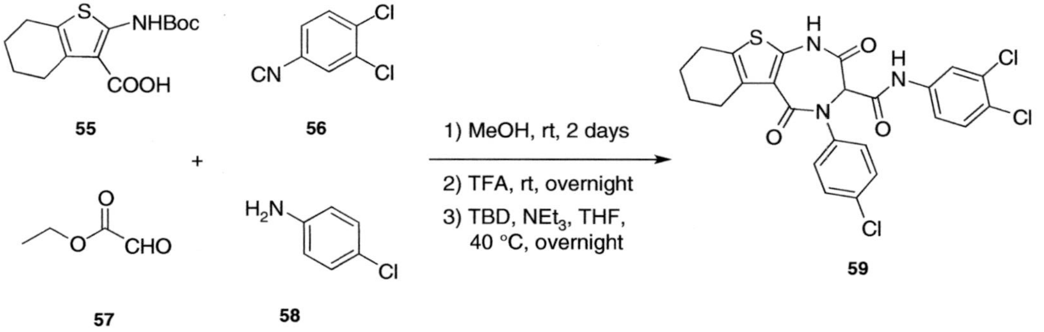

Scheme 3.9 Gewald thienodiazepine-2,5-dione synthesis. Source: Huang et al. [86].

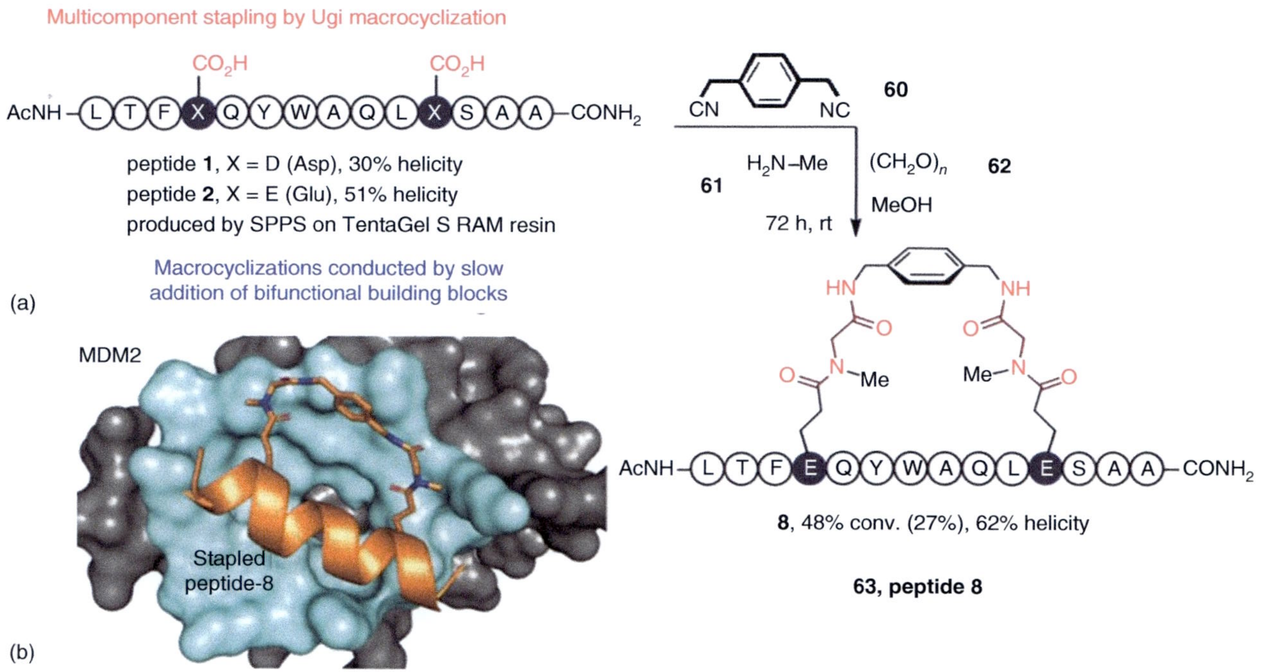

Figure 3.6 MCR peptide stapling. (a) Solid-phase synthesis of stapled peptide 8, **63**; (b) Cocrystal structure of Ugi-peptide 8 and MDM2 (PDB ID 6T2D). The buried surface area (BSA) of the peptide 8 is depicted as a cyan surface.

3.2 Scaffolds and the Chemical Space of MCR

The scaffold is an important theoretical concept in medicinal chemistry which is very helpful to rationalize SAR and optimize compounds activities. A classical definition of the scaffold is "a molecular core to which functional groups are attached." [92] Bemis and Murcko scaffolds, cyclic skeletons, and other reductions are graph theoretical constructs describing different topologies of groups of molecules foremost useful for computational analysis [93, 94]. An alternative definition of a scaffold is "the smallest atomic denominator and its connectivity as resulting from a reaction or reaction sequence using starting materials with common functional groups." [52] This definition of a scaffold is practically connected to their synthesis and is used in the following. The usefulness of a scaffold in medicinal chemistry is the notion that a core structure with its surrounding function groups is responsible for the biological activity and the observed SAR. The bioisosteric replacement of the core structure, while keeping the surrounding ligand sphere, was coined "scaffold hopping" and is of great importance in medicinal chemistry to enter favorable drug-like properties or to secure intellectual property. The scaffold has thus major implications for the definition of a chemical space around an intellectual IP.

The Gewald-3CR (G-3CR) is one of the very mild reactions of elemental sulfur in organic chemistry. In the GW-3CR, elemental sulfur reacts with a-methylene aldehydes or ketones and α-methylene α-acidic nitriles under basic conditions. The G-3CR already proved its use in medicinal chemistry with accessing the block buster drug olanzapine. An example of a recent compound **64** is the pan-serotype inhibitor of the dengue virus NS4B protein, NITD-688, showing favorable pharmacokinetics and efficacy in preclinical animal models [95]. Due to the 2-amino thiophene moiety with a further reactive substituent in 3-position, the primary GW-3CR scaffold is a valuable hub in organic chemistry leading to a plethora of additional products (Scheme 3.10).

Inhibition of angiogenesis is a promising strategy for the treatment of cancer. Compound **65** had a considerable activity against prostate cancer cell line, PC3 (IC_{50} = 0.97 μM) [96]. Meanwhile, the cardiac troponin I interacting kinase (TNNI3K) exacerbates ischemia-reperfusion (IR) injury via oxidative stress, thereby promoting cardiomyocyte death. Some compounds with thienopyrimidine scaffold exhibited sub-micromolar TNNI3K inhibitory capacity and good kinase selectivity, as well as cytoprotective activity [97].

Thieno[2,3-b]pyridine amines can be synthesized by Gewald reaction. Compound **66** not only showed the high activity toward acetylcholinesterase and butyrylcholinesterase but also performed good inhibitory activity toward amyloid-beta (βA) aggregation and β-secretase 1, neuroprotectivity, and cytotoxicity against HepG2 cells [98]. Gewald chemistry could be used to synthesize the compound **67**. A novel Ni-catalyzed denitrogenative cross-coupling between **67** and cyclopropanols to yield **68** was reported. This neoteric reactivity allows for the convenient synthesis of β-(*o*-amido)aryl ketones **68** from readily available starting materials with good yields and general substrates [99]. Many natural or synthetic chalcones have potential anti-tumor activity. Chalcone analogues containing a thieno[2,3-d]pyrimidin-2-yl group **69** can also be synthesized via Gewald chemistry. This chalcone series

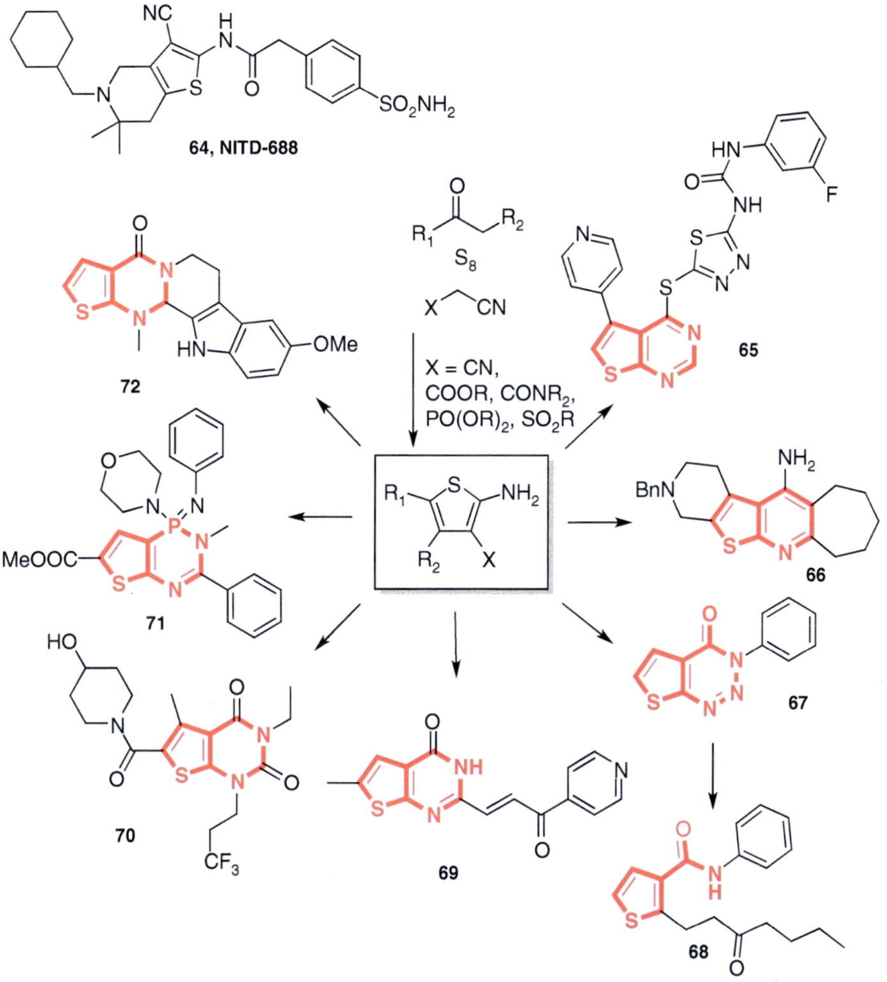

Scheme 3.10 NITD688 as a potent Dengue virus inhibitor and the Gewald chemistry hub.

could perform good inhibitory activity against HCT-116 cells and induce cell apoptosis via the mitochondrial death pathway [100]. Bayer Pharmaceuticals found a novel non-xanthine (thienouracil) antagonist of the A_{2B} adenosine receptor from high throughput screening. After the structure optimization work, compound **70** was described and showed the best activity [101]. G-3CR product could be used in the fused phospha-pyrimidines synthesis as a starting point. Two general approaches are described to synthesize compound **71** [102]. The generation of natural product-inspired scaffold diversity is an effective but challenging strategy to investigate the broader chemical space and identify promising drug leads. From the evodiamine, compound **72** as an evodiamine-inspired novel scaffolds was designed and synthesized. From the G-3CR product as the synthesis starting point, compound 68 could been synthesized very easily (3 steps) with moderate yields [103].

3.2.1 Marketed and Clinical Stage Drugs

Many marketed or clinical stage drugs comprise fragments which can be assembled advantageously using MCR chemistry. Its use can potentially shorten the time to discover and optimize a lead. Moreover, it can potentially shorten the synthesis pathway of the API, saving time and money, and leading to an overall more sustainable synthesis. A survey of the current pharmacopeia, e.g. Drug Bank, reveals that ~10% contains MC fragments [104]. Surprisingly, however, only very few drugs indeed were discovered and produced by MCR. Thus, there is a considerable potential for MCR chemistry not only to save time, resources, and money but also to establish alternative intellectual property. In the following, examples of advantageous use of MCR chemistry in the discovery and/or production of drugs are discussed.

Phenserine **77** is a synthetic drug which has been investigated as a medication to treat the Alzheimer's disease (AD), as the drug exhibits significant neuroprotective and neurotrophic effects. The intermediate **76** can be synthesized by a Pd-induced MCR of *o*-iodobenzacylamide **75**, benzylalcohol **74**, and carbon monoxide **73** (Scheme 3.11).

Tadalafil **83** is also called the "weekend pill" [105]. It is a medication used to treat erectile dysfunction (ED), benign prostatic hyperplasia (BPH), and pulmonary arterial hypertension. Tadalafil can be synthesized starting from Boc-protected (D)-tryptophan **78** and piperonal **82** via Ugi-4CR reaction and Pictet–Spengler reaction (Scheme 3.12) [106].

Bestatin **87** is a natural product isolated from *Streptomyces olivoreticuli*. It acts as a broad spectrum competitive reversible protease inhibitor of aminopeptidase B, leukotriene A4 hydrolase, aminopeptidase M/N, oxytocinase/vasopressinase, and leukotriene D4 hydrolase. It is studied in clinical trials for use in the treatment of

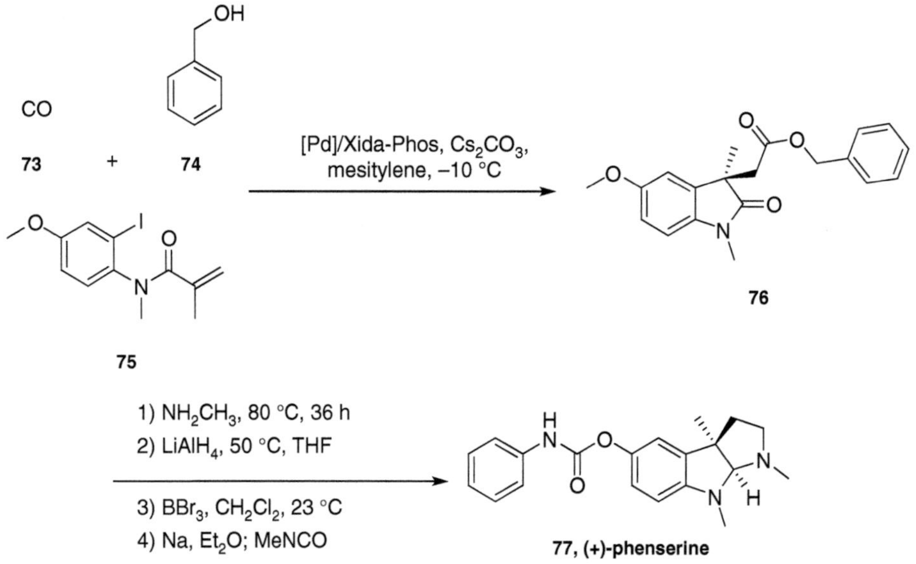

Scheme 3.11 Phenserine synthesis by MCR.

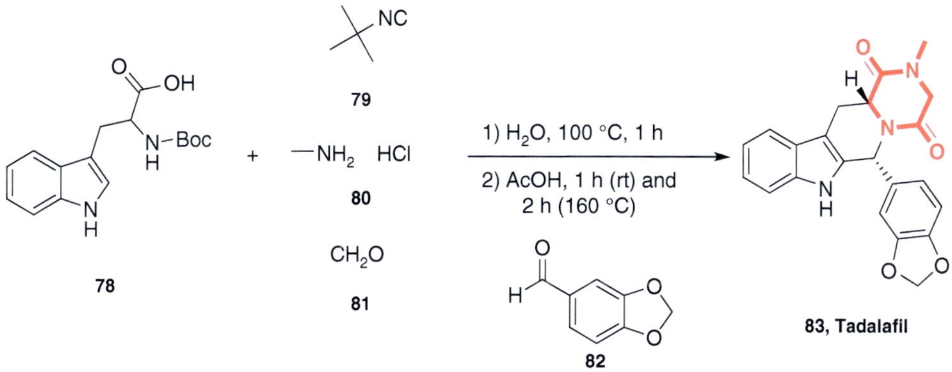

Scheme 3.12 U-4CR for Tadalafil synthesis. Source: Adapted from Jida et al. [106].

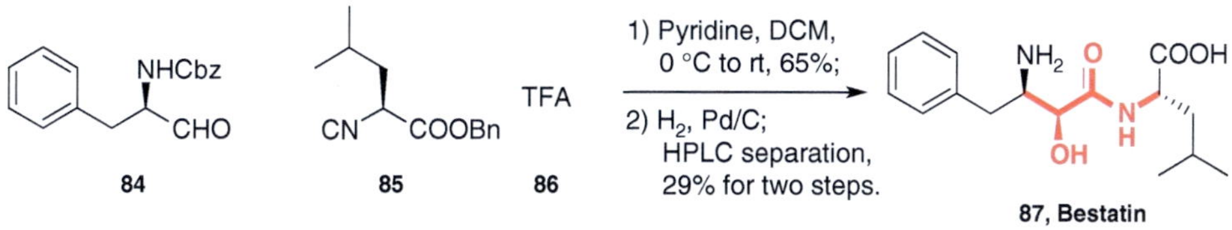

Scheme 3.13 MCR synthesis for Bestatin.

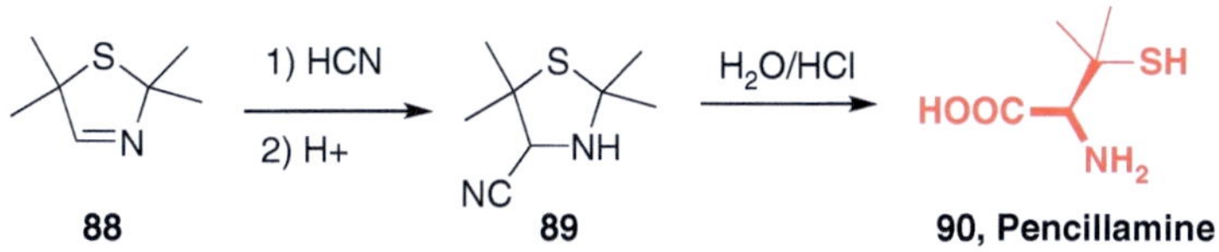

Scheme 3.14 Racemic penicillamine can be synthesized by an Asinger-3CR followed by a Strecker reaction and hydrolysis.

acute myelocytic leukemia and hypercholesterolemia. Bestatin is also used in the construction of PROTAC as it binds the E3 ligase cIAP1 [107]. The molecule with 3 adjacent stereocenters has not only been synthesized by sequential multistep syntheses [108–110] but can also be prepared by a Passerini 3-component reaction [111]. Reaction of *N*-α-Cbz-D-Phe-H **84** and the leucine-derived isocyanide **85** with TFA in pyridine produced adduct in 65% yield as a 1.5 : 1 mixture of diastereomers at the new hydroxy center with retained configuration at the original aldehyde and isonitrile chiral centers. Separation of the α-hydroxy diastereomers, hydrogenolysis, and HPLC separation afforded bestatin **87** in satisfactory overall yield (Scheme 3.13).

Penicillamine **90** is a chelating agent used for the removal of excess copper in patients with Wilson's disease. It is also used to reduce cystine excretion in cystinuria and to treat patients with severe, active rheumatoid arthritis unresponsive to conventional therapy. Penicillamine can be synthesized by a combination of the Asinger and Strecker MCR in a racemic form (Scheme 3.14) [112].

Ivosidenib **96** is the first approved drug which was discovered by Ugi MCR chemistry and is also produced using the same chemistry [113–116]. It inhibits

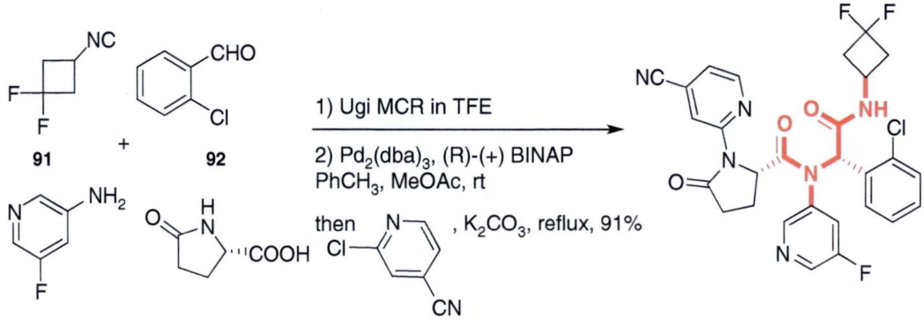

91 92 93 94 95 96, Ivosidenib

Scheme 3.15 MCR synthesis of ivosidenib.

isocitrate dehydrogenase-1 (IDH1), which is mutated in several forms of cancer. It is a medication for the treatment of acute myeloid leukemia (AML) with IDH1 mutations. A process described by TEVA, resulting in the API, is as follows. The isocyanide **91** is produced from the cyclobutylamine by formylation and dehydration with $POCl_3$/TEA and then dissolved in 2,2,2-trifluoroethanol (TFE). The Ugi reaction was performed on the preformed imine and the acid **94** and resulted in a mixture of the two diastereomers. An efficient workup and crystallization procedure was developed for the isolation of desired diastereomer Ugi product. Acid/base extraction and treatment with piperidine allowed for isolation of pure Ugi product in >99.9% chiral purity and 27–31% yield from amine **93** without the need of any chiral chromatography. In a second step, 2-chloroisonicotinonitrile **95** was selectively coupled onto the γ-butyrolactame by a Buchwald–Hartwig procedure followed by crystallization from EtOAc and *n*-heptane gave ivosidenib in 91% yield (Scheme 3.15).

Telaprevir **97** is a initially marketed HCV NS3/4A serine protease inhibitor and was co-developed by Vertex Pharmaceuticals and Johnson & Johnson [117]. Recently, Telaprevir and its close relative Boceprevir have been shown to potent inhibitor SARS-CoV-2 main protease exhibiting also strong inhibition of cellular viral replication [118]. Especially, Boceprevir seems to be interesting as a candidate of repurposing [119]. Telaprevir as all other HCV NS3/4A serine protease inhibitors belongs to the class of peptidomimetics, which are mimicking the endogenous substrate and are complex compounds with high molecular weight and full stereochemistry. Not surprisingly, the original synthesis was lengthy, expensive, and complex, involving for example chiral chromatographic separation [120]. Even the described GMP synthesis of the drug exceeded well 20 linear steps and features low overall yields. An elegant total synthesis of Telaprevir was described featuring a Passerini and diastereoselective Ugi reaction [121]. Through the incorporation of two MCRs, the overall step number and time to product could be cut in more than half featuring a much better overall yield (~30%, Scheme 3.16). The *N*-terminal α-ketoamide moiety which is important for the covalent addition to the HCV NS3/4A serine protease was synthesized by a Passerini reaction (P-3CR). Toward this end, commercial (*S*)-2-amino-1-pentanol (**98**) was transformed to aldehyde **99** by *N*-formylation and IBX oxidation and *in situ* subsequently reacted in the P-3CR. The one-pot Dess–Martin oxidation/Passerini reaction of **99** furnished **102** in 60%

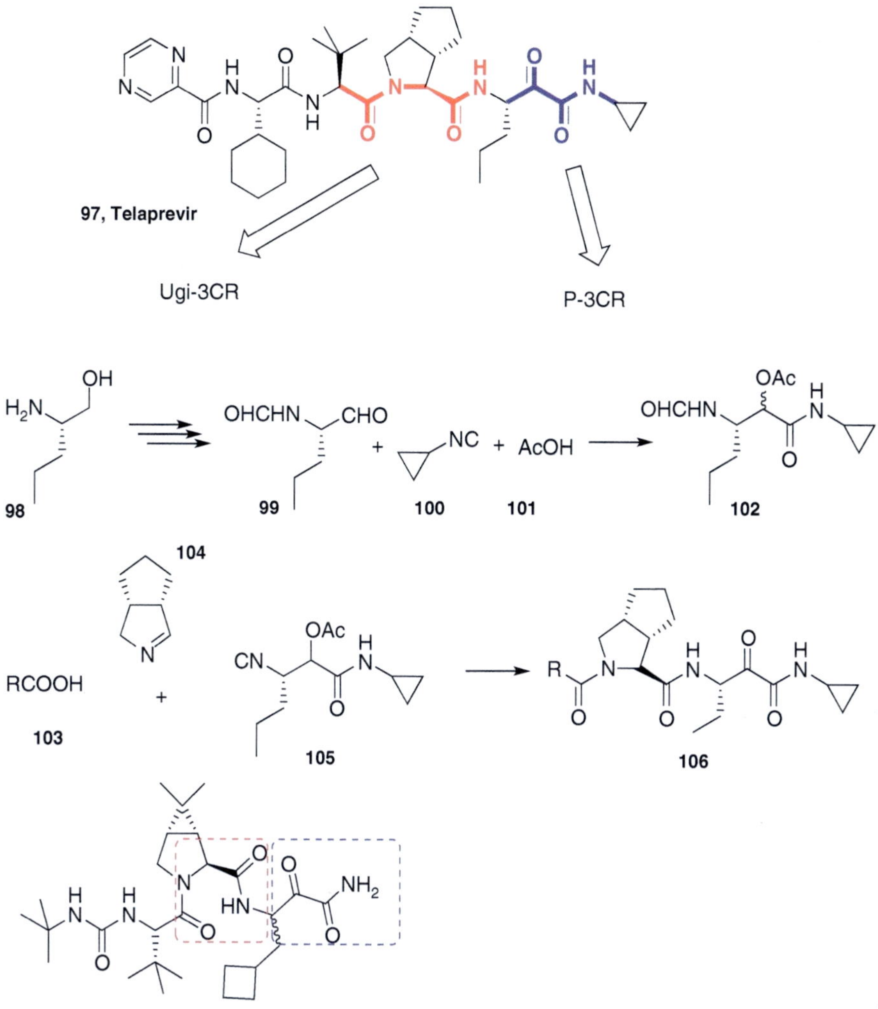

Scheme 3.16 HCV inhibitors Telaprevir and Boceprevir and MCR synthesis of Telaprevir.

overall yield as a 78 : 22 mixture of diastereomers. Dehydration of the formamide using triphosgene and NMM in DCM at, −30 °C yielded 87% of the isocyanide **105** with no racemization of the C3 stereocenter. The second MCR facilitating the overall Telaprevir total synthesis is a highly diastereoselective Ugi reaction performed in DCM, of the cyclic **104**, the Passerini reaction derived isocyanide **105** and the *C*-terminal carboxylic acid **103** in 76% yield. The cyclic Schiff base **104** was produced enantiomerically pure by the enzymatic desymmetrization of meso-pyrrolidines by means of a monoamine oxidase-N (MAO-N) from *Aspergillus niger* optimized by directed evolution [122]. The synthesis is scalable and has been performed on a GMP scale to produce 1000 kg of drug. This Telaprevir synthesis is an example *par excellence* for the usefulness of MCR chemistry in medicinal chemistry. It is conceivable that the close relative Boceprevir can be synthesized by an analogous MCR route.

Praziquantel **113** (PZQ) is an antihelminthic and is the pharmacological mainstream to treat schistosomiasis [123]. Schistosomiasis also called Bilharzia is a neglected tropical disease of poverty that leads to chronic severe morbidity, including abdominal pain, diarrhea, blood in the stool, liver enlargement, accumulation of fluid in the peritoneal cavity, and hypertension of the abdominal blood vessels and bladder cancer. 240 Million people are infected with Schistosomiasis, ~200,000 die yearly, and more than 700 million people live in endemic areas. Hence, affordable drugs are of highest importance. The disease is acquired in standing fresh water when parasitic blood flukes, known as schistosomes penetrate through the human skin. A tetrahydroisoquinoline derivative, PZQ, is an effective one-time-use treatment to kill the adult form of the schistosomes. It is well tolerated, and no signs of resistance have been observed, despite treatment of millions of patients in sub-Sahara Africa. The classical synthesis of PZQ involves a sequential five-step synthesis including a Reissert reaction using large excess of toxic KCN [124]. Alternatively, PZQ can also be synthesized by a short convergent sequence involving a Ugi reaction followed by a Pictet–Spengler reaction (Scheme 3.17). Reaction of phenylethylamine derived isocyanide **108** with formaldehyde **110**, aminoacetaldehyde dimethyl acetal **109,** and cyclohexyl carboxylic acid **111** yields PZQ in one pot in 86% yield [125]. The subsequent Pictet–Spengler reaction was reported with 75% yield. The synthesis has been scaled to several tens of gram. A mole scale synthesis of the phenylethyl isocyanide has been described using a dry work-up with increased safety and producing less reaction waste resulting in a highly reduced environmental footprint [60]. Variations of carboxylic acids and oxo components yielded a library of derivatives, none of which was more active then PZQ in a schistosoma killing assay [126]. Ketocarboxylic acids used in the same

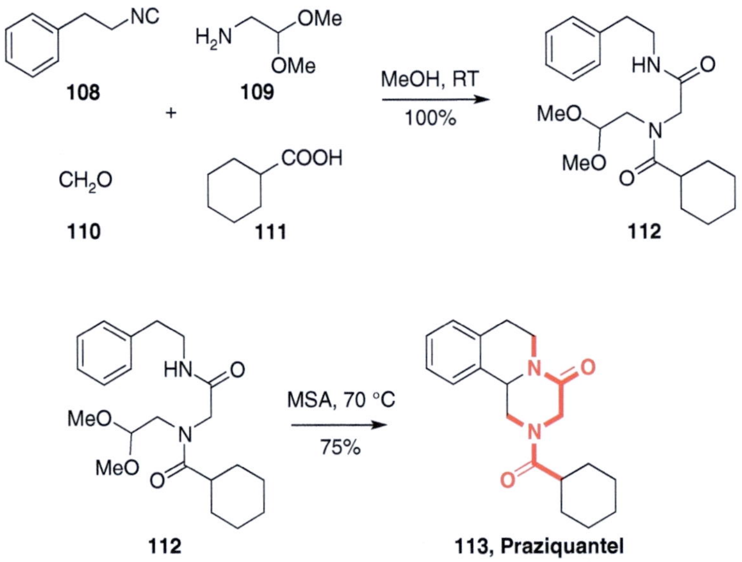

Scheme 3.17 MCR synthesis of Praziquantel.

scheme give unusual polycyclic scaffolds [127]. Praziquantel due to its important for the worlds health and its safety and efficacy is on the WHO model list of essential medication. Recently, the billionth PZQ tablet was donated to the WHO by the producer Merck KG.

Atorvastatin **120** for many years was the best-selling drug ever. It is a statin, used for the prevention of cardiovascular diseases originating from abnormal blood lipid levels. The mode-of-action of Atorvastatin involves the inhibition of the NADH-dependent liver enzyme HMG-CoA reductase, which is the rate-controlling enzyme of the mevalonate pathway, the metabolic pathway that produces cholesterol and other isoprenoids. The different industrial atorvastatin syntheses proceed via the Paal–Knorr route consisting of between 6 and 11 steps excluding the synthesis of amine **115** which is commercially available. The MCR approach features a key Ugi reaction followed by a 3 + 2 cycloaddition through a Munchnone intermediate (Scheme 3.18) [128]. The reaction at rt of *p*-fluorobenzaldehyde **114**, the suitably functionalized, commercially available amine **115**,(29,39–41) the convertible isocyanide **116**,(42–45) and isobutyric acid **117** in TFE afforded the Ugi adduct (U-4C) **118** in 40% yield (performed on a 10 mmole scale). The regioselective [3 + 2] cycloaddition of **118** with the *N*,3-diphenylpropiolamide **119**

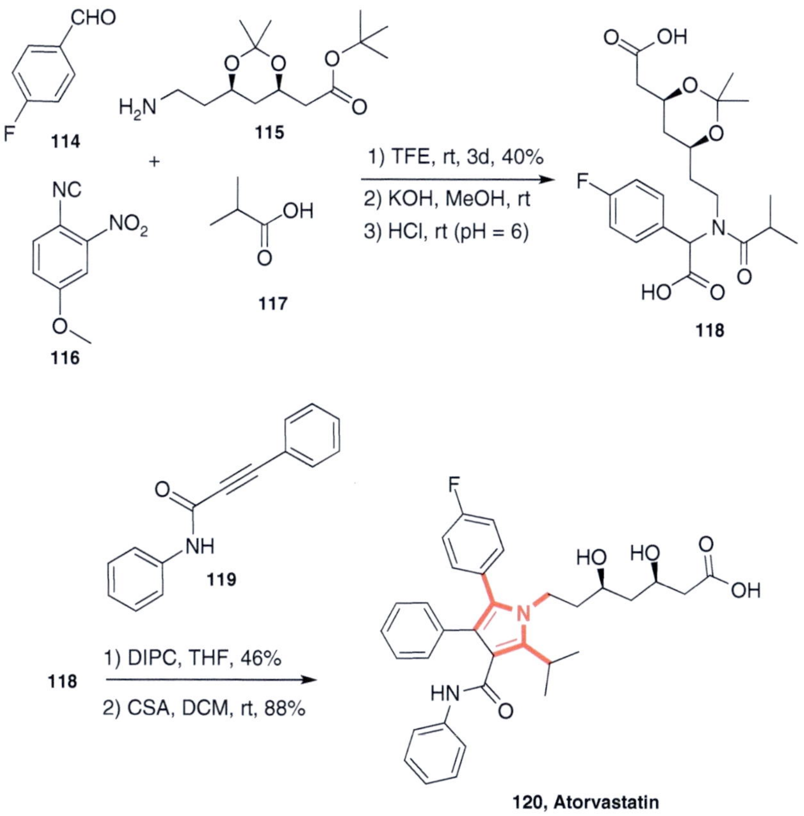

Scheme 3.18 MCR route to Atorvastatin. Source: Zarganes-Tzitzikas et al. [128].

and *N*,*N*′-diisopropylcarbodiimide (DIPC) in THF, yielding the advanced intermediate **120** in 46% yield which can be readily converted by acidic deprotection with 10-camphorsulfonic acid (CSA) to atorvastatin. Amongst all described atorvastatin syntheses, the MCR strategy can be ranked among the most competitive one with just four steps.

Carfentanil **126** is a synthetic opioid analgesic introduced by Paul Janssen in the 1970s. It is a controlled substance in many countries due to its widespread use as an illegal narcotic. The mode of action is revealed by u-opioid receptor binding, which is >10 times higher than morphine. Moreover, carfentanil easily crosses the blood–brain barrier. Many different synthesis routes toward carfentanil are known. The Ugi route involves secondary amide to ester transformation in the presence of a tertiary amide. This is a demanding transformation since tertiary amide is more reactive then secondary. For this purpose, so-called convertible isocyanides were developed. Convertible isocyanides are called convertible because once they reacted in the MCR, they can be selectively transformed into another functional group. Already in 1962, Ugi described the use of cyclohexenyl isocyanide in the U-4CR and subsequent conversion of the product to a primary amide. Later, the "child" was named by Armstrong. Clearly, the carfentanil skeleton can also assembled by a Ugi MCR [129–132]. The Ugi 4CR of **121**, commercial ketone **123**, aniline **122**, and propionic acid **124** followed by methanolysis yields carfentanil **126** (Scheme 3.19). The U-4CR afforded **125** in excellent yield (92%). Acidic methanolysis then gave carfentanil **126** in near-quantitative yield (98%). The Ugi route was used to discover novel analgesics with better side effect profile. The synthesis and pharmacological characterization of novel carfentanil amide analogues by receptor binding and analgesia assays revealed **127**, which showed moderate analgesic affinity and some respiratory depression, but produced no physical dependence or inhibition of GI transit in mouse models [131]. **128** has a mixed μOR/δOR receptor binding profile.

Clopidogrel **135** is an antiplatelet medication to reduce the risk of heart disease and stroke. It is a prodrug which is activated by different CYPs and then covalently binds and inactivates the P2Y12 subtype of ADP receptor. **135** has been synthesized in racemic form using two MCRs [133]. The Petasis-3CR of (2-chlorophenyl)boronic acid **129**, 2-oxoacetic acid **130,** and 4,5,6,7-tetrahydrothieno[3,2-c]pyridine **131** can help to yield the clopidogrel. **135** can also be synthesized using the Ugi-4CR employing the convertible isocyanide cyclohexenylisocyanide **132**, aldehyde **133,** and amine **134** (Scheme 3.20).

Xylocaine **140** is a local anesthetic. Xylocaine and similar caines were first synthesized by Ugi using his U-3CR. 1,3,5-Triazinanes were introduced as formaldimine surrogates in the Ugi reaction yielding i.a. xylocaine [134]. Pt/TiO_2 catalysts under visible/near-UV light irradiation were used to synthesize xylocaine from the isocyanide **138**, diethylamine **136,** and formaldehyde **137** (Scheme 3.21) [135].

Quinapril **144** is an ACE inhibitor and works by decreasing renin-angiotensin-aldosterone system activity. It is used to treat high blood pressure, heart failure, and diabetic kidney disease. Its synthesis was described using Ugi MCR (Scheme 3.22) [136]. $TiCl_4$ catalyst was used to accomplish the U-3CR product of acetaldehyde **141**, ethyl-2-amino 4-phenyl butanoate **142,** and benzyl-2-isocyano-3-phenyl propionate **143**. The major diastereomer was separated by crystallization as the

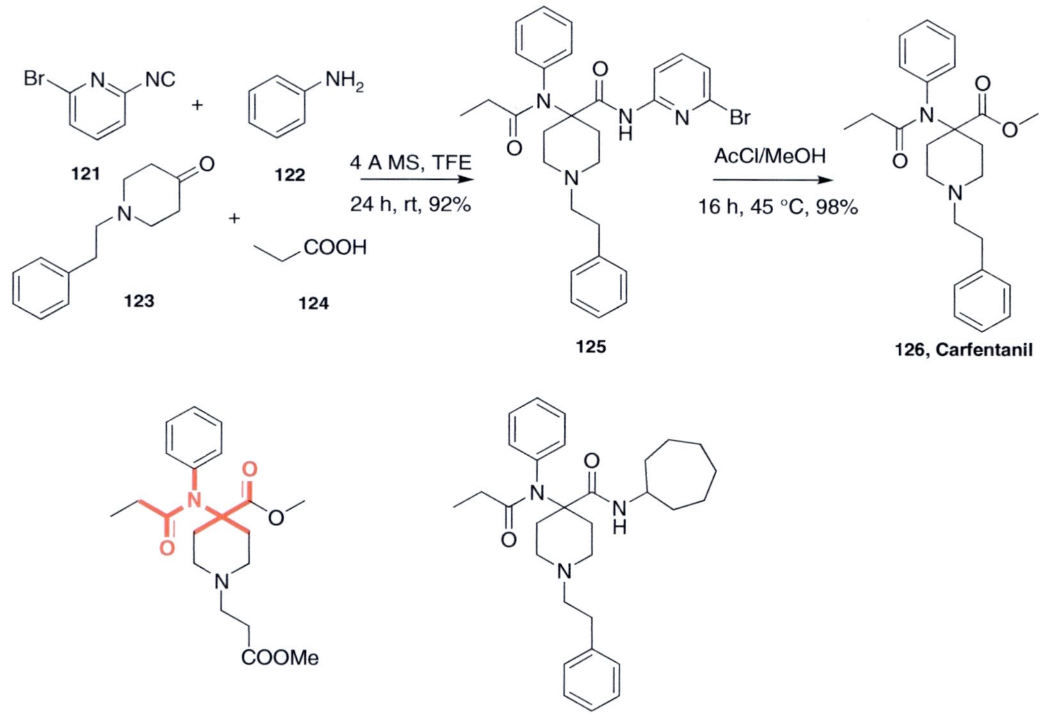

Scheme 3.19 A MCR synthesis of carfentanil.

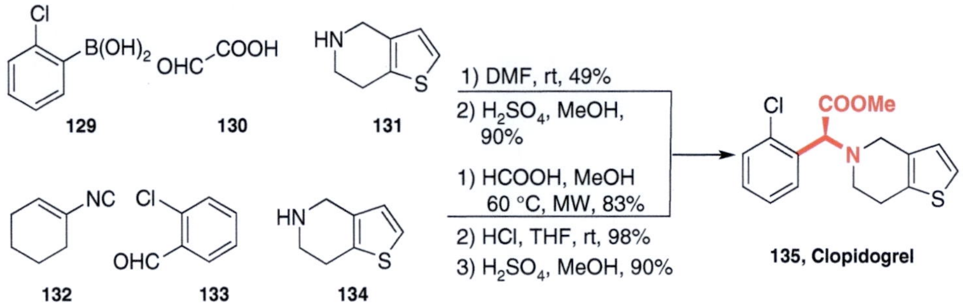

Scheme 3.20 Two MCR syntheses of racemic clopidogrel.

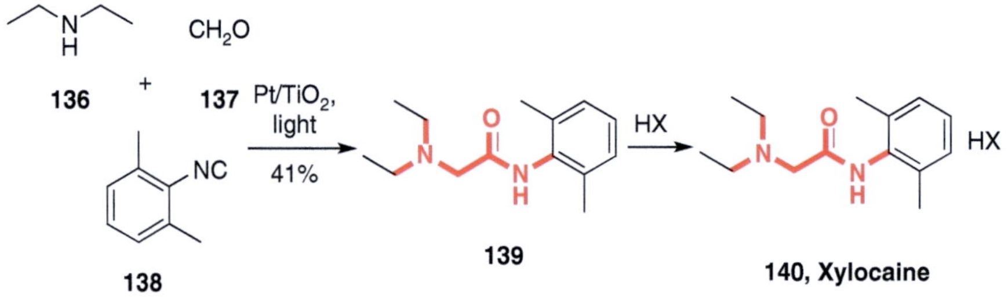

Scheme 3.21 Synthesis of xylocaine by U-3CR.

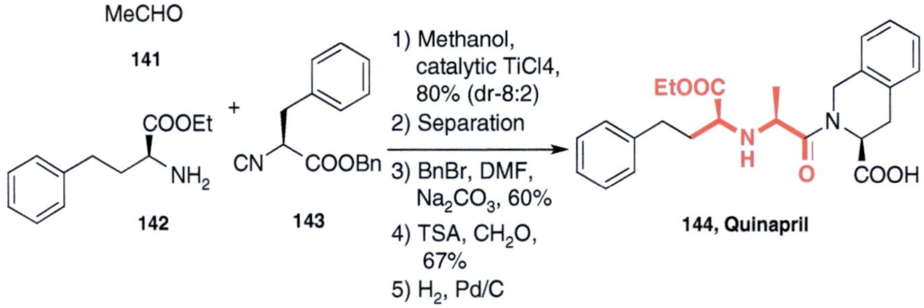

Scheme 3.22 U-3CR synthesis of quinapril. Source: Based on Borase et al. [136].

maleate salt. After benzylation of the secondary amine, the tetrahydroisoquinoline was formed through a Pictet–Spengler reaction using *p*-toluene sulfonic acid monohydrate, paraformaldehyde in toluene. Debenzylation with Pd/C yielded quinapril.

Racetams are a class of nootropic or anticonvulsant drugs that share a pyrrolidone nucleus. Using Ugi MCR of γ-amino butyric acids and the corresponding aldehydes and isocyanides, the approved piracetam **148**, etiracetam, (*S*)-levetiracetam **149**, nefiracetam **150**, and pramiracetam **151** as well as a library of other highly substituted novel racetams were synthesized (Scheme 3.23) [137].

Lacosamide is an important drug for the treatment of epilepsy. A scalable version of the Ugi reaction was used to synthesize (*R*)-Lacosamide **156** [138]. To induce

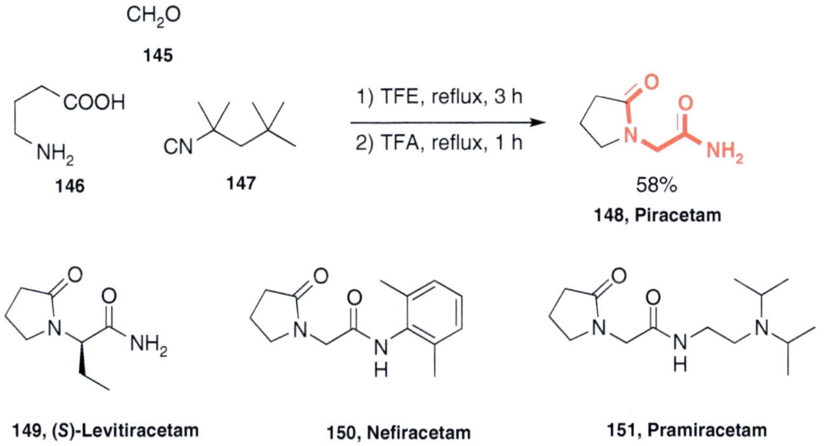

Scheme 3.23 MCR synthesis of piracetam and structure of other synthesized racetams. Source: Cioc et al. [137].

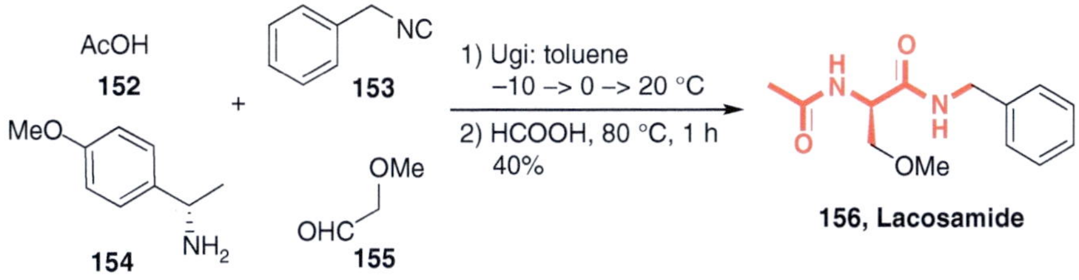

Scheme 3.24 2-Step (*R*)-lacosamide synthesis via Ugi MCR.

the stereocenter of lacosamide the chiral auxiliary (*S*)-4-methoxy phenethylamine **154** was chosen to yield the two diastereomers in 70 : 30, in favor of the desired (*S*,*R*)-diastereomer. They could be separated by crystallization. Cleavage of the auxiliar was accomplished by the use of formic acid at 80 °C. The API was synthesized on a 50 kg scale in 40% overall yield, meeting all required purity specifications (Scheme 3.24).

Olanzapine **165** is an atypical antipsychotic used to treat schizophrenia and bipolar disorder. It has a higher affinity for 5-HT2A serotonin receptors than D2 dopamine receptors, which is a common property of most atypical antipsychotics. As a generic drug, many different synthesis pathways have been described, including a flow synthesis [139]. A key element in the structure is the 2-amino thiophene which is synthesized using the Gewald reaction (GW-3CR) of acetonitrile derivatives, sulfur and a-methylene ketones or aldehydes (Scheme 3.25). In this particular case, propionic aldehyde **157** and malondinitrile **159** are reacted with sulfur **158** in DMF. The Gewald reaction is one of the few reactions in organic chemistry where elemental sulfur reacts at mild room temperature conditions [13, 140]. Furthermore, the 2-aminothiophene fragment is present in numerous drugs.

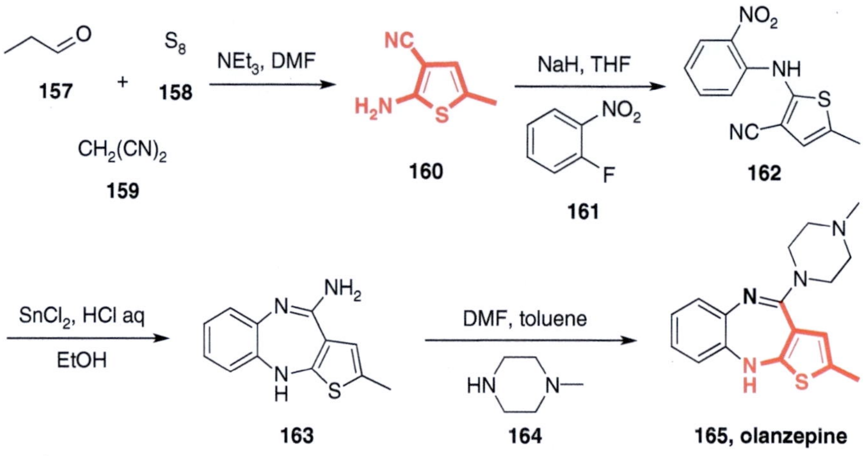

Scheme 3.25 Olanzapine synthesis involving a Gewald MCR fragment.

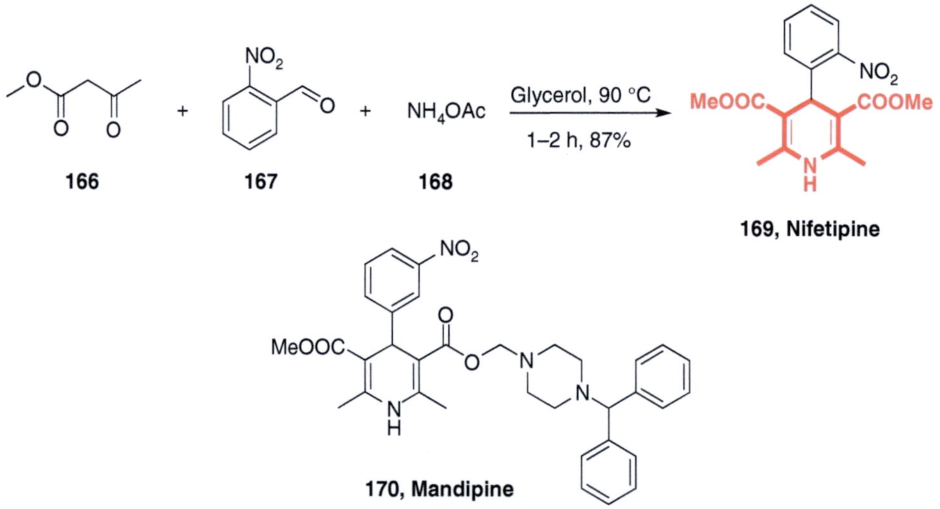

Scheme 3.26 Nifedipine synthesis using the Hantzsch MCR and structures of other dihydropyridine-based drugs. Source: Based on Sohal et al. [141].

Nifedipine **169** is another classic drug. It is a calcium channel blocker medication used to manage angina, high blood pressure, Raynaud's phenomenon, and premature labor. Dihydropyridines are synthesized by the Hantzsch reaction (Scheme 3.26) [141]. Several other drugs **170** based on the dihydropyridine scaffold are now approved.

Almorexant **176** is an orexin antagonist, functioning as a competitive receptor antagonist of the OX1 and OX2 orexin GPCR receptors, which was developed by the pharmaceutical companies Actelion and GSK for the treatment of insomnia [142]. Development of the drug was abandoned in January 2011 due to undisclosed issues pertaining to almorexant's safety profile. The U-3CR (performed

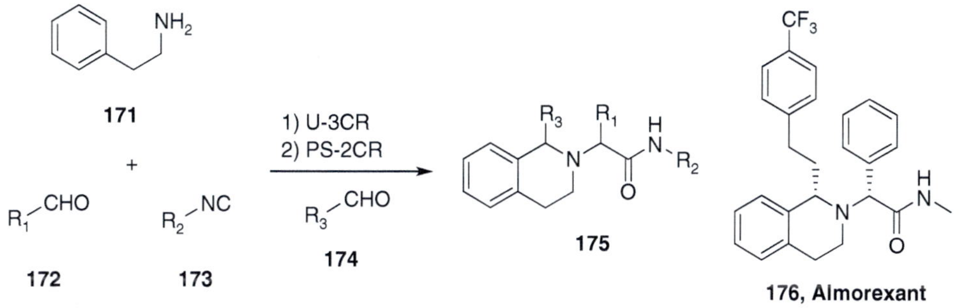

Scheme 3.27 Almorexant synthesis via the U-3CR.

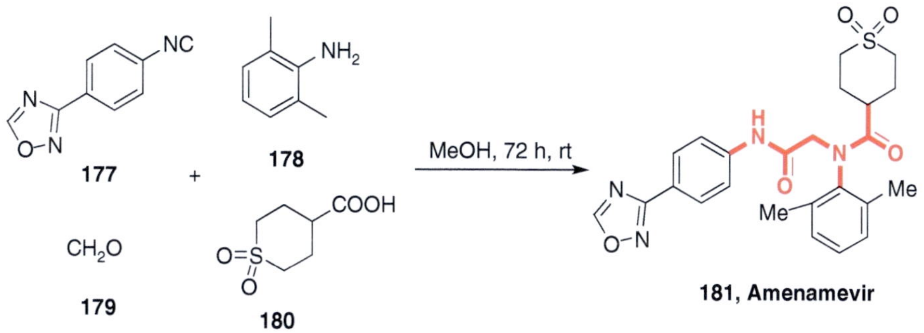

Scheme 3.28 Ugi reaction synthesis of antiviral Amenamevir.

by 2-phenylethan-1-amine **171**, aldehyde **172,** and isocyanide **173**) followed by a Pictet–Spengler reaction was functional in the discovery of Almorexant (Scheme 3.27). However, the synthesis pathway was abandoned in favor of an enantioselective one.

Amenamevir **181**, approved in Japan, is an antiviral drug used for the treatment of shingles (herpes zoster). It acts as an inhibitor of helicase–primase [143, 144]. Clearly, Amenamevir can be assembled as a product of a simple U-4CR of formaldehyde **179**, 2,6-dimethyl aniline **178**, carboxylic acid **180**, and *p*-oxadiazole phenylisocyanide **177**. Even taking the isocyanide building block synthesis into account, it outperforms the published multi-step sequential synthesis (Scheme 3.28).

3.3 Some Biopharmaceutical Application of MCR

Antibody drug conjugate (ADC) is an established, rapidly growing drug modality, mostly with applications in cancer but recently also in CNS disorders. An issue in ADC is the selective coupling of the linker and warhead on specific substituents of the antibody [145]. It is well established that the site-specific conjugation and the number of warheads per antibody have a profound influence on the PKPD properties of the ADC. This is often accomplished by disulfide bridge reduction and random reaction with Michael acceptors. Recently, genetically modified mAbs can be generated with non-proteinogenic amino acids in specific positions which can selectively react.

A novel site-selective strategy was recently introduced by employing Ugi- or Passerini MCR to conjugate different payloads. Native acid side chain amino acids (aspartate/glutamate) neighboring lysine work as acid and amine component in the Ugi reaction can be selectively modified with an oxo and isocyanide component for the conjugation process. This strategy is in sharp contrast to most conjugation techniques which target a single amino acid residue in the mAb. Specifically, Trastuzumab was coupled with different aldehydes and tert-butyl isocyanide to yield mAb with different ratios of payloads. Aldehydes included TAMRA aldehyde **186**, azide-containing aldehyde **188**, or glycosyl aldehyde **187**. The azide side chain was further derivatized by click chemistry, for example with a dolastatin derivative (Scheme 3.29).

While the number of modification sites on trastuzumab is too high to call this strategy site-specific, the *N*-terminal modification of both chains coupled with the excellent aspartate/glutamate selectivity (Figure 3.7).

Tubulysins are a class of highly cyctotoxic natural products isolated from *Myxobacteria* sp. [146] They act by binding to the cytoskeleton and preventing the necessary tubulin dynamic [147, 148]. They are exceedingly potent apoptosis inducers up to femtomolar concentrations in different cancer cell lines. Their small therapeutic window prevents their direct use as anticancer agents. However, tubulysins have been clinically evaluated as conjugates to folate and monoclonal antibodies [149–151]. Interestingly, Tubulysins have been conjugated very flexible at three different parts of the molecule: *C*-terminal, *N*-terminal, and on the aromatic part of the TubuPhe moiety. Despite multiple attempts to produce tubulysins by fermentation, the yields are too low and the separation of the different family members makes this process not viable for large-scale production. Multiple tubulysin total syntheses have been described, most of them lengthy and low yielding. A total synthesis of N^{14}-desacetoxytubulysin **190** using a Passerini reaction indeed is amongst the most convergent, shortest, and highest yielding (Scheme 3.30) [152]. This synthesis features less than 20 steps, with an overall yield approaching 30%, while being scalable. A key step involves the Passerini reaction of configurationally stable chiral cysteine-derived isocyanide, dipeptide-derived carboxylic acid, and chiral b-amino aldehyde **191**. The reaction was performed in the presence of chiral chiral ligand x together with $ZnBr_2$, to afford **195a** as the correct diastereomer as a major product (diastereoselectivity 92 : 8) in acceptable yield (71%). The Passerini reaction is included into the so-called PADAM [153, 154] (Passerini-deprotection-migration) sequence of *N*-protected aldehydes which after deprotection undergo a smooth transacylation.

3.3.1 Computational Methods of MCR Chemical Space Screening

The unique MCR feature of a very large chemical space at the same time is also a burden. While $>10^{12}$ compounds in principle are accessible by MCR no finite way to synthesize or screen, this numbers will be available anytime soon. Therefore,

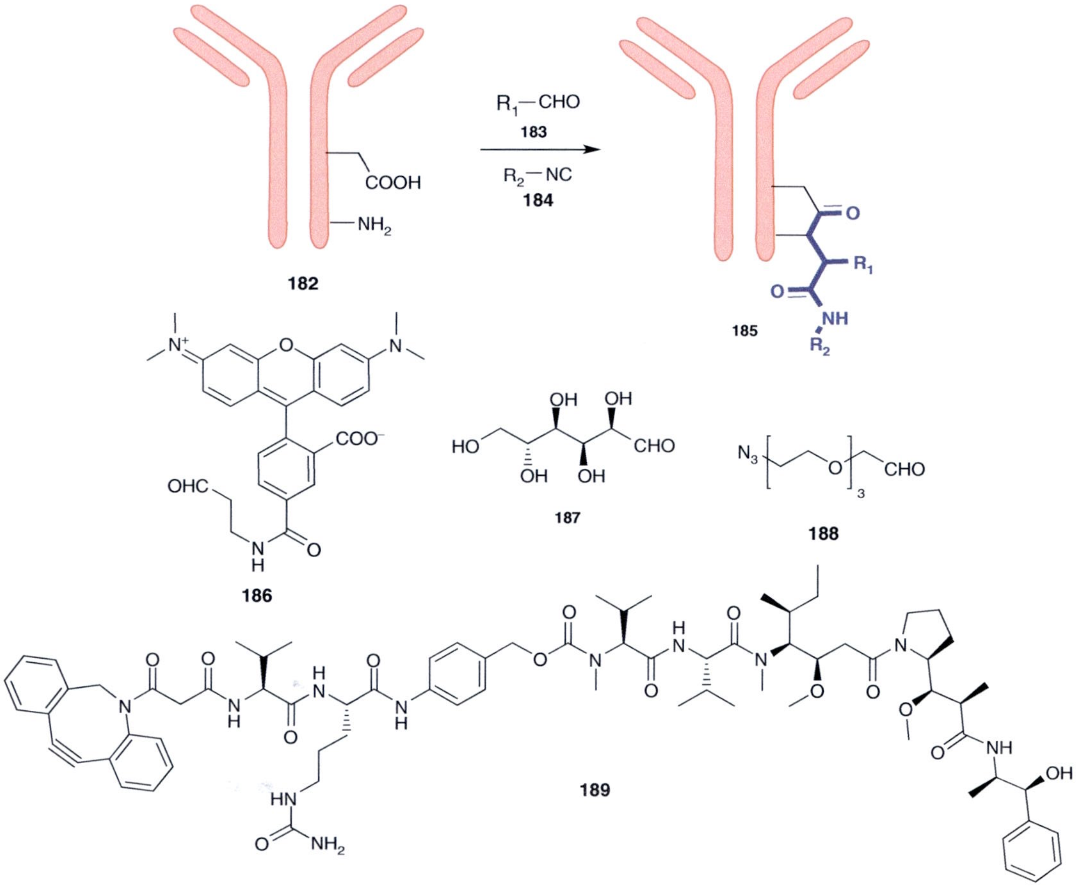

Scheme 3.29 Synthesis of antibody drug conjugates (ADC) using the U-4CR.

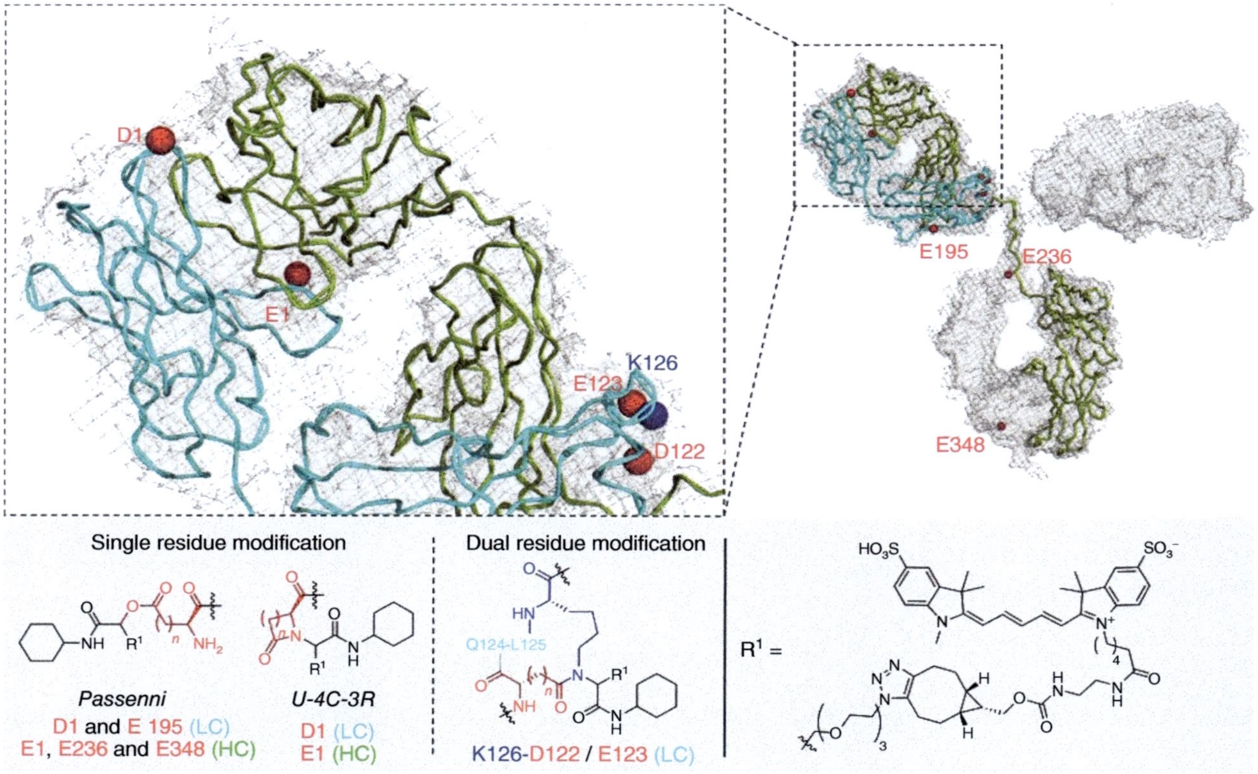

Figure 3.7 IMCR protein conjugation.

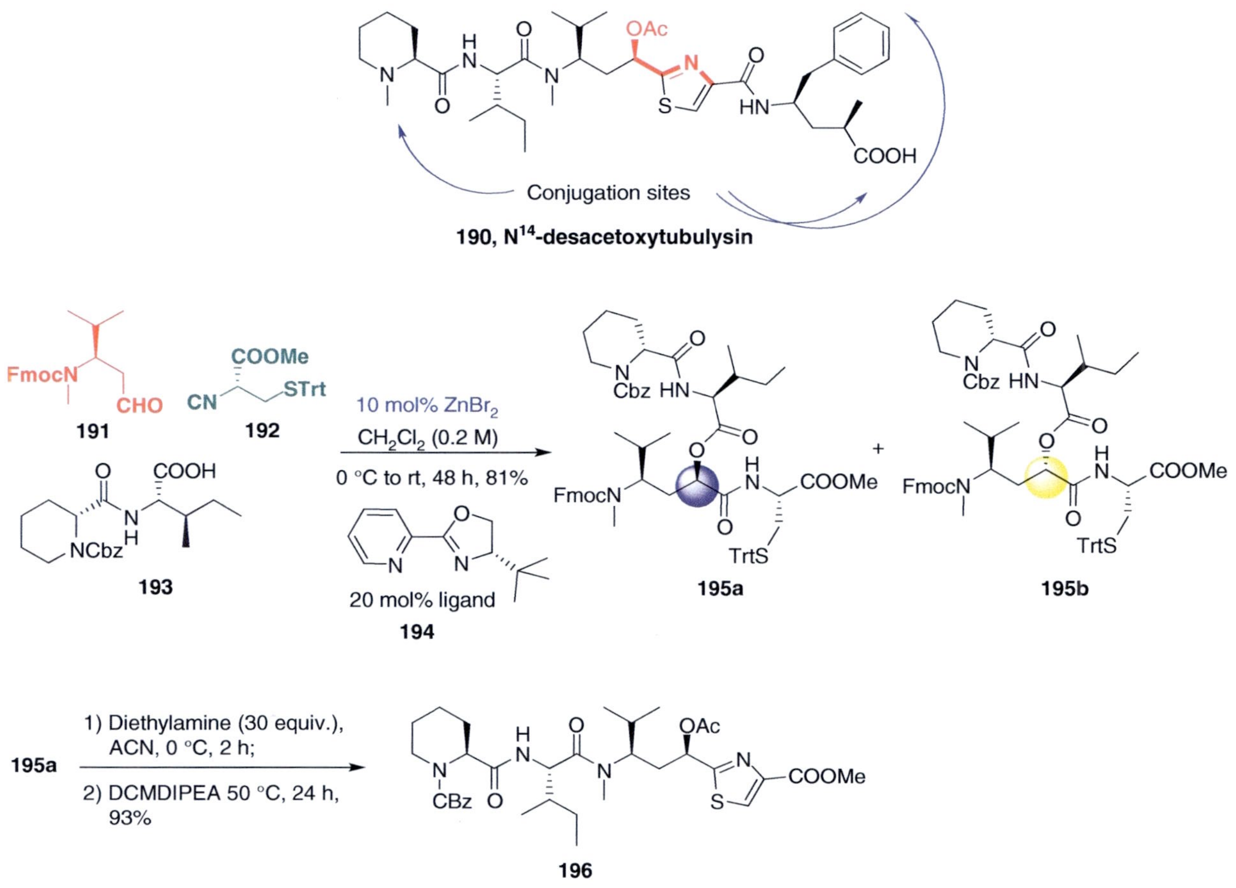

Scheme 3.30 Stereoselective, convergent, low step, and high yielding N^{14}-desacetoxytubulysin synthesis featuring a Passerini reaction. Source: Vishwanatha et al. [152].

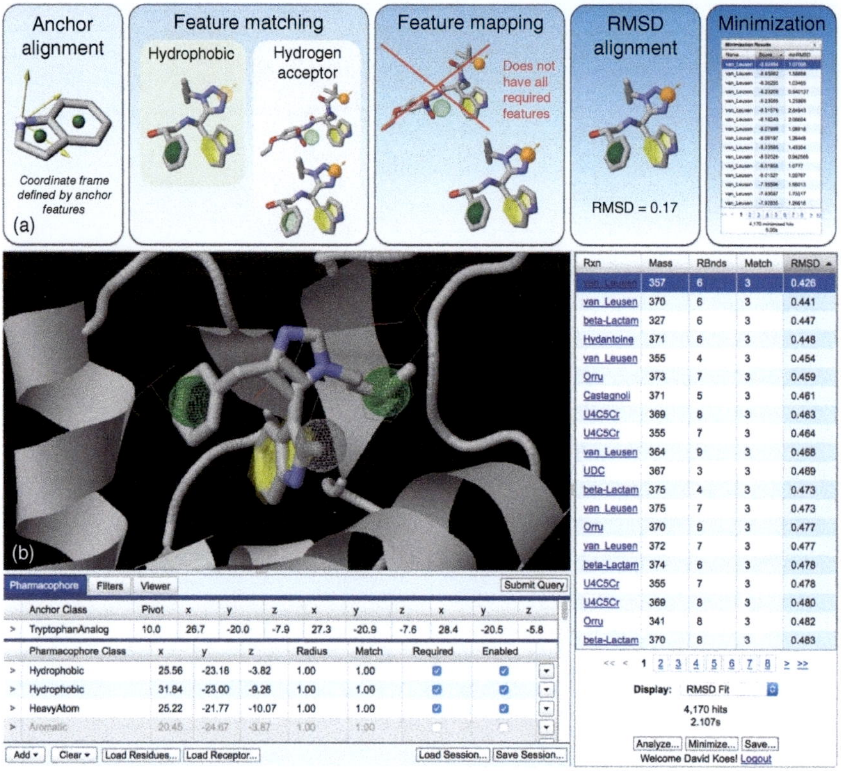

Figure 3.8 The ANCHOR.QUERY VS platform to screen >31 million MCR molecules. (a) The ANCHOR.QUERY computational workflow. (b) The ANCHOR.QUERY web-based interface for pharmacophore search. An example query derived from the p53/MDM2 PPI is shown (PDB ID 1YCR). Three key residues of p53 (thin wireframe) are shown with the MDM2 receptor (cartoon), the pharmacophore query (spheres and yellow anchor indicator), and a matching compound based on the van Leusen imidazole MCR from the tryptophan-biased library (sticks).

computational ways to narrow down the almost infinite chemical space are very useful. The ANCHOR.QUERY is a specialized pharmacophore search technology that brings interactive virtual screening of novel protein–protein inhibitors to the desktop [155, 156]. It is a freely accessible interactive webpage (http://anchorquery.csb.pitt.edu/) useful to screen >30 million amino acids mimics for drug-like MCR accessible compounds (Figure 3.8). ANCHOR.QUERY leverages the concept of anchors, amino acid residues that bury a large amount of solvent accessible surface area at the protein–protein interface. Every compound in our MCR accessible virtual library contains an anchor analog, a functional group that is a chemical mimic of a specific amino acid. ANCHOR.QUERY™ pharmacophore queries always include an anchor feature in addition to the standard hydrophobic, ionic, and hydrogen bond donors. All non-anchor features are stored relative to a coordinate system defined by the anchor in an efficient spatial index. Pharmacophore searches scale relative to the breadth and complexity of the query, not the size of the database. As a result, full 3D pharmacophore searches can be executed over millions of explicit conformations in a matter of seconds.

ANCHOR.QUERY was successfully applied to the discovery of multiple p53/MDM2 antagonists [75, 83, 157] and PDK1 allosteric inhibitors [158].

3.4 Conclusion

In our small review, we hope to have convinced the medicinal chemist that MCR is an interesting synthetic technology which can have advantages over traditional sequential multi-step approaches. Survey of contemporary literature suggests that MCR is heavily underused. Chemists are encouraged to think out-of-the-box of traditional chemistry and to try new pathways less walked. MCR chemistry is solid and reliably producing projected target compounds. MCR chemistry is useful at the discovery stage as well as for the production of APIs. Like few other reactions, MCR has an outstanding functional group compatibility which can be used for combining MCR with a very large number of other organic reactions to result in a sheer indefinite scaffold diversity. The building blocks used in MCR are archetypical chemicals available in numerous variations. Thus, an unprecedented large chemical space is accessible, with respect to number of steps. MCR chemistry can also be advantageously combined with modern approaches in artificial intelligence and automation of chemistry.

20 years ago Barry Sharpless stated: "Click chemistry, promised access to novel chemical space by empowering combinatorial library synthesis with a 'few good reactions.' These click reactions fulfilled key criteria (broad scope, quantitative yield, abundant starting material, mild reaction conditions, and high chemoselectivity), keeping the focus on molecules that would be easy to make, yet structurally diverse." [159]. We strongly believe that MCR chemistry is an even greater treasure trove that can help to focus on "form follows function."

References

1 Ugi, I., Domling, A., and Horl, W. (1994). Multicomponent reactions in organic-chemistry. *Endeavour* 18 (3): 115–122.

2 Domling, A. (1998). Isocyanide based multi component reactions in combinatorial chemistry. *Comb. Chem. High Throughput Screen.* 1 (1): 1–22.

3 Domling, A. (2000). The discovery of new isocyanide-based multi-component reactions. *Curr. Opin. Chem. Biol.* 4 (3): 318–323.

4 Domling, A. and Ugi, I. (2000). Multicomponent reactions with isocyanides. *Angew. Chem., Int. Ed.* 39 (18): 3168–3210.

5 Domling, A. (2006). Recent developments in isocyanide based multicomponent reactions in applied chemistry. *Chem. Rev.* 106 (1): 17–89.

6 Domling, A., Wang, W., and Wang, K. (2012). Chemistry and biology of multicomponent reactions. *Chem. Rev.* 112 (6): 3083–3135.

7 Zarganes-Tzitzikas, T. and Domling, A. (2014). Modern multicomponent reactions for better drug syntheses. *Org. Chem. Front.* 1 (7): 834–837.

8 Ruijter, E., Scheffelaar, R., and Orru, R.V.A. (2011). Multicomponent reaction design in the quest for molecular complexity and diversity. *Angew. Chem. Int. Ed.* 50 (28): 6234–6246.

9 Neochoritis, C.G., Zhao, T., and Domling, A. (2019). Tetrazoles via multicomponent reactions. *Chem. Rev.* 119 (3): 1970–2042.

10 Boltjes, A. and Dömling, A. (2019). The groebke-blackburn-bienayme reaction. *Eur. J. Org. Chem.* 2019 (42): 7007–7049.

11 Abdelraheem, E.M.M., Shaabani, S., and Domling, A. (2018). Artificial macrocycles. *Synlett* 29 (9): 1136–1151.

12 Zarganes-Tzitzikas, T., Chandgude, A.L., and Domling, A. (2015). Multicomponent reactions, union of MCRs and beyond. *Chem. Rec.* 15 (5): 981–996.

13 Huang, Y.J. and Domling, A. (2011). The Gewald multicomponent reaction. *Mol. Divers.* 15 (1): 3–33.

14 Domling, A. (2002). Recent advances in isocyanide-based multicomponent chemistry. *Curr. Opin. Chem. Biol.* 6 (3): 306–313.

15 Ugi, I., Domling, A., Gruber, B., and Almstetter, M. (1997). Multicomponent reactions and their libraries - A new approach to preparative organic chemistry. *Croat. Chem. Acta* 70 (2): 631–647.

16 Tripolitsiotis, N.P., Thomaidi, M., and Neochoritis, C.G. (2020). The ugi three-component reaction; a valuable tool in modern organic synthesis. *Eur. J. Org. Chem.* 2020 (42): 6525–6554.

17 Ugi, I., Werner, B., and Domling, A. (2003). The chemistry of isocyanides, their multicomponent reactions and their libraries. *Molecules* 8 (1): 53–66.

18 Banfi, L.; Riva, R. (2004). The Passerini Reaction. Organic reactions 65: 1–140.

19 Olyaei, A. and Sadeghpour, M. (2019). Recent advances in the synthesis and synthetic applications of Betti base (aminoalkylnaphthol) and bis-Betti base derivatives. *RSC Adv.* 9 (32): 18467–18497.

20 Serafini, M., Murgia, I., Giustiniano, M. et al. (2021). The 115 years old multicomponent Bargellini reaction: perspectives and new applications. *Molecules* 26 (3): 558.

21 Shumaila, A.M.A. and Al-Thulaia, A.A.N. (2019). Mini-review on the synthesis of Biginelli analogs using greener heterogeneous catalysis: recent strategies with the support or direct catalyzing of inorganic catalysts. *Synth. Commun.* 49 (13): 1613–1632.

22 Blanco-Urgoiti, J., Añorbe, L., Pérez-Serrano, L. et al. (2004). The Pauson–Khand reaction, a powerful synthetic tool for the synthesis of complex molecules. *Chem. Soc. Rev.* 33 (1): 32–42.

23 Ricker, J.D. and Geary, L.M. (2017). Recent advances in the Pauson-Khand reaction. *Top. Catal.* 60 (8): 609–619.

24 Ingate, S.T. and Marco-Contellers, J. (1998). The asymmetric Pauson-Khand reaction. a review. *Org. Prep. Proced. Int.* 30 (2): 121–143.

25 Shibata, T. (2006). Recent advances in the catalytic Pauson–Khand-type reaction. *Adv. Synth. Catal.* 348 (16–17): 2328–2336.

26 Geis, O. and Schmalz, H.-G. (1998). New developments in the Pauson–Khand reaction. *Angew. Chem. Int. Ed.* 37 (7): 911–914.
27 Candeias, N.R., Montalbano, F., Cal, P.M.S.D., and Gois, P.M.P. (2010). Boronic acids and esters in the petasis-borono mannich multicomponent reaction. *Chem. Rev.* 110 (10): 6169–6193.
28 Wu, P., Givskov, M., and Nielsen, T.E. (2019). Reactivity and synthetic applications of multicomponent petasis reactions. *Chem. Rev.* 119 (20): 11245–11290.
29 Yu Tao, L.H., Xinyan, W., and Jun, Y. (2012). Progress in petasis reaction. *Chinese J. Org. Chem.* 32 (10): 1836–1845.
30 Cherkasov, R.A. and Galkin, V.I. (1998). The Kabachnik–Fields reaction: synthetic potential and the problem of the mechanism. *Russ. Chem. Rev.* 67 (10): 857–882.
31 Keglevich, G. and Bálint, E. (2012). The Kabachnik-Fields reaction: mechanism and synthetic use. *Molecules* 17 (11): 12821–12835.
32 Grieco, P.A., Yoshida, K., and Garner, P. (1983). Aqueous intermolecular Diels-Alder chemistry: reactions of diene carboxylates with dienophiles in water at ambient temperature. *J. Org. Chem.* 48 (18): 3137–3139.
33 Cao, M.-H., Green, N.J., and Xu, S.-Z. (2017). Application of the aza-Diels–Alder reaction in the synthesis of natural products. *Org. Biomol. Chem.* 15 (15): 3105–3129.
34 Vinogradov, M.G., Turova, O.V., and Zlotin, S.G. (2021). Catalytic asymmetric Aza-Diels-Alder reaction: pivotal milestones and recent applications to synthesis of nitrogen-containing heterocycles. *Adv. Synth. Catal.* 363: 1466–1526.
35 Ghashghaei, O., Masdeu, C., Alonso, C. et al. (2018). Recent advances of the Povarov reaction in medicinal chemistry. *Drug Discov. Today Technol.* 29: 71–79.
36 Krasavin, M. and Dar'in, D. (2016). Current diversity of cyclic anhydrides for the Castagnoli–Cushman-type formal cycloaddition reactions: prospects and challenges. *Tetrahedron Lett.* 57 (15): 1635–1640.
37 González-López, M. and Shaw, J.T. (2009). Cyclic anhydrides in formal cycloadditions and multicomponent reactions. *Chem. Rev.* 109 (1): 164–189.
38 Stout, D.M. and Meyers, A.I. (1982). Recent advances in the chemistry of dihydropyridines. *Chem. Rev.* 82 (2): 223–243.
39 Zai-Qun, L. (2015). Two neglected multicomponent reactions: asinger and groebke reaction for constructing thiazolines and imidazolines. *Curr. Org. Synth.* 12 (1): 20–60.
40 Weyesa, A. and Mulugeta, E. (2020). Recent advances in the synthesis of biologically and pharmaceutically active quinoline and its analogues: a review. *RSC Adv.* 10 (35): 20784–20793.
41 Garbarino, S., Ravelli, D., Protti, S., and Basso, A. (2016). Photoinduced multicomponent reactions. *Angew. Chem. Int. Ed.* 55 (50): 15476–15484.
42 Lennine, R.M. and Wender, A.S. (2016). Ionic liquid in multicomponent reactions: a brief review. *Curr. Green Chem.* 3 (2): 120–132.
43 Moni, L., Banfi, L., Riva, R., and Basso, A. (2016). External-oxidant-based multicomponent reactions. *Synthesis* 48 (23): 4050–4059.

44 Sharma, U.K., Ranjan, P., Van der Eycken, E.V., and You, S.-L. (2020). Sequential and direct multicomponent reaction (MCR)-based dearomatization strategies. *Chem. Soc. Rev.* 49 (23): 8721–8748.

45 Fairoosa, J., Neetha, M., and Anilkumar, G. (2021). Recent developments and perspectives in the copper-catalyzed multicomponent synthesis of heterocycles. *RSC Adv.* 11 (6): 3452–3469.

46 Hulme, C.; Gore, V. (2003). Multi-component Reactions: Emerging Chemistry in Drug Discovery 'From Xylocain to Crixivan'. Current medicinal chemistry. 10(1): 51–80.

47 Wenyi, Z. and Fen-Er, C. (2012). One-pot synthesis and its practical application in pharmaceutical industry. *Curr. Org. Synth.* 9 (6): 873–897.

48 Banfi, L., Basso, A., Moni, L., and Riva, R. (2014). The alternative route to enantiopure multicomponent reaction products: biocatalytic or organocatalytic enantioselective production of inputs for multicomponent reactions. *Eur. J. Org. Chem.* 2014 (10): 2005–2015.

49 Wang, J., Qin, A., and Tang, B.Z. (2020). Multicomponent polymerizations involving green monomers. *Macromol. Rapid Commun.*: 42 (6) 2000547.

50 Reguera, L., Méndez, Y., Humpierre, A.R. et al. (2018). Multicomponent reactions in ligation and bioconjugation chemistry. *Acc. Chem. Res.* 51 (6): 1475–1486.

51 Lamberth, C. (2020). Multicomponent reactions in crop protection chemistry. *Bioorg. Med. Chem.* 28 (10): 115471.

52 Dömling, A. and Huang, Y. (2010). Piperazine scaffolds via isocyanide-based multicomponent reactions. *Synthesis* 2010 (17): 2859–2883.

53 Huang, Y., Khoury, K., Chanas, T., and Dömling, A. (2012). Multicomponent synthesis of diverse 1, 4-benzodiazepine scaffolds. Organic Letters. 14(23): 5916–5919.

54 Gerencsér, J., Dormán, G., and Darvas, F. (2006). Meldrum's acid in multicomponent reactions: applications to combinatorial and diversity-oriented synthesis. *QSAR & Comb. Sci.* 25 (5-6): 439–448.

55 Martinez de Marigorta, E., de los Santos, J.M., Ochoa de Retana, A.M. et al. (2018). Multicomponent reactions in the synthesis of γ-lactams. *Synthesis* 50 (23): 4539–4554.

56 Biesen, L. and Müller, T.J.J. (2021). Multicomponent and one-pot syntheses of quinoxalines. *Adv. Synth. Catal.* 363 (4): 980–1006.

57 Reguera, L. and Rivera, D.G. (2019). Multicomponent reaction toolbox for peptide macrocyclization and stapling. *Chem. Rev.* 119 (17): 9836–9860.

58 Abdelraheem, E.M.M., Shaabani, S., and Dömling, A. (2018). Macrocycles: MCR synthesis and applications in drug discovery. *Drug Discov. Today Technol.* 29: 11–17.

59 Plowright, A.T., Johnstone, C., Kihlberg, J. et al. (2012). Hypothesis driven drug design: improving quality and effectiveness of the design-make-test-analyse cycle. *Drug Discov. Today* 17 (1): 56–62.

60 Patil, P., Ahmadian-Moghaddam, M., and Dömling, A. (2020). Isocyanide 2.0. *Green Chem.* 22 (20): 6902–6911.

61 Neochoritis, C.G., Zhang, J., and Domling, A. (2015). Leuckart-wallach approach to sugar isocyanides and its IMCRs. *Synthesis-Stuttgart* 47 (16): 2407–2413.

62 Neochoritis, C.G., Zarganes-Tzitzikas, T., Stotani, S. et al. (2015). Leuckart-wallach route toward isocyanides and some applications. *ACS Comb. Sci.* 17 (9): 493–499.

63 Cioc, R.C., Ruijter, E., and Orru, R.V.A. (2014). Multicomponent reactions: advanced tools for sustainable organic synthesis. *Green Chem.* 16 (6): 2958–2975.

64 Mironov, M.A., Ivantsova, M.N., and Mokrushin, V.S. (2003). Ugi reaction in aqueous solutions: a simple protocol for libraries production. *Mol. Divers.* 6 (3): 193–197.

65 Cannalire, R., Pelliccia, S., Sancineto, L. et al. (2021). Visible light photocatalysis in the late-stage functionalization of pharmaceutically relevant compounds. *Chem. Soc. Rev.* 50: 766–897.

66 Neochoritis, C., Estrada-Ortiz, N., Khoury, K., and Domling, A. (2014). p53-MDM2 and MDMX antagonists. In: *Annual Reports in Medicinal Chemistry*, vol. 49 (ed. M.C. Desai), 167–187. Elsevier, Amsterdam, The Netherlands.

67 Popowicz, G.M., Domling, A., and Holak, T.A. (2011). The structure-based design of Mdm2/Mdmx-p53 inhibitors gets serious. *Angew. Chem., Int. Ed.* 50 (12): 2680–2688.

68 Khoury, K., Popowicz, G.M., Holak, T.A., and Domling, A. (2011). The p53-MDM2/MDMX axis - A chemotype perspective. *Medchemcomm* 2 (4): 246–260.

69 Wang, W.J., Groves, M.R., and Domling, A. (2018). Artificial macrocycles as IL-17A/IL-17RA antagonists. *Medchemcomm* 9 (1): 22–26.

70 Kussie, P.H., Gorina, S., Marechal, V. et al. (1996). Structure of the MDM2 oncoprotein bound to the p53 tumor suppressor transactivation domain. *Science (New York, N.Y.)* 274 (5289): 948–953.

71 Sisko, J., Kassick, A.J., Mellinger, M. et al. (2000). An investigation of imidazole and oxazole syntheses using aryl-substituted TosMIC reagents1. *J. Org. Chem.* 65 (5): 1516–1524.

72 Popowicz, G.M., Czarna, A., Wolf, S. et al. (2010). Structures of low molecular weight inhibitors bound to MDMX and MDM2 reveal new approaches for p53-MDMX/MDM2 antagonist drug discovery. *Cell Cycle* 9 (6): 1104–1111.

73 Czarna, A., Beck, B., Srivastava, S. et al. (2010). Robust generation of lead compounds for protein-protein interactions by computational and MCR chemistry: p53/Hdm2 antagonists. *Angew. Chem., Int. Ed.* 49 (31): 5352–5356.

74 Furet, P., Chène, P., De Pover, A. et al. (2012). The central valine concept provides an entry in a new class of non peptide inhibitors of the p53-MDM2 interaction. *Bioorg. Med. Chem. Lett.* 22 (10): 3498–3502.

75 Czarna, A., Beck, B., Srivastava, S. et al. (2010). Robust generation of lead compounds for protein-protein interactions by computational and MCR chemistry: p53/Hdm2 antagonists. *Angew. Chem. Int. Ed. Eng.* 49 (31): 5352–5356.

76 Raboisson, P., Marugán, J.J., Schubert, C. et al. (2005). Structure-based design, synthesis, and biological evaluation of novel 1, 4-diazepines as HDM2 antagonists. *Bioorg. Med. Chem. Lett.* 15 (7): 1857–1861.

77 Parks, D.J., LaFrance, L.V., Calvo, R.R. et al. (2006). Enhanced pharmacokinetic properties of 1, 4-benzodiazepine-2, 5-dione antagonists of the HDM2-p53 protein–protein interaction through structure-based drug design. *Bioorg. Med. Chem. Lett.* 16 (12): 3310–3314.

78 Grasberger, B.L., Lu, T., Schubert, C. et al. (2005). Discovery and cocrystal structure of benzodiazepinedione HDM2 antagonists that activate p53 in cells. *J. Med. Chem.* 48 (4): 909–912.

79 Wang, W., Cao, H.P., Wolf, S. et al. (2013). Benzimidazole-2-one: a novel anchoring principle for antagonizing p53-Mdm2. *Bioorg. Med. Chem.* 21 (14): 3982–3995.

80 Huang, Y.J., Wolf, S., Koes, D. et al. (2012). Exhaustive fluorine scanning toward potent p53-Mdm2 antagonists. *ChemMedChem* 7 (1): 49–52.

81 Huang, Y.J., Wolf, S., Beck, B. et al. (2014). Discovery of highly potent p53-MDM2 antagonists and structural basis for anti-acute myeloid leukemia activities. *ACS Chem. Biol.* 9 (3): 802–811.

82 Skalniak, L., Twarda-Clapa, A., Neochoritis, C.G. et al. (2019). A fluorinated indole-based MDM2 antagonist selectively inhibits the growth of p53(wt) osteosarcoma cells. *FEBS J.* 286 (7): 1360–1374.

83 Shaabani, S., Neochoritis, C.G., Twarda-Clapa, A. et al. (2017). Scaffold hopping via ANCHOR.QUERY: β-lactams as potent p53-MDM2 antagonists. *MedChemComm* 8 (5): 1046–1052.

84 Estrada-Ortiz, N., Neochoritis, C.G., Twarda-Clapa, A. et al. (2017). Artificial macrocycles as potent p53-MDM2 inhibitors. *ACS Med. Chem. Lett.* 8 (10): 1025–1030.

85 Wang, K., Nguyen, K., Huang, Y., and Dömling, A. (2009). Cyanoacetamide multicomponent reaction (I): parallel synthesis of cyanoacetamides. *J. Comb. Chem.* 11 (5): 920–927.

86 Huang, Y.J., Wolf, S., Bista, M. et al. (2010). 1,4-Thienodiazepine-2,5-diones via MCR (I): synthesis, virtual space and p53-Mdm2 activity. *Chem. Biol. Drug Des.* 76 (2): 116–129.

87 Ricardo, M.G., Ali, A.M., Plewka, J. et al. (2020). Multicomponent peptide stapling as a diversity-driven tool for the development of inhibitors of protein–protein interactions. *Angew. Chem. Int. Ed.* 59 (13): 5235–5241.

88 Goodnow, R.A., Dumelin, C.E., and Keefe, A.D. (2017). DNA-encoded chemistry: enabling the deeper sampling of chemical space. *Nat. Rev. Drug Discov.* 16 (2): 131–147.

89 Favalli, N., Bassi, G., Scheuermann, J., and Neri, D. (2018). DNA-encoded chemical libraries – achievements and remaining challenges. *FEBS Lett.* 592 (12): 2168–2180.

90 Kunig, V.B.K., Ehrt, C., Dömling, A., and Brunschweiger, A. (2019). Isocyanide multicomponent reactions on solid-phase-coupled DNA oligonucleotides for encoded library synthesis. *Org. Lett.* 21 (18): 7238–7243.

91 Kunig, V.B.K., Potowski, M., Akbarzadeh, M. et al. (2020). TEAD-YAP interaction inhibitors and MDM2 binders from DNA-encoded indole-focused Ugi peptidomimetics. *Angew. Chem.*, Int. Ed. 132: 20518–20522

92 Hu, Y., Stumpfe, D., and Bajorath, J. (2016). Computational exploration of molecular scaffolds in medicinal chemistry. *J. Med. Chem.* 59 (9): 4062–4076.

93 Bemis, G.W. and Murcko, M.A. (1996). The properties of known drugs. 1. Molecular frameworks. *J. Med. Chem.* 39 (15): 2887–2893.

94 Xu, Y.-j. and Johnson, M. (2001). Algorithm for naming molecular equivalence classes represented by labeled pseudographs. *J. Chem. Inf. Comput. Sci.* 41 (1): 181–185.

95 Moquin, S.A., Simon, O., Karuna, R. et al. (2021). NITD-688, a pan-serotype inhibitor of the dengue virus NS4B protein, shows favorable pharmacokinetics and efficacy in preclinical animal models. *Sci. Transl. Med.* 13 (579): eabb2181.

96 Faraji, A., Oghabi Bakhshaiesh, T., Hasanvand, Z. et al. (2021). Design, synthesis and evaluation of novel thienopyrimidine-based agents bearing diaryl urea functionality as potential inhibitors of angiogenesis. *Eur. J. Med. Chem.* 209: 112942.

97 Pang, H., Wang, N., Chai, J. et al. (2020). Discovery of novel TNNI3K inhibitor suppresses pyroptosis and apoptosis in murine myocardial infarction injury. *Eur. J. Med. Chem.* 197: 112314.

98 Saeedi, M., Safavi, M., Allahabadi, E. et al. (2020). Thieno[2,3-b]pyridine amines: synthesis and evaluation of tacrine analogs against biological activities related to Alzheimer's disease. *Arch. Pharm.* 353 (10): 2000101.

99 Li, J., Zheng, Y., Huang, M., and Li, W. (2020). Ni-catalyzed denitrogenative cross-coupling of benzotriazinones and cyclopropanols: an easy access to functionalized β-aryl ketones. *Org. Lett.* 22 (13): 5020–5024.

100 Wang, F.-C., Peng, B., Cao, S.-L. et al. (2020). Synthesis and cytotoxic activity of chalcone analogues containing a thieno[2,3-d]pyrimidin-2-yl group as the A-ring or B-ring. *Bioorg. Chem.* 94: 103346.

101 Härter, M., Kalthof, B., Delbeck, M. et al. (2019). Novel non-xanthine antagonist of the A2B adenosine receptor: From HTS hit to lead structure. *Eur. J. Med. Chem.* 163: 763–778.

102 Volochnyuk, D.M., Kovaleva, S.A., Chernega, A.N. et al. (2006). Novel approaches to fused phospha-pyrimidines. *Synthesis* 2006 (10): 1613–1624.

103 Wang, S., Fang, K., Dong, G. et al. (2015). Scaffold diversity inspired by the natural product evodiamine: discovery of highly potent and multitargeting antitumor agents. *J. Med. Chem.* 58 (16): 6678–6696.

104 Berman, H.M., Westbrook, J., Feng, Z. et al. (2000). The protein data bank. *Nucleic Acids Res.* 28 (1): 235–242.

105 Daugan, A., Grondin, P., Ruault, C. et al. (2003). The discovery of tadalafil: a novel and highly selective PDE5 inhibitor. 2: 2,3,6,7,12,12a-hexahydropyrazino[1′,2′:1,6]pyrido[3,4-b]indole-1,4-dione analogues. *J. Med. Chem.* 46 (21): 4533–4542.

106 Jida, M. and Ballet, S. (2018). An efficient one-pot synthesis of chiral N-protected 3-substituted (Diketo)piperazines via Ugi-4CR/De-Boc/cyclization process. *ChemistrySelect* 3 (4): 1027–1031.

107 Itoh, Y., Ishikawa, M., Naito, M., and Hashimoto, Y. (2010). Protein knockdown using methyl bestatin–ligand hybrid molecules: design and synthesis of inducers of ubiquitination-mediated degradation of cellular retinoic acid-binding proteins. *J. Am. Chem. Soc.* 132 (16): 5820–5826.

108 Nemoto, H., Ma, R., Suzuki, I., and Shibuya, M. (2000). A new one-pot method for the synthesis of α-siloxyamides from aldehydes or ketones and its application to the synthesis of (−)-bestatin. *Org. Lett.* 2 (26): 4245–4247.

109 Richter, A. and Hedberg, C. (2010). A practical and convenient synthesis of the protease inhibitor epibestatin. *Synthesis* 2010 (12): 2039–2042.

110 George, S., Suryavanshi, G.S., and Sudalai, A. (2008). A short enantioselective synthesis of (−)-bestatin via l-proline-catalyzed α-amination of an aldehyde. *Tetrahedron Lett.* 49 (48): 6791–6793.

111 Semple, J.E., Owens, T.D., Nguyen, K., and Levy, O.E. (2000). New synthetic technology for efficient construction of α-hydroxy-β-amino amides via the passerini reaction1. *Org. Lett.* 2 (18): 2769–2772.

112 Weigert, W.M., Offermanns, H., and Degussa, P.S. (1975). D-Penicillamine—Production and Properties. *Angew. Chem. Int. Ed. Engl.* 14 (5): 330–336.

113 DiNardo, C.D., Stein, E.M., de Botton, S. et al. (2018). Durable remissions with ivosidenib in IDH1-mutated relapsed or refractory AML. *N. Engl. J. Med.* 378 (25): 2386–2398.

114 Fan, B., Dai, D., DiNardo, C.D. et al. (2020). Clinical pharmacokinetics and pharmacodynamics of ivosidenib in patients with advanced hematologic malignancies with an IDH1 mutation. *Cancer Chemother. Pharmacol.* 85 (5): 959–968.

115 Popovici-Muller, J., Lemieux, R.M., Artin, E. et al. (2018). Discovery of AG-120 (ivosidenib): a first-in-class mutant IDH1 inhibitor for the treatment of IDH1 mutant cancers. *ACS Med. Chem. Lett.* 9 (4): 300–305.

116 Roboz, G.J., DiNardo, C.D., Stein, E.M. et al. (2020). Ivosidenib induces deep durable remissions in patients with newly diagnosed IDH1-mutant acute myeloid leukemia. *Blood* 135 (7): 463–471.

117 Kim, J.J., Culley, C.M., and Mohammad, R.A. (2012). Telaprevir: an oral protease inhibitor for hepatitis C virus infection. *Am. J. Health-Syst. Pharm.* 69 (1): 19–33.

118 Fu, L., Ye, F., Feng, Y. et al. (2020). Both Boceprevir and GC376 efficaciously inhibit SARS-CoV-2 by targeting its main protease. *Nat. Commun.* 11 (1): 4417.

119 Oerlemans, R., Ruiz-Moreno, A.J., Cong, Y. et al. (2021). Repurposing the HCV NS3–4A protease drug boceprevir as COVID-19 therapeutics. *RSC Med. Chem.* 12: 370–379.

120 Yip, Y., Victor, F., Lamar, J. et al. (2004). Discovery of a novel bicycloproline P2 bearing peptidyl α-ketoamide LY514962 as HCV protease inhibitor. *Bioorg. Med. Chem. Lett.* 14 (1): 251–256.

121 Znabet, A., Polak, M.M., Janssen, E. et al. (2010). A highly efficient synthesis of telaprevir by strategic use of biocatalysis and multicomponent reactions. *Chem. Commun.* 46 (42): 7918–7920.

122 Köhler, V., Bailey, K.R., Znabet, A. et al. (2010). Enantioselective biocatalytic oxidative desymmetrization of substituted pyrrolidines. *Angew. Chem. Int. Ed.* 49 (12): 2182–2184.

123 Dömling, A. and Khoury, K. (2010). Praziquantel and schistosomiasis. *ChemMedChem* 5 (9): 1420–1434.

124 Seubert, J., Pohlke, R., and Loebich, F. (1977). Synthesis and properties of praziquantel, a novel broad spectrum anthelmintic with excellent activity against Schistosomes and Cestodes. *Experientia* 33 (8): 1036–1037.

125 Cao, H., Liu, H., and Dömling, A. (2010). Efficient multicomponent reaction synthesis of the schistosomiasis drug praziquantel. *Chem. Eur. J.* 16 (41): 12296–12298.

126 Liu, H., William, S., Herdtweck, E. et al. (2012). MCR synthesis of praziquantel derivatives. *Chem. Biol. Drug Des.* 79 (4): 470–477.

127 Wang, W., Ollio, S., Herdtweck, E., and Dömling, A. (2011). Polycyclic compounds by Ugi–Pictet–Spengler sequence. *J. Org. Chem.* 76 (2): 637–644.

128 Zarganes-Tzitzikas, T., Neochoritis, C.G., and Dömling, A. (2019). Atorvastatin (Lipitor) by MCR. *ACS Med. Chem. Lett.* 10 (3): 389–392.

129 Malaquin, S., Jida, M., Gesquiere, J.-C. et al. (2010). Ugi reaction for the synthesis of 4-aminopiperidine-4-carboxylic acid derivatives. Application to the synthesis of carfentanil and remifentanil. *Tetrahedron Lett.* 51 (22): 2983–2985.

130 de Castro, S., Camarasa, M.-J., Balzarini, J., and Velázquez, S. (2014). Discovery and SAR studies of a novel class of cytotoxic 1,4-disubstituted piperidines via Ugi reaction. *Eur. J. Med. Chem.* 83: 174–189.

131 Váradi, A., Palmer, T.C., Haselton, N. et al. (2015). Synthesis of carfentanil amide opioids using the Ugi multicomponent reaction. *ACS Chem. Neurosci.* 6 (9): 1570–1577.

132 van der Heijden, G., Jong, J.A.W., Ruijter, E., and Orru, R.V.A. (2016). 2-Bromo-6-isocyanopyridine as a universal convertible isocyanide for multicomponent chemistry. *Org. Lett.* 18 (5): 984–987.

133 Kalinski, C., Lemoine, H., Schmidt, J. et al. (2008). Multicomponent reactions as a powerful tool for generic drug synthesis. *Synthesis* 2008 (24): 4007–4011.

134 Golubev, P., Guranova, N., and Krasavin, M. (2020). 1,3,5-Triazinanes as formaldimine surrogates in the Ugi reaction. *Eur. J. Org. Chem.* 2020 (29): 4517–4520.

135 Adolph, C.M., Werth, J., Selvaraj, R. et al. (2017). Dehydrogenative transformations of imines using a heterogeneous photocatalyst. *J. Org. Chem.* 82 (11): 5959–5965.

136 Borase, B.B., Godbole, H.M., Singh, G.P. et al. (2020). Application of Ugi three component reaction for the synthesis of quinapril hydrochloride. *Synth. Commun.* 50 (1): 48–55.

137 Cioc, R.C., Schaepkens van Riempst, L., Schuckman, P. et al. (2017). Ugi four-center three-component reaction as a direct approach to racetams. *Synthesis* 49 (07): 1664–1674.

138 Wehlan, H., Oehme, J., Schäfer, A., and Rossen, K. (2015). Development of scalable conditions for the Ugi reaction—application to the synthesis of (R)-lacosamide. *Org. Process Res. Dev.* 19 (12): 1980–1986.

139 Hartwig, J., Ceylan, S., Kupracz, L. et al. (2013). Heating under high-frequency inductive conditions: application to the continuous synthesis of the neuroleptícum olanzapine (Zyprexa). *Angew. Chem. Int. Ed. (English)* 52 (37): 9813–9817.

140 Gewald, K., Schinke, E., and Böttcher, H. (1966). Heterocyclen aus CH-aciden nitrilen, VIII. 2-amino-thiophene aus methylenaktiven nitrilen, carbonylverbindungen und schwefel. *Chem. Ber.* 99 (1): 94–100.

141 Sohal, H.S., Goyal, A., Sharma, R., and Khare, R. (2014). One-pot, multicomponent synthesis of symmetrical Hantzsch 1,4-dihydropyridine derivatives using glycerol as clean and green solvent. *Eur. J. Chem.* 5 (1) https://doi.org/10.5155/eurjchem.5.1.171-175.943.

142 Boss, C. (2020). 20 years of medicinal chemistry - always look at the bright side (of life). *Chimia* 74 (7): 549–560.

143 Kawashima, M., Nemoto, O., Honda, M. et al. (2017). The study, i., Amenamevir, a novel helicase–primase inhibitor, for treatment of herpes zoster: a randomized, double-blind, valaciclovir-controlled phase 3 study. *J. Dermatol.* 44 (11): 1219–1227.

144 Yajima, M., Yamada, H., Takemoto, M. et al. (2017). Profile of anti-herpetic action of ASP2151 (amenamevir) as a helicase-primase inhibitor. *Antivir. Res.* 139: 95–101.

145 Beck, A., Goetsch, L., Dumontet, C., and Corvaïa, N. (2017). Strategies and challenges for the next generation of antibody–drug conjugates. *Nat. Rev. Drug Discov.* 16 (5): 315–337.

146 Steinmetz, H., Glaser, N., Herdtweck, E. et al. (2004). Isolation, crystal and solution structure determination, and biosynthesis of tubulysins—powerful inhibitors of tubulin polymerization from myxobacteria. *Angew. Chem. Int. Ed.* 43 (37): 4888–4892.

147 Kaur, G., Hollingshead, M., Holbeck, S. et al. (2006). Biological evaluation of tubulysin A: a potential anticancer and antiangiogenic natural product. *Biochem. J.* 396 (2): 235–242.

148 Khalil, M.W., Sasse, F., Lünsdorf, H. et al. (2006). Mechanism of action of tubulysin, an antimitotic peptide from myxobacteria. *ChemBioChem* 7 (4): 678–683.

149 Reddy, J.A., Dorton, R., Dawson, A. et al. (2009). In vivo structural activity and optimization studies of folate–tubulysin conjugates. *Mol. Pharm.* 6 (5): 1518–1525.

150 Tumey, L.N., Leverett, C.A., Vetelino, B. et al. (2016). Optimization of tubulysin antibody–drug conjugates: a case study in addressing ADC metabolism. *ACS Med. Chem. Lett.* 7 (11): 977–982.

151 Thurston, D.E., Jackson, P.J. et al. Contents (2019). *Cytotoxic Payloads for Antibody–Drug Conjugates*, P011–P021. The Royal Society of Chemistry.

152 Vishwanatha, T.M., Giepmans, B., Goda, S.K., and Dömling, A. (2020). Tubulysin synthesis featuring stereoselective catalysis and highly convergent multicomponent assembly. *Org. Lett.* 22 (14): 5396–5400.

153 Banfi, L., Basso, A., Guanti, G., and Riva, R. (2003). Passerini reaction--amine deprotection--acyl migration (PADAM): a convenient strategy for the solid-phase preparation of peptidomimetic compounds. *Mol. Divers.* 6 (3-4): 227–235.

154 Owens, T.D. and Semple, J.E. (2001). Atom-economical synthesis of the N(10)-C(17) fragment of cyclotheonamides via a novel Passerini reaction-deprotection-acyl migration strategy. *Org. Lett.* 3 (21): 3301–3304.

155 Koes, D., Khoury, K., Huang, Y. et al. (2012). Enabling large-scale design, synthesis and validation of small molecule protein-protein antagonists. *PLoS One* 7 (3): e32839.

156 Koes, D.R., Dömling, A., and Camacho, C.J. (2018). AnchorQuery: rapid online virtual screening for small-molecule protein-protein interaction inhibitors. *Protein Sci.* 27 (1): 229–232.

157 Surmiak, E., Neochoritis, C.G., Musielak, B. et al. (2017). Rational design and synthesis of 1,5-disubstituted tetrazoles as potent inhibitors of the MDM2-p53 interaction. *Eur. J. Med. Chem.* 126: 384–407.

158 Kroon, E., Schulze, J.O., Süß, E. et al. (2015). Discovery of a potent allosteric kinase modulator by combining computational and synthetic methods. *Angew. Chem. Int. Ed. Engl.* 54 (47): 13933–13936.

159 Kolb, H.C., Finn, M.G., and Sharpless, K.B. (2001). Click chemistry: diverse chemical function from a few good reactions. *Angew. Chem. Int. Ed.* 40 (11): 2004–2021.

4

Solid-Phase Heterocycle Synthesis Using Multicomponent Reactions

Leonardo G. Ceballos[1,2], *Daylin F. Pacheco*[3], *Bernhard Westermann*[2], *and Daniel G. Rivera*[1,3]

[1] *University of Havana, Laboratory of Synthetic and Biomolecular Chemistry, Faculty of Chemistry, Zapata and G, Havana, 10400, Cuba*
[2] *Leibniz Institute of Plant Biochemistry, Department of Bioorganic Chemistry, Weinberg 3, Halle/Saale 06120, Germany*
[3] *University of Havana, Center for Natural Products Research Zapata and G, Havana 10400, Cuba*

In memory of Éric Marsault (1971–2021)

4.1 Introduction

Solid-phase synthesis (SPS) represents one of the greatest technological advances of synthetic chemistry, with areas such as automated synthesis of peptides, nucleic acids, and polysaccharides proving the notable success of this methodology. Whereas SPS has been mostly dedicated to produce large biopolymers by iterative coupling cycles, the combinatorial synthesis of small molecules also benefitted significantly from the development of this technology, as very diverse libraries of compounds – mostly heterocycles – have been prepared using solid-supported, diversity-generating procedures. Among those processes capable to generate both skeletal and appendage diversity with low synthetic cost are multicomponent reactions (MCRs) [1, 2], which are ideally designed to construct complex heterocycles and peptidomimetics in short reaction sequences. However, the field of MCRs has been dominated by solution-phase synthesis, while the adaptation of new multicomponent procedures into solid-phase has been less exploited.

Nevertheless, the reproducibility, speed, usually high efficiency, and the few number of purification steps required to obtain and decorate medicinally privileged scaffolds are key factors for seeking an effective translation of MCRs into solid phase, eventually automated, protocols. Two general reviews have been published [3, 4] covering the field of solid-phase MCRs from the 1990s until the middle of the 2000s, while a very recent one describes multicomponent protocols used to assemble and derivatize biopolymers [5]. However, no focused review has been devoted exclusively

Multicomponent Reactions towards Heterocycles: Concepts and Applications, First Edition.
Edited by Erik V. Van der Eycken and Upendra K. Sharma.

to heterocycle synthesis using solid-phase MCRs, despite heterocycles perhaps being the most targeted class of compounds of this type of reaction. This chapter describes solid-phase procedures in which the MCR is the key step to assemble either the heterocycle skeleton or the polyfunctional moiety that is later cyclized to render the heterocyclic core.

MCRs are convergent processes that incorporate three or more reactants into a final product, with formation of several covalent bonds in a time-resolved manner and often involving rearrangements and ring-closing steps [1–4]. A well-accepted requisite of a MCR is that the incorporation of all reacting building blocks into the final skeleton must take place in one pot, a concept arising from the classic solution phase synthesis. However, in SPS, it is common to space out the addition of reactants using intermediate washing procedures. The addition of reactants one after each other, and not all at once, is frequently a requirement in SPS and not a choice as in solution phase. As a result, in the context of this chapter, we assume the MCRs follow all those processes that have been traditionally considered – mostly named reactions – despite the fact that there might be a washing step in-between the addition of one component and another. In principle, there is not a one-pot process in SPS if the solid-supported substrate is retained on the resin in the form of a stable, isolatable, compound. Thus, the reaction sequence could be considered equivalent to a one-pot procedure. An example of this is the imine formation, a crucial, almost ubiquitous step in many MCRs that is often done before the addition of other reactants [1–4]. In SPS, the imine formation is typically followed by a washing step to remove the excess of either the carbonyl or amino component, and the imine is treated with further components to complete the reaction. In this review, such a process, and others similar are considered as MCRs, since the imine is never isolated for testing or characterization. Insights into challenges associated with this type of multicomponent SPS are provided, along with details of which resins or other polymeric supports that are typically employed. The chapter is divided by the ring size of the heterocycles formed, and by the type of solid or soluble support used in the synthesis.

4.2 Synthesis of Five-Membered Ring Heterocycles

1,3-Dipolar cycloadditions are probably the most versatile reactions for the construction of structurally diverse five-membered heterocyclic rings, and such transformations have proven efficiency in SPS. Haap et al. [6] reported the condensation of aldehydes (**1**) with hydroxylamines (**2**) and trapping the resulting nitrones (**3**) with various dipolarophiles (**4**) to produce isoxazolidines (**5**) (Scheme 4.1a). Three different reaction variants – each dealing with one of the three components attached to 2-chlorotrityl or Rink amide resins – were evaluated concerning versatility and yield. The best results were obtained when the hydroxylamine component was attached to the solid support (**2a–d**) achieving excellent yields (up to 95%). In general, three- or four-substituted and strongly electron-deficient or electron-rich alkenes gave the poorer reaction outcome or no reaction at all. However, in the case of the aldehyde (**1**), no conclusion regarding the yields and the nature of the substituent relationship could be stated.

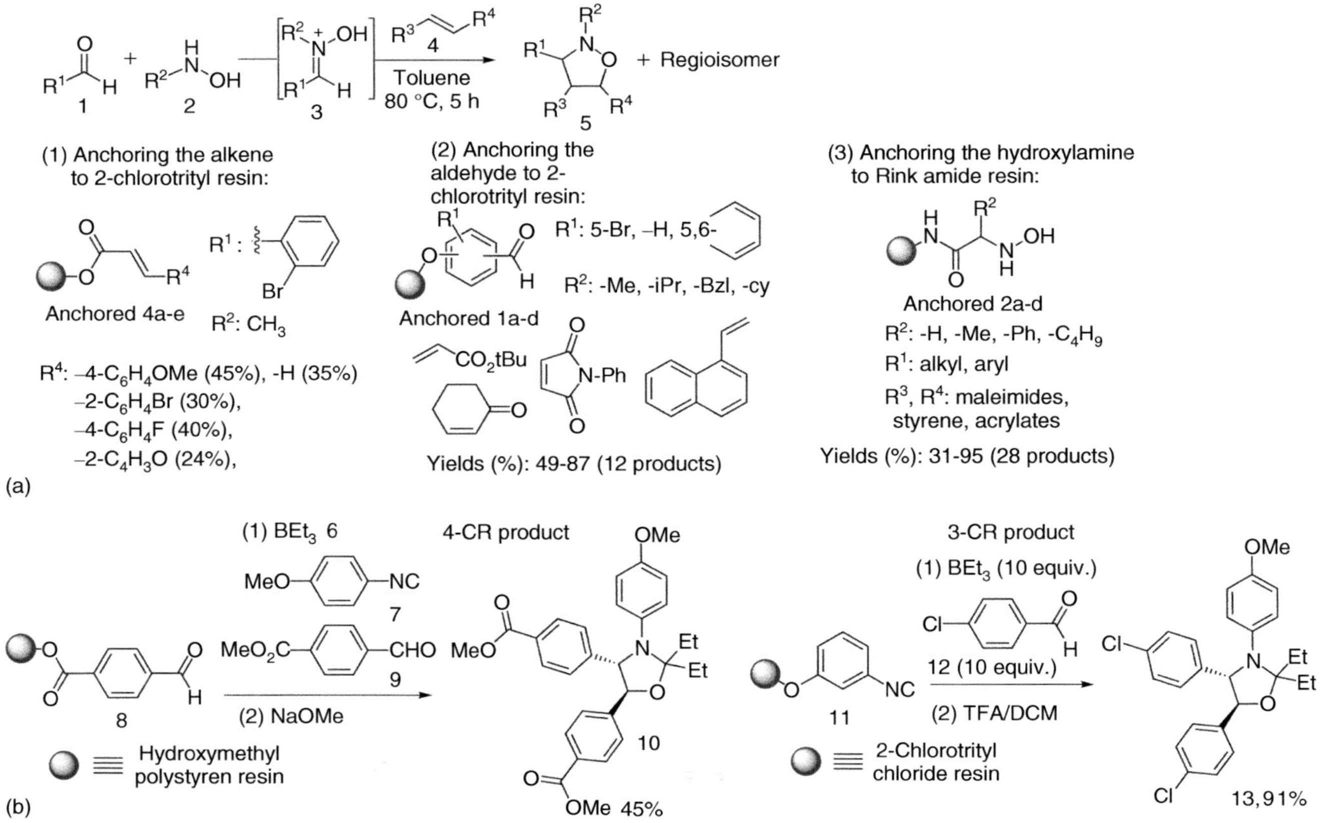

Scheme 4.1 Applications of solid-phase 1,3-dipolar cycloadditions in the synthesis of (a) isoxazolidines and (b) oxazolidines.

More recently, Lavilla and coworkers [7] expanded a previous study dedicated to the on-resin preparation of oxazolidines (**10**) from alkylboranes (**6**), isocyanides (**7**), and aldehydes (**8** and **9**) [8]. In their updated reaction scope, the authors intercepted the dipolar component, an azomethine ylide generated from the condensation of the anchored aldehyde (**8**), the isocyanide (**7**) and the borane (**6**), with another aldehyde molecule (**9**, same or different) (Scheme 4.1b). In the first study, this modification was carried out by immobilizing 4-formylbenzoic acid onto hydroxymethyl polystyrene resin. To prove the formation of the azomethine ylide in the solid support, the authors just mixed the resin with the borane and the isocyanide resulting in an aziridine that was obtained from this three-component reaction (3-CR). In a subsequent experiment, a new aldehyde was added to the reaction mixture and the desired oxazolidine was obtained in a four-component reaction (4-CR). Additionally, the isocyanide component (**11**) was anchored to 2-chlorotrityl chloride resin and the oxazolidine (**13**) was also obtained in excellent yield in another 3-CR.

The traditional Gewald-3CR has also been implemented in SPS. In 2001, Castanedo and Sutherlin [9] functionalized the Argo-Gel Wang resin through esterification with cyanoacetic acid and the modified resin (**14**) was reacted with an oxo component, elemental sulfur, and morpholine for eight hours in refluxing ethanol. Several aldehydes and ketones (**15a–h**) were employed during the study and in many cases, excellent yields and purities were achieved. The obtained thiophenes (**16**) were acylated before the cleavage. To demonstrate the utility of the reaction, the authors performed the synthesis of a known protein tyrosine phosphatase 1B (**17**, PTP1B) inhibitor and an adenosine receptor A3 inhibitor (**18**) (Scheme 4.2a). The results of conducting the Gewald-3CR in solid phase under microwave (MW) irradiation was reported by Hoener et al. [12] Even when the oxo components were very similar, the MW-assisted approach provided higher yields and purities within just one hour.

The synthesis of thiazolidinones, another class of important sulfur-containing heterocycles, has also been reported in solid phase through a 3-CR that employs an amine, an aldehyde, and mercaptosuccinic acid [13]. In the strategy used by Munson et al. [13] to prepare a library of more than 1000 thiazolidinones (**23**), diamines (**19**) were linked to the Wang resin and the free amino group was reacted with aldehydes (**20**) to prepare an imine (**21**), which was trapped *in situ* by the aforementioned bisnucleophile **22** (Scheme 4.2b). On average, purity of the library members was greater than 65%, thus proving the usefulness of this straightforward method for the synthesis of these complex molecules bearing a pendant amine suitable for scaffold diversification. Another library of thiazolidinones was also prepared by Gupta et al. [14] using the resin loaded with a succinic acid anhydride/5-hydroxyfurfural-derived linker.

In 2000, Heck and Dömling reported the discovery of a new MCR for the synthesis of 4-carboxy-2-acylaminomethylthiazoles (**28**) [15]. The reaction consisted of a one-pot condensation of a thiocarboxylic acid (**27**), an aldehyde (**26**), an amine (**24**), and 3-(*N,N*-dimethylamino)-2-isocyanoacrylate (**25**). Later, Dömling and coworkers reported the extension of this reaction to the solid phase, using either Rink amide resin as the amino component [16, 17] or anchoring

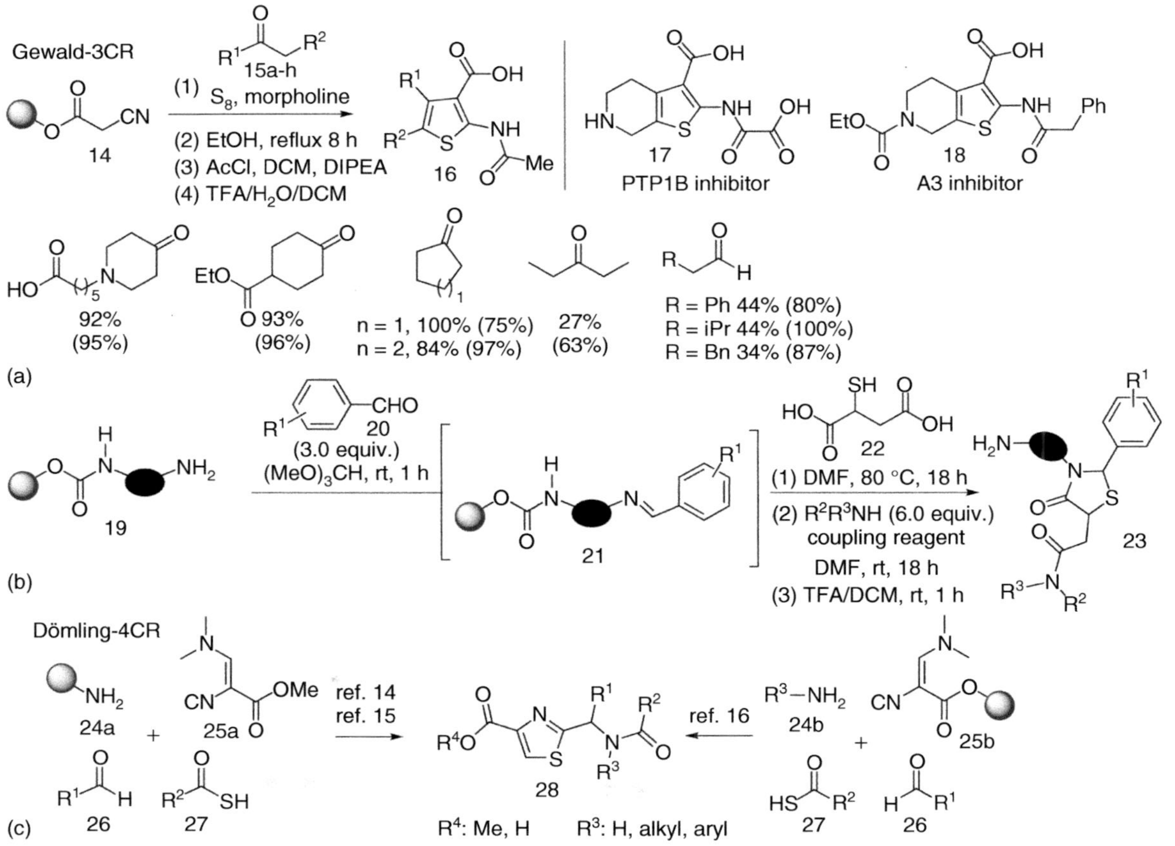

Scheme 4.2 Solid-phase synthesis of sulfur-containing five-membered ring heterocycles such as thiophene (a), thiazolidin-4-one. Source: Based on Lusch and Tallarico [10]. (b) and thiazole (c) by MCRs. Source: Based on Henkel et al. [11].

3-(*N,N*-dimethylamino)-2-isocyanoacrylate to HMBA resin (**25b**) (Scheme 4.2c) [11]. In the former study, the authors employed a variety of aldehydes and two different thiocarboxylic acids and as a result 24 disubstituted thiazoles were obtained after TFA/DCM cleavage with poor to excellent purities (31–96%) [16]. In the latter, the same thiocarboxylic acids were combined with several aldehydes and amines which made possible the preparation of 25 new thiazole-containing scaffolds [11].

4.3 Synthesis of Six-Membered Ring Heterocycles

Thus far, one of the most exploited reactions for the SPS of six-membered rings is the Biginelli reaction. The fact that the dihydropyrimidine (DHPM) moiety is found in many active compounds due to its pharmacological versatility has been, from our perspective, the driving force to continuously expand the scope of this reaction. The first report of the solid-phase Biginelli reaction (Scheme 4.3) was published in 1995 by Wipf and Cunningham [18]. The use of a γ-aminobutyric acid (GABA) urea derivative (**29**) linked to Wang resin gave the authors the chance to synthesize 10 DHPM derivatives (**30**) in good to excellent yields (67–98%). Impressively, the yields were higher than those obtained when the reaction was conducted in solution (54–86%). A macrobead-anchored urea was also used by Lusch and Tallarico to conduct split-and-pool Biginelli synthesis of a DHPM library using a variety of aryl aldehydes and β-ketoesters in the second step [10].

In the following years, Kappe's group dedicated a significant effort to study this MCR in solid phase. In 2001, they reported two orthogonal on-resin methods to conduct the Biginelli condensation using Wang resin-supported β-ketoesters [19]. In the first one (Scheme 4.3b), the transient *N*-acyliminium ion, formed from the aldehyde (**32**) and the (thio)urea component (**33**), was condensed with the resin-bound β-ketoester (**31**) under acid catalysis to assemble the DHPMs (**34**), which were released from resin to obtain DHPM-5-carboxylic acid derivatives **35** in

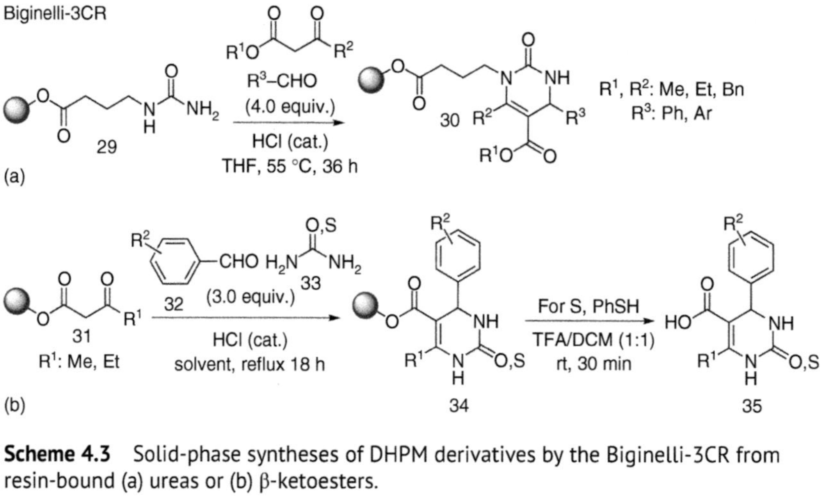

Scheme 4.3 Solid-phase syntheses of DHPM derivatives by the Biginelli-3CR from resin-bound (a) ureas or (b) β-ketoesters.

moderate to good overall yields (39–82%) and high purities (>95%) after standard cleavage. In an alternative method (not shown), β-ketoesters were condensed with aldehydes and *O*-methylisourea under mild basic conditions, followed by hydrolysis of the *O*-methylisourea to obtain the desired DHPM in moderate yields (30–66%) and high purities (>95%).

Another publication from the same group [20] reported the post-condensation derivatization of DHPM leading to furo[3,4-*d*]pyrimidines (**38**), pyrrolo[3,4-*d*]-pyrimidines (**39**), and pyrimido[4,5-*d*]pyridazines (**40**) using three different traceless cyclative-cleavage strategies. The study comprised the on-resin Biginelli 3-CR from hydroxymethyl polystyrene resin 4-chloroacetoacetate (**36**) using 12 different aldehydes followed by substitution reactions of the chlorine either with primary amines, hydrazines, or cyclative cleavage to obtain bicyclic DHPM libraries (Scheme 4.4). Finally, it is worth mentioning a related MCR comprising the condensation of two equivalents of benzaldehyde and one of urea with a resin-bound β-keto-γ-lactam reported by Byk and Kabha (not shown) [21]. This MCR allowed obtaining, in a diastereospecific manner, a spiro-bicyclic β-keto-γ-lactam containing three stereogenic centers, while the solution phase reaction resulted in a much lower stereoselectivity.

The Hantzsch 3-CR is another process that has been employed on solid supports to generate 4-aryl-1,4-dihydropyridine libraries. In a report by Breitenbucher and Figliozzi [22], a compound library was prepared utilizing the aforementioned reaction and based on previous experiences of other groups [23, 24]. Thus, the 1,4-dihydropyridine moiety was assembled on Wang resin functionalized with a β-ketoester (**41**) upon treatment with eight aminocrotonates (**42**) and several aryl aldehydes in DMF, generating 192 resin-bound 1,4-dihydropyridine derivatives (**43–46**) which were finally cleaved (Scheme 4.5) and obtained with an HPLC average purity of 81%. This procedure was complemented with a post-MCR diversification of the resin-bound 4-aryl-1,4-dihydropyridine providing several more derivatives, only with a minor contamination with Knoevenagel reaction product (c. 15%).

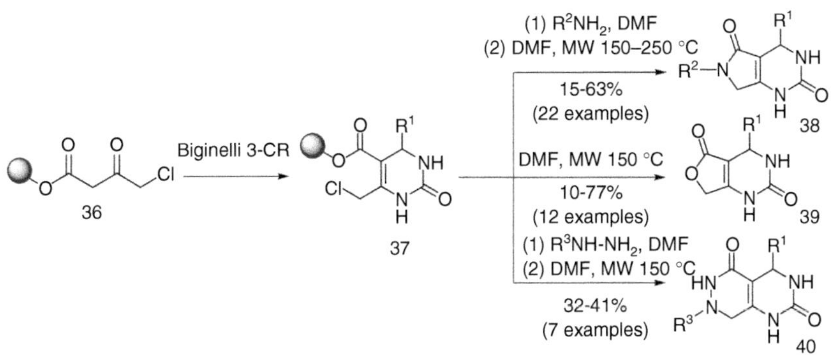

Scheme 4.4 Synthesis of furo[3,4-*d*]pyrimidines, pyrrolo[3,4-*d*]-pyrimidines, and pyrimido[4,5-*d*]pyridazines by derivatization of resin-bound Biginelli-3CR products.

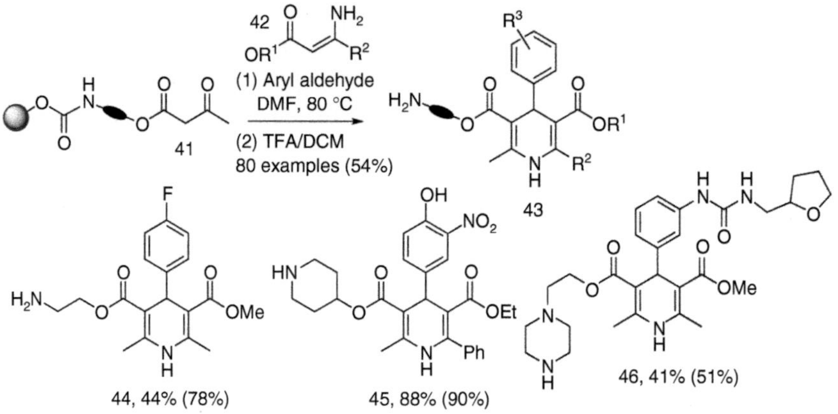

Scheme 4.5 Synthesis of functionalized 1,4-dihydropyridine by the Hantzsch 3-CR.

Several other methods have been developed for the synthesis of nitrogen-containing heterocycles. For example, in a report published by Wang and Wilson [25], an anchored benzaldehyde (**47**) was mixed with varied primary and secondary amines and the *in situ* generated imines (**48**) were reacted with Danishefsky's diene (**49**) to form 2,3-dihydro-4-pyridones (**50**) through a tandem Mannich–Michael process catalyzed by a Lewis acid (Scheme 4.6a). The small library of heterocycles was obtained in moderate to excellent yields (60–90%). In addition, a multicomponent aza-Diels–Alder reaction (aza-DA) was employed by Zhang et al. [26] to synthesize a related family of six-membered heterocycles (**53**) by reaction of an anchored preformed imine – derived from amine (**51**) and acting as dienophile – with a set of dienes (Scheme 4.6b). The aza-DA reaction was Lewis acid-catalyzed and could be run in the presence of water when formalin was employed to form intermediate **52**.

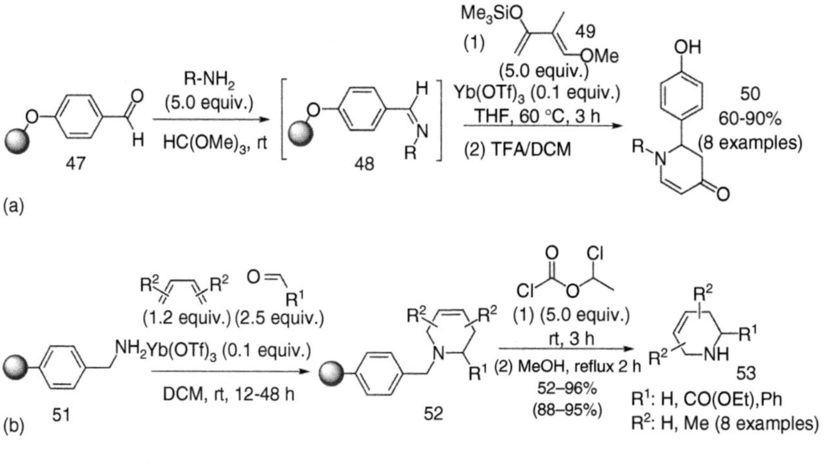

Scheme 4.6 SPS of 2,3-dihydro-4-pyridones (a) and tetrahydropyridines (b) by aza-Diels–Alder reaction.

4.4 Synthesis of Fused Heterocyclic Ring Systems

In some cases, a MCR condensation may lead not only to one heterocycle ring but to a system of two or more fused rings. Among, the first reports are three papers published by Armstrong and coworkers describing the synthesis of complex molecules comprising up to four condensed cycles by a related multicomponent aza-DA reaction, namely the Grieco-3CR [27–29]. In their first report, a Wang resin-anchored aniline (**54**) was condensed in a combinatorial manner with various aldehydes and alkenes (Scheme 4.7) in the presence of catalytic amounts of TFA, to obtain 40 tetrahydroquinoline derivatives (**56–59**). In general, the yield of isolated products was moderate to good (53–92%), with the only detectable impurity corresponding to Schiff base. The yields for aromatic aldehydes bearing electron-withdrawing groups (EWGs) was higher (about 80–90%) than those obtained when aromatic aldehydes were substituted by electron-donating groups (EDGs), while aliphatic aldehydes did not afford the desired tetrahydroquinolines. In a subsequent study [28], the aldehyde or the alkene were the components anchored to the resin, while different aniline derivatives and alkenes were employed in excess to render compounds in moderate to good yields (60–87%). Again, the use of anilines substituted by EWG gave the best yields (72–87%) and highest purities (>80%).

A related Povarov-type [30] multicomponent aza-DA reaction was implemented by Lavilla and coworkers [31] to produce pyrido-fused tetrahydroquinolines in both solution and solid phase. In this class of Lewis acid-catalyzed MCR, a *formal* aza-DA reaction takes places between an aryl imine – usually derived from an aniline bearing EDG – and, for example, an enol ether or enamine as dienophile component. It has been suggested that the reaction proceeds via an actual aza-DA or a Mannich-type pathway [30]. As shown in Scheme 4.8, the authors employed a glyoxylate-functionalized Merrifield resin (**60**) as oxo-component, *p*-methyl aniline as aryl amine component and *N*-methyl-3-cyano-1,4-dihydropyridine as

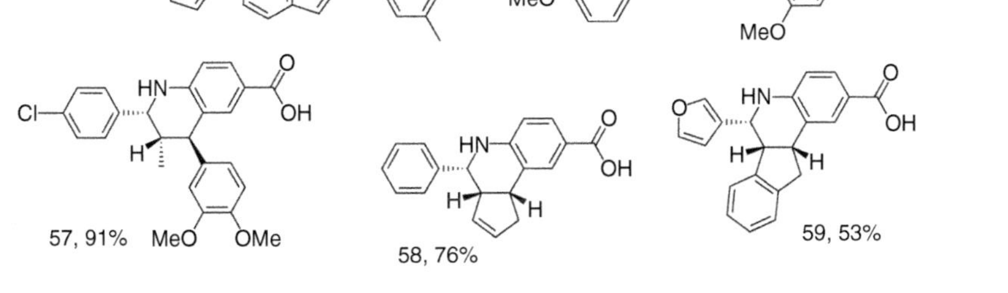

Scheme 4.7 SPS of tetrahydroquinoline derivatives by the Grieco-3CR with a resin-anchored amine.

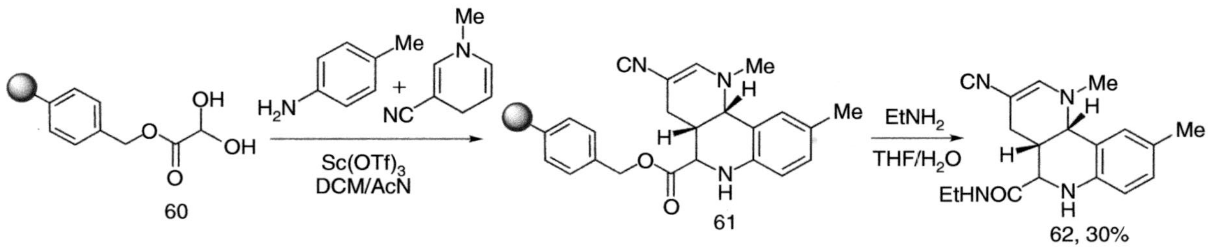

Scheme 4.8 SPS of pyrido-fused tetrahydroquinolines by an aza-Diels–Alder reaction of a resin-anchored aldehyde.

electron-rich dienophile. After preformation of the aryl imine – acting as azadiene partner – the multicomponent aza-DA reaction proved efficiency in providing the fused tetrahydroquinoline (**61**) under $Sc(OTf)_3$-catalzyed conditions. Final release from the Merrifield resin under a variety of basic conditions led to varied carboxylic acid derivatives (**62**). In parallel, attachment of the aniline to Wang resin enabled the preparation of similar tetrahydroquinolines using ethyl glyoxylate and the above-mentioned dihydropyridine as the oxo and dienophile components, respectively, in this case with TFA-mediated cleavage of the product from the Wang resin (not shown) [31].

Multicomponent 1,3-dipolar cycloadditions are suitable processes for assembling very complex fused heterocyclic systems. For example, Bickneil et al. [32] used a pyridinium salt (**63**) anchored to Wang resin in a Tsuge-type reaction with *N*-substituted maleimides and the subsequent reaction of the initially formed cycloadduct with nitrile oxides to form **64**, which resulted in the final tricyclic product **65** upon acidic cleavage from the resin (Scheme 4.9). From a mechanistic point of view, the product is supposed to be formed via isoxazoline fragmentation to give an iminium intermediate which undergoes oxidation driven by the recovery of aromaticity in the resulting pyridinium salt and final hydrolysis of the oxime. In total, 96 products were obtained in an automated synthesizer.

In 2003, Hall and coworkers [33] reported the first three-component aza-[4+2]/allylboration reaction to access polysubstituted α-hydroxyalkyl piperidines in a highly diastereocontrolled fashion from maleimides (**66**), 4-boronohydrazonodienes (**67**), and aldehydes (**68**). After a very detailed study of this MCR in the solution phase, the authors decided to apply it in the solid phase, immobilizing the dienophile maleimide component (**66**) to SASRIN resin (Scheme 4.10a). Conducting the reaction in solid-phase shortened reaction times from 72 to 14 hours, while keeping comparable yields, good purities (80–90%), and maintaining high levels of stereocontrol in the preparation of these densely functionalized nitrogen heterocycles (**69**). More recently, the same group reported another variation of the current reaction where the immobilized component was the aldehyde [34].

In a related approach also with resin-anchored maleimides, Byk and coworkers reported a novel 4-CR based on the condensation of a β-keto lactam (**70**), an aldehyde (**71**), an isocyanide (**72**), and a dienophile (**73**), whose product formation, a 3-substituted isoindolinone, is catalyzed by a Lewis acid [35]. This reaction comprises a Knoevenagel condensation between the first two components, and this intermediate reacts with the isocyanide in a [4+1] cycloaddition to produce a fused bicycle [5+5] furan, which is too reactive to be isolated and reacts in the presence

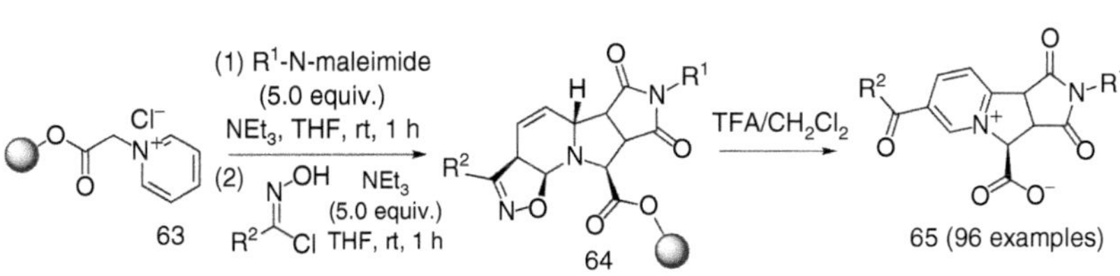

Scheme 4.9 Synthesis of zwitterionic tricyclic compounds by solid-phase Tsuge reaction.

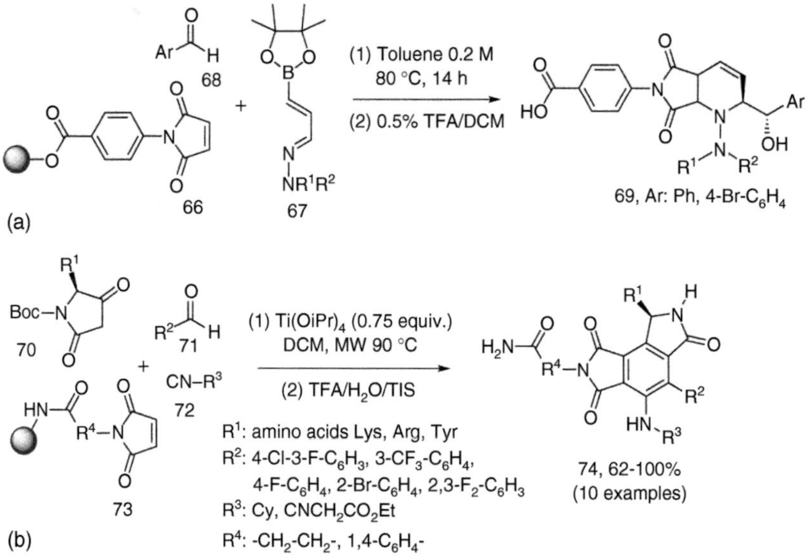

Scheme 4.10 Synthesis of polysubstituted fused heterocycles from resin-anchored maleimides by three (a) and four (b) component reactions.

of the dienophile to generate a Diels–Alder intermediate, which undergoes oxygen elimination and aromatization. Recently, this reaction was extended to the solid phase [36], and two strategies were applied during its adaptation. Both used the Rink amide resin as the solid support, but one with the aldehyde (not shown) and the other with the dienophile (**73**) anchored to the resin (Scheme 4.10b). The latter method delivered cleaner crude products since the dienophile can only react with the *isocyanide-aldehyde-β-keto lactam* adduct, generating fewer side products. In total, 10 different 3-substituted isoindolinones (**74**) were obtained in moderate to excellent yields (62–100%).

4.5 Synthesis of Heterocycles on Solid-Supported Amino Acids

Several examples of MCRs on solid-supported amino acids, peptides, and oligonucleotides have emerged in the last decades. As a recent review [5] has covered most of the examples being conducted on peptides and deoxyribonucleic acid (DNA), here we focus mostly on those heterocycle-forming methods implemented on solid-supported amino acids, but their extension to nucleic acids and peptides is also commented aiming at motivating further inputs into the production of heterocycle–biomolecule hybrids.

First, some of the procedures discussed above have been also implemented with resin-bound peptides. For example, the method reported by Byk's group [36] on the Lewis acid-catalyzed 4-CR between a β-keto lactam, an aldehyde, an isocyanide, and a dienophile was employed in the preparation of cell-penetrating peptides showing fluorescent properties as a result of the fused isoindolinone moiety. In a model peptide, a 14-amino-acid amphipathic α-helical peptide was functionalized with a maleimide group at the *N*-terminus and next derivatized with the 4-CR to render the fluorescent heterocycle–peptide hybrids in acceptable overall yields.

Similarly, an application of the Grieco aza-DA 3-CR in the derivatization of resin-anchored amino acids was provided by Kiselyov et al. [29] The authors employed acylated 4-aminophenylalanine, immobilized on Wang resin, in the above-mentioned reactions with cyclopentadiene and other electron-rich olefins as well as a variety of aldehydes, mostly benzaldehyde derivatives. Using $Yb(OTf)_3$ as catalyst under optimized conditions, a library of 50 products was constructed with yields ranging from moderate to excellent (62–92%), with the best efficiency resulting for benzaldehydes bearing EWGs. A similar example of the utilization of the Grieco-3CR – in this case it could also be also considered as the Povarov-3CR – in the solid-phase derivatization of amino acids was reported by Demaude et al. [37] As depicted in Scheme 4.11, the authors implemented the aza-DA 3-CR between resin-bound 4-aminophenylalanine (**75**), an aromatic aldehyde and phenyl vinyl thioether to produce a thiophenyl-substituted tetrahydroquinoline product (**76**). Subsequent oxidation with *m*-chloroperoxybenzoic acid (*m*CPBA) and pyrolysis of the resulting sulfoxide provided a small library of quinolone-functionalized amino acids (**77**) after conventional cleavage from the resin, albeit in poor to moderate yields. Albericio and coworkers [38] also implemented a Lewis acid-catalyzed reaction between the amino group of the Lys side chain, a nitrostyrene, and a 1,3-dicarbonyl compound, under MW irradiation, to produce pyrrole–peptide conjugates, a report that has been compiled in a recent review [5].

The Ugi-4CR is among the MCRs most widely employed in SPS, although it does not lead to cyclic compounds typically, unless bifunctional components are used. However, variants of the Ugi-4CR including hydrazoic and isocyanic acids as one of the components lead to the formation of tetrazoles and hydantoinimides, respectively [1]. Constabel and Ugi employed a solid-phase sequence of two consecutive Ugi reactions to produce peptidomimetics bearing tetrazoles and hydantoinimides [39]. As shown in Scheme 4.12, the first Ugi-4CR was performed on free amino groups of AM RAM or Tentagel S RAM resins using isobutyraldehyde, Fmoc-Gly-OH, and *tert*-butyl isocyanide to assemble a peptidic moiety (**78**) in high

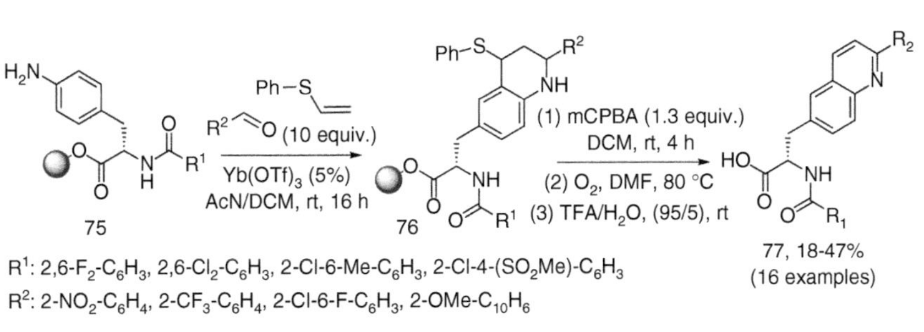

Scheme 4.11 Synthesis of quinoline derivatives by the Grieco-3CR with *p*-amino-derived phenylalanines.

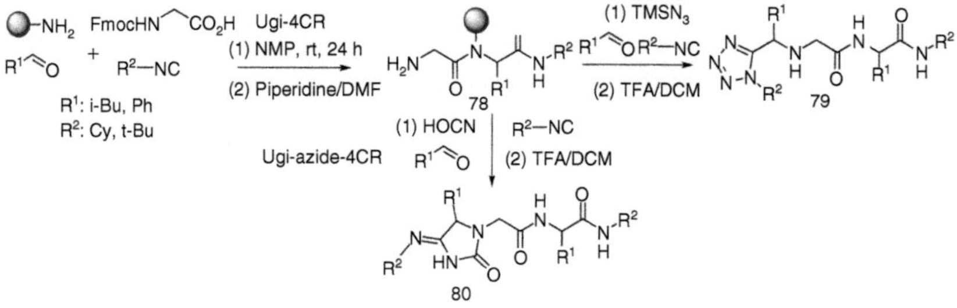

Scheme 4.12 Synthesis of tetrazoles and hydantoinimides by two consecutive on-resin Ugi-4CRs.

yield and purity. After the Fmoc removal, either the hydantoinimide-forming variant of the Ugi-4CR with isocyanic acid, aldehydes, and isocyanides or the tetrazole-forming Ugi-azide-4CR using trimethylsilyl azide (as surrogate of hydrazoic acid) and the same two components, were performed to produce a library of heterocycle–peptidomimetic hybrids **79** and **80** in moderate to good yield and good purity.

Additionally, further applications of Ugi-type-4CRs for the construction of heterocycles were reported by Liu and Nefzi [40], and Yudin and coworkers [41]. In both cases, the MCR took place intramolecularly leading to the formation of six-membered ring heterocycles. The first example [40] comprised the modification of MBHA resin-linked Fmoc-cysteine (**81**) with bromoacetic acid to render a bifunctional amino acid that underwent an intramolecular Ugi-4CR with several cyclic ketones and isocyanides to obtain 12 different *N*-cycloakyl thiomorpholinones (**82**) in moderate to good yields (62–81%) and excellent (81–93%) purities (Scheme 4.13a). In the second example [41], various amino acids were attached to the 4-(4-formyl-3-methoxyphenoxy)ethyl (*FMP*) *resin via reductive amination to* render the resin-linked bifunctional components (**83**) bearing a secondary

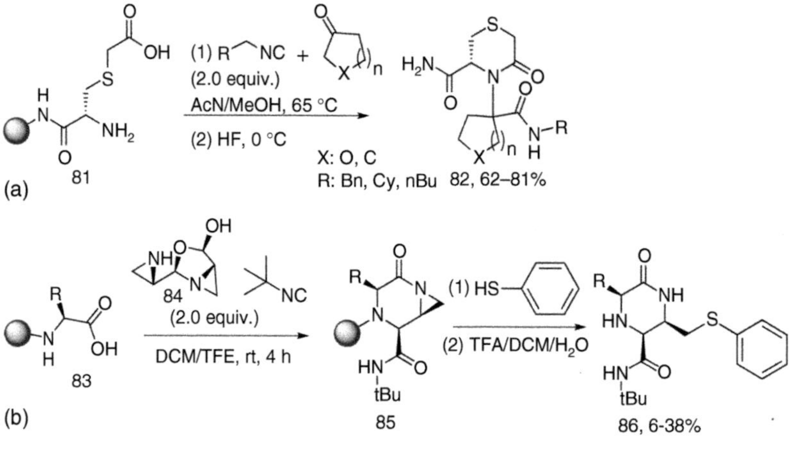

Scheme 4.13 Synthesis of six-membered heterocycle such as thiomorpholinones (a) and piperazinones (b) derived from α-amino acids.

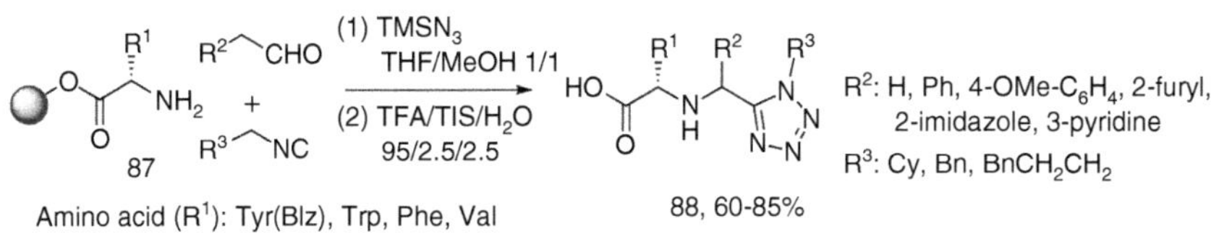

Scheme 4.14 Synthesis of tetrazole-peptidomimetics as M1-aminopeptidase inhibitors.

amine and a carboxylic acid, which upon condensation with *t*-butyl isocyanide and (*S*)-aziridine aldehyde dimer (**84**) led to the formation of bicyclic aziridine products (**85**) (Scheme 4.13b). The authors tried to cleave and isolate the product; however, these efforts were unsuccessful due to the instability of such compounds. Nevertheless, they opted for reacting the bicyclic aziridines (**85**) with a series of nucleophiles such as thiophenol, thiobenzoic acid, and sodium azide, resulting in an efficient and stereoselective on-resin method for the assembly and further derivatization of functionalized piperazinones (**86**).

A more recent publication by Méndez et al. [42] describes the SPS of tetrazole-peptidomimetics (**88**) as inhibitors of the *Escherichia coli* neutral M1-aminopeptidase (ePepN). The on-resin methodology employed was previously developed by the same group of Rivera and coworkers [43] to conjugate lipids to peptides by means – among other methods – of the tetrazole-forming Ugi-4CR (not shown). As depicted in Scheme 4.14, the same on-resin Ugi-4CR with Wang resin-anchored amino acids (**87**) was employed to construct a library of 21 peptidomimetics (**88**) by the tetrazole-forming method. The motivation behind the design of these inhibitors was the metal-chelating capacity of tetrazoles, which if suitably functionalized with enzyme-active site matching functionalities derived from the amino acid, could inhibit the ePepN enzymatic activity via chelation of the Zn^{2+} cation. The overall solid-phase protocol comprised the deprotection of the resin-bound amino acids by Fmoc removal, followed by the on-resin Ugi-4CR for 72 hours and final cleavage from the resin. The protocol rendered a wide set of compounds for biological evaluation that resulted in the discovery of new and potent inhibitors [42].

4.6 Solid-Phase Multicomponent Construction of DNA-Encoded Heterocycle Libraries

In the DNA-encoded screening technology [44], a library of molecules is produced by combinatorial synthetic tools in a way that each compound is anchored to an encoding DNA sequence to enable a subsequent separation, amplification, and sequencing of the encoding DNA tag [45]. The library construction can be done in aqueous solution phase using diversity-generating – usually split and pool – methodologies that should be tolerant to the double-stranded DNA structure to avoid its damage. The second alternative is to carry out the library synthesis on DNA oligonucleotide chains linked to a solid support, which allows utilizing standard solid-phase conditions including the use of protecting groups DNA structure, metal catalysis, and nonpolar organic solvents. In recent years, the scaffold diversity generation capacity of

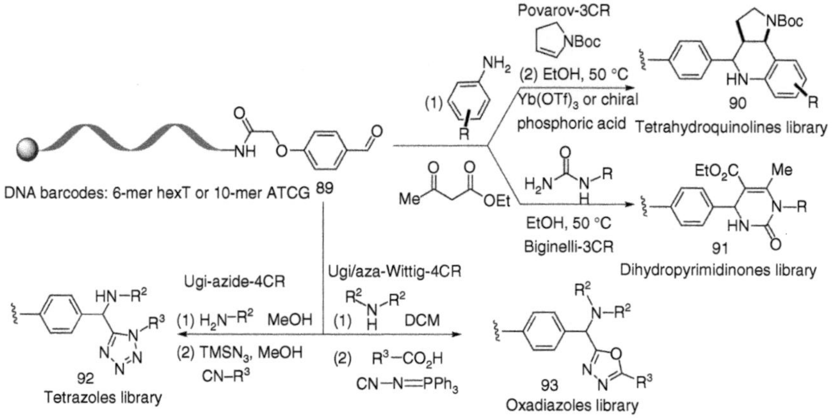

Scheme 4.15 On-resin construction of DNA-encoded heterocycle libraries by MCRs.

MCRs has been used for the constructions of DNA-encoded libraries [46–49]. In this sense, Brunschweiger and coworkers [46–49] have implemented a variety of on-resin MCRs for the preparation of DNA-coupled heterocycles starting from aldehyde- and amino-functionalized DNA barcodes anchored to controlled pore glass (CPG) beads, a type of solid support that combine rigidity and high surface as well as chemical and solvent stability.

Typically, the DNA oligonucleotide chains are synthesized on CPG beads by standard phosphoramidite chemistry, followed by installation of the MCR-reactive functionality. As shown in Scheme 4.15, the incorporation of an aldehyde handle (**89**) in varied solid-supported DNA barcodes enabled the implementation of on-resin MCRs such as the metal-mediated Castagnoli–Cushman reaction [46], metal- and organocatalyzed versions of the Povarov-3CR to render tetrahydroquinolines (**90**) [47], as well as the Biginelli-3CR reaction to furnish DHPMs (**91**) [47]. In cooperation with Dömling's group [48], the same team also exploited the versatility of isocyanide-based MCRs[35] to produce DNA-coupled heterocycles such as tetrazoles (**92**) and oxadiazoles (**93**) (Scheme 4.15). Also, libraries of benzimidazoles and peptidomimetics were produced employing the Groebke–Blackburn–Bienaymé-3CR and the Ugi-4CR, respectively [48]. DNA functionalization with primary and secondary amines was also used to implement the Ag(I)-mediated 1,3-dipolar cycloaddition upon treatment with varied aldehydes and dipolarophiles [46] and construction of Petasis-3CR-derived libraries [49], respectively (not shown). Considering the power of the DNA-encoded library for discovery purposes, we foresee new applications of solid-supported MCR chemistry with such technologies. A recent report on the discovery of inhibitors of protein–protein interactions using MCR chemistry on DNA oligonucleotides has been recently published [50].

4.7 Miscellaneous Supports for Multicomponent Synthesis of Heterocycles

Cellulose is a biopolymer that is typically used in the macroarray technology to address several of the drawbacks of combinatorial libraries generated on synthetic

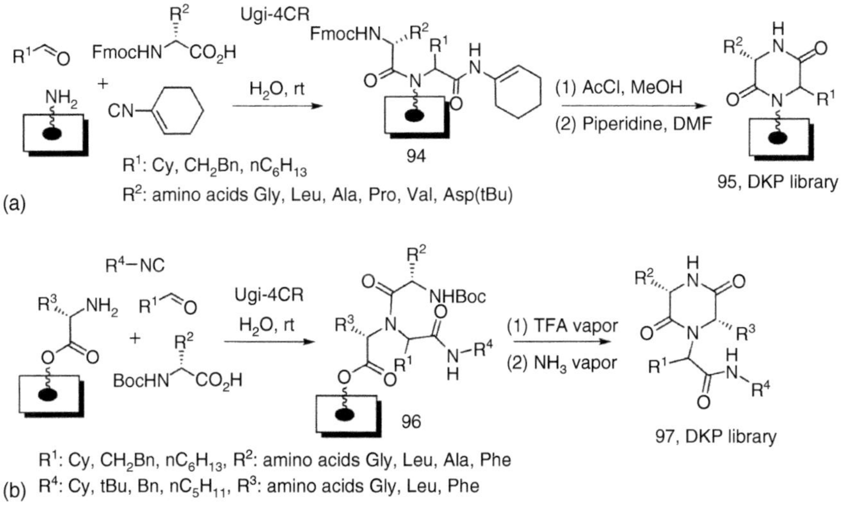

Scheme 4.16 On-cellulose synthesis of DKP libraries by Ugi/cyclization strategies using cleavable (a) and non-cleavable (b) isocyanides.

polymers such as polystyrene beads. Among the advantages of cellulose as solid support are its robustness, inexpensive character, and amenability to conduct biological screening with the compound still bound to the solid support. Blackwell and coworkers pioneered the field of macroarray heterocycle assembly using MCRs [51, 52]. As depicted in Scheme 4.16, this group developed two macroarray synthesis strategies for the construction of libraries of diketopiperazines (DKPs) using the Ugi-4CR. The first one used the convertible 1-cyclohexenyl isocyanide for the assembly of the Ugi-derived dipeptide skeleton (**94**) on an amino-derived cellulose support, followed by methanolysis and cyclization to render the cellulose-supported DKP (**95**) [51]. The second approach also relied on the Ugi-4CR to assemble a cellulose-supported peptide (**96**) followed by a deprotection-cyclization–cleavage procedure to produce DPKs (**97**). Variation of the isocyanide, amino acid, and carbonyl components allowed generating a ~400-member library in excellent purities and yields [52]. Analysis of the DKP library revealed a small set of strong quorum sensing antagonists in *Vibrio fischeri*. These reports pave the way for further development in macroarray library generation using MCRs.

As mentioned before, conducting an organic reaction in the solid phase has many advantages compared with conventional methods. However, this does not mean that we must stop the liquid phase if we want to keep those benefits. The use of soluble polymers or ionic liquids might offer many of those advantages. Using soluble polymers supporting the starting materials could allow following the progress of a given reaction by TLC, UV, IR, or even NMR techniques, while it also obviates the need of purification along the sequence since the excess of reagents is removed through precipitation steps as well. Soluble polymers are cheaper and have higher loading than their solid counterparts, and the former may skip the slow and nonlinear kinetics due to poor diffusion and difficulties to access active sites that is frequently found for the latter. These difficulties can be solved with soluble supports based on dendritic polyalcohols [53], such as polyethylene glycols (PEGs), or soluble polystyrenes

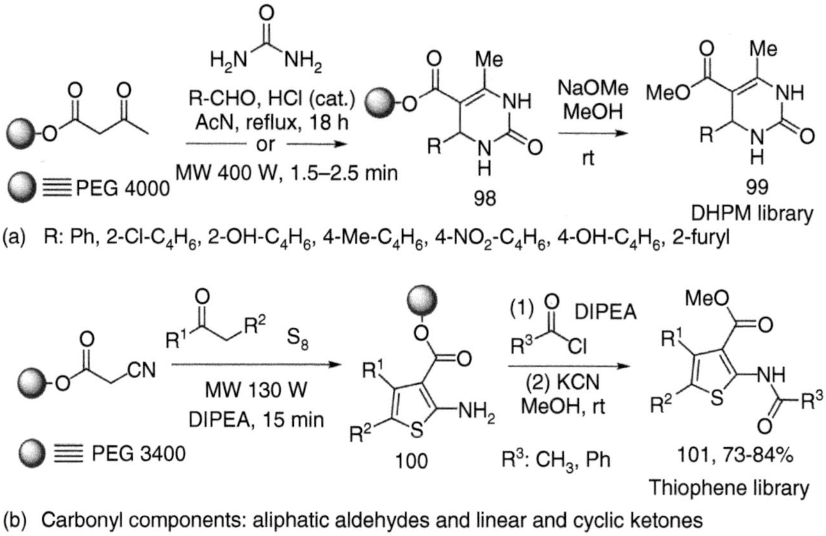

Scheme 4.17 Synthesis of heterocycles by the Biginelli-3CR (a) and the Gewald-3CR (b) using PEG-supported starting materials.

(JandaJels), just to mention the most important ones [54–56]. They allowed keeping liquid-phase properties by fine-tuning solvent conditions and linker structures. PEGs, for example, have been used successfully in the synthesis of biopolymers.

Among the first reports on the synthesis of heterocycles immobilized to soluble supports is the procedure developed by Xia and Wang [57] in 2002 for the synthesis of 3,4-dihydropyrimidin-2(1*H*)-ones (DHPM) by the Biginelli-3CR. As shown in Scheme 4.17a, the acetoacetate moiety supported in PEG-4000 was subjected to the MCR to produce (**99**) either in refluxing ACN during 18 hours (70–94% yield) or under neat conditions with MW irradiation for 1.5–2.5 minutes (70–88% yield). The very good yields obtained were compared with those resulting from the classical solution-phase Biginelli-3CR using the same substrates, and the former resulted to be 2–3 times better.

The application of the Gewald-3CR using nitrile acetates anchored to PEG-3400 was reported by Zhang et al. in 2004 [58]. Addition of elemental sulfur, drops of diisopropylethylamine (DIPEA), and several ketones made it possible to obtain tetrasubstituted thiophene derivatives (**101**). Reactions were carried out under MW irradiation and the products acylated afterward using acetyl or benzoyl chloride (Scheme 4.17b). Cleavage from the support was performed using MeOH in the presence of catalytic amounts of potassium cyanide at room temperature overnight. A small library of 24 compounds was obtained in moderate to excellent yields (38–95%). In addition, other relevant inputs include the PEG-supported assembly of Povarov-3CR products and their derivatization by a [1,3]-sigmatropic rearrangement [59].

Although it is beyond the scope of this chapter, non-polymeric supports such as ionic liquids have also been used in heterocycle synthesis with great success. These supports have attracted attention due to their intriguing physical and chemical properties [60], such as high thermal and chemical stabilities, negligible vapor pressure, non-flammability, and high loading capacity. Some of the applications of these materials in multicomponent heterocycle synthesis include the preparation of ionic-liquid-supported thiazolidines [61], 2-amino thiophenes by the Gewald-3CR [62], DHPMs by the Biginelli-3CR [63], and tetrahydroquinolines by the Grieco-3CR [64].

4.8 Conclusions

Solid-supported MCRs offer a variety of advantages over traditional solution phase synthesis that need to be considered for drug discovery purposes. The rapid generation, in a cost-effective manner, of medium and large compound, highly diversified libraries is indeed one of the advantages, but so is the possibility to conduct multiple synthetic steps to assemble complex heterocycles without the need of purifications of intermediates. Unfortunately, there hasn't been a lot of growth in this field over the last decade, which is evidenced by the reducing number of publications dealing with solid-supported synthesis of heterocycles using MCRs. Nevertheless, this chapter has shown that solid-phase protocols offer a similar or sometimes even higher synthetic scope than solution-phase procedures and are well suited for creating a level of molecular complexity using consecutive reactions steps. Not only the classic polystyrene and PEG-based resins can be used, but also soluble polymers and ionic liquids share some of the benefits of having one of the reactants anchored to facilitate long pathways and avoid complex purification procedures. Perhaps the new research into this field should be directed toward implementing asymmetric versions of MCRs in the solid phase, thus enabling the generation of enantio-enriched products more suitable for biological evaluation. The field of MCRs can benefit a lot from further developments of this technology, which – as already proven for peptides – can be used not only for discovery purpose but also for the large-scale production of complex skeletons using iterative and efficient on-resin protocols.

References

1 Dömling, A. and Ugi, I. (2000). Multicomponent reactions with isocyanides. *Angew. Chem. Int. Ed.* 39 (18): 3168–3210. https://doi.org/10.1002/1521-3773(20000915)39:18<3168.

2 Zhu, J., Wang, Q., and Wang, M. (eds.) (2014). *Multicomponent Reactions in Organic Synthesis*. Wiley.

3 Dax, S.L., McNally, J.J., and Youngman, M.A. (1999). Multi-component methodologies in solid-phase organic synthesis. *Curr. Med. Chem.* 6 (3): 255–270.

4 Banfi, L., Guanti, G., Riva, R. et al. (2007). Multicomponent reactions in solid-phase synthesis. *Curr. Opin. Drug. Discov. Devel.* 10 (6): 704–714.

5 Rivera, D.G., Ricardo, M.G., Vasco, A.V. et al. (2021). On-resin multicomponent protocols for biopolymers assembly and derivatization. *Nat. Protoc.* 16: 561–578. https://doi.org/10.1038/s41596-020-00445-6.

6 Haap, W.J., Kaiser, D., Walk, T.B. et al. (1998). Solid phase synthesis of diverse isoxazolidines via 1,3 dipolar cycloaddition. *Tetrahedron* 54 (15): 3705–3724. https://doi.org/10.1016/S0040-4020(98)00099-4.

7 Kielland, N., Catti, F., Bello, D. et al. (2010). Boron- based dipolar multicomponent reactions: simple generation of substituted aziridines, oxazolidines and pyrrolidines. *Chem Eur J* 16 (26): 7904–7915. https://doi.org/10.1002/chem.201000349.

8 Hesse, G., Witte, H., and Gulden, W. (1965). Preparation of 1,3-oxazolidines from isonitriles and aldehydes by a multicomponent reaction. *Angew. Chem. Int. Ed. Engl.* 4 (7): 569–569. https://doi.org/10.1002/anie.196505961.

9 Castanedo, G.M. and Sutherlin, D.P. (2001). Synthesis of tetrasubstituted thiophenes on solid-support using the Gewald reaction. *Tetrahedron Lett.* 42 (41): 7181–7184. https://doi.org/10.1016/S0040-4039(01)01470-8.

10 Lusch, M.J. and Tallarico, J.A. (2004). Demonstration of the feasibility of a direct solid-phase split-pool biginelli synthesis of 3,4-dihydropyrimidinones. *Org. Lett.* 6 (19): 3237–3240. https://doi.org/10.1021/ol048946r.

11 Henkel, B., Westner, B., and Dömling, A. (2003). Polymer-bound 3-*N,N*-(dimethylamino)-2-isocyanoacrylate for the synthesis of thiazoles via a multicomponent reaction. *Synlett* 2003 (15): 2410–2412. https://doi.org/10.1055/s-2003-43331.

12 Hoener, A.P.F., Henkel, B., and Gauvin, J.C. (2003). Novel one-pot microwave assisted Gewald synthesis of 2- acyl amino thiophenes on solid support. *Synlett* 2003 (1): 63–66. https://doi.org/10.1055/s-2003-36229.

13 Munson, M.C., Cook, A.W., Josey, J.A. et al. (1998). An efficient high-speed synthetic route to amino-substituted thiazolidinone libraries. *Tetrahedron Lett.* 39 (40): 7223–7226. https://doi.org/10.1016/S0040-4039(98)01592-5.

14 Gupta, P., Singh, S.K., Pathak, A. et al. (2002). Template-directed approach to solid-phase combinatorial synthesis of furan-based libraries. *Tetrahedron* 58 (52): 10469–10474. https://doi.org/10.1016/S0040-4020(02)01396-0.

15 Heck, S. and Dömling, A. (2000). A versatile multi-component one-pot thiazole synthesis. *Synlett* 2000 (3): 424–426. https://doi.org/10.1055/s-2000-6517.

16 Henkel, B., Sax, M., and Dömling, A. (2003). A new and efficient multicomponent solid-phase synthesis of 2-acylaminomethylthiazoles. *Tetrahedron Lett.* 44 (18): 3679–3682. https://doi.org/10.1016/S0040-4039(03)00662-2.

17 Kolb, J., Beck, B., Almstetter, M. et al. (2003). New MCRs: the first 4-component reaction leading to 2,4-disubstituted thiazoles. *Mol. Divers.* 6 (3–4): 297–313. https://doi.org/10.1023/B:MODI.0000006827.35029.e4.

18 Wipf, P. and Cunningham, A. (1995). A solid phase protocol of the Biginelli dihydropyrimidine synthesis suitable for combinatorial chemistry. *Tetrahedron Lett.* 36 (43): 7819–7822. https://doi.org/10.1016/0040-4039(95)01660-A.

19 Valverde, M.G., Dallinger, D., and Kappe, C.O. (2001). Solid-phase synthesis of dihydropyrimidones via *N*-acyliminium ion-based α-ureidoalkylations. *Synlett* 2001 (6): 741–744. https://doi.org/10.1055/s-2001-14608.

20 Pérez, R., Beryozkina, T., Zbruyev, O.I. et al. (2002). Traceless solid-phase synthesis of bicyclic dihydropyrimidones using multidirectional cyclization cleavage. *J. Comb. Chem.* 4 (5): 501–510. https://doi.org/10.1021/cc0200181.

21 Byk, G. and Kabha, E. (2006). A solid-supported stereoselective multicomponent reaction: one-pot generation of three asymmetric carbons. *Synlett* 2006 (5): 747–748. https://doi.org/10.1055/s-2006-932498.

22 Breitenbucher, J.G. and Figliozzi, G. (2000). Solid-phase synthesis of 4-aryl-1,4-dihydropyridines via the Hantzsch three component condensation. *Tetrahedron Lett.* 41 (22): 4311–4315. https://doi.org/10.1016/S0040-4039(00)00660-2.

23 Gordeev, M.F., Patel, D.V., and Gordon, E.M. (1996). Approaches to combinatorial synthesis of heterocycles: a solid-phase synthesis of 1,4-dihydropyridines. *J. Org. Chem.* 61 (3): 924–928. https://doi.org/10.1021/jo951706s.

24 Gordeev, M.F., Patel, D.V., England, B.P. et al. (1998). Combinatorial synthesis and screening of a chemical library of 1,4-dihydropyridine calcium channel blockers. *Bioorg. Med. Chem. Lett.* 6 (7): 883–889. https://doi.org/10.1016/S0968-0896(98)00048-0.

25 Wang, Y. and Wilson, S.R. (1997). Solid phase synthesis of 2,3-dihydro-4-pyridones: reaction of Danishefsky's diene with polymer-bound imines. *Tetrahedron Lett.* 38 (23): 4021–4024. https://doi.org/10.1016/S0040-4039(97)00824-1.

26 Zhang, W., Xie, W., Fang, J. et al. (1999). Ytterbium(III) trifluoromethanesulfonate catalyzed solid phase aza Diels–Alder reaction and subsequent facile adduct release. *Tetrahedron Lett.* 40 (45): 7929–7933. https://doi.org/10.1016/S0040-4039(99)01587-7.

27 Kiselyov, A.S. and Armstrong, R.W. (1997). Solid support synthesis of tetrahydroquinolines via the Grieco three component condensation. *Tetrahedron Lett.* 38 (35): 6163–6166. https://doi.org/10.1016/S0040-4039(97)01425-1.

28 Kiselyov, A.S., Smith, L., and Virgilio, A. (1998). Immobilized aldehydes and olefins in the solid support synthesis of tetrahydroquinolines via a three component condensation. *Tetrahedron* 54 (28): 7987–7996. https://doi.org/10.1016/S0040-4020(98)00446-3.

29 Kiselyov, A.S., Smith, L., and Armstrong, R.W. (1998). Solid support synthesis of polysubstituted tetrahydroquinolines via three-component condensation catalyzed by $Yb(OTf)_3$. *Tetrahedron* 54 (20): 5089–5096. https://doi.org/10.1016/S0040-4020(98)00248-8.

30 Ghashghaei, O., Masdeu, C., Alonso, C. et al. (2018). Recent advances of the Povarov reaction in medicinal chemistry. *Drug Discov. Today Technol.* 29: 71–79. https://doi.org/10.1016/j.ddtec.2018.08.004.

31 Carranco, I., Díaz, J.L., Jiménez, O. et al. (2005). Multicomponent reactions with dihydroazines: efficient synthesis of a diverse set of pyrido-fused tetrahydroquinolines. *J. Comb. Chem.* 7 (1): 33–41. https://doi.org/10.1021/cc049877a.

32 Bickneil, A.J., Hird, N.W., and Readshaw, S.A. (1998). Efficient robotic synthesis. Multi-component preparation of a tricyclic template by solid phase Tsuge reaction. *Tetrahedron Lett.* 39 (32): 5869–5872. https://doi.org/10.1016/S0040-4039(98)01185-X.

33 Touré, B.B., Hoveyda, H.R., Tailor, J. et al. (2003). A three-component reaction for diversity-oriented synthesis of polysubstituted piperidines: solution and solid-phase optimization of the first tandem aza[4+2]/allylboration. *Chem Eur J* 9 (2): 466–474. https://doi.org/10.1002/chem.200390049.

34 Ulaczyk-Lesanko, A., Pelletier, E., Lee, M. et al. (2007). Optimization of three- and four-component reactions for polysubstituted piperidines: application to the synthesis and preliminary biological screening of a prototype library. *J. Comb. Chem.* 9 (4): 695–703. https://doi.org/10.1021/cc0700344.

35 Gelman, M., Massarano, T., Lavi, R. et al. (2018). A new multicomponent reaction MCR4 for the synthesis of analogs of staurosporine. *Curr. Org. Chem.* 22 (5): 505–517. https://doi.org/10.2174/1385272821666170817110101.

36 Massarano, T., Mazir, A., Lavi, R. et al. (2020). Solid-phase multicomponent synthesis of 3- substituted isoindolinones generates new cell-penetrating probes as drug carriers. *ChemMed Chem* 15 (10): 833–838. https://doi.org/10.1002/cmdc.201900656.

37 Demaude, T., Knerr, L., and Pasau, P. (2004). New synthetic pathway to diverse 2-substituted quinolines based on a multicomponent reaction: solution-phase and solid-phase applications. *J. Comb. Chem.* 6 (5): 768–775. https://doi.org/10.1021/cc049937c.

38 Jad, Y.E., Gudimella, S.K., Govender, T. et al. (2018). Solid- phase synthesis of pyrrole derivatives through a multicomponent reaction involving Lys containing peptides. *ACS Comb. Sci.* 20 (4): 187–191. https://doi.org/10.1021/acscombsci.8b00006.

39 Constabel, F. and Ugi, I. (2001). Repetitive Ugi reactions. *Tetrahedron* 57 (27): 5785–5789. https://doi.org/10.1016/S0040-4020(01)00516-6.

40 Liu, Z. and Nefzi, A. (2012). Ugi four-center three-component reaction for the parallel solid-phase synthesis of *N*-substituted thiomopholinones. *Tetrahedron Lett.* 53 (9): 1013–1014. https://doi.org/10.1016/j.tetlet.2011.12.074.

41 Treder, A.P., Tremblay, M.C., Yudin, A.K. et al. (2014). Solid- phase synthesis of piperazinones via disrupted ugi condensation. *Org. Lett.* 16 (17): 4674–4677. https://doi.org/10.1021/ol5023118.

42 Méndez, Y., De Armas, G., Pérez, I. et al. (2019). Discovery of potent and selective inhibitors of the *Escherichia coli* M1-aminopeptidase via multicomponent solid-phase synthesis of tetrazole-peptidomimetics. *Eur. J. Med. Chem.* 163 (3): 481–499. https://doi.org/10.1016/j.ejmech.2018.11.074.

43 Morales, F.E., Garay, H.E., Muñoz, D.F. et al. (2015). Aminocatalysis-mediated on-resin Ugi reactions: application in the solid-phase synthesis of *N*-substituted and tetrazolo lipopeptides and peptidosteroids. *Org. Lett.* 17 (11): 2728–2731. https://doi.org/10.1021/acs.orglett.5b01147.

44 Brenner, S. and Lerner, R.A. (1992). Encoded combinatorial chemistry. *Proc. Natl. Acad. Sci. USA* 89 (12): 5381–5383. https://doi.org/10.1073/pnas.89.12.5381.

45 Dickson, P. and Kodadek, T. (2019). Chemical composition of DNA-encoded libraries, past present and future. *Org. Biomol. Chem.* 17: 4676–4688. https://doi.org/10.1039/C9OB00581A.

46 Potowski, M., Kunig, V.B.K., Losch, F. et al. (2019). Synthesis of DNA-coupled isoquinolones and pyrrolidines by solid phase ytterbium- and silver-mediated imine chemistry. *Med. Chem. Commun.* 10: 1082–1093. https://doi.org/10.1039/C9MD00042A.

47 Potowski, M., Losch, F., Wünnemann, E. et al. (2019). Screening of metal ions and organocatalysts on solid support-coupled DNA oligonucleotides guides design of DNA-encoded reactions. *Chem. Sci.* 10 (45): 10481–10492. https://doi.org/10.1039/C9SC04708E.

48 Kunig, V.B.K., Ehrt, C., Dömling, A. et al. (2019). Isocyanide multicomponent reactions on solid-phase-coupled DNA oligonucleotides for encoded library synthesis. *Org. Lett.* 21 (18): 7238–7243. https://doi.org/10.1021/acs.orglett.9b02448.

49 Potowski, M., Esken, R., and Brunschweiger, A. (2020). Translation of the copper/bipyridine-promoted Petasis reaction to solid-phase coupled DNA for encoded library synthesis. *Bioorg. Med. Chem.* 28 (9): 115441–115448. https://doi.org/10.1016/j.bmc.2020.11.5441.

50 Kunig, V.B.K., Potowski, M., Akbarzadeh, M. et al. (2020). TEAD–YAP interaction inhibitors and MDM2 binders from DNA- encoded indole-focused Ugi peptidomimetics. *Angew. Chem. Int. Ed.* 59 (46): 20338–20342. https://doi.org/10.1002/anie.202006280.

51 Lin, Q. and Blackwell, H.E. (2006). Rapid synthesis of diketopiperazine macroarrays via Ugi four-component reactions on planar solid supports. *Chem. Commun.*: 2884–2886. https://doi.org/10.1039/B604329A.

52 Campbell, J. and Blackwell, H.E. (2009). Efficient construction of diketopiperazine macroarrays through a cyclative-cleavage strategy and their evaluation as luminescence inhibitors in the bacterial symbiont *Vibrio fischeri*. *J. Comb. Chem.* 11 (6): 1094–1099. https://doi.org/10.1021/cc900115x.

53 Roller, S., Siegers, C., and Haag, R. (2004). Dendritic polyglycerol as a high-loading support for parallel multistep synthesis of GABA lactam analogues. *Tetrahedron* 60 (39): 8711–8720. https://doi.org/10.1016/j.tet.2004.05.108.

54 Bergbreiter, D.E. (2002). Using soluble polymers to recover catalysts and ligands. *Chem. Rev.* 102 (10): 3345–3384. https://doi.org/10.1021/cr010343v.

55 Gravert, D.J. and Janda, K.D. (1997). Organic synthesis on soluble polymer supports: liquid-phase methodologies. *Chem. Rev.* 97 (2): 489–510. https://doi.org/10.1021/cr960064l.

56 Toy, P.H. and Janda, K.D. (2000). Soluble polymer-supported organic synthesis. *Acc. Chem. Res.* 33 (8): 546–554. https://doi.org/10.1021/ar990140h.

57 Xia, M. and Wang, Y. (2002). Soluble polymer-supported synthesis of Biginelli compounds. *Tetrahedron Lett.* 43 (43): 7703–7705. https://doi.org/10.1016/S0040-4039(02)01850-6.

58 Zhang, H., Yang, G., Chen, J. et al. (2004). Synthesis of thiophene derivatives on soluble polymer-support using Gewald reaction. *Synthesis* 2004 (18): 3055–3059. https://doi.org/10.1055/s-2004-834895.

59 Chen, C.H., Yello, G.S., Lin, P.T. et al. (2011). Base-catalyzed Povarov reaction: an unusual [1,3] sigmatropic rearrangement to dihydropyrimidobenzimidazoles. *Org. Lett.* 13 (19): 5120–5123. https://doi.org/10.1021/ol201985p.

60 (a) Mamantov, G. and Mamantov, C.B. (1985). Advances in molten salt chemistry. *Berichte der Bunsengesellschaft für physikalische Chemie* 5, 89 (10): 1121–1121. https://doi.org/10.1002/bbpc.19850891025; (b) Hussey, C.L. (1988). Room temperature haloaluminate ionic liquids. Novel solvents for transition metal solution chemistry. *Pure Appl. Chem.* 60 (12): 1763–1772. https://doi.org/10.1351/pac198860121763. (c) Mamantov, G. and Marassi, R. (1987). *Molten Salt Chemistry an Introduction and Selected Applications*. Dordrecht: Springer https://doi.org/10.1007/978-94-009-3863-2.

61 Fraga-Dubreuil, J. and Bazureau, J.P. (2003). Efficient combination of task-specific ionic liquid and microwave dielectric heating applied to one-pot three component synthesis of a small library of 4-thiazolidinones. *Tetrahedron* 59 (32): 6121–6130. https://doi.org/10.1016/S0040-4020(03)00954-2.

62 Hu, Y., Wei, P., Huang, H. et al. (2006). Synthesis of 2-aminothiophenes on ionic liquid phase support using the Gewald reaction. *Synth. Commun.* 36 (11): 1543–1548. https://doi.org/10.1080/00397910600588819.

63 Legeay, J.C., Eynde, J.J.V., Toupet, L. et al. (2007, 2007). A three-component condensation protocol based on ionic liquid phase bound acetoacetate for the synthesis of Biginelli 3,4-dihydropyrimidine-2(1*H*)-ones. *ARKIVOC* 3: 13–28. ISSN 1424-6376.

64 Hassine, F., Gmouh, S., Pucheault, M. et al. (2007). Task specific onium salts and ionic liquids as soluble supports in *Grieco*'s multicomponent synthesis of tetrahydroquinolines. *Monatsh. Chem.* 138: 1167–1174. https://doi.org/10.1007/s00706-007-0708-0.

5

Green Synthesis of Heterocycles Via MCRs

Wei Zhang

University of Massachusetts Boston, Center for Green Chemistry and Department of Chemistry, 100 William T Morrissey Blvd, Boston, MA 02125, USA

5.1 Introduction

Reaction efficiency is one of the most important aspects in green synthesis. Multicomponent reactions (MCRs) with three or more components have inherent advantages of pot, atom, and step economy (PASE) to form multiple new bonds in a simple operation [1–3]. MCRs integrate most parts of reactants to the product structures for high atom economy, eliminating intermediate purification step for low waste disposal, and performing simple operation for saving time and resources. Similar to one-pot synthesis and cascade reactions, MCRs are an attractive and active topic in the development of new methodologies for organic synthesis and catalysis [4–6]. The greenness and effectiveness of MCRs could be quantitatively evaluated by a series of green metrics analysis including atom economy, atom efficiency, carbon efficiency, reaction mass efficiency, process mass intensity, solvent and water intensity, and E factor [7–9].

There are only limited numbers of review articles on MCR-based green synthesis [10, 11]. Many aspects of MCRs are presented in other chapters of this book. This chapter only covers MCRs with enhanced efficiency in the synthesis of heterocyclic compounds. These methods are: (i) high-order MCRs; (ii) consecutive MCRs; (iii) MCRs followed by cyclization; and (iv) MCRs followed by cycloaddition or annulation. In addition to efficiency, these MCRs and post-MCR modifications significantly increase the diversity and complexity of heterocyclic compounds which may have biological activities and valuable in drug discovery research. Due to the page limit, only selected examples are presented in this chapter to illustrate the MCR-based green synthesis of heterocycles. Other green aspects including using water or bio-based solvents; using alternative energies such as photo, microwave, and ultrasound for reactions; employing renewable substrates as starting materials; and recovery and recycling of reagents and catalysts are mentioned wherever appropriate, but they are not the major focuses of this chapter. Interested readers

Multicomponent Reactions towards Heterocycles: Concepts and Applications, First Edition.
Edited by Erik V. Van der Eycken and Upendra K. Sharma.

can reference related books and review articles on green organic synthesis and catalysis [12, 13].

5.2 High-Order MCRs

Reactions with single operational step and involving five or more components are called high-order MCRs [14]. Compared to popular three-component reactions (3CRs), high-order MCRs are more efficient and the product structures are more complicated. But because of sophistic reaction mechanisms and competitive side-reactions, the number of high-order MCRs is limited and the development of such new reactions are not an easy task. In many cases, the high-order MCRs are discovered by accident instead of by design.

Presented in this section are high-order MCRs including 5CRs, 6CRs, and 7CRs. If two or more of the same molecules are involved in the MCRs, they are called pseudo-MCRs. Using 5CRs as examples (Scheme 5.1), in addition to 5CR of five different components A + B + C + D + E (Scheme 5.1, I), other types of reactions such as 2A + B + C + D (Scheme 5.1, II), 2A + 2B + C (Scheme 5.1, III), 3A + B + C, and 4A + B are pseudo-5CRs. It is noteworthy that reactants used for MCRs could have more than one reaction center. In these cases, the intramolecular reactions are involved and thus the number of reaction centers is higher than the components. For example, a five-center and four-component reaction (5C4CR) (Scheme 5.1, IV) is also a high-order MCR. It is also noteworthy that MCRs only have a single operational step in which all the components are charged to the reaction vessel at same time. If components are introduced stepwisely at different reaction stages or under different conditions, then they should be called one-pot reactions instead of MCRs.

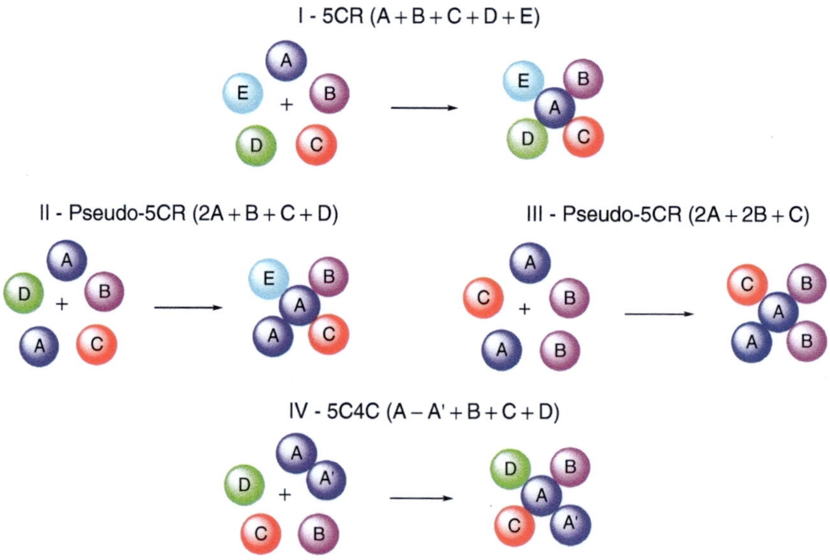

Scheme 5.1 Conceptual figures for representative 5CRs.

Shown in Scheme 5.2 are two A + B + C + D + E-type 5CRs of *N*-propargyl isatins, malononitrile, benzyl bromides, sodium azide with 5,5-dimethyl-1,3-cyclohexa nedione or 4-hydroxycarbazole in the synthesis of triazole-tethered spirochromeno-carbazoles **1** or **2** using reusable cellulose-supported CuI nanoparticles (Cell-CuI NPs) as a catalyst [15, 16]. In the synthesis of **1**, the condensation of *N*-propargyl isatins, malononitrile, and 1,3-cyclohexanedione gives **3** which undergoes cyclization to form spirochromenes **4** and then click reaction with benzylic azides to afford products **1**. The Cell-CuI NP catalyst can be recovered for reuse. Synthesized compounds from this work were submitted for biological screening on human cancer cell lines. The most active compound showed antiproliferative activity at IC_{50} 3.78 μM.

Another A + B + C + D + E kind 5CR involving phenylacetonitriles, aryl aldehydes, hydrazine, ammonium acetate, and ethyl acetoacetate has been reported for the synthesis of 4,7-dihydro-1*H*-pyrazolo[3,4-b]pyridin-6-amines **5** using EtOH as a green solvent (Scheme 5.3) [17]. Under the catalysis of meglumine, the reaction of intermediates **6** and **7** and then with NH_3 affords products after dehydrative cyclization.

A pseudo-5CR of 2A + B + C + D kind has been developed using 2 equiv of β-nitrostyrenes and 1 equiv each of aryl aldehydes, ammonium acetate, and Meldrum's acid in EtOH for diastereoselective synthesis of a small compound library of functionalized piperidines **8** which has pharmacology interest (Scheme 5.4) [18]. The reaction of two intermediates generated separately from the condensation of β-nitrostyrenes with Meldrum's acid and aryl aldehydes with NH_4OAc affords intermediates **9** which react with aryl aldehydes again followed by base-promoted cyclization to give products **8**.

Shown in Scheme 5.5 is another pseudo-5CR of 2 equiv of indene-1,3-dione and 1 equiv each *o*-benzenediamines, indene-1,2,3-trione, and anilines under the catalysis of *p*-TSA to give highly condensed polycyclic system **10** [19]. The reaction process involves iminization–aromatization reaction of indene-1,2,3-trione with *o*-benzenediamine followed by condensation with two indene-1,3-dione give **11** and then with anilines to afford products **10** after cyclization and tautomerization.

Pseudo-5CRs of 2A + 2B + C is more popular than other kind of 5CRs. A method for the synthesis of pyrazole- and pyrazolone-fused dihydrooxepines **12** is established through the reaction of 1 equiv of acetophenones and 2 equiv each of phenylhydrazines and ethyl acetoacetate (Scheme 5.6) [20]. Condensation of intermediates **13** with 2 equiv of **14** affords compounds **15** which then undergo I_2-promoted cyclization to afford the products.

The 2A + 2B + C kind pseudo-5CRs shown in Scheme 5.7 involves double Huisgen [3 + 2] cycloadditions of azomethine ylides using 1 equiv of *α*-aminoester and 2 equiv each of aldehydes and maleimides for diastereoselective synthesis of fused-pyrrolidine compounds **16** [21]. Azomethine ylides generated from *α*-aminoester and aldehydes undergo [3 + 2] cycloaddition with maleimides to form adducts **17** which then react with second aldehydes and maleimides for another [3 + 2] cycloaddition to give products **16**. This reaction can be conducted using amino acids to replace amino esters for decarboxylative double [3 + 2] cycloadditions [22].

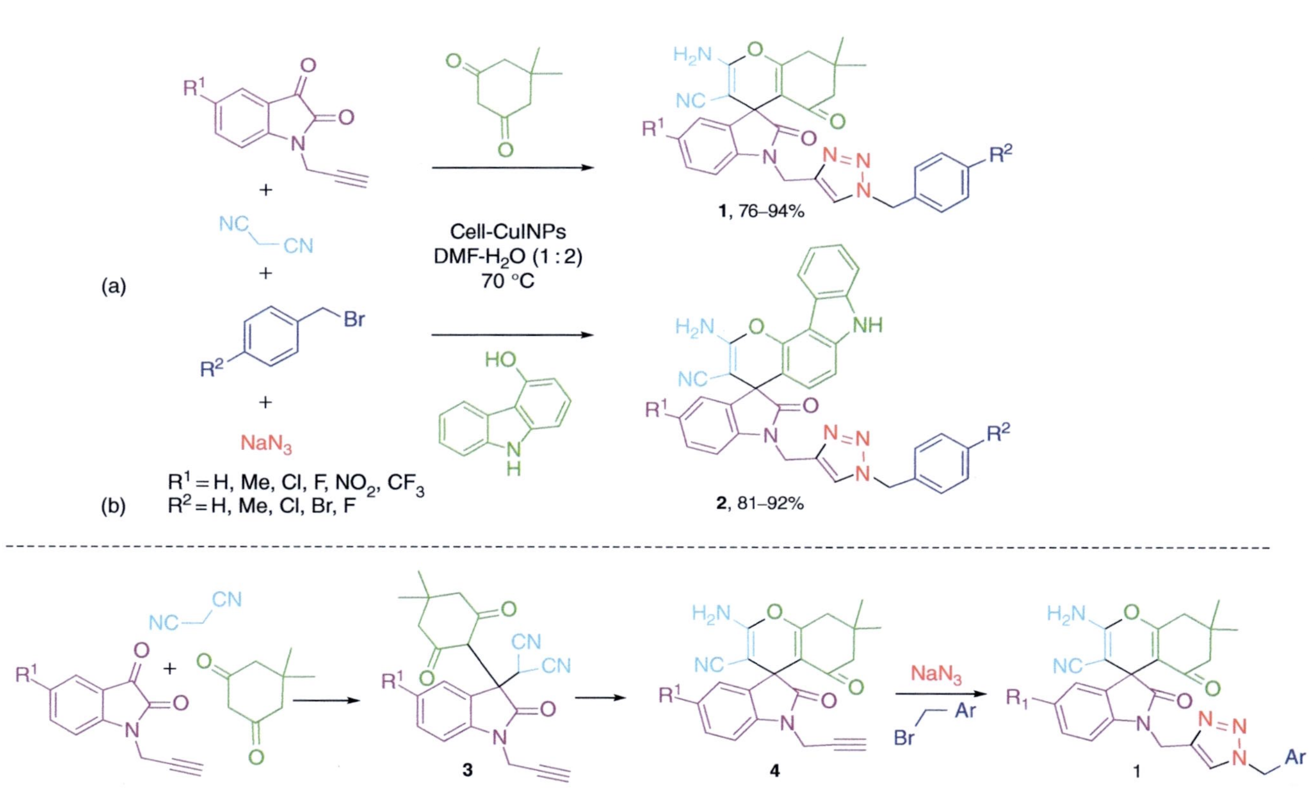

Scheme 5.2 5CRs for triazole-tethered spirochromenocarbazoles.

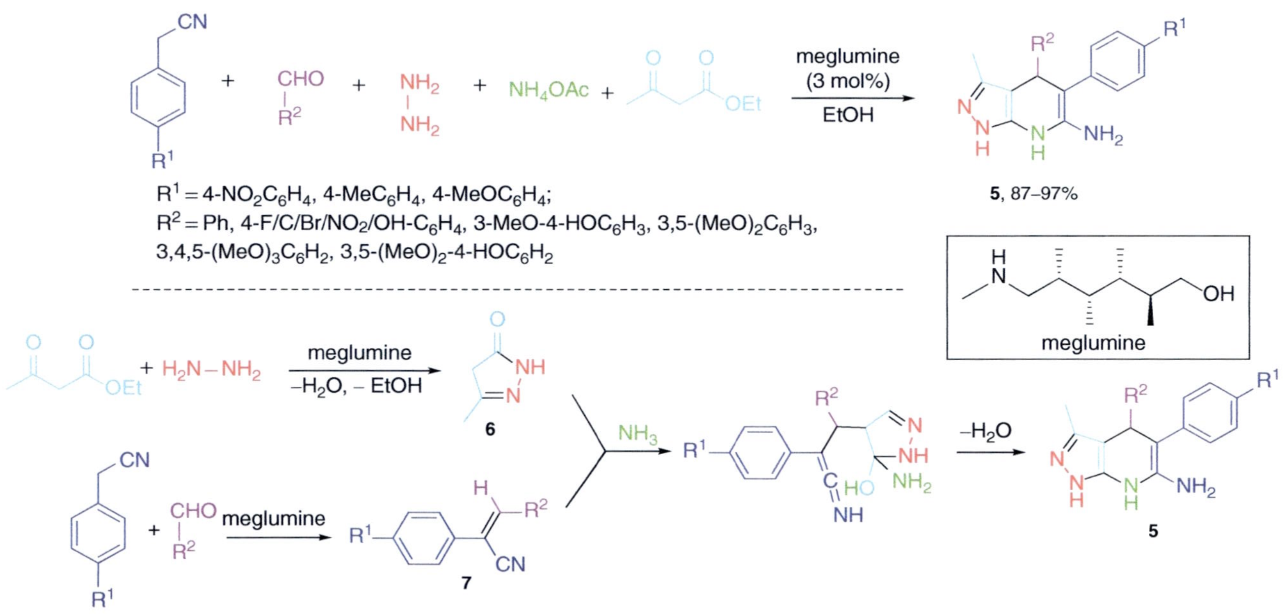

Scheme 5.3 5CR for dihydropyrazolopyridines. Source: Govindaraju et al. [17].

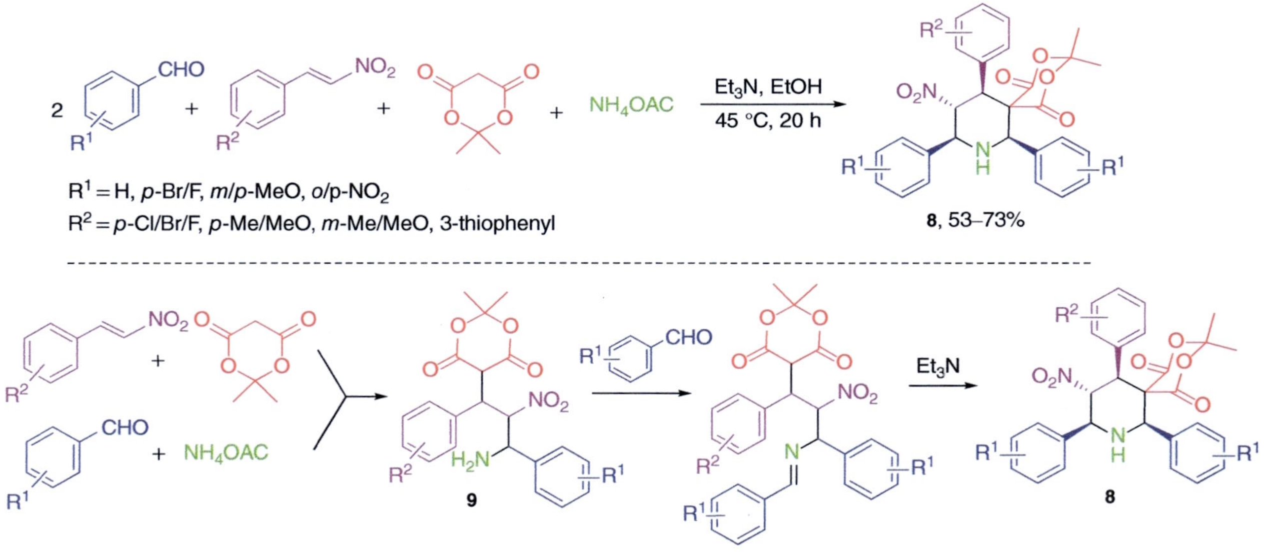

Scheme 5.4 Pseudo-5CR for spiro-piperidines bearing Meldrum's acid. Source: Li et al. [18].

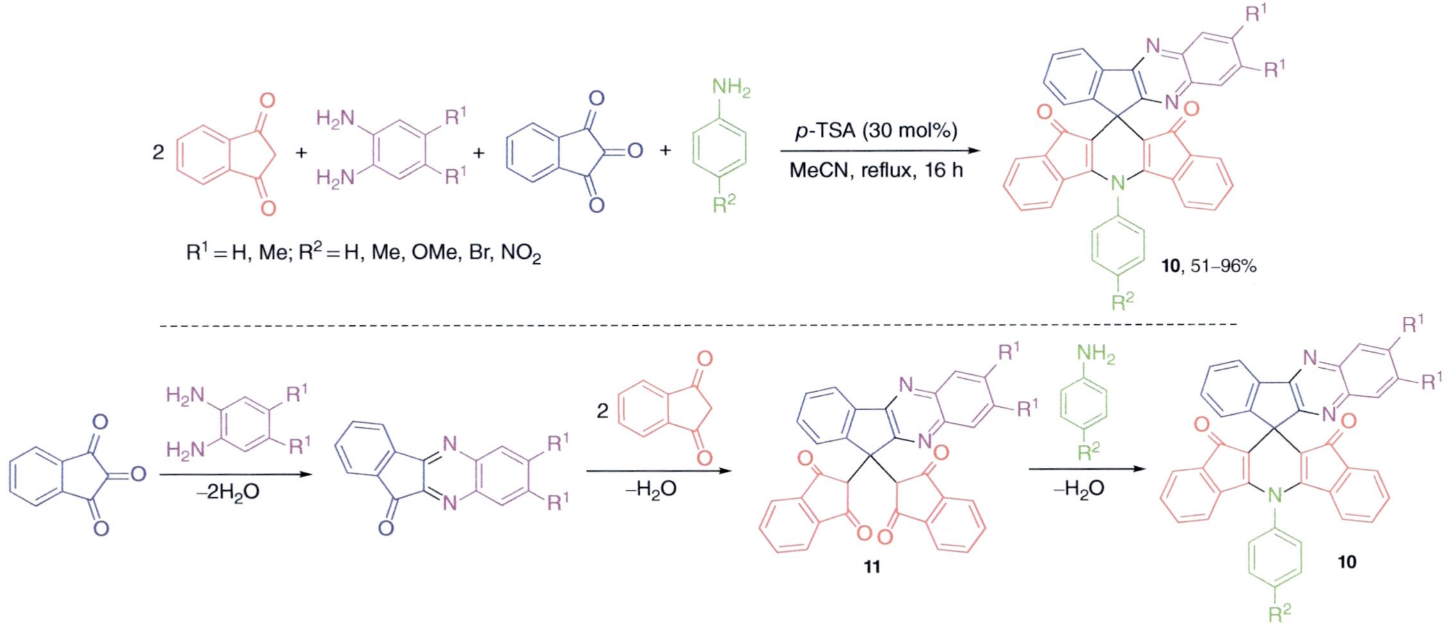

Scheme 5.5 Pseudo-5CR for polyheterocycles.

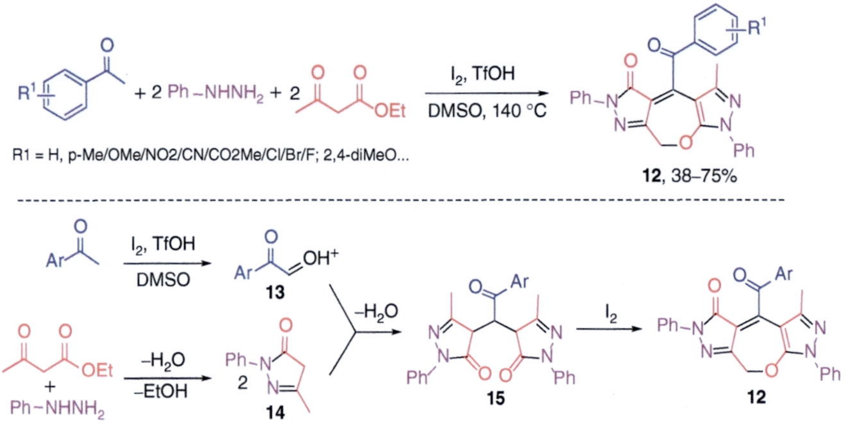

Scheme 5.6 Pseudo-5CR for pyrazole- and pyrazolone-fused dihydrooxepines. Source: Based on Zhao et al. [20].

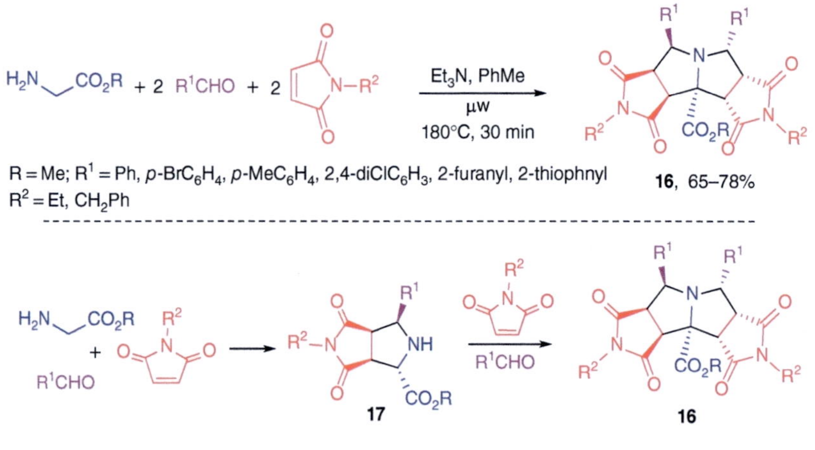

Scheme 5.7 Pseudo-5CR for pyrrolidine-based polycyclic compounds.

The 3A + 2B kind pseudo-5CRs are rare in literature. Presented in Scheme 5.8 is an example of such reaction using 3 equiv of isatins and 2 equiv of β-keto amides for the synthesis of oxoindolines bearing two quinoline and pyrrolodione-fused rings **18** [23]. The reaction process involves the condensation of isatins with 2 equiv of β-keto amides to form **19** which react with 2 equiv of isatins followed by ring opening of isatins to form **20** and then amidative cyclization and aromatization to afford products **18**.

An interesting 4A + B kind pseudo-5CR of 4 equiv of 1,3-indanedione and 1 equiv of benzaldehydes for the synthesis of highly condensed ring system **21** is shown in Scheme 5.9 [24]. Et_3N-catalyzed cyclotrimerization of 1,3-indanedione results active cyclic diene **22** which undergoes diastereoselective Diels–Alder reaction with 2-aryllidene-1,3-indanediones **23** generated in situ from indanedione and benzaldehydes to form polycyclic structure **21** with a core of bicyclo[2.2.2]octane.

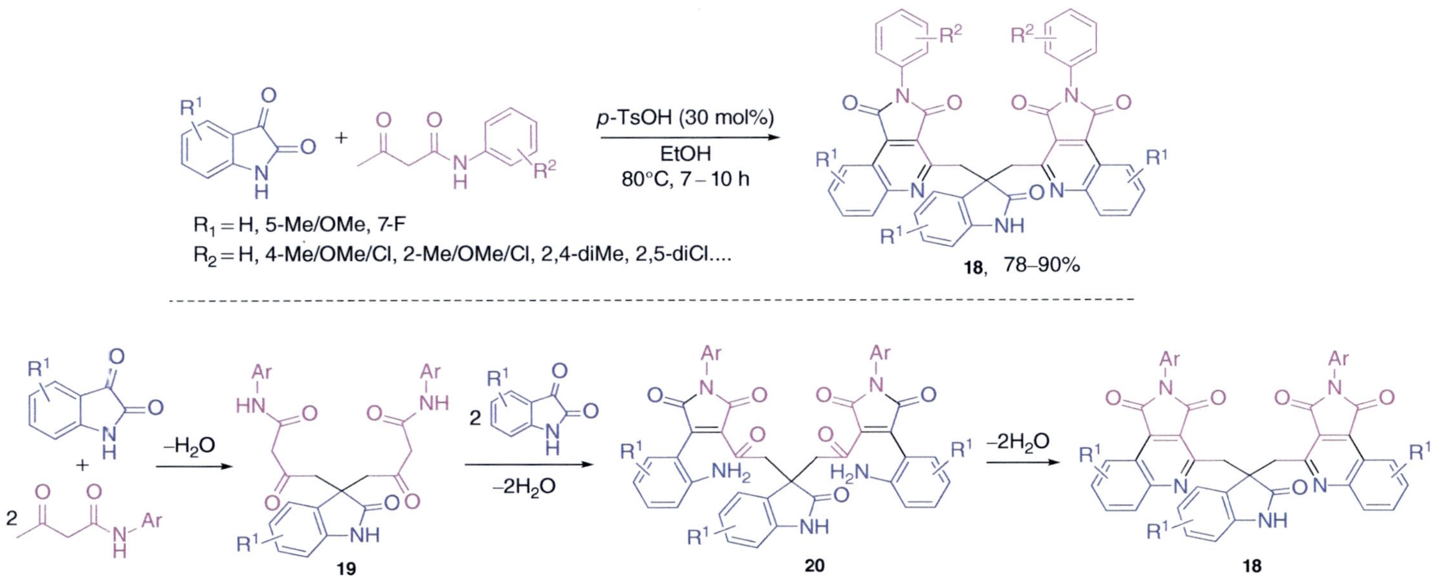

Scheme 5.8 Pseudo-5CR for heterocycles bearing oxoindoline, quinoline, and pyrrolodione rings.

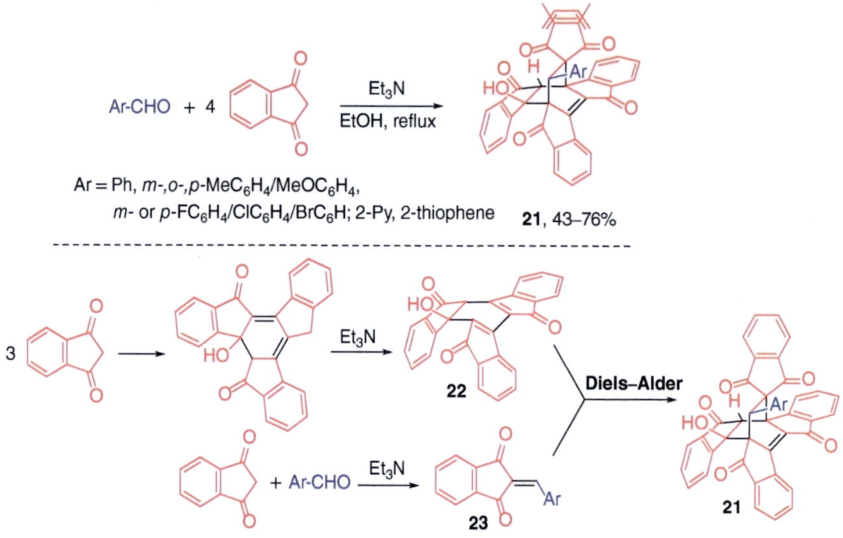

Scheme 5.9 Pseudo-5CR for bicyclo[2.2.2]octane-centered polycycles. Source: Cao et al. [24].

If a reactant for MCR has more than one reaction center, then the MCR has to indicate both the centers and components. For example, the reaction in Scheme 5.10 is a 5C4CR, in which malononitrile is a two-centered reactant [25]. The Knoevenagel condensation of barbituric acid followed by Michael addition with malononitrile afford **25**. Cyclization and sequential dehydration and aromatization give pyridopyrimidine products **24**. Some of the product analogs were found to have good anti-inflammatory activity.

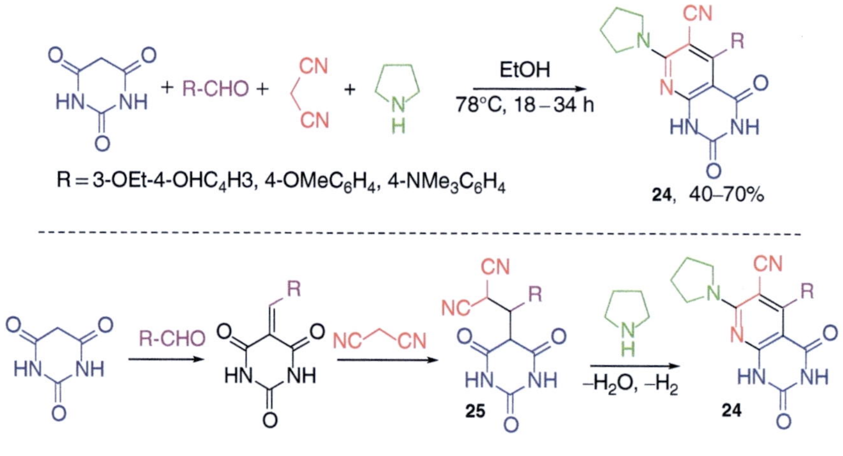

Scheme 5.10 5C4CR for pyridopyrimidines.

Shown in Scheme 5.11 is a double Huisgen 5C3CR in which a two-centered reactant **26** is used in 2 equiv. The reaction process involves two intermolecular [3 + 2] cycloadditions of azomethine ylides, first to form pyrrolidine intermediates **28** which then react with another equiv of **26** to form products **27** [26]. This one-pot reaction is conducted under microwave heating in 10 minute to generate four new rings and seven stereocenters in a diastereoselective fashion.

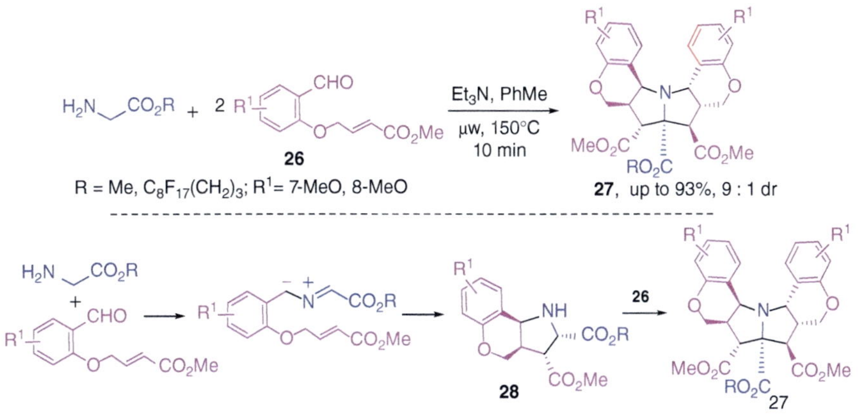

Scheme 5.11 5C3CR for polyheterocycles.

A 6CR of six different reactants (A + B + C + D + E + F) is hard to find in literature. Presented in Scheme 5.12 is a single-operational reaction involving Knoevenagel/Ugi/click reactions of six different molecules for the synthesis of triazole- and coumarin-containing carboxamides **29** [27]. The Knoevenagel reaction of 2-hydroxy aldehydes and Meldrum's acid gives intermediates **30**, while the reaction of propargyl aldehydes with anilines gives intermediates **31**. The Ugi-4C3CR of these two intermediates gives **32** after rearrangement. The last step click reaction gives final products **29**.

Shown in Scheme 5.13 is a 2A + B + C + D kind pseudo-6CR using 2 equiv of isocyanides and 1 equiv each of primary amines, carbon disulfide, maleic anhydride, and dialkyl acetylenedicarboxylates for the synthesis of a heterocyclic products **33** containing rhodanine and furan rings [28]. Intermediates **34** and **35** generated from corresponding starting materials are coupled to form **36** which them undergo Mumm rearrangement for **37** followed by [4 + 1] cycloaddition with another equiv of isocyanides to form products **33**. The reaction is conducted under sonication radiation and using H_2O as a solvent.

A 2A + 2B + C + D-type azido-Ugi pseudo-6CR using 2 equiv each of isocyanides and $TMSN_3$ and 1 equiv each of cyclic imines (pyrrolines) and formaldehyde for the synthesis of bistetrazole-containing cyclic amines **38** is shown in Scheme 5.14 [29]. The reaction is carried out in MeOH addition of 2 drops of Et_3N. The reaction process initiates with the azido-Ugi reaction of pyrrolines with isocyanides and HN_3 (derived from $TMSN_3$ and MeOH) to form monotetrazole intermediates **39** which undergo second azido-Ugi reaction with aldehydes, isocyanides and HN_3 to give final products **38**. In addition to pyrrolines, other cyclic imines could be used for the synthesis.

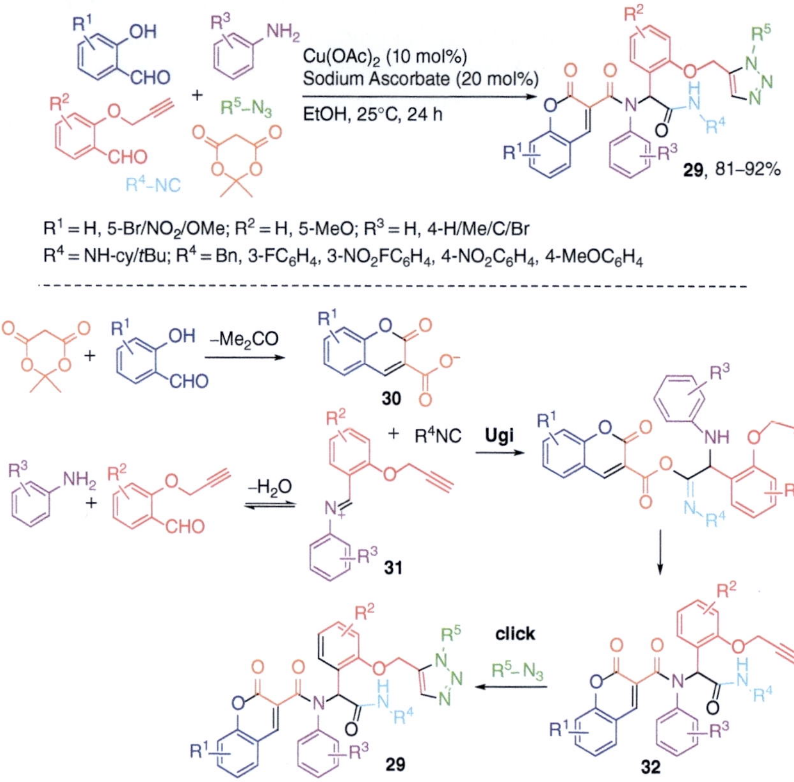

Scheme 5.12 7CR for heterocycles involving Knoevenagel/Ugi/click reactions.

Shown in Scheme 5.15 is another 2A + 2B + C + D kind pseudo-6CR with 2 equiv each of glutaraldehyde and malononitrile and 1 equiv each of cyclic 1,3-diketones and primary amines for the synthesis of highly condensed ring systems **40** [30]. Intermediate **41** generated from the reaction of glutaraldehyde and 2 equiv of malononitrile reacts with 1,3-diketones for **42** which further react with glutaraldehyde and primary amines to form products **40**. The cyclic 1,3-ketones can be replaced with acyclic ketones or 4-hydroxycoumarins for making different products.

The highest order MCRs covered in this chapter are 7CRs. It is hard to find 7CRs of seven different molecules. Shown in Scheme 5.16 is a 2A + 2B + 2C + D kind pseudo-7CR employing 2 equiv each of 4-hydroxycoumarin, β-ketoesters, and phenylglyoxal monohydrate and 1 equiv of *p*-phenylenediamine in the synthesis of C_2 symmetric compounds **43** [31]. This reaction process is developed from a 4CR of 4-hydroxycoumarin, β-ketoesters, and phenylglyoxal monohydrate and primary amines by replacing the primary amines with *p*-phenylenediamine. The last step of the pseudo-7CR is the reaction of intermediates **44** with 2 equiv of **45** to form products **43**.

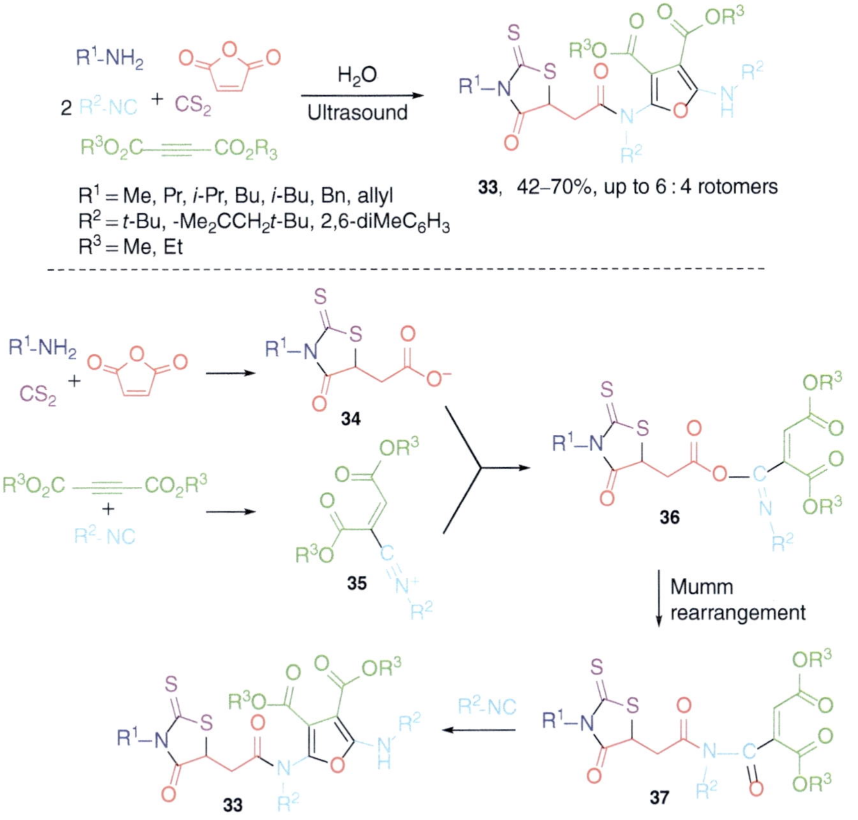

Scheme 5.13 Pseudo-6CR for rhodanine- and furan-containing compounds.

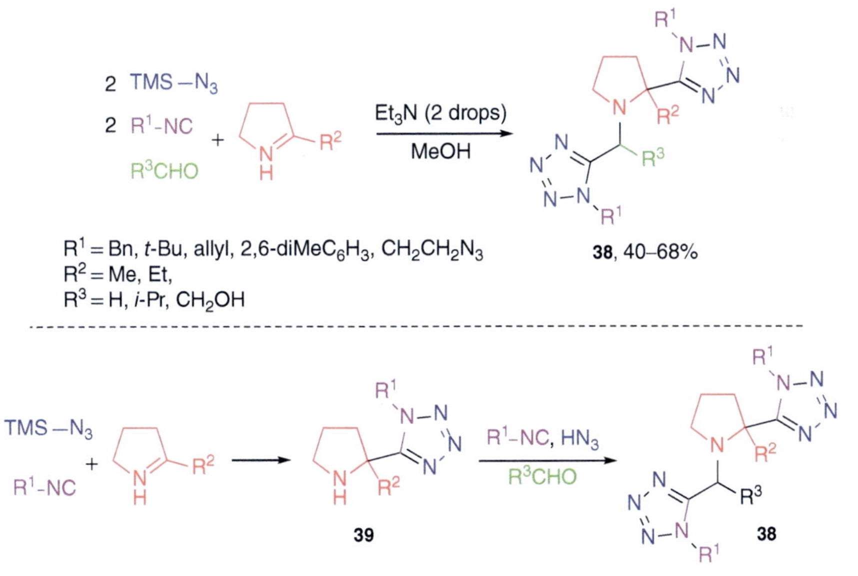

Scheme 5.14 Pseudo-6CR for bistetrazole-containing cyclic amines. Source: Kutovaya et al. [29].

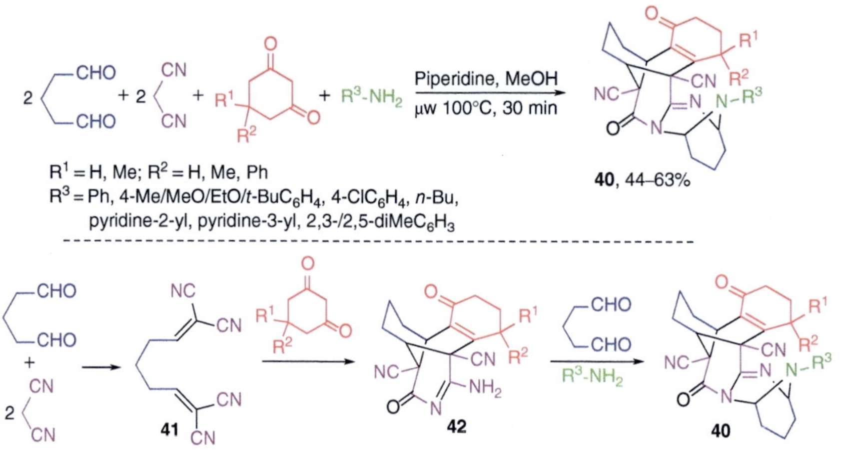

Scheme 5.15 Pseudo-6CR for highly condensed heterocycles.

5.3 Consecutive MCRs

The consecutive MCRs, also called sequential or multiple MCRs, are to combine two or more MCRs (Scheme 5.17) [32]. It is a stepwise reaction process, in which the unreacted functional group (A′) from the first MCR is used for the second MCR. Consecutive MCRs make the synthesis more powerful in efficiency, and the products have significantly high structural diversity and molecular complexity.

A double Groebke–Blackburn–Bienaymé (GBB) reaction sequence using 2,4-diaminopyrimidine as a two-centered component has been reported for the synthesis of substituted diimidazopyrimidines (Scheme 5.18) [33]. Under catalysis of $Yb(OTf)_3$, the reaction of aminoazine, aldehyde, and isocyanide affords first GBB products **46**, which then undergo second GBB reaction with a different set of aldehyde and isocyanide under the catalysis of *p*-toluenesulfonic acid (PTSA) to afford products **47**. Other bifunctional components, such as 2,4-diaminoquinazoline and 3,6-diaminopyridazine, could be used for the synthesis of corresponding products.

Sequential Biginelli and Ugi-4CR reactions for the synthesis of tetrahydropyrimidinone amides **48** are shown in Scheme 5.19 [35]. In the reaction process, the Biginelli reaction of ureidoalkyl acids with aldehydes and benzyl acetoacetates affords tetrahydropyrimidinone carboxylic acids **49**. The adducts are then used for the Ugi-4CR with amines, aldehydes, and isocyanides to form products **48**. The first MCR is conducted at room temp, while the second one is under microwave heating.

Sequential intermolecular and intramolecular Huisgen reactions of azomethine ylides for the synthesis of pyrrolidine-based polycyclic compounds **50** are shown in Scheme 5.19 [34]. In the reaction process, the intermolecular [3 + 2] cycloaddition of benzaldehydes, glycine ester, and *N*-ethylmaleimides affords products **51** which then undergo intramolecular [3 + 2] cycloaddition with *O*-allyl or *O*-alkynyl salicylaldehydes to afford products **40** diastereoselectively (Scheme 5.20).

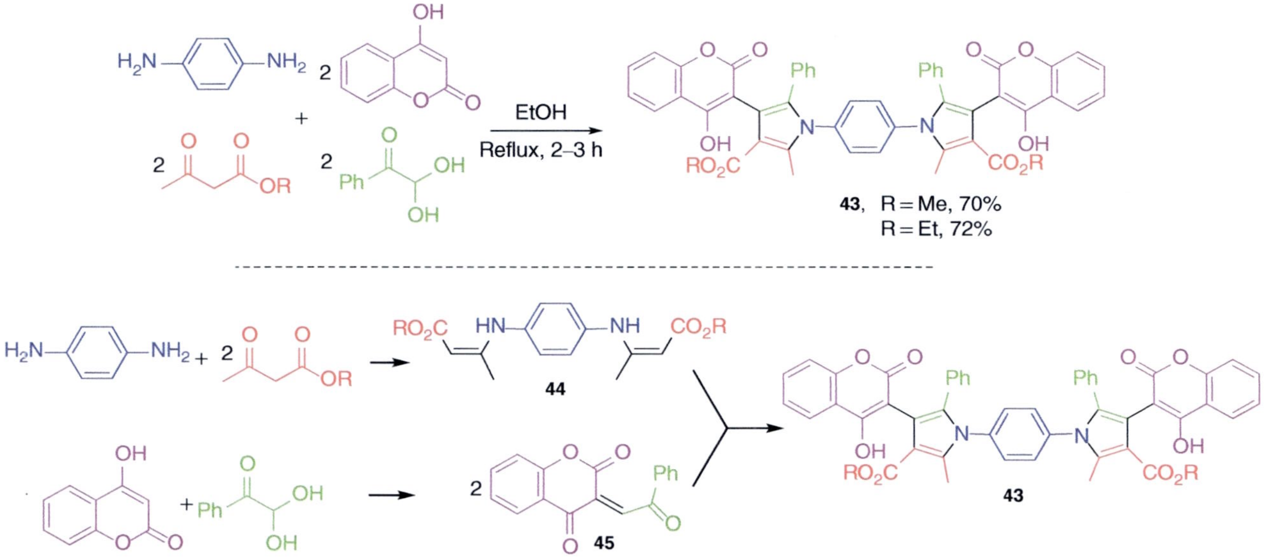

Scheme 5.16 Pseudo-7CR for C_2-symmetric heterocycles.

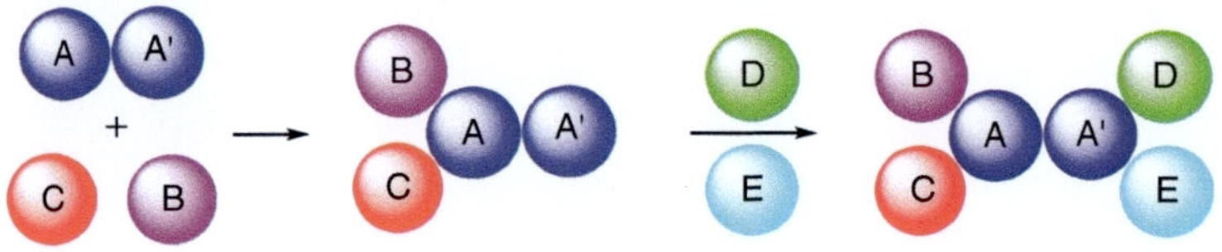

Scheme 5.17 Conceptual figure of consecutive MCR. Source: Zhi et al. [32].

Sequential 3CR and GBB reactions for the synthesis of imidazotriazoles **52** have been reported (Scheme 5.21) [36]. The 3CR of orthoesters, aryl hydrazine, and cyanamide for 5-aminotriazoles **53** is followed by the $Sc(OTf)_3$-promoted GBB reaction with aryl aldehydes and isonitriles to afford imidazotriazole products **52**. The first MCR was conducted under microwave heating, while the second one was at room temp.

Double Ugi reactions for the synthesis of tetrazolopiperazinones **54** are shown in Scheme 5.22 [37]. ω-Carboxyl α-aminotetrazoles **55** are synthesized from the first Ugi-4CR of $TMSN_3$, isocyanides, aldehydes, and tritylamine followed by the treatment with HCl. The second Ugi-4C3CR of two-centered components **55** with isocyanides and aldehydes affords products **54** as a mixture of diastereomers.

The combination of Ugi-4C3CR and Pictet–Spengler reactions for the synthesis of polycyclic compounds **56** is shown in Scheme 5.23 [38]. In this sequence, 2-isocyanoethylindole, carbonyl-protected aminoacetaldehyde, and two-centered 2-formyl benzoic acid are used for the Ugi-4C3CR reaction. Adduct **57** is treated with formic acid to free the carbonyl group for the sequential intramolecular PS reaction to form indole alkaloids **56** as a mixture of two diastereomers. Other polycyclic indole alkaloid-type compounds could be prepared by using different cyclic keto carboxylic acids as starting materials.

A triple Ugi-MCR sequence for the synthesis of cyclic pentapeptoids **58** is shown in Scheme 5.24 [39]. After ester hydrolysis of the adduct from the first Ugi-4CR, the resulted acid **59** is used for the second Ugi-4CR reaction to give ester **60** which undergoes ester hydrolysis and Cbz deprotection to give a two-centered reactant for Ugi-4C3CR with *t*-butyl isocyanide and paraformaldehyde to form cyclopeptoid **58** after the treatment with TFA. A similar reaction sequence to combining four Ugi-4CRs has also been developed for the synthesis of cyclic heptapeptoid [40].

Another triple MCR sequence involving Ugi-4CR, Passerini, and Ugi-4C3CR has been developed for the synthesis of cyclic depsipeptoids **61** (Scheme 5.25) [41]. The first Ugi-4CR reaction of isocyanide, paraformaldehyde, benzylamine, and *N*-Boc-glycine under microwave heating followed by hydrolysis provides acid **62**. It is used for the Passerini reaction with *i*-butyraldehyde and *t*-butyl isocyanoacetate to give amino acid salt **63** after the treatment with TFA. Ugi-4C3CR of **63** with paraformaldehyde and an isocyanide affords cyclic pentadepsipeptoids **61**. Other analogues shown in Scheme 5.24 could be prepared by this method.

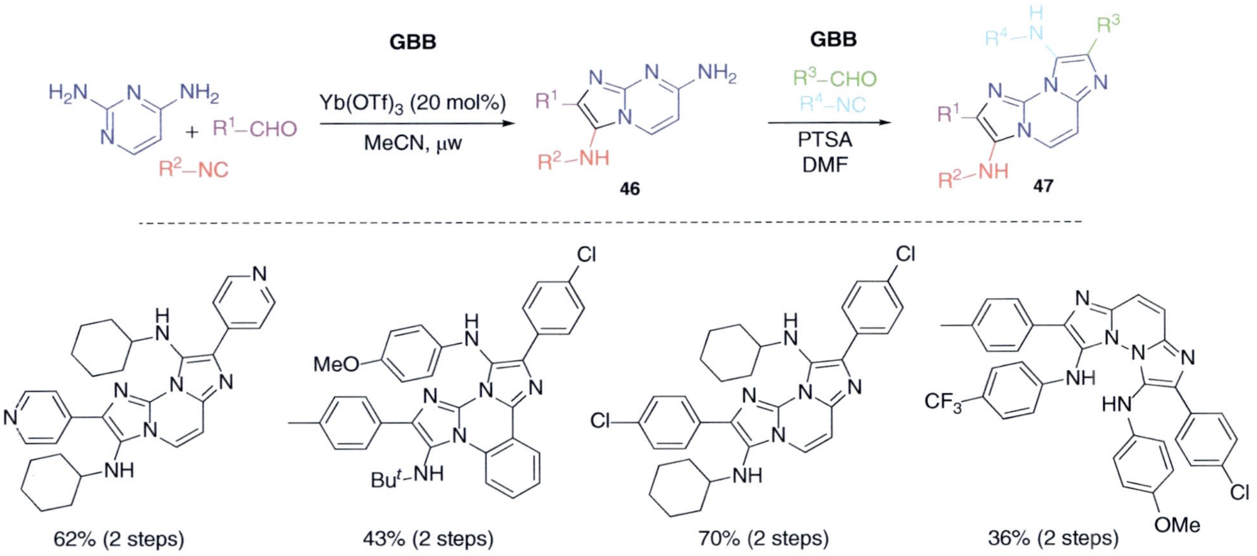

Scheme 5.18 Double GBB reactions for diimidazopyrimidines. Source: Based on Ghashghaei et al. [33].

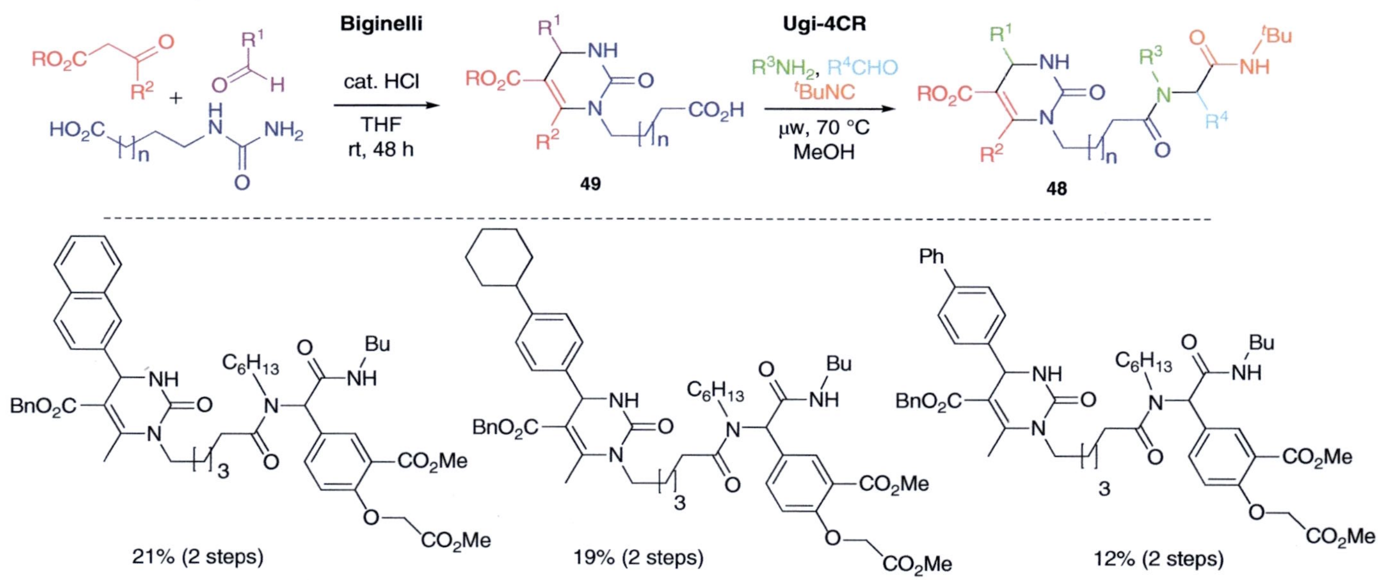

Scheme 5.19 Biginelli and Ugi-4CR reactions for tetrahydropyrimidinone amides. Source: Based on Werner et al. [34].

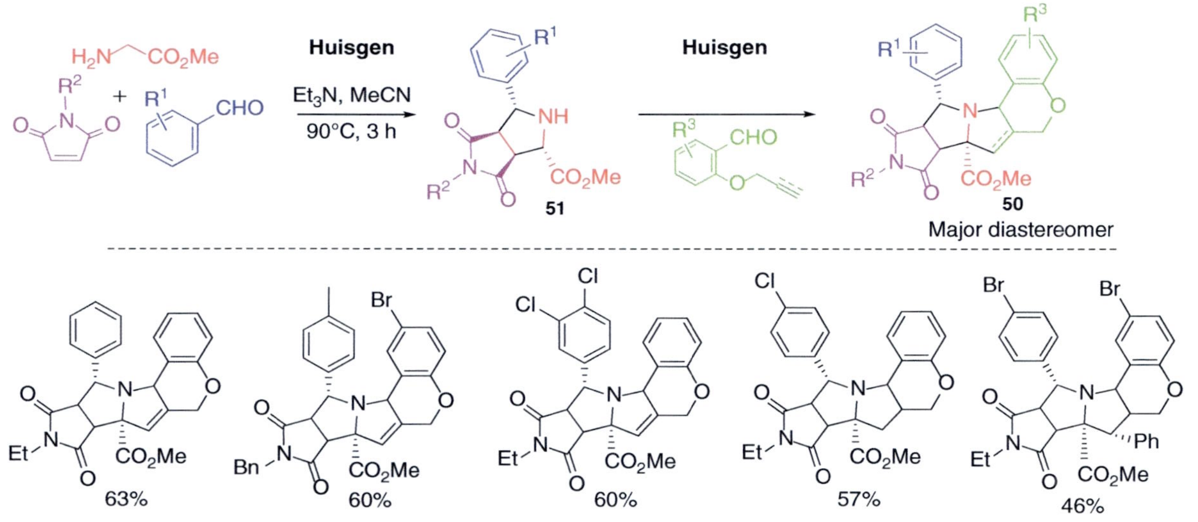

Scheme 5.20 Double inter- and intramolecular Huisgen reactions for polyheterocycles.

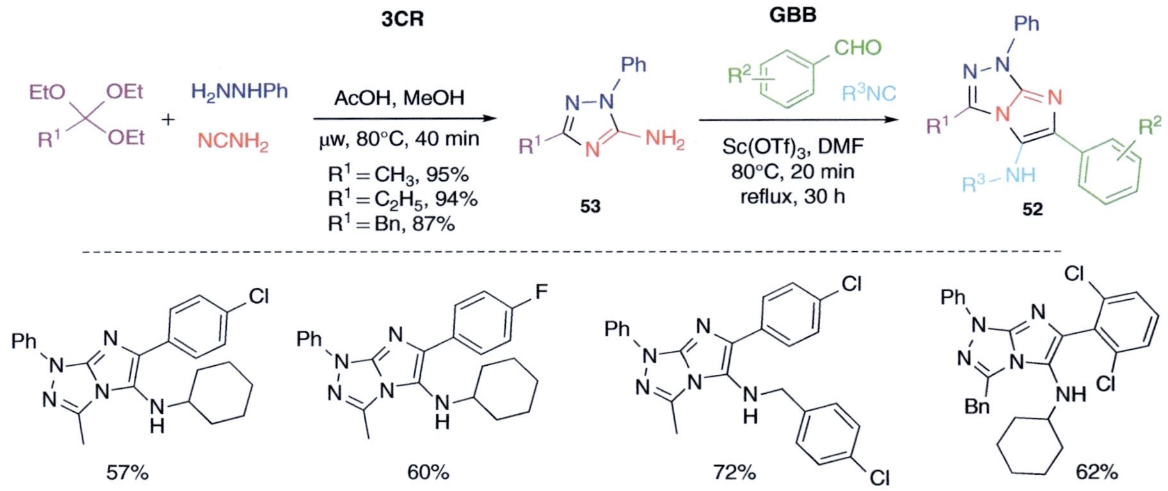

Scheme 5.21 3CR and GBB reactions for imidazotriazoles. Source: Aouali et al. [36].

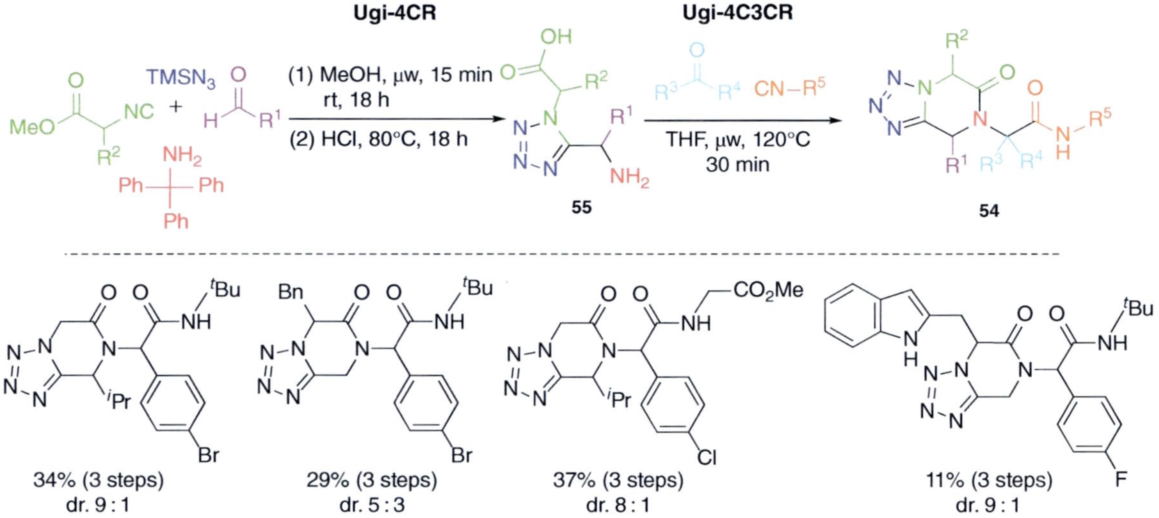

Scheme 5.22 Ugi-4C4CR and Ugi-4C3CR reactions for tetrazolopiperazinones. Source: Based on Zarganes-Tzitzikas et al. [37].

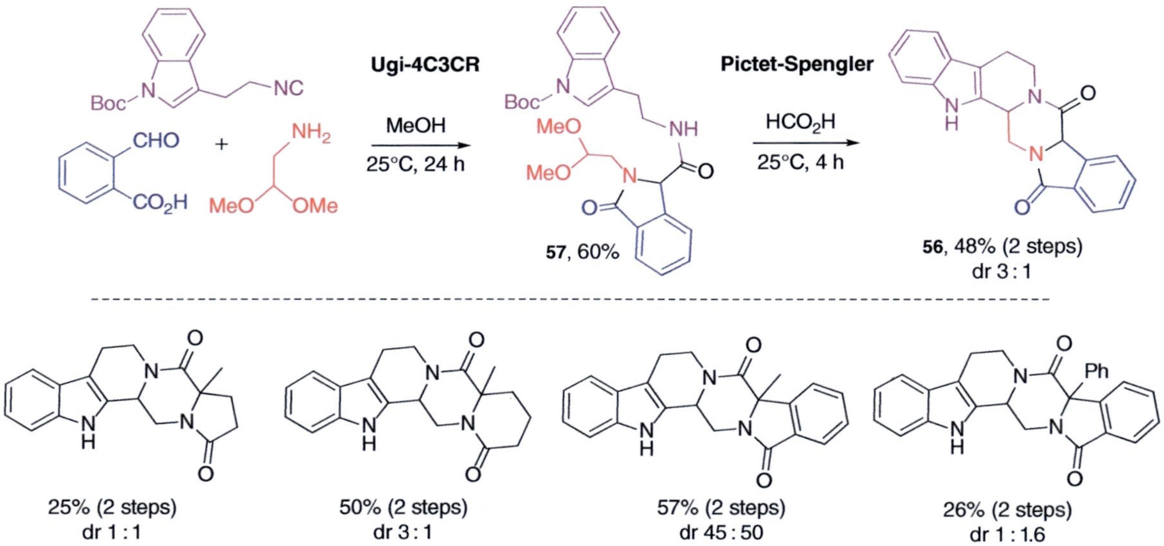

Scheme 5.23 Ugi-4C3CR and Pictet–Spengler reactions for polycyclic indole alkaloids. Source: Based on Wang et al. [38].

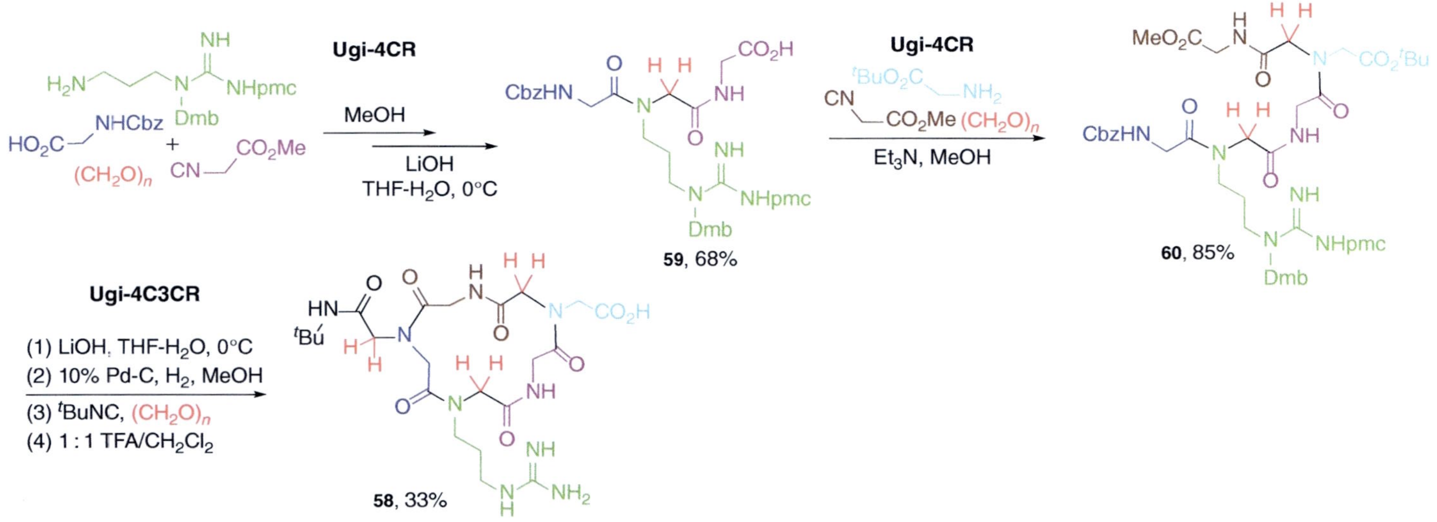

Scheme 5.24 Two Ugi-4CRs and a Ugi-4C3CR reactions for cyclopeptoid. Source: Based on Vercillo et al. [39].

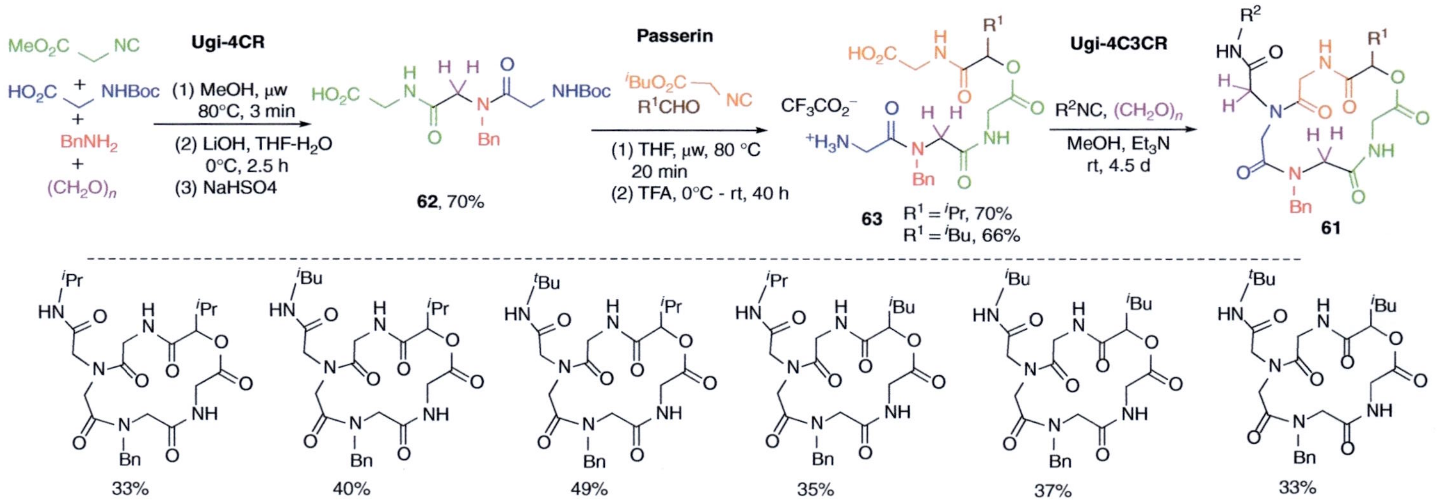

Scheme 5.25 Sequential Ugi-4CR, Passerini, and Ugi-4C3CR reactions for pentadepsipeptoids. Source: Based on Barreto et al. [41].

5.4 MCRs Followed by Cyclization Reactions

The MCRs are synthetically efficient, but the reported numbers of MCRs are limited. Also not all the MCRs directly lead to the formation of desired heterocyclic compounds. Thus, conducting post-MCR modification is a good strategy for the construction of new rings with increased structural diversity and complexity [3]. Highlighted in this session are selective examples of MCRs followed by cyclization reactions (Scheme 5.26). The reaction processes could be in one-pot or involving intermediate separations.

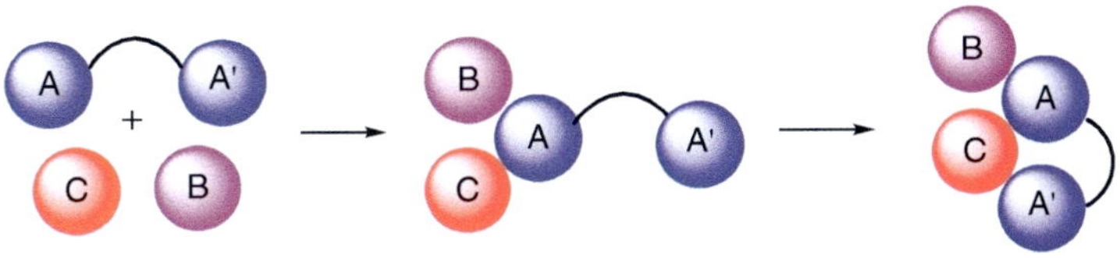

Scheme 5.26 Conceptual figure of MCR and cyclization.

Using Huisgen [3 + 2] cycloaddition products **64** as common intermediates to access three different scaffolds through cyclization to form 5-, 6-, and 7-membered rings **65**, **66**, and **67** are shown in Scheme 5.27 [42]. Fluorous tags (R_f) were employed in the parallel synthesis of these three compound libraries. In the synthesis of hydantoin-fused compounds **65**, [3 + 2] adducts **64** are reacted with arylisocyanates in the presence of a catalytic amount of DMAP to afford ureas. Microwave-assisted cyclization affords hydantoin-fused products **65** as a 420-compound library [43]. In the synthesis of piperazinedione-fused compounds **66**, intermediates **64** are treated with chloroacetyl chloride to afford *N*-acylated products, then react with R^4NH_2 followed by microwave-promoted cyclization to afford piperazinediones **66** as a 90-compound library [44]. In the synthesis of benzodiazepine-fused tricyclic compounds **67**, *N*-acylation of **64** with 2-nitrobenzoyl chloride, reduction of the nitro group with zinc dust in acetic acid, and then DBU-promoted cyclization afford piperazinediones **67** as a 60-compound library [43].

Intramolecular Huisgen [3+2] cycloaddition product **68** was used for post-condensation modifications in the synthesis of heterocycles **69** and **70** (Scheme 5.28). In the synthesis of hydantoin-fused compound **69**, [3 + 2] adduct **68** is reacted with arylisocyanate to afford a urea followed by microwave-assisted cyclization to give products **69** as mixtures of six diastereomers [45]. In the synthesis of piperazinedione-fused compound **70**, intermediates **68** are reacted with chloroacetyl chloride followed by cyclization to afford piperazinedione **70** as mixtures of four diastereomers [46].

Shown in Scheme 5.29 is another Huisgen [3 + 2] cycloaddition-initiated synthesis of heterocyclic compounds **71** [47]. The azido group of [3 + 2] adducts undergo second [3 + 2] cycloaddition with another equiv of maleimides followed by decomposition of triazoles to give **72**. Aminomethylation with formaldehyde and cyclization affords products **71**.

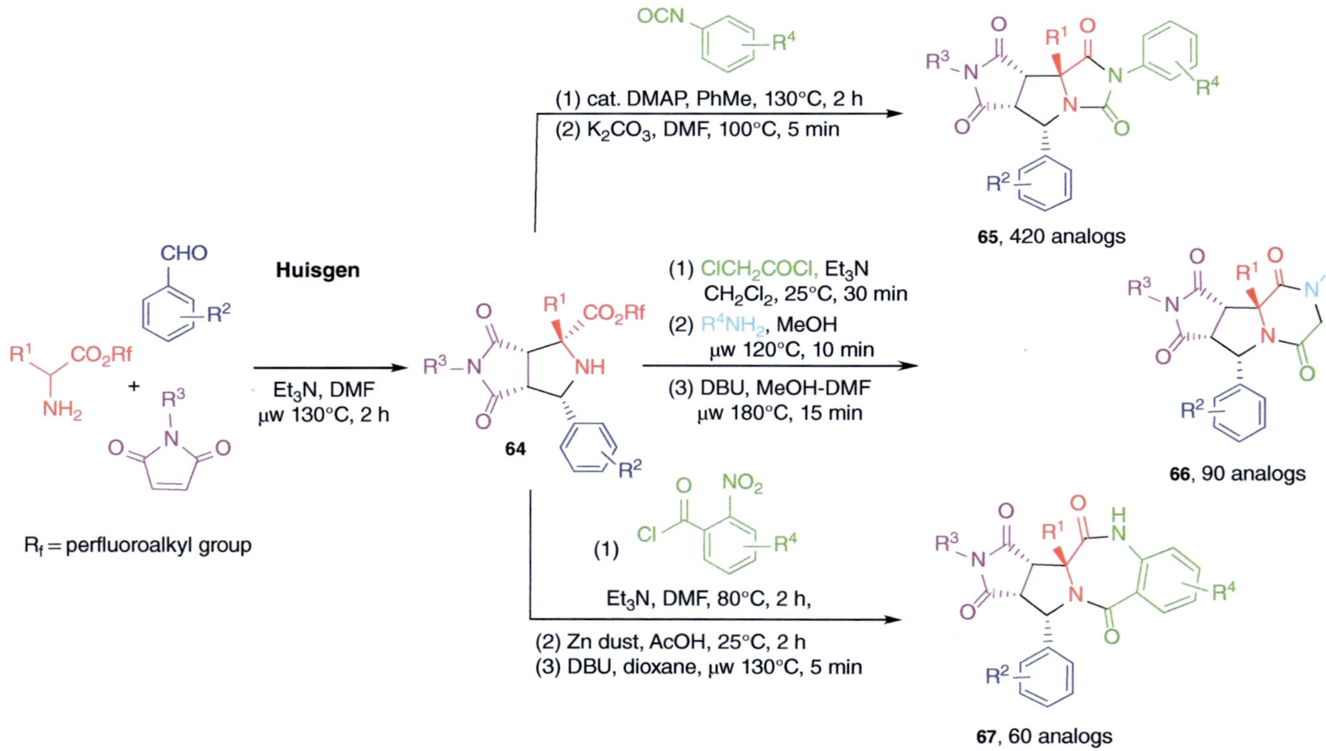

Scheme 5.27 Huisgen reaction and post-condensation modifications for three different libraries. Source: Zhang et al. [42].

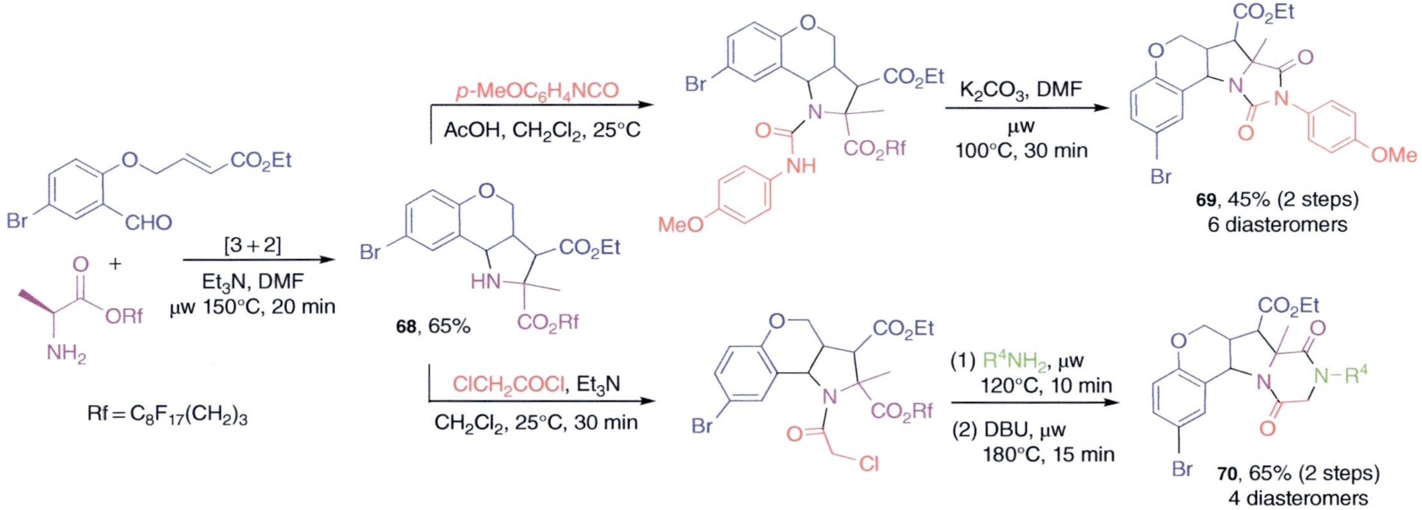

Scheme 5.28 Intramolecular [3 + 2] and post-condensation modifications.

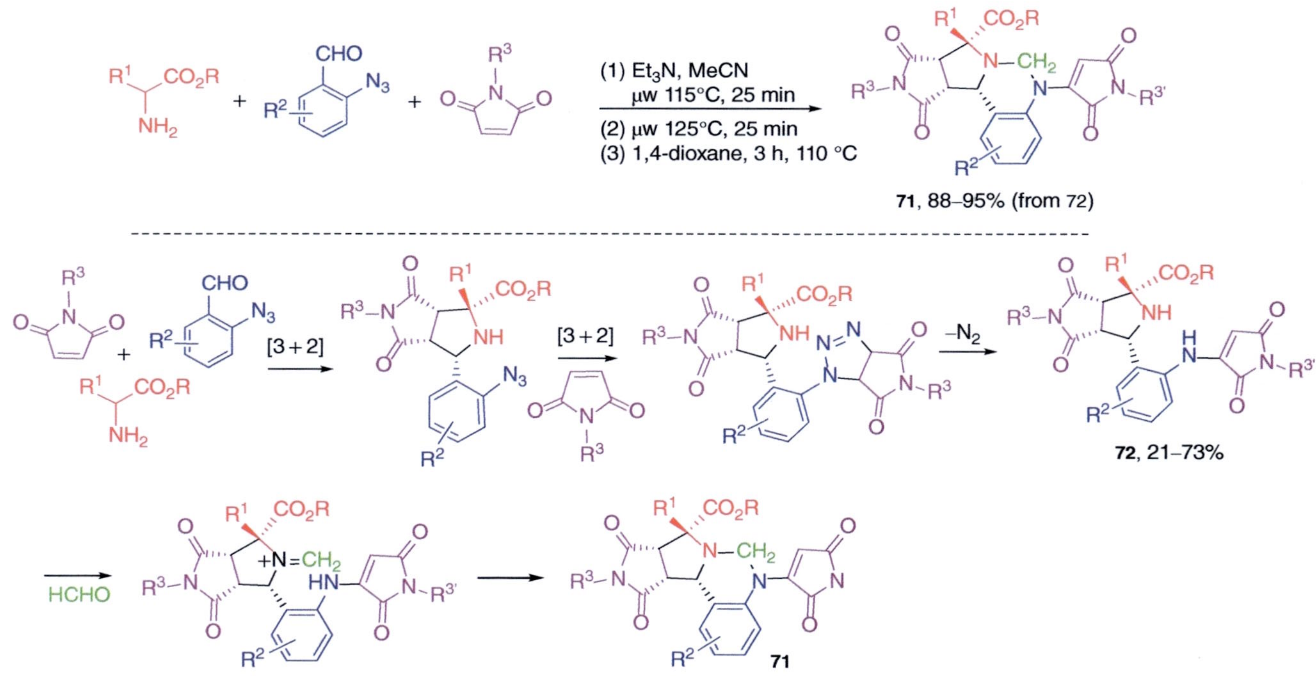

Scheme 5.29 Huisgen and cyclization reactions for heterocycles.

A reaction sequence has been developed by conducting the Huisgen [3 + 2] cycloaddition of azomethine ylides for **73**, *N*-acylation of **74,** and then CuI-catalyzed CF_3 radical cascade addition and cyclization for diastereoselective synthesis of fused-tetrahydrobenzodiazepin-3-ones **75** (Scheme 5.30) [48]. The combination of Huisgen reaction and cascade radical transformation is a good approach for assembling heterocycles.

A one-pot and three-step synthesis has been developed for diastereoselective synthesis of tetrahydropyrrolodiazepinones **76** through the combination of Huisgen reaction for **77**, *N*-acylation for **78,** and Staudinger/aza-Wittig reactions for the products (Scheme 5.31) [49]. Green chemistry metrics analysis of the reaction process was conducted. By replacing animo esters with amino acids, this reaction sequence could be modified by having the decarboxylative [3 + 2] cycloaddition as the initial step [50].

The Huisgen [3 + 2] cycloaddition intermediates **79** generated from the reaction of 2-bromobenzaldehydes, maleimides, and amino esters are treated with *N*-allylation for **80** and intramolecular Heck reactions to form hexahydropyrroloisoquinolines **81** (Scheme 5.32) [51]. EtOH and MeCN are used as preferable green solvents for the reactions. When using alkynyl instead of allyl bromides to react with the [3 + 2] cycloaddition products, the scope of this reaction is extended for cyclative Heck alkyne hydroarylation at the last step [52].

One-pot reactions for the synthesis of polycyclic spiroindolines **82** are shown in Scheme 5.33 [53]. The reaction process involves Ugi-4CR for **83** and then de-Boc double cyclization for **82**. Resulted products were submitted for screening tests on several difficult-to-inhibit cancer cell lines including Huh7. The Ugi-4CR is conducted at room temp, while the double cyclization reaction was performed under microwave heating.

A method for the synthesis of natural product-like polycyclic azepinoindoles **84** has been developed (Scheme 5.34) [54]. The reaction process involves an Ugi-4CR using readily available starting materials for **85** followed by Au-catalyzed dearomatization and Michael addition reactions to afford products in chemo-, regio-, and diastereoselective manners.

A reaction sequence involves Ugi-4CR for **86,** and 6-*exo*-dig cyclization of *N*-substituted 2-alkynamides for the synthesis of a library of functionalized morpholinone glycoconjugates **87** is shown in Scheme 5.35 [55]. The reactions are carried out under mild conditions using regeneratable Ph_3P as a catalyst.

A reaction sequence of consecutive Asinger and Ugi-4C3CR followed by a post-Cu-catalyzed cyclization for making bridged-dihydrothiazolines **88** is shown in Scheme 5.36 [56]. In this reaction process, Asinger products **89** are used for the sequential Ugi-4C3CR with isocyanides and *E*-bromoacrylic acid to afford bisamides **90**. The last step of Cu-catalyzed cyclative rearrangement of give tricyclic ketone products **88**.

Shown in Scheme 5.37 is a reaction sequence of Asinger followed by Ugi-4C3CR and then cyclization reaction for making thiazolo- and oxazolo-annulated benzodiazepinediones **91** [57]. NaSH or NaOH is used for the Asinger to afford thiazolines or oxazolines **92** for the Ugi-4C3CR with allyl isocyanides and 2-fluoro-benzoic acids to form **93** followed by LDA-promoted cyclization to give products **91**.

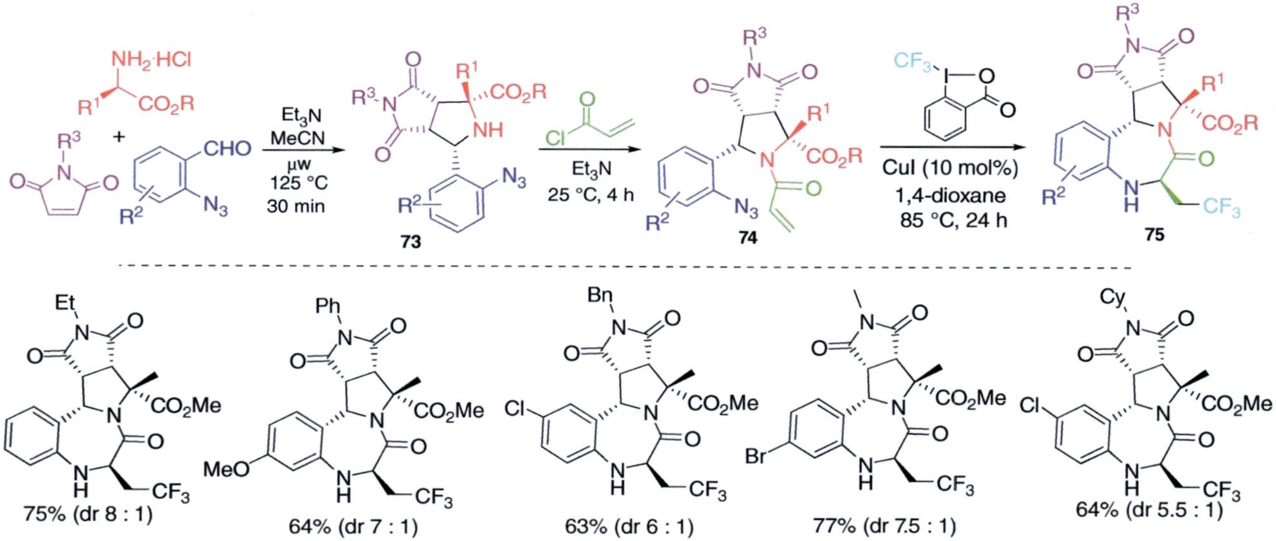

Scheme 5.30 Huisgen and radical reactions for CF_3-substituted heterocycles. Source: Muthengi et al. [48].

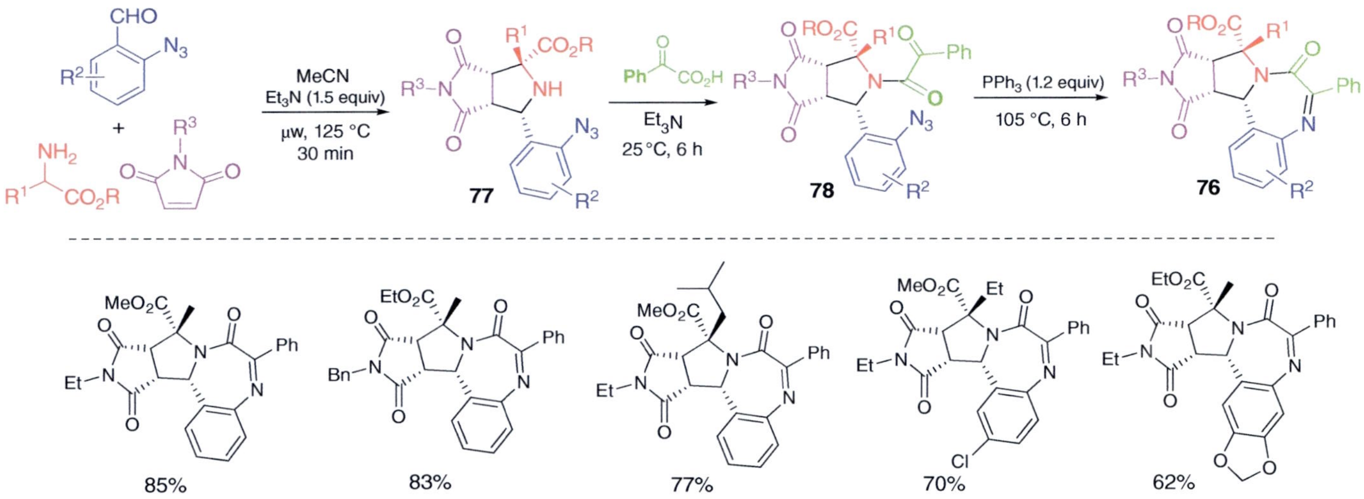

Scheme 5.31 Huisgen and Staudinger/aza-Wittig reactions for tetrahydropyrrolodiazepinones. Source: Ma et al. [49].

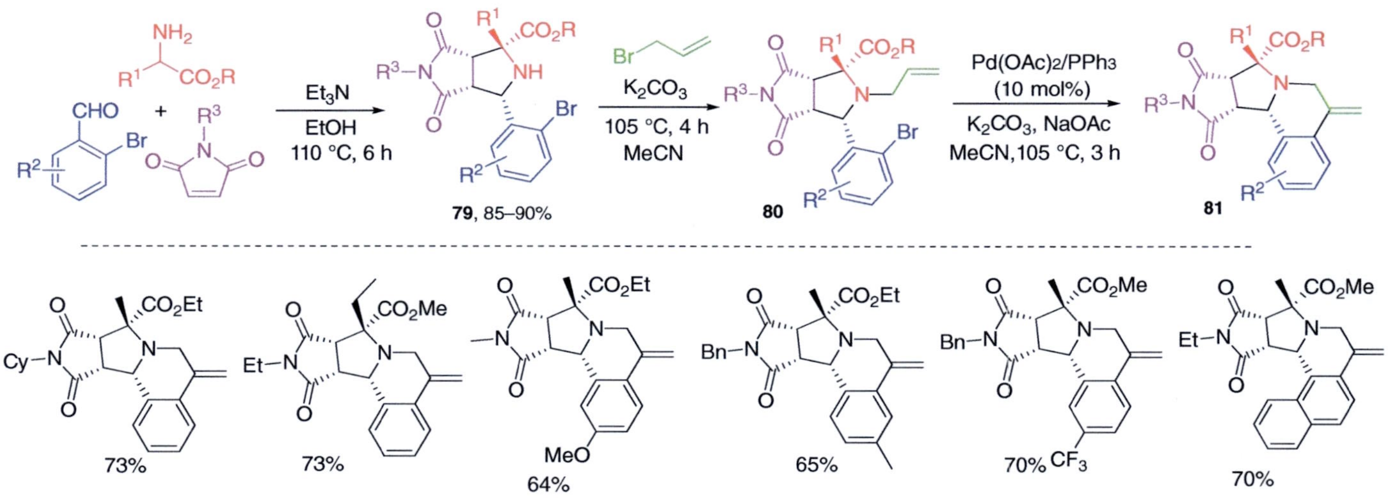

Scheme 5.32 Huisgen and Heck reactions for hexahydropyrroloisoquinolines. Source: Ma et al. [51].

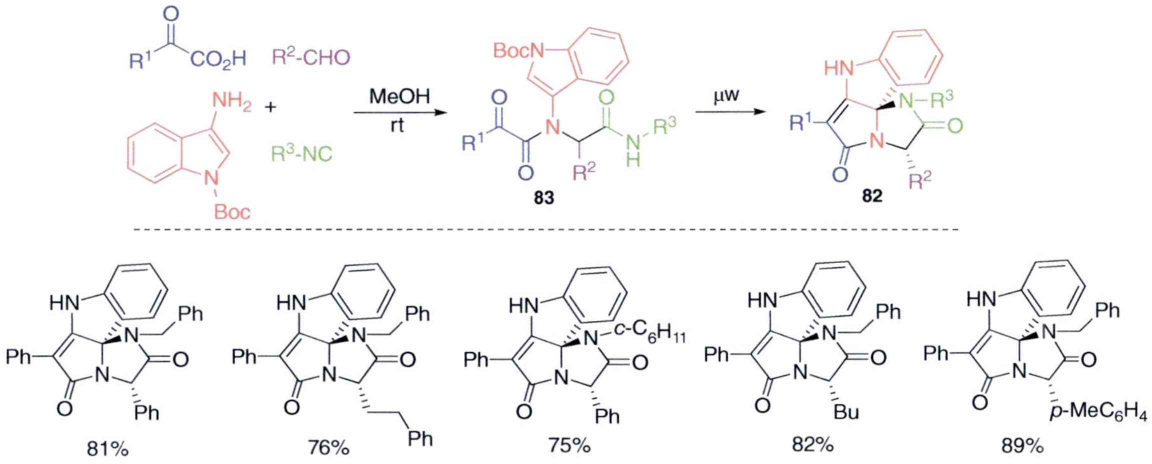

Scheme 5.33 Ugi-4CR and cyclization for spiroindolines. Source: Based on Xu et al. [53].

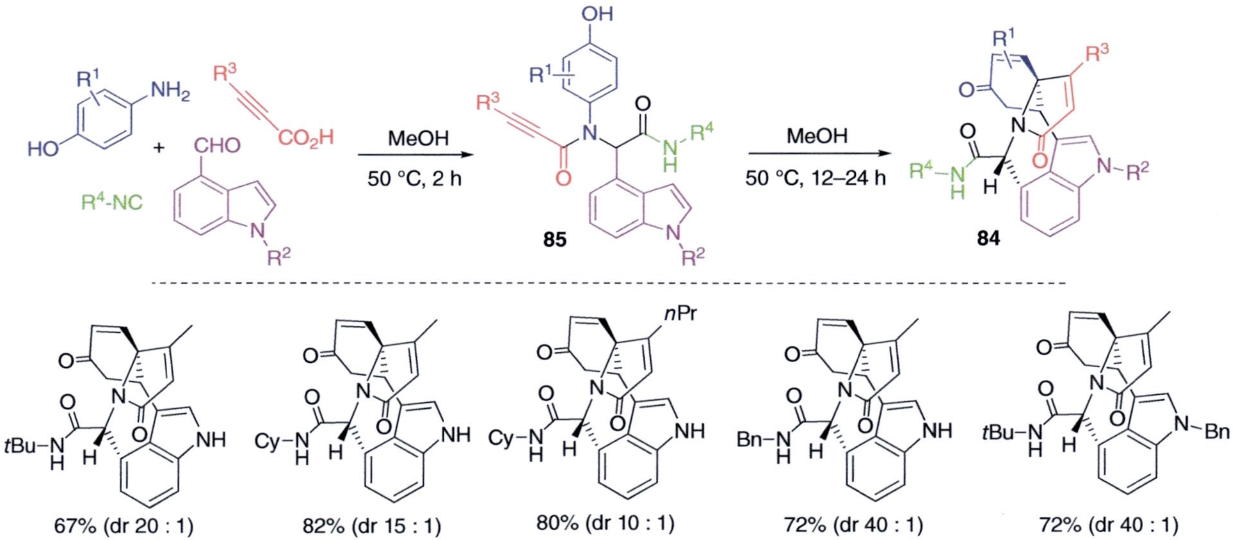

Scheme 5.34 Ugi-4CR and Au-catalyzed cyclization for polycyclic azepinoindoles. Source: Based on He et al. [54].

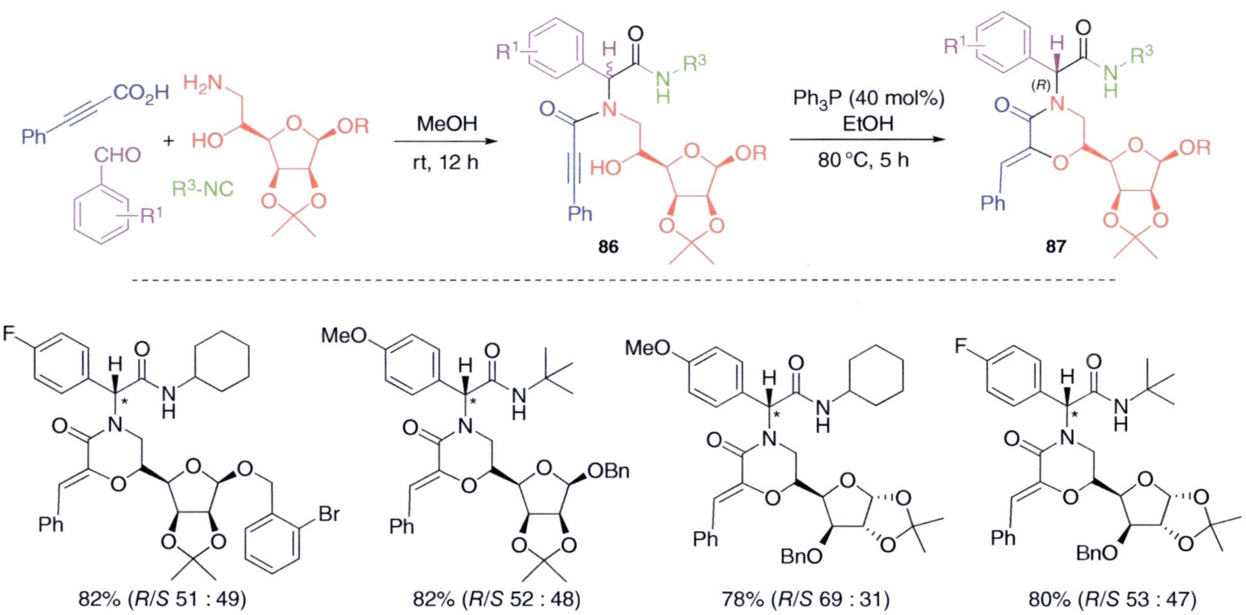

Scheme 5.35 Ugi-4CR and 6-*exo*-dig cyclization for morpholinone glycoconjugates. Source: Kumar et al. [55].

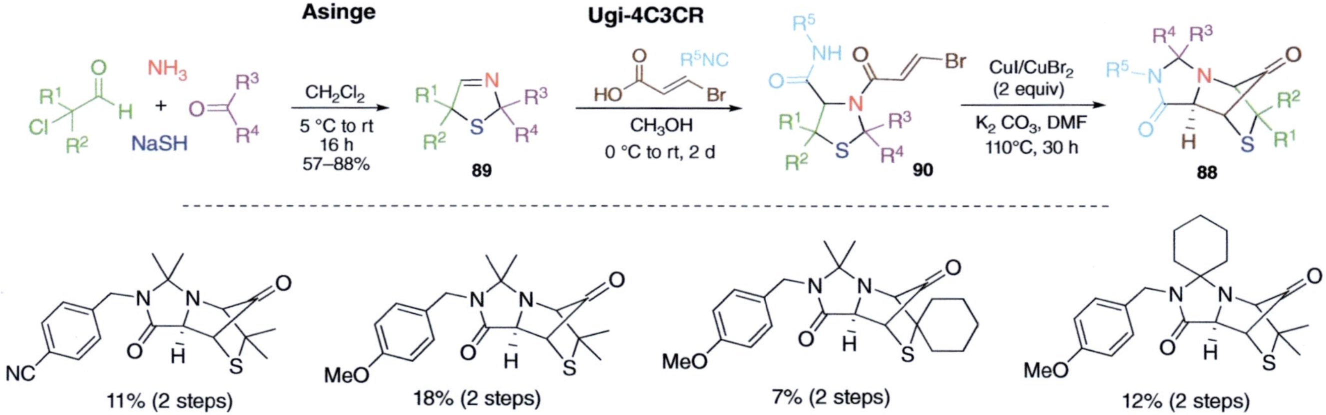

Scheme 5.36 Asinger, Ugi-4C3CR, and cyclization reactions for heterocycles. Source: Based on Kröger et al. [56].

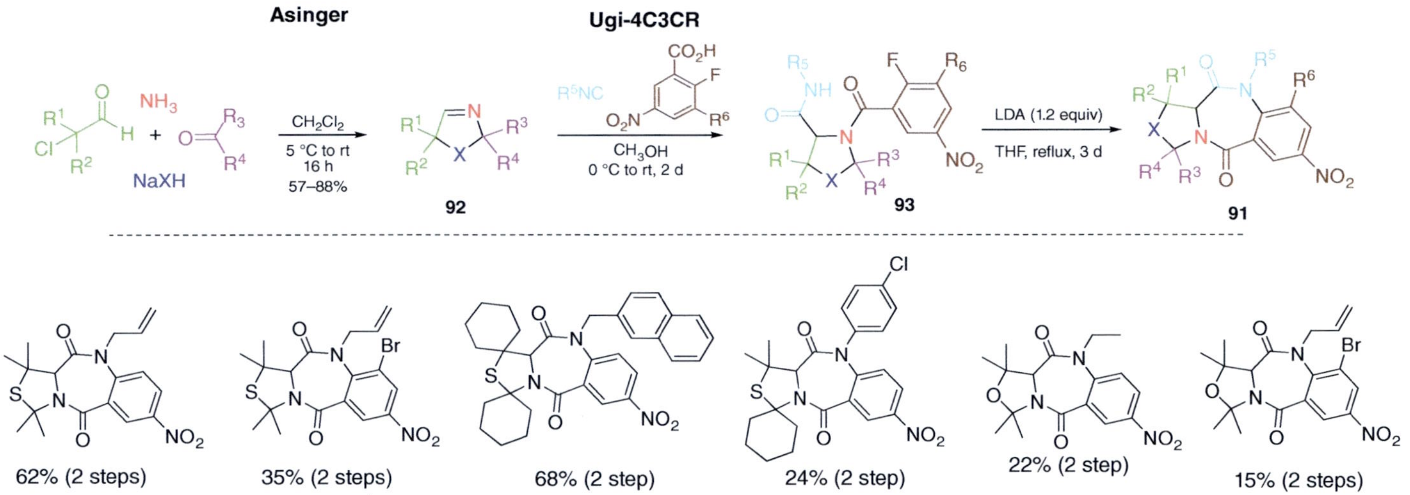

Scheme 5.37 Asinger and Ugi-4C3CR reactions for benzodiazepinediones.

5.5 MCRs Followed by Cycloaddition or Annulation Reactions

Cycloaddition is even more powerful than cyclization in increasing molecular diversity and complexity. Combination of MCRs and cycloadditions or annulation has been demonstrated in the synthesis of polyheterocycles, many of them are in diastereoselective fashions (Scheme 5.38) [3].

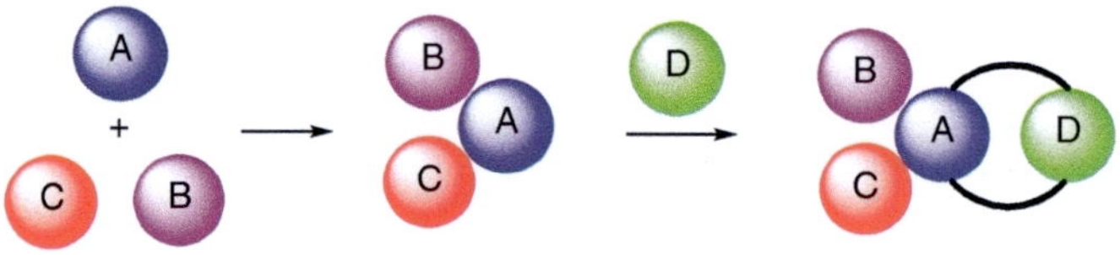

Scheme 5.38 Conceptual figure for MCR and cycloaddition. Source: Based on Zhang and Yi [3].

The combination of intermolecular Huisgen [3 + 2] cycloaddition and intramolecular [4 + 2] cycloaddition reactions has been developed for making heterocyclic scaffold **94** (Scheme 5.39) [58]. After diastereoselective [3 + 2] cycloaddition of amino esters, 2-furanylaldehydes and maleimides, the resulting intermediates **95** are *N*-acylated with acid chlorides followed by intramolecular [4 + 2] cycloaddition to afford polycyclic compounds **33**. In this reaction sequence, *N*-acylation and intramolecular Diels–Alder cycloaddition were conducted in one-pot to give final products containing up to eight stereocenters as single diastereomers.

A one-pot synthesis involving Huisgen [3 + 2] cycloaddition reaction of an azomethine ylides fpr **96**, propargylation for **97,** and azide-alkyne click reaction has been developed for diastereoselective synthesis of novel triazolobenzodiazepine-containing polycyclic compounds **98** (Scheme 5.40) [59]. In this intergrated reaction process, the click reaction is conducted under Cu catalyst-free conditions. The scope of this reactions has been extended by using amino acids to replace amino esters for the initial Huisgen [3 + 2] cycloaddition [60].

A two-step synthesis involving Huisgen [3 + 2] cycloaddition of azomethine ylides for **99** followed by double S_N2 reaction-based [5 + *n*] annulation has been developed for modular synthesis of dihydrobenzoxazines **100**, dihydrobenzooxazinones **101**, tetrahydrobenzoxazepines **102,** and tetrahydrobenzoxazocines **103** (Scheme 5.41) [61, 62]. The [3 + 2] cycloaddition adducts bearing –NH and –OH groups are used for the [5 + *n*] annulation with one-, two-, and three-carbon electrophiles to give different ring-size scaffolds. Green chemistry metrics analysis of the synthetic processes provides favorable results.

A two-step synthesis involving a 4CR of 2-hydroxybenzldehydes, amines, isonitriles, and TMS azide for **104** followed by [5 + 1] annulation with 1,1′-carbonyldiimidazole (CDI) for multisubstituted oxazinanones **105** has been reported (Scheme 5.42) [63]. This method is applicable for the synthesis of oxazolidinones and oxazepanones derivatives which demonstrates the feasibility in combination of MCR and annulation reactions to access different heterocyclic scaffolds.

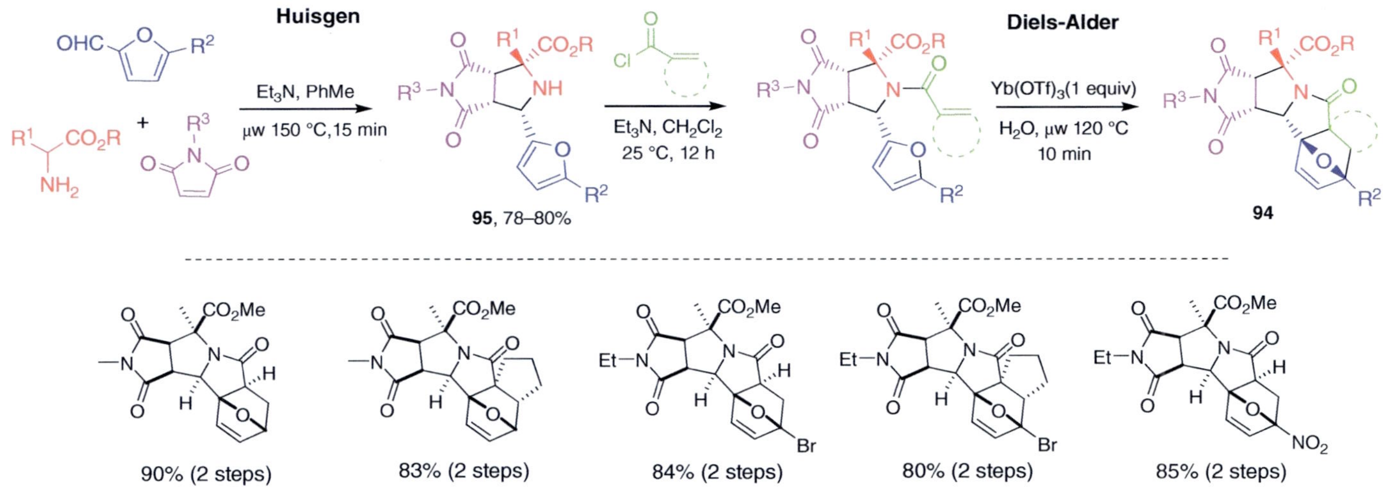

Scheme 5.39 Huisgen and Diels–Alder reactions for polyheterocycles. Source: Lu et al. [58].

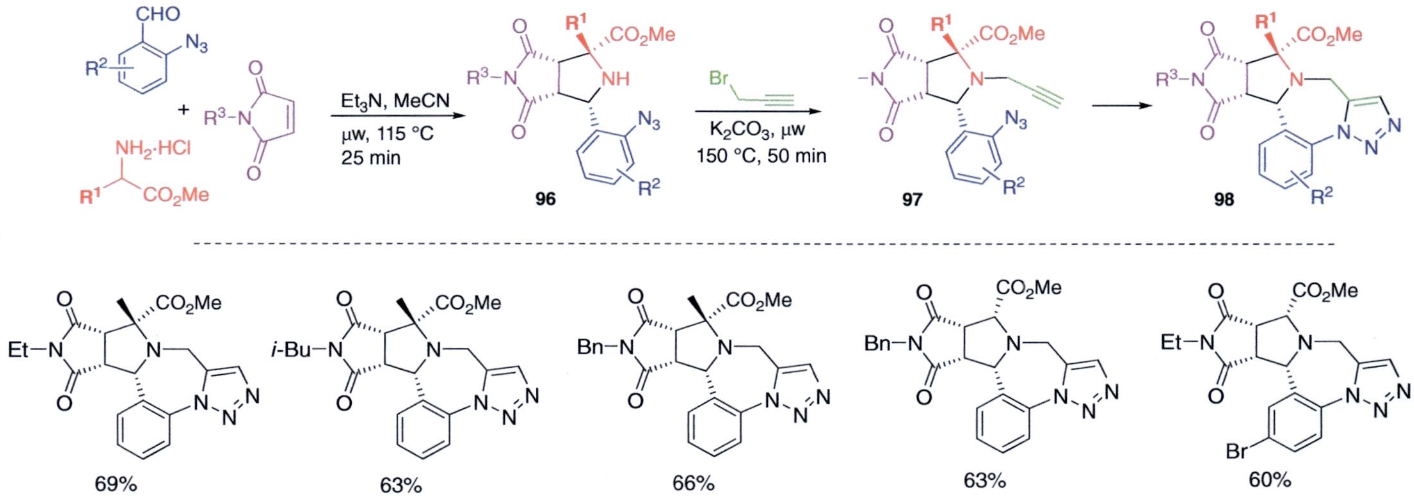

Scheme 5.40 One-pot Huisgen and click reactions for heterocycles. Source: Zhang et al. [59].

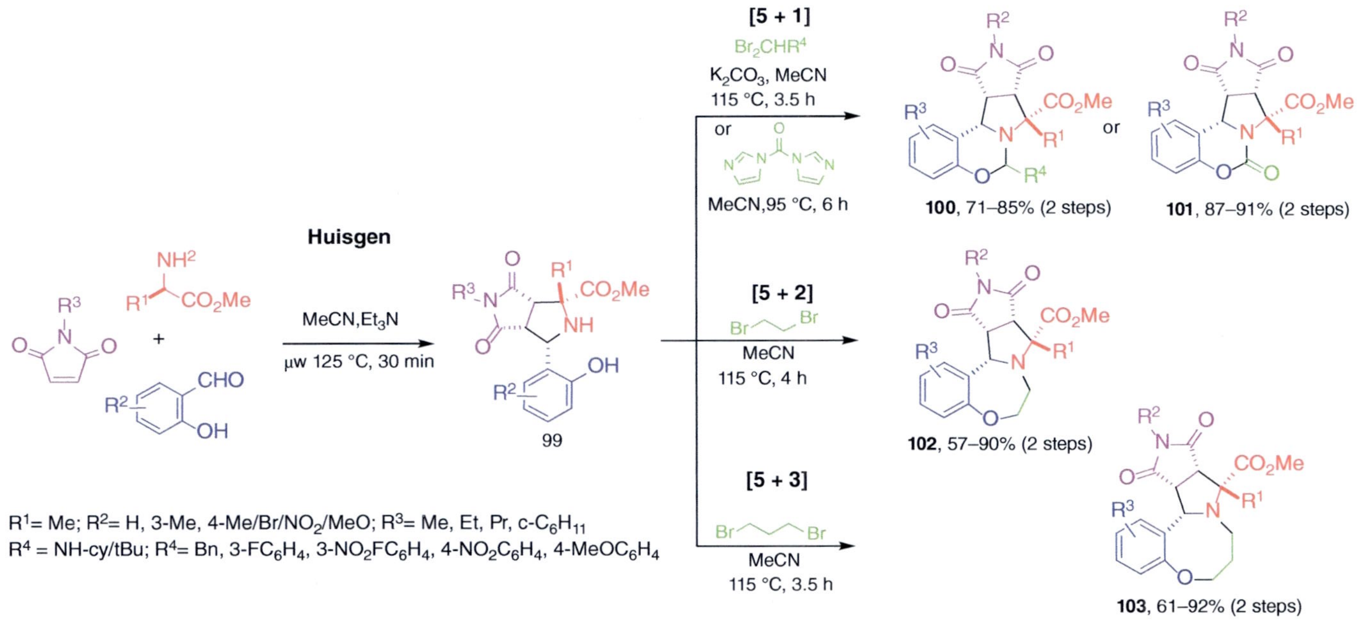

Scheme 5.41 Huisgen and [5 + *n*] annulation reactions for heterocycles. Source: Adapted from Muthengi et al. [61]; Zhang et al. [62].

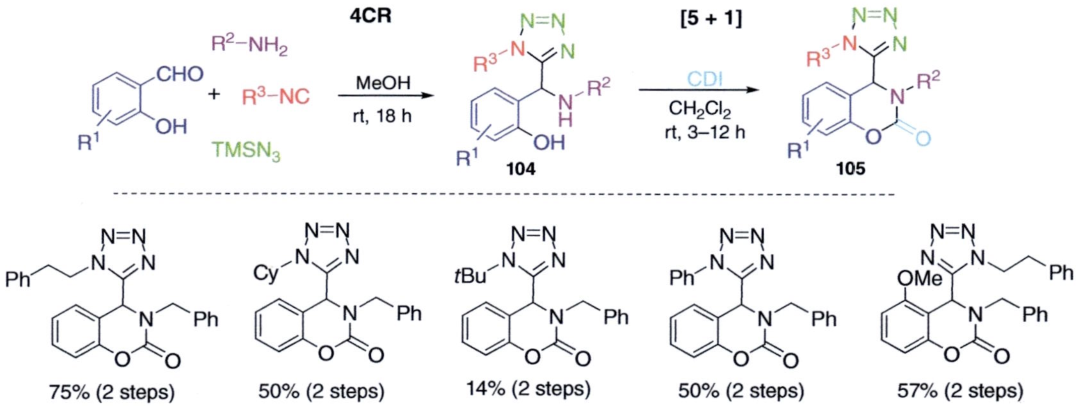

Scheme 5.42 4CR and [5 + 1] annulation for oxazinanones. Source: Based on Li et al. [63].

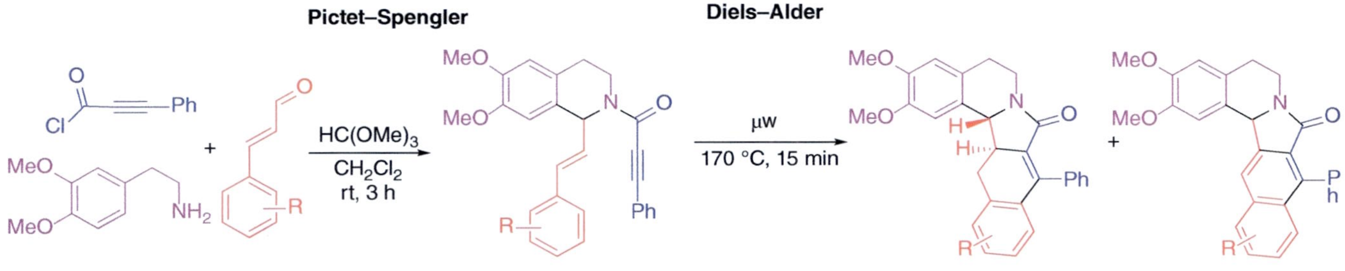

R = H, 2-MeO, 2-NO_2, 2-CF_3

106, 43–85%

107a

107b

up to 100%
107a/107b 1:3–10:1

Scheme 5.43 Pictet–Spengler and Diels–Alder reactions pentacyclic alkaloids. Source: Ruijter et al. [64].

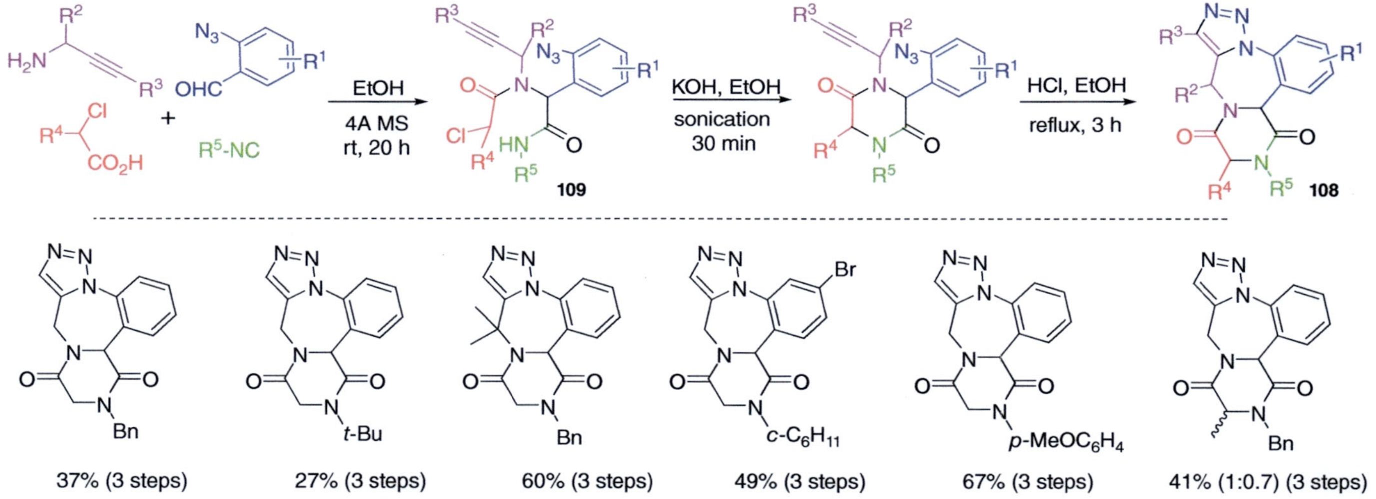

Scheme 5.44 Ugi-4CR and post modifications for polyheterocycles. Source: Based on Li et al. [63].

Several structurally diverse penta- and hexacyclic alkaloid-type compounds have been prepared through a two-step synthesis using readily available starting materials (Scheme 5.43) [64]. The *N*-acyliminium Pictet–Spengler reaction of electron-rich β-arylethylamines, cinnamaldehyde derivatives, and alkynoyl chlorides provides intermediates **106** for an intramolecular Diels–Alder cycloaddition to afford diverse polycyclic alkaloid-like compounds **107** through the reliable and efficient reaction sequence.

A straightforward three-step synthesis to assemble triazolobenzodiazepine-fused diketopiperazines **108** has been developed (Scheme 5.44) [65]. The initial Ugi-4CR products **109** bearing the necessary functionalities for sequential base-induced cyclization and azide–alkyne cycloaddition reactions afford highly condensed cyclic frameworks **108**. EtOH is used as a green solvent for this three-step synthesis.

5.6 Conclusion and Outlook

This chapter highlights the MCR-based PASE synthesis, including high-order MCRs, consecutive MCRs, and MCRs followed by post-condensation modifications such as cyclization and cycloaddition reactions. Those methods have intrinsic efficiency for being simple in operation, short in reaction time, saving energy, and reducing waste. The examples shown in this chapter demonstrates the great achievement of organic chemists in the development of MCR-based synthesis of heterocycles. Integration of MCRs with other reaction processes, such as radical cascade reactions, photoredox catalysis, transition metal- and organocatalysis, and electrochemical reactions can further empower the capability of MCRs. There is no doubt that MCR-related methods will continuously play important roles in the synthesis of complex heterocycles which have biological interest and other utilities.

References

1 Clarke, P.A., Santos, S., and Martin, W.H.C. (2007). *Green Chem.* 9: 438–440.

2 Hayashi, Y. (2016). *Chem. Sci.* 7: 866–880.

3 Zhang, W. and Yi, W.-B. (2019). *Pot, Atom, and Step Economy (PASE) Synthesis.* Cham: Springer.

4 Zhu, J., Wang, Q., and Wang, M.-X. (eds.) (2015). *Multicomponent Reactions in Organic Synthesis.* Weinheim: Wiley-VCH.

5 Herrera, R.P. and Marques-Lopez, E. (eds.) (2015). *Multicomponent Reactions – Concepts and Applications for Design and Synthesis.* Hoboken: Wiley.

6 Ameta, K.L. and Dandia, A. (2017). *Multicomponent Reactions – Synthesis of Bioactive Heterocycles.* London: CRC Press.

7 Constable, D. and Jimenez-Gonzalez, C. (eds.) (2018). *Handbook of Green Chemistry – Green Metrics*, vol. 11. Weinheim: Wiley-VCH.

8 Dicks, A.P. and Hent, A. (2015). *Green Chemistry Metrics Metrics – A Guide to Determining and Evaluating Process Greenness.* Cham: Springer.

9 Roschangar, F., Zhou, Y., Constable, D.J.C. et al. (2018). *Green Chem.* 20: 2206–2211.

10 Cioc, R.C., Ruijter, E., and Orru, R.V.A. (2014). *Green Chem.* 16: 2958–2975.

11 Huang, Y., Yazbak, A., and Dömling, A. (2012). *Green Techniques for Organic Synthesis and Medicinal Chemistry*, 1e, 497–522. Chichester: Wiley.

12 Beach, E.S. and Kundu, S. (eds.) (2010). *Handbook of Green Chemistry – Tools for Green Chemistry*, vol. 10. Weinheim: Wiley-VCH.

13 Zhang, W. and Cue, B.W. (eds.) (2018). *Green Techniques for Organic Synthesis and Medicinal Chemistry*, 2e. Chichester: Wiley.

14 Brauch, S., van Berkel, S.S., and Westermann, B. *Chem. Soc. Rev.* 42: 4948–4962.

15 Chavan, P.V., Pandit, K.S., Desai, U.V. et al. (2017). *Res. Chem. Intermed.* 43: 5675–5690.

16 Chavan, P.V., Desai, U.V., Wadgaonkar, P.P. et al. (2019). *Bioorg. Chem.* 85: 475–486.

17 Govindaraju, S., Tabassum, S., Khan, R.-U.-R., and Pasha, M.A. (2017). *Chin. Chem. Lett.* 28: 437–441.

18 Li, Y., Xue, Z., Ye, W. et al. (2014). *ACS Comb. Sci.* 16: 113–119.

19 Amanpour, T., Bazgir, A., Ardekani, A.M., and Ghahremanzadeh, R. (2014). *Monatsh. Chem.* 145: 627–632.

20 Zhao, P., Wu, X., Geng, X. et al. (2018). *Tetrahedron* 74: 4323–4330.

21 Lu, Q., Song, G., Jasinski, J.P. et al. (2012). *Green Chem.* 14: 3010–3012.

22 Zhang, X., Qiu, W., Evans, J. et al. (2019). *Org. Lett.* 21: 2176–2179.

23 Mao, K., Dai, L., Liu, Y., and Rong, L. (2019). *J. Heterocycl. Chem.* 56: 2111–2120.

24 Cao, J., Shi, R.-G., Sun, J. et al. (2020). *J. Org. Chem.* 85: 2168–2179.

25 Firoozi, N., Torres, G.M., and Arndtsen, B.A. (2016). *J. Org. Chem.* 81: 11145–11152.

26 Zhang, W., Lu, Y., and Geib, S.J. (2005). *Org. Lett.* 7: 2269–2272.

27 Shaabani, S., Shaabani, A., and Ng, S.W. (2014). *ACS Comb. Sci.* 16: 176–183.

28 Hooshmand, S.E., Ghadari, R., Mohammadian, R. et al. (2019). *ChemistrySelect* 4: 11893–11898.

29 Kutovaya, I.V., Zarezin, D.P., Shmatova, O.I., and Nenajdenko, V.G. (2019). *Eur. J. Org. Chem.*: 2675–2681.

30 Fu, L., Feng, X., Zhang, J.-J. et al. (2015). *Green Chem.* 17: 1535–1545.

31 Karamthulla, S., Pal, S., Khan, N., and Choudhury, L.H. (2014). *Synlett* 25: 1926–1936.

32 Zhi, S., Ma, X., and Zhang, W. (2019). *Org. Biomol. Chem.* 17: 7632–7650.

33 Ghashghaei, O., Caputo, S., Sintes, M. et al. (2018). *Chem. Eur. J.* 24: 14513–14521.

34 Zhang, X., Qiu, W., Ma, X. et al. (2018). *J. Org. Chem.* 83: 13536–13542.

35 Werner, S., Turner, D.M., Lyon, M.A. et al. (2006). *Synlett* 14: 2334–2338.

36 Aouali, M., Mhalla, D., Allouche, F. et al. (2015). *Med. Chem. Res.* 24: 2732–2741.

37 Zarganes-Tzitzikas, T., Patil, P., Khoury, K. et al. (2015). *Eur. J. Org. Chem.*: 51–55.

38 Wang, W., Herdtweck, E., and Dömling, A. (2010). *Chem. Commun.* 46: 770–772.

39 Vercillo, O.E., Andrade, C.K.Z., and Wessjohann, L.A. (2008). *Org. Lett.* 10: 205–208.

40 Barreto, A.F.S. and Andrade, C.K.Z. (2018). *Tetrahedron* 74: 6861–6865.

41 Barreto, A.F.S., Vercillo, O.E., Wessjohann, L.A., and Andrade, C.K.Z. (2014). *Beilstein J. Org. Chem.* 10: 1017–1022.

42 Zhang, W., Lu, Y., Chen, C.H.-T. et al. (2006). *Eur. J. Org. Chem.*: 2055–2059.

43 Zhang, W., Lu, Y., Chen, C.H.-T. et al. (2006). *J. Comb. Chem.* 8: 687–695.

44 Werner, S., Nielsen, S.D., Wipf, P. et al. (2019). *J. Comb. Chem.* 11: 452–459.

45 Lu, Y., Geib, S.J., Damodaran, K. et al. (2010). *Chem. Commun.* 46: 7578–7580.

46 Zhang, W. (2013). *Chem. Lett.* 42: 676–681.

47 Zhang, X.F., Pham, K., Liu, S. et al. (2016). *Beilstein J. Org. Chem.* 12: 2204–2210.

48 Muthengi, A., Erickson, J., Muriph, R.E., and Zhang, W. (2019). *J. Org. Chem.* 84: 5927–5935.

49 Ma, X., Zhang, X., Awad, J.M. et al. (2019). *Green Chem.* 21: 4489–4494.

50 Ma, X., Zhang, X., Awad, J.M. et al. (2020). *Tetrahedron Lett.* 61: 151392.

51 Ma, X., Meng, S., Zhang, X. et al. (2020). *Beilstein J. Org. Chem.* 16: 1225–1233.

52 Yang, G., Liu, Y., Wang, L. et al. (2021). *Tetrahedron Lett.* 62: 152859.

53 Xu, Z.-G., Li, S.-Q., Meng, J.-P. et al. (2018). *Chem. Eur. J.* 24: 6732–6736.

54 He, Y., Song, L., Liu, C. et al. (2020). *J. Org. Chem.* 85: 15092–15103.

55 Kumar, H., Prajapati, G., Dubey, A. et al. (2020). *Org. Lett.* 22: 9258–9262.

56 Kröger, D., Franz, M., Schmidtmann, M., and Martens, J. (2015). *Org. Lett.* 17: 5866–5869.

57 Kröger, D., Stalling, T., and Martens, J. (2016). *Synthesis* 48: 4189–4198.

58 Lu, Q., Huang, X., Song, G. et al. (2013). *ACS Comb. Sci.* 15: 350–355.

59 Zhang, X., Zhi, S., Wang, W. et al. (2016). *Green Chem.* 18: 2642–2646.

60 Ma, X., Zhang, X., Qiu, W. et al. (2019). *Molecules* 24: 601.

61 Muthengi, A., Zhang, X., Dhawan, G. et al. (2018). *Green Chem.* 20: 3134–3139.

62 Zhang, W., Zhang, X., Ma, X., and Zhang, W. (2018). *Tetrahedron Lett.* 59: 3845–3847.

63 Li, J., Lorenzo, V.D., Patil, P. et al. (2020). *ACS Comb. Sci.* 22: 356–360.

64 Ruijter, E., Garcia-Hartjes, J., Hoffmann, F. et al. (2010). *Synlett*: 2485–2489.

65 Vroemans, R., Bamba, F., Winters, J. et al. (2018). *Beilstein J. Org. Chem.* 14: 626–633.

6

The Use of Flow Chemistry in the Multicomponent Synthesis of Heterocycles

Chiara Lambruschini, Lisa Moni, and Andrea Basso

Università degli Studi di Genova, Dipartimento di Chimica e Chimica Industriale, Via Dodecaneso 31, Genova, Italy

6.1 Introduction

The combination of flow processing with multicomponent reactions has demonstrated to be a valuable approach towards an efficient and sustainable chemical synthesis. Reactions performed in continuous flow systems are characterized by high interfacial area per volume and take advantage of the rapid heat and mass transmission, thus giving excellent safety control in the case of hazardous reagents and/or highly exothermic reactions. In addition, they can be performed with reduced reagent consumption and can be easily scaled up simply by running the reaction for longer times or by using more reactors in parallel, thus avoiding re-optimization of the reaction conditions. The use of flow chemistry positively affects yields, selectivity and reaction times, and the integration of on-line detection modules allows real-time monitoring of the process. In the case of photochemical or gas–liquid heterogeneous reactions, the flow approach has additional advantages, such as the homogeneous penetration of light or the intimate interaction between the two phases [1, 2]. When coming to multicomponent reactions, another advantage over batch reactions is the possibility to perform the addition of the components at different stages, in a telescoped manner, still falling within the definition of a multicomponent condensation. A few years ago, the flow-assisted synthesis of heterocycles via multicomponent reactions was reviewed by De Borggraeve, with the main focus on the different types of heterocylces [3]. The reason for a new chapter on this subject, apart from including the relevant literature of the last few years, is to focus mainly on the different approaches that can be used to perform heterocycle multicomponent syntheses in continuous flow and on the advantages of such approaches over batch synthesis. In particular, we have divided this chapter in four different sections, and, apart from the "classical" flow synthesis, we have included three sub-chapters dedicated to the use of (i) hazardous reagents, (ii) special conditions, and (iii) telescoped systems, respectively.

Multicomponent Reactions towards Heterocycles: Concepts and Applications, First Edition.
Edited by Erik V. Van der Eycken and Upendra K. Sharma.

6.2 Multicomponent Reactions Under Standard Flow Conditions

Generally, the results obtained in flow are equal or superior than those in batch, since the use of continuous flow systems may overcome some limitations that affect several organic reactions. In this section, examples in which classical batch reactions are upgraded to flow chemistry are reported and in particular the advantages of using flow over batch are highlighted.

In 2006, Stevens reported the employment of a microreactor for the continuous synthesis of tetrasubstituted imidazoles **5** using a modified Radziszewski reaction, in which an α-diketone **1**, an aldehyde **2**, and an amine **3** react with ammonia (ammonium acetate **4** is usually used as source of NH_3) (Scheme 6.1) [1]. This four multicomponent reaction is already known in batch [4], but the need of MW heating hampers a scale-up to multigram productions of the desired imidazoles. The use of a microreactor has the inherent advantages of an extremely high mixing rate and a very narrow temperature profile due to a greater surface-to-volume ratio. The choice of the optimal solvent mixture was a key point. The ideal solvent should have a b.p. 10 °C below the employed temperature, otherwise the gas-bubble formation results in unequal residence times, and should be able to dissolve the reactants, since handling of solids in this type of microreactor is disfavored. The authors were not able to identify a single solvent fulfilling these needs, therefore they opted for a mixture of two solvents: *N*-methyl-2-pyrrolidone (NMP) and *n*-pentanol that were used separately in the two inlets of the system. The presence of acetic acid (10 equiv) was another key point, because it suppresses the formation of the Radziszewski product (R^3 = H), which is the main side product formed by the incorporation of two moles of NH_3 instead of the primary amine. After a careful optimization of the reaction parameters (i.e. temperature, solvents, and additive), the authors were able to obtain similar conversions to the batch reaction.

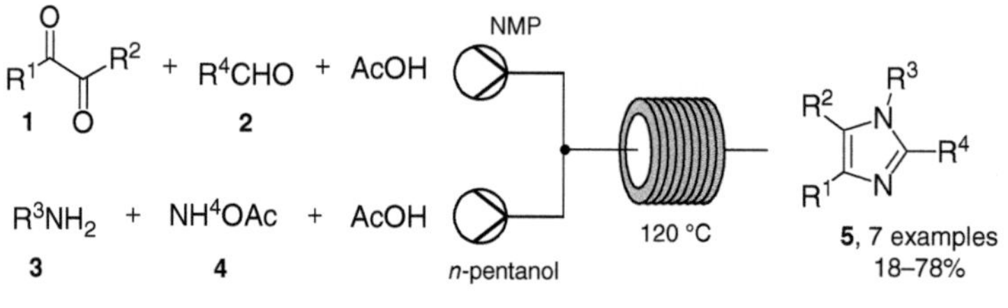

Scheme 6.1 Synthesis of tetrasubstituted imidazoles **5** using a modified Radziszewski reaction. Source: Based on Wegner et al. [1].

1,4-Dihydropyridines **9** are commonly prepared by the Hantzsch three component reaction starting from an aldehyde **6**, an acetoacetate **7**, and ammonium acetate **8**. Typically, this transformation requires long reaction times (hours) to obtain full conversion in batch. Hessel reported this synthesis under superheated conditions by the use of a microreactor equipped with a back pressure regulator to avoid gas formation (Scheme 6.2) [5]. Such superheated reaction mixtures are not easy to process on a batch scale. The small volume of the microreactor (200 μl) and the short residence time (six minutes) minimizes the exposure of the reaction mixture in the hot zone and ensures a limited formation of by-products under such harsh conditions.

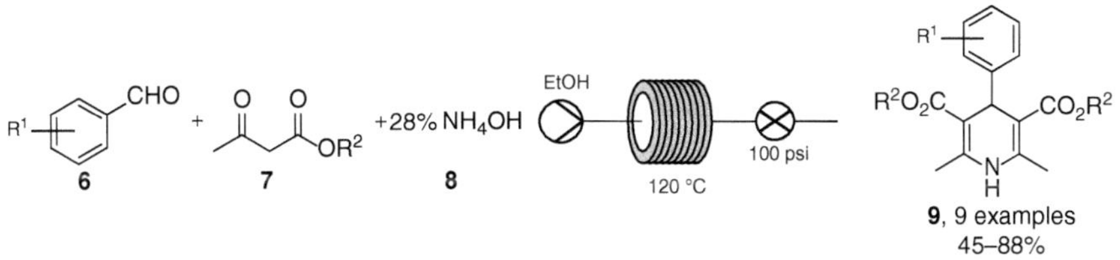

Scheme 6.2 Hantzsch synthesis of 1,4-dihydropyridines **9** under superheated conditions. Source: Baraldi et al. [5].

The Groebke–Blackburn–Bienaymé three-component reaction involves the interaction of a 2-aminoazine (such as **10**), an aldehyde **11**, and an isocyanide **12** under acidic conditions to form a fused imidazole ring. One of the most employed source of the amine is 2-aminopyrimidine **10** leading to aminoimidazo[1,2-*a*]pyrimidines **13**. Under conventional batch conditions, in addition to the expected 3-amino substituted product (thermodynamic product), the regioisomeric 2-amino compound (kinetic product) is also formed in various amounts depending on the nature of the other components and on the reaction conditions. Chen and coworkers showed that using a continuous flow process the formation of the undesired regioisomer can be almost suppressed (>1%) [6]. After a careful optimization (temperature, residence time, type of catalyst, and catalyst loading), the authors found the optimal parameters (50 minutes at 80 °C) that can assure a remarkable improved selectivity and shorter reaction times than the batch reaction (Scheme 6.3). Moreover, the successful formation of 3-aminoimidazo[1,2-*a*]pyrimidines employing aliphatic aldehydes, albeit in low yield, is unprecedented under batch conditions.

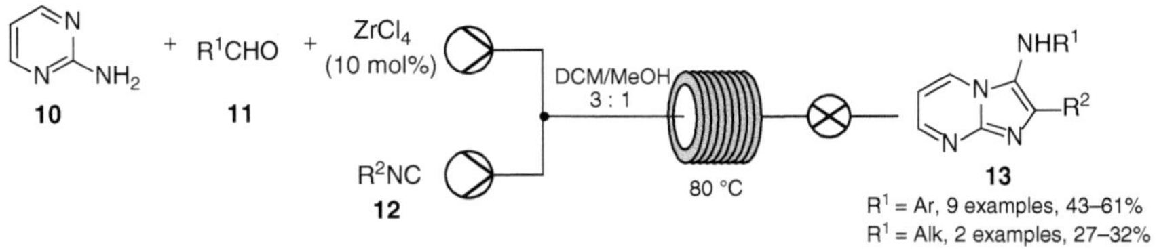

Scheme 6.3 Regioselective synthesis of aminoimidazo[1,2-*a*]pyrimidines via the Groebke-Blackburn-Bienaymé MCR.

In 2013, Stevens and coworkers reported the preparation of a peculiar tricyclic system through the condensation of 6-amminouracil **14** with an excess of benzyl amine **15** and formaldehyde **16**. The reaction was optimized in batch and required microwave heating, however the synthesis of **17** was scaled out using a microreactor heated at 90 °C (Scheme 6.4) [7].

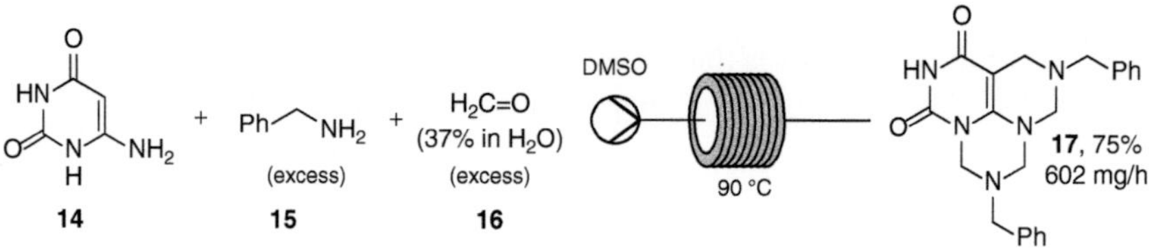

Scheme 6.4 Flow synthesis of substituted tetrahydropentaazaphenalene-1,3-dione **17**. Source: Based on García et al. [7].

The three-component Huisgen 1,4-dipolar cycloaddition allows the construction of six-membered heterocycles through a [4+2] cycloaddition of a 1,4-dipole, generated from an isoquinoline **18** and an acetylenedicarboxylate **19**, and a dipolarophile. In batch, the transformation is plagued by a side reaction between the 1,4-dipolar intermediated and another molecule of the alkyne leading to the cycloaddition by-product like **22**. In 2012, Lei et al. developed the 1,4-dipolar cycloaddition using aryl trifluoromethyl ketones **20** as dipolarophiles [8]. A preliminary study in batch showed the formation of the desired product **21** (41%), but, as expected, a significant amount of the side product **22** (34%) was also present. The translation to flow microreactor led to a complete suppression of the by-product and a library of **21** was obtained in high yields (Scheme 6.5). The dramatic effect of the microflow system is attributed to the extremely rapid and efficient mixing of reagents.

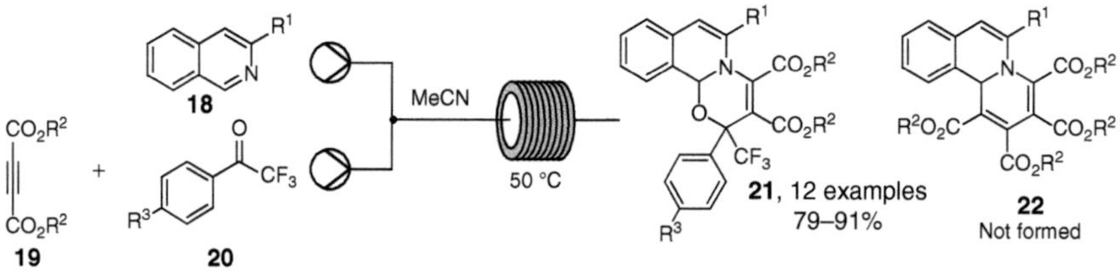

Scheme 6.5 Highly selective Huisgen 1,4-dipolar cycloaddition.

Gonzalez and coworkers developed the continuous flow synthesis of 2-aminochromenes **26–27** from aromatic aldehydes **23**, malononitrile **24**, and α- or β-naphthol **25** in the presence of a basic catalyst (DBU) [9]. After the Knoevenagel condensation of the aldehyde with the active methylene compound, the resulting arylidenemalononitrile reacts with naphthol to form the C—C bond. The C-alkylated intermediate undergoes an intramolecular cyclization (C—O bond forming event) and an isomerization to yield the final product. The authors used the ThalesNano H-Cube apparatus as flow system equipped with an inert titanium cartridge, since in this reaction there is no need of an heterogeneous catalyst (Scheme 6.6). After optimization, a library of 2-aminochromenes was obtained in high yield and short time (residence time: 50 s).

The flow synthesis of 5-(thiazol-2-yl)-3,4-dihydropyrimidin-2(1*H*)-ones **33** was reported by Cosford and coworkers in two steps, without isolation of the intermediate **30** thanks to an automated system [10]. The first step is the Hantzsch thiazole

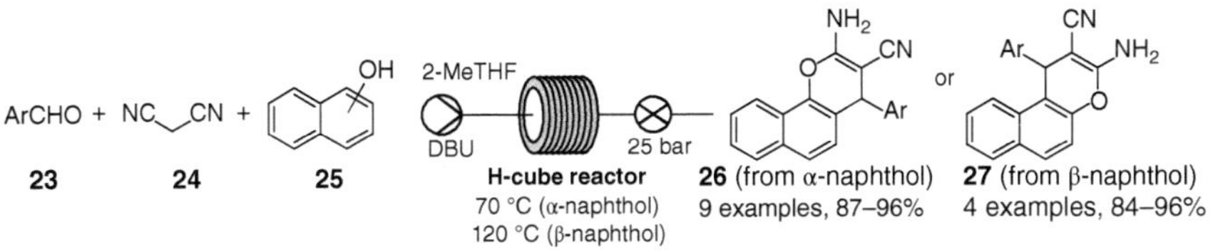

Scheme 6.6 Multicomponent synthesis of 2-aminochromenes **26–27**.

synthesis via condensation of ketal-protected thioamide **28** and α-bromoketones **29**. HBr and water are generated *in situ* and used to remove the protecting group and to catalyze the subsequent Biginelli three-component reaction (Scheme 6.7). This flow protocol allows for a very rapid construction of libraries of such interesting compounds.

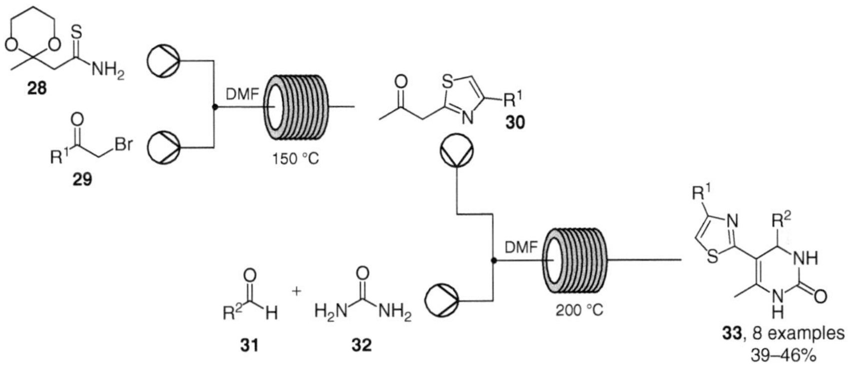

Scheme 6.7 Sequential Hantzsch/Biginelli continuous flow synthesis of **33**.

Guo and coworkers developed a novel three component reaction of 1,7-enynes **34** with sulfinic acids **36** and diselenides **35** yielding highly substituted 3,4-dihydroquinolin-2(1*H*)-ones **37**, through a sulfonyl radical-induced 6-*exo*-dig cyclization in the presence of an oxidant (*tert*-butyl hydroperoxide, TBHP) [11]. After batch optimization, the authors moved to the microflow set-up obtaining the desired product in higher yields and shorter times (Scheme 6.8).

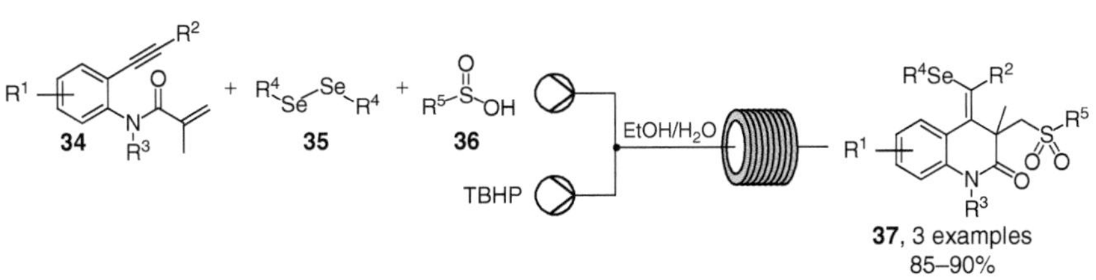

Scheme 6.8 Radical multicomponent selenosulfonation for the synthesis of **37**.

While the examples discussed so far highlight the beneficial effects of flow conditions on the yield, selectivity and time, the following ones demonstrate how easily flow processes can be combined with in-line purification or reaction monitoring, thus also allowing a rapid isolation of products and optimization of the conditions.

The successful combination of continuous flow and the three component Povarov reaction for the rapid generation of a library of tricyclic tetrahydroquinolines **41** was developed in 2016 by Gioiello and coworkers [12]. This reaction is a formal cycloaddition of an electron-rich dienophile **40** and an *N*-arylimine heterodyne, generated *in situ* from an aromatic aldehyde **38** and an aniline **39**, but it is believed that the mechanism proceeds step-wise. Since the reaction is catalyzed by acids, the authors added 5% of PEG300 as co-solvent to prevent the precipitation of the aniline hydrochloride within the lines. The flow system was equipped by an in-line work-up (liquid–liquid extraction) and an automatic chromatographic separation allowing the purification of the reaction mixture and each compound was isolated diastereomerically pure, with few exceptions (Scheme 6.9).

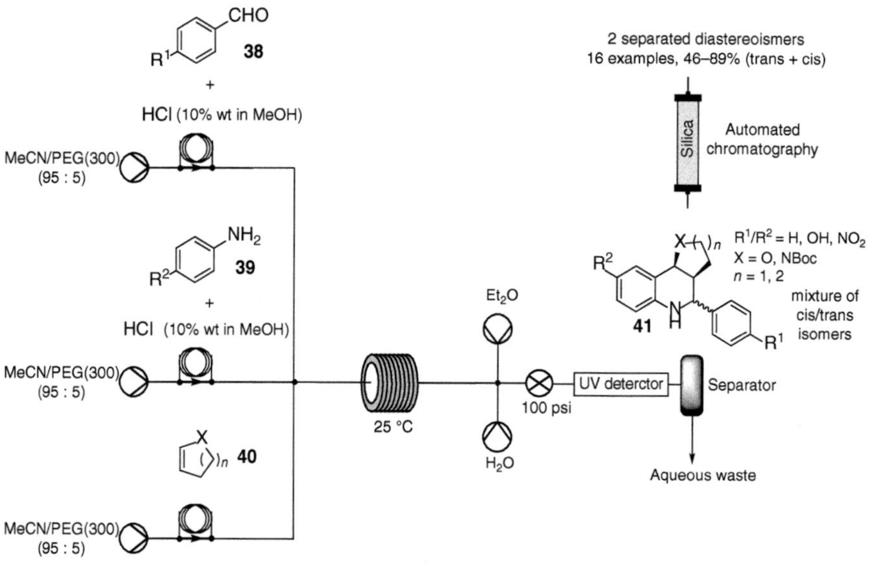

Scheme 6.9 Highly automated setup for the synthesis and purification of diastereomerically pure tricyclic tetrahydroquinolines **41**.

Martha et al. developed the high-throughput optimization of the synthesis of a substituted 2-imidazoline **45** via a three components reaction between benzylamine **42**, acetone **43**, and *p*-nitrobenzylisocyanide **44** (an isocyanide having α-acidic protons) in the presence of a ferrocene-based Lewis acid catalyst [13]. The system combines an automated continuous flow reactor with an in-line atmospheric-pressure chemical ionization mass spectrometer (APCI-MS) detector (Scheme 6.10). The typical optimization of reaction parameters (i.e. reactants ratio, temperature, and catalyst loading) was carried out automatically changing the relative flow rate of the reactant or the concentration of the injected model catalyst. Then the authors performed a high-throughput screening of 17 ferrocenes in six solvents in a very straightforward manner thanks to the high degree of automation.

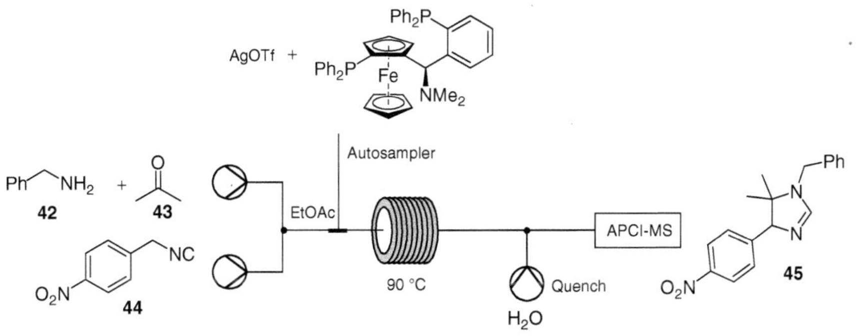

Scheme 6.10 High-throughput reaction optimization and catalyst screening by combining continuous-flow and mass spectrometry detection.

6.3 Multicomponent Reactions with Hazardous Reagents

One of the most important advantage of microreactors is the possibility of using toxic reagents in a safe way. Moreover, the use of flow reactors reduces the interaction of the operators with toxic reagents/intermediates.

In this context, Stevens reported the successful application of a microreactor for the safe handling of the highly toxic hydrogen cyanide, generated *in situ* from KCN and AcOH, which were pumped separately into the microreactor. HCN is one of the three components needed in this modified Strecker multicomponent reaction that leads to 3-amino-4-(arylamino)-1*H*-isochromen-1-ones **48** (Scheme 6.11) [14]. The mechanism proceeds through a Strecker reaction between 2-formylbenzoic acid **46**, an aromatic amine **47** and potassium cyanide, followed by a intramolecular

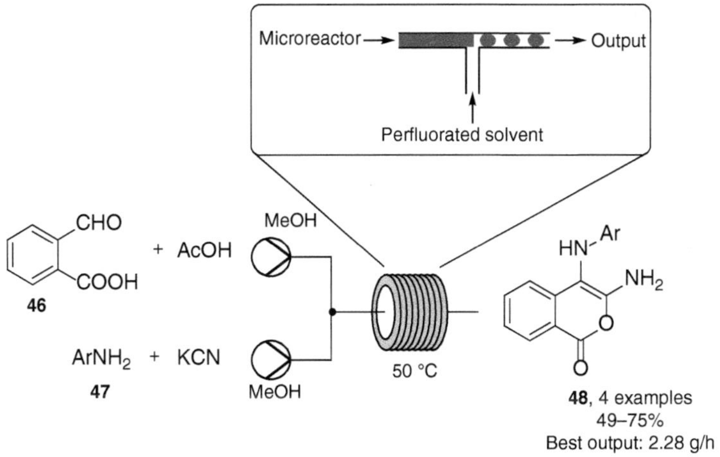

Scheme 6.11 Safe handling of HCN in a modified Strecker reaction. Source: Adapted from Acke et al. [14].

nucleophilic addition of the carboxylic group to the nitrile. Finally, the resulting *O*-acyl imidate tautomerizes to the final isochromenone. The choice of solvent and temperature is limited by the limited solubility of KCN and the authors found out that MeOH was the best solvent. One of the main problems faced during the optimization was the crystallization of the product at the end of the tubing. They solved the issue pumping an immiscible perfluorinated solvent between the microreactor and the residence time unit. This method creates plugs of the reaction mixture, separated from each other by the immiscible solvent. In such way, the perfluorinated solvent drives the product further through the system, even if there is crystallization in the plugs itself. Although the yields are lower using the microreactor system than in batch, the small dimensions of the microreactor (2 ml) ensures that only minor amounts of the toxic reagent are formed and even if a qualitative HCN spot-test was positive, a slower coloration was detected in the case of the microreactor setup than in batch conditions.

The Bucherer–Bergs reaction is the most popular method for the synthesis of hydanthoins **52** starting from aldehydes or ketones. The three components reaction is carried out by simply heating a mixture of the carbonyl compound **49**, KCN and ammonium carbonate (which thermally decomposes into NH_3 and CO_2) in $EtOH/H_2O$. However in batch the reaction presents several limitations: (i) long reaction time (several hours or even days), (ii) loss of NH_3 and CO_2 due to their volatility, (iii) limited solubility of many starting materials in such polar solvents, (iv) safety hazard on large scale due to the presence of KCN. In 2016, Kappe and coworkers reported the continuous flow variant of the Bucherer–Bergs reaction with results far superior than the batch process [15]. The flow system allows a safe handling of KCN and the reaction can be carried out keeping the gaseous reagents in solution working at high pressure thanks to the back pressure regulator (Scheme 6.12). The presence of two independent in-lets allows the use a segmented flow pattern of a biphasic liquid-liquid solvent mixture, composed by any organic solvent for the carbonyl compounds and water for the inorganic reagents. The increased interfacial area and the absence of the headspace in flow lead to higher yield (91%) than the biphasic batch reaction (40%).

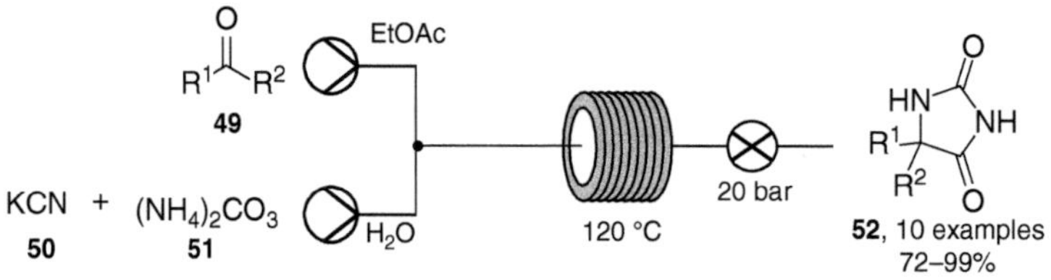

Scheme 6.12 The Bucherer–Bergs reaction for the synthesis of hydanthoins.

One of the most notorious "click reaction" is the two-component Cu(I)-catalyzed alkyne azide cyclization (CuAAC), that proceeds through a 1,3-dipolar cycloaddition yielding 1,4-disubstituted 1,2,3-triazoles **56** with high regioselectivity. Zhang and coworkers translated the one-pot version to a flow system, avoiding the

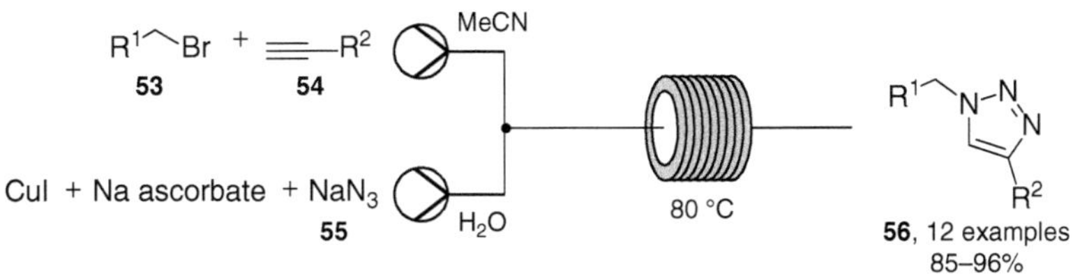

Scheme 6.13 Continuous flow synthesis of 1,4-disubstituted triazoles by CuAAC.

isolation and handling of the potentially unstable organic azides [16]. A tandem alkylation/CuACC resulted in a three-component reaction between **53**, **54**, and **55**. After optimization of the parameters, a library of triazoles **56** was obtained in high yield, in short time and without the formation of by-products thanks to the excellent mixing and mass transfer in micro flow system (Scheme 6.13). The coupling of an automated continuous flow system to the CuAAC can potentially accelerate the production of products containing this widespread heterocycle in a combinatorial fashion.

N-unsubstituted 4-aryl-1*H*-1,2,3-triazoles **60** can be prepared by cyclization of nitroalkenes with azides. However, the highly toxic and explosive hydrazoic acid may be generated during the process and nitroalkenes may rise safety issues. Zhang and coworkers reported the optimization of the two components process both in batch and in flow, demonstrating the superior performance of the flow reactor thanks to the possibility of using the high temperature (190 °C) that benefits this cycloaddition [17]. Moreover, nitroalkenes can be prepared via Henry reaction followed by dehydration. The authors optimized the one-pot three-component flow protocol starting from benzaldehydes **57**, nitromethane **58**, and sodium azide proving that there is no need to prepare nitroalkenes in advance (Scheme 6.14).

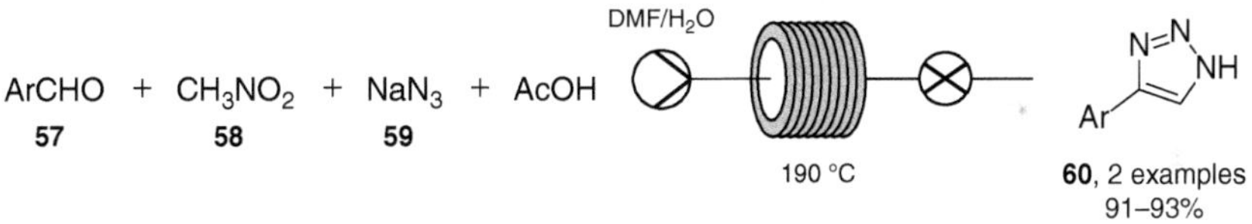

Scheme 6.14 Three component synthesis of *N*-unsubstituted 4-aryl-1*H*-1,2,3-triazoles **60**.

6.4 Multicomponent Reactions Under Special Conditions

The beneficial effects of continuous flow approaches in comparison with classical batch conditions have been thoroughly discussed in the previous paragraphs. Improvements are observed in terms of yields, selectivity, and kinetics; factors that render flow technology an efficient solution to the multicomponent synthesis of heterocycles can be further increased when special conditions are applied. In this section the use of microwave heating, of catalytically active reactors, of heterogeneous systems and of light irradiation is discussed, showing how this area of research is lively and constantly evolving.

6.4.1 Reactions with Microwave or Inductive Heating

The limited geometry of microwave (MW) devices is a major drawback for the scale-up of reactions, and has therefore limited the industrial applications of this technology. To overcome this problem, the use of flow chemistry (microwave-assisted continuous flow organic synthesis, MACOS) has been proposed at the end of last century, combining the sample-handling and scaling-up advantages of flow with the rate-enhancing features of microwave heating; the relevant literature, up to 2007, has been reviewed by Baxendale et al. [18].

Organ and coworker [19] have used the MACOS approach to prepare medicinally relevant, heterocyclic compounds in a MCR format. Specifically, reaction of various aldehydes **61** with dimedone **62** and pyrazole **63** has been used to form quinolinones **64** (Scheme 6.15). The three components were introduced into a 1180 μm glass capillary, placed into the microwave oven, through separate feeds, in equal concentrations and at the same rate. On one hand, this greatly expanded the combinatorial efficiency of the system for library generation, but on the other hand it required careful optimization of capillary diameter, microwave power, flow-rate, reaction concentration and solvent, in order to avoid laminar flow, and therefore, poor mixing and conversions.

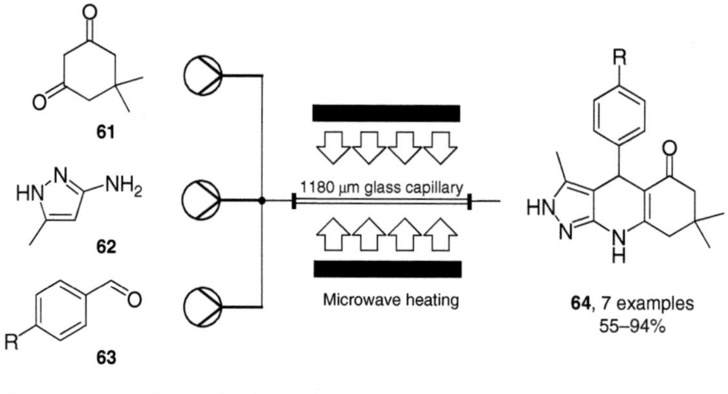

Scheme 6.15 Synthesis of quinolinones **64** through a three component reaction in a MW heated capillary.

Although microchannel devices usually produce small amounts of product, compound **64** was collected in about 300 mg total amount after only 30 minutes of operation. Using the same approach, tetrasubstituted furans were prepared from aldehydes, cyclohexyl isocyanide, and dimethyl acetylenedicarboxylate.

Bàlint et al. have used the Kabachnik–Fields reaction to obtain α-aryl-α-aminophosphonates from primary amines, benzaldehyde derivatives and dialkyl phosphites [20]. The use of a continuous flow microwave reactor allowed to synthetize the target compounds in high (c. 90%) yields without any catalyst in alcoholic solvent. Compared to traditional batch reactions, shorter reaction times and lower excess of reagents were employed. The authors further extended this methodology to the synthesis of isoindolin-1-one-3-phosphonates **68** by the three-component condensation of 2-formylbenzoic acid **65**, aliphatic primary amines **66**, and diethyl phosphite **67** (Scheme 6.16) [21]. A dual pump, connected to a commercially available continuous flow cell, was used to feed the reactor with two independent lines. The mixture was pumped in the reactor with a flow rate of 0.70–0.15 ml/minute, corresponding to residence times of 10–45 minutes, the mixture leaving the reactor was cooled down to 25 °C using a spiral-like cooler and a back pressure regulator (operating at 17 bar) was used to avoid solvent boiling. Change in flow rate only slightly improved the overall yield, while better improvement was obtained by using excess amine and phosphite, with conversions close to 100%.

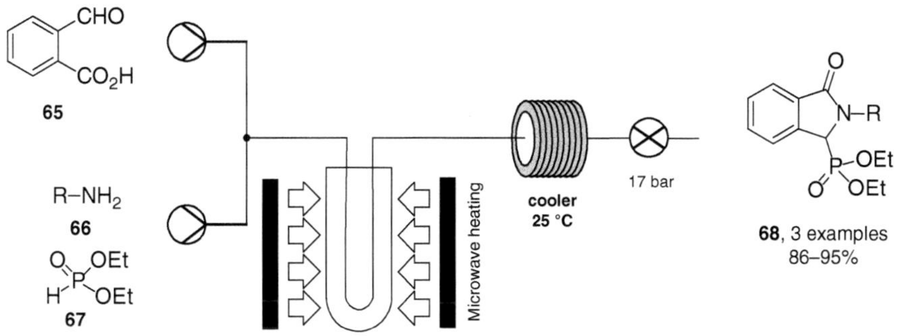

Scheme 6.16 Synthesis of isoindolin-1-one-3-phosphonates **68** with a continuous flow microwave reactor. Source: Based on Tajti et al. [21].

Larhed and coworkers developed a three-component MW-flow reaction using a silicon carbide (SiC) tubular reactor able to efficiently absorb microwaves [22]. The reaction consisted of the condensation of isatoic anhydride **69**, an isatin derivative **70** and aniline **71** at 160 °C, yielding spiro-oxindole dihydroquinazolinones **72**, known as insulin-regulated aminopeptidase (IRAP) inhibitors, in good to moderate yields and short residence times (168 s) (Scheme 6.17).

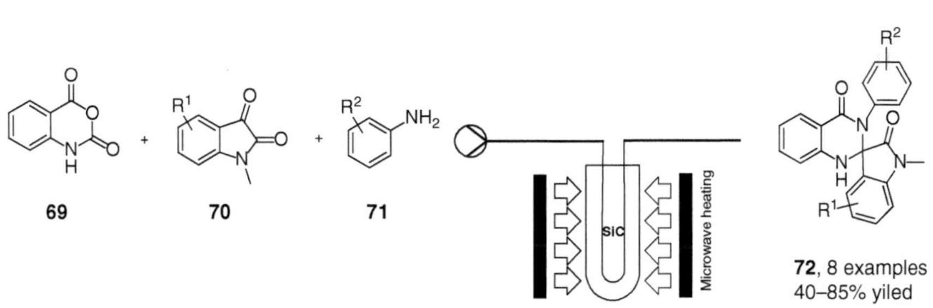

Scheme 6.17 Synthesis of spiro-oxindole dihydroquinazolinones **72** using a SiC reactor heated with microwaves.

The paper published by Li and coworkers [23] in 2010, still related to the use of microwaves, anticipates the next paragraph, as it is dealing with catalytically active reactors. Although this is not strictly an example of MCR synthesis of heterocycles, it does represent a case of multicomponent functionalization of heterocycles and therefore the authors considered it consistent with this chapter An A^3-coupling between aldehydes **73**, amines **74**, and alkynes **75** is performed using MACOS, but the reactor is coated with thin films of Cu or Au. A dramatic improvement of the reaction kinetics is observed, either due to direct metal catalysis or to the extremely high temperatures (up to 950 °C) reached by the thin film, when microwaved. Although the A^3 multicomponent reaction does not generate a heterocyclic system, various cyclic amines were employed in this study, thus affording *N*-functionalized *N*-heterocycles **76**. The scope of the reaction was investigated feeding the reactor with a mixture of the three reagents, however a test experiment was carried out also using three independent lines, in order to improve combinatorial efficiency and library synthesis. Both set-ups afforded the same final outcome, yielding *N*-propargylated cyclic amines **76** in moderate to good yields (Scheme 6.18).

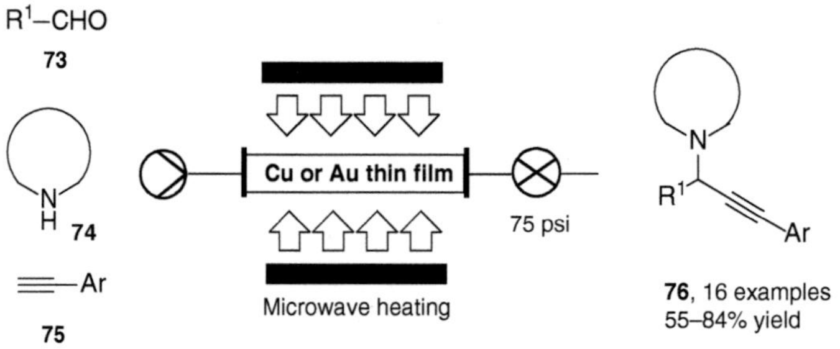

Scheme 6.18 Synthesis of propargylated cyclic amines **76** with a copper or gold coated microwave reactor.

The use of microwaves for Hantzsch dihydropyridine synthesis is a well-established methodology, affording the desired products in high yields and purities. The use of a continuous flow approach was independently reported by Bagley et al. [24] and Baxendale et al. [25] in 2013, starting from propargyl or aromatic aldehydes respectively. Yields were slightly lower than in batch mode, but the use of highly concentrated solutions and short residence times were a clear advantage in comparison with existing methodologies.

In conclusion of this section, it is worth mentioning the work by Kirschning and coworkers, although not strictly related to microwaves [26]. Steel beads or silica gel coated superparamagnetic nanoparticles have been used as fixed-bed material for flow reactors to efficiently heat reaction mixtures in an inductive field. Inductive heating relies on magnetic inductive hyperthermia, also termed superparamagnetism, according to the Neel relaxation principle: when magnetic nanoparticles are exposed to a changing magnetic field, their loss of magnetic hysteresis creates heat. The scope and limitations of this technology have been investigated, suggesting that inductive heating can be compared to MACOS with respect to rate acceleration. In addition, nanoparticles have several advantages when incorporated into flow devices, such as a simple setup (nanoparticles are incorporated inside a flow reactor made of glass and encased with an inductor) and minimal safety issues; moreover, only the superparamagnetic particles are the initial source for heating, which occurs inside the reactor, and they can be superheated (above 500 °C). Among many different reactions performed under flow conditions by using inductively heated reactors, multicomponent Biginelli, Mannich, and Petasis reactions were successfully performed. Some years later, the same authors reported the design of a steel flow reactor that could be inductively heated in bulk without the need for steel beads or superparamagnetic nanoparticles [27]. This type of heating is really fast and the desired temperature can be reached within 30 s. The author tested the system under high temperature/high pressure condition in some selected transformation using a water/toluene solvent mixture. Enones **77** react with benzaldehydes **78** and benzyl amines **79** in a multicomponent fashion yielding acetyl dihydropyridines **80** through a formal [1+2+1+2] cycloaddition process. Batch conditions require an acid catalyst (Sc(OTf)$_3$) and the reaction lasts 2–20 days in DCM. Under aqueous high temperature flow conditions dihydropyridine **80** can be prepared in high yield at 180 °C in the absence of a Lewis acid and with residence times below eight minutes (Scheme 6.19).

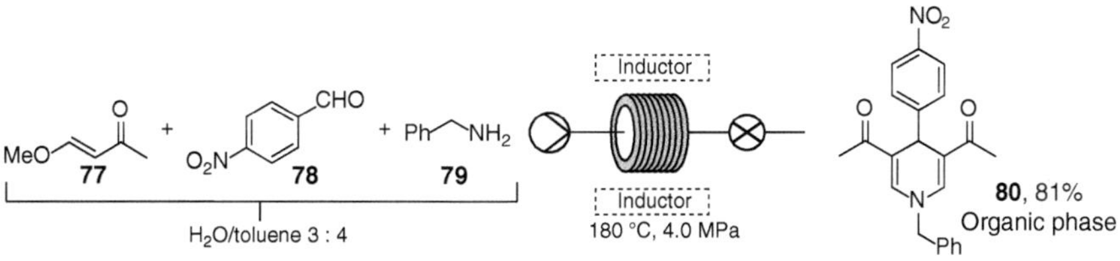

Scheme 6.19 High temperature/high pressure synthesis of dihydropyridine **80** in a water/toluene mixture.

6.4.2 Reactions with Active Packed-Bed Columns

The work published by Ley in 2010 regarding the flow synthesis of pyrrolidines and pyrroles clearly highlights how packed-bed columns can dramatically improve the efficiency of multicomponent reactions, when loaded with reagents, catalysts, or scavengers [28]. According to the general procedure for the obtainment of the pyrrolidine ring via the cycloaddition between nitrostyrenes and azomethine ylides, the author elaborated a three-component process in which stock solutions of β-nitrostyrene **81**, glycine methylester hydrochloride **82**, and aldehyde **83** (with triethylamine), all dissolved in acetonitrile, were introduced into a flow reactor and directed into a heated (60–80 °C) glass column filled with anhydrous sodium sulfate (Scheme 6.20). This first step was required to promote the imine formation between **82** and **83**, leaving the nitrostyrene derivative untouched. Upon exiting the drying column, the reaction mixture was directed into a coil heated at 80–100 °C, where the desired cycloaddition between **81** and the in situ formed stabilized azomethine ylide took place. In-line work-up with a polymer-supported benzylamine scavenger (QP-BZA) removed residual starting materials, thus affording the desired tetra-substituted pyrrolidine **84** in yields higher than 70% as a mixture of stereoisomers. Further elaboration of the product could also be introduced in the streamline, with a packed-bed column loaded with activated MnO_2. The oxidant, at elevated temperatures (85–110 °C) could afford clean conversion to the corresponding pyrroles **85**.

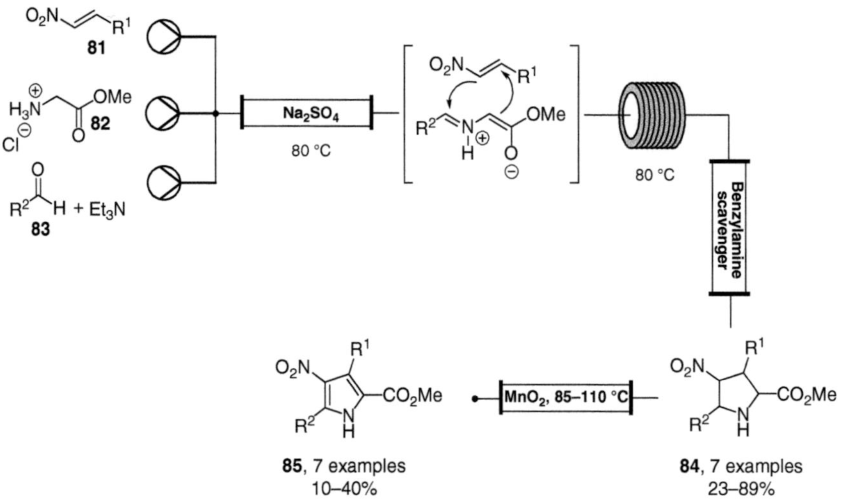

Scheme 6.20 In-line synthesis of pyrrolidines **84** and pyrroles **85** with the aid of packed-bed columns loaded with reagents and scavengers.

The nitro-Mannich/lactamization reaction between imines and nitropropanoates affords pyrrolidin-2-ones **89** and can be favored by the presence of acids [29]. Vaccaro has proposed a continuous flow diastereoselective multicomponent synthesis of **89** from benzaldehyde **86**, benzylamine **87**, and methyl 3-nitropropanoate **88** employing a packed-bed reactor filled with a solid acid catalyst, Aquivion® perfluorosulfonic acid (PFSA) (Scheme 6.21) [30]. Remarkable features of this catalyst are a superacid character, comparable with pure sulfuric acid, and a high

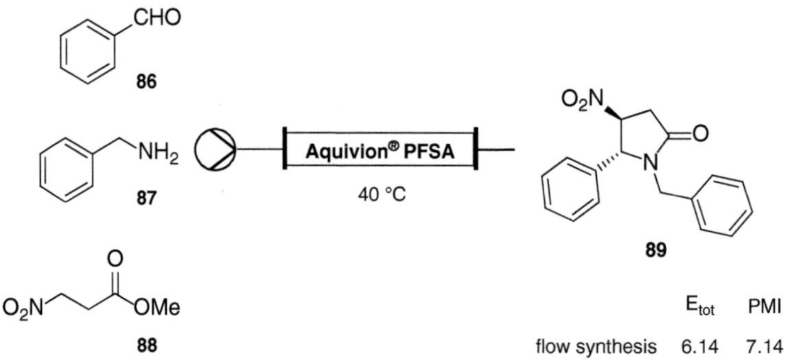

	E_{tot}	PMI
flow synthesis	6.14	7.14
batch synthesis	29.42	30.42

Scheme 6.21 Synthesis of pyrrolidin-2-one **89** using a packed-bed reactor filled with a superacid catalyst. Source: Bianchi et al. [30].

chemical inertness of the perfluorinated polymeric backbone, that can resist to strong acids and bases, as well as to oxidative and reductive environments. The approach displayed advantages in terms of green metrics (E-values and process mass intensity, PMI) compared to the same reaction performed in batch conditions, and the reduction of wastes was greater than 99% compared to not-optimized syntheses previously reported in the literature. On the other hand, energy-dispersive x-ray spectroscopy (EDS) and scanning electron microscopy (SEM) analyses of the recovered catalyst showed that the physical structure was preserved, but that a superficial encrustation was formed and regularly increased after each run. This was attributed to the precipitation of product **89**, poorly soluble in the reaction environment.

Rodrigues and Neto have reported the multicomponent Biginelli reaction using a stable and robust zinc-based coordination polymer (Zn-CP) as heterogeneous catalysts in a packed-bed column [31]. A model reaction with benzaldehyde **90**, urea **91**, and ethyl acetoacetate **92** under continuous flow conditions yielded excellent yields of Biginelli product **93** for residence time of 34 minutes at 70 °C or 17 minutes at 100 °C, while a batch process reaction at 100 °C for five hours gave almost the same results (Scheme 6.22). The recyclability of Zn-CP was reported for three cycles without loss of catalytic activity. Bioactive compounds oxomonastrol, monastrol, enastron, dimethylenastron, and piperastrol were synthesized with excellent yields with this approach.

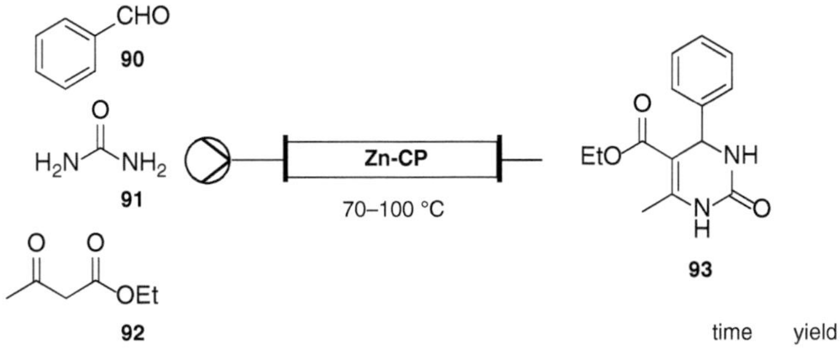

	time	yield
flow synthesis (70 °C)	34 min	96%
flow synthesis (100 °C)	17 min	99%
batch synthesis (100 °C)	5 h	91%

Scheme 6.22 Continuous flow Biginelli reaction using a Zn-based coordination polymer.

In 2018, Tian reported the preparation of N-heterocyclic carbene (NHC)-stabilized silver nanoparticles (NPs) supported on polyacrylonitrile fiber (PANF) [32]. In general, NPs exhibit higher catalytic activity than their bulk counterparts, but aggregation and formation of a deactivated bulk mass is a major drawback, especially when the catalyst has to be reused several times in a flow reactor. The formation of coordination bonds between NHCs and the surface of metal NPs can be exploited to retain their unique nanoscale properties, while the support on polymeric fibers can further improve their stability, regeneration, and separation. The AgNPs have been tested in a three-component reaction between alkynes, haloalkanes, and cyclic amines and a synergistic effect of the AgNPs and of the PANF support was observed in the alkyne deprotonation, that occurred with a remarkable turnover number (3500). A stock solution of alkyne **96**, dihalomethane **95** (5 equiv), cyclic amine **94** (1.2 equiv) in the presence of DBU as a base was pumped into a silicone column loaded with PANF-NHC@Ag at 25 °C, giving a yield of *N*-propargylated cyclic amine **97** of 95% and a productivity of 57 mmol/h (Scheme 6.23). The system was operated for 24 hours without interruption.

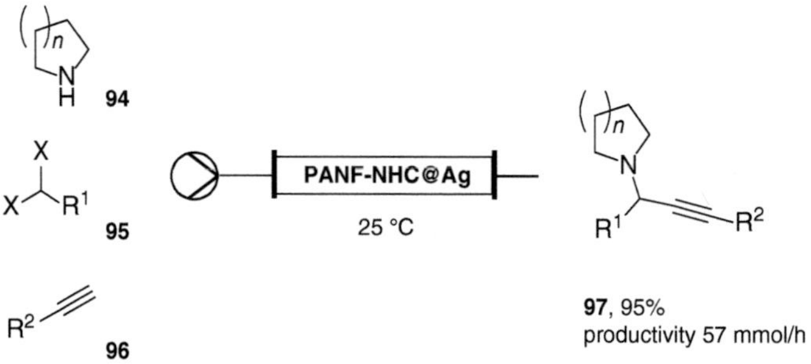

Scheme 6.23 Multicomponent functionalization of cyclic amines with haloalkanes and alkynes using a packed-bed reactor loaded with silver nanoparticles.

Finally, an A3 coupling between aminothiazoles **98**, alkynes **99**, and aromatic aldehydes **100** has been reported by Volkova and coworkers [33]. Imidazo[2,1-*b*]thiazoles have been obtained in batch and under continuous flow with the aid of Cu catalysis. The multicomponent reaction was first optimized in batch, showing a broad scope. Continuous flow synthesis with heterogeneous Cu catalyst was then investigated using a steel column filled with CuI/SiO_2 and heated at 200 °C. The improved heat transfer reduced side reactions and a back-pressure regulator was introduced to eliminate solvent boiling, resulting in substantial enhancement of reaction kinetics. Solubility issues were addressed in order to avoid column clogging, and to this purpose anisole was used in place of toluene as the solvent of choice. Overall, a general improvement of yields was observed, and compounds **102** and **103** were obtained in 59% and 54% yield respectively, where the batch synthesis failed to afford them (Scheme 6.24). It should be noted, however, that the flow synthesis was performed using preformed imines **101**. Unfortunately, no attempts were made mixing the three reagents directly in the flow reactor.

6.4.3 Reactions Under Other Conditions

Traditionally, photochemical and heterogeneous gas/liquid reactions largely benefit from continuous flow approaches. In the first case, the small dimensions of the reaction coils allow an optimal and uniform irradiation of the solution, eliminating the problems of light attenuation observed in batch reactions, especially during

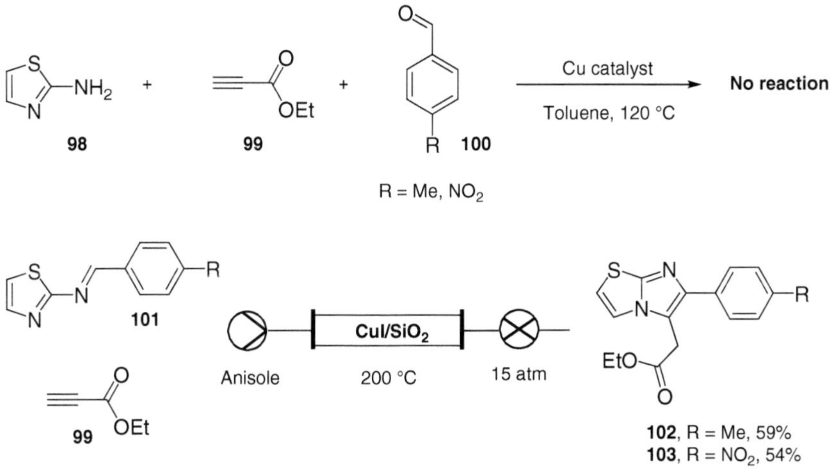

Scheme 6.24 Synthesis of imidazo[2,1-*b*]thiazoles with Cu-based catalytically active reactor.

scaling-up. In the case of gas/liquid reactions, on the other hand, the formation of regular segmented flows ensures an intimate contact between the reagents present in the two phases. It is therefore surprising that only two heterocyclic molecules obtained through a multicomponent approach have been found in the literature.

The first molecule has been reported by Basso and coworkers and was synthetized in a three-component reaction from diazonium salt **104**, styrene derivative **105** and acetonitrile, used as the solvent [34]. $Ru(bpy)_3Cl_2$ was used as photoredox catalyst, able to reduce, upon blue light activation, **104** to an aryl radical; the aryl radical in turn reacted with **105** and the resulting benzyl radical restored the catalyst, generating carbocation **106**. The carbocation could be directly trapped intramolecularly by the carboxylic functionality displayed by **105**, yielding **108** (path a), or could be firstly trapped by acetonitrile, yielding **107** and in turn benzoxazepinone **109** (path b). To improve the selectivity towards the multicomponent adduct **109**, the reaction was performed in batch and under continuous flow conditions, using an in-house made Booker-Millburn reactor, with external illumination, but no improvement was observed (Scheme 6.25).

Radical stannylcarbonylation has been used by Ryu and coworkers to obtain lactam **111** from 1,6-azaenyne **110** [35]. Thermally induced radical carbonylations using tin hydride usually require pressurized CO conditions to avoid premature quenching of radicals by tin hydride, and the use of a segmented flow resulted optimal to this aim. A stream of CO gas was mixed with a toluene solution containing tributyltin hydride, radical initiator V-65 (2,2′-azobis(2,4-dimethylvaleronitrile) and azaenyne **110** in a T-shaped micromixer. This biphasic (gas–liquid) mixture was then fed into a stainless-steel reactor, heated at 80 °C using an oil bath. A back-pressure control valve was connected to regulate and maintain the pressure of the reactor system (Scheme 6.26). The product was obtained in 85% yield with complete Z selectivity.

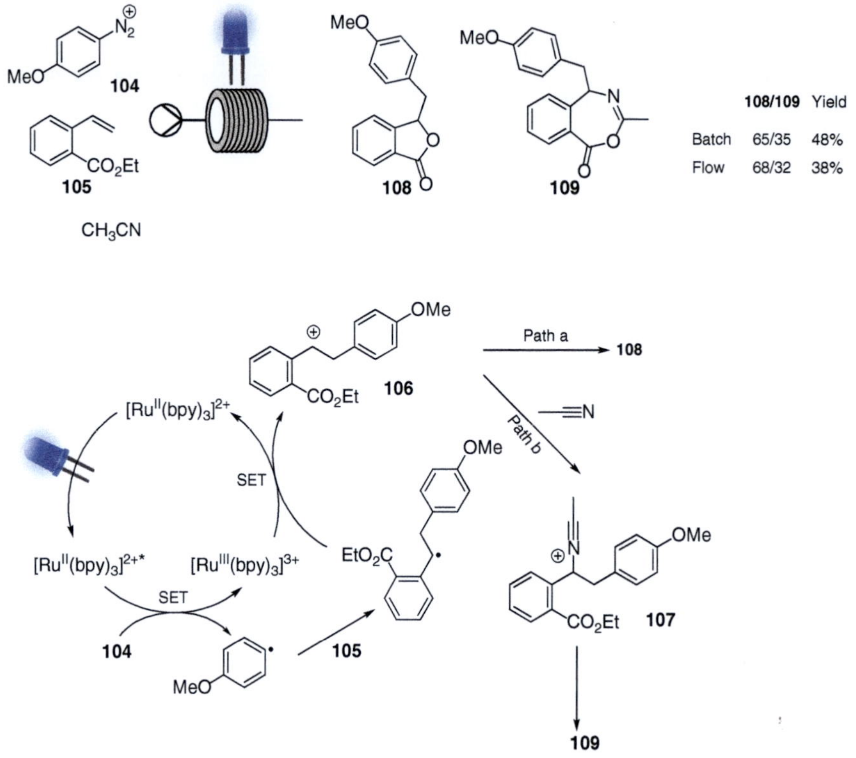

	108/109	Yield
Batch	65/35	48%
Flow	68/32	38%

Scheme 6.25 Synthesis of benzoxazepinone **109** via a photoredox multicomponent process in flow.

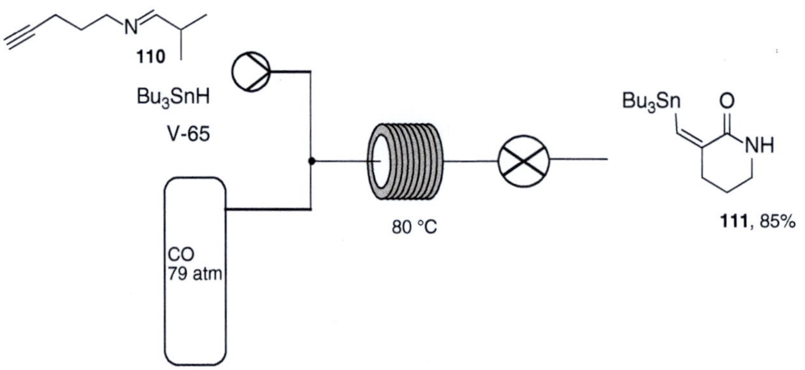

Scheme 6.26 Synthesis of stannylated lactam **111** via a three component reaction with carbon monoxide.

6.5 Telescoped Reactions

By definition, multicomponent reactions require all regents to be added from the beginning in the same reaction environment. When moving to flow chemistry, however, the definition is not so stringent, and a process in which various reagents are added at different stages can also be considered multicomponent, provided that all are incorporated into the final product and that the additions take place within the same stream-line, without interruptions to isolate or purify reaction intermediates. The "telescoping" of multiple synthetic steps into a single continuous process provides an efficient method for the production of heterocyclic compounds, with considerable decrease of waste production and without the need of exhaustive workup procedures. Furthermore, the amount of reagents/solvents is reduced compared to batch multistep synthesis, the screening of reaction conditions is facilitated and the rapid generation of focused compound libraries is appealing to pharmaceutical companies, which could benefit from cost reduction and rapid drug manufacturing. For these reasons telescoped reactions have also been incorporated in this chapter and the most relevant examples are reported below.

An interesting study was recently carried out by Tallarek and coworkers on a Knowenagel–Michael domino reaction to give **115**, employing an aminopropylated monolith cartridge. The reaction of various aldehydes **112**, malononitrile **113** and dimedone **114** was first studied as a one pot reaction, and then as a sequential process, with the delayed addition of dimedone in the reaction flow (Scheme 6.27). The latter design proved to give better results in terms of turnover frequencies (TOF) of the catalyst. It was demonstrated, in fact, that the amine catalyst played a dual role: as a primary amine in the aldimine formation in the Knoevenagel condensation and as a base for the deprotonation of dimedone prior to its nucleophilic attack onto the Knovenagel product. The delayed addition of dimedone on the cartridge avoided that this competed with the aldehyde for the active sites of the heterogeneous catalyst [36].

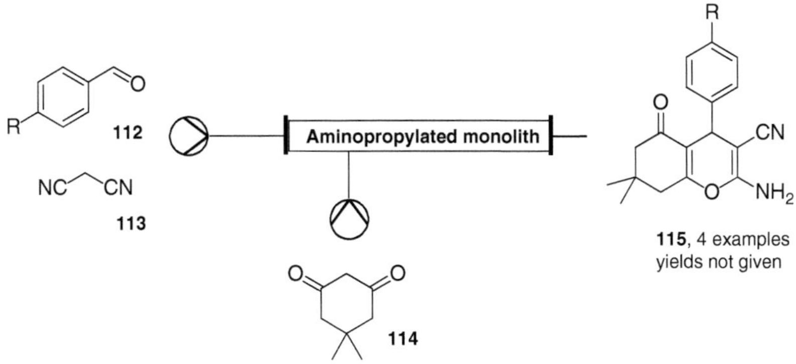

Scheme 6.27 Three component synthesis of **115** using an aminopropylated monolith cartridge.

The synthesis of 2,3,5-trisubstituted tetrahydrofurans **119** can be accomplished by a sequential vinylogous Mukaiyama aldol reaction (VMAR) between bis(silyl) dienediolate **116** and an aldehyde **117**, followed in situ by a Prins-type cyclization of the resulting silyl enol ether with a second aldehyde **118**. In batch the reaction proceeds with high diastereoselection (97 : 3), but when two different aldehydes are employed the method shows a limitation in terms of selectivity, in fact a significant amount of **120** is formed. Schneider solved the selectivity issue using flow chemistry, employing a smaller reactor, cooled at 0 °C, for the Mukaiyama step and a 10 times bigger one for the Prins cyclization [37]. After a careful optimization of the parameters of both steps (equivalents, concentration, flow rate, residence time, volume of the reactor) the amount of the side product was reduced by 50% (Scheme 6.28). The high yield and selectivity reached with the flow synthesis remain unchanged on a larger scale, on the contrary in batch the yield of **119** drops by 80% and the amount of **120** increases by 300%.

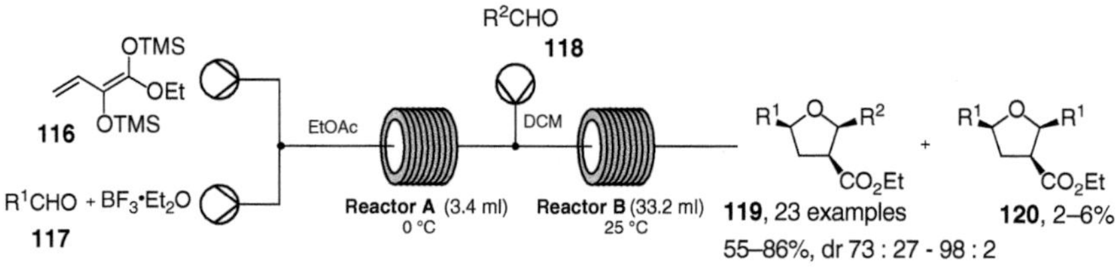

Scheme 6.28 Selective synthesis of highly substituted tetrahydrofurans **119**.

Climent and coworkers have reported the one-pot synthesis of 2,3-dihydro-1,5-benzothiazepine **124** using fixed-bed reactors loaded with solid acid and base catalysts [38]. A Claisen–Schmidt condensation between benzaldehyde **121** and acetophenone **122** was achieved in a stainless-steel reactor filled with rehydrated Al–Mg mixed oxide (CAT 1) at 50 °C; the outcome of the reaction, in which the conversion was complete and the selectivity towards chalcone was 99%, was fed directly together with 2-aminothiophenol **123**, into a second fixed-bed reactor that contained MCM-41 (CAT 2), heated at 100 °C (Scheme 6.29). MCM-41, an aluminosilicate, was selected after a catalyst screening thanks to its weak acidity and its large pore diameter, able to prevent diffusional problems, strong product adsorption, and pore blockage. The yield in **124** was 95% and remained constant after 10 hours of operation.

Pastre and coworkers reported the application of continuous flow to prepare 3-thio-1,2,4-triazole derivatives [39]. A telescoped process was developed, based on the condensation of hydrazides **125** and isothiocyanates **126** to afford a thiosemicarbazide **127**, which was then cyclized under basic conditions, provided by a stream of a sodium hydroxide solution. The obtained 1,2,4-triazole-3-thiol **128** was further alkylated with benzyl/alkyl halides **129**, thus affording compounds **130** with three diversity inputs (Scheme 6.30). The single reactions were first tested in solution, however, during translation into the continuous flow system, a solvent change to a mixture of MeCN/DMF (8 : 2) was necessary to keep all reagents and products

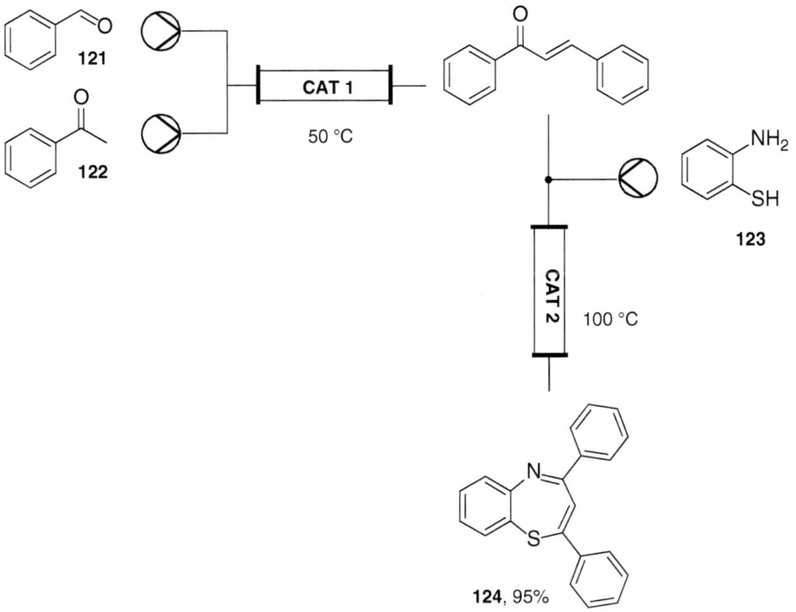

Scheme 6.29 Synthesis of 2,3-dihydro-1,5-benzothiazepine **124** through a telescoped three component process.

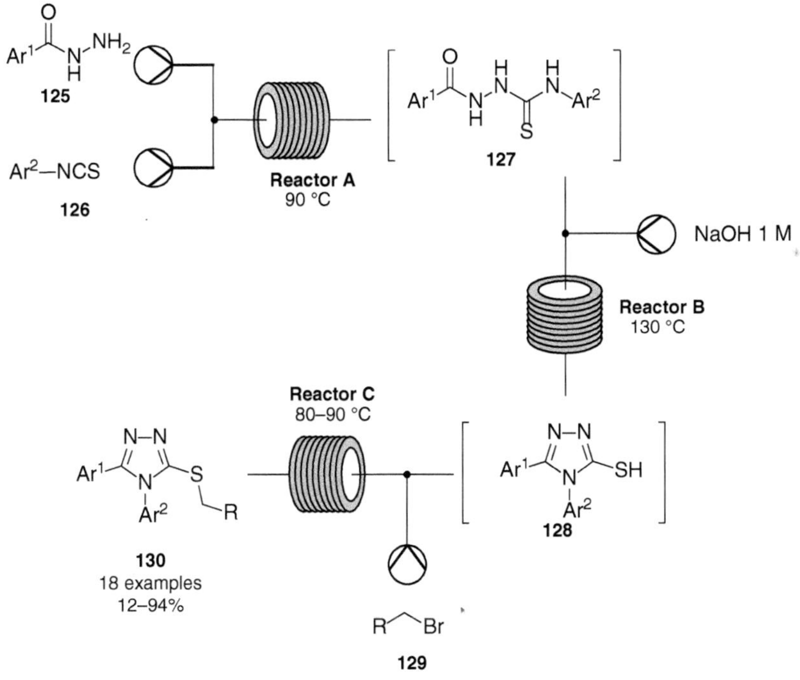

Scheme 6.30 Continuous flow synthesis of 3-thio-1,2,4-triazole derivatives **130**.

soluble inside the flow system. The flow platform enabled the fast generation of compound libraries and the scale-up of selected compounds.

Recently, liquid–liquid microextraction and microfluidic distillation have been implemented in continuous-flow stream-lines to allow separations and purifications in a continuous fashion. As an example we describe the synthesis of imidazo[1,2-*a*]pyridin-2-yl-1,2,4-oxadiazoles reported by Cosford and coworkers [40]. The use of polar, high-boiling solvents is often necessary in flow chemistry to prevent the formation of insoluble intermediates in the reactor. This, however, represents a challenge when large amounts of solvents have to be removed during purification and isolation of the products, especially during scale-up. The synthesis of 1,2,4-oxadiazoles was achieved from arylnitriles **131**, carboxylic acid derivatives and aminopyridines in a multistep process (Scheme 6.31). First, bromoacetic acid **132** and aminopyridine **133** were combined in a glass reactor, heated at 100 °C, to yield imidazo[1,2-*a*]pyridine-2-carboxylic acid **134**; the acid exiting the first reactor was combined with EDC/HOBt/DIPEA (1 : 1 : 1) in a T-mixer and then loaded into reactor C (heated at 150 °C) together with amidoxime **135**, in turn obtained

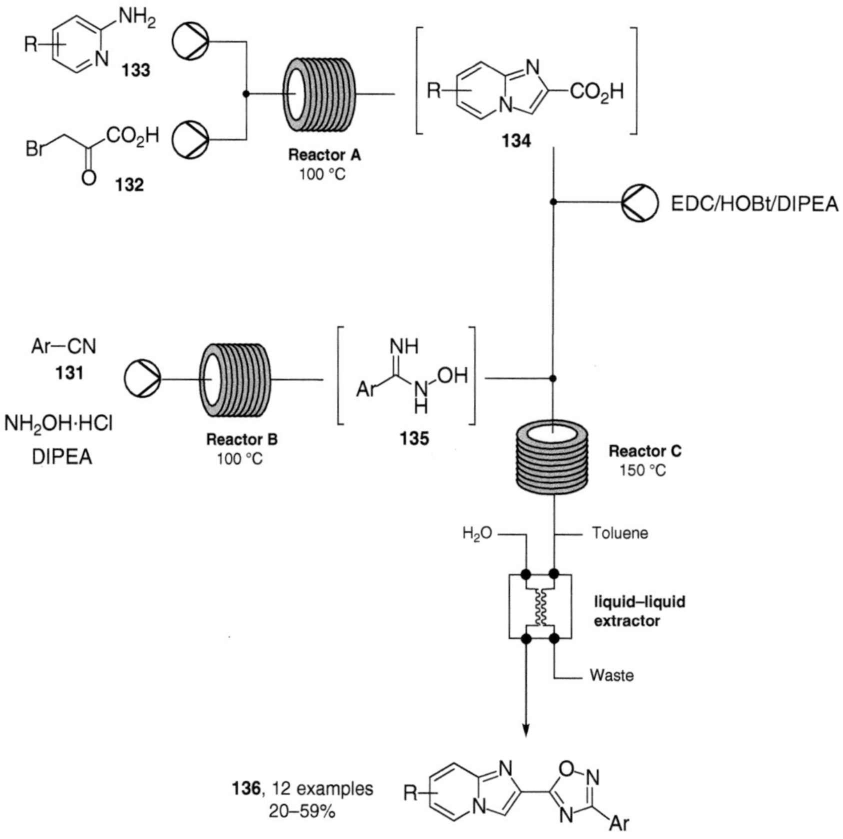

Scheme 6.31 Telescoped synthesis of oxadiazoles **136** using a liquid–liquid microextractor.

in reactor B from arylnitrile **131** and hydroxylamine in the presence of a base. A microfluidic liquid–liquid extraction module was introduced at the end of the streamline to remove the high boiling point and water soluble solvent (dimethylacetamide, DMA). Before entering the phase separation device, the mixture exiting reactor C was mixed with toluene and water using an external pump: this efficiently removed DMA and allowed isolation of the desired array of oxadiazoles **136** in the toluene phase, with a remarkable productivity of ≈0.5 g/h.

How many steps can be combined in a stream-line process? Jamison and Jensen in 2017 published the longest linear sequence of reactions telescoped in continuous flow without any workup interruption, known to that date [41]. A sequence of six chemical reactions in five distinct reactors was developed for the flow synthesis of Ciprofloxacin (Scheme 6.32): the first step was the acylation of vinylogous carbamate **138** with **137**, achieved in reactor A, followed by displacement of the dimethylamino group of **139** with cyclopropylamine **140** in reactor B. At this stage it was found that the presence of dimethylamine could interfere with the subsequent cyclization/nucleophilic substitution of **141**, therefore acetyl chloride was pumped together with the reaction mixture in reactor C, resulting in complete conversion of dimethylamine in DMA. Reactor D, heated at 180 °C, allowed conversion of **141** into **143**, mediated by piperazine **142**. Reactor E, finally, allowed the hydrolysis of the ethyl ester affording Ciprofloxacin sodium salt **144**. Sequential offline acidification and filtration afforded ciprofloxacin hydrochloride **145**. The overall yield of the seven-step sequence was 60%, comparable to batch and semi-batch syntheses, but with a total residence time of only 9 minutes, compared to the 24 hours required in batch. Moreover, no separation of intermediates was required throughout the synthesis.

6.6 Conclusions

The numerous examples illustrated in this chapter have highlighted the enormous potential of multicomponent approaches performed under flow conditions. The results obtained in terms of selectivity and yield are sometimes much better than the ones obtained in batch, while in other cases they are comparable. However, it is undeniable that flow approaches still allow a significant reduction in reaction times, as well as easy scale-up and simpler purification of the products. Moreover, various authors have demonstrated the improvement of green metrics (such as *E*-values and process mass intensity). It is important to underline that the success of continuous flow approaches is also due to a significant improvement in the technologies available, which allows the reactions to be conducted at high temperatures, for example with the aid of microwaves, at high pressures, with the use of back pressure regulators, with real-time monitoring of processes, through the coupling with analytical tools. As pointed out in the chapter, it is surprising that there are (still) very few examples of photochemical or gas/liquid heterogeneous phase reactions, but it is worth noting that these approaches are indeed widely used, but this chapter has investigated only a very limited portion of the literature, involving the use of multicomponent reactions to synthesize heterocycles.

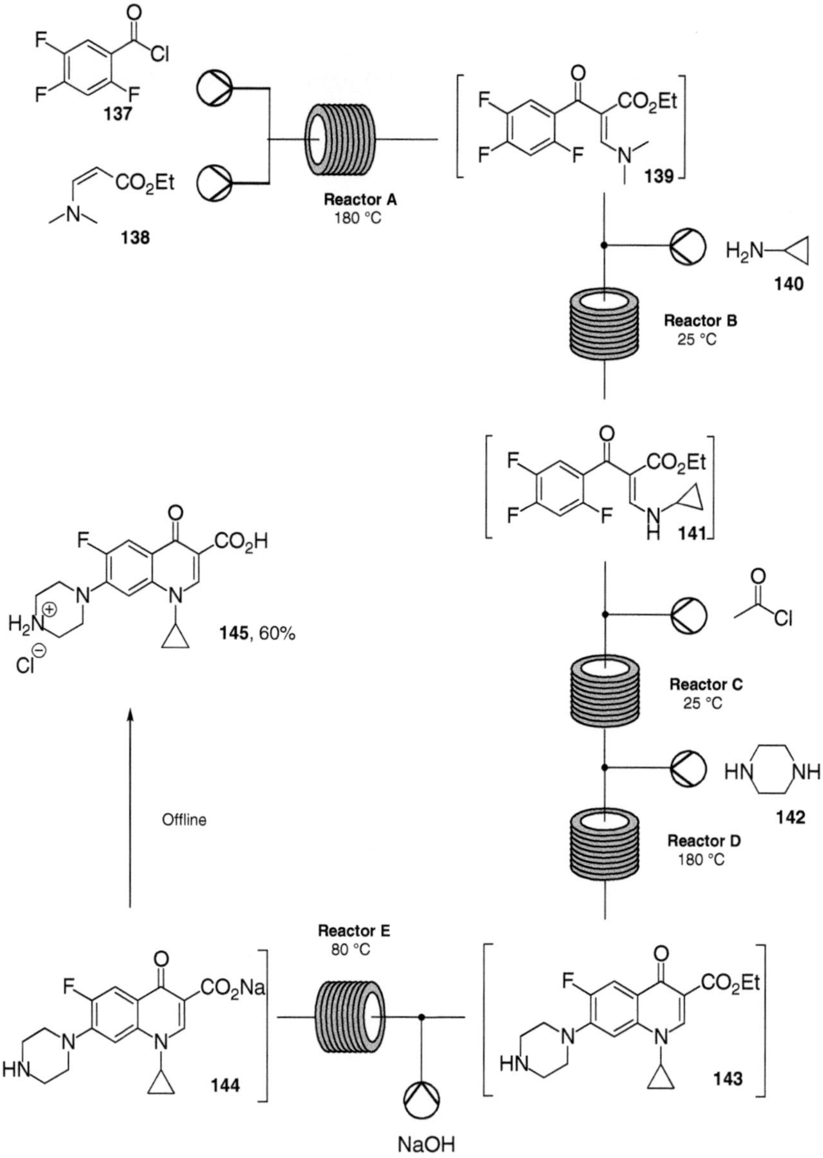

Scheme 6.32 Telescoped synthesis of Ciprofloxacin **145** in a sequence of six chemical reactions in five distinct reactors.

As we have witnessed that multicomponent reactions over the last few decades have gone from being a purely academic field to being widely used also by industry, so it is reasonable to expect that the combination of multicomponent reactions and flow chemistry will become more and more the prerogative of the industrial realm, not only for scale up but also in drug discovery processes.

References

1 Wegner, J., Ceylan, S., and Kirschning, A. (2011). Ten key issues in modern flow chemistry. *Chem. Commun.* 47: 4583–4592.

2 Plutschack, M.B., Pieber, B., Gilmore, K., and Seeberger, P.H. (2017). The Hitchhiker's guide to flow chemistry parallel. *Chem. Rev.* 117: 11796–11893.

3 Van Mileghem, S., Veryser, C., and De Borggraeve, W.M. (2018). *Flow-Assisted Synthesis of Heterocycles via Multicomponent Reactions*, vol. 56, 133–159. Springer International Publishing AG https://doi.org/10.1007/7081_2018_23.

4 Acke, D.R.J., Orru, R.V.A., and Stevens, C.V. (2006). Continuous synthesis of tri- and tetrasubstituted imidazoles via a multicomponent reaction under microreactor conditions. *QSAR Comb. Sci.* 25: 474–483.

5 Baraldi, P.T., Noel, T., Wang, Q., and Hessel, V. (2014). The accelerated preparation of 1,4-dihydropyridines using microflow reactors. *Tetrahedron Lett.* 55: 2090–2092.

6 Butler, A.J.E., Thompson, M.J., Maydom, P.J. et al. (2014). Regioselective synthesis of 3-aminoimidazo[1,2-a]-pyrimidines under cntinuous flow conditions. *J. Org. Chem.* 79: 10196–10202.

7 García, D., Jatczak, M., Muylaert, K. et al. (2013). Straightforward microwave-assisted synthesis of 5,8-disubstituted 5,6,8,9-tetrahydro-4H, 7H-2,5,6a,8,9a-pentaazaphenalene-1,3-diones. *Eur. J. Org. Chem.* 2013: 1732–1739.

8 Lei, M., Tian, W., Hu, R.J. et al. (2012). An efficient synthesis of 2-(trifluoromethyl)-2H-1,3 oxazino 2,3-a isoquinolines via a three-component cascade approach using a continuous-flow microreactor. *Synthesis-Stuttgart* 44: 2519–2526.

9 Vaddula, B.R., Yalla, S., and Gonzalez, M.A. (2015). An efficient and more sustainable one-step continuous-flow multicomponent synthesis approach to chromene derivatives. *J. Flow Chem.* 5: 172–177.

10 Pagano, N., Herath, A., and Cosford, N.D.P. (2011). An automated process for a sequential heterocycle/multicomponent reaction: multistep continuous flow synthesis of 5-(thiazol-2-yl)-3,4-dihydropyrimidin-2(1H)-ones. *J. Flow Chem.* 1: 28–U63.

11 Qiu, J.-K., Shan, C., Wang, D.-C. et al. (2017). Metal-free radical-triggered selenosulfonation of 1,7-enynes for the rapid synthesis of 3,4-dihydroquinolin-2(1H)-ones in batch and flow. *Adv. Synth. Catal.* 359: 4332–4339.

12 Cerra, B., Mostarda, S., Custodi, C. et al. (2016). Integrating multicomponent flow synthesis and computational approaches for the generation of a tetrahydroquinoline compound based library. *MedChemComm* 7: 439–446.

13 Martha, C.T., Heemskerk, A., Hoogendoorn, J.C. et al. (2009). High-throughput reaction optimisation and activity screening of ferrocene-based Lewis acid-catalyst complexes by using continuous-flow reaction detection mass spectrometry. *Chem. Eur. J.* 15: 7368–7375.

14 Acke, D.R.J. and Stevens, C.V. (2007). A HCN-based reaction under microreactor conditions: industrially feasible and continuous synthesis of 3,4-diamino-1H-isochromen-1-ones. *Green Chem.* 9: 386.

15 Monteiro, J.L., Pieber, B., Corrêa, A.G., and Kappe, C.O. (2016). Continuous synthesis of hydantoins: intensifying the Bucherer–Bergs reaction. *Synlett* 27: 83–87.

16 Lei, M., Hu, R., Wang, Y., and Zhang, H. (2014). Synthesis of 1,4-disubstituted-1,2,3-trizazoles via click reaction in micro flow reactor. *Heterocycles* 88: 1511–1517.

17 Li, D., Liu, L., Tian, Y. et al. (2017). A flow strategy for the rapid, safe and scalable synthesis of N-H 1, 2, 3-triazoles via acetic acid mediated cycloaddition between nitroalkene and NaN_3. *Tetrahedron* 73: 3959–3965.

18 Baxendale, I.R., Hayward, J.J., and Ley, S.V. (2007). Microwave reactions under continuous flow conditions. *Comb. Chem. High Throughput Screen.* 10: 802–836.

19 Bremner, W.S. and Organ, M.G. (2007). Multicomponent reactions to form heterocycles by microwave-assisted continuous flow organic synthesis. *J. Comb. Chem.* 9: 14–16.

20 Bálint, E., Tajti, Á., Ladányi-Pára, K. et al. (2019). Continuous flow synthesis of α-aryl-α-aminophosphonates. *Pure Appl. Chem.* 91: 67–76.

21 Tajti, A., Toth, N., Ravai, B. et al. (2020). Study on the microwave-assisted batch and continuous flow synthesis of N-alkyl-isoindolin-1-one-3-phosphonates by a special Kabachnik-Fields condensation. *Molecules* 25: 3307. https://doi.org/10.3390/molecules25143307.

22 Engen, K., Sävmarker, J., Rosenström, U. et al. (2014). Microwave heated flow synthesis of spiro-oxindole dihydroquinazolinone based IRAP inhibitors. *Org. Process. Res. Dev.* 18: 1582–1588.

23 Shore, G., Yoo, W.J., Li, C.J., and Organ, M.G. (2010). Propargyl amine synthesis catalysed by gold and copper thin films by using microwave-assisted continuous-flow organic synthesis (MACOS). *Chemistry* 16: 126–133.

24 Bagley, M.C., Fusillo, V., Jenkins, R.L. et al. (2013). One-step synthesis of pyridines and dihydropyridines in a continuous flow microwave reactor. *Beilstein J. Org. Chem.* 9: 1957–1968.

25 Baxendale, I.R., Hornung, C., Ley, S.V. et al. (2013). Flow microwave technology and microreactors in synthesis. *Aust. J. Chem.* 66: 131.

26 Ceylan, S., Coutable, L., Wegner, J., and Kirschning, A. (2011). Inductive heating with magnetic materials inside flow reactors. *Chem. Eur. J.* 17: 1884–1893.

27 Hartwig, J. and Kirschning, A. (2016). Flow synthesis in hot water: synthesis of the atypical antipsychotic iloperidone. *Chem. Eur. J.* 22: 3044–3052.

28 Ley, V., Baumann, M., Baxendale, R.I. et al. (2010). Synthesis of highly substituted nitropyrrolidines, nitropyrrolizines and nitropyrroles via multicomponent-multistep sequences within a flow reactor. *Heterocycles* 82: 1297.

29 Pelletier, S.M., Ray, P.C., and Dixon, D.J. (2009). Nitro-Mannich/lactamization cascades for the direct stereoselective synthesis of pyrrolidin-2-ones. *Org. Lett.* 11: 4512–4515.

30 Bianchi, L., Ballerini, E., Curini, M. et al. (2015). Aquivion PFSA as a novel solid and reusable acid catalyst in the synthesis of 2-pyrrolidin-2-ones in flow. *ACS Sustain. Chem. Eng.* 3: 1873–1880.

31 Silva, G.C.O., Correa, J.R., Rodrigues, M.O. et al. (2015). The Biginelli reaction under batch and continuous flow conditions: catalysis, mechanism and antitumoral activity. *RSC Adv.* 5: 48506–48515.

32 Cao, J. and Tian, H. (2018). N-Heterocyclic carbene-protected Ag nanoparticles immobilized on polyacrylonitrile fiber as efficient catalysts for a three-component coupling reaction. *Chem. Asian J.* 13: 1561–1569.

33 Rassokhina, I.V., Tikhonova, T.A., Kobylskoy, S.G. et al. (2017). Synthesis of imidazo[2,1-b]thiazoles via copper-catalyzed A(3)-coupling in batch and continuous flow. *J. Org. Chem.* 82: 9682–9692.

34 Anselmo, M., Moni, L., Ismail, H. et al. (2017). Photocatalyzed synthesis of isochromanones and isobenzofuranones under batch and flow conditions. *Beilstein J. Org. Chem.* 13: 1456–1462.

35 Fukuyama, T., Rahman, M.T., Kamata, N., and Ryu, I. (2009). Radical carbonylations using a continuous microflow system. *Beilstein J. Org. Chem.* 5: 34.

36 Haas, C.P. and Tallarek, U. (2019). Kinetics studies on a multicomponent Knoevenagel-Michael Domino reaction by an automated flow reactor. *ChemistryOpen* 8: 606–614.

37 Hoffmeyer, P. and Schneider, C. (2019). Continuous flow synthesis of highly substituted tetrahydrofurans. *Eur. J. Org. Chem.* 2019: 5326–5333.

38 Climent, M.J., Corma, A., Iborra, S., and Marti, L. (2014). Solid catalysts for multistep reactions: one-pot synthesis of 2,3-dihydro-1,5-benzothiazepines with solid acid and base catalysts. *ChemSusChem* 7: 1177–1185.

39 Damião, M.C.F.C.B., Galaverna, R., Kozikowski, A.P. et al. (2017). Telescoped continuous flow generation of a library of highly substituted 3-thio-1,2,4-triazoles. *React. Chem. Eng.* 2: 896–907.

40 Herath, A. and Cosford, N.D. (2017). Continuous-flow synthesis of highly functionalized imidazo-oxadiazoles facilitated by microfluidic extraction. *Beilstein J. Org. Chem.* 13: 239–246.

41 Lin, H., Dai, C., Jamison, T.F., and Jensen, K.F. (2017). A rapid total synthesis of Ciprofloxacin hydrochloride in continuous flow. *Angew. Chem. Int. Ed. Engl.* 56: 8870–8873.

7

C–H Functionalization as an Imperative Tool Toward Multicomponent Synthesis and Modification of Heterocycles

Alexey A. Festa and Leonid G. Voskressensky

Organic Chemistry Department, Science Faculty, Peoples' Friendship University of Russia (RUDN University), Miklukho-Maklaya st., 6, Moscow, 117198, Russia

7.1 Introduction

Functionalization of carbon–hydrogen bonds, the most abundant bonds in organic compounds, is challenging yet synthetically valuable. With no need to initially replace hydrogen with a reactive functional group, new opportunities for step- and atom-economical syntheses are opening. Given that multicomponent reactions (MCRs) also afford such opportunities, combining C–H functionalization with MCRs should be synergistic. Common methods to functionalize C—H bonds are based on transition metal insertion (C–H activation), free-radical formation (for instance, via hydrogen atom abstraction), electrochemical processes, or photoinduced processes. All these approaches inevitably generate reactive intermediates that can undergo the desired two-component transformations, but are also under development as excellent transmitters that involve a third component. In so doing, one attains maximum molecular diversity and complexity in a single step. This chapter analyzes a merging of C–H functionalization and MCRs, toward synthesis and modification of heterocyclic compounds. For more general reviews on C–H functionalization, refer to [1–3], and for recent reviews focused on other aspects of the field, refer to [4–11].

The scope of the chapter is limited to works where functionalization of nonactivated C—H bonds takes place. Because of mechanistic differences, we categorize the material as either transition-metal-involved or transition-metal-free. We divide these categories into reactions where one forms a new heterocyclic entity (synthesis of a heterocycle), or one functionalizes a heterocycle without formation of a new cycle (modification). We survey papers published from 2003 to December of 2020.

A promising development in the MCR field pertains to post-MCR C–H functionalization strategies; we do not cover these reactions. See 12–15 for selected examples.

Multicomponent Reactions towards Heterocycles: Concepts and Applications, First Edition.
Edited by Erik V. Van der Eycken and Upendra K. Sharma.

7.2 Transition-metal-involved C–H Functionalization

7.2.1 Multicomponent Synthesis of Heterocycles Through C–H Functionalization

In 2009, Zhu and coworkers [16] reported a Pd-catalyzed, sequential, three-component reaction toward synthesis of oxindoles **1** (Scheme 7.1). In the first step, the reaction between aryl bromide **2** and phenylpropiolamide **3** took place, followed by addition of aryl iodide. Xantphos is ordinarily a ligand of choice for *N*-arylation but inhibited the carbopalladation step. The ratio of Pd source and phosphorous ligand was proposed to be the key for the ability of the catalytic system to promote both arylation steps. The reaction worked smoothly for aryl bromides, bearing electron-withdrawing groups in the para-position. However, ortho or meta nitro-substituted aryl bromides did not participate in the reaction.

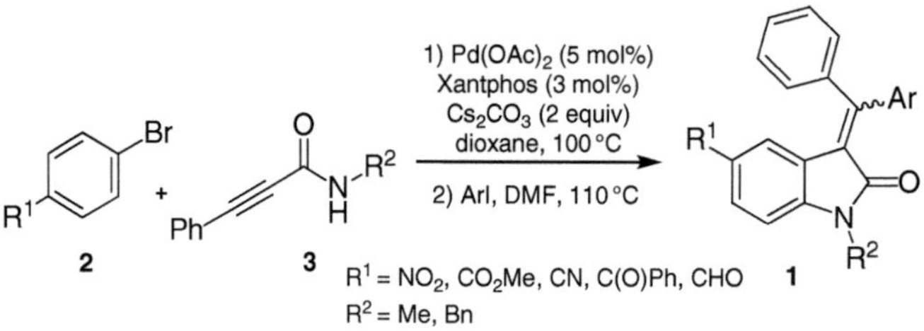

Scheme 7.1 Sequential three-component synthesis of oxindoles.

The transformation starts with *in situ* generated Pd(0) oxidative addition toward the C—Br bond and transmetalation of the resulting species with phenylpropiolamide anion, producing intermediate **4** (Scheme 7.2). Subsequently, carbopalladation of a triple bond is followed by C–H-insertion, delivering a palladacycle **5**, reductive elimination from which the target oxindole product **1** is yielded and the Pd(0) catalyst is recycled. Interestingly, the stereochemistry of the products does not perfectly correspond to the mechanism; the carbopalladation step is *syn* selective, whereas an *E*/*Z*-mixture forms. The authors explain this observation via isomerization of the intermediate prior to C–H activation.

In 2015, Hwang and coworkers [17] reported three-component synthesis of indoles **6** (Scheme 7.3). The reaction between anilines, acetylenes, and benzoquinones was catalyzed by CuCl (5 mol%) and carried out in MeOH under irradiation with a blue LED. The reaction proceeded through photo-irradiation of *in situ* generated copper acetylenide, oxidation of the Cu(I)-center to Cu(II), and recombination of the benzoquinone radical anion with Cu(II)-acetylenide. The resulting alkyne was reacted with aniline, followed by intramolecular Friedel–Crafts cyclization.

Lan, Li, and coworkers [18] reported a redox-neutral Rh(III)-catalyzed transformation of *N*-sulfonyl ketoimines **7** into isoindoles **8** (Scheme 7.4). Heating sulfonyl derivatives with an excess of acrylate esters in a sealed tube in DCE/AcOH under a nitrogen atmosphere delivered isondoles **8** (Scheme 7.4). Alternatively, the use

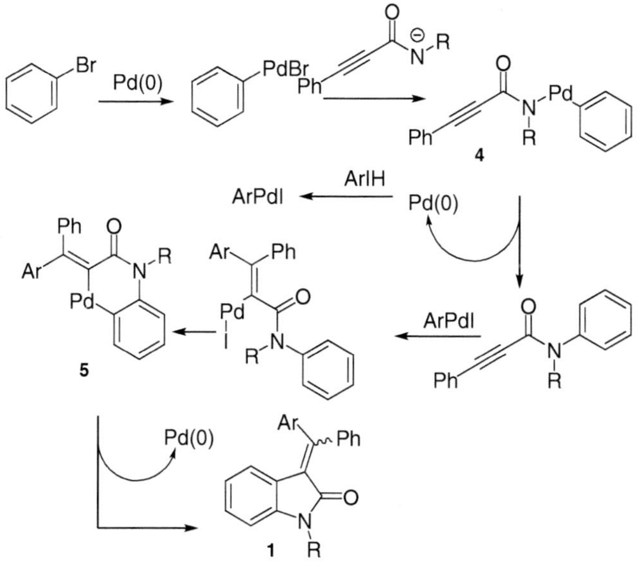

Scheme 7.2 Mechanistic proposal for oxindole synthesis.

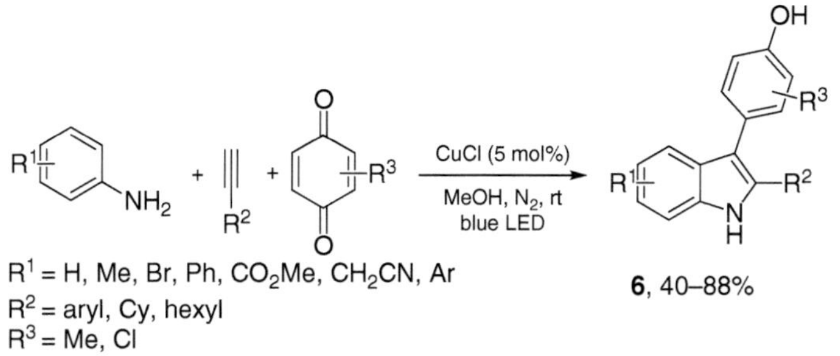

Scheme 7.3 Copper-catalyzed three-component synthesis of indoles.

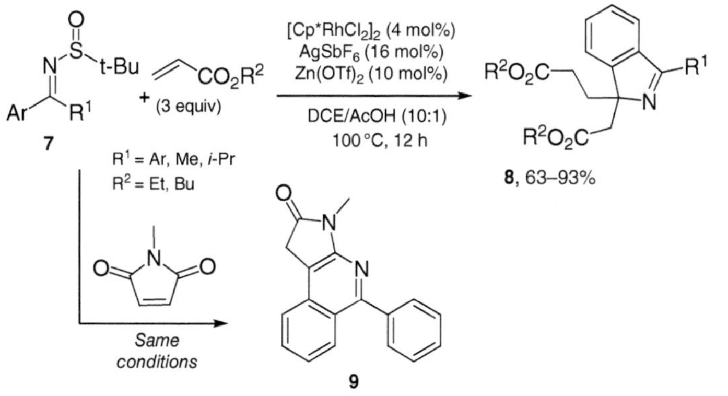

Scheme 7.4 Transformations of N-sulfonyl ketoimines.

of maleimides instead of acrylate facilitated formation of isoquinolines **9** in a two-component process. The reactions were catalyzed by $[Cp^*RhCl_2]_2$ (4 mol%), $AgSbF_6$ (16 mol%), and $Zn(OTf)_2$ (10 mol%). First, the *N*-sulfonyl imine group acted as a directing group, and then cleavage of the N—S bond delivered the final product **8**.

In 2016, Cheng and coworkers [19] reported the transformation of 2-phenylpyridines **10** into isoindolium salts **11** via a pseudo-three-component reaction (Scheme 7.5). They determined that the complex $[Cp^*Rh(MeCN)_3](BF_4)_2$ was the optimal catalyst and used water as the solvent. The reaction was successful in a pseudo-four-component reaction of complex polyheterocycle **12** [20]. In this case, silver acetate was an oxidant and a source of one carbon atom.

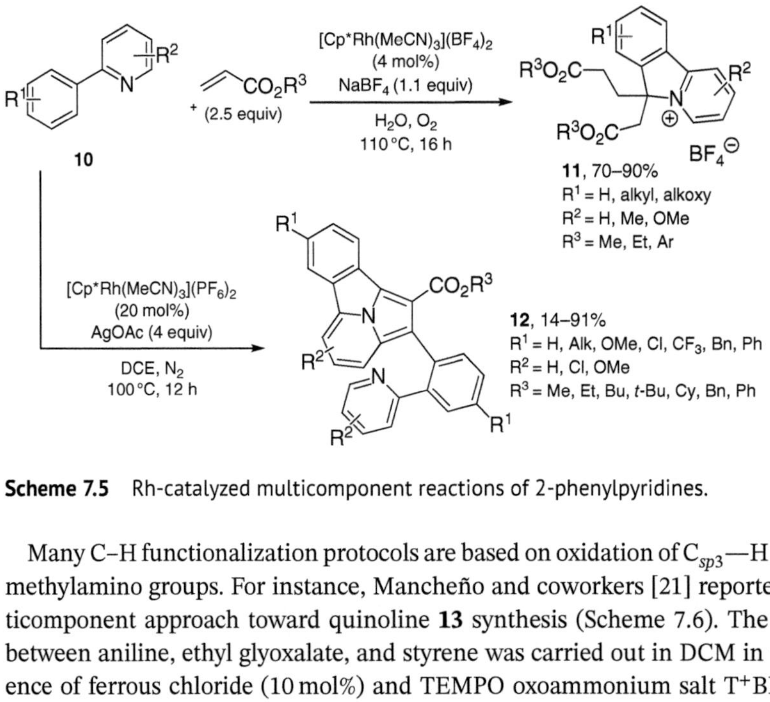

Scheme 7.5 Rh-catalyzed multicomponent reactions of 2-phenylpyridines.

Many C–H functionalization protocols are based on oxidation of C_{sp3}—H bonds in methylamino groups. For instance, Mancheño and coworkers [21] reported a multicomponent approach toward quinoline **13** synthesis (Scheme 7.6). The reaction between aniline, ethyl glyoxalate, and styrene was carried out in DCM in the presence of ferrous chloride (10 mol%) and TEMPO oxoammonium salt T^+BF4^-. The iminium intermediate **14** participated in a Povarov reaction, followed by dehydrogenation to furnish the resulting heterocycle. Although no organometallic intermediate is involved, the process might be considered as a formal CH-functionalization of a styrene. The scope of suitable anilines was wide, including electron-donating and electron-withdrawing substituents. When meta-substituted aniline was used, a mixture of 5- and 7-methyl quinolines was obtained. Different styrenes smoothly worked in the reaction, although 4-CF_3-substituted styrene gave the target product with a lower 55% yield. Octene-1 and norbornene (NBE) were also used for the reaction and successfully gave quinolines in moderate 36–50% yields.

One can achieve an effective three-component synthesis of quinolines **15** with the help of rhodium catalysis (Scheme 7.7) [22]. Anilines were reacted with alkynes and paraformaldehyde, the latter of which acted as a CO source. Under the action of

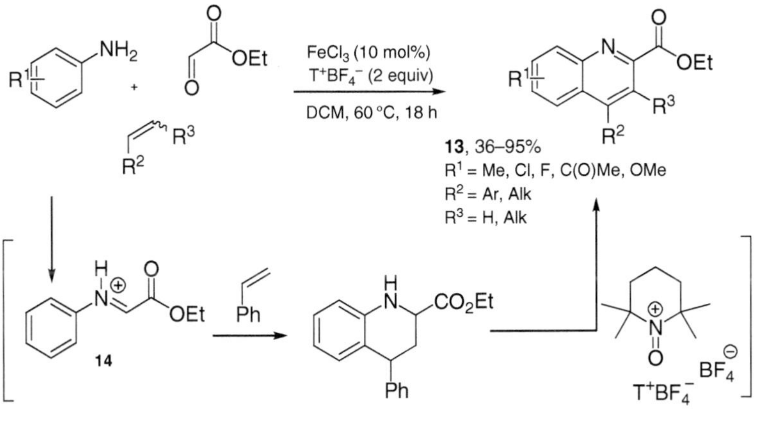

Scheme 7.6 TEMPO oxoammonium salt-mediated three-component synthesis of quinolines.

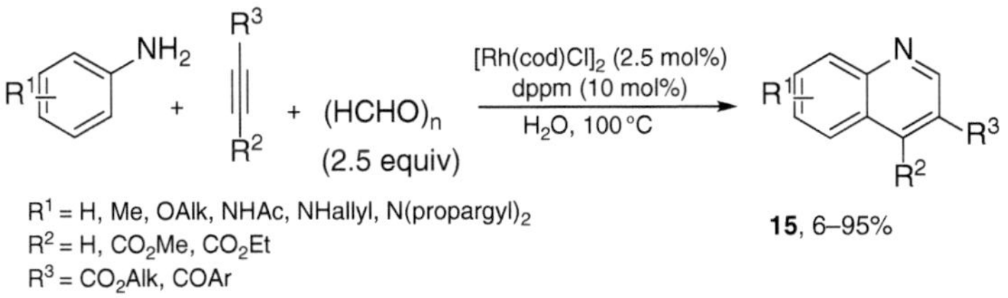

Scheme 7.7 Rh-catalyzed synthesis of quinolines from anilines, alkynes and paraformaldehyde.

Rh(I), paraformaldehyde underwent dehydrogenation, producing CO, which was trapped by the rhodium complex with the aniline and alkyne. Further CO insertion, protodemetalation, and elimination of H_2O delivered the final compounds in 6–95% yields.

Aryl ketoximes **16** are suitable substrates for a pseudo-four-component synthesis of isoquinolines **17** (Scheme 7.8) [23]. The reaction of 3,4-dihydronaphthalen-1(2*H*)-one *O*-methyl oxime with diphenyl acetylene (3 equiv) was carried out in the presence of rhodium catalyst $[Cp^*RhCl_2]_2$. The aryl moiety of ketoximes was able

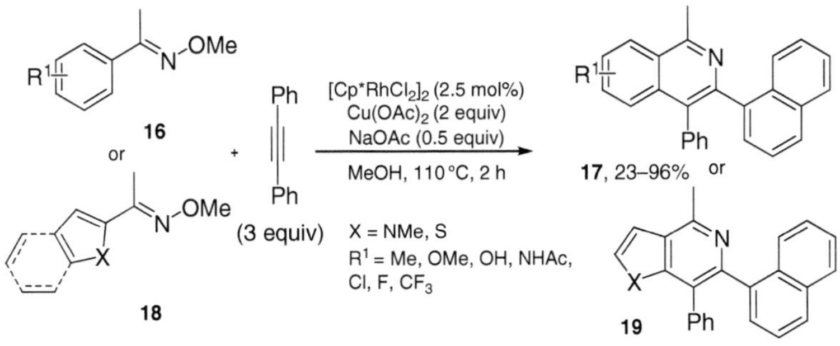

Scheme 7.8 Multicomponent reactions of O-methyl oximes.

to bear various functionalities: methyl, methoxy, hydroxy, acetylamino, chloro, fluoro, or trifluoromethyl groups. The use of heterocyclic ketoximes **18** was possible as suggested by using pyrrole, thiophene, and indole substrates. The resulting isoquinolines **17** were obtained in 23–96% yields, with the lowest 23% yield for *ortho*-methyl-substituted aryl ketoxime. The *O*-methyl oxime directing group was subsequently exploited for pyridine-annulation to porphyrins [24].

Kapur and coworkers [25] reported a three-component synthesis of isoquinolinones **20** via Ru(II)-catalyzed carbene migratory insertion (Scheme 7.9). They reacted oxazolinyl-substituted arenes **21** with diazomalonates and cesium acetate in the presence of $[RuCl_2(p\text{-cymene})]_2]$ (5 mol%) and $AgSbF_6$ (35 mol%) in refluxing DCE for 10–12 hours. A proton source was needed for the transformation to proceed efficiently, and the addition of $(PhO)_2P(O)OH$ was optimal. Various aryl oxazolines and diazomalonates were suitable for the reaction conditions, delivering target products with no considerable effect of the substitution on the reaction yield. When meta-substituted phenyl oxazoline was used, a mixture of regioisomers formed. Use of CsOPiv or CsF as nucleophile instead of acetate failed to facilitate the corresponding reaction, whereas CsOBz provided the target product. The reaction proceeds through an oxazolinyl-directed C–H activation, trapping of diazomalonate, migratory insertion, and formation of intermediate **22**. The latter is capable of intramolecular nucleophilic cyclization, followed by nucleophilic ring opening with an acetate anion.

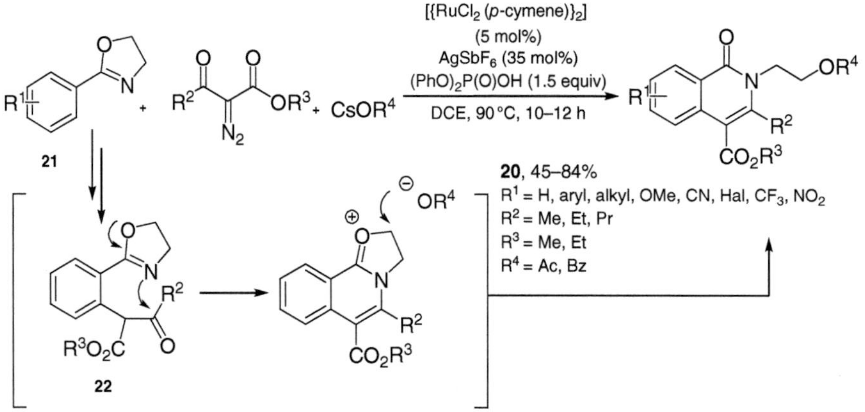

Scheme 7.9 Synthesis of isoquinolinones from oxazolines, diazomalonates and cesium carboxylates.

In 2020, Zhou, Yi, and coworkers [26] reported a divergent approach toward quinolones **23** (Scheme 7.10). Methyleneoxetanones **24** and *N*-acetyloxy amides **25** were subjected to a Rh(III)-catalyzed sequence of C–H activation/Lossen rearrangement reactions. The reactants were combined in MeOH at rt for 12 hours in the presence of [Cp*RhCl2]2 (2.5 mol%), potassium pivalate (1 equiv), and acetic acid (1 equiv). Alcohol participated in the reaction as a third component. Amides with electron-donating groups reacted better than amines with electron-withdrawing substituents. The oxetanones could bear various aliphatic

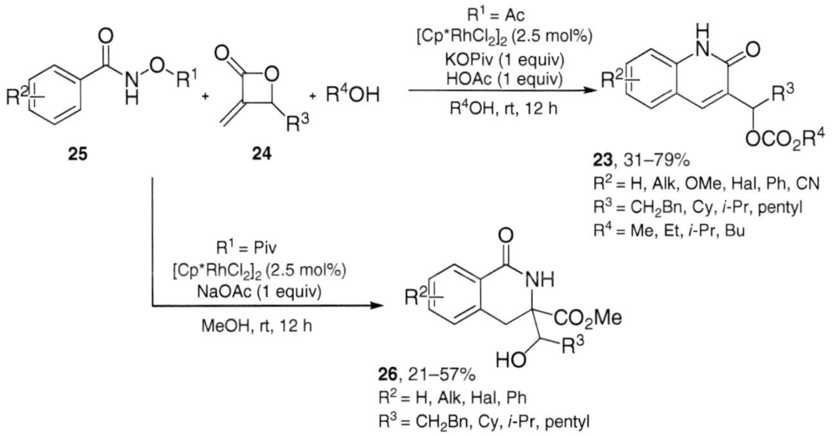

Scheme 7.10 Divergent reactivity of N-acetyloxy amides towards methyleneoxetanones.

substituents. Primary aliphatic alcohols were suitable reaction media and reactants, generating 2-quinolones **23** in 31–79% yields. Interestingly, when *N*-pivaloyloxy amides were used as a substrate, another pathway became a major reaction route, giving dihydroquinolones **26**. Although the scope of this route was comparatively wide, the yields in general were lower.

A carbonylative reaction of anilines with benzyne performed under dual Pd-catalysis delivered phenanthridinones **27** (Scheme 7.11) [27]. As a source of benzyne, *ortho*-trimethylsilyl phenyltriflate **28** was used. The most efficient generation of the reactive species took place in the presence of copper fluoride. Recycling of the Pd(II)-catalyst was achieved with stoichiometric quantities of copper acetate. The utility of the methodology was demonstrated on synthesis of a series of *crinasiadine*-type natural products.

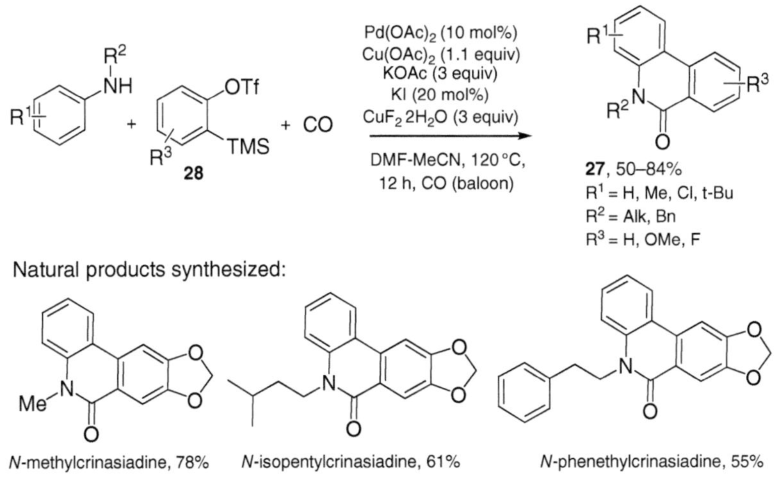

Scheme 7.11 A carbonylative interaction of anilines with benzynes.

You and coworkers [28] used *in situ* generated *N*-vinylpyridinium cations in Rh-catalyzed synthesis of polycyclic azinium salts **29** (Scheme 7.12). Pyridine and alkyne in 1,2-dichloroethane participated in a three-component interaction, catalyzed by 5 mol% $[Cp^*RhCl_2]_2$ in the presence of $Cu(OAc)_2$ (2 equiv) and potassium acetate (3 equiv). The reaction was performed for 24 hours at 150 °C under a nitrogen atmosphere. The transformation was general, hence one can employ various alkynes (e.g. unsymmetrical) and diverse azines. *N*-methylimidazole and *N*-methylbenzimidazole can undergo the present transformation, giving corresponding imidazole-fused heterocycles.

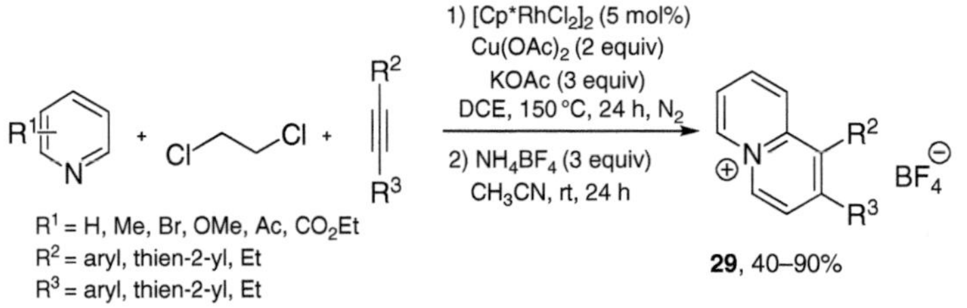

Scheme 7.12 Rh-catalyzed synthesis of polycyclic azinium salts.

Wu, Jiang, and coworkers [29] reported a versatile synthesis of pyrimidines **30** through an effective Pd-catalyzed pseudo-four-component reaction (Scheme 7.13). Reaction of amidine, styrene was performed and DMF was used both as a solvent and reactant at 120 °C for 24 hours in the presence of Pd(II) trifluoroacetate (5 mol%), Xantphos ligand (4,5-bis(diphenylphosphino)-9,9-dimethylxanthene (5 mol%), and *tert*-butyl hydroperoxide (TBHP, 3 equiv). As a result, 5-formamido pyrimidines **30** were obtained in 43–71% yields. Amidines could bear aryl moieties with methyl, methoxy, or halogen groups. Alkyl-substituted amidines were also tolerated. One can smoothly use nicotinamidine and isonicotinamidine in this transformation. Various styrenes, vinylnaphtalene, vinylpyridine, and 4-methyl-5-vinylthiazole gave desired products with moderate-to-good 47–71% yields, demonstrating the generality of the approach. One can also employ various substituted formamides. In this case, one unit of formamide was incorporated in the final product, whereas another unit of formaldehyde acted as a C1 synthon. The reaction pathway involves formation of an enamide **31**, a product of the interaction of DMF and styrene via acyl radical formation by TBHP. Then, enamide **31** coordinated with the Pd(II) species and subsequently underwent addition with the amidine, β-hydride elimination of palladium, and condensation with a C1 synthon **32**.

In 2020, Müller and coworkers [30] reported a consecutive, Pd-catalyzed, three-component reaction toward preparation of pyrazoles **33** (Scheme 7.14). In this one-pot methodology, a synthesis of aryl enals/enones was first attained through a Heck reaction of α,β-unsaturated carbonyl derivatives with aryl bromides. Further addition of tosylhydrazides and concentrated sulfuric acid (1 equiv), followed by heating with sodium hydroxide under microwave irradiation, delivered target heterocycles in 12–98% yields. A wide variety of aryl bromides smoothly reacted, favoring electron-withdrawing substitution to electron-rich substrates. One can incorporate

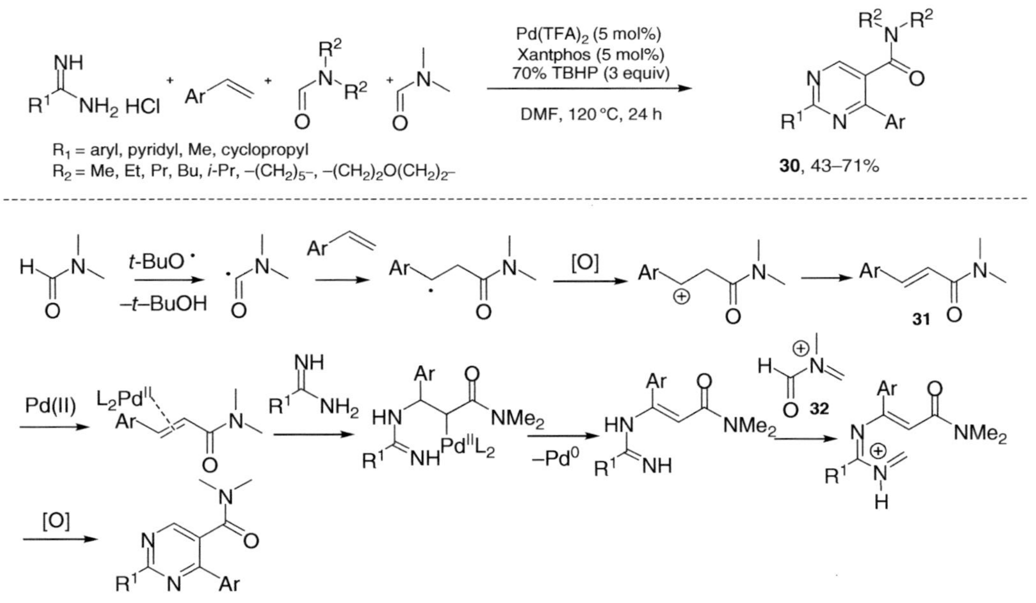

Scheme 7.13 Pseudo-four-component synthesis of pyrimidines.

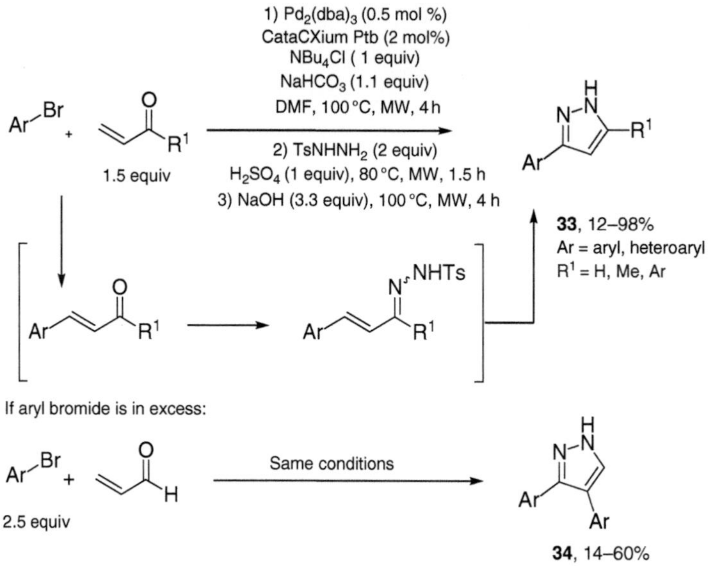

Scheme 7.14 A consecutive Pd-catalyzed thee-component synthesis of pyrazoles.

isoquinoline, thiophene, and dibenzothiophene bromides in the final pyrazoles in moderate yields. Interestingly, boronic ester-containing bromides successfully facilitated the desired molecules. One can use acrolein, methyl vinyl ketone, or aryl vinyl ketones as an alkene. When excess quantities of aryl bromides were used, double Heck-cyclization occurred, delivering nonaromatic 3,3-diaryl pyrazoles **34**, which further rearranged into stable aromatic 3,4-disubstituted isomers.

In 2019, Li and coworkers reported an intriguing route to perfluoroalkylated isoxazoles **35** via N—O bond formation (Scheme 7.15) [31]. Perfluoroalkyl iodides and styrenes were reacted with sodium azide and *tert*-butylhydroperoxide under Co(II)-catalysis in the presence of DABCO. The reaction was carried out in hexane/THF for 12 hours under a nitrogen atmosphere. The reaction started with Co(II)-induced formation of perfluoroalkyl radical, which added to a styrene double bond. Oxidation of the latter derived carbonyl derivative **36**. Two fluorine atoms in **36** were sacrificial, with one being eliminated as HF, and the other substituted by an azide anion. The resulted intermediate **37** underwent intramolecular cyclization with an extrusion of N_2 to furnish the final isoxazole. The scope of styrenes was wide, including complex alkenes with various functional groups. Vinyl heterocycles were also well tolerated, whereas internal and aliphatic alkenes were not suitable reaction partners. Acrylate derivatives did not undergo the transformation either. The perfluoroalkyl moiety was varied, revealing the possibility to chemoselectively generate isoxazoles with CF_2Br- or CF_2Cl-groups.

The approach was successfully expanded with the use of Togni's reagent instead of perfluoroalkyl iodide. In this case, synthesis of azide-substituted isoxazoles **38** occurred with a CF_3 group from Togni's reagent acting as a C1-synthon (Scheme 7.16).

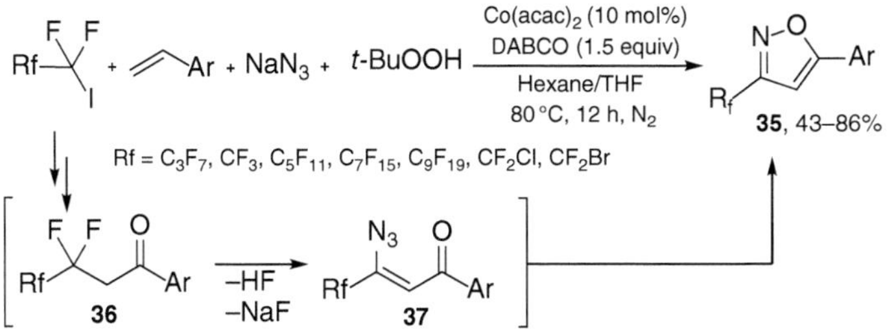

Scheme 7.15 A four-component synthesis of fluorinated isoxazoles.

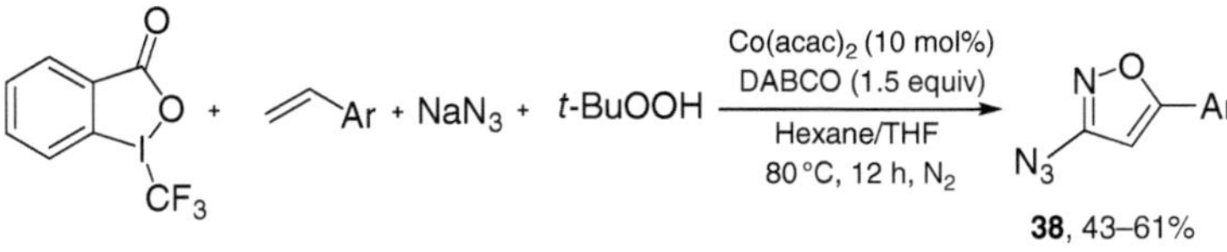

Scheme 7.16 Synthesis of isoxazolylazides employing Togni's reagent as a C1-donor.

Inn 2018, Li, Lu, Liu, and coworkers [32] reported an oxidative MCR toward synthesis of imidazo-fused heterocycles **39** (Scheme 7.17). Terminal alkynes, and allyl or benzyl bromides, were reacted with 2-amino-substituted heterocycles or amidines. The reaction was performed in the presence of copper (II) triflate (10 mol%), lithium carbonate (30 mol%), and TEMPO (30 mol%) in toluene at 100 °C for 12 hours under an oxygen atmosphere. Amino heterocycles afforded fused imidazoles, whereas amidines gave imidazoles **40**. The reaction proceeded through alkylation of an amino group, followed by oxidation to the imine, and copper-catalyzed addition of acetylenide. Then, copper-catalyzed intramolecular cyclization and oxidation took place to deliver the final products.

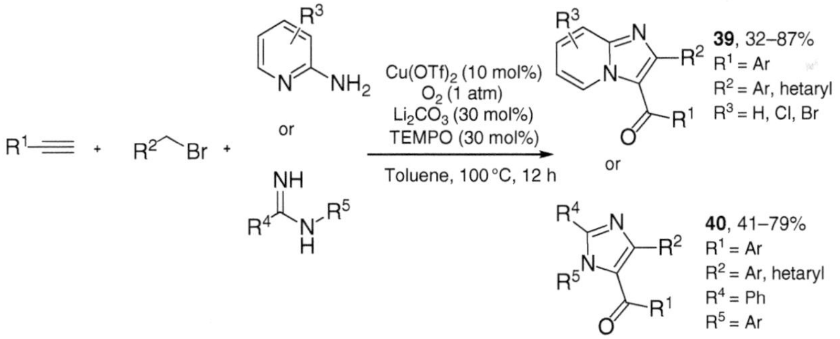

Scheme 7.17 Three-component reactions of alkynes, bromides and 2-amino-substituted heterocycles or amidines.

Another oxidative domino reaction employed Mn(III) or permanganate to facilitate extremely diverse formation of fused imidazoles [33]. Alternatively, a three-component reaction of pyridin-2-ones, acetophenones, and *O*-tosylhydroxylamine in ionic liquid led to arylimidazo[1,2-*a*]pyridines under Cu(II)-catalysis.

In 2013, Jiang, Wang, and coworkers [34] reported formation of valuable condensed benzazepines **41** in a pseudo-three-component manner (Scheme 7.18). They reacted isatins with alkynes (5 equiv) under catalysis with $Pd(OAc)_2$ (10 mol%) in the presence of stoichiometric oxidant AgOAc in a nitrogen atmosphere in MeCN/dioxane (1 : 1). The resulting benzazepines **41** were isolated in 55–96% yields. One can employ a wide scope of isatins, bearing various alkyl, halogen, or methoxy groups. When *N*-methyl or 7-chloroisatin was tested, no desired transformation took place, upholding the need of N—H and C(7)—H bonds. Interestingly, when naphthalene-based isatin **42** was employed, the reaction pathway changed and a series of unexpected pyrroloindoline-3-ones **43** was prepared under the same conditions. Regarding the alkyne component, one can use symmetrical and unsymmetrical alkynes. Addition to unsymmetrical alkynes followed Markovnikov's rule when the alkynes were strongly electron-rich or electron-deficient. In other cases, the majority of the product was governed by steric hindrance. Heteroaryl, alkyl, or ester-substituted alkynes were tolerated.

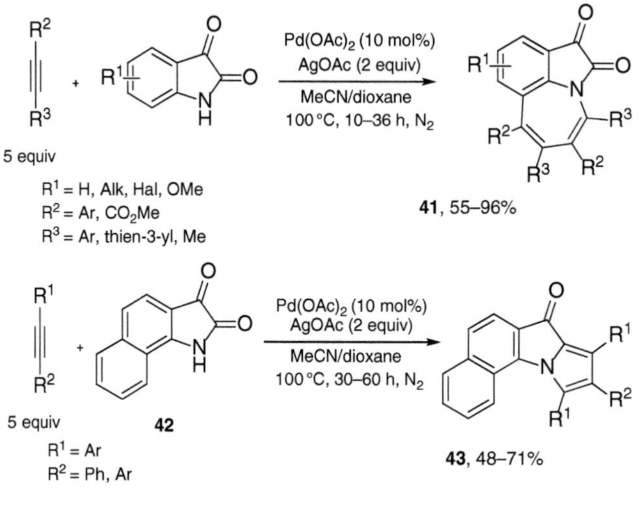

Scheme 7.18 Multicomponent Pd-catalyzed reactions of isatins and benzoisatins with alkynes.

The following mechanism was proposed for the transformation (Scheme 7.19). First, palladation of isatin takes place, giving intermediate **44**, capable of *syn*-addition to diphenylacetylene. The resulting vinyl palladium intermediate **45** reacts with another molecule of alkyne, followed by formation of palladacycloheptadiene **46**. Reductive elimination of Pd(0) delivers target product **41**, whereas the oxidant recycles Pd(0) to active Pd(II) species.

In 2013, Qian et al. [35] reported an efficient Cu-catalyzed synthesis of polycyclic molecules **47** (Scheme 7.20). They reacted bis(2-bromophenyl)methanone, alkyne, and 2-azidoacetamide in toluene in the presence of copper iodide, *N*,*N*'-dimethylethane-1,2-diamine (DMEDA), and cesium carbonate at elevated temperature for 60 hours. The resulting dibenzotriazolonaphtyridinones **47** were

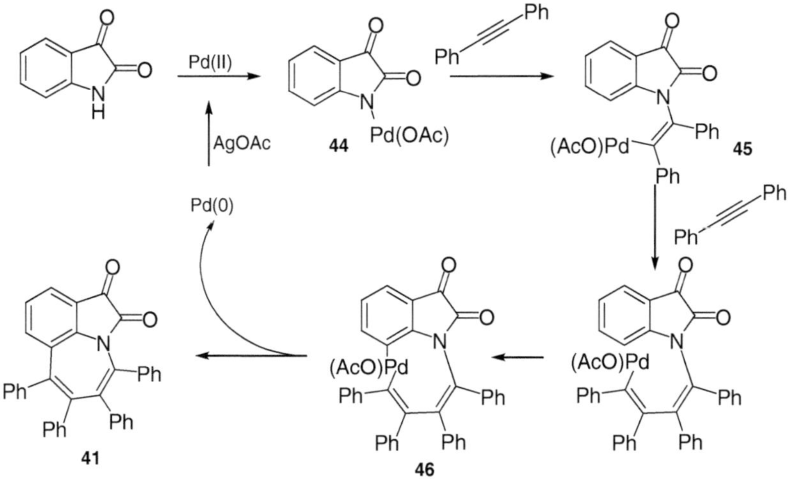

Scheme 7.19 Proposed mechanism for the synthesis of annulated benzazepines.

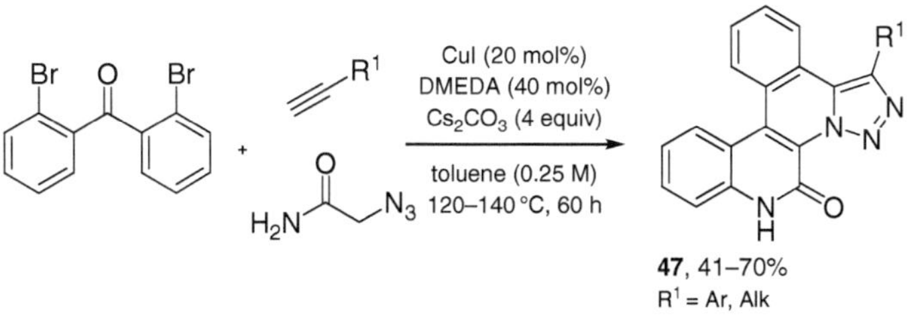

Scheme 7.20 Copper-catalyzed reaction of bis(2-bromophenyl)methanone, alkynes and 2-azidoacetamide.

prepared in 41–70% yields as an outcome of copper-catalyzed azide–alkyne cycloaddition, Goldberg intermolecular amidation, Camps cyclization, followed by a copper-catalyzed intramolecular C–H arylation reaction sequence.

Chuang, Cheng, and coworkers [36] reported an excellent method for preparing polyaromatic compounds based on multicomponent, rhodium-catalyzed, multiple C–H activation (Scheme 7.21). A pseudo-four-component interaction of (*Z*)-*N*-hydroxybenzamidines **48** with alkynes (3 equiv) was promoted by the $[Cp^*Rh(CH_3CN)_3](SbF_6)_2$ complex and needed $Cu(OAc)_2$ as an oxidant. The reaction was performed in *tert*-amyl alcohol for 16 hours and delivered benzo(isoquinolino)naphtyridines **49** with very good 60–86% yields. Various para-substituted benzamidines were tolerated, bearing electron-donating or electron-withdrawing groups. When meta-substituted substrate was tested, a mixture of regioisomers was obtained in a 60 : 40 ratio. Various substituted symmetric diarylalkynes worked smoothly. Interestingly, transformation of aliphatic alkynes was also possible, but the process required 48 hours to reach completion and resulted in a double C–H activation sequence, affording corresponding alkyl-substituted naphtyridines **50**. One can also synthesize analogous polycycles by electrochemistry [37].

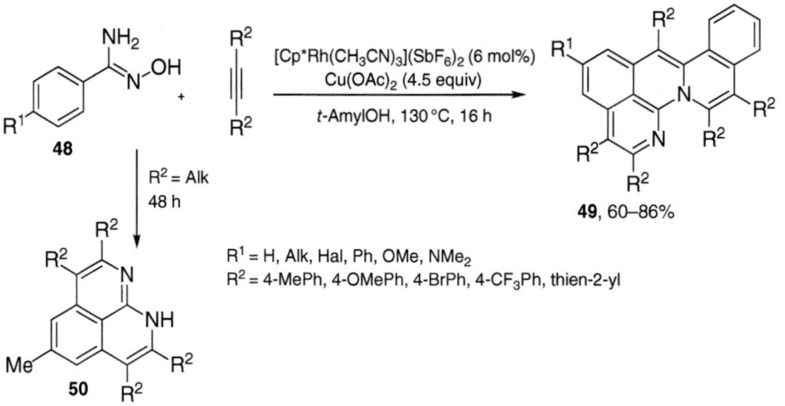

Scheme 7.21 Rh-catalyzed preparation of polyannulated heterocycles.

The authors proposed the following mechanism for the described transformation (Scheme 7.22). In the first step, nitrogen of the benzamidine coordinates to the Rh metal center and participates in ortho C–H activation. The rhodacycle **51** undergoes a coordinative insertion into the rhodium–carbon bond, followed by reductive elimination of Rh(III). Next, rhodium coordinates with the amine nitrogen and activates the C—H bond to provide rhodacycle **52**. Subsequent alkyne insertion, coordination of Rh to nitrogen, and C–H activation, followed by insertion of another unit of alkyne, complete the sequence.

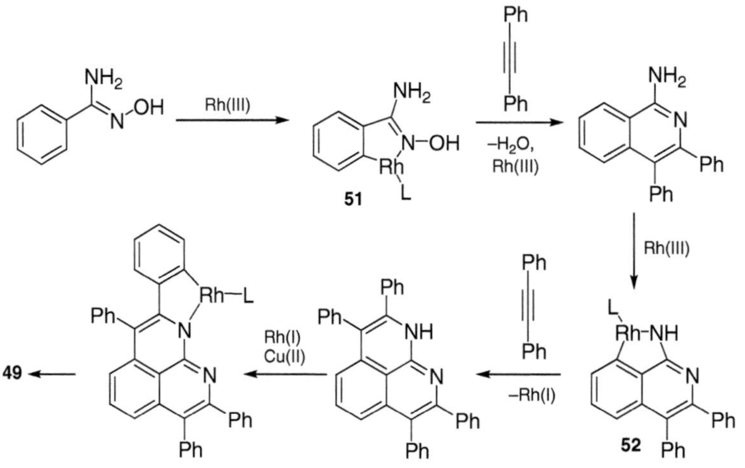

Scheme 7.22 Proposed mechanism for polyannulation of alkynes to benzamidines.

One can use benzamides [38] or benzonitriles [39] in analogous two- or three-fold alkyne annulations, toward polycyclic molecules with both nitrogen- and oxygen-containing heterocycles. Multiple annulation of alkynes to *N*-heterocyclic carbenes [40–42] delivers nitrogen-containing polycycles. One can synthesize oxygen polycycles through multiple alkyne annulations to naphthoquinones [43], naphthaldehydes [44], or benzoylacetonitriles [45]. Rhodium-catalyzed double annulation of diazo compounds to 2-phenylpyridine-3-carbonitriles [46] or benzoylacetonitriles [47, 48] is also feasible.

In 2016, Lautens and coworkers [49] reported an extremely effective route toward triazole-condensed benzazepines **53** (Scheme 7.23). The key to the success of a dual catalytic system is the use of Pd catalyst, precomplexed with phosphine ligand; i.e. the Herrmann–Beller palladacycle (HBP). Otherwise, phosphine coordinated copper iodide, inhibiting formation of the triazole.

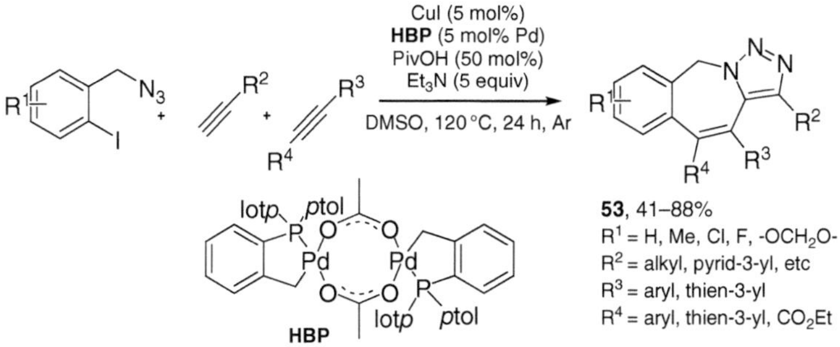

Scheme 7.23 Multicomponent synthesis of triazolobenzazepines.

Qian, Dong, and coworkers [50] reported a rhodium-catalyzed, three-component reaction of cyclic *N*-sulfonyl ketimines **54**, alkynes, and aldehydes, furnishing complex polycyclic molecules **55** (Scheme 7.24). To achieve the best yields, they refluxed the reactants in DCE for 40 hours under an argon atmosphere in the presence of 2.5 mol% $[Cp^*RhCl_2]_2$ catalyst, $AgSbF_6$ (10 mol%), and di-*tert*-butyldicarbonate (1 equiv) as additives. Interestingly, other anhydrides such as Ac_2O or TFAA could also facilitate the process, but less effectively. The scope of suitable aldehydes included both aromatic and aliphatic aldehydes, except for pyridine-2-carbaldehyde, which failed to give the desired product. Use of *N*-sulfonyl ketimines with various aryl substituents at C(3) was possible. Diarylacetylenes were most commonly used, but a smooth reaction was also demonstrated on butylphenylacetylene (R^3 = Bu).

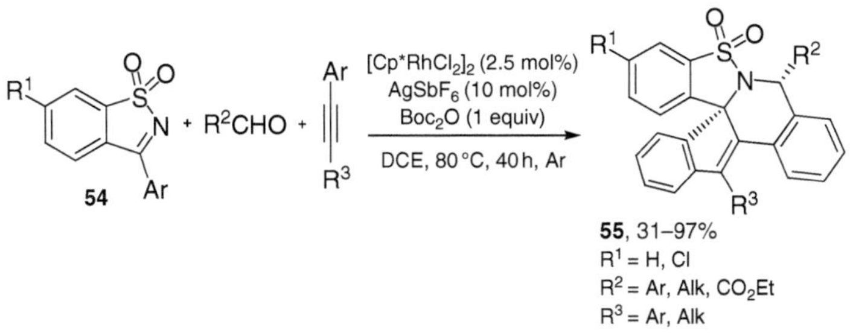

Scheme 7.24 Three-component reaction of cyclic N-sulfonyl ketimines, aldehydes, and alkynes.

In accordance with control experiments and literature reports, Scheme 7.25 shows the following mechanism for this domino transformation. First, the activated

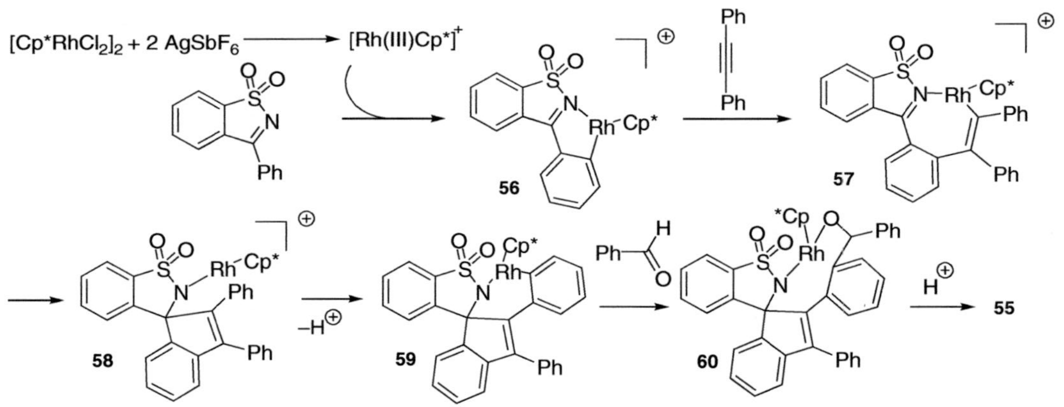

Scheme 7.25 Proposed mechanism for a three-component reaction of cyclic N-sulfonyl ketimines, aldehydes, and alkynes.

rhodium species coordinates with an imine nitrogen and undergoes C–H activation to form rhodacycle **55**. Then, insertion of an alkyne gives intermediate **56**, where C=N insertion generates intermediate **57**. The adjacent aryl group participates in a second C–H activation step, giving complex **58**. Insertion onto an aldehyde C=O bond gives intermediate **59**, which readily transforms into the final product, and rhodium active species are liberated for the next catalytic cycle. An alternative pathway suggests acylation of intermediate **60** with Boc_2O, followed by nucleophilic substitution to form the final product (not shown).

In 2013, Jiang and coworkers [51] reported an unexpected three-component synthesis of indolines **61**. Heating acetanilides **62** with propenol for 16 hours in methanol in the presence of 2 mol% $[Cp^*RhCl_2]_2$ catalyst, $AgSbF_6$ (10 mol%), and copper (II) acetate (2 equiv) delivered indolines **61** in 57–63% yields (Scheme 7.26). This transformation was successful in preparing indolines with methyl, bromo, and methoxy groups in various positions of the benzene ring.

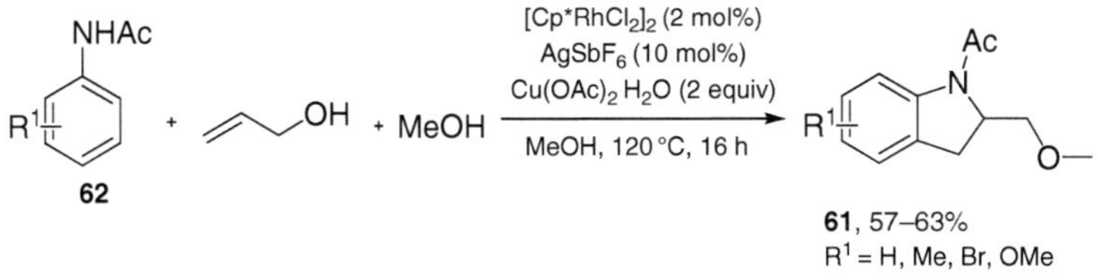

Scheme 7.26 Three-component synthesis of indolines.

In 2017, Huang and coworkers [52] elegantly expanded the methodology to a four-component synthesis of 2-alkoxytetrahydroquinolines **62** (Scheme 7.27). Anilines, carboxylic acid anhydrides, and propenol were heated in alcohol (the fourth component) in the presence of 2.5 mol% $[Cp^*RhCl_2]_2$ catalyst, $AgSbF_6$ (10 mol%), pivalic acid (2 equiv), and copper (II) acetate (2 equiv) under a nitrogen atmosphere to give quinolines **62** in 40–91% yields. The reaction was compatible

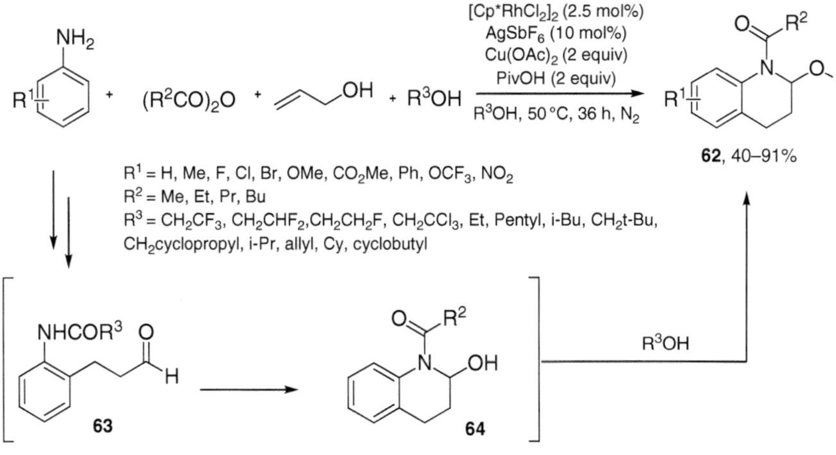

Scheme 7.27 Four-component synthesis of 2-alkoxytetrahydroquinolines.

with a broad scope of anilines, aliphatic anhydrides, and primary and secondary alcohols. Initially, acylation of aniline and ortho-directed oxidative alkylation generated key intermediate **63**, capable of intramolecular *semi*-aminal formation. Interaction of *semi*-aminal **64** with the alcohol delivered the final product.

In 2003, Lautens and coworkers [53] reported one of the first multicomponent syntheses of oxygen-containing heterocycles **65**, based on C–H functionalization (Scheme 7.28). The methodology is based on the Catellani reaction, but ortho positions of aryl iodide are functionalized with two different alkyl halides. One installs an alkyl halide on the aryl iodide prior to transformation and uses the other alkyl halide in excess. Annulation of five-membered rings is the most effective (85%), but one also can synthesize six- (62%) or seven-membered (69%) rings through this procedure. The reaction proceeds in the presence of $Pd(OAc)_2$ (10 mol%), PPh_3 (20 mol%), NBE (5 equiv), and Cs_2CO_3 (5 equiv). The key step of the reaction is formation of a palladacycle from aryl iodide, Pd(0), and NBE, which readily participate in oxidative addition with the external alkyl halide.

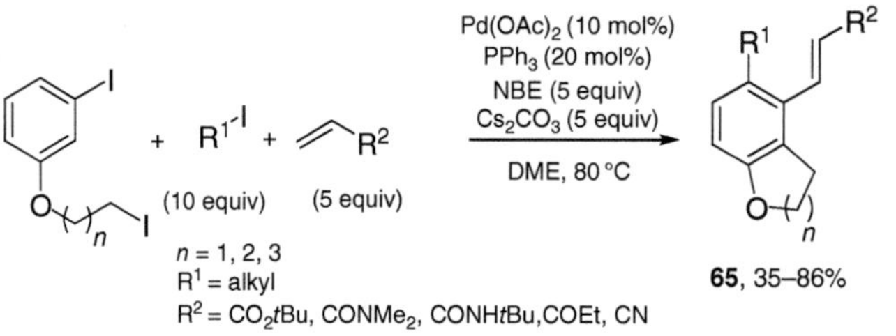

Scheme 7.28 Catellani reaction in the synthesis of oxygen-containing heterocycles.

In 2016, Gogoi and coworkers [54] reported a multicomponent transformation of isatins, based on C–H activation. In this work, isatins reacted with alkynes and carboxylic acids in the presence of 2.5 mol% $[\{RuCl_2(p\text{-cymene})\}_2]$ and CsOAc (20 mol%) to give 8-amido coumarins **66** in 51–85% yields (Scheme 7.29). No external oxidant was needed. Various alkynes were suitable. For instance, not only diaryl- or dialkylacetylenes worked, but also mixed substrates such as methylphenylacetylene, ethylphenylacetylene, or methoxymethylphenylacetylene smoothly generated the desired products. The more sterically hindered or electron-rich groups occupied the C(3) of coumarin. Terminal alkynes were incompatible with the reaction conditions.

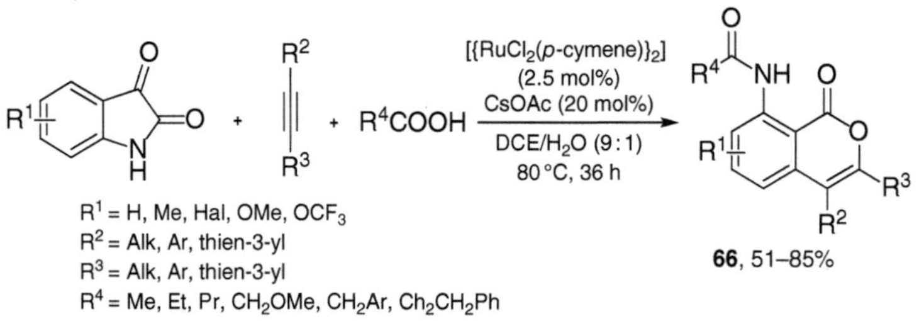

Scheme 7.29 Three-component reaction of isatins, alkynes and carboxylic acids.

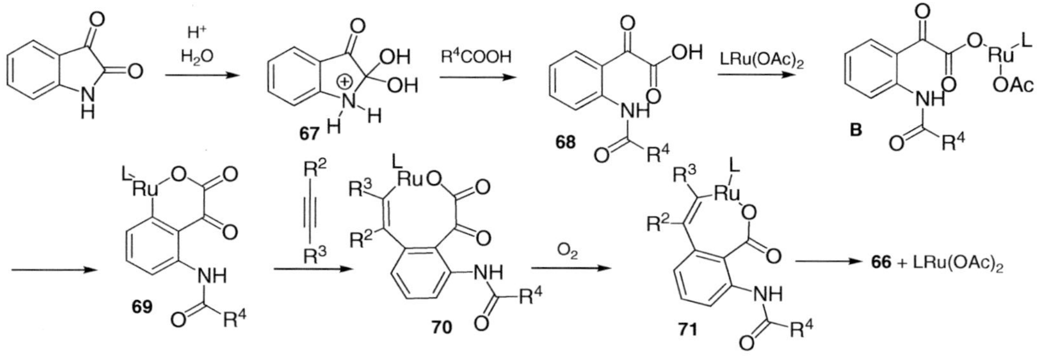

Scheme 7.30 Proposed mechanism for a three-component reaction of isatins, alkynes and carboxylic acids.

The scope of isatins was large: one can use methyl, methoxy, trifluoromethoxy, fluoro, chloro, or bromo-substituted substrates. Regarding the carboxylic acid counterpart, one can employ acetic, propanoic, butanoic, 2-methoxyacetic, or arylacetic acids. Aromatic acids were not reactive and failed to give the desired coumarins.

The reaction proceeds through an initial formation of 2-amido α-oxo phenylacetic acid **68**, through an intermediate **67** (Scheme 7.30). Then, Ru(II) acetate forms a complex with **68**, which undergoes C–H activation to give intermediate **69**. Subsequent alkyne insertion into **69** generates intermediate **70**. Further CO_2 release and ruthenium oxidation give complex **71**. Reductive elimination completes the catalytic cycle and gives the desired coumarin product.

One can synthesize 2-iminocoumarins **72** through a three-component Rh(III)-catalyzed reaction of thiocarbamates **73**, alkynes, and sulfonamides (Scheme 7.31) [55]. The reactants were heated in a sealed tube in the presence of $[Cp^*RhCl_2]_2$ (2.5 mol%), $AgSbF_6$ (10 mol%), and $Cu(OAc)_2$ (1 equiv) in DCE at 120 °C for 12 hours to deliver target 2-iminocoumarins **72** in 22–88% yields (Scheme 7.31). The thiocarbamate moiety can possess a wide variety of functional groups of electron-donating or -withdrawing nature. One can employ sulfonamides with alkyl groups, giving corresponding derivatives in 36% yield. When a 4-nitrophenyl sulfonamide was used, a diminished yield of 22% was obtained. Acetylenes with aromatic or aliphatic substituents were tolerated.

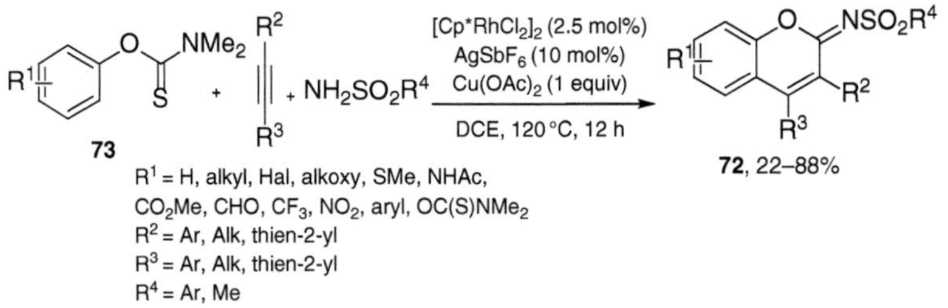

Scheme 7.31 Rh-catalyzed synthesis of iminocoumarins.

In 2020, Nishii, Miura, and coworkers [56] synthesized benzo[b]thiophenes **74** through a Rh(III)-catalyzed interaction of arylboronic acids, alkynes, and elemental sulfur (Scheme 7.32). Optimizing the reaction conditions revealed that one can achieve the best yield via a $[Cp^*Rh(CH_3CN)_3](SbF_6)_2$ (4 mol%) complex as the catalyst and silver acetate (3 equiv) as an oxidant. Mixing DMF with $PhCF_3$ improved the yield, probably due to the higher solubility of sulfur in the latter. Alkynes bearing aryl groups with electron-donating moieties gave higher yields than alkynes with electron-withdrawing substituents. When butylphenylacetylene was employed, an 88 : 12 mixture of regiosomeric benzothiophenes (R^2 = Bu or R^3 = Bu) was synthesized. For superior reproducibility, the scope of aryboronic acids was evaluated via boroxines. Para-substituted boroxines gave products with a group at C(6), and meta-substituted boroxines delivered groups at the C(5) position, activating the more sterically accessible C—H bond. The reaction failed with *ortho*-Br boroxine.

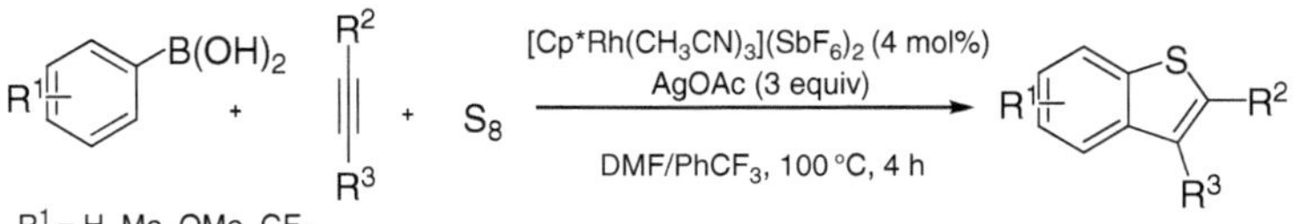

R^1 = H, Me, OMe, CF_3
R^2 = aryl, thien-2-yl
R^3 = aryl, thien-2-yl, Bu

74, 36–78%

Scheme 7.32 Three-component synthesis of benzo[b]thiophenes.

7.3 Transition-metal-involved C–H Functionalization

7.3.1 Multicomponent C–H Functionalization of Heterocycles

7.3.1.1 C(sp^2)-H Functionalization

An impressive enantioselective three-component functionalization of indoles with aryldiazoacetates and imines was achieved via dual rhodium and chiral phosphoric acid catalysis by Hu and coworkers [57]. The reaction was performed in toluene, rhodium (II) acetate and chiral phosphorous acid **PA** were identified as catalysts of choice to produce the desired compounds **75** in excellent yields and stereoselectivities (Scheme 7.33). It was found that the addition of molecular sieves was crucial to obtain high enantioselectivities. Indoles with methyl, methoxy, and bromo groups in the benzene ring were tolerated. Substitution in the diazo component had no significant effect on the reaction yield and selectivity either. Electron-rich imines gave poor enantioselectivity in comparison with electron-neutral or electron-withdrawing moieties. The key intermediate of this transformation was believed to be a zwitterionic compound **76**, undergoing an addition to a PA-activated imine. Later, rhodium catalysis was proved unsuitable for analogous three-component functionalization of pyrroles, producing either C–H insertion derivatives or poisoning the catalyst (for *N*-unprotected pyrroles). An allyl Pd(II) complex [{PdCl(η^3-C_3H_5)}$_2$] was identified as the best catalyst for the desired three-component reaction [58]. The stereodivergent outcome was achieved through

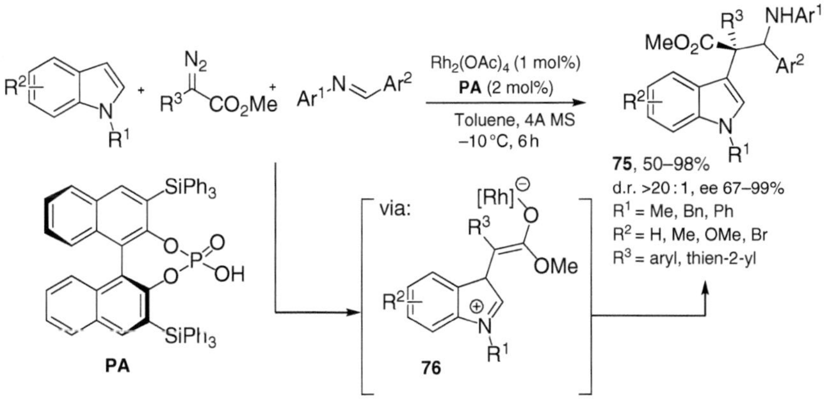

Scheme 7.33 Three-component functionalization of indoles with aryldiazoacetates and imines.

the use of various phosphorous acids. Thus, tri-iso-propyl-substituted phosphorous acid (R^4 = 2,4,6-*i*-PrC_6H_4) was promoting a syn-selective reaction, while the use of triphenylsilyl-derived catalyst (R^4 = SiPh_3), combined with L-tartaric acid, was favoring *anti*-selective process.

The Hu's work had been granted an excellent expansion by Gong and coworkers in 2014 [59]. The idea was to trap a zwitterionic intermediate analogous to **76** with a nitroalkene, instead of an imine. Indeed, an interaction between indoles, diazooxindoles, and nitroalkenes successfully furnished 3,3′-bisindoles **77** in a highly stereoselective manner under the catalysis with ruthenium (II) complex and chiral squaramide derivative (Scheme 7.34). The reaction was performed in DCM/Et_2O mixture and target molecules were prepared with 64–98% yields and up to 99% ee. The methodology was used for the synthesis of intermediate in (−)-folicanthine alkaloid.

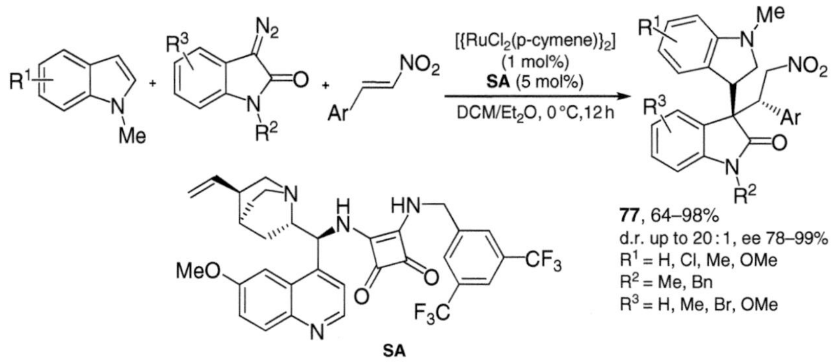

Scheme 7.34 Three-component functionalization of indoles with aryldiazoacetates and nitroalkenes.

An annulation of a cyclopentane ring to an indole in a diastereoselective fashion became the next development in the field [60]. Firstly, the three-component Rh(II)-catalyzed interaction of indole, aryldiazoacetate, and unsaturated α-keto ester delivered enol **78** as a result of Michael addition of rhodium intermediate **76** to the latter (Scheme 7.35). In the next step, the enol **78** was subjected to a

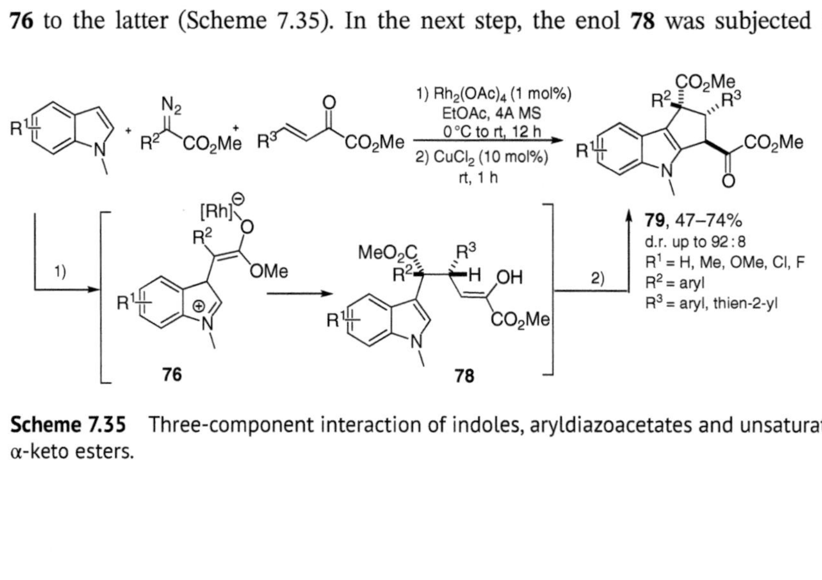

Scheme 7.35 Three-component interaction of indoles, aryldiazoacetates and unsaturated α-keto esters.

copper(II)-catalyzed dehydrogenative coupling, furnishing cyclopenta[b]indoles **79** in 47–74% yields. Various *N*-methylindoles were tolerated, as well as keto esters. Diazoacetates were limited to aryl substituted (R^2 = Ar), as far as the reaction with ethyl diazoacetate failed. When 1 equiv of $CuCl_2$ was used under oxygen-free conditions, the target product was obtained with 50% yield, supporting the idea of air being the terminal oxidant, while Cu(II) was converted into Cu(I), but not Cu(0) in the course of the reaction. This fact was also upheld by X-ray photoelectron spectroscopy.

A carbonylative strategy toward directed arylation of electron-rich heterocycles was developed by Beller and coworkers [61]. Pyridine was used as a directing group, and a series of pyrrolyl, furyl, and thienyl-based substrates and their benzannulated analogs were treated with aryl iodides in the presence of $[RuCl_2(cod)]_n$, KOAc (20 mol%), $NaHCO_3$ (2 equiv) under atmosphere of CO (30 bar) (Scheme 7.36). The carbonylated products **80** were isolated with 10–91% yields. In case of C(2)-selective arylation (pyridine moiety at C(3)), the best result of 91% was achieved for 2-iodotoluene, and in general, ortho-substituted iodoarenes were giving superior yields. Unexpectedly, when more substituents were introduced into iodoarene, the yields were substantially diminished. For instance, 2-iodomesitilene delivered only trace amounts of the corresponding product. In case of C(3)-selective arylation (pyridine moiety at C(2)), the ortho-effect was not so pronounced and the respectful molecules were synthesized in moderate yields. In general, furans were reacting worse than thiophenes. Moreover, anisole failed to give the desired product with the furan substrate. *N*-Pyridyl or *N*-pyrimidyl pyrroles or indoles could be arylated at C(2) under analogous conditions as well, delivering desired aryl heteroaryl ketones **81** with moderate-to-good yields 45–75%. Strongly chelating substrates, such as

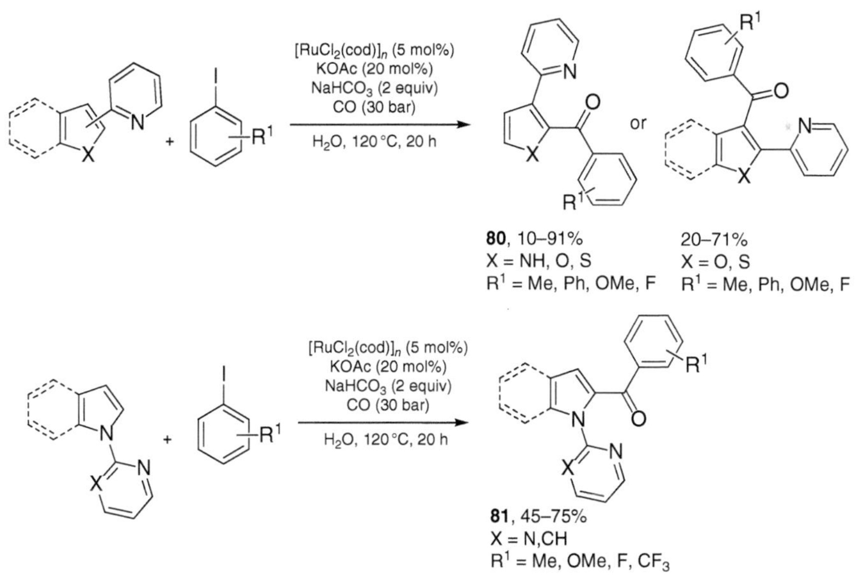

Scheme 7.36 A carbonylative functionalization of pyridyl-substituted heterocycles.

2-(1*H*-pyrazol-1-yl)pyridine and 2-(1*H*-imidazol-1-yl)pyridine, failed to undergo the process.

Later, aminocarbonylation of pyridyl-substituted heterocycles was developed by Driver and coworkers [62]. This palladium-catalyzed three-component reaction employed nitroarenes as the nitrogen source and $Mo(CO)_6$ as both reductant and source of CO (Scheme 7.37). The transformation was carried out in DCE at 120 °C for 12 hours. It was shown that the addition of PivOH (0.5 equiv) was necessary to promote the C—H bond cleavage. The reaction was tested on *N*-pyridin-2-yl pyrroles and indoles, and it was also successful with (thien-2-yl)pyridines. The nitroarene component was evaluated in the reaction with 2-phenylpyridine, and it was found that substrates with electron-withdrawing substituents were superior to the nitroarenes with 4-methoxy or 4-methyl moieties.

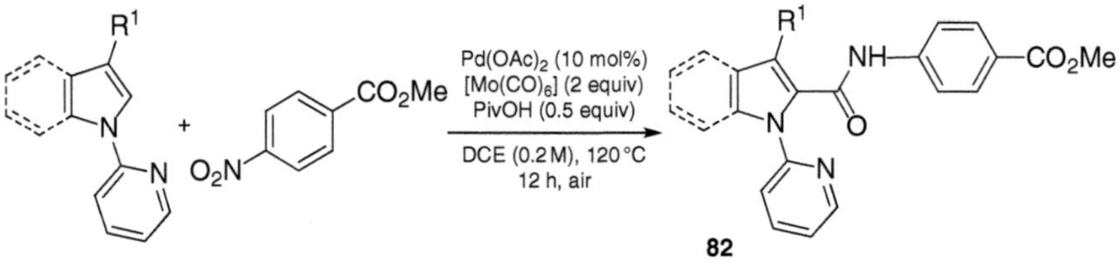

82

Scheme 7.37 Aminocarbonylation of pyridyl-substituted heterocycles.

A one-pot installation of an 8-aminoquinoline directing group followed by a C–H activation step was successfully realized by Wan an coworkers in 2015 [63]. For this transformation to proceed, 8-aminoquinoline, benzoyl chloride, and aryl iodide (3 equiv) were refluxed in xylene for 12 hours in the presence of $Pd(OAc)_2$ (3 mol%) and K_2CO_3 (2 equiv) (Scheme 7.38). As a result, diarylated benzamides **83** were synthesized in 48–95% yields. Lower yields were obtained for 4-acetyliodobenzene, while employment of other aryl iodides or para-substituted benzoyl chlorides led

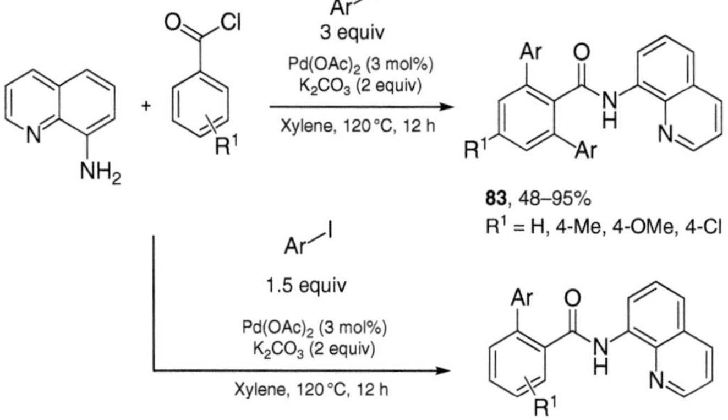

Scheme 7.38 Three-component reaction of 8-aminoquinoline, benzoyl chlorides, and aryl iodides.

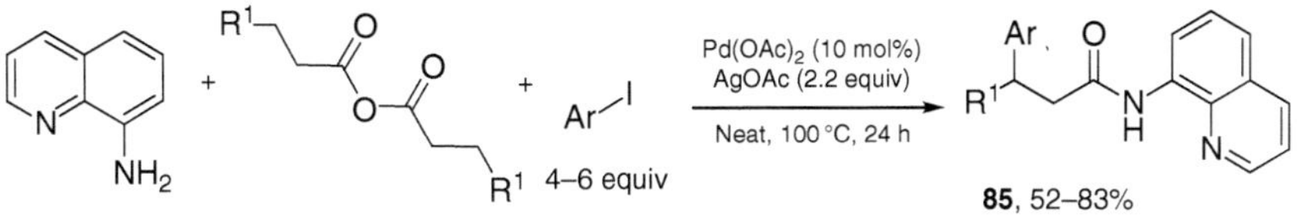

Scheme 7.39 Three-component reaction of 8-aminoquinoline, anhydrides of aliphatic carboxylic acids, and aryl iodides.

to a smooth conversion. When meta- or ortho-substituted benzoyl chlorides were used, monoarylation was becoming a favored process, giving products **84**.

Interestingly, when anhydrides of aliphatic carboxylic acids were used in combination with 8-aminoquinoline and aryl iodide, C(sp^3)-H functionalization toward **85** took place (Scheme 7.39) [64]. Silver salt could be replaced with K_2CO_3 or KOAc, as it was demonstrated in several examples. In case of propionic anhydride (R^1 = H), the bis β-arylated product was formed in majority.

Amides of 8-aminoquinolines could be modified at C(5) through a remote C–H functionalization strategy [65]. Thus, when 8-aminoquinoline derivatives, styrenes, and difluoromethyl bromides were subjected to Ru catalysis with [Ru(*p*-cymene)Cl_2]$_2$ (10 mol%), 5-alkylated quinolines **86** were produced in moderate-to-good yields (Scheme 7.40). The reaction mixture was heated in PhCl for 24 hours in the presence of phosphine ligand (P(4-$C_6H_4CF_3$)$_3$, 40 mol%), *N*-acetylvaline (30 mol%), and PivOK (3 equiv). The choice of the ligand was critical for the success of the reaction, hence PPh_3, PCy_3, Ruphos, and Dppe did not facilitate the process. Various amides could be transformed into products with moderate yields, but the lowest yields were for ortho-substituted aryl in R^1. When isoquinoline ring had a group in C(2) or (C3), the yields were also lower. Unexpectedly, when quinoline with a methoxy group at C(6) was employed, the target C(5)-functionalized product was obtained with 68% yield. In addition to 2-bromo-2,2-difluoroester and 2-bromo-2,2-difluoroamides, perfluoroalkane and bromo ethyl carboxylates worked well in the transformation.

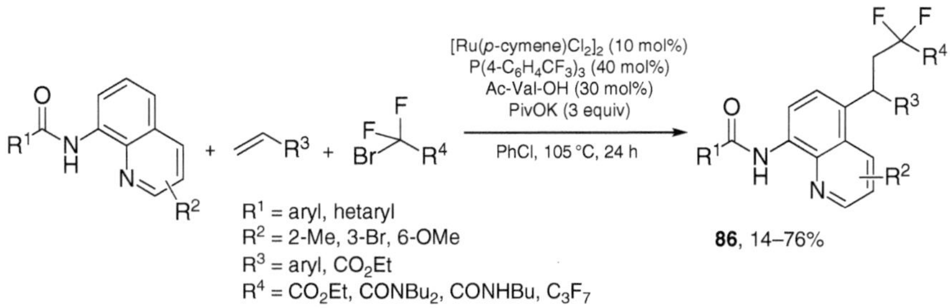

Scheme 7.40 Ru-catalyzed synthesis of difluoromethyl-substituted aminoquinolines.

An efficient three-component 1,4-carboamination of conjugated dienes with arenes and dioxazolones was developed by Glorius and coworkers [66]. A solitary example of heterocyclic C–H functionalization was demonstrated on an indole. Stirring an indole with an amide directing group with the diene and *tert*-butyl-substituted oxazolone under the action of $[Cp^*RhCl_2]_2$ (5 mol%), $AgBF_4$ (20 mol%), and AgOAc (50 mol%) in acetone furnished the alkylated indole **87** with 74% yield (Scheme 7.41). The regioselectivity was excellent, and 1,2-carboamination product was not noticed. Oxazolone acted as an available and cheap nitrogen source.

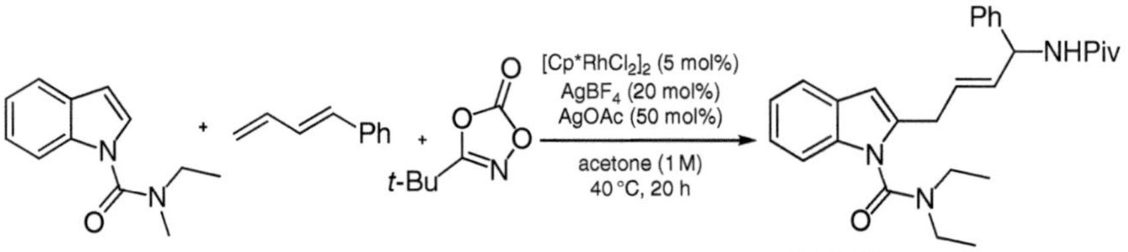

Scheme 7.41 Three-component 1,4-carboamination of conjugated dienes.

A mild method for Friedel–Crafts alkylation of π-excessive heterocycles was developed by Xu and coworkers [67]. The alkylating agent, *N*-alkenoxypyridinium salt, was generated *in situ* from *N*-pyridine oxide **88** and alkyne (Scheme 7.42). The reaction was carried under Au(I)-catalysis in 1,1,1,3,3,3-hexafuoroisopropanol (HFIP) at rt, then heteroarene was added, and the mixture was heated for 4–6 hours. Pyrroles, furans, and thiophenes were alkylated at C(2), while indoles were functionalized at C(3).

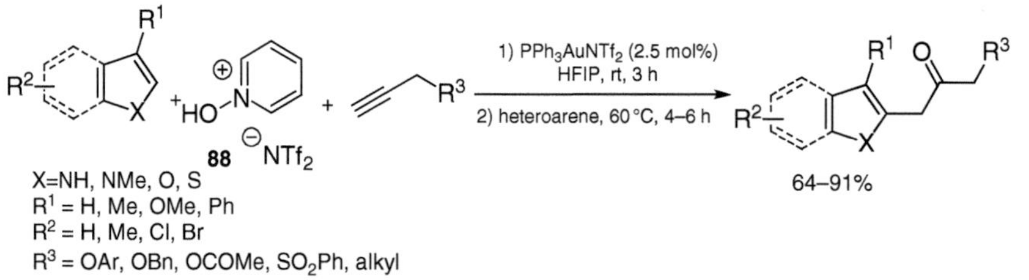

Scheme 7.42 Friedel–Crafts alkylation of π-excessive heterocycles.

Pyridine was an effective directing group for cobalt(III)-catalyzed three-component functionalization of isoquinolinones [68]. Stirring *N*-pyridin-2-ylisoquinolinone **89**, ethyl vinyl ketone, and an aldehyde in the presence of $[Cp^*Co(C_6H_6)][B(C_6F_5)_4]_2$ (10 mol%), LiOAc (20 mol%) in dioxane at rt for 20 hours delivered alkylated heterocycles **90** with good and very good yields and high diastereoselectivity (Scheme 7.43). The reaction worked successfully with both aromatic and aliphatic aldehydes. Later, conjugated dienes were used for the functionalization of arenes and aldehydes, and the individual example of heteroarene was a thiophene substrate [69]. Interestingly, heterocyclic ketones – azetidinones, oxetanones, and isatins, were successfully reacting with arenes and dienes under similar conditions as a carbonyl component [70].

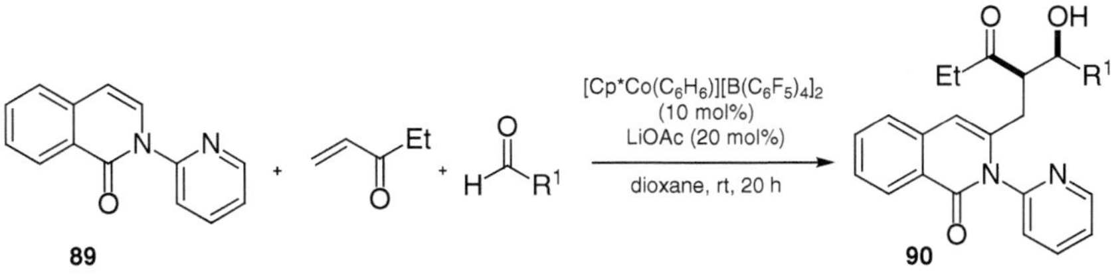

R^1 = Ph, 82%, d.r. 95 : 5
R^1 = Cy, 67%, d.r. 98 : 2

Scheme 7.43 Co(III)-catalyzed three-component functionalization of isoquinolinones.

A pseudo three-component functionalization of benzobisthiazoles **91** was developed in 2014 [71]. The reaction of benzobisthiazole with excess amounts of (het)aryl bromide was co-catalyzed by Pd(II) acetate (1 mol%) and copper (II) acetate (20 mol%) (Scheme 7.44). Heating the reactants in DMF in the presence of PPh_3 (0.5 equiv) and K_2CO_3 (2 equiv) for 18 hours allowed the synthesis of double arylation products **92** with 15–81% yields. Electron-neutral bromo arenes were converted into the product with very good yields. The employment of electron-rich substrates such as bromoanisole or 2-bromothiophene resulted in lower 38–49% yields. In its turn, the electron-poor bromides were converting smoothly. Interestingly, 2-bromo and 3-bromopyridine were transformed in 66–58% yields, respectively.

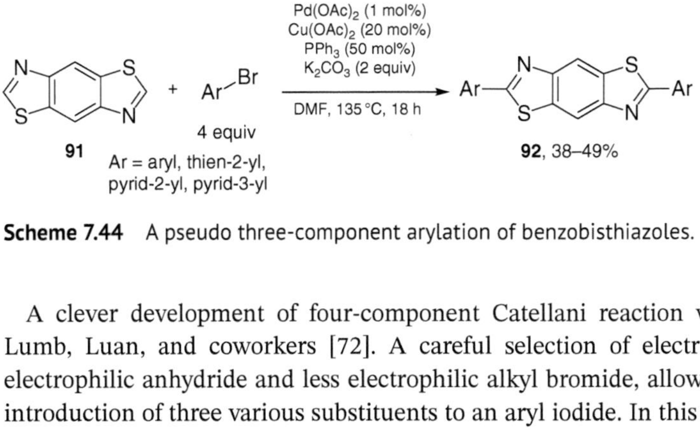

Scheme 7.44 A pseudo three-component arylation of benzobisthiazoles.

A clever development of four-component Catellani reaction was realized by Lumb, Luan, and coworkers [72]. A careful selection of electrophiles – more electrophilic anhydride and less electrophilic alkyl bromide, allowed the selective introduction of three various substituents to an aryl iodide. In this transformation, aryl iodide underwent an oxidative addition with Pd (0), followed by an addition of NBE and C–H activation step. Further sequential reaction with benzoic anhydride, alkyl bromide, and an alkene delivered the desired products **93**. A solitary example of heterocyclic C–H functionalization, employing this methodology, was demonstrated on 3-iodothiophene (Scheme 7.45). Another example of thiophene functionalization was achieved via rhodium-catalyzed three-component reaction of 2-(thien-2-yl)pyridine, benzaldehyde dimethylacetal, and tosyl amide by Le Gall, Presset, and coworkers in 2019 [73].

Functionalization of 1,3-azoles was achieved through their three-component reaction with hetaryl bromides and norbornadiene (NBD) [74]. The reactants were

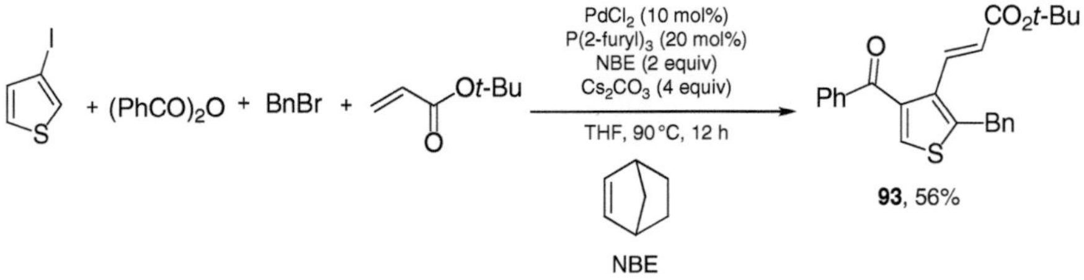

Scheme 7.45 Catellani reaction in functionalization of thiophene.

heated in dioxane for 16 hours. As a catalyst, $Pd(OAc)_2$ (5 mol%) was used, and DPEPhos (7.5 mol%) as a ligand (Scheme 7.46). The addition of TBAI (0.5 equiv) improved the reactivity and allowed the use of hetaryl bromides instead of iodides. Brominated pyrazoles, thiophenes, and *N*-acetylindole were tolerated. Indole derivative was the only substrate to give product as a trans-isomer. As for the azole component, the reaction works on benzoxazoles, benzimidazole, oxazoles, thiazoles, and benzothiazole. Benzimidazole and benzo(thiazole) were less reactive in C–H functionalization step, and the addition of CuI (50 mol%) was needed to promote these reactions. Various bridged bicyclic alkenes could be also used.

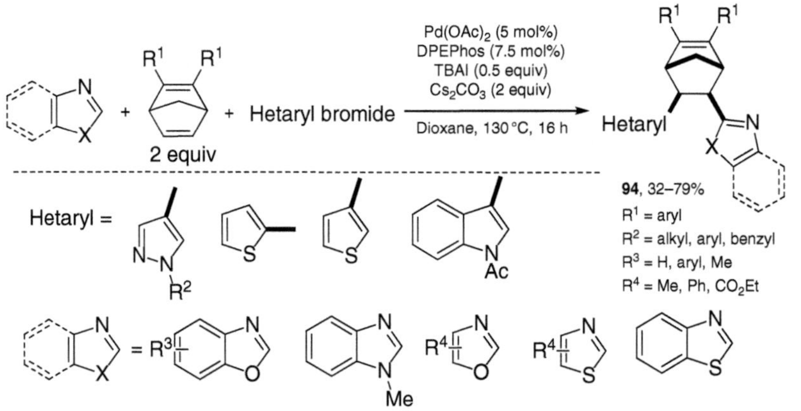

Scheme 7.46 Three-component functionalization of 1,3-azoles.

The reaction of 2-iodopyrazole with NBD could be accompanied by spontaneous retro-Diels–Alder reaction, and polycyclic molecules **95** were formed. The addition of pivalic acid was facilitating this reaction pathway. When NBE was used with this substrate, a cyclic bis-pyrazole **96** was obtained (Scheme 7.47). The analogous transformations were observed for 2-iodo-*N*-methylindole, 3-bromothiophene, and 4-bromo-2-methylthiophene.

Functionalization of 4-nitropyrazoles via a sequential three-component reaction was achieved by Iaroshenko, Langer, and coworkers [75]. The Pd-catalyzed diarylation of 4-nitropyrazole took place with the excess amounts of CuI (up to 6 equiv).

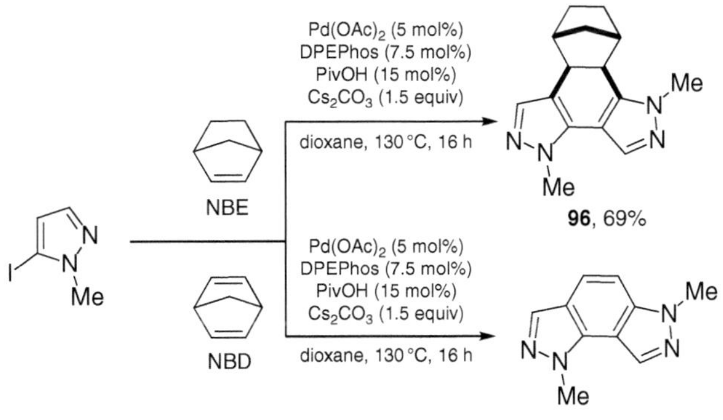

Scheme 7.47 Synthesis of polycyclic pyrazole-containing molecules.

7.3.1.2 C(sp^3)-H Functionalization

For the functionalization of six-membered nitrogen-containing heterocycles, reactions of tetrahydroisoquinolines (THIQ) are the most studied. The general strategy is the *in situ* formation of iminium ions and the addition of various nucleophiles to them: alkynes [76], dialkyl phosphonates [77], diazoesters [78] (Scheme 7.48).

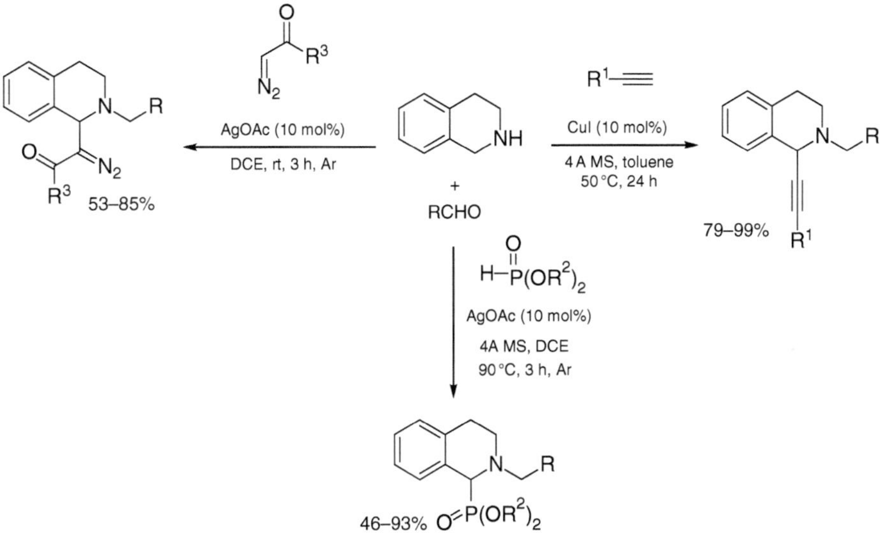

Scheme 7.48 Functionalization of tetrahydroisoquinolines with alkynes, dialkyl phosphonates, diazoesters. Source: Adapted from Zheng et al. [76]; Hu et al. [77]; Huang et al. [78].

Iminium-ion generation under visible-light photocatalysis was compatible with Ugi-type reaction [79]. Combination of *N*-aryltetrahydroisoquinoline with isocyanides and carboxylic acids efficiently furnished C(1)-substituted isoquinolines **97** through excited $Ru(bpy)_3Cl_2$-catalyzed oxidation of tertiary amine (Scheme 7.49). The use of CH_3CN as a solvent was found crucial for the success of the reaction, which worked in a broad scope of reactants.

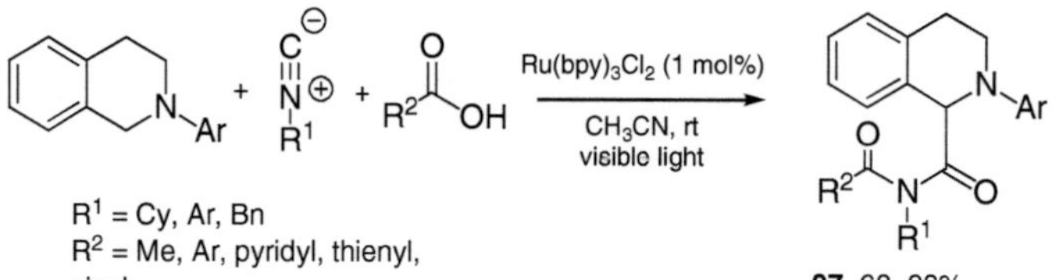

97, 68–98%

Scheme 7.49 Ugi-type reaction of tetrahydroisoquinolines.

For the functionalization of saturated oxygen-containing heterocycles, Luo and coworkers developed a Pd-catalyzed three-component reaction of cyclic ethers with *N*-tosylhydrazones and aryl halides [80]. After the oxidative addition of Pd(0) to an aryl bromide, the *in situ* generated diazo derivative formed a carbene with Pd complex, which could undergo a migratory insertion to provide an alkyl complex, prone to β-hydride elimination. Thus formed 1,1-diphenylethylene reacted with the radical, formed from DTBP and ether. Further disproportionation of the resulted radical derived a mixture of saturated and unsaturated compounds. When the mixture was subjected to H_2 reduction on Pd/C, a saturated derivative **98** was becoming the only product (Scheme 7.50). Various aryl halogenides could be employed, and the both electron-deficient and electron-rich substrates reacted smoothly. Furyl bromide could be also used, delivering target molecule with 78% yield. The reaction failed with 2-bromopyridine, giving desired product in trace amounts. Except for 1,4-dioxane, THF, tetrahydropyran, and dioxolane were tested. The latter furnished a mixture of C(2) and C(4)-functionalized products.

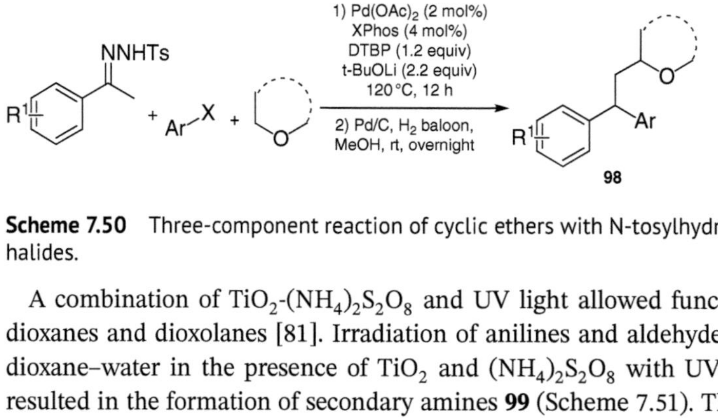

98

Scheme 7.50 Three-component reaction of cyclic ethers with N-tosylhydrazones and aryl halides.

A combination of TiO_2-$(NH_4)_2S_2O_8$ and UV light allowed functionalization of dioxanes and dioxolanes [81]. Irradiation of anilines and aldehydes in mixture of dioxane–water in the presence of TiO_2 and $(NH_4)_2S_2O_8$ with UV LED (365 nm) resulted in the formation of secondary amines **99** (Scheme 7.51). The reaction was

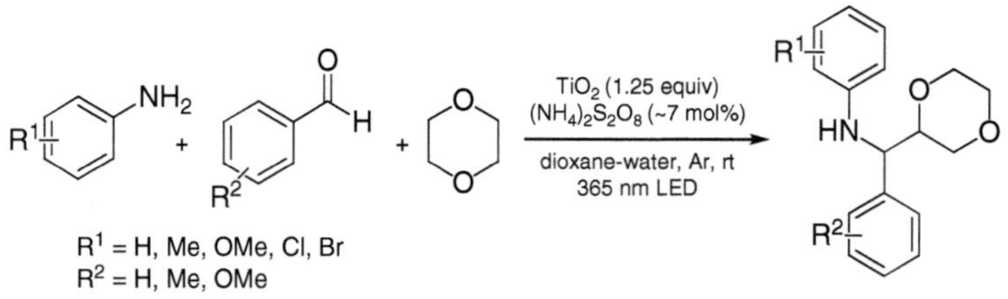

99, 46–90%

Scheme 7.51 Functionalization of dioxanes and dioxolanes.

successful with TiO_2 only, but the addition of peroxodisulfate could improve the yield from 66% to 86% yield. This transformation proceeded through the formation of α-oxo radicals and its addition to imine.

7.4 Transition-metal-free C–H Functionalization

In this part of the chapter, contemporary examples of heterocycle synthesis or modification under transition-metal-free conditions are summarized. The functionalization of unactivated C—H bonds is reviewed in this survey, while employment of activated substrates such as CH-acids (malonates, nitroalkanes, alkylpyridine, terminal alkynes, etc.) is mostly omitted. For selected examples of the latter in heterocyclic synthesis, see 82–89. For heterocyclic modification, see 90–94.

7.4.1 Multicomponent Synthesis of Heterocycles Through C–H-functionalization

A TBHP-mediated synthesis of indolizines **100** was developed by Ablajan and coworkers in 2020 [95]. Initially, acetophenone was converted into α-iodoacetophenone **101**, which subsequently reacted with pyridine to give a pyridinium salt **102** (Scheme 7.52). Methyl radical was generated from DMSO and *tert*-butoxy radical. The interaction of the latter with pyridinium radical was eventually giving intermediate **103**. Cycloaddition of **103** and pyridinium ylide was followed with pyridine elimination. The oxidation of **104** was delivering the target molecule. The reaction was tolerant to a broad scope of ketones. Notably, heteroaromatic

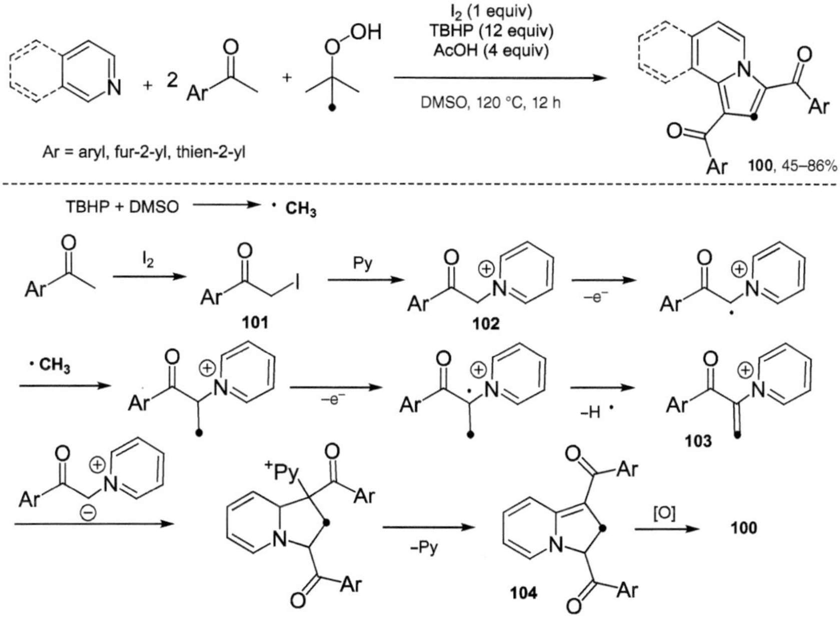

Scheme 7.52 Multicomponent synthesis of indolizines, mediated by TBHP.

methyl ketones could be used. Thus, furyl and thienyl-substituted indolizines were obtained. Isoquinoline could be used instead of pyridine, generating corresponding benzoindolizine with 45% yield.

An effective synthesis of polysubstituted pyrroles **105** was developed in our group recently [96]. In the first step, 2-imidazolines were reacting with terminal alkynes bearing an electron-withdrawing group (ester or ketone, 2 equiv) in xylene at rt, delivering imidazolidines **106** with *N*-vinylpropargylamine fragment (Scheme 7.53). Then, the reaction mixture was refluxed on air to produce 2-benzoyl pyrroles. The reaction worked smoothly with a broad scope of 2-arylimidazolines. Various substituents R^1 on N(1) were also tolerated. When aliphatic group was on nitrogen (R^1), the second step was becoming sluggish. To overcome this limitation, an addition of acylating reagent was needed. According to the control experiments and literature reports, in the second step an adduct A was undergoing a domino aza-Claisen/transannular nucleophilic addition/oxidative ring cleavage/acylation reaction sequence.

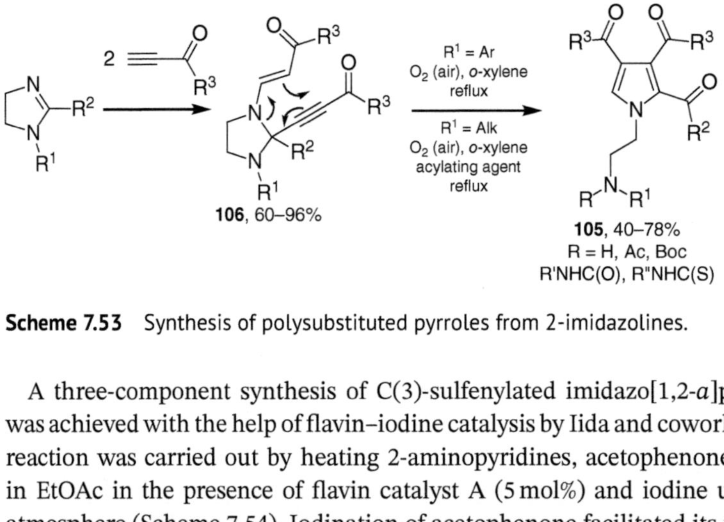

Scheme 7.53 Synthesis of polysubstituted pyrroles from 2-imidazolines.

A three-component synthesis of C(3)-sulfenylated imidazo[1,2-*a*]pyridines **107** was achieved with the help of flavin–iodine catalysis by Iida and coworkers [97]. The reaction was carried out by heating 2-aminopyridines, acetophenones, and thiols in EtOAc in the presence of flavin catalyst A (5 mol%) and iodine under oxygen atmosphere (Scheme 7.54). Iodination of acetophenone facilitated its reaction with aminopyridine to give imidazopyridine. The iodide is recycled with the help of flavin and oxygen. Then, thiol was oxidized to disulfide, which was transformed

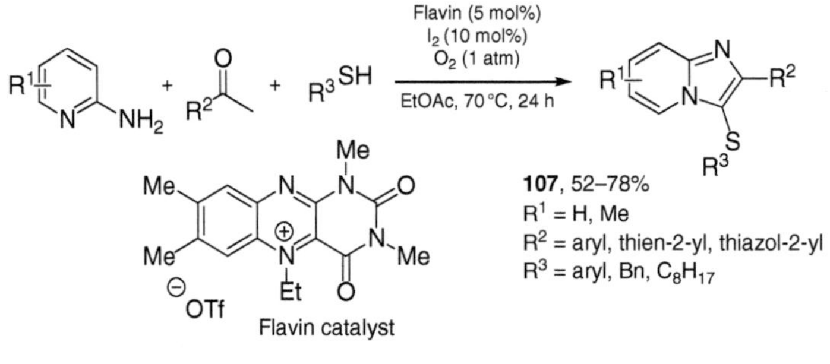

Scheme 7.54 Synthesis of imidazo[1,2-a]pyridines from 2-aminopyridines, acetophenones, and thiols.

into sulfenyl iodide (RSI) in the presence of iodine. The latter sulfenylated the imidazopyridine to complete the sequence. The reaction worked with both aromatic and aliphatic thiols. 2-acetylthiophene and 2-acetylthiazole were also smoothly converted into the desired products.

Visible-light-promoted three-component reaction of 2-aminopyridines, benzylamines, and *tert*-butylisocyanide led to the formation of 3-aminoimidazo[1,2-*a*] pyridines **108** with excellent yields 92–97% [98]. Simple white LED was used as a light source. A xanthene dye Eosin Y was employed as a photocatalyst, mediating synthesis of intermediate **109** through a hydrogen atom abstraction from benzyl amine (Scheme 7.55). The interaction of **109** with the isocyanide delivered the target product. Benzylamines with electron-withdrawing substituents were reacting faster, but electron-rich substrates were also reacting with excellent yields. Other isocyanides were not tested.

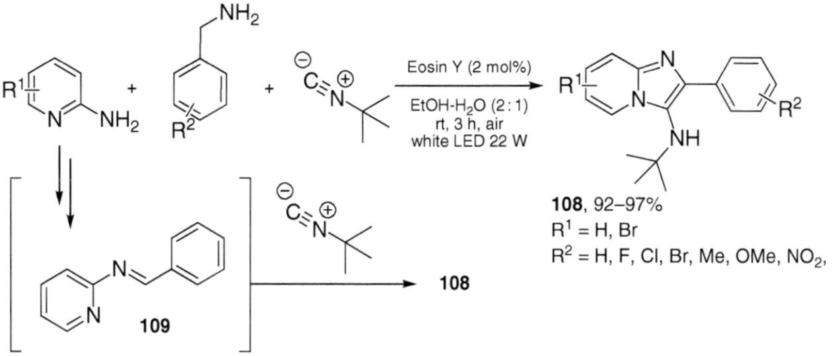

Scheme 7.55 Visible-light-promoted three-component synthesis of 3-aminoimidazo [1,2-a]pyridines.

An electrochemical transformation of methyl aryl ketones, benzyl amines, and amines allowed a preparation of imidazoles **110** [99]. The plausible mechanism was as follows: iodide oxidation to iodine radical, formation of α-iodomethylaryl ketone, formation of intermediate **111**, capable of anodic oxidation to aminal **112**, and final aromatization (Scheme 7.56). The transformation was of general character. Various amines could be employed. Interestingly, allylamine and propargyl amine were suitable with the reaction conditions, giving products with 63–71% yields, respectively.

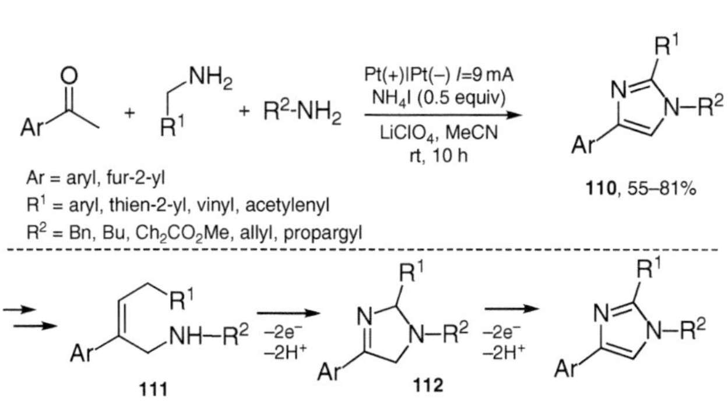

Scheme 7.56 An electrochemical preparation of imidazoles.

Fused triazines **113** were effectively synthesized from 2-aminobenzimidazoles, aromatic aldehydes (2 equiv), and ammonium iodide in a multicomponent manner [100]. Under optimized conditions, benzimidazo[1,2-*a*]-1,3,5-triazines **113** were obtained with 36–90% yields (Scheme 7.57). The addition of DMSO and molecular sieves increased the yield of the model reaction to 79%. Electronic effects were not making impact on the reaction yield, and different aldehydes worked smoothly. Lower yields were obtained in the case of more sterically hindered naphthaldehydes. Noteworthy, heteroaromatic aldehydes (pyridyl and thienyl) could be employed. The reactions with aliphatic aldehydes failed. Interestingly, indazole-3-amine gave the desired product **114**, while indole-2-amine was transformed to an unexpected compound **115**.

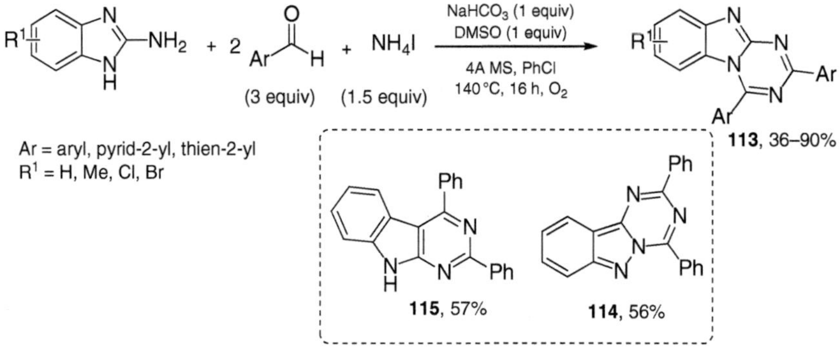

Scheme 7.57 Multicomponent synthesis of condensed triazines.

Deb, Baruah, and coworkers [101] reported a green methodology for the synthesis of 1,3-oxazines **116**, as an alternative to copper-catalyzed process [102]. Naphtols, aldehydes, and tetrahydroisoquinoline were refluxed in H_2O in the oxygen atmosphere. Firstly, the reaction proceeded as a three-component process, delivering intermediate **117** (Scheme 7.58). Further oxidation of **117** to an iminium ion was followed by the fragmentation to quinone methide **118** and imine **119**. These fragments were believed to undergo [4 + 2]-cycloaddition with each other to furnish

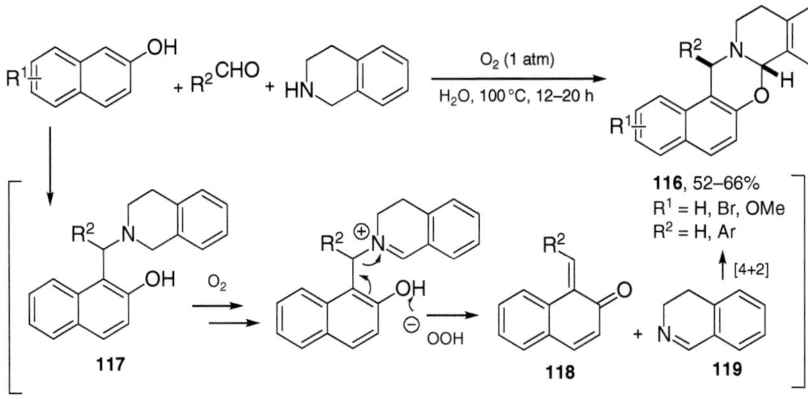

Scheme 7.58 Synthesis of 1,3-oxazines under aerobic conditions.

the final oxazine. Various aromatic aldehydes (R^2 = Ar) and formaldehyde (R^2 = H) were tolerated, but aliphatic aldehydes were not reactive. The reaction also failed with non-benzylic amines taken instead of tetrahydroisoquinoline.

7.4.2 Multicomponent C–H Functionalization of Heterocycles

In 2008, Itami, Li, and coworkers [103] reported one of the first multicomponent transition-metal-free functionalizations of a heterocycle. Pyridine-*N*-oxides were heated in cycloalkanes in the presence of di-*tert*-butyl peroxide (3 equiv) to 135 °C for 15 hours in air (Scheme 7.59). The reaction proceeded with di-functionalization of pyridine-*N*-oxide or tri-functionalization, when R^1 was not H; mixtures of di- and tri-alkylated products **120** formed. One can use various alkanes, such as norbornane or dioxane.

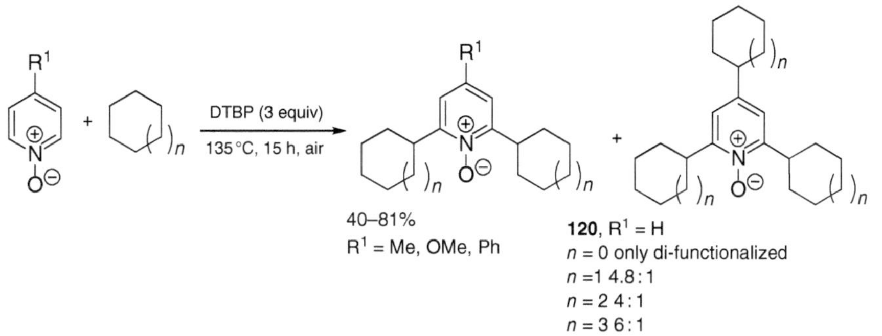

Scheme 7.59 Transition-metal-free functionalizations of pyridine-N-oxides.

Subsequently, Antonchick and Burgmann [104] functionalized azines with alkanes. The reaction was performed with a large excess of cyclohexane in DCM at rt (Scheme 7.60). Bis(trifluoroacetoxy)iodobenzene (PIFA, 4 equiv) was used as the oxidant and sodium azide (6 equiv) as an additive. The pseudo three-component reaction works for quinoline, phenantroline, 5-nitroisoquinoline, quinoxaline, quinazoline, phthazlazine, pyridopyrazine, or pyrimidine (four-component reaction). In general, the alkyl radical favored more electron-deficient positions, suggesting its nucleophilic character. The following mechanism was proposed. Ligand exchange in PIFA was followed by decomposition, releasing two radicals. Azide radical **121** abstracted hydrogen from an alkane to give an alkyl radical. Subsequent addition of this radical to protonated azine, followed by oxidation of a cation radical with $PhIOCOCF_3$, delivered the protonated form of the desired compound **122**.

Three-component coupling of THIQ, aromatic aldehydes, and aminopyrazoles was reported in 2020 [105]. The reactants were heated neat for six hours in the presence of L-proline (20 mol%), producing modified tetrahydroquinolines **123** in 67–90% yields (Scheme 7.61). Various aldehydes are suitable for the reaction conditions, except for aliphatic aldehydes. One can also perform the reaction on an indole, smoothly generating the desired products **124** in very good 80–86% yields. Only *N*-unsubstituted indoles react. Despite using a chiral catalyst, the resulting compounds are racemic. One can replace the proline catalyst with other acids

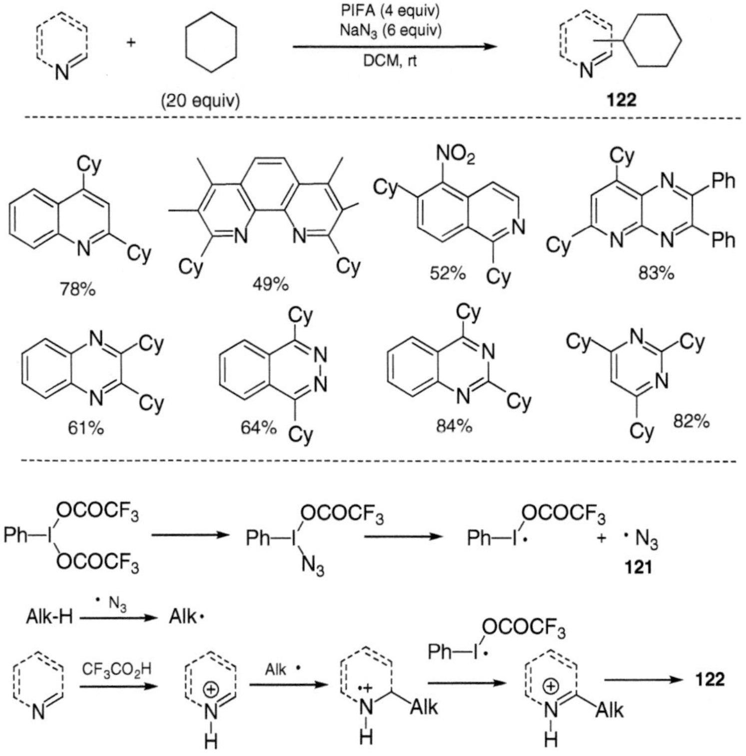

Scheme 7.60 Functionalization of azines with alkanes.

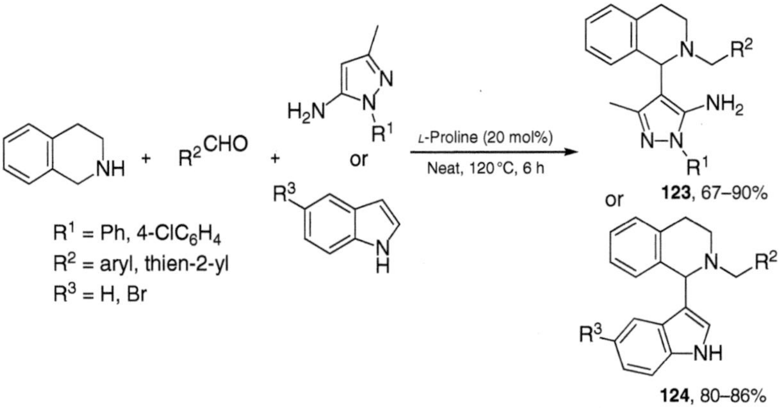

Scheme 7.61 Transition-metal-free functionalization of tetrahydroisoquinolines.

such as benzoic or acetic acid, but doing so decreased the yield to 32% and 27%, respectively.

One can functionalize pyrroles in a diaseteroselective manner through a three-component photoinduced process, reported in 2019 by Melchiorre and coworkers [106]. In a typical reaction, a mixture of pyrrole, alkyl chloride, and maleimide was irradiated with a blue LED in the presence of dithiocarbamate catalyst **125** (20 mol%;

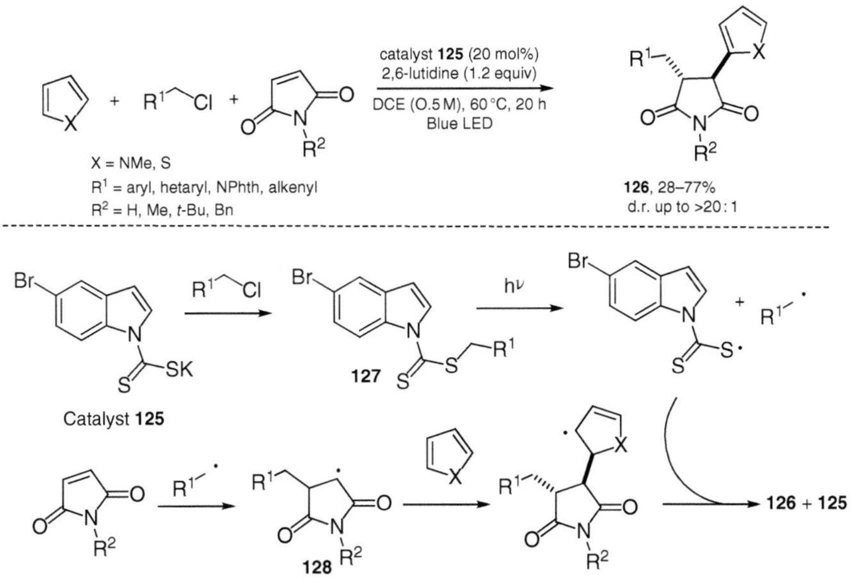

Scheme 7.62 Visible-light-promoted functionalization of pyrroles and thiophenes.

Scheme 7.62). DCE was used as a solvent and 2,6-lutidine as a base. Difunctionalized with alkyl and pyrrolyl moieties maleimides **126** were isolated in 28–77% yields and excellent diastereoselectivities, with d.r. > 20 : 1. Alkyl chloride, the radical precursor, can bear a broad scope of heterocyclic moieties: triazole, pyrazole, isoxazole, benzothiazole, thiazole, furan, and thiophene. Substituents such as phthalimide and alkenyl were also tolerated. Because all the radicals generated were stabilized, one cannot use simple alkanes. Instead of *N*-methylpyrrole as a heterocyclic component, one can employ thiophene. Interestingly, one can also use *N*-unsubstituted maleimides as well as *N*-benzyl and *N*-*tert*-butyl maleimides.

The mechanistic considerations are the following. Alkyl chloride is redox-inactive and does not participate in visible-light-mediated reactions by itself. Otherwise, when dithiocarbamate catalyst is alkylated with alkyl chloride, the resulting intermediate **127** might undergo photoinduced C—S bond cleavage, generating two radicals. The alkyl radical readily reacts with maleimide, giving radical **128**. Further addition to the hetaryl moiety is followed by aromatization through oxidation with dithiocarbamate radical. This act also recycles the catalyst.

One can achieve pyridine C–H functionalization through its reaction with benzyne [107]. The reaction between pyridine, *ortho*-silyl phenyltriflate (benzyne precursor), and isatin proceeds under reflux in THF in the presence of KF (2 equiv) and [18]crown-6 (2 equiv; Scheme 7.63). The initially formed zwitterion **129** underwent conversion into pyridylidene intermediate **130**. This intermediate was trapped with a non-enolizable and highly electrophilic *N*-methyl isatin. Further aryl migration delivered product **131** of formal C–H functionalization in moderate-to-good yields.

Researchers recently reported a series of transition-metal-free C(3)–H functionalizations of quinoxalinones. Studer and coworkers [108] developed a three-component reaction of quinaxolinones, alkenes, and perfluoroalkyliodides.

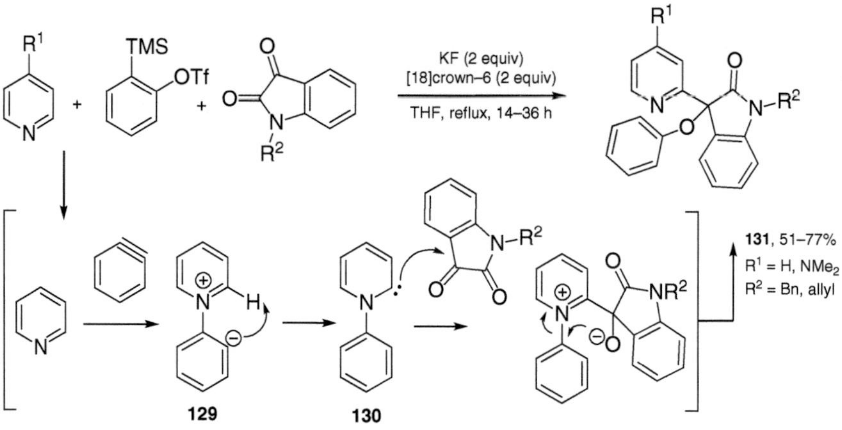

Scheme 7.63 Functionalization of pyridine with benzyne.

The reaction was performed under irradiation with visible light in DMP in the presence of DBU under an argon atmosphere. The alkylated quinoxalinones **132** were obtained in 34–98% yields (Scheme 7.64). The reaction had a broad scope with respect to the alkene component. Aliphatic electron-rich alkenes reacted smoothly, whereas styrene gave the desired product with the lowest 34% yield. In accordance with mechanistic studies, the reaction is an electron-catalyzed transformation. It is initiated by photoinduced decomposition of an R_fI/DBU complex, leading to formation of perfluoroalkyl radical. The radical chain is sustained by the reduction of R_fI with a quinoxalinyl radical anion.

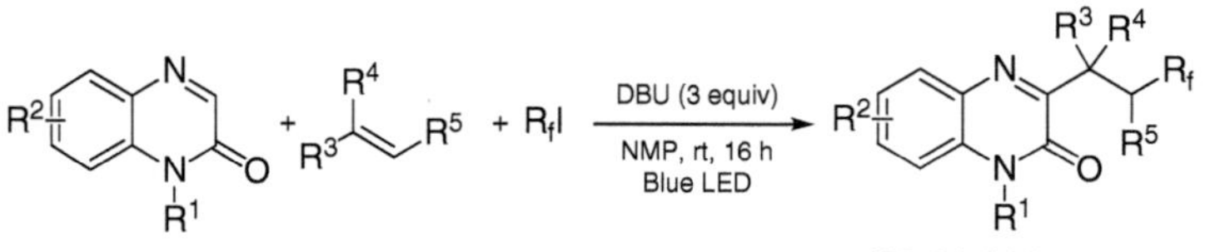

Scheme 7.64 Three-component functionalization of quinoxalinones with alkenes and perfluoroalkyliodides.

One can perform similar reactions with various free-radical precursors. Koley and coworkers [109] reported a three-component reaction of quinaxolines, styrenes, and sulfinic acids. Free radicals were generated from sulfinic acids by the action of TBHP. Xu, Zhu, Zhang, and coworkers [110] employed $TMSN_3$ as a free-radical precursor to obtain corresponding alkylazides of quinaxolinones. Generation of $N_3\bullet$ was achieved via PIFA (refer to the alkylation of pyridine for the mechanism of azide radical generation). Lv, Yue, Wei, and coworkers [111] used Langlois' reagent for an analogous three-component transformation and employed potassium peroxodisulfate as an oxidant.

Li, Zhang, and coworkers [112] reported formation of quinaxoline oximes through a three-component process between quinaxolinones, ketones, and *tert*-butyl nitrite (Scheme 7.65). The reaction took place in the presence of methanesulfonic acid

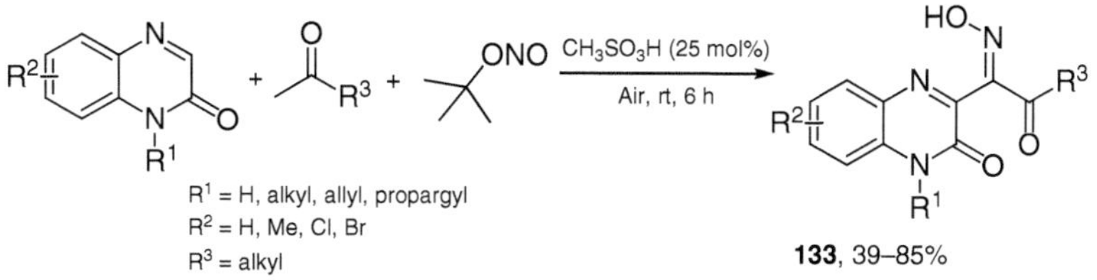

Scheme 7.65 Three-component functionalization of quinoxalinones with ketones tert-butyl nitrite.

(25 mol%). Oximes **133** were formed as a single (*E*)-isomer, which was 1.2 kcal/mol more stable than the (*Z*)-isomer, in accordance with DFT studies. Alkyl substituents on *N*(1) of the tested quinoxaline were broad in scope. Allyl and propargyl groups were well tolerated. Various substituents in the benzene ring were also compatible with the reaction conditions. Regarding the ketone partner, one can smoothly use methyl alkyl ketones, whereas acetophenones are not suitable for the reaction.

References

1 Hartwig, J.F. (2016). Evolution of C–H bond functionalization from methane to methodology. *J. Am. Chem. Soc.* 138: 2–24.

2 Gensch, T., Hopkinson, M.N., Glorius, F., and Wencel-Delord, J. (2016). Mild metal-catalyzed C—H activation: examples and concepts. *Chem. Soc. Rev.* 45: 2900–2936.

3 Roudesly, F., Oble, J., and Poli, G. (2017). Chemical metal-catalyzed C H activation/functionalization: the fundamentals. *Journal Mol. Catal. A, Chem.* 426: 275–296.

4 Colby, D.A., Bergman, R.G., and Ellman, J.A. (2010). Rhodium-catalyzed C–C bond formation via heteroatom-directed C–H bond activation. *Chem. Rev.* 110: 624–655.

5 Shul'Pin, G.B. (2016). New trends in oxidative functionalization of carbon–hydrogen bonds: a review. *Catalysts* 6: 50.

6 Wan, J.P., Gan, L., and Liu, Y. (2017). Transition metal-catalyzed C–H bond functionalization in multicomponent reactions: a tool toward molecular diversity. *Org. Biomol. Chem.* 15: 9031–9043.

7 Song, L. and Van der Eycken, E.V. (2020). Transition metal-catalyzed intermolecular cascade C–H activation/annulation processes for the synthesis of polycycles. *Chem. – A Eur. J.* 27 (1): 121–144.

8 Uygur, M. and García Mancheño, O. (2019). Visible light-mediated organophotocatalyzed C—H bond functionalization reactions. *Org. Biomol. Chem.* 17: 5475–5489.

9 Jiao, K.J., Xing, Y.K., Yang, Q.L. et al. (2020). Site-selective C–H functionalization via synergistic use of electrochemistry and transition metal catalysis. *Acc. Chem. Res.* 53: 300–310.

10 Batra, A. and Singh, K.N. (2020). Recent developments in transition metal-free cross-dehydrogenative coupling reactions for C–C bond formation. *European J. Org. Chem.* 2020: 6676–6703.

11 Kang, E., Kim, H.T., and Joo, J.M. (2020). Transition-metal-catalyzed C–H functionalization of pyrazoles. *Org. Biomol. Chem.* 18: 6192–6210.

12 Bonnaterre, F., Bois-choussy, M., and Zhu, J. (2008). Synthesis of dihydrophenanthridines by a sequence of Ugi-4CR and palladium-catalyzed intramolecular C—H functionalization. *Beilstein J. Org. Chem.* 6: 1–6.

13 Balalaie, S. (2014). Sequential ugi four-component reaction (4-CR)/C–H activation using (diacetoxyiodo) benzene for the synthesis of 3- (diphenylmethylidene)-2, 3-dihydro-1 H -indol-2-ones. *Helv. Chim. Acta* 97: 1555–1563.

14 Ojeda, G.M., Ranjan, P., Fedoseev, P. et al. (2019). Combining the ugi-azide multicomponent reaction and rhodium(III)-catalyzed annulation for the synthesis of tetrazole-isoquinolone/pyridone hybrids. *Beilstein J. Org. Chem.* 15: 2447–2457.

15 Saha, D., Kharbanda, A., Essien, N. et al. (2019). Intramolecular cyclization of imidazo[1,2-: A] pyridines via a silver mediated/palladium catalyzed C–H activation strategy. *Org. Chem. Front.* 6: 2234–2239.

16 Pinto, A., Neuville, L., and Zhu, J. (2009). Palladium-catalyzed domino N-arylation/carbopalladation/C – H functionalization: three-component synthesis of 3-(diarylmethylene) oxindoles. *Tetrahedron Lett.* 50: 3602–3605.

17 Sagadevan, A., Ragupathi, A., and Hwang, K.C. (2015). Photoinduced copper-catalyzed regioselective synthesis of indoles: three-component coupling of arylamines, terminal alkynes, and quinones. *Angew. Chemie – Int. Ed.* 54: 13896–13901.

18 Wang, Q., Li, Y., Qi, Z. et al. (2016). Rhodium(III)-catalyzed annulation between N-sulfinyl ketoimines and activated olefins: C–H activation assisted by an oxidizing N–S bond. *ACS Catal.* 6: 1971–1980.

19 Upadhyay, N.S., Jayakumar, J., and Cheng, C.-H. (2016). Rhodium-catalyzed regioselective synthesis of isoindolium salts from 2-arylpyridines and alkenes in aqueous medium under oxygen. *Adv. Synth. Catal.* 358: 3381–3386.

20 Liu, C., Wu, S., Sun, W. et al. (2020). Rapid and efficient construction of indolizino[3,4,5-ab]isoindole skeletons by a rhodium-catalyzed tandem reaction. *Asian J. Org. Chem.* 9: 68–72.

21 Rohlmann, R., Stopka, T., Richter, H., and Mancheño, O.G. (2013). Iron-catalyzed oxidative tandem reactions with TEMPO oxoammonium salts: synthesis of dihydroquinazolines and quinolines. *J. Org. Chem.* 78 (12): 6050–6064.

22 Midya, S.P., Sahoo, M.K., Landge, V.G. et al. (2015). Reversed reactivity of anilines with alkynes in the rhodium-catalysed C–H activation/carbonylation tandem. *Nat. Commun.* 6: 8591.

23 Liu, B., Hu, F., and Shi, B.F. (2014). Synthesis of sterically congested polycyclic aromatic hydrocarbons: rhodium(III)-catalyzed cascade oxidative annulation of aryl ketoximes with diphenylacetylene by sequential cleavage of multiple C–H bonds. *Adv. Synth. Catal.* 356: 2688–2696.

24 Li, C., Zhu, L., Liang, W. et al. (2019). An unusual [4 + 2] fusion strategy to forge: meso-N/O -heteroarene-fused (quinoidal) porphyrins with intense near-infrared Q-bands. *Chem. Sci.* 10: 7274–7280.

25 Kumar, G.S., Khot, N.P., and Kapur, M. (2019). Oxazolinyl-assisted Ru(II)-catalyzed C–H functionalization based on carbene migratory insertion: a one-pot three-component cascade cyclization. *Adv. Synth. Catal.* 361: 73–78.

26 Bian, M., Mawjuda, H., Gao, H. et al. (2020). Lossen rearrangement vs C–N reductive elimination enabled by Rh(III)-catalyzed C–H activation/selective lactone ring-opening: chemodivergent synthesis of quinolinones and dihydroisoquinolinones. *Org. Lett.* 22: 9677–9682.

27 Feng, M., Tang, B., Xu, H., and Jiang, X. (2016). Collective synthesis of phenanthridinone through C–H activation involving a Pd-catalyzed aryne multicomponent reaction. *Org. Lett.* 18: 4352–4355.

28 Wang, Z., Yin, J., Zhou, F. et al. (2019). Multicomponent reactions of pyridines to give ring-fused pyridiniums: in situ activation strategy using 1,2-dichloroethane as a vinyl equivalent. *Angew. Chemie – Int. Ed.* 58: 254–258.

29 Guo, W., Liao, J., Liu, D. et al. (2017). A four-component reaction strategy for pyrimidine carboxamide synthesis. *Angew. Chemie – Int. Ed.* 56: 1289–1293.

30 Stephan, M., Panther, J., Wilbert, F. et al. (2020). Heck reactions of acrolein or enones and aryl bromides – synthesis of 3-aryl propenals or propenones and consecutive application in multicomponent pyrazole syntheses. *European J. Org. Chem.* 2020: 2086–2092.

31 Chen, Y., Li, L., He, X., and Li, Z. (2019). Four-component reactions for the synthesis of perfluoroalkyl isoxazoles. *ACS Catal.* 9: 9098–9102.

32 Li, X., Wang, T., Lu, Y.J. et al. (2018). Copper-catalyzed oxidative multicomponent reaction: synthesis of imidazo fused heterocycles with molecular oxygen. *Org. Biomol. Chem.* 16: 7143–7151.

33 Storozhenko, O.A., Festa, A.A., Ndoutoume, D.R.B. et al. (2018). Mn-mediated sequential three-component domino knoevenagel/cyclization/michael addition/oxidative cyclization reaction towards annulated imidazo[1,2-a]pyridines. *Beilstein J. Org. Chem.* 14: 3078–3087.

34 Wang, L., Huang, J., Peng, S. et al. (2013). Palladium-catalyzed oxidative cycloaddition through C–H/N–H activation: access to benzazepines. *Angew. Chem. Int. Ed. Engl.* 52: 1768–1772.

35 Qian, W., Wang, H., and Allen, J. (2013). Copper-catalyzed domino cycloaddition/C–N coupling/cyclization/(C–H arylation): an efficient three-component synthesis of nitrogen polyheterocycles. *Angew. Chemie – Int. Ed.* 52: 10992–10996.

36 Jayakumar, J., Parthasarathy, K., Chen, Y.H. et al. (2014). One-pot synthesis of highly substituted polyheteroaromatic compounds by rhodium(III)-catalyzed multiple C–H activation and annulation. *Angew. Chemie – Int. Ed.* 53: 9889–9892.

37 Kong, W.J., Shen, Z., Finger, L.H., and Ackermann, L. (2020). Electrochemical access to aza-polycyclic aromatic hydrocarbons: rhoda-electrocatalyzed domino alkyne annulations. *Angew. Chemie – Int. Ed.* 59: 5551–5556.

38 Li, J., Liu, J., Yin, J. et al. (2019). Double ortho -C–H activation/annulation of benzamides with aryl alkynes: a route to double-helical polycyclic heteroaromatics. 84 (23): 15697–15705.

39 Yin, J., Zhou, F., Zhu, L. et al. (2018). Annulation cascade of arylnitriles with alkynes to stable delocalized PAH carbocations: via intramolecular rhodium migration. *Chem. Sci.* 9: 5488–5493.

40 Ghorai, D. and Choudhury, J. (2015). Rhodium(III)-N-heterocyclic carbene-driven cascade C–H activation catalysis. *ACS Catal.* 5: 2692–2696.

41 Ge, Q., Li, B., Song, H., and Wang, B. (2015). Rhodium(iii)-catalyzed cascade oxidative annulation reactions of aryl imidazolium salts with alkynes involving multiple C–H bond activation. *Org. Biomol. Chem.* 13: 7695–7710.

42 Ge, Q., Hu, Y., Li, B., and Wang, B. (2016). Synthesis of conjugated polycyclic quinoliniums by rhodium(III)-catalyzed multiple C–H activation and annulation of arylpyridiniums with alkynes. *Org. Lett.* 18: 2483–2486.

43 Wang, J., Qin, D., Lan, J. et al. (2015). Rh-catalysed direct cyclisation of 1,4-naphthoquinone and 9,10-phenanthraquinone with alkyne: facile access to 1,8-dioxapyrenes and 1,12-dioxaperylenes as orange and red-emitting luminophores. *Chem. Commun.* 51: 6337–6339.

44 Yin, J., Tan, M., Wu, D. et al. (2017). Synthesis of phenalenyl-fused pyrylium cations: divergent C–H activation/annulation reaction sequence of naphthalene aldehydes with alkynes. *Angew. Chem. Int. Ed. Engl.* 56: 13094–13098.

45 Tan, X., Liu, B., Li, X. et al. (2012). Rhodium-catalyzed cascade oxidative annulation leading to substituted naphtho[1,8-bc]pyrans by sequential cleavage of C(sp^2)-H/C(sp^3)-H and C(sp^2)-H/O-H bonds. *J. Am. Chem. Soc.* 134: 16163–16166.

46 Zhang, B., Li, B., Zhang, X., and Fan, X. (2017). Synthesis of naphthoquinolizinones through Rh(III)-catalyzed double C(sp2)-H bond carbenoid insertion and annulation of 2-aryl-3-cyanopyridines with α-diazo carbonyl compounds. *Org. Lett.* 19: 2294–2297.

47 Yan, K., Li, B., and Wang, B. (2018). Iridium-catalyzed tandem cyclization of benzoylacetonitriles with diazo compounds leading to substituted naphtho[1,8-bc]pyrans by sequential C–H functionalization. *Adv. Synth. Catal.* 360: 2272–2279.

48 Fang, F., Zhang, C., Zhou, C. et al. (2018). Rh(III)-catalyzed C–H activation of benzoylacetonitriles and tandem cyclization with diazo compounds to substituted benzo [de] chromenes. *Org. Lett.* 20: 1720–1724.

49 Qureshi, Z., Kim, J.Y., Bruun, T. et al. (2016). Pd-catalyzed synthesis of fully decorated polycyclic triazoles: introducing C–H functionalization to multicomponent multicatalytic reactions ((MC)2R). *ACS Catal.* 6: 4946–4952.

50 Huang, J.R., Song, Q., Zhu, Y.Q. et al. (2014). Rhodium(III)-catalyzed three-component reaction of imines, alkynes, and aldehydes through C–H activation. *Chem. – A Eur. J.* 20: 16882–16886.

51 Huang, L., Wang, Q., Qi, J. et al. (2013). Rh(iii)-catalyzed ortho-oxidative alkylation of unactivated arenes with allylic alcohols. *Chem. Sci.* 4: 2665–2669.

52 Bai, P., Li, Y.Q., and Huang, Z.Z. (2017). A four-component cascade C–H functionalization/cyclization/nucleophilic substitution reaction to construct α-functionalized tetrahydroquinolines by the strategy of *in situ* directing group formation. *Org. Lett.* 19: 1374–1377.

53 Pache, S. and Lautens, M. (2003). Palladium-catalyzed sequential alkylation–alkenylation reactions: new three-component coupling leading to oxacycles. *Org. Lett.* 5: 4827–4830.

54 Kaishap, P.P., Sarma, B., and Gogoi, S. (2016). The amide C–N bond of isatins as the directing group and the internal oxidant in Ru-catalyzed C–H activation and annulation reactions: access. *Chem. Commun.* 52: 9809–9812.

55 Chen, Z., Jin, S., Jiang, W. et al. (2020). Multicomponent synthesis of iminocoumarins via rhodium-catalyzed C–H bond activation. *J. Org. Chem.* 85: 11006–11013.

56 Moon, S., Kato, M., Nishii, Y., and Miura, M. (2020). Synthesis of benzo[b]thiophenes through rhodium-catalyzed three-component reaction using elemental sulfur. *Adv. Synth. Catal.* 362: 1669–1673.

57 Qiu, H., Li, M., Jiang, L.Q. et al. (2012). Highly enantioselective trapping of zwitterionic intermediates by imines. *Nat. Chem.* 4: 733–738.

58 Zhang, D., Qiu, H., Jiang, L. et al. (2013). Enantioselective palladium(II) phosphate catalyzed three-component reactions of pyrrole, diazoesters, and imines. *Angew. Chemie – Int. Ed.* 52: 13356–13360.

59 Chen, D.F., Zhao, F., Hu, Y., and Gong, L.Z. (2014). C–H functionalization/asymmetric michael addition cascade enabled by relay catalysis: metal carbenoid used for C–C bond formation. *Angew. Chemie – Int. Ed.* 53: 10763–10767.

60 Jiang, L., Jin, W., and Hu, W. (2016). Double C–H functionalization of indoles via three-component reactions/CuCl2-catalyzed aerobic dehydrogenative coupling for the synthesis of polyfunctional cyclopenta[b]indoles. *ACS Catal.* 6: 6146–6150.

61 Pospech, J., Tlili, A., Spannenberg, A. et al. (2014). Regioselective ruthenium-catalyzed carbonylative direct arylation of five-membered and condensed heterocycles. *Chemistry* 20: 3135–3141.

62 Zhou, F., Wang, D.S., Guan, X., and Driver, T.G. (2017). Nitroarenes as the nitrogen source in intermolecular palladium-catalyzed aryl C–H bond aminocarbonylation reactions. *Angew. Chemie – Int. Ed.* 56: 4530–4534.

63 Liu, Y., Zhang, Y., Huang, M., and Wan, J.P. (2015). Step economical synthesis of o-aryl benzamides via C–H activation relayed by the in situ installation of directing group: a multicomponent method. *RSC Adv.* 5: 46192–46196.

64 Dalal, A., Singh, P., and Babu, S.A. (2019). One-pot, solvent-free Pd(II)-catalyzed direct β–C–H arylation of carboxamides involving anhydrides as substrates via *in situ* installation of directing group. *Tetrahedron* 75: 1246–1257.

65 Shi, W.Y., Ding, Y.N., Liu, C. et al. (2020). Three-component ruthenium-catalyzed remote C–H functionalization of 8-aminoquinoline amides. *Chem. Commun.* 56: 12729–12732.
66 Pinkert, T., Wegner, T., Mondal, S., and Glorius, F. (2019). Intermolecular 1,4-carboamination of conjugated dienes enabled by Cp*RhIII-catalyzed C–H activation. *Angew. Chemie – Int. Ed.* 58: 15041–15045.
67 Zhai, R.L., Xue, Y.S., Liang, T. et al. (2018). Regioselective arene and heteroarene functionalization: N-alkenoxypyridinium salts as electrophilic alkylating agents for the synthesis of α-aryl/α-heteroaryl ketones. *J. Org. Chem.* 83: 10051–10059.
68 Boerth, J.A., Hummel, J.R., and Ellman, J.A. (2016). Highly stereoselective cobalt(III)-catalyzed three-component C–H bond addition cascade. *Angew. Chemie – Int. Ed.* 55: 12650–12654.
69 Boerth, J.A., Maity, S., Williams, S.K. et al. (2018). Selective and synergistic cobalt(iii)-catalysed three-component C–H bond addition to dienes and aldehydes. *Nat. Catal.* 1: 673–679.
70 Shen, Z., Li, C., Mercado, B.Q., and Ellman, J.A. (2020). Cobalt(III)-catalyzed diastereoselective three-component C–H bond addition to butadiene and activated ketones. *Synthesis* 52: 1239–1246.
71 Bon, J.L., Feng, D., Marder, S.R., and Blakey, S.B. (2014). A C–H functionalization protocol for the direct synthesis of benzobisthiazole derivatives. *J. Org. Chem.* 79: 7766–7771.
72 Wang, J., Qin, C., Lumb, J.P., and Luan, X. (2020). Regioselective synthesis of polyfunctional arenes by a 4-component catellani reaction. *Chem* 6: 2097–2109.
73 Xavier, T., Rayapin, C., Le Gall, E., and Presset, M. (2019). Multicomponent aromatic and benzylic mannich reactions through C–H bond activation. *Chemistry* 25: 13824–13828.
74 Jeong, S., Kim, E., Kim, M. et al. (2020). Divergent strategies for the π-extension of heteroaryl halides using norbornadiene as an acetylene synthon. *Org. Lett.* 22: 9670–9676.
75 Iaroshenko, V.O., Gevorgyan, A., Davydova, O. et al. (2014). Regioselective and guided C–H activation of 4-nitropyrazoles. *J. Org. Chem.* 79: 2906–2915.
76 Zheng, Q.H., Meng, W., Jiang, G.J., and Yu, Z.X. (2013). CuI-catalyzed C1-alkynylation of tetrahydroisoquinolines (THIQs) by A 3 reaction with tunable iminium ions. *Org. Lett.* 15: 5928–5931.
77 Hu, G., Chen, W., Ma, D. et al. (2016). Silver-catalyzed, aldehyde-induced α–C–H functionalization of tetrahydroisoquinolines with concurrent C–P bond formation/N-alkylation. *J. Org. Chem.* 81: 1704–1711.
78 Huang, J., Li, L., Xiao, T. et al. (2016). Redox-neutral C1 functionalization of unprotected tetrahydroisoquinolines with diazo carbonyl compounds. *Asian J. Org. Chem.* 5: 1204–1207.
79 Chen, Y. and Feng, G. (2015). Visible light mediated sp3 C—H bond functionalization of N-aryl-1,2,3,4-tetrahydroisoquinolines via Ugi-type three-component reaction. *Org. Biomol. Chem.* 13: 4260–4265.

80 Wan, Z.j., Yuan, X.f., Xiao, T. et al. (2019). A Pd-catalyzed three-component reaction and hydrogenation strategy to prepare 2-functionalized cyclic ethers involving a radical initiating C(sp3)-H activation. *Asian J. Org. Chem.* 8: 2196–2200.

81 Zhang, L., Deng, Y., and Shi, F. (2013). Light promoted aqueous phase amine synthesis via three-component coupling reactions. *Tetrahedron Lett.* 54: 5217–5219.

82 Yaragorla, S., Dada, R., and Singh, G. (2016). Alkaline-earth-catalyzed sp3 C–H functionalization of methyl azaarenes and its use in a one-pot four-component synthesis of azaarenyl benzylpyrazolones. *Synlett* 27: 912–918.

83 Paul, K., Jalal, S., Kundal, S., and Jana, U. (2016). Synthesis of fused dibenzofuran derivatives via palladium-catalyzed domino C–C bond formation and iron-catalyzed cycloisomerization/aromatization. *J. Org. Chem.* 81: 1164–1174.

84 Yang, J., Zhou, P., Hu, B. et al. (2018). TBAI/$K_2S_2O_8$ – promoted multicomponent domino reaction of aryl methyl ketones, enaminones, and indoles: a facile access to multisubstituted 3-indolyl-pyrroles. *ChemistrySelect* 3: 11938–11942.

85 Jana, A., Bhaumick, P., Panday, A.K. et al. (2019). I2/DMSO mediated multicomponent reaction for the synthesis of 2-arylbenzo[d] imidazo[2,1- b] thiazole derivatives. *Org. Biomol. Chem.* 17: 5316–5330.

86 Chen, M.N., Di, J.Q., Li, J.M. et al. (2020). Eosin Y-catalyzed one-pot synthesis of spiro[4H-pyran-oxindole] under visible light irradiation. *Tetrahedron* 76: 131059.

87 Zi, Q.X., Yang, C.L., Li, K. et al. (2020). Multicomponent cascade reaction by metal-free aerobic oxidation for synthesis of highly functionalized 2-amino-4-coumarinyl-5-arylpyrroles. *J. Org. Chem.* 85: 327–338.

88 Sokolova, E.A., Festa, A.A., Golantsov, N.E. et al. (2019). Highly fluorescent pyrido[2,3-b]indolizine-10-carbonitriles through pseudo three-component reactions of N-(cyanomethyl)pyridinium salts. *European J. Org. Chem.* 2019: 6770–6775.

89 Sokolova, E.A., Festa, A.A., Subramani, K. et al. (2020). Microwave-assisted synthesis of fluorescent pyrido[2,3-b]indolizines from alkylpyridinium salts and enaminones. *Molecules* 25: 1–12.

90 Kumar, A., Gupta, G., and Srivastava, S. (2011). Synthesis of new class of alkyl azarene pyridinium zwitterions via iodine mediated sp^3 C–H bond activation. *Org. Lett.* 13: 6366–6369.

91 Sivakumar, S., Kanchithalaivan, S., and Kumar, R.R. (2013). A one-pot three-component domino protocol for the synthesis of penta-substituted 4H-pyrans. *RSC Adv.* 3: 13357–13364.

92 Pathan, M.Y., Chavan, S.S., Shaikh, T.M.Y. et al. (2017). Facile one-pot multi-component synthesis of spirooxindoles and 3, 3'-disubstituted oxindoles via sp3C-H activation/functionalization of azaarenes. *ChemistrySelect* 2: 9147–9152.

93 Singh, G. and Yaragorla, S. (2017). Highly efficient one-pot tandem Friedlander annulation and chemo-selective Csp3-H functionalization under calcium catalysis. *RSC Adv.* 7: 18874–18882.

94 Panigrahi, A., Muniraj, N., and Prabhu, K.R. (2020). Iodine-promoted one-pot multicomponent chemoselective reaction for C—C/C—N and C-C/C–S bond formation using thiols. *European J. Org. Chem.* 2020: 5780–5784.

95 Zhang, Q., Ablajan, K., Wang, B. et al. (2020). One-pot synthesis of indolizines using TBHP as the methylene source under metal-free condition. *European J. Org. Chem.* 2020: 262–266.

96 Golantsov, N.E., Golubenkova, A.S., Festa, A.A. et al. (2020). A domino route toward polysubstituted pyrroles from 2-imidazolines and electron-deficient alkynes. *Org. Lett.* 22: 4726–4731.

97 Okai, H., Tanimoto, K., Ohkado, R., and Iida, H. (2020). Multicomponent synthesis of imidazo[1,2- a]pyridines: aerobic oxidative formation of C–N and C–S bonds by flavin-iodine-coupled organocatalysis. *Org. Lett.* 22: 8002–8006.

98 Singh, H.K., Kamal, A., Kumari, S. et al. (2020). Eosin Y-catalyzed synthesis of 3-aminoimidazo[1,2- a]pyridines via the HAT process under visible light through formation of the C–N bond. *ACS Omega* 5: 29854–29863.

99 Zeng, L., Li, J., Gao, J. et al. (2020). An electrochemical oxidative multicomponent cascade annulation of ketones and amines used to produce imidazoles. *Green Chem.* 22: 3416–3420.

100 Chen, J., Sun, Z., Xiao, F., and Deng, G. (2020). Base-promoted aerobic oxidative synthesis of fused 1,3,5-triazines under metal-free conditions. *Green Chem.* 22: 6778–6782.

101 Deb, M.L., Pegu, C.D., Borpatra, P.J. et al. (2017). Catalyst-free multi-component cascade C–H-functionalization in water using molecular oxygen: an approach to 1,3-oxazines. *Green Chem.* 19: 4036–4042.

102 Gupta, K.S.V., Ramana, D.V., Vinayak, B. et al. (2016). Copper-catalyzed regio and diastereoselective three component C–N, C–C and C–O bond forming reaction: oxidative sp^3 C–H functionalization. *New J. Chem.* 40: 6389–6395.

103 Deng, G., Ueda, K., Yanagisawa, S. et al. (2009). Coupling of nitrogen heteroaromatics and alkanes without transition metals: a new oxidative cross-coupling at C–H/C–H bonds. *Chem. – A Eur. J.* 15: 333–337.

104 Antonchick, A.P. and Burgmann, L. Direct selective oxidative cross-coupling of simple alkanes with heteroarenes. *Angew. Chemie – Int. Ed.* 52 (11): 3267–3271.

105 Rahman, I., Deka, B., Thakuria, R. et al. (2020). L-proline-catalyzed regioselective C1 arylation of tetrahydroisoquinolines through a multicomponent reaction under solvent-free conditions. *Org. Biomol. Chem.* 18: 6514–6518.

106 Cuadros, S., Horwitz, M.A., Schweitzer-Chaput, B., and Melchiorre, P. (2019). A visible-light mediated three-component radical process using dithiocarbamate anion catalysis. *Chem. Sci.* 10: 5484–5488.

107 Nawaz, F., Mohanan, K., Charles, L. et al. (2013). Temporary intramolecular generation of pyridine carbenes in metal-free three-component C–H bond functionalisation/aryl-transfer reactions. *Chem. – A Eur. J.* 19: 17578–17583.

108 Zheng, D. and Studer, A. (2019). Photoinitiated three-component α-perfluoroalkyl-β-heteroarylation of unactivated alkenes via electron catalysis. *Org. Lett.* 21: 325–329.

109 Sekhar Dutta, H., Ahmad, A., Khan, A.A. et al. (2019). Metal free benzylation and alkylation of quinoxalin-2(1H)-ones with alkenes triggered by sulfonyl radical generated from sulfinic acids. *Adv. Synth. Catal.* 361: 5534–5539.

110 Shen, J., Xu, J., Huang, L. et al. (2020). Hypervalent iodine(III)-promoted rapid cascade reaction of quinoxalinones with unactivated alkenes and TMSN3. *Adv. Synth. Catal.* 362: 230–241.

111 Meng, N., Wang, L., Liu, Q. et al. (2020). Metal-free trifluoroalkylation of quinoxalin-2(1 H)-ones with unactivated alkenes and langlois' reagent. *J. Org. Chem.* 85: 6888–6896.

112 Xu, J., Yang, H., He, L. et al. (2021). Synthesis of (E)-quinoxalinone oximes through a multicomponent reaction under mild conditions. *Org. Lett.* 23: 195–201.

8

Multicomponent-Switched Reactions in Synthesis of Heterocycles

Valentyn A. Chebanov[1,2], Serhiy M. Desenko[1], Victoria V. Lipson[1,2,3], and Nikolay Yu. Gorobets[1]

[1] *State Scientific Institution "Institute for Single Crystals" NAS of Ukraine, Department of Organic and Bioorganic Chemistry, Nauky ave., 60, Kharkiv, 61072, Ukraine*
[2] *Karazin Kharkiv National University, Chemistry Faculty, Svobody Sq., 4, Kharkiv, 61022, Ukraine*
[3] *State Institution "V. Ya. Danilevsky Institute for Endocrine Pathology Problems" NAMS of Ukraine, Department of Medicinal Chemistry, Alchevskikh St., 10, Kharkiv, 61002, Ukraine*

Multicomponent reaction (MCR) is an important class of organic reactions and a powerful tool in modern synthesis allowing one-pot treatment of three and more compounds and leading to the simultaneous formation of several chemical bonds between molecules of reagents (usually not in a synchronous manner but in cascade one) [1]. There are MCRs when all reagents in one vessel are different as well as the cases when one of the reagents may supply two or more fragments of its molecules into the final compound structure (with or without chemodifferentiation) and, therefore, the number of starting compounds can be less than three. Another variation is when a reaction is carried out with sequential addition of the reactants without changing the reaction conditions (sequential MCR) or with their changing from step-to-step (consecutive MCR) [2].

Among other organic reactions, MCRs were discovered one of the first at the end of the nineteenth centuries – the beginning of the twentieth centuries – reactions of Strecker [3], Mannich [4], Hantzsch [5], Doebner [6], Biginelli [7, 8], etc. Due to their great efficiency in the generation of chemical complexity and diversity, the renaissance of such reactions refers to the violent growth of combinatorial chemistry about three decades ago and continues now. Application of MCRs is rather wide and concerns classical target-oriented organic synthesis [1, 9], synthesis of heterocyclic compounds [10–12] as well as diversity-oriented synthesis [13–16], medicinal chemistry and new drugs search [17–20], green synthesis [21, 22], material science [23–25], and other.

There are several rational design strategies for discovering novel MCRs [26–29] to increase capabilities, diversity of final products, and efficiency of multicomponent approaches in general. Among them Single Reactant Replacement, Modular Reaction Sequences, Condition-based Divergence Strategy, Combination of MCRs, Post-cyclization Strategy, and others.

Multicomponent Reactions towards Heterocycles: Concepts and Applications, First Edition. Edited by Erik V. Van der Eycken and Upendra K. Sharma.

In most cases, it is assumed that multicomponent and step-by-step approaches yielded the same final compounds and their comparison is based on such advantages as reducing the number of stages and amounts of solvents, shorter reaction times, atom economy, ease of generating libraries of compounds with high diversity, simplicity of automation, etc. However, there is another alternative – to use MCRs as a tool to switch the direction of the transformation and, therefore, to obtain different final compounds. In this case, the increased number of reagents itself becomes the reaction direction differentiation factor, and, thus, we deal with multicomponent-switched reactions – the processes in which MCR switching effect acts as the main control tool of the reaction's selectivity (Scheme 8.1).

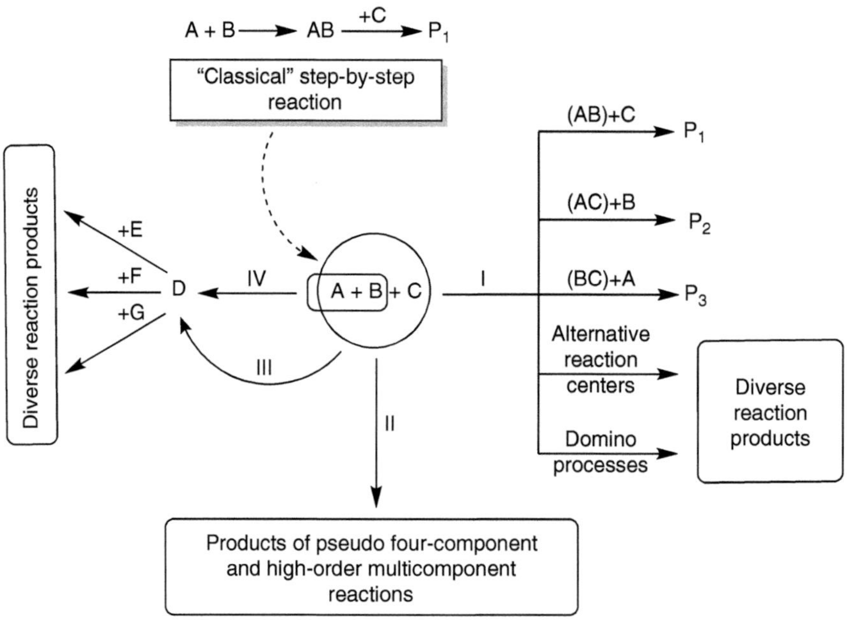

Scheme 8.1 Multicomponent switching effect.

This possibility is based on the fact that commutative law may not be performed for chemical reactions and $A+B+C \neq (A+B)+C$. Indeed, the "classical" procedure for the treatment of three compounds A, B, and C is a step-by-step procedure including the reaction of two components A and B, isolation of intermediate AB, and then its reaction with component C that gives final product P_1 (Scheme 8.1). In case of the multicomponent procedure (Scheme 8.1, Pathway I), when compound AB is replaced by its precursors A and B, the reaction may proceed either via the same cascade of stages (AB) + C or via alternative cascades: (AC) + B and (BC) + A yielding different final products P_2 and P_3.

Due to the presence of alternative reaction centers in a polyfunctional reactant, the MCRs may involve these centers and, thereby, gives novel reaction products; the same situation is possible when some additional cascade processes like rearrangements or post-cyclizations are possible after the completing MCR that allows more reaction products as well. There are also possible proceeding pseudo-four-component reactions and even high-order MCRs between three reactants (Scheme 8.1, Pathway II).

Such divergent character of multicomponent treatments may be a problem that leads to the formation of complicated mixtures of several final compounds; however, it also gives a possibility to switch their course between several alternatives by variation of the reaction conditions or by structure modification of the reagents [14, 15, 30].

Another option to tune the reaction directions consists in the approach that based on the fixation of an unstable but high reactive intermediate D of MCR (Scheme 8.1, Pathway III) or two-component reaction (Scheme 8.1, Pathway IV) and then carrying out its treatment (two- or multicomponent) under different specific conditions that give novel diverse final compounds.

One of the first and simplest examples of the multicomponent-switched reaction was described by Desenko and coworkers in 1990 [31, 32]. It was established that cyclocondensation of 3-amino-1,2,4-triazoles **3** with cyclohexalidencyclohexanones **1**, which had been obtained by reaction of two molecules of cyclohexanone **2**, yielded 4,7-dihydro-1,2,4-triasolo[1,5-*a*]pyrimidines **4** (Scheme 8.2, Pathway **a**). On the one hand, its pseudo-three-component analog using two equivalents of cyclohexanone gave 4,5-dihydroregioisomer **5** with different location of substituents in pyrimidine ring (Pathway **b**).

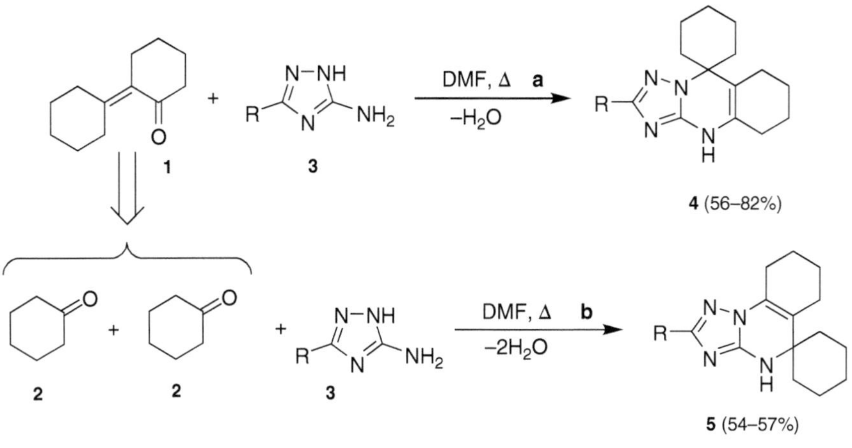

Scheme 8.2 The first example of the multicomponent-switched reaction.

A similar situation was observed for the condensation of 3-amino-1,2,4-triazoles **3** with β-dypnone **6** (two-component reaction, Scheme 8.3, Pathway **a**) or with two equivalents of acetophenone **7** (pseudo-three-component reaction, Scheme 8.3, Pathway **b**) [33–35]. But there are two differences in comparison with the reaction of cyclohexanone (Scheme 8.2). Firstly, the compounds **10** as the minor product of a side-three-component reaction with the participation of DMF (solvent) was formed [33] (Scheme 8.3, Pathway **b4**). Secondly, both 4,7- and 4,5-dihydroisomers

8 and **9** were isolated in the presence of $ZnCl_2$ as a catalyst (Scheme 8.3, Pathway **b4**) [34] while in case of catalysis by acetic or mineral acids [33] as well as under the promotion by chlorotrimethylsilane [35] the pseudo-three-component reaction gave only one regioisomer – heterocycles **9**.

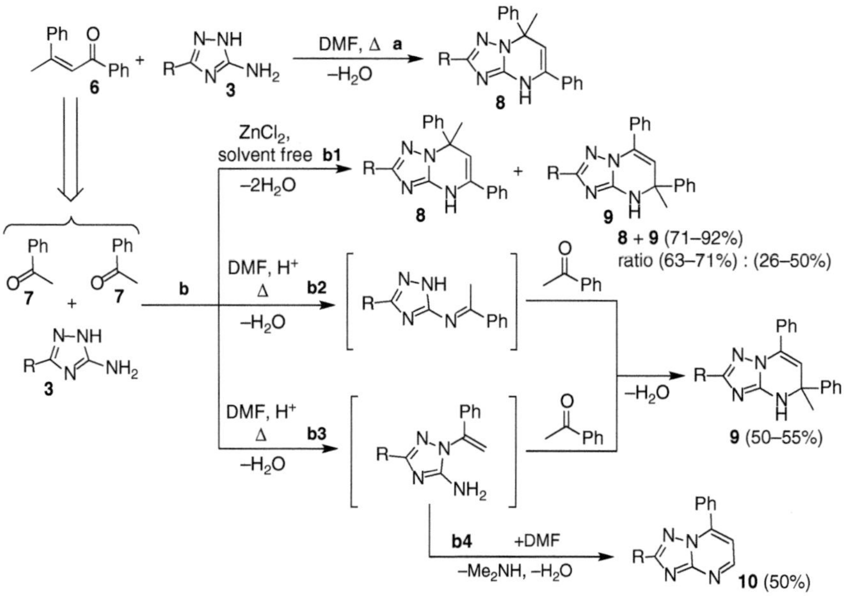

Scheme 8.3 Multicomponent-switched reaction of 3-amino-1,2,4-triazoles with β-dypnone or its precursor.

The formation of compounds **8** in the pseudo-three-component reaction of aminotriazoles with acetophenone corresponds to the "classical" behavior of α,β-unsaturated carbonyls because the generation of the latter occurs *in situ*. Taking into account that self-condensation of two molecules of ketones cannot be the first step of the formation of dihydrotriazolopyrimidines **9** (the reaction **a** in Scheme 8.3 always leads to heterocycles **8**), in publications [33–35] two alternative mechanisms were suggested. According to the authors of [34, 35], the key intermediate of the treatment was azomethine (Scheme 8.3, Pathway **b2**), while in the publication [33] the formation of enamines was proposed (Pathway **b3**). In our opinion, mechanism **b3** is more probable for the formation of compound **9** because the product of side reaction **10** corresponds to the cyclization of the enamine intermediate but not azomethine.

Other authors [36] inspired by the publication of Wermann and Hartmann [34] carried out a similar MCR with the participation of 5-amino-3-methylpyrazole **11** instead of 3-amino-1,2,4-triazoles **3** and found that under $ZnCl_2$ catalysis the reaction of amine **11** with dypnone **6** gave dihydropyrazolepyrimidine **12** (Scheme 8.4, Pathway **a**), while pseudo-three-component heterocyclization involving two equivalents of acetophenone yielded isomeric heterocyclic compounds **13** (Pathway **b**).

In this case, Bellec and Lhommet offered the mechanism including the preliminary formation of azomethine; however, no evidence of it was proposed in [36].

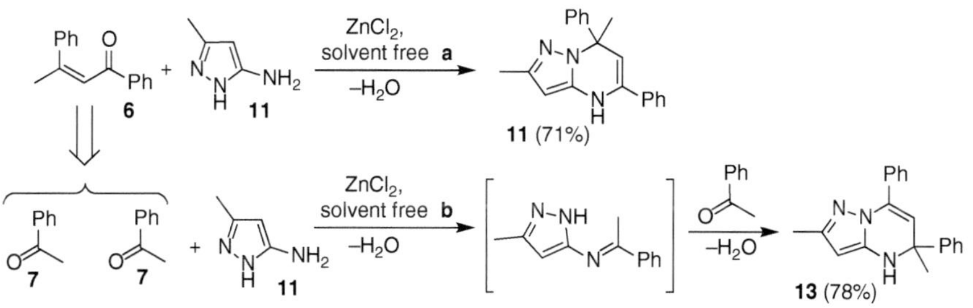

Scheme 8.4 Multicomponent-switched reaction of 5-amino-3-methylpyrazole with β-dypnone or its precursor.

One of the widespread approach to the synthesis of 5,7-diarylsubstituted 4,7-dihydro-1,2,4-triazolo[1,5-*a*]pyrimidines **17** is the reaction of 3-aminotriazoles **3** with chalcones **14** (Scheme 8.5, Pathway **a**) [37–43]. However, in contrast to the pseudo-three-component heterocyclizations involving only acetophenones (Scheme 8.3), the starting from the synthetic precursors of chalcones, aromatic aldehydes and acetophenones, does not change the direction of the reaction and the three-component cyclocondensation yielded the same products **17** as in case of the two-component interaction (Scheme 8.5, Pathway **b**) [44–46].

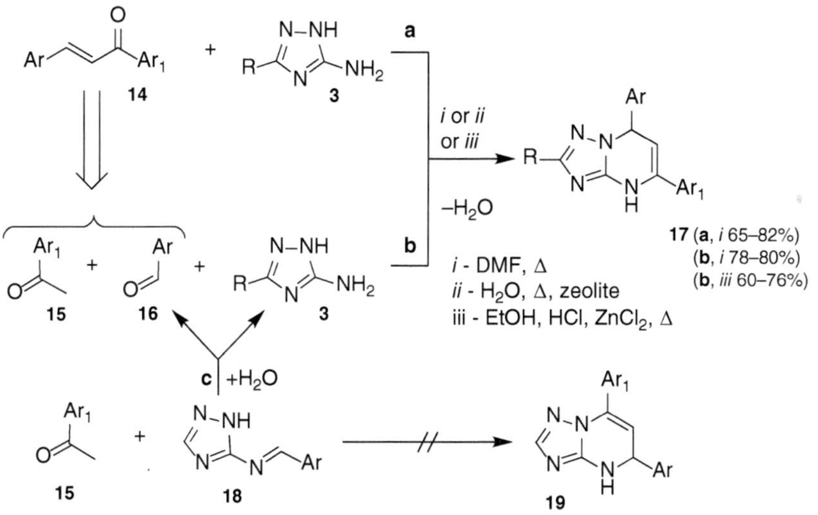

Scheme 8.5 MCR switching effect does not work.

It was shown [44, 45] that multicomponent procedure cannot be considered as an independent synthetic method in this case, instead, the cyclocondensation proceeded via the initial formation of α,β-unsaturated ketones like **14**.

It is also interesting that the same compounds **17** were isolated in the condensation of acetophenones **15** with azomethines **18,** while no trace of regioisomeric compounds **19** was found. Thus, the reason of this is the decomposition of imine **18** into starting aldehydes **16** and aminotriazole **3** during the reaction (Scheme 8.5, Pathway **c**).

Different behavior was observed in the case of arylidenacetones **20** – their reactions with 3-amino-1,2,4-triazole **3** led to the formation of dihydrotriazolopyrimidi **21** (Scheme 8.6, Pathway **a**) [41, 47–49], while heterocyclization involving aldehyde **16** and acetone **22** proceeded in another course yielding tetrahydrotriazolopyrimidines **23** (Pathway **b**) with the opposite location of substituents (methyl group and aryl) in pyrimidine ring [50, 51].

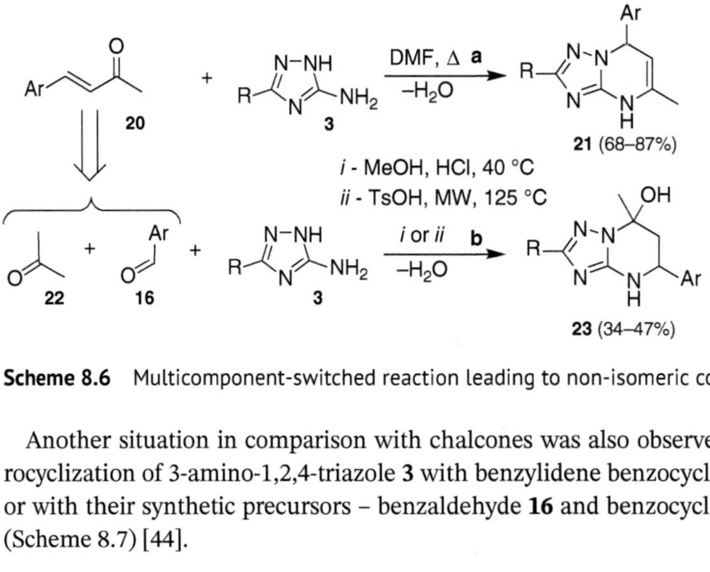

Scheme 8.6 Multicomponent-switched reaction leading to non-isomeric compounds.

Another situation in comparison with chalcones was also observed in the heterocyclization of 3-amino-1,2,4-triazole **3** with benzylidene benzocycloalkanones **24** or with their synthetic precursors – benzaldehyde **16** and benzocycloalkanones **26** (Scheme 8.7) [44].

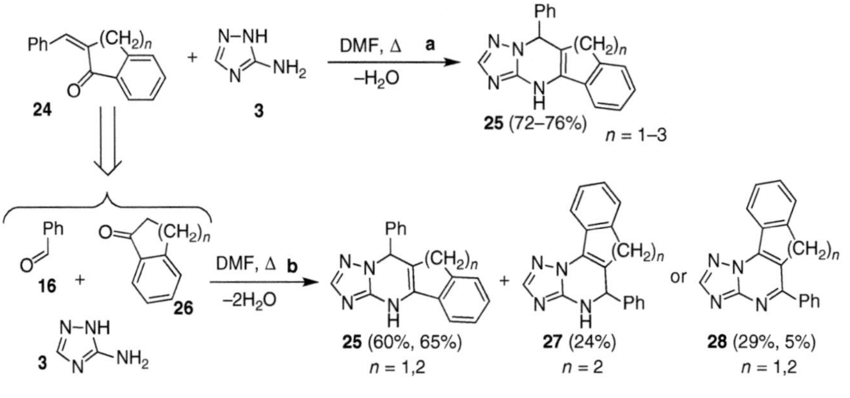

Scheme 8.7 Multicomponent-switched reaction as a side process.

For the two-component heterocyclization, the sole reaction products were compounds **25** (Scheme 8.7, Pathway **a**), while for three-component treatment in addition to the major products, isomeric compounds **27** ($n = 2$) and/or products of their dehydrogenation **28** ($n = 1,2$) were also isolated (here and in most cases the oxidation of dihydro azoloazines occurs in the reaction mixture as a redox process with the participation of oxygen of air or one of the intermediates). Thus, in this case, the multicomponent-switched heterocyclization is observed; however, the new direction is only a side process.

In publications [52–54], a similar situation is described – the reaction of 3-aminotriazole either with arylidencyclohexanones [52] or with their precursors – aldehyde and cyclohexanone [53, 54] – gave different final heterocyclic compounds of the linear structure of the tricyclic skeleton, in the first case, and with angular structure, in the letter case. However, the selectivity of these MCRs was also very low that led to the formation of mixtures of several final compounds.

Thus, above we described one type of multicomponent-switched reactions in which the control of the direction of heterocyclizations was carried out by increasing the number of reagents (e.g. transition from two-component reactions to MCRs). However, there is another type of such reaction when their course is tuned by the application of Condition-based Divergence Strategy [14, 15, 27].

As it has been already mentioned, there are several reasons for the divergent character of MCRs (Scheme 8.1): (i) not a single-stage synchronous character of MCRs but the possibility of several cascades of elementary stages giving different final products (for compounds A,B, and C these cascades are (AB) + C, (AC) + B, (BC) + A); (ii) existence of several alternative reaction centers in the reactants and, therefore, the possibility of the parallel competing reactions; (iii) proceeding rearrangements, post-cyclizations, etc; and (iv) proceeding more complex processes, e.g. pseudo-high-order MCRs. Condition-based Divergence is one of the strategies that allows MCR to be switched selectively from one direction to another.

It is worth to note that this strategy is applicable not only for MCRs – a two-component reaction can also be effectively redirected between several courses [14, 55]. For instance, heterocyclization of aryliden pyruvic acids **29** with 5-amino-*N*-aryl-1*H*-pyrazole-4-carboxamides **30** depending on the conditions applied can yield either pyrazolopyrimidine carboxylic acids **31** or furanones **32** (Scheme 8.8, Pathways **a** and **b**, correspondingly) [56].

However, there are other interesting cases when the switching of two-component heterocyclizations applies together with the multicomponent-switched strategy that gives an additional possibility to increase molecular diversity of the final heterocycles. Thus, the reaction of 5-aminopyrazole **33** with aryliden pyruvic acids **29** (Scheme 8.8) can lead to the formation of both dihydropyrazolopyrimidines **34** (Pathway **c**) [57] and pyrazolopyridines **35** (Pathway **d**) [58], while using the three-component reaction involving synthetic precursors of aryliden pyruvic acids, aromatic aldehydes **16** and pyruvic acid **36**, yields positional isomers of the latter, compounds **37** (Pathway **e**) [58].

A similar example was also described in another publication of Chebanov and coworkers [59]. It was established that the cyclocondensation of 3,4-disubstituted 5-aminopyrazoles **38** with alkyl esters of arylpyruvic acids **39** depending on the reaction conditions can proceed in two directions: the reaction in acetic acid under

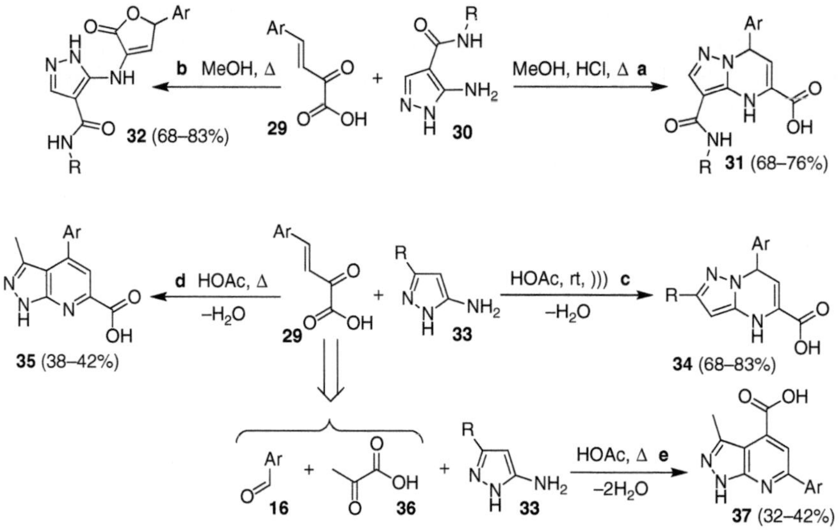

Scheme 8.8 More functions, more direction: the use of arylidene pyruvic acids and their precursors.

ultrasonication at room temperature yielded alkyl 6,7-dihydropyrazolo[1,5-*a*]pyrimidine-5-carboxylate **40** containing OH group in position 6 (Scheme 8.9, Pathway **a**). The same reaction under refluxing gave alkyl 6-hydroxypyrazolo[1,5-*a*]pyrimidine-5-carboxylates **41** (Pathway **b**).

Further, it was shown that three-component reaction of 5-aminopyrazoles **38** with synthetic precursors of arylidenpyruvic acid esters **39** (aromatic aldehydes **16** and pyruvic acid esters **42**) can also be switched between two different courses giving either alkyl 7-hydroxy-4,5,6,7-tetrahydropyrazolo[1,5-*a*]pyrimidine-7-carboxylates **43** (acetic acid, room temperature, ultrasonication, Pathway **c**) or heteroaromatized pyrazolopyrimidines **41** (acetic acid, refluxing, Pathway **d**).

Interesting results concerning tuning the selectivity of the similar reactions involving 3-amino-5-methylisoxazole were published by Morozova et al. [60]. It was shown that heterocyclization of arylidenpyruvic esters **39** and amine **44** under wide range of the reaction conditions led to the mixture of final compounds – derivatives of furanone **45** and pyrrolone **46** (Scheme 8.10, Pathway **a**). However, changing the unsaturated acids **29** with their potassium salts (R_1 = K) or with ethyl esters **39** (R_1 = Et) gave the possibility to obtain both of these compounds selectively. Thus, the reaction involving the potassium salts in acetic acid at room temperature yielded 3-(isoxazol-3-ylamino)furan-2-ones **45** (Pathway **b**) while in the case of the ethyl esters the heterocyclization in acetonitrile in the presence of scandium triflate at −15 °C proceeded toward the formation of 1-(isoxazol-3-yl)-3-(isoxazol-3-ylamino)-1,5-dihydro-pyrrol-2-ones **46** (Pathway **c**).

Unexpectedly, the three-component reaction of synthetic precursors of acids **39** (aromatic aldehydes **16** and derivatives of pyruvic acid **47**) with 3-amino-5-methylisoxazole **44** in all cases gave only the mixture of both possible compounds (Pathway **d**). In turn, the third type of final compounds, 3-hydroxy-1-(isoxazol-3-yl)-1,

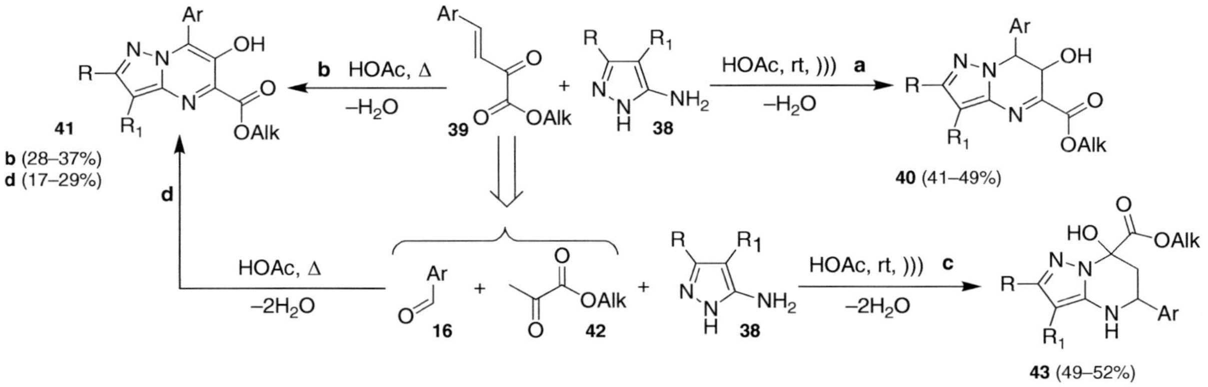

Scheme 8.9 Switched reaction pathways of arylidene pyruvic acid esters or their precursors.

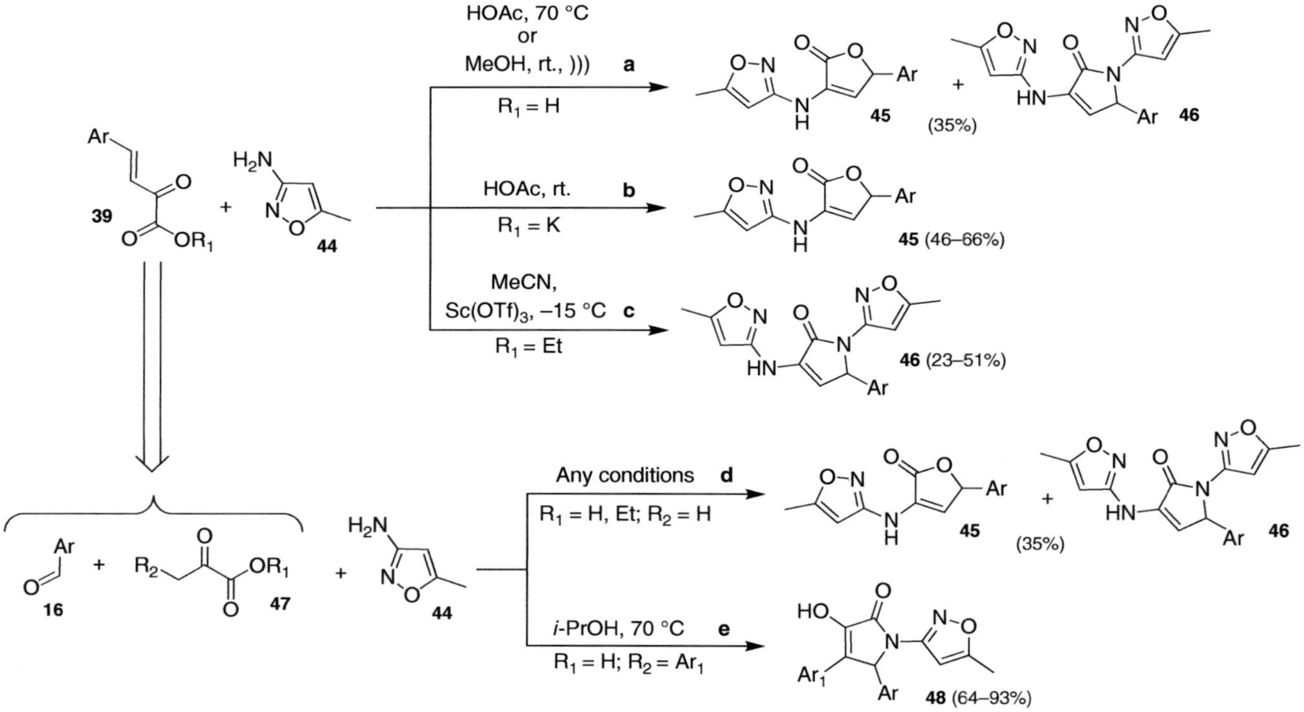

Scheme 8.10 Multicomponent-switched reaction of 3-amino-5-methylisoxazole and arylidene pyruvic acids or their precursors.

5-dihydropyrrol-2-ones **48**, was isolated in the multicomponent condensation involving arylpyruvic acid (Pathway **e**).

One of the widespread and effective methods to tune the selectivity of MCRs is a variation of properties of the reaction medium – solvent and catalysts sometimes along with other parameters. For instance, two very similar examples of switchable MCRs of this type were published by Tu et al. [61] and Liu et al. [62]. In both cases, CCN binucleophile (either 2,6-diaminopyrimidin-4-one or 5-amino-3-methyl-1-phenylpyrazole) reacted with aldehydes **16** and 4-hydroxychromen-2-one **49**; depending on the reaction conditions, different final products were selectively obtained (Scheme 8.11). In the case of 2,6-diaminopyrimidin-4-one, the reaction under MW irradiation in DMF yielded tetracyclic compounds **51** (Pathway **a**), while the same three-component treatment in the mixture of acetic acid/DMF (1 : 4) gave other final heterocycles **52** (Pathway **b**) [61].

For 5-amino-3-methyl-1-phenylpyrazole, the three-component reaction with the same reagents can also proceed in similar pathways **c** and **e** (Scheme 8.11) giving either dihydrochromenopyrazolopyridinone heterocycles **53** (in the mixture of acetic acid/acetonitrile (1 : 5) at reflux) or tetrahydropyrazolopyridinones **55** (in the mixture of acetic acid/ethanol (1 : 20) at reflux) [62]. The third direction with the formation of heteroaromatized chromenopyrazolopyridinone **54** (Pathway **d**) was observed also in the mixture acetic acid/acetonitrile (1 : 5), however, at the temperature 160 °C along with increasing the reaction time from 6 to 10 hours.

In all these cases, the three-component reaction proceeds through the formation of intermediate **50** (Scheme 8.11) which then can transform into tetracyclic final products **51** or **53** (**54**) (the reaction involve carbonyl group) or into heterocycles **52** or **55** (the reaction involve lactone fragment).

Tu and colleagues in another publication [63] demonstrated the switching of MCR involving aromatic aldehydes **16**, 5-amino-3-methyl-1-phenylpyrazole **56**, and cyclopentanone **57** depending on the type of the solvent/catalyst system (Scheme 8.12).

There are two courses possible: in acetic acid under MW heating at 120 °C the reaction proceeds through the formation of azomethine **58** ant its further cyclization with cyclopentanone giving 3,6,7,8-tetrahydrocyclopenta[*d*]pyrazolo[3,4-*b*]pyridine **59** (Pathway **a**); in the mixture DMF/NaOH with the same all other conditions on the first of the reaction diarylidencyclopentanone **60** forms and then reacts with aminopyrazole yielding 1,5,6,7-tetrahydrocyclopenta[b]pyrazolo[4,3-*e*]pyridines **61** (Pathway **b**). Thus, the variation of the acid–base properties of the catalysts allows the authors to redirect the treatment due to the changing the first stage of the reaction – acidic medium facilitates the formation of the imine, while basic conditions promote the Claisen–Schmidt reaction involving aldehyde and methylene active cyclopentanone.

Lipson and coworkers studied in detail MCRs involving aminoazoles **3**, aromatic aldehydes **16**, and Meldrum's acid **62** and found that these cyclocondensations were rather sensitive to the variation of the solvent/catalyst system; however, the selectivity of some directions were low [64–68]. Thus, 3-amino-1,2,4-triazole **3** (R_1 = H) in this three-component reaction gave either triazolopyrimidin-5-ones

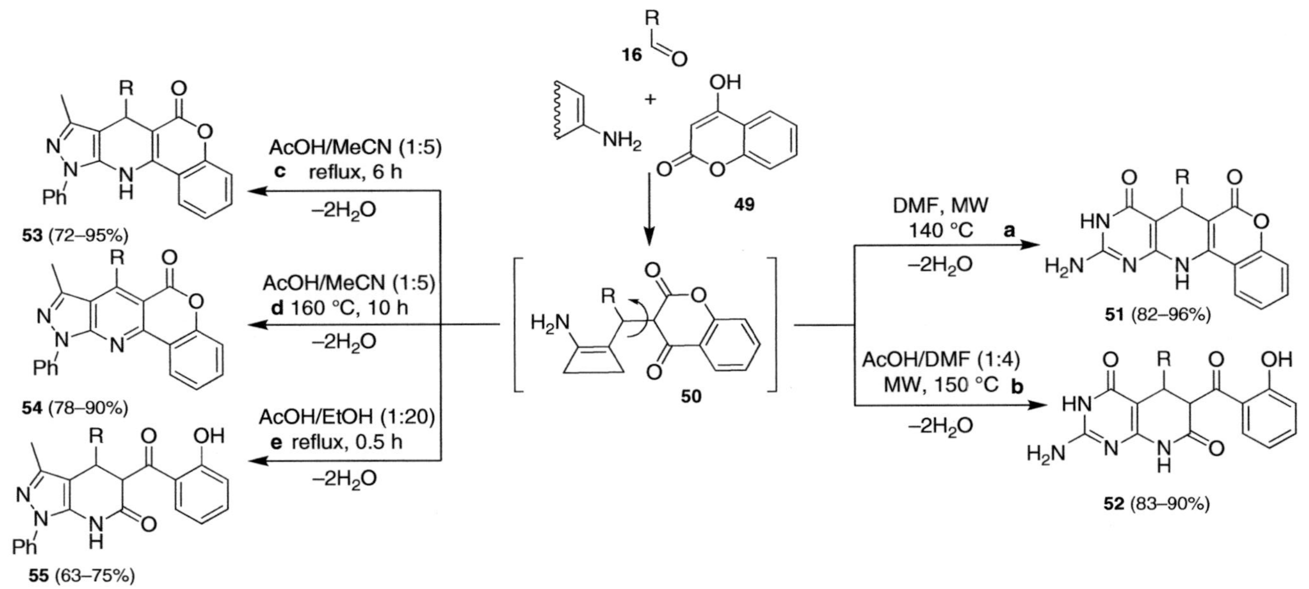

Scheme 8.11 Tuning the selectivity of MCR by a variation of solvent.

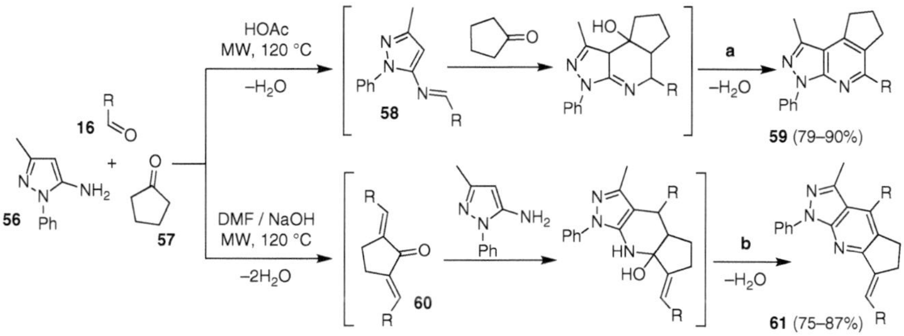

Scheme 8.12 Tuning the selectivity of MCR by a variation of solvent/catalyst system.

63 (Scheme 8.13, Pathway **a**) or noncyclic compounds **64** together with *N*-(1,2,4-triazol-5-yl)acetamide (Pathway **b**) [65].

Reactions of 5-amino-3-methylthio-1,2,4-triazole (R_1 = SMe) and 3,5-diamino-1,2,4-triazole ($R_1 = NH_2$) in all cases yielded heterocyclic compounds, but their structures are dependent on the reaction conditions. For the first aminotriazole (R_1 = SMe) treatment in ethyl acetate/pyridine led to the formation of a mixture of two compounds – triazolopyrimidin-5-ones **65** and isomeric triazolopyrimidin-7-ones **66** (Pathway **c**), while the condensation in boiling DMF gave only one isomer – heterocycles **65** (Pathway **d**) [67]. The reaction involving aminotriazole with $R_1 = NH_2$ was different and allowing formation of only triazolopyrimidin-7-ones **67** (refluxing in MeOH or *i*-PrOH, Pathway **e**), while the treatment in boiling DMF was unselective and yielded two isomers again **68**, **69** (Pathway **f**) [68].

The heterocyclization of 2-aminobenzimidazole **70** with Meldrum's acid **62** and aromatic aldehydes **16** was strictly switched between two directions [66] – the reaction in boiling DMF or in boiling nitrobenzene yielded positional isomers **71** and **72**, correspondingly (Scheme 8.13, Pathways **g** and **h**). The selective formation of benzo[4,5]imidazo[1,2-*a*]pyrimidin-4(1*H*)-one **71** in higher yields was described by Sheibani et al. [69] in the same reaction in boiling ethanol in the presence of L-proline, while isomeric pyrimidin-2-one derivatives **72** were obtained by Yao et al. [70] in ionic liquid [bmim$^+$][BF_4^-].

A different situation was observed when Meldrum's acid was introduced in MCRs with aldehydes **16** and 5-aminopyraziles or isoxazole **73** (Scheme 8.14) [71]. For X = NPh and O, the three-component reaction in aqueous medium under MW irradiation yielded spiro{pyrazolo[1,3]dioxanopyridine}-4,6-diones **74** (Pathway **a**), while in DMF or acetic acid the final products were pyrazolo[3,4-*b*]pyridines **75** (Pathway **b**). Interesting, that for X = NH and NMe the latter heterocycles were obtained under all conditions including boiling in MeOH, MeCN, DMF, $C_6H_5NO_2$ [72], or MW synthesis in water (Pathway **c**).

Chebanov and coworkers [73] developed other efficient and reproduceable procedures to switch such MCRs between these two directions by the example of 5-amino-3-methylisoxazole treatment (Scheme 8.15).

This MCR of 5-amino-3-methylisoxazole **76**, aromatic aldehydes **16**, and Meldrum's acid **62** under ultrasonication of the starting materials in EtOH at room

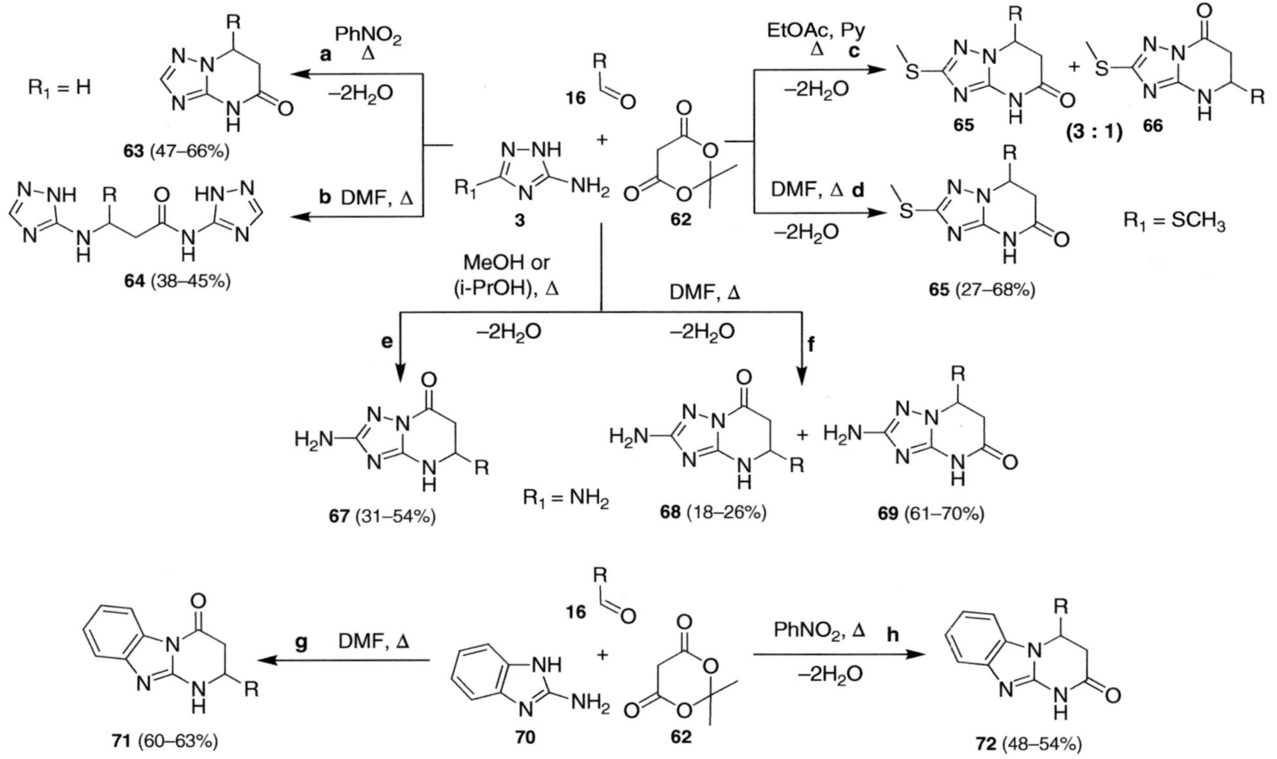

Scheme 8.13 Tuning the selectivity of MCR of Meldrum's acid by a variation of solvent.

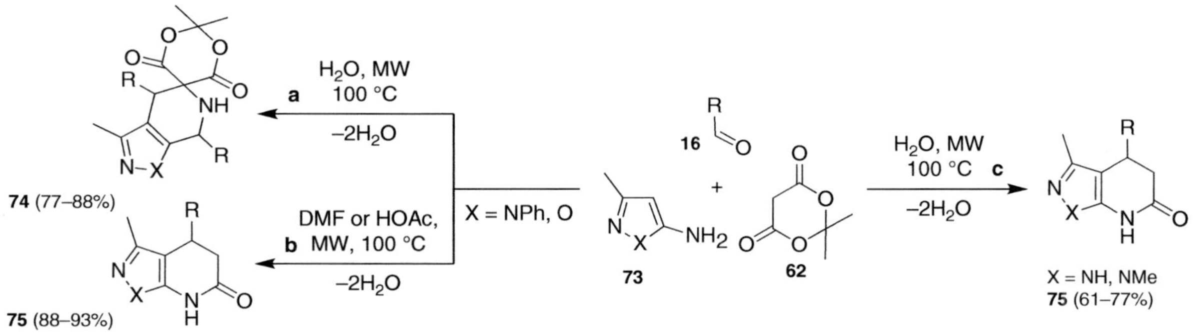

Scheme 8.14 Influence of binucleophile structure on the direction of MCR.

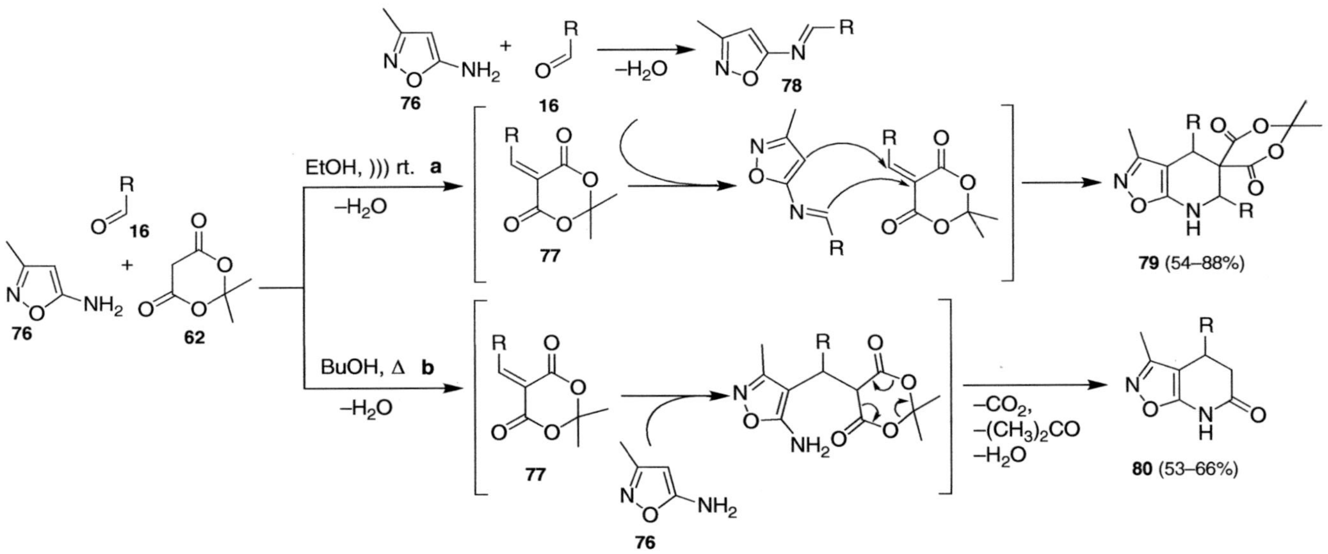

Scheme 8.15 Switching of MCR between 5-amino-3-methylisoxazole, aldehydes, and Meldrum's acid.

temperature resulted in the formation of spiroheterocycles **79** (Scheme 8.15, Pathway **a**), while the refluxing of the reaction mixture in *n*-BuOH gave dihydroisoxazolopyridinones **80** (Pathway **b**).

The plausible mechanisms for both directions begin from the formation of the arylidene derivatives of Meldrum's acid **77**, but then there are two alternative sequences: pathway **b** occurring under heating supposes the nucleophilic addition of the CH reaction center of aminoisoxazole **76** to the double bond of unsaturated compound **77**. Further elimination of acetone and CO_2 results in unstable ketene which condenses in the pyridone **80**. In contrast, the formation of the spiro-substance **79** observing at room temperature presumably proceeds as hetero-Diels–Alder reaction between arylidene **77** and azomethine **78** forming *in situ* (Pathway **a**).

Among the MCRs involving aminoazoles, methylene active compounds, and aldehydes, the condensation with the participation of glyoxals occupies a special place. These processes are characterized by a diversity of possible directions; however, among the numerous examples of such reactions, there are only a few of them allow switching their course [14].

For instance, it was found that the three-component reaction of amine **70**, cyclohexane-1,3-dione **81**, and arylglyoxal hydrates **82**, regardless of the process conditions always gave enolates **83** at the first stage (Scheme 8.16) [74]. Subsequent degradation of these compounds led to the formation of intermediates **84** which then cyclized in acetic acid mostly into imidazoquinazolinones **85** (the interaction of amino group with carbonyl group of the cyclohexanedione moiety), while in DMF at higher temperature their predominant conversion into compounds **86** was observed (the interaction of amino group with the carbonyl of the arylglyoxal moiety).

MCRs of arylglyoxals **82** with cyclohexanediones **87** and 5-aminopyrazoles **88** can also be switched between several directions depending on the reaction conditions and the structure of aminoazole. In the case of R_1 = CONHAr, the treatment via the formation of intermediate **89** (arylglyoxal reacted with exocyclic NH_2 group of aminopyrazole and active methylene center of cyclohexanedione) yielded 5,7,8,9-tetrahydropyrazolo[1,5-*a*]quinazolin-6-ones **90** (Scheme 8.17, Pathway **a**) [75].

The same reaction involving 4-unsabstituted 5-aminopyrazoles **88** (R_1 = H) gave either 1,4,6,7,8,9-hexahydropyrazolo[3,4-*b*]quinolin-5-ones **92** (under refluxing in ethanol, Pathway **b**) [75] or, under MW irradiation in DMSO-Et_2N at 120 °C, products of their subsequent heterocyclization with the second molecule of arylglyoxal – 3,9-dihydropyrazolo[3,4-*b*]pyrrolo[3,2,1-*ij*]quinolin-5-ones **94** (Pathway **c**) [76]. The last two reactions proceed via the preliminary formation of intermediate **91** – in contrast to Pathway **a** in this case arylglyoxal reacts with the CH reaction center of aminopyrazole instead of its exocyclic NH_2 group.

It is also interesting that MCRs involving arylglyoxals can be rather sensitive to the structure of the reagents that allows to switch them selectively between several courses depending on the type of the substituent or on the type of the binucleophile. For instance, Jiang et al. [77] in the reaction of 5-aminopyrazoles **95** and anilines

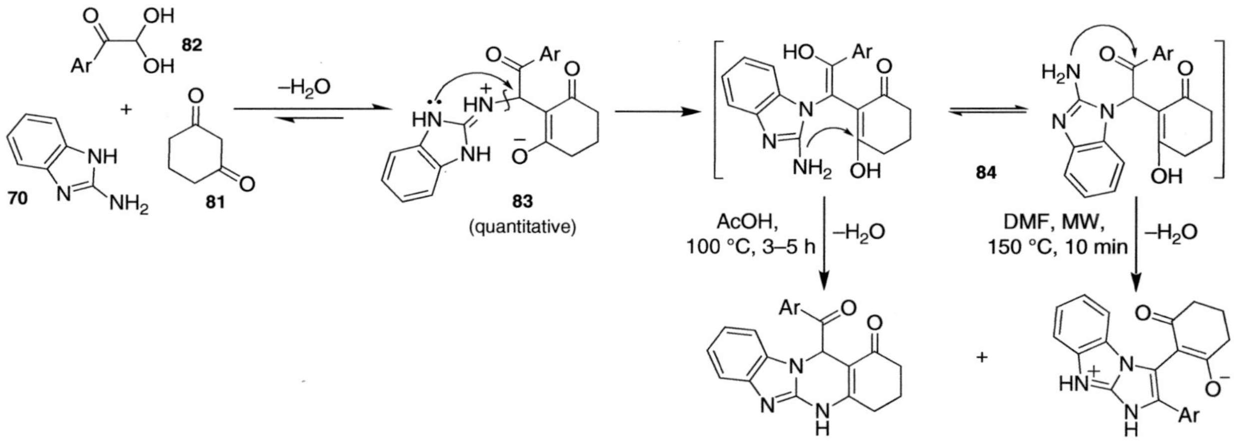

Scheme 8.16 Tuning the selectivity of MCR by a variation of solvent and activation type.

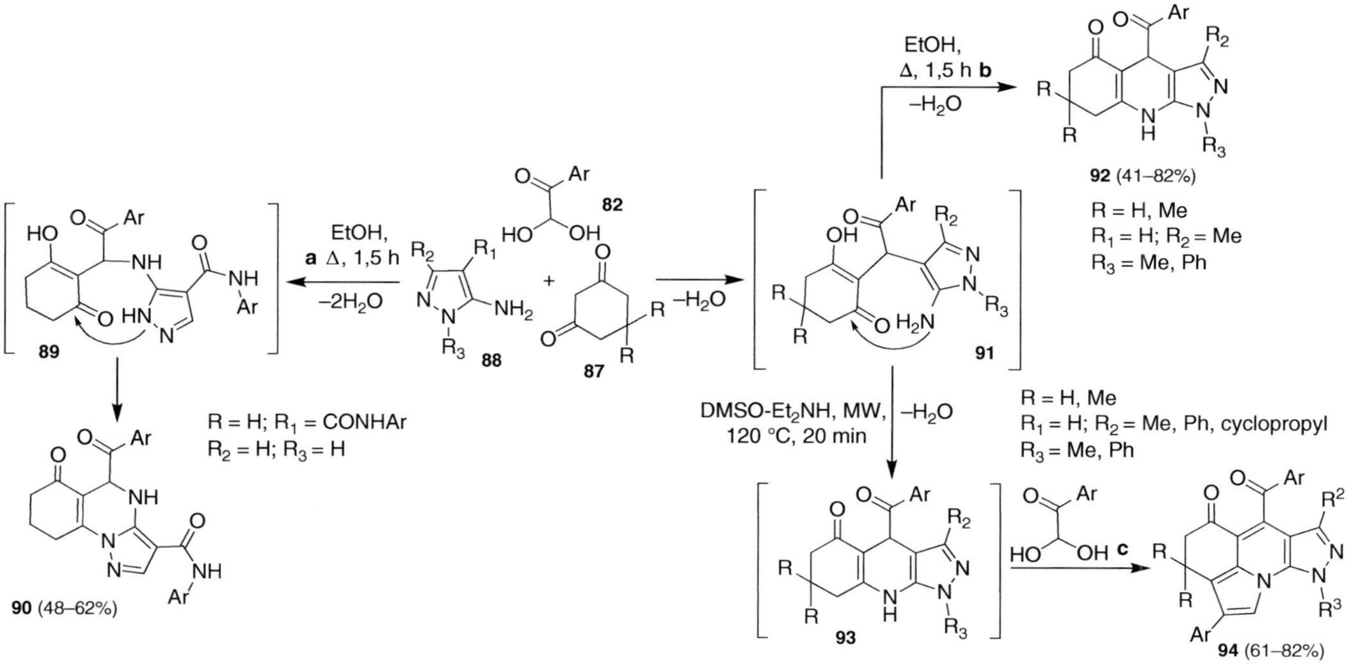

Scheme 8.17 Switching of MCR of arylglyoxals, cyclohexanediones, and 5-aminopyrazoles.

96 with two equivalents of arylglyoxals **82** (Scheme 8.18) described the isolation of either azepinoindoles **99** or pyrazolopyridines **100** under the same conditions. In both cases, the treatment passed via the formation of two intermediates with the participation of arylglyoxal – compound **97** and two different protonated forms of iminone (**98** and **98**). The further attack of imino group of the intermediate **97** on the carbon of either C=N or C=O group allowed to proceed the reaction through several elementary stages in the Pathway **a** (R_2 = Ph, R_3 = 4-Me) or the Pathway **b**, correspondingly.

In turn, Mishra and colleagues showed [78] that MCRs of arylglyoxals with 4-hydroxycoumarin and some amines led to the formation of both pyridine cycle (in case of aminopyrazole or aminoisoxazole, compounds **101**, Scheme 8.18) and pyrrole cycle (in case of aminocoumarine or 6-aminouracil, compounds **102** and **103**).

There are also possible pseudo-MCRs with participation of only two components, arylglyoxals and aminoazoles, which selectively give diverse types of final heterocycles depending on the reagents structure [79, 80].

Pyruvic acid and its derivatives are also an interesting object for the application in switchable MCRs that was shown in several publications of Chebanov and colleagues. In some cases, the reactions of pyruvic acids with binucleophiles and carbonyl compounds can proceed nonselectively but, on the other hand, there are a lot of examples of successful tuning their selectivity.

In publication [81], it was established that three-component reaction of pyruvic acid **36**, aldehydes **16**, and 3-amino-1,2,4-triazole **3** in boiling DMF was not selective and gave the mixture of two compounds–triazolopyrimidine carboxylic acids **104** and **105**, forming with the participation of alternative endocyclic reaction centers of aminotriazole (Scheme 8.19, Pathway **a**). The authors did not develop synthetic procedures allowing obtaining both of these heterocycles selectively; however, 4,7-dihydro[1,2,4]triazolo[1,5-*a*]pyrimidine-5-carboxylic acids **106** were isolated as sole products of the reaction carried out in acetic acid under refluxing (Pathway **c**) [81, 82].

The three-component reaction of the same reagents in acetic acid at 65 °C changed the direction and yielded other heterocyclic compounds–7-hydroxy-4,5,6,7-tetrahydro[1,2,4]triazolo[1,5-*a*]pyrimidine-7-carboxylic acids **107** (Pathway **b**) [82]. An additional interesting observation was reported in this publication – heterocycles **107** can be transformed into compounds **106** via disintegration into the starting materials under refluxing in acetic acid (Pathway **d**).

Cyclocondensations of pyruvic acids are sensitive not only to the reaction conditions but also to the structure of reagents. For instance, the three-component reaction of 3-amino-1,2,4-triazole and aldehydes with arylpyruvic acids **108** instead of pyruvic acid **36** under boiling in acetic acid gave triazolylpyrrolones **109** (Pathway **e**) but not heterocyclic carboxylic acids like **106** (Pathway **c**) [82, 83]. Changing heating to ultrasonication at room temperature switched the reaction course again toward tetrahydrotriazolopyrimidine carboxylic acids **110** (Pathway **f**) as it had been observed for the heterocyclizations involving pyruvic acid at 65 °C (Pathway **b**). Heterocycles **110** can be transformed into pyrrolones **109** by refluxing in acetic

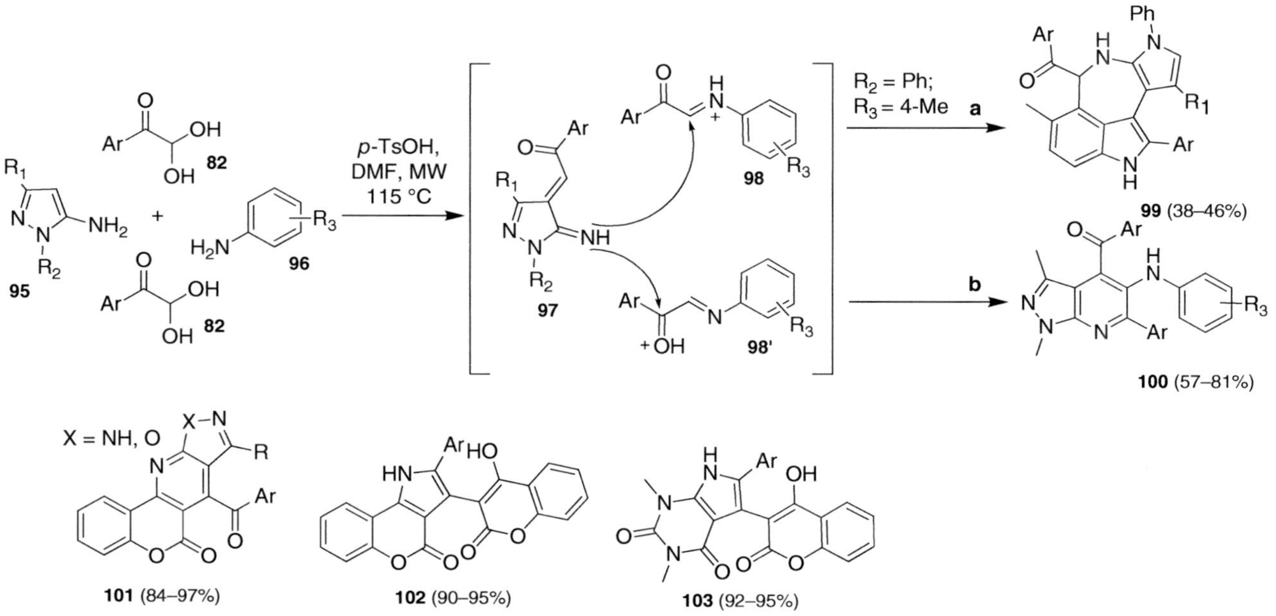

Scheme 8.18 Other examples of switching of MCR involving arylglyoxals.

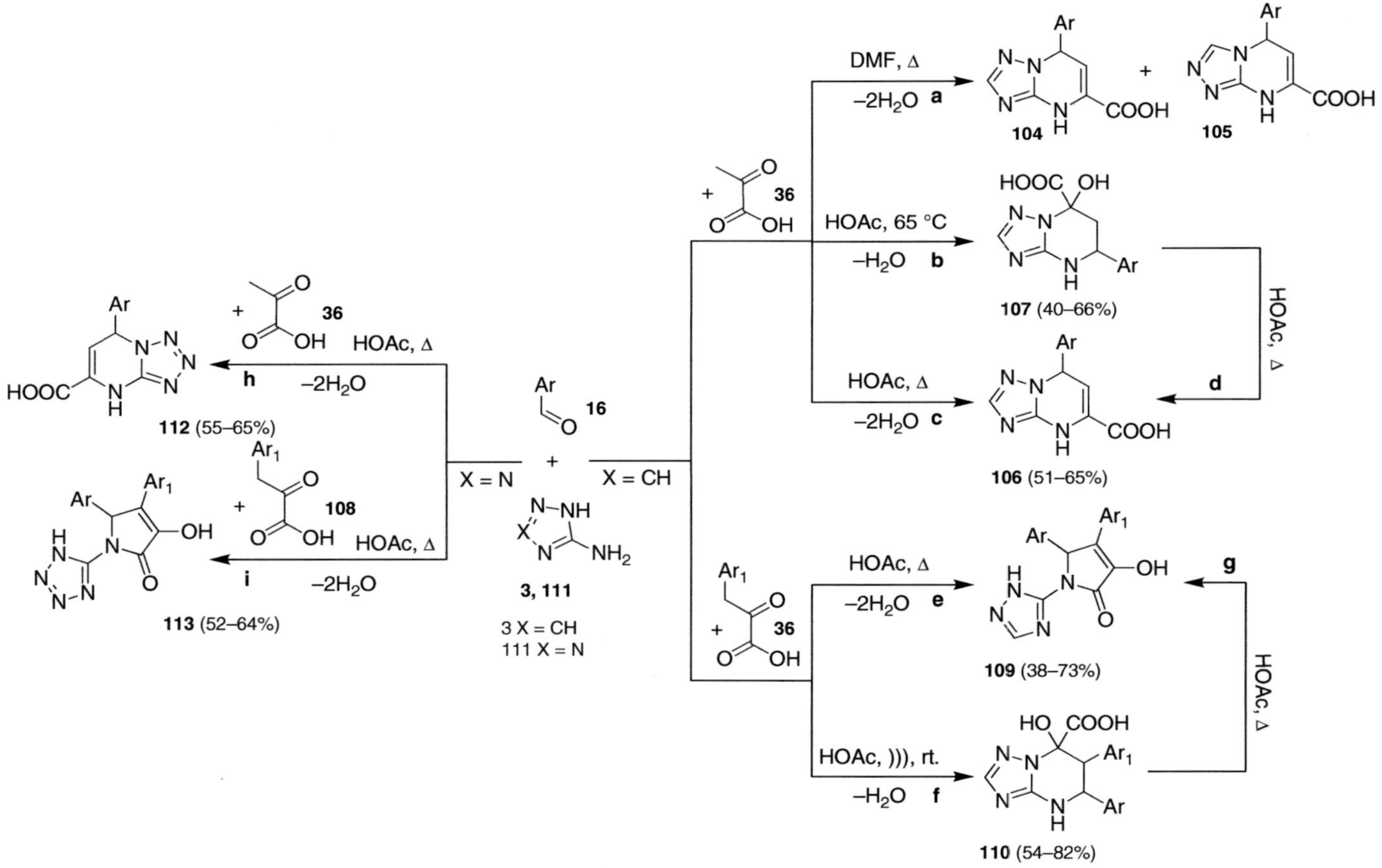

Scheme 8.19 Tuning the selectivity of MCR involving pyruvic acids.

acid as well (Pathway **g**). Similar behavior was described for MCRs involving some 4-substituted 5-aminopyrazoles [58, 82, 83].

The three-component reaction of 5-aminotetrazole **111** and aromatic aldehydes **16** also depends on the structure of pyruvic acids **36** [81, 83]. It was found that MCR involving pyruvic acid **36** in boiling acetic acid gave "classical" for such reactions 4,7-dihydrotetrazolo[1,5-*a*]pyrimidine-5-carboxylic acid **112** (Scheme 8.20, Pathway **h**) while the heterocyclization of arylpyruvic acids **108** yielded under the same conditions 3-hydroxy-1-(tetrazol-5-yl)-1,5-dihydropyrrol-2-one **113** (Pathway **i**).

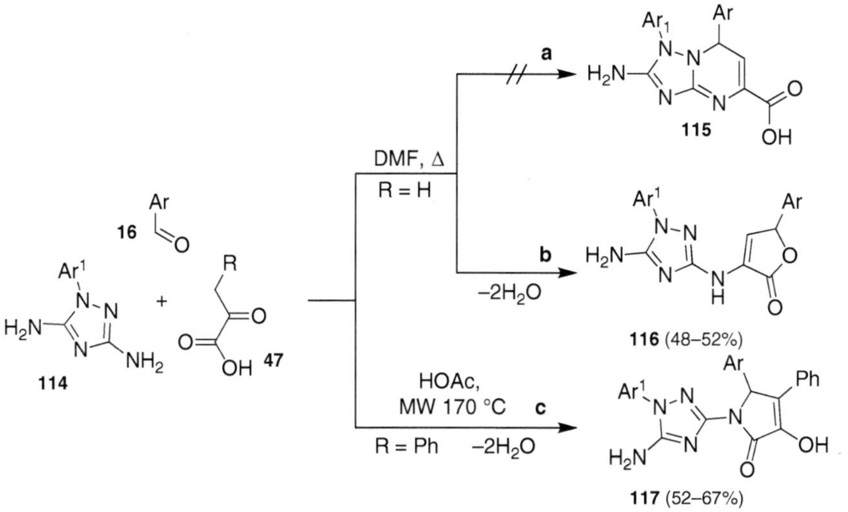

Scheme 8.20 Influence of binucleophile structure on the direction of MCR involving pyruvic acids.

In some cases, the structure of binucleophile can also affect the course of the heterocyclizations of such type.

As it was demonstrated by Sakhno et al. [83, 84], an introduction of aryl substituent in position 1 of 3-amino-1,2,4-triazol **114** (Scheme 8.20) change the direction of its reaction with pyruvic acid **47** (R = H) and aldehydes **16** from the formation of triazolopyrimidine carboxylic acids like **115** (Pathway **a**) toward 3-((1,2,4-triazol-3-yl)amino)furan-2(5*H*)-ones **116** (Pathway **b**). According to the authors hypothesis, the alternative direction of the cyclization is associated with a loss of aromaticity by the aminoazole fragment during proceeding the reaction in Pathway **a** [84].

The reaction of aminoazole **114** and aldehydes **16** with phenylpyruvic acid **47** (R = Ph) in acetic acid under conventional or microwave heating yielded triazolylpyrrolones **117** (Pathway c) as it had been observed for 1-unsubstituted 3-amino-1,2,4-triazoles (Scheme 8.19, Pathway **e**).

In some examples described above, the significant difference of the reaction temperature was the main factor that affected the reaction behavior. Indeed, numerous organic reactions may proceed either under kinetic or under thermodynamic control [85] and one of the powerful and very effective tools to switch them between these pathways is the variation of the reaction temperature. One of the classical examples, included in most of organic textbooks, is the application of temperature regime to control the selectivity of naphthalene sulfonation proceeding at the α-position under 80 °C and at the β-position under 160 °C [86]. Another well-known process of such type is an addition of HBr to 1,3-butadiene to form either 3-bromo-1-butene (at −80 °C) or 1-bromo-2-butene (at 40 °C) as the major products [87]. It is important that the formation of kinetically controlled compounds should be reversible and, thereby, under heating them can be converted into thermodynamically controlled compounds.

For MCR, the variation of the reaction temperature to direct the treatment toward kinetic or thermodynamic pathway is another effective tool applied for swathing the reaction selectivity. Chen and colleagues showed [88] that MCR of some 5-amino-1,2,4-triazols **3**, aromatic aldehydes **16,** and derivatives of acetoacetic esters **118** in water in the presence of *p*-TsOH at 80 °C gave a mixture of two heterocycles **119** and **120** (Scheme 8.21, Pathway **a**). In the further publication of another scientific group, it was demonstrated that such type of the reaction gave good perspectives to switch them between several routes by changing of the reaction parameters including the key one – the temperature regime.

First of all, in most cases, similar three-component reactions under heating in numerous solvents yield only one heterocyclic system – 4,7-dihydro[1,2,4]triazolo[1,5-*a*]pyrimidine-6-carboxylates **121** (Scheme 8.21, Pathway **b**) [14, 15, 89]. In contrast, Gorobets and coworkers [89] found that carrying out the same treatments at the lower temperature (in EtOH/HCl at 40 °C or in water at room temperature) changed their course to the formation of another type of final heterocyclic compounds – 5-aryl-7-hydroxy-4,5,6,7-tetrahydro[1,2,4]triazolo[1,5-*a*]pyrimidine-6-carboxylates **122** (Pathway **c**) with different location of substituents in pyrimidine ring. The structure of these compounds was ultimately proved by NMR and X-ray data. Moreover, it was shown that these products under higher temperature underwent rearrangement and dehydration into the previously known thermodynamic control products **121** (Pathway **d**).

It is interesting that two groups, Shaabani et al. [90] and Koryakova et al. [91], studied very similar reactions and established the formation of other final compounds – 7-aryl-5-hydroxy derivative of tetrahydro[1,2,4]triazolo[1,5-*a*]pyrimidine-6-carboxylates **123** (Pathway **e**), e.g. intermediate of the "classical" 4,7-dihydro[1,2,4]triazolo[1,5-*a*]pyrimidine-6-carboxylates **121**. However, there is no strong evidence of the formation of such compounds in the corresponding publications.

Chebanov and colleagues described a similar reaction with the participation of acetoacetamides **124**, aromatic aldehydes **16**, and two aminoazoles – 3-amino-1,2,4-triazole **125** or substituted 5-aminopyrazoles **126** [92, 93]. It was found, that under ultrasonication in ethanol at room temperature, only kinetically controlled direction occurred with the selective formation of the tetrahydrotriazolo- or tetrahydropyrazolopyrimidines **127** and **129** (Scheme 8.22, Pathways **a** and **c**). The same reaction either in refluxing DMF or under microwave irradiation at 120 °C in ethanol proceeded as thermodynamically controlled step sequences in

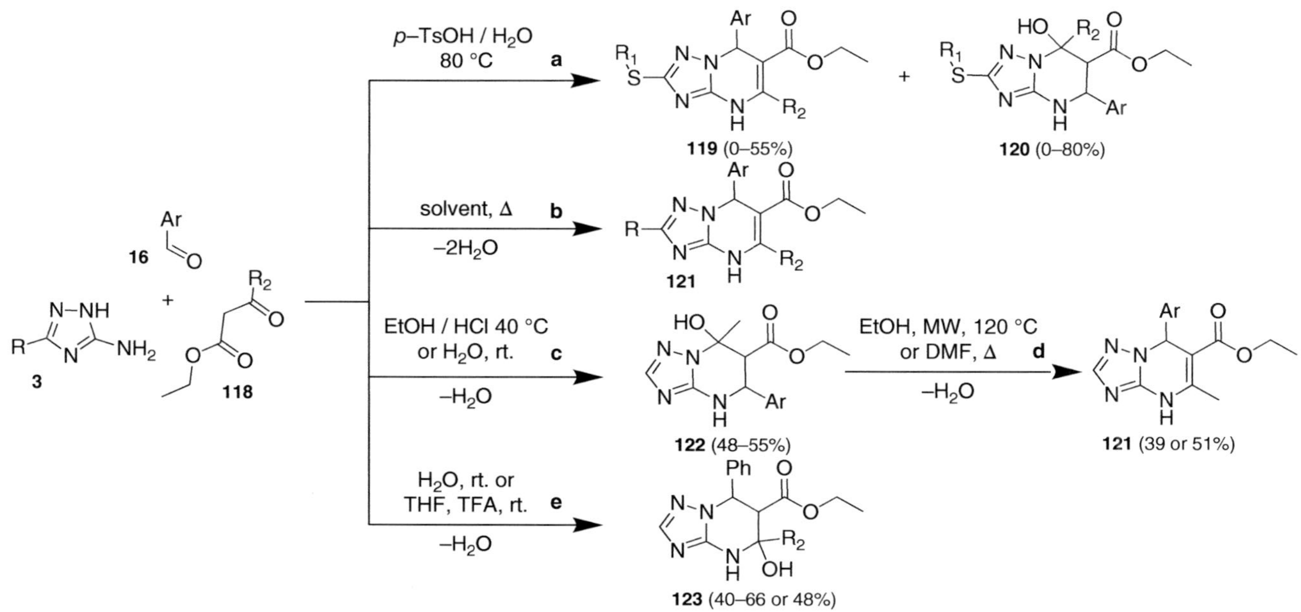

Scheme 8.21 Kinetic control *vs* thermodynamic control in MCRs of 5-amino-1,2,4-triazols, aromatic aldehydes, and acetoacetic esters.

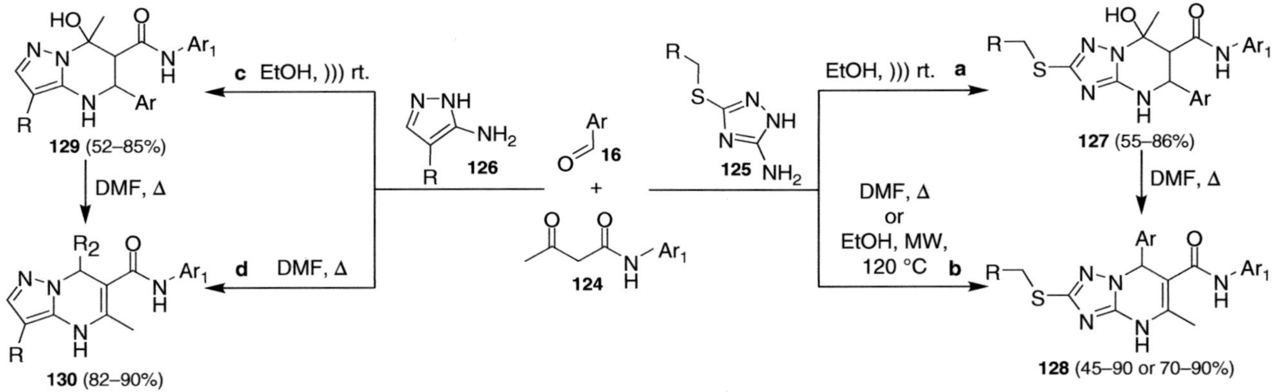

Scheme 8.22 Kinetic control *vs* thermodynamic control in MCRs involving acetoacetamides, aldehydes, and aminoazoles.

both cases and yielded dihydrotriazolo- and dihydropyrazolopyrimidines **128** and **130** via skeletal rearrangement (Pathways **b** and **d**).

As it was observed in the previous example (compounds **127**, **129** and **128**, **130**, Scheme 8.22), kinetically and thermodynamically controlled products have the different location of substituents (aryl and methyl groups) in partially hydrogenated pyrimidine ring. However, under refluxing in DMF, the first compounds can be easily converted into the latter ones. According to the data presented in the publications [93], such conversion included decomposition of the tetrahydropyrimidine ring, the formation of the starting materials, and their reaction via thermodynamically controlled pathway.

An additional possible pathway for such reactions was described by Gadara and Ladva [94]. The reaction was carried out in a small amount of DMF as solvent at 120–130 °C (practically solid-phase synthesis). According to the authors, the heterocyclization yielded 5,8-dihydro[1,2,4]triazolo[4,3-*a*]pyrimidine-6-carboxamides **132** (Scheme 8.23) which were formed with the participation of NH_2 group and endocyclic 4-NH reaction center of aminotriazole instead of 2-NH one. However, no evidence of the formation of this type of final product is provided in the article while spectral and other data are in good accordance with the structure of usual products of similar treatments – heterocycles **133** having triazolo[1,5-*a*]pyrimidine skeleton. It should be noted that there are only several proven examples of obtaining triazolo[4,3-*a*]pyrimidines in MCRs involving 3-amino-1,2,4-triazole as a 1,3-binuclephile and highly reactive carbonyl compounds [89].

As it was found by Muravyova et al. [95], the temperature is also the main factor that influences the structure of final heterocyclic compounds in the three-component reaction of barbituric acids **134**, 5-aminopyrazoles **95**, and aromatic aldehydes **16** (Scheme 8.24). Under refluxing in DMF or under microwave irradiation depending on the nature of the substituent R_1, the treatment led to the formation of either pyrazolopyridopyrimidines **139** ($R_1 \neq H$, Scheme 8.24, Pathway **a**), or their dihydro derivatives **138** ($R_1 = H$). The treatment of the same reagents under ultrasonication at room temperature switched toward the four-component heterocyclization yielding spiroheterocycles **140** (Pathway **b**).

It was proven [95] that in both cases, these MCRs proceed through the formation of arylidenebarbituric acid **135**. On the next step under high temperature, it reacts with 5-aminopyrazole to form the second intermediate which is cyclized into pyrazolopyridopyrimidines **138** or **139**. On the other hand, at room temperature under ultrasonication, the reaction includes also the formation of imine **137** between 5-aminopyrazole and aldehyde, which then reacts with arylidenebarbituric acid according to the mechanism of a formal aza-Diels–Alder reaction.

The variation of temperature becomes a much more effective tool to switch MCRs between several courses by its combination with other methods of selectivity tuning. For instance, it is known that the three-component reaction of 5-amipyrazoles, aldehydes, and derivatives of 1,3-cyclohexanedione in most cases yielded only one type of reaction product based on pyrazolo[3,4-*b*]quinolin-5-one skeleton (compounds **141**, Scheme 8.25, see, for example, [96–99]). However, this treatment under conventional thermal conditions sometimes leads to the

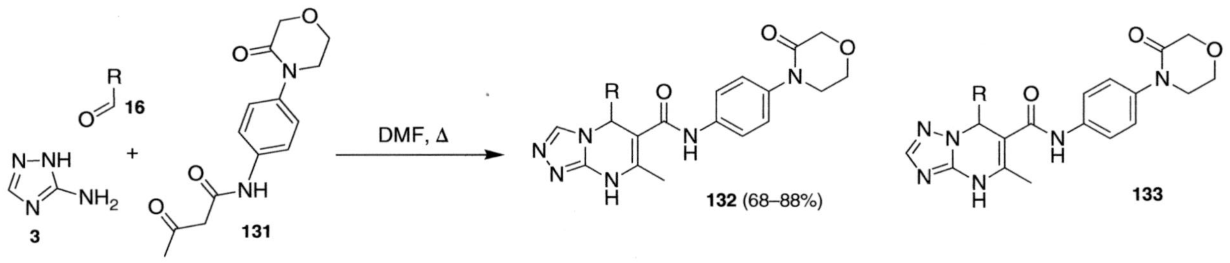

Scheme 8.23 Unproven pathway of MCRs involving acetoacetamides, aldehydes, and aminoazoles.

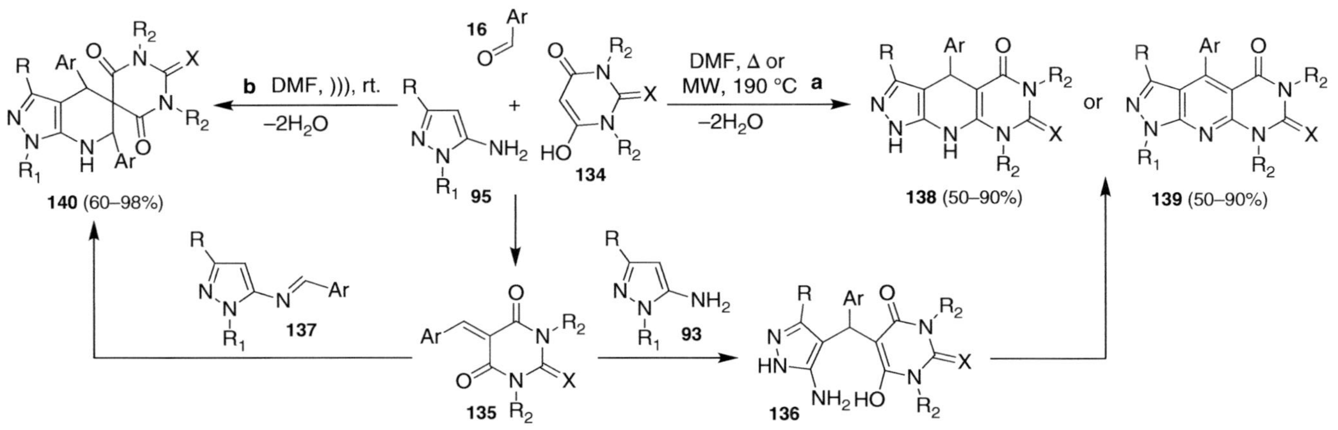

Scheme 8.24 Kinetic control *vs* thermodynamic control in three-component reaction of barbituric acids, 5-aminopyrazoles, and aromatic aldehydes.

formation of the mixtures containing both pyrazolo[3,4-*b*]quinolin-5-one **141** and pyrazolo[5,1-*b*]quinazolin-8-one **142** (Scheme 8.25, Pathway **a**) [100, 101]. Such situations give a good chance to tune the selectivity of this MCR.

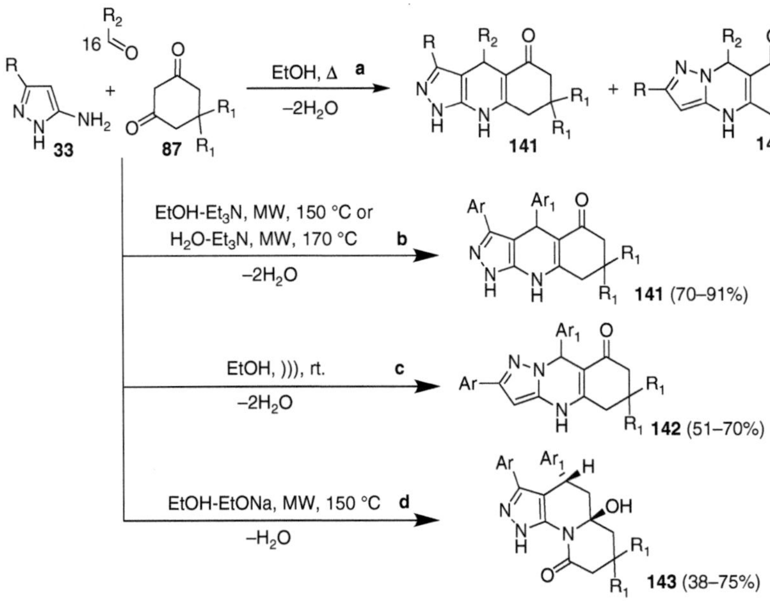

Scheme 8.25 From one set of the starting materials to three different final compounds by variation of solvent/catalyst/activation types.

Indeed, the application of nonclassical methods of activation, the variation of temperature, and the addition of certain catalysts allowed to direct this treatment to three routs giving different heterocyclic systems. The microwave-assisted reaction of the reagents in EtOH containing catalytic amounts of Et_3N at 150 °C [101] or in water at 170 °C in the presence of triethylamine [102] enabled pyrazolopyridines **141** to be obtained selectively in high yields via a thermodynamically controlled pathway (Scheme 8.25, Pathway **b**). In contrast, at room temperature under ultrasonication of the equimolar mixture of the starting materials in EtOH without any catalyst, this MCR proceeded as a kinetically controlled reaction with the formation of the dihydropyrimidine ring (compound **142**, Pathway **c**) [101].

The third type of heterocyclic compounds, pyrazolo[4,3-*c*]quinolizin-9-one **143** (Pathway **d**), was selectively obtained from the same starting materials on adding strong bases (EtONa or *t*BuOK) to the reaction medium in a microwave field at 150 °C [101, 103]. In this particular case, the application of microwave activation, together with strong bases and high temperature, is the key factor for the selectivity of the reaction, that is connected with the simultaneous acting of several factors including the very high rate of self-heating of the highly polar alcohol–alcoholate system, which may not be achieved under the conventional heating [101].

The variation of temperature mode with the help of nonclassical methods of activation and special catalytic systems was also applied to switch between three routs the analogous MCR involving 5-aminopyrazole, containing a carboxamide group in position 4 [104].

It was established [104] that the three-component treatment of 5-aminopyrazoles **30**, aromatic aldehydes **16**, and 1,3-cyclohexanedione **87** (Scheme 8.26) in boiling DMF or in a microwave reactor (DMF, 150 °C) always yielded pyrazolopyrimidines **144** (Pathway **a**).

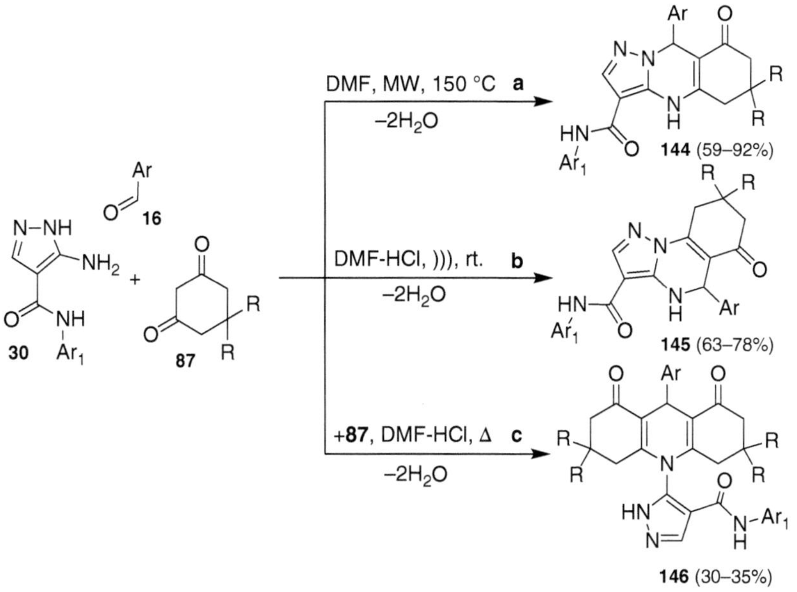

Scheme 8.26 Switching of MCR of 5-aminopyrazoles, aromatic aldehydes, and 1,3-cyclohexanediones.

In contrast, the application of DMF–HCl mixture as reaction media under ultrasonication switched the direction of the MCR toward the positional isomeric heterocycles **145** (Pathway **b**). Under very similar conditions but at a higher temperature and in the presence of the second equivalent of cyclic diketone, the reaction changed the course again yielding 3,4,6,7,9,10-hexahydroacridine-1,8-diones **146** (Pathway **c**).

An additional possibility to diverse structures of the final compounds with multicomponent-switched reactions is domino intramolecular post-cyclization processes. For the application of this approach, the starting materials should contain functional groups that are able to take part in such intramolecular processes or the corresponding reaction centers have to be formed in the MCR.

One of the classical starting reagents applying in the MCRs following with intramolecular heterocyclizations is *ortho*-substituted aromatic aldehydes, e.g. salicylic aldehyde containing additional nucleophilic reaction center, the hydroxyl group, in the *ortho*-position to their formyl function. This creates additional heterocyclization pathways where the reactivity of this group can be involved (one of the recent reviews in the field was published by Heravi et al. [105]).

Surprisingly, the use of this potential of salicylic aldehydes for the creation of the switchable MCRs is at the initial stage of development today [106, 107]. Thus, Gorobets and coworkers [50, 108] described a transformation of salicylic aldehyde **147** in the reaction with 3-amino-1,2,4-triazole **3** and acetone **22** that under mild reaction conditions proceeds similarly to the ordinary aromatic aldehydes containing no additional reaction center in the molecules with the formation of tetrahydropyrimidines **148**, but in this case, much lower temperature is required for the reaction (Scheme 8.27, Pathway **a**). On the other hand, the oxygen bridge closure is achieved under evaluated temperature [50, 109] allowing the synthesis of a triazolobenzoxadiazocine library **149** with variable substituents R_1 and R_2 (Pathway **b**).

Two-component modification of such reaction starting from salicylidenacetone **150** and several aminoazoles was independently studied by Svetlik et al. [110]. It is shown that the reaction with 5-amino-3-methyl-1*H*-pyrazole **33** (Pathway **c**) afforded pyrazolobenzoxadiazocine derivatives **151**. Also, with the use of 2-aminotetrazole, 2-aminobenzimidazole, and 2-amino-2-thiazoline within this methodology, the corresponding bridges **152–154** were obtained. Skeletons **151–154** (Pathway **c**) principally differ from class **149** (Pathway **b**) by the regiodirection of the tetrahydropyrimidine ring closure: aldehyde is attached to the exocyclic aminogroup in the aminoazloe within compound **149**, but to the endocyclic one for **152–154**, which is an obvious consequence of different mechanisms of these condensations.

The direction of similar MCRs involving salicylic aldehydes **147**, pyruvic acids **47** (R_1 = H), and 5-aminopyrazoles **155** (Scheme 8.28) depends on the temperature regime that allows tuning their selectivity toward two heterocyclic compounds, one of which is oxygen-bridged as well [111]. This three-component treatment can proceed either under thermodynamic control (refluxing in acetic acid or MW heating) giving pyrazolo[3,4-*b*]pyridines **156** (Pathway **a**) or under kinetic control (ultrasonication at room temperature) following with intramolecular cyclization and yielding pyrazolobenzoxazocines **157** (Pathway **b**).

In the case of arylpyruvic acids (R_1 = Ar_1), the MCR under ultrasonication yielded tetrahydropyrazolopyrimidines **158** instead of expected oxygen-bridged pyridine derivatives like **157** (Scheme 8.28, Pathway **c**).

There are three alternative reaction directions described for such MCRs with the use of acetoacetic esters **159**. Similarly, to the previous examples, Pathway **a** (Scheme 8.29) leads to tetrahydropyrimidines **160** [50] that does not differ from the reactivity of aromatic aldehydes behavior under the analogous conditions (Scheme 8.21, Pathway **c** and Scheme 8.22, Pathway **a**). However, Světlík and Kettmann [112] using slightly modified reaction conditions managed to obtain a product of pseudo-four-component reaction **161** with its minor diastereomer **162** in the case of parent salicylic aldehyde and two other byproducts **163** and **164** identified in the reaction mixtures for the cases of other substituted salicylic aldehydes (Scheme 8.29, Pathway **b**). Finally, as it was shown by Magedov's group [113], the application of boiling acetic acid in the presence of piperidine results in the formation of four-cyclic condensed chromeno[3,4-*e*][1,2,4]triazolo[1,5-*a*]pyrimidin-6-one **165**.

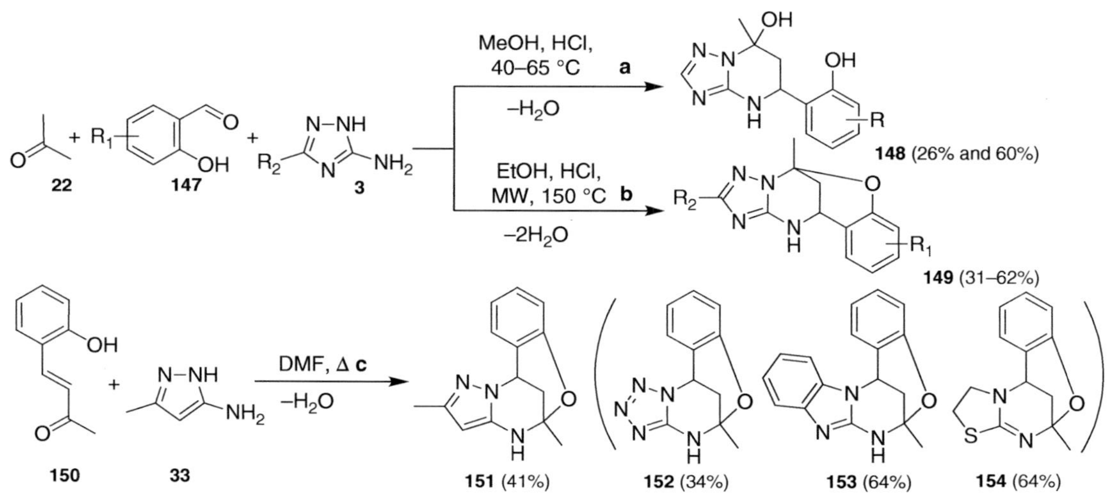

Scheme 8.27 Switching of MCR involving salicylic aldehydes with post-cyclization.

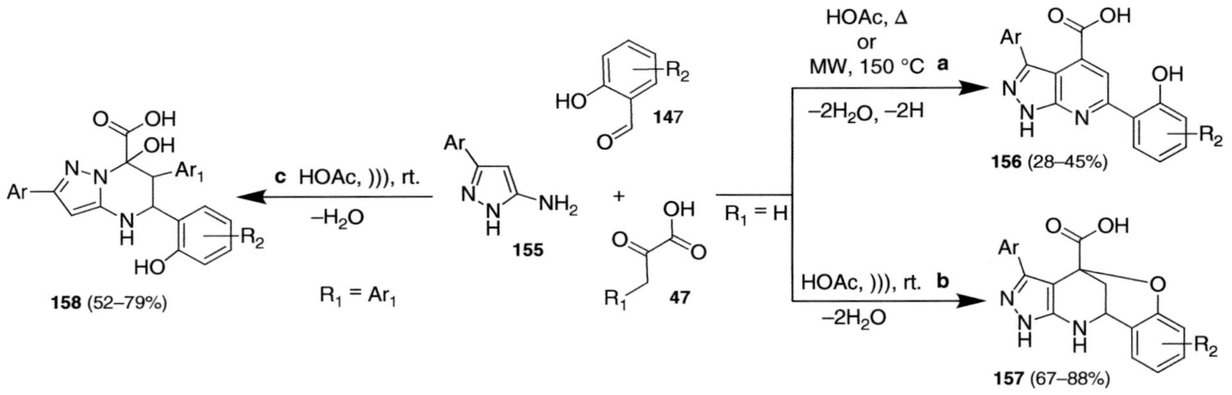

Scheme 8.28 Switching of MCR involving salicylic aldehydes, pyruvic acids, and 5-aminopyrazoles.

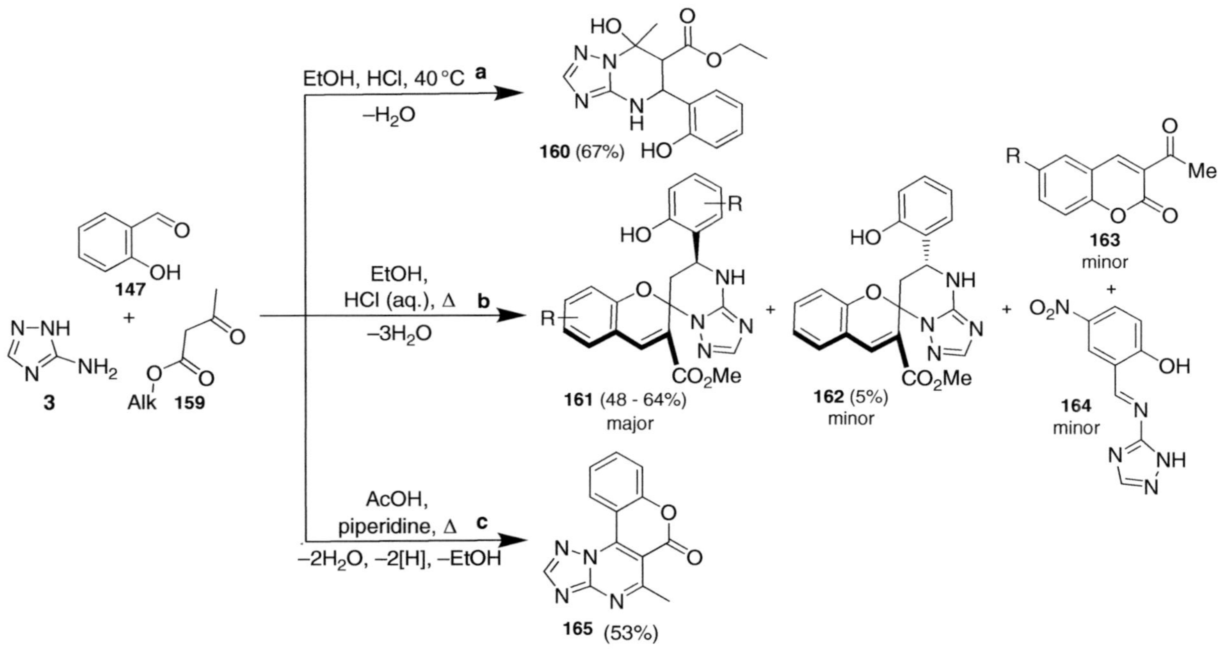

Scheme 8.29 Diverse directions of MCRs involving salicylic aldehydes and acetoacetic esters.

The approach illustrated in Scheme 8.29, Pathway **a**, was further developed [114] using 3-acetyldihydrofuran-2(3*H*)-one **166**, as a cyclic derivative of acetoacetates (Scheme 8.30, Pathway **a**). This uncatalyzed reaction is carried out in water at room temperature resulting in the stereospecific formation of spiro-conjugated heterocyclic system **168**. The stereoselectivity of the product formation was lost, when benzaldehydes **169** without *ortho*-OR_1 group were involved in this transformation (R_2 = H, Me or Et). Světlík's route (Scheme 8.30, Pathway **b**) has received an interesting development [115], where aminoazoles **170** (2-aminobenzothiazoles or 2-amino-5-methylthiazole) were involved in the similar MCR. This resulted in the formation of spiroketals isolated as a mixture of two diasteriomers **171** and **172**, in which the first one dominates in most cases (Scheme 8.30, Pathway **b**).

Presumably, such redirection of the reaction pathway is due to lower nucleophilicity of endocyclic thiazole nitrogen atom compared to that in 3-aminotriazole, thus instead of this nitrogen atom, the OH group of salicylic aldehyde is capable to realize its reactivity to lock the spiro-moiety.

Similar to Pathway **c** (Scheme 8.29), MCRs involving 3-aminopyrazol-5-ones **173** or 5-amino-1-phenyl-3-methylpyrazole **56** (Scheme 8.31, Pathways **a** and **b**) lead to the isolation of polycondensed chromene derivatives **175** and **176**, correspondingly, via oxidative heterocyclization [113]. However, as it was shown elsewhere [110], the isolation of non-oxidized product **177** is also possible. It is formed along with an admixture of oxidized derivative **178** as a result of the similar MCR using starting aminopyrazole **33** in boiling ethanol (Pathway **c**).

It should be noted that in Scheme 8.31, the aminopyrazole derivatives **33**, **56**, and **173** act as 1,3-*C*,*N*-binucleophile (see also Schemes 8.12–8.14), in contrast, in Scheme 8.27 both exo- and endocyclic aminogroups of aminopyrazole **33** take part in the heterocyclization.

Three-component reaction of salicylaldehyde **147**, 5-amino-3-methylisoxazole **76** and *N*-(2-methoxyphenyl)-3-oxobutanamide **179** was studied in [116] under different reaction conditions.

It was observed that depending on the activation method and kind of catalyst, three types of the final heterocyclic systems can be selectively obtained (Scheme 8.32): the treatment of the starting materials at room temperature using mechanical stirring or under ultrasonication gives only chroman **180** (Pathway **a**); the same reaction in the presence of ytterbium triflate leads exclusively to isoxazolopyridine **181** (Pathway **b**); the application of ultrasonic activation together with ytterbium triflate further intramolecular cyclization with the formation of isoxazolobenzoxazocine **182** is promoted. Application of other starting amides of type **179** showed that the bridges **182** are formed only when 2-alkoxyl substituent is presented in the aromatic ring. The unusual benzoxazocine ring closure at room temperature in this case is rationalized by the formation of an intermediate complex with ytterbium cation stabilized by additional chelation with the alkoxyl group [116].

On the other hand, MCR of the same aminoisoxazole **76** and salicylic aldehydes **169** with the derivatives of 1,3-cyclohexanedione **87** under similar conditions showed different behavior without the formation of any oxygen-bridged heterocyclic compound [117]. In particular, the authors observed that this treatment under

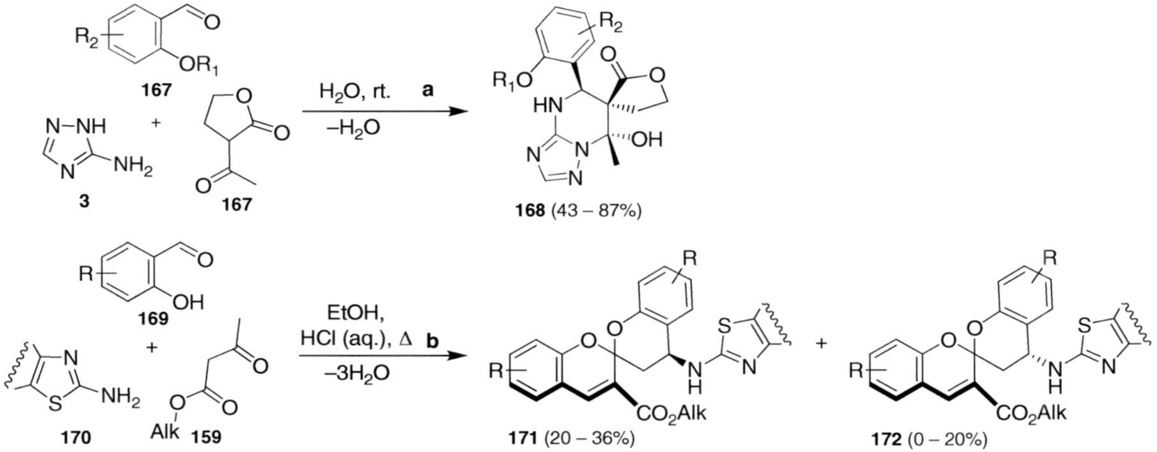

Scheme 8.30 Stereochemical features of MCRs involving some aminoazoles and salicylic aldehydes.

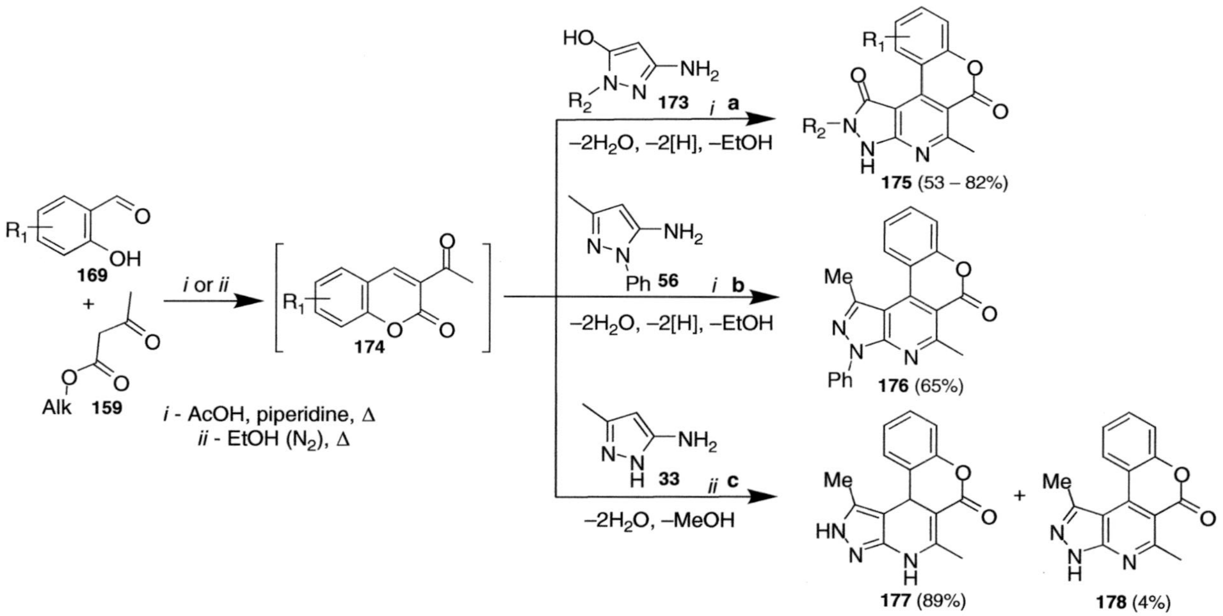

Scheme 8.31 Formation of polycondensed chromenes in MCRs involving salicylic aldehydes.

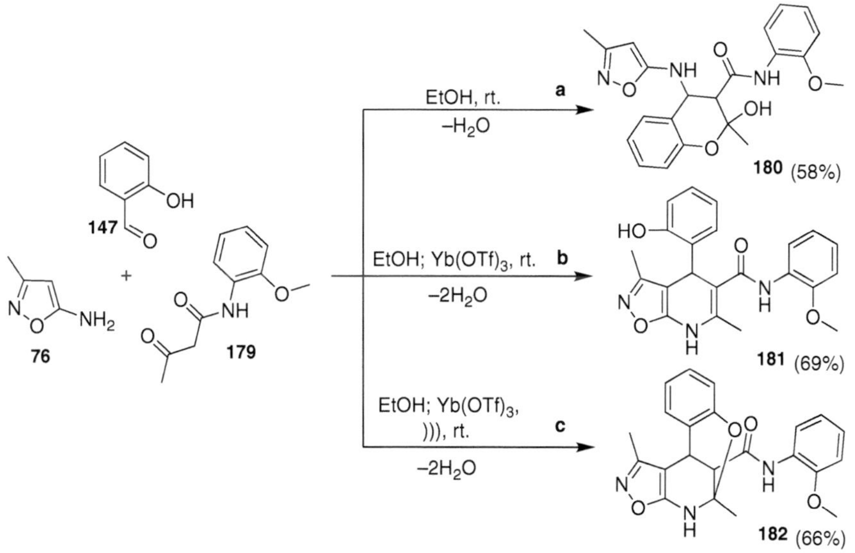

Scheme 8.32 Tuning the selectivity of MCR of salicylaldehyde, 5-amino-3-methylisoxazole, and N-(2-methoxyphenyl)-3-oxobutanamide by variation of solvent/catalyst/activation types.

MW or thermal heating in DMF or water gave the complicated mixtures containing isoxazoloquinolinones **183** and tetrahydroxanthenones **184** and **185** (Scheme 8.33, Pathway **a**).

Variation of the reaction conditions allowed tuning the selectivity of the heterocyclization: in the presence of Brønsted or Lewis acids heating of the starting materials in DMF or H_2O led to the formation of isoxazoloquinolinones **183** (Pathway **b**), while ultrasonication in EtOH-Et_3N gave mixtures of both tetrahydroxanthenones **184** and **185**. Compound **185**, in turn, may be selectively obtained by the treatment of the reagents in boiling DMF when the second equivalent of salicylic aldehyde is added (Pathway **c**).

Thus, the combination of such polyfunctional building blocks as different aminoazoles with salicylic aldehydes showing versatile reaction ability has a great potential for the development of novel multidirectional MCRs that are certainly not limited by the currently existing examples in the literature.

Modular reaction sequences (the term has been introduced by Orru and coworkers [118]) use a versatile reactive intermediate (D, see Scheme 8.1) which can be trapped converted further in the same pot to create different complex scaffolds in a minimal number of steps. This concept has a great potential allowing the combination of not only different reactants but also different reactions under different reaction conditions within one reactor. The development of research in this area is partially illustrated in several reviews [26, 27, 107, 119, 120], here we will overview an example that utilizes the reactivity of polyfunctional 2-(2-cyanovinyl)-3-oxo-cyclohex-1-enes as the common intermediates in the synthesis of a series of heterocyclic systems.

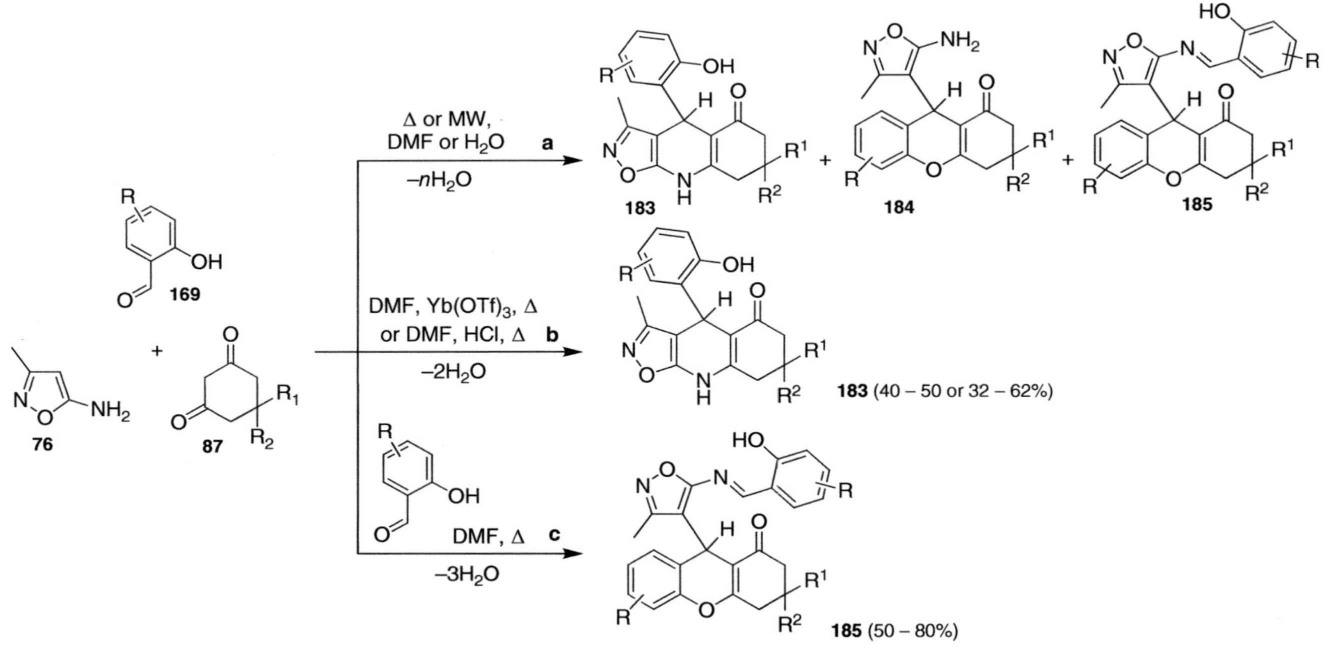

Scheme 8.33 Switching of MCR 5-amino-3-methylisoxazole, salicylic aldehydes, and 1,3-cyclohexanediones.

The reaction of 1,3-dicarbonyl compound **186** with DMFDMA **187** at the beginning of the sequence occurs at room temperature during five minutes for most 1,3-cyclohexanone derivatives (Scheme 8.34). Acyclic 1,3-dicarbonyl compounds require higher temperature giving high conversion into the desired enamine **188**, as the first two-component intermediate in the sequence. Due to the solvent-free format of the first stage, the second step can be carried out in the same pot using a solvent suitable for the next intermediate formation and for its further transformation.

It has been shown [121] that cyclic and acyclic enimines **188** react at high temperature with a series of active methylene nitriles **184** in the presence of catalytic amounts of piperidine giving 2-pyridone derivatives **191** (Scheme 8.34, Pathway **a**). The 2-(2-cyanovinyl)-3-oxo-cyclohex-1-enes **190** are the second three-component intermediates in these sequences, and they were isolated at room temperature in high yields with different substituents and cations and characterized as an individual compound [122]. This made it possible to investigate pathways of their transformations to choose certain conditions for their desired selective transformation and further apply these conditions in stepwise or one-pot formats. In this way, trapping of enoles **190** and processing the cyclization in the presence of water and two equivalents of piperidine under evaluated temperature [123] results in the formation of *N*1-substituted 2,5-dioxo-1,2,5,6,7,8-hexahydro-3-quinolinecarbonitriles **192** (Pathway **b**). The switching between two pathways is controlled by the basicity of the reaction mixture, where more basic conditions activate the amide group involving it into heterocyclization.

However, the increased nucleophilicity of R_3 function may also favor such switching that is shown by the following reactions of enoles **190** bearing the hydrazide group (R_3). In water at boiling (Pathway **c**) or in AcOH at room temperature (Pathway **d**), in both cases, the hydrazide group is involved in two different types of heterocyclizations (products **193** and **194**, correspondingly, [124]). Interestingly, the benzimidazolyl fragment appeared to be so reactive (Pathway **e**), that it was not possible to stop the reaction on the enolate formation stage or involve the cyano group in the cyclization. Thus, under all the conditions applied, the reaction products were only derivatives **195** [125].

The application of additional reagents makes the variability of transformations even more diverse. Water in the presence of a strong acid (Pathway **f**, [126]) results in the formation of 2,5-dioxo-5,6,7,8-tetrahydro-2*H*-chromene-3-carboxamides **196**, whereas (hetero)aromatic amines in AcOH [121, 122] or arylhydrozine hydrochlorides in water [124] (Pathway **g**) provide *N*1-substituted 2,5-dioxo-1,2,5,6,7,8-hexahydro-3-quinolinecarboxamide **197**. For the transformation **190** → **197,** the starting enolates **190** were isolated from the reaction mixture [122].

However, as it was shown using 2-(thiazol-2-yl)acetonitriles as**189**, the one-pot protocol is also possible with solvent replacement: isopropanol, which is needed at the stage of the salt formation, can be evaporated off and replaced with acetic acid, which is required for the final stage of the interaction (Pathway **g**). The synthesis of 3-(1,3-thiazol-2-yl)-7,8-dihydroquinoline-2,5(1*H*,6*H*)-diones **189** (Pathway **i**) can be

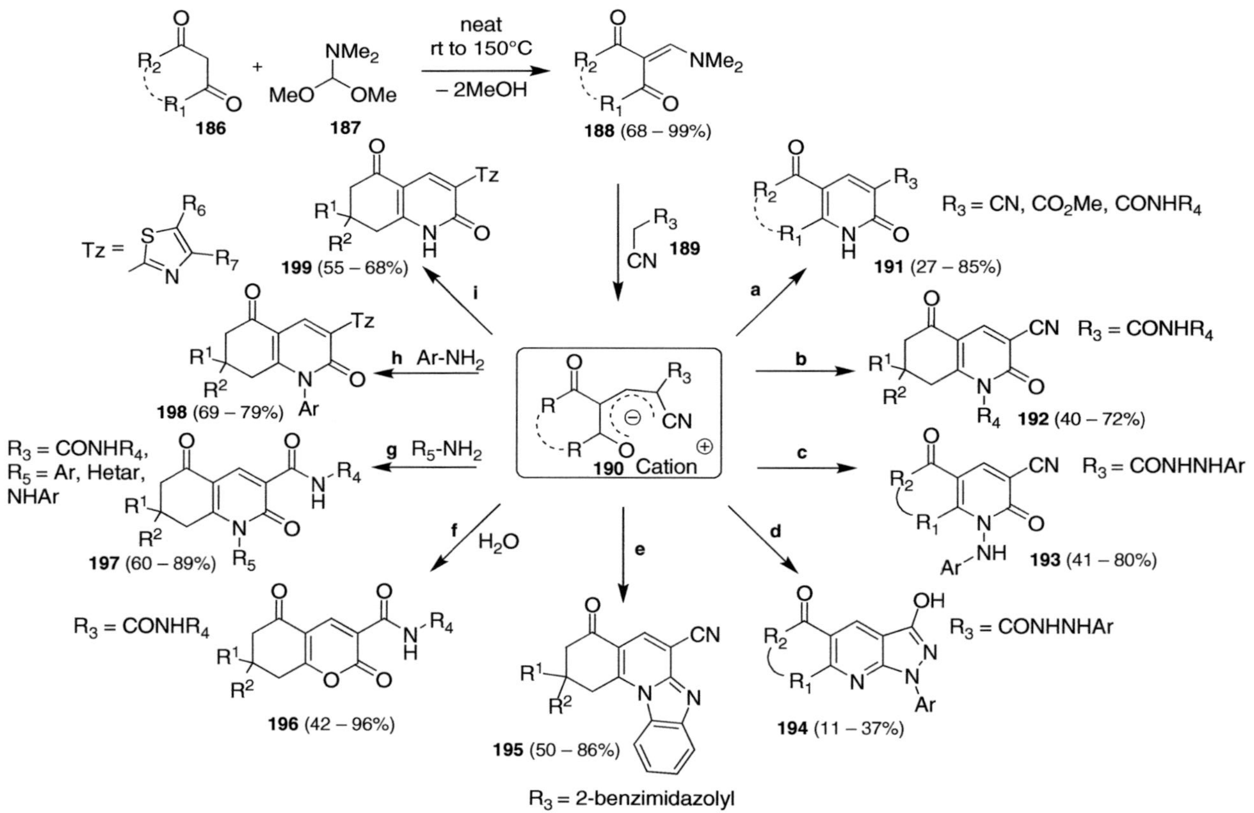

Scheme 8.34 Examples of modular reaction sequences *via* stabilized enolates.

performed using different conditions also with and without isolation of the enolates **190** [127].

The presence of cyclic fragment seems to be crucial for the stabilization of the enolates **190**, since, in the absence of it (Scheme 8.35), such intermediates tend to react with *in situ* generated diethylamine that attacks postulated iminopyran intermediate [128] providing amides **202** [129, 130]. Cyclization of these amides into 2-pyridones **203** is accomplished under more vigorous conditions, thus modeling the final stage of an ANRORC mechanism [121, 125] proposed for such 2-pyridone ring closure (Pathways **a** and **i**, Scheme 8.34).

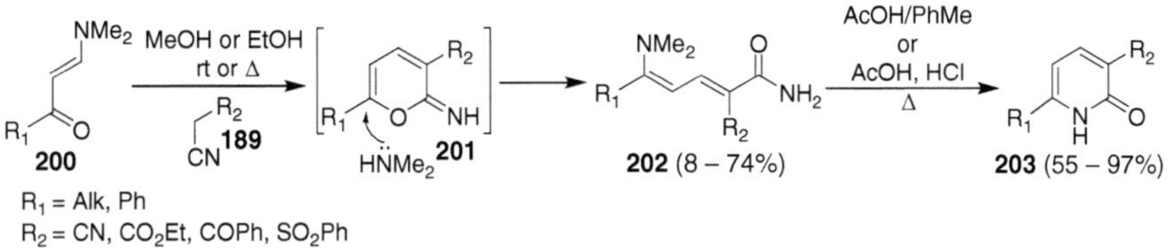

Scheme 8.35 Modular reaction sequences are impossible for unstabilized intermediates.

Thus, MCR switching effect creates new possibilities to control the reaction's selectivity and to switch MCRs between several courses. First of all, the increasing number of the reagents leads to the drastic accretion of the diversity of molecules in the reaction mixture (both starting materials and stable/unstable intermediates) and, therefore, to the strict increasing the probable directions of the interaction that gives additional capability to influence the transformation's proceeding by different external factors. On the other hand, the MCR switching effect can be itself a control tool – a replacement of one reagent by its synthetic precursors (or *vice versa* a replacement of sequential procedure with one-pot one) often leads to the switching the interaction course. These two factors assure a high potential of multicomponent-switched heterocyclizations for solving numerous tasks of synthetic chemistry and open up the possibilities for efficient synthesis of diverse heterocyclic scaffolds from a limited set of starting materials.

References

1 Zhu, J., Wang, Q., and Wang, M.-X. (eds.) (2014). *Multicomponent Reactions in Organic Synthesis*. Weinheim, Germany: Wiley-VCH Verlag GmbH & Co. KGaA.

2 Müller, T.J.J. (ed.) (2014). *Multicomponent Reactions 1. General Discussion and Reactions Involving a Carbonyl Compound as Electrophilic Component*. Stuttgart: Georg Thieme Verlag KG.

3 Strecker, A. (1850). Ueber die künstliche Bildung der Milchsäure und einen neuen, dem Glycocoll homologen Körper. *Ann. der Chemie und Pharm.* 75 (1): 27–45.

4 Mannich, C. and Krösche, W. (1912). Ueber ein Kondensationsprodukt aus Formaldehyd, Ammoniak und Antipyrin. *Arch. Pharm. (Weinheim).* 250 (1): 647–667.

5 Hantzsch, A. (1881). Condensationsprodukte aus Aldehydammoniak und ketonartigen Verbindungen. *Ber. Dtsch. Chem. Ges.* 14 (2): 1637–1638.
6 Döbner, O. (1887). Ueber α-Alkylcinchoninsäuren und α-Alkylchinoline. *Justus Liebig's Ann. der Chemie* 242 (3): 265–289.
7 Biginelli, P. (1891). Ueber Aldehyduramide des Acetessigäthers. *Ber. Dtsch. Chem. Ges.* 24 (1): 1317–1319.
8 Biginelli, P. (1891). Ueber Aldehyduramide des Acetessigäthers. II. *Ber. Dtsch. Chem. Ges.* 24 (2): 2962–2967.
9 Zhi, S., Ma, X., and Zhang, W. (2019). Consecutive multicomponent reactions for the synthesis of complex molecules. *Org. Biomol. Chem.* 17 (33): 7632–7650.
10 Bhat, I.S. (2020). One-pot construction of bis-heterocycles through isocyanide based multicomponent reactions. *ChemistrySelect* 5 (27): 8040–8061.
11 Mohammadkhani, L. and Heravi, M.M. (2019). Synthesis of various N-heterocycles using the ugi four-center three-component reaction. *ChemistrySelect* 4 (34): 10187–10196.
12 Neochoritis, C.G., Zhao, T., and Dömling, A. (2019). Tetrazoles via multicomponent reactions. *Chem. Rev.* 119 (3): 1970–2042.
13 Palanivel, L. and Gnanasambandam, V. (2020). Diversity oriented multi-component reaction (DOS–MCR) approach to access natural product analogues: regio- and chemo-selective synthesis of polyheterocyclic scaffolds via one-pot cascade reactions. *Org. Biomol. Chem.* 18 (16): 3082–3092.
14 Murlykina, M.V., Morozova, A.D., Zviagin, I.M. et al. (2018). Aminoazole-based diversity-oriented synthesis of heterocycles. *Front. Chem.* 6 (article 527): 1–43.
15 Chebanov, V.A. and Desenko, S.M. (2012). Multicomponent heterocyclization reactions with controlled selectivity. *Chem. Heterocycl. Compd.* 48 (4): 566–583.
16 Devi, N., Rawal, R.K., and Singh, V. (2015). Diversity-oriented synthesis of fused-imidazole derivatives via Groebke–Blackburn–Bienayme reaction: a review. *Tetrahedron* 71 (2): 183–232.
17 Insuasty, D., Castillo, J., Becerra, D. et al. (2020). Synthesis of biologically active molecules through multicomponent reactions. *Molecules* 25 (3): 1–71.
18 Graebin, C.S., Ribeiro, F.V., Rogério, K.R., and Kümmerle, A.E. (2019). Multicomponent Reactions for the Synthesis of Bioactive Compounds: A Review. *Curr. Org. Synth.* 16 (6): 855–899.
19 Muñoz-Torrero, D., Lavilla, R., Pérez-Areales, F.J., and Ghashghaei, O. (2019). Multicomponent reactions: a mighty journey partner for infectious tropical disease drug discovery. In: *Medicinal Chemistry Approaches to Malaria and Other Tropical Diseases* (ed. C. Kelly), 181–217. Elsevier.
20 Ruijter, E. and Orru, R.V.A. (2013). Multicomponent reactions – opportunities for the pharmaceutical industry. *Drug Discov. Today Technol.* 10 (1): e15–e20.
21 Hussain, A. and Dar, B.A. (2020). Environmentally benign organic synthesis. In: *Applications of Nanotechnology for Green Synthesis. Nanotechnology in the Life Sciences* (ed. A.A. Inamuddin), 125–144. Springer.
22 Nainwal, L.M., Tasneem, S., Akhtar, W. et al. (2019). Green recipes to quinoline: a review. *Eur. J. Med. Chem.* 164: 121–170.

23 Javanbakht, S. and Shaabani, A. (2020). Multicomponent reactions-based modified/functionalized materials in the biomedical platforms. *ACS Appl. Bio Mater.* 3 (1): 156–174.

24 Afshari, R. and Shaabani, A. (2018). Materials functionalization with multicomponent reactions: state of the art. *ACS Comb. Sci.* 20: 499–528.

25 Levi, L. and Müller, T.J.J. (2016). Multicomponent syntheses of functional chromophores. *Chem. Soc. Rev.* 45 (10): 2825–2846.

26 Ruijter, E., Scheffelaar, R., and Orru, R.V.A. (2011). Multicomponent reaction design in the quest for molecular complexity and diversity. *Angew. Chem. Int. Ed. Eng.* 50 (28): 6234–6246.

27 van der Heijden, G., Ruijter, E., and Orru, R.V.A. (2013). Efficiency, diversity, and complexity with multicomponent reactions. *Synlett* 24 (06): 666–685.

28 Ganem, B. (2009). Strategies for innovation in multicomponent reaction design. *Acc. Chem. Res.* 42 (3): 463–472.

29 Weber, L., Illgen, K., and Almstetter, M. (1999). Discovery of new multi component reactions with combinatorial methods. *Synlett* (3): 366–374.

30 Chebanov, V.A., Gura, K.A., and Desenko, S.M. (2010). Aminoazoles as key reagents in multicomponent heterocyclizations. In: *Topics in Heterocyclic Chemistry* (eds. R.V.A. Orru and E. Ruijeter), 41–84. Berlin Heidelberg: Springer.

31 Desenko, S.M., Orlov, V.D., and Estrada, K. (1990). Formation of derivatives of 1,2,4-triazoloquinazolines in the reactions of 3-amino-1,2,4-triazoles with cyclohexanone. *Chem. Heterocycl. Compd.* 26 (7): 839–840.

32 Chernyshev, V.M., Khoroshkin, D.A., Sokolov, A.N. et al. (2008). Synthesis, structure and some reactions of 4a',5′,6′,7′,8′,8a'-hexahydro-4′ H -spiro[cyclohexane-1,9′-[1,2,4]triazolo[5,1- b]-quinazolines]. *J. Heterocyclic Chem.* 45 (5): 1419–1427.

33 Desenko, S.M., Orlov, V.D., Getmanskii, N.V. et al. (1992). Aromatic substituted 1,2,4-triazolo[1,5-a]pyrimidine. *Dokl. Akad. Nauk SSSR Ser. Khim.* 324: 801–804.

34 Wermann, K. and Hartmann, M. (1991). Synthesis of dihydro-1,2,4-triazolo [1,5- a]pyrimidines. *Synthesis (Stuttg).* 03: 189–191.

35 Ryabukhin, S.V., Naumchik, V.S., Grygorenko, O.O., and Tolmachev, A.A. (2012). Chlorotrimethylsilane-promoted condensation of ketones and aminoazoles. *J. Heterocyclic Chem.* 49 (5): 1147–1150.

36 Bellec, C. and Lhommet, G. (1995). Chemical and electrochemical reduction of some pyrazolo[1,5- a]pyrimidines. *J. Heterocyclic Chem.* 32 (6): 1793–1800.

37 Chebanov, V.A. and Desenko, S.M. (2006). Dihydroazines based on α,β-unsaturated ketones reactions. *Curr. Org. Chem.* 10: 297–317.

38 Chebanov, V.A., Desenko, S.M., and Gurley, T.W. (2008). *Azaheterocycles Based on α,β-Unsaturated Carbonyls.* Berlin Heidelberg: Springer.

39 Yu, W., Goddard, C., Clearfield, E. et al. (2011). Design, synthesis, and biological evaluation of triazolo-pyrimidine derivatives as novel inhibitors of hepatitis B virus surface antigen (HBsAg) secretion. *J. Med. Chem.* 54 (16): 5660–5670.

40 Umar, T., Gusain, S., Raza, M.K. et al. (2019). Naphthalene-triazolopyrimidine hybrid compounds as potential multifunctional anti-Alzheimer's agents. *Bioorg. Med. Chem.* 27 (14): 3156–3166.

41 Desenko, S.M., Orlov, V.D., Lipson, V.V. et al. (1994). Synthesis and pharmacological activity of N-pyrimidinylsuccinaminic acids and succinimides. *Pharm. Chem. J.* 28 (3): 154–157.

42 Orlov, V.D., Desenko, S.M., Potekhin, K.A., and Struchkov, Y.T. (1988). Cyclocondensation of α,β-unsaturated ketones with 3-amino-1,2,4-triazole. *Chem. Heterocycl. Compd.* 24 (2): 192–196.

43 Desenko, S.M., Orlov, V.D., and Beskorovainyi, V.I. (1991). Tautomerism of dihydro-1,2,4-triazolo[1,5-a]pyrimidines containing 2-hydroxy- and 4-dimethylaminoaryl substituents. *Chem. Heterocycl. Compd.* 27 (2): 201–204.

44 Desenko, S.M., Orlov, V.D., Getmanskii, N.V. et al. (1993). Three component condensation of 3-amino-1,2,4-triazole with carbonyl compounds. A new synthesis of 1,2,4-triazolo[1,5-a]pyrimidines. *Chem. Heterocycl. Compd.* 29 (4): 406–410.

45 Hassaneen, H.M.E. and Farghaly, T.A. (2015). A simple, convenient, one-pot synthesis of dihydro-azolopyrimidines, DFT calculation, and NMR determination by using H-ferrierite zeolite as catalyst. *J. Heterocyclic Chem.* 52 (4): 1154–1161.

46 Shahi, M.N., Perveen, N., Khan, M.N. et al. (2016). Triazolopyrimidines: synthesis of Aryl-1,2,4-triazolo[1,5-a]pyrimidines by Doebner-Miller reaction and their antibacterial activity. *Asian J. Chem.* 28 (12): 2665–2669.

47 Desenko, S.M., Orlov, V.D., and Lipson, V.V. (1990). Chemical conversions of 5,7-disubstituted dihydro-1,2,4-triazolo[1,5-a]pyrimidines. *Chem. Heterocycl. Compd.* 26 (12): 1362–1366.

48 Lipson, V.V., Desenko, S.M., Orlov, V.D. et al. (2000). Cyclocondensation of 3-amino-1,2,4-triazoles with esters of substituted cinnamic acids and aromatic unsaturated ketones. *Chem. Heterocycl. Compd.* 36: 1329–1335.

49 Massari, S., Nannetti, G., Desantis, J. et al. (2015). A broad anti-influenza hybrid small molecule that potently disrupts the interaction of polymerase acidic protein–basic protein 1 (PA-PB1) subunits. *J. Med. Chem.* 58 (9): 3830–3842.

50 Gorobets, N.Y., Sedash, Y.V., Ostras, K.S. et al. (2010). Unexpected alternative direction of a Biginelli-like multicomponent reaction with 3-amino-1,2,4-triazole as the urea component. *Tetrahedron Lett.* 51 (16): 2095–2098.

51 Komykhov, S.A., Bondarenko, A.A., Musatov, V.I. et al. (2017). (5S,7R)-5-Aryl-7-methyl-4,5,6,7-tetrahydro-[1, 2, 4]triazolo[1,5-a]pyrimidin-7-ols as products of three-component condensation. *Chem. Heterocycl. Compd.* 53 (3): 378–380.

52 Desenko, S.M., Estrada, K., Orlov, V.D., and Ponomarev, O.A. (1991). Carbocycle annelated 1,2,4-triazolo[1,5-a]pyrimidines. *Chem. Heterocycl. Compd.* 27 (1): 88–91.

53 Vasil'kova, N.O., Anis'kov, A.A., and Kriven'ko, A.P. (2015). Three-component cyclocondensation of 5-nitrothiophene-2-carbaldehyde with cyclohexanone and

C-aminotri(tetr)azoles. Formation paths of isomeric hexahydrotriazoloquinazolines. *Russ. J. Org. Chem.* 51 (12): 1766–1769.

54 Vasil'kova, N.O., Anis'kov, A.A., Sorokin, V.V., and Kriven'ko, A.P. (2018). Three-component condensation of 1,2,4-triazol-3-amine with cyclohexanone and ortho-substituted benzaldehydes. *Russ. J. Org. Chem.* 54 (6): 933–937.

55 Beletskaya, I.P., Nájera, C., and Yus, M. (2020). Chemodivergent reactions. *Chem. Soc. Rev.* 49 (19): 7101–7166.

56 Sakhno, Y.I., Shishkina, S.V., Shishkin, O.V. et al. (2010). Diversity oriented heterocyclizations of pyruvic acids, aldehydes and 5-amino-N-aryl-1H-pyrazole-4-carboxamides: catalytic and temperature control of chemoselectivity. *Mol. Divers.* 9: 523–531.

57 Chebanov, V.A., Sakhno, Y.I., and Desenko, S.M. (2012). High regioselective ultrasonic-assisted synthesis of 2,7-diaryl-4,7-dihydropyrazolo [1,5-a]pyrimidine-5-carboxylic acids. *Ultrason. Sonochem.* 19 (3): 707–709.

58 Chebanov, V.A., Sakhno, Y.I., Desenko, S.M. et al. (2007). Cyclocondensation reactions of 5-aminopyrazoles, pyruvic acids and aldehydes. Multicomponent approaches to pyrazolopyridines and related products. *Tetrahedron* 63 (5): 1229–1242.

59 Sakhno, Y.I., Kozyryev, A.V., Desenko, S.M. et al. (2018). Features of two- and multicomponent heterocyclization reactions involving 3,4-disubstituted 5-aminopyrazoles and alkyl pyruvates. *Tetrahedron* 74: 564–571.

60 Morozova, A.D., Muravyova, E.A., Shishkina, S.V. et al. (2019). Features of 3-amino-5-methylisoxazole in heterocyclizations involving pyruvic acids. *Chem. Heterocycl. Compd.* 55 (1): 78–89.

61 Tu, S., Li, C., Shi, F. et al. (2008). An efficient chemoselective synthesis of pyrido[2,3-d]pyrimidine derivatives under microwave irradiation. *Synthesis (Stuttg).* 2008 (3): 369–376.

62 Liu, M., Yin, G., Zhu, C., and Yao, C. (2016). Selective synthesis of new tetracyclic coumarin-fused pyrazolo[3,4-b]pyridines and pyrazolo[3,4-b]pyridin-6 (7 H)-ones. *J. Heterocyclic Chem.* 53 (5): 1617–1625.

63 Wang, S.-L., Liu, Y.-P., Xu, B.-H. et al. (2011). Microwave-assisted chemoselective reaction: a divergent synthesis of pyrazolopyridine derivatives with different substituted patterns. *Tetrahedron* 67 (48): 9417–9425.

64 Lipson, V.V. and Gorobets, N.Y. (2009). One hundred years of Meldrum's acid: advances in the synthesis of pyridine and pyrimidine derivatives. *Mol. Divers.* 13 (4): 399–419.

65 Lipson, V.V., Orlov, V.D., Desenko, S.M. et al. (1999). Reaction of arylidene derivatives of Meldrum's acid with 3-amino-1,2,4-triazole. *Chem. Heterocycl. Compd.* 35 (5): 595–599.

66 Lipson, V.V., Orlov, V.D., Desenko, S.M. et al. (2000). 1,2,3,4-Tetrahydropyrimido[1,2-a]benzimidazol-2- and -4-ones. *Chem. Heterocycl. Compd.* 36: 1039–1043.

67 Lipson, V.V., Desenko, S.M., Borodina, V.V. et al. (2005). 2-Methylthio-4,5,6,7-tetrahydro-1,2,4-triazolo[1,5-a]pyrimidin-5- and -7-Ones. *Chem. Heterocycl. Compd.* 41 (2): 216–220.

68 Lipson, V.V., Borodina, V.V., and Shirobokova, M.G. (2005). Cyclocondensation of 3,5-diamino-1,2,4-triazole with benzaldehydes and Meldrum's acid. *Ukr. Khimicheskii Zhurnal* 71: 95–99.

69 Sheibani, H., Saidi, K., and Lakaei, M. (2012). Three-component one-pot synthesis of 4-aryl-2,3-dihydropyrimido[1,2-a]benzimidazol-2-ones catalyzed by L-proline. *J. Heterocyclic Chem.* 49 (6): 1386–1390.

70 Yao, C., Lei, S., Wang, C. et al. (2009). Three-component synthesis of 4-aryl-1 H -pyrimido[1,2-a]benzimidazole derivatives in ionic liquid. *J. Heterocyclic Chem.* 47 (1): 26–32.

71 Ma, N., Jiang, B., Zhang, G. et al. (2010). New multicomponent domino reactions (MDRs) in water: highly chemo-, regio- and stereoselective synthesis of spiro{[1,3]dioxanopyridine}-4,6-diones and pyrazolo[3,4-b]pyridines. *Green Chem.* 12 (8): 1357–1361.

72 Lipson, V.V., Desenko, S.M., Shirobokova, M.G. et al. (2006). Heterocyclizations of 5-methylpyrazol-3-amine with unsaturated arylaliphatic carboxylic acid derivatives. *Russ. J. Org. Chem.* 42 (7): 1022–1027.

73 Morozova, A.D., Muravyova, E.A., Shishkina, S.V. et al. (2017). Diversity-oriented multicomponent heterocyclizations involving derivatives of 3(5)-aminoisoxazole, aldehydes and meldrum's or N,N′-dimethylbarbituric acid. *J. Heterocyclic Chem.* 54 (2): 932–943.

74 Petrova, O.N., Zamigajlo, L.L., Ostras, K.S. et al. (2015). Multicomponent reaction of 2-aminobenzimidazole, arylglyoxals, and 1,3-cyclohexanedione. *Chem. Heterocycl. Compd.* 51 (4): 310–319.

75 Petrova, O.N., Zamigailo, L.L., Shirobokova, M.G. et al. (2013). Cyclocondensation of 3(5)-aminopyrazoles with arylglyoxals and cyclohexane-1,3-diones. *Chem. Heterocycl. Compd.* 49 (7): 955–967.

76 Wang, J.-J., Feng, X., Xun, Z. et al. (2015). Multicomponent strategy to pyrazolo[3,4-e]indolizine derivatives under microwave irradiation. *J. Org. Chem.* 80 (16): 8435–8442.

77 Jiang, B., Ye, Q., Fan, W. et al. (2014). Four-component strategy for selective synthesis of azepino[5,4,3-cd]indoles and pyrazolo[3,4-b]pyridines. *Chem. Commun.* 50 (46): 6108–6111.

78 Mishra, R., Panday, A.K., Choudhury, L.H. et al. (2017). Multicomponent reactions of arylglyoxal, 4-hydroxycoumarin, and cyclic 1,3-C,N-binucleophiles: binucleophile-directed synthesis of fused five- and six-membered N-heterocycles. *Eur. J. Org. Chem.* 2017 (19): 2789–2800.

79 Fan, W., Ye, Q., Xu, H.-W. et al. (2013). Novel double [3 + 2 + 1] heteroannulation for forming unprecedented dipyrazolo-fused 2,6-naphthyridines. *Org. Lett.* 15 (9): 2258–2261.

80 Jiang, B., Fan, W., Sun, M.-Y. et al. (2014). Domino reaction of arylglyoxals with pyrazol-5-amines: selective access to pyrazolo-fused 1,7-naphthyridines, 1,3-diazocanes, and pyrroles. *J. Org. Chem.* 79 (11): 5258–5268.

81 Chebanov, V.A., Sakhno, Y.I., Desenko, S.M. et al. (2005). Three-component procedure for the synthesis of 5-aryl-5,8-dihydroazolo[1,5-a]pyrimidine-7-carboxylic acids. *Synthesis (Stuttg).* 15: 2597–2601.

82 Murlykina, M.V., Sakhno, Y.I., Desenko, S.M. et al. (2015). Study of the chemoselectivity of multicomponent heterocyclizations involving 3-amino-1,2,4-triazole and pyruvic acids as key reagents, and biological activity of the reaction products. *Eur. J. Org. Chem.* 2015 (20): 4481–4492.

83 Sakhno, Y.I., Desenko, S.M., Shishkina, S.V. et al. (2008). Multicomponent cyclocondensation reactions of aminoazoles, arylpyruvic acids and aldehydes with controlled chemoselectivity. *Tetrahedron* 64 (49): 11041–11049.

84 Sakhno, Y.I., Desenko, S.M., Shishkina, S.V. et al. (2011). Unusual direction of cyclocondensation of 1-(4-Chlorophenyl)-3,5-diamino-1,2,4-triazole, pyruvic acid, and aldehydes. *Synthesis (Stuttg).* 07: 1120–1124.

85 Carey, F.A. and Sundberg, R.J. (2007). *Advanced Organic Chemistry. Part A: Structure and Mechanisms.* New York: Springer.

86 Bruice, P.Y. (2014). *Organic Chemistry*. Pearson Education, Upper Saddle River.

87 Wade, L.G. (2012). *Organic Chemistry*. Glenview: Pearson Education.

88 Chen, Q., Jiang, L.-L., Chen, C.-N., and Yang, G.-F. (2009). The first example of a regioselective Biginelli-like reaction based on 3-alkylthio-5-amino-1,2,4-triazole. *J. Heterocyclic Chem.* 46 (2): 139–148.

89 Sedash, Y.V., Gorobets, N.Y., Chebanov, V.A. et al. (2012). Dotting the i's in three-component Biginelli-like condensations using 3-amino-1,2,4-triazole as a 1,3-binucleophile. *RSC Adv.* 2 (17): 6719–6728.

90 Shaabani, A., Rahmati, A., Rezayan, A.H. et al. (2007). Clean synthesis in water: uncatalyzed three-component condensation reaction of 3-amino-1, 2,4-triazole or 2-aminobenzimidazole with aldehyde in the presence of activated CH-acids. *QSAR Comb. Sci.* 26 (9): 973–979.

91 Koryakova, O.V., Titova, Y.A., Murashkevich, A.N., and Fedorova, O.V. (2018). IR-Spectroscopic study of adsorption of aminoazoles on oxide catalysts of Biginelli reaction. *Kinet. Catal.* 59 (2): 188–195.

92 Chebanov, V.A., Muravyova, E.A., Desenko, S.M. et al. (2006). Microwave-assisted three-component synthesis of 7-aryl-2-alkylthio-4, 7-dihydro-1, 2, 4-triazolo -pyrimidine-6-carboxamides and their selective reduction. *J. Comb. Chem.* 8: 427–434.

93 Muravyova, E.A., Desenko, S.M., Rudenko, R.V. et al. (2011). Switchable selectivity in multicomponent heterocyclizations of acetoacetamides, aldehydes, and 3-amino-1,2,4-triazoles/5-aminopyrazoles. *Tetrahedron* 67 (48): 9389–9400.

94 Gadara, S.A. and Ladva, K.D. (2020). Solid phase synthesis and antimicrobial activity of novel triazolo[1,5-a]pyrimidine derivatives. *Asian J. Chem.* 32 (9): 2298–2302.

95 Muravyova, E.A., Shishkina, S.V., Musatov, V.I. et al. (2009). Chemoselectivity of multicomponent condensations of barbituric acids, 5-aminopyrazoles, and aldehydes. *Synthesis (Stuttg).*: 1375–1385.

96 Khumalo, M.R., Maddila, S.N., Maddila, S., and Jonnalagadda, S.B. (2019). A multicomponent, facile and catalyst-free microwave-assisted protocol for the synthesis of pyrazolo-[3,4-b]-quinolines under green conditions. *RSC Adv.* 9 (53): 30768–30772.

97 Wang, H.-Y. and Shi, D.-Q. (2012). Three-component one-pot synthesis of pyrazolo[3,4-b]quinolin-5(6H)-one derivatives in aqueous media. *J. Heterocyclic Chem.* 49 (1): 212–216.

98 Patel, D.M., Sharma, M.G., Vala, R.M. et al. (2019). Hydroxyl alkyl ammonium ionic liquid assisted green and one-pot regioselective access to functionalized pyrazolodihydropyridine core and their pharmacological evaluation. *Bioorg. Chem.* 86: 137–150.

99 Sharma, M.G., Vala, R.M., and Patel, H.M. (2020). Pyridine-2-carboxylic acid as an effectual catalyst for rapid multi-component synthesis of pyrazolo[3,4-b]quinolinones. *RSC Adv.* 10 (58): 35499–35504.

100 Drizin, I., Holladay, M.W., Yi, L. et al. (2002). Structure-activity studies for a novel series of tricyclic dihydropyrimidines as KATP channel openers (KCOs). *Bioorg. Med. Chem. Lett.* 12 (11): 1481–1484.

101 Chebanov, V.A., Saraev, V.E., Desenko, S.M. et al. (2008). Tuning of chemo- and regioselectivities in multicomponent condensations of 5-aminopyrazoles, dimedone, and aldehydes. *J. Org. Chem.* 73 (13): 5110–5118.

102 Andriushchenko, A.Y., Desenko, S.M., Chernenko, V.N., and Chebanov, V.A. (2011). Green and efficient synthesis of pyrazolo[3,4-b]quinolin-5-ones derivatives by microwave-assisted multicomponent reaction in hot water medium. *J. Heterocyclic Chem.* 48: 365–367.

103 Chebanov, V.A., Saraev, V.E., Desenko, S.M. et al. (2007). One-pot, multicomponent route to pyrazoloquinolizinones. *Org. Lett.* 9 (9): 1691–1694.

104 Chebanov, V.A., Saraev, V.E., Shishkina, S.V. et al. (2012). Controlled switching of multicomponent heterocyclizations of 5-amino-N-arylpyrazole-4-carboxamides, 1,3-cyclohexanediones, and aldehydes. *Eur. J. Org. Chem.*: 5515–5524.

105 Heravi, M.M., Zadsirjan, V., Mollaiye, M. et al. (2018). Salicylaldehydes as privileged synthons in multicomponent reactions. *Russ. Chem. Rev.* 87 (6): 553–585.

106 Shi, Y.-L. and Shi, M. (2007). The synthesis of chromenes, chromanes, coumarins and related heterocycles via tandem reactions of salicylic aldehydes or salicylic imines with α,β-unsaturated compounds. *Org. Biomol. Chem.* 5 (10): 1499–1504.

107 Ibarra, I.A., Islas-Jácome, A., and González-Zamora, E. (2018). Synthesis of polyheterocycles via multicomponent reactions. *Org. Biomol. Chem.* 16 (9): 1402–1418.

108 Kondratiuk, M., Gorobets, N.Y., Sedash, Y.V. et al. (2016). 5-(5-Bromo-2-hydroxy-3-methoxyphenyl)-7-methyl-4,5,6,7-tetrahydro[1,2,4]triazolo[1,5-a]pyrimidin-7-ol. *Molbank* 2016 (2): M898.

109 Gümüş, M.K., Gorobets, N.Y., Sedash, Y.V. et al. (2017). A modified Biginelli reaction toward oxygen-bridged tetrahydropyrimidines fused with substituted 1,2,4-triazole ring. *Chem. Heterocycl. Compd.* 53 (11): 1261–1267.

110 Svetlik, J., Veizerová, L., Mayer, T.U., and Catarinella, M. (2010). Monastrol analogs: a synthesis of pyrazolopyridine, benzopyranopyrazolopyridine, and oxygen-bridged azolopyrimidine derivatives and their biological screening. *Bioorg. Med. Chem. Lett.* 20 (14): 4073–4076.

111 Murlykina, M.V., Sakhno, Y.I., Desenko, S.M. et al. (2013). Features of switchable multicomponent heterocyclizations of salicylic aldehydes and 5-aminopyrazoles with pyruvic acids and antimicrobial activity of the reaction. *Tetrahedron* 69: 9261–9269.

112 Světlík, J. and Kettmann, V. (2011). The chameleon-like behaviour of 3-amino-1,2,4-triazole in the Biginelli reaction: unexpected formation of a novel spiroheterocyclic system. *Tetrahedron Lett.* 52 (10): 1062–1066.

113 Frolova, L., Malik, I., Uglinskii, P. et al. (2011). Multicomponent synthesis of 2,3-dihydrochromeno[4,3-d]pyrazolo[3,4-b]pyridine-1,6-diones: a novel heterocyclic scaffold with antibacterial activity. *Tetrahedron Lett.* 52 (49): 6643–6645.

114 Gümüş, M.K., Gorobets, N.Y., Sedash, Y.V. et al. (2017). Rapid formation of chemical complexity via a modified Biginelli reaction leading to dihydrofuran-2(3H)-one spiro-derivatives of triazolo[1,5-a]pyrimidine. *Tetrahedron Lett.* 58 (35): 3446–3448.

115 Světlík, J., Prónayová, N., Švorc, L., and Frecer, V. (2014). A complicated path of salicylaldehyde through the Biginelli reaction: a case of unexpected spiroketalization. *Tetrahedron* 70 (44): 8354–8360.

116 Tkachenko, V.V., Muravyova, E.A., Desenko, S.M. et al. (2014). The unexpected influence of aryl substituents in N -aryl-3-oxobutanamides on the behavior of their multicomponent reactions with 5-amino-3-methylisoxazole and salicylaldehyde. *Beilstein J. Org. Chem.* 10: 3019–3030.

117 Muravyova, E.A., Tkachenko, V.V., Desenko, S.M. et al. (2013). Behavior of 5-amino-3-methylisoxazole in multicomponent heterocyclizations with carbonyl compounds under thermal heating and non-classical conditions. *ARKIVOC* (3): 338–371.

118 Paravidino, M., Scheffelaar, R., Schmitz, R.F. et al. (2007). A flexible six-component reaction to access constrained depsipeptides based on a dihydropyridinone core. *J. Org. Chem.* 72 (26): 10239–10242.

119 Groenendaal, B., Ruijter, E., and Orru, R.V.A. (2008). 1-Azadienes in cycloaddition and multicomponent reactions towards N-heterocycles. *Chem. Commun.* (43): 5474–5489.

120 Malinakova, H. Recent advances in the discovery and design of multicomponent reactions for the generation of small-molecule libraries. *Reports Org. Chem.* 5: 75–90.

121 Gorobets, N.Y., Yousefi, B.H., Belaj, F., and Kappe, C.O. (2004). Rapid microwave-assisted solution phase synthesis of substituted 2-pyridone libraries. *Tetrahedron* 60 (39): 8633–8644.

122 Yermolayev, S.A., Gorobets, N.Y., Lukinova, E.V. et al. (2008). An efficient synthesis of N1-substituted 2,5-dioxo-1,2,5,6,7,8-hexahydro-3-quinolinecarboxamide via enolate salts. *Tetrahedron* 64 (20): 4649–4655.

123 Yermolayev, S.A., Gorobets, N.Y., and Desenko, S.M. (2009). Rapid three-step one-pot microwave-assisted synthesis of 2,5-dioxo-1,2,5,6,7,8-hexahydro-3-quinolinecarbonitrile library. *J. Comb. Chem.* 11 (1): 44–46.

124 Yermolayev, S.A., Gorobets, N.Y., Shishkin, O.V. et al. (2011). Pathways for cyclizations of hydrazine-derived 2-(2-cyanovinyl)-3-oxo-cyclohex-1-ene enolates. *Tetrahedron* 67 (16): 2934–2941.

125 Vodolazhenko, M.A., Mykhailenko, A.E., Gorobets, N.Y., and Desenko, S.M. (2017). One-pot synthesis of benzo[4,5]imidazo[1,2-a]pyridine derivatives in aqueous conditions. *J. Heterocyclic Chem.* 54 (1): 753–757.

126 Vodolazhenko, M.A., Gorobets, N.Y., Yermolayev, S.A. et al. (2012). Application of stable fused dienolates for diversity oriented synthesis of 2,5-dioxo-5,6,7,8-tetrahydro-2H-chromene-3-carboxamides. *RSC Adv.* 2 (3): 1106–1111.

127 Dzhavakhishvili, S.G., Gorobets, N.Y., Chernenko, V.N. et al. (2008). Synthesis of novel 3-(1,3-thiazol-2-yl)-7,8-dihydroquinoline-2,5(1H,6H)-diones. *Russ. Chem. Bull.* 57 (2): 422–427.

128 Vodolazhenko, M.A., Gorobets, N.Y., Zhikol, O.A. et al. (2016). A quantum chemical approach towards understanding stability and tautomerism of 2-imino-2H-pyran derivatives. *RSC Adv.* 6 (57): 52201–52211.

129 Bondavalli, F., Bruno, O., Presti, E.L. et al. (1999). An efficient synthesis of functionalized 2-pyridones by direct route or via amide/enolate ammonium salt intermediates. *Synthesis (Stuttg).* 7: 1169–1174.

130 Gorobets, N.Y., Sedash, Y.V., Shishkina, S.V. et al. (2009). Structure of the intermediate in the synthesis of 6-methyl-2-oxo-1, 2-dihydropyridine-3-carbonitrile. *ARKIVOC, (xiii)*: 23–30.

9

Recent Applications of Multicomponent Reactions Toward Heterocyclic Drug Discovery

Nathan Bedard[1], Alessandra Fistrovich[1], Kevin Schofield[1], Arthur Shaw[1], and Christopher Hulme[1,2]

[1] *The University of Arizona, College of Science, Department of Chemistry & Biochemistry, Tucson, AZ 85721, USA*
[2] *The University of Arizona, College of Pharmacy, Department of Pharm./Tox., Tucson, AZ 85721, USA*

9.1 Introduction

Modern-day strategies for drug enhancement incorporate fragment-, lead-, or drug-like properties that push the efficiency of medicinal chemistry operations toward superior physicochemical properties, potency, and selectivity [1]. Heterocyclic structures have risen in this domain as they play a vital role in many drugs and natural products [2]. To make these structures, medicinal chemists must incorporate multicomponent reactions (MCRs) as they offer increased atom and time efficiency while providing a high molecular diversity. This chapter highlights recent advancements in heterocyclic MCRs when it comes to U.S. Food and Drug Administration (FDA)-approved drugs, drug intermediates, natural products, and modern trends of modified reactions to help readers obtain useful knowledge toward designing new reaction strategies regarding pharmaceuticals.

9.2 Multicomponent Reactions

MCRs are generally defined as chemical reactions that incorporate three or more reactants in a single pot that generates one product in which every reactant has contributed most (or all) of its atoms [1]. There are a vast number of advantages that MCRs provide which include commonly high yields, cost-effectiveness, high atom efficiency, high molecular diversity and complexity, and mild conditions. One of the downsides includes the unpleasant smell of more volatile isocyanides. Another downside, not essentially intrinsic to MCRs, is the ignorance of the class of reactions by many chemists that may lead to biased views against their use [1]. However, this can be combatted by increasing the overall knowledge of the benefits of MCRs and carefully conveying this to the chain of command.

Multicomponent Reactions towards Heterocycles: Concepts and Applications, First Edition.
Edited by Erik V. Van der Eycken and Upendra K. Sharma.

The first documented MCR was the Strecker reaction in 1850 [3] that produced α-amino nitriles that α-amino acids originate from. The consecutive steps of the Strecker reaction are common to most MCRs in which a carbonyl condenses with an amine and nucleophilic addition ensuing. This general reactivity order led to many more MCRs being discovered which have been covered in the recent decades [4–7].

9.3 The Ugi Reaction

In 1959, Ivar Ugi reported a reaction involving a carbonyl, amine, isocyanide, and a carboxylic acid to form a bis-amide (Scheme 9.1) and is known today as the Ugi four-component reaction (4CR) [8]. The most widely accepted mechanism for this transformation was proposed by Ugi in the 1960s [9, 10]. The mechanism for an Ugi 4CR starts with the condensation of the amine with the carbonyl of the ketone/aldehyde. The isocyanide component is proposed to act bivalently where the isocyanide adds to the Schiff base, while the carboxylate adds to the isocyanide forming the imidoyl intermediate. This undergoes a Mumm rearrangement where the amine attacks the ester forming an α-acetamido carboxamide as the final product (Scheme 9.2).

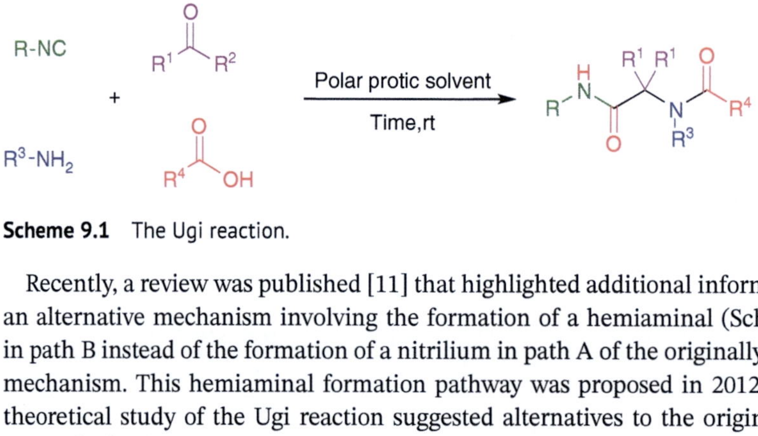

Scheme 9.1 The Ugi reaction.

Recently, a review was published [11] that highlighted additional information for an alternative mechanism involving the formation of a hemiaminal (Scheme 9.3) in path B instead of the formation of a nitrilium in path A of the originally reported mechanism. This hemiaminal formation pathway was proposed in 2012 after the theoretical study of the Ugi reaction suggested alternatives to the original mechanism [12]. This mechanism was later debunked by the investigation of the Ugi 4CR intermediates by electrospray ionization mass spectrometry (ESI-MS[/MS]) that confirmed the formation of the nitrilium ion and finding no hemiaminal intermediate [13].

The Ugi reaction has been expanded to suit a whole range of diverse substrates. Primary and secondary amines work well, as well as hydroxylamines and aminophosphonic acids [11–13]. Azides (Ugi-azide) [8], cyanates, thiocyanates, water, hydrogen selenide, hydrazone, *N*-hydroxyimide, hydrazoic acid, thioacetic acid, phenol (Ugi-Smiles), thiophenol, squaric acid, and hydroxycoumarin can be used in place of carboxylic acid [8–19, 14–23, 176–178]. With substrate variation almost exhausted, research groups have been going after enantioselective Ugi reactions. Past efforts attempted to use chiral isocyanides or amino acids to give chiral

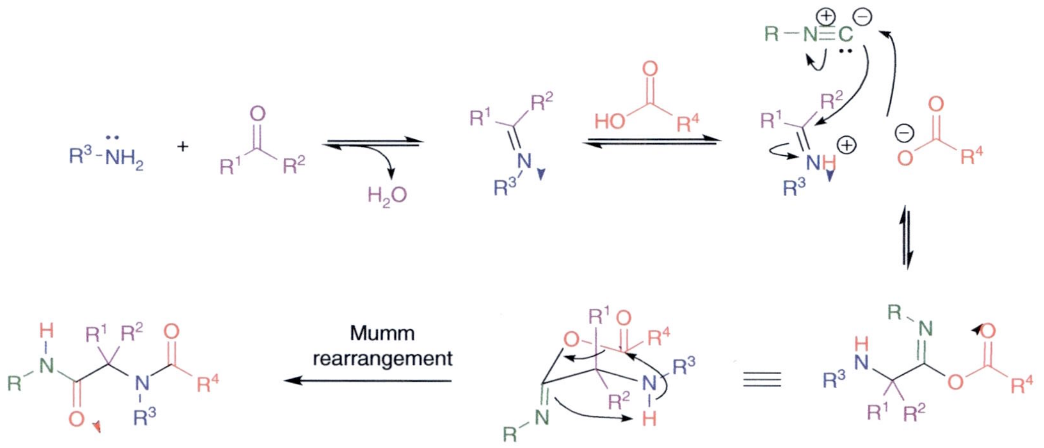

Scheme 9.2 Proposed mechanism for an Ugi 4CR. Source: Based on Moni et al. [48].

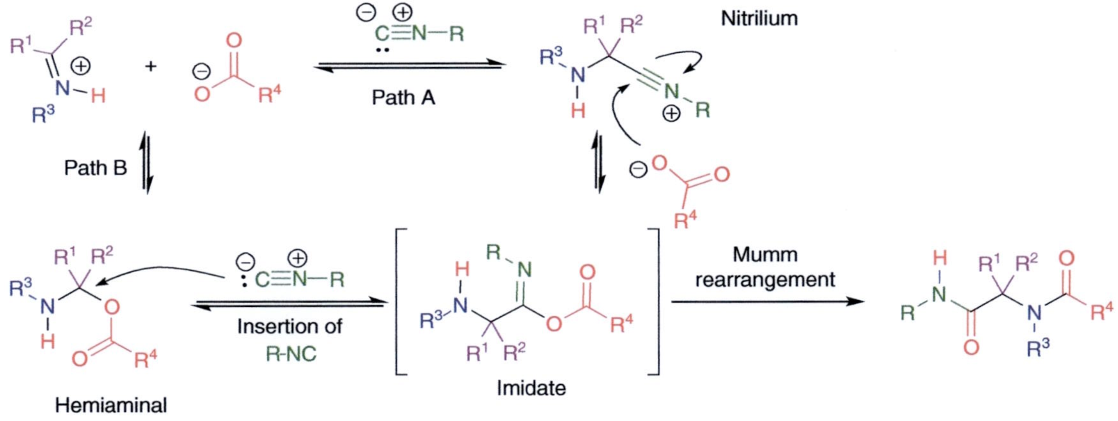

Scheme 9.3 Alternative mechanism for an Ugi 4CR.

Ugi 4CR products; however, this methodology often afforded low enantioselectivity. In 2018, Tan and coworkers discovered that using chiral phosphoric acid ligand **CPA6** (Figure 9.1) gave chiral Ugi 4CR products up to 90% yield and 92% ee [21]. All previous acid component listed above is summarized (Figure 9.1) below.

For the Ugi 4CR, typically all reagents are added simultaneously; however, pre-formation of the Schiff base by addition of the carbonyl and amine components is also common. Polar protic solvents are used with methanol, ethanol, and 2,2,2-trifluoroethanol (TFE) being the most common [25, 26], and the molarity of solvent used is often around 0.1 M. The reactions are commonly run at room temperature for 24 hours; however, the time can vary drastically based on the reaction rate. The Ugi reaction time can be reduced by microwave-assisted synthesis, and the selectivity, purity, and yield can also be increased efficiently [27–29].

The true power of the Ugi reaction is being able to design its substrates to give heterocyclic drug-like products in two to three steps. These types of processes are called post-Ugi transformations. Some of which include Ugi-Heck [30–33], Ugi-de-boc-cyclization (UDC) [34–37], and Ugi-oxidation [38–41]. A general schematic for an Ugi 4CR and post-modification is represented below (Scheme 9.4) in which Peshkov and coworkers demonstrated the utility of the Ugi reaction by deploying specific functional groups which are then utilized in a post-Ugi cascade to give tetracyclic spiroindolines [36].

A 5-endo trig oxidative radical cyclization of benzylamine-derived Ugi three-component reaction (3CR) products rapidly affords imidazolidinones with three diversity elements. This MCR–oxidation methodology further showcases manipulation of the diverse elements in MCR products via oxidative radical cyclizations, which generates highly decorated privileged heterocycles (Scheme 9.5) [38].

9.3.1 The Ugi Reaction Used in Natural Product Synthesis

Ustiloxins are a family of cyclic peptides isolated from the fungus *Ustilaginoidea virens*. This fungus causes the disease False smut in rice crops which reduces both yield and grain quality [43, 44]. Hutton and coworkers synthesized cyclic peptide natural product, ustiloxin D in six steps from a known β-hydroxydopa derivative which is rapidly converted to macrocycle ustiloxin D via UDC methodology (Scheme 9.6) [37]. The reaction of ammonia, *N*-Boc-valine, and isonitrile with the β-hydroxydopa derivative affords a branched tripeptide, and subsequent global deprotection and cyclization afford ustiloxin D. Use of this ammonia-Ugi MCR was more step-economical and higher-yielding than the previous route reported by Joullié [38].

9.3.2 The Ugi Reaction in FDA-approved Drugs and Drug Candidates

Deploying 2-bromo-6-isocycanopyridine 1 as a convertible isonitrile in conjunction with the ketone 2 affords the Ugi 4CR adduct 3 where subsequent methanolysis gives the FDA-approved drug carfentanil 4 in 90% yield over two steps (Scheme 9.7) [47].

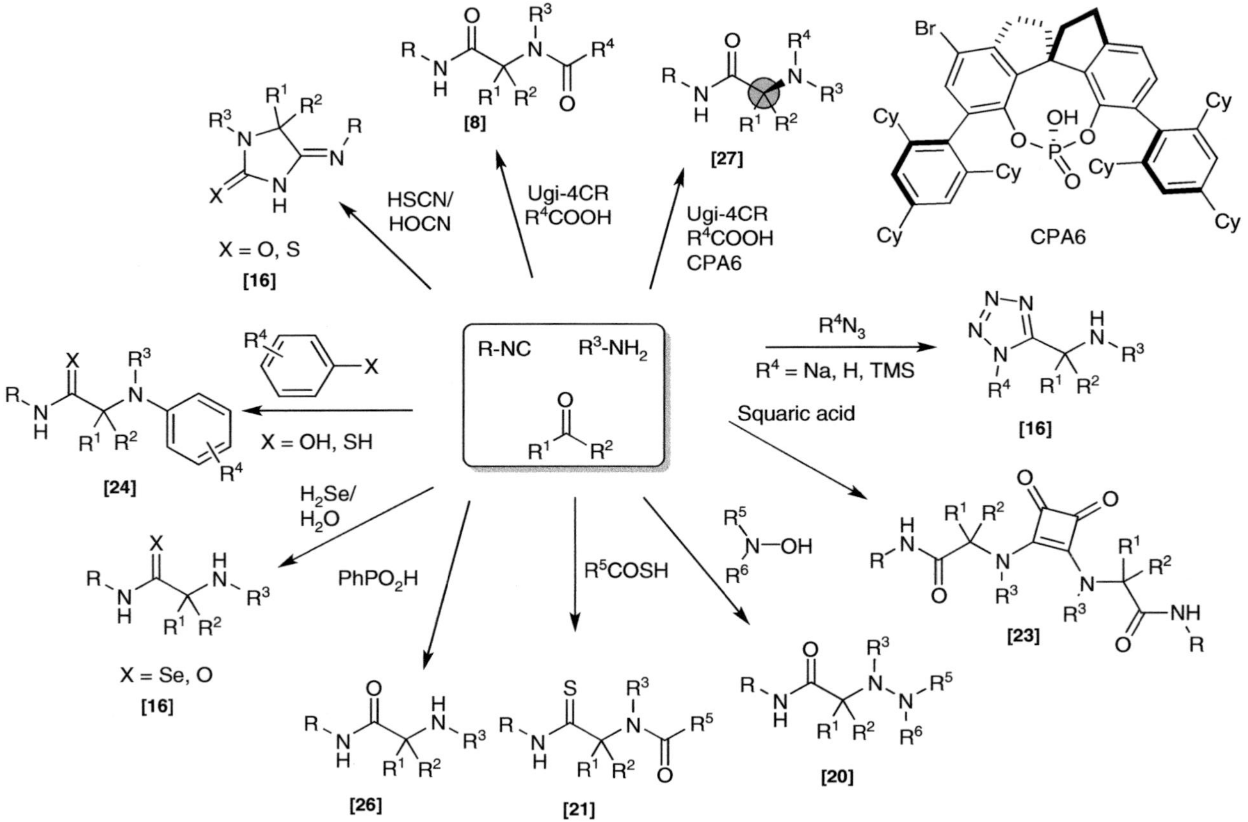

Figure 9.1 Manipulation of the acid component in Ugi 4CR.

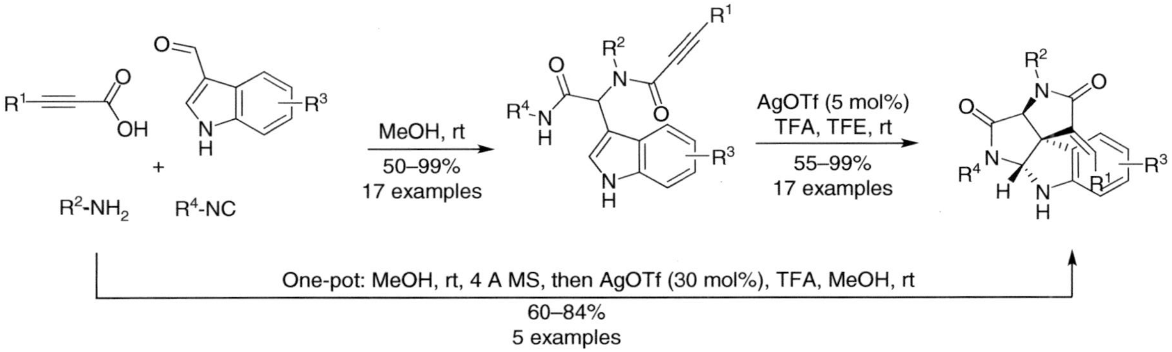

Scheme 9.4 Post-Ugi cascade to tetracyclic spiroindolines.

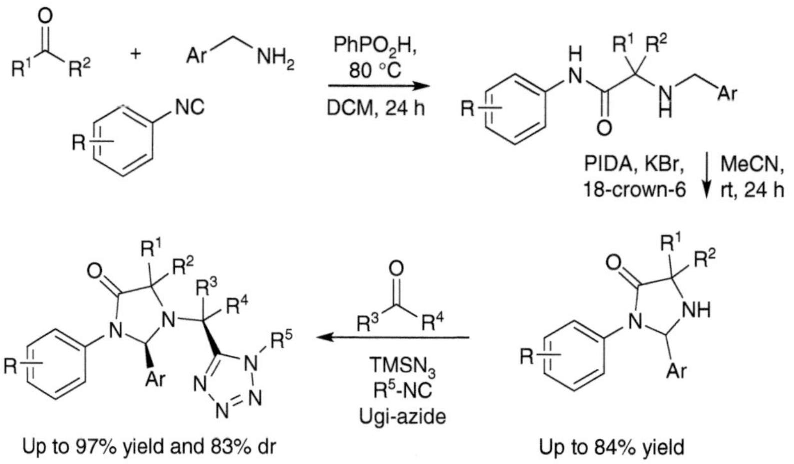

Scheme 9.5 5-endo-trig oxidative radical cyclization of benzylamine-derived Ugi product.

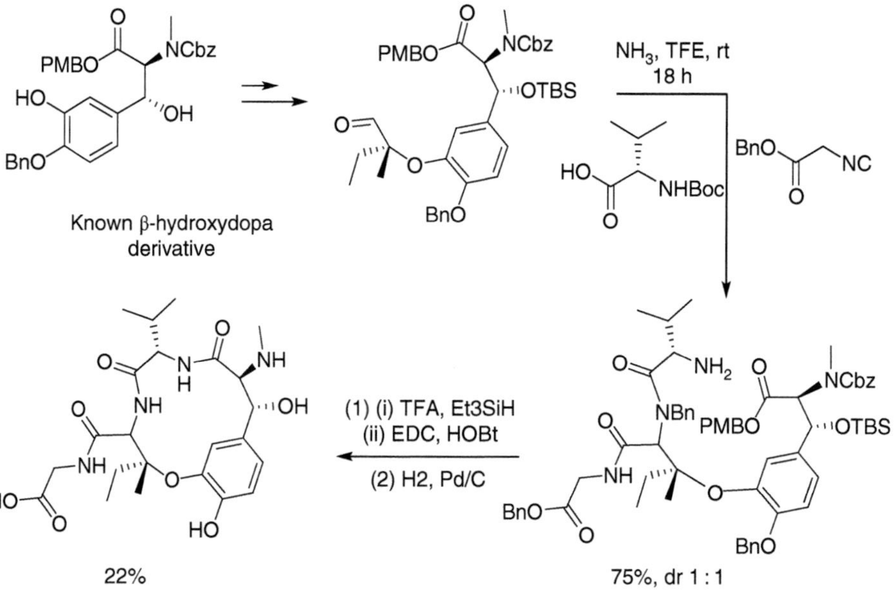

Scheme 9.6 Synthesis of ustiloxin D utilizing an Ugi 4CR.

Majumdar and coworkers produced potent carfentanil amide opioids **5–7** that showed high affinity for mu-opioid receptors (MOR) and delta-opioid receptors (DOR) in one step utilizing an Ugi 4CR, while **7** showed no visible signs of physical dependence or constipation in mice and produced less respiratory depression than morphine (Figure 9.2) [39]. Through this methodology, mass libraries of opioids can be produced.

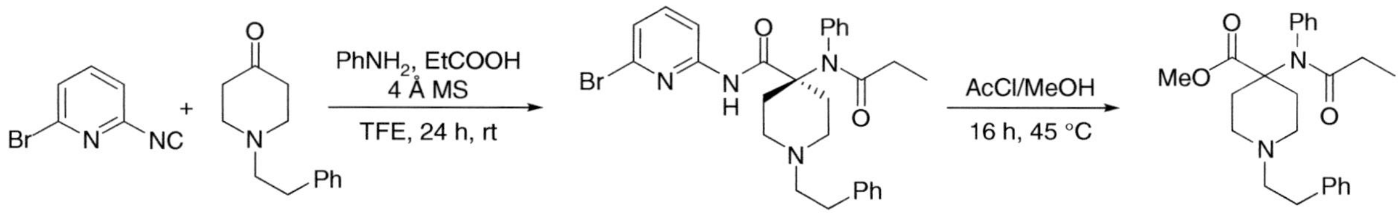

1 **2** **3**, 92% Carfentanil **4**, 98%

Scheme 9.7 A two-step synthesis of Carfentanil via an Ugi 4CR.

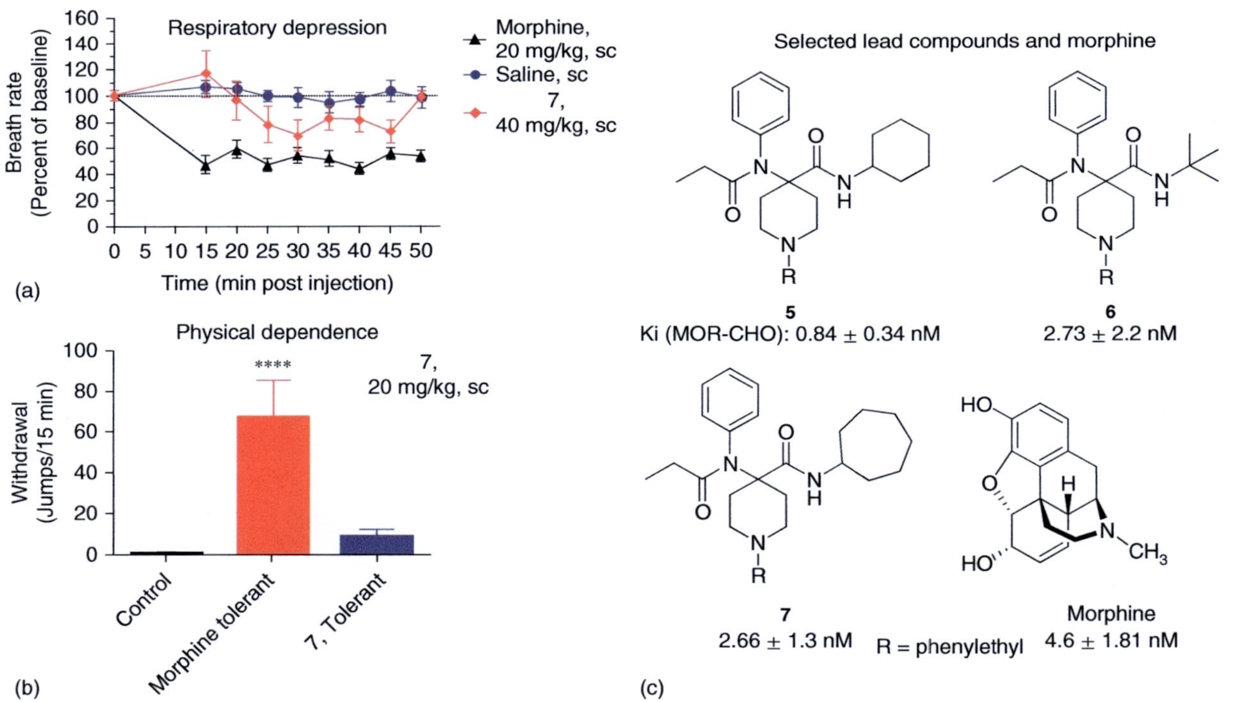

Figure 9.2 (a) Sensitivity of **7** to opioid antagonist: groups of mice ($n = 10$) received a fixed dose of **7** (15 mg/kg) alone or with β-FNA or NTI. Tail flick analgesia was measured 30 min after dosing. (b) Physical dependence. Groups of mice (n greater than or equal to 10) received either morphine (10 mg/kg) or **7** (1 mg/kg) until they showed complete tolerance. Naloxone precipitated a profound withdrawal syndrome in animals treated with morphine, while mice administered with **7** displayed no significant difference from controls. (c) Selected lead compounds in comparison to morphine [48].

9.3.2.1 Synthesis of Lipitor Using Ugi 4CR

Lipitor belongs to the drug class of statins, lipid-lowering drugs for the prevention of cardiovascular disease (CVD). Lipitor has attracted much attention synthetically as it is one of the best-selling medications of all time. Previous routes include Paal–Knorr, Hantzsch, and Munchnone which all require between 5 and 7 steps synthetically [49–51]. The issue with these routes is that there are a greater number of steps to produce the starting materials and Lipitor, while the Hantzsch route produces the lactone of Lipitor. Using the Ugi 4CR methodology, Dömling and coworkers were able to synthesize Lipitor in just four steps (Scheme 9.8) [43]. Another highlight of this synthetic route is that by diversifying the aldehyde, amine, acid, and/or isocyanide component, one can produce a large library of highly diverse substituted bioactive pyrroles, something that is not as feasible with the previous routes mentioned. Therefore, this synthetic route is among the most competitive.

9.3.2.2 Synthesis of Ivosidenib Utilizing Ugi 4CR

Agios Pharmaceuticals first published their patent on Ivosidenib, an IDH1 inhibitor, in 2015 (TIBSOVO®) and was rapidly approved by the FDA on 2 May 2019 [53]. Ivosidenib is the first oral, targeted therapy for adult patients with acute myeloid leukemia (AML) with a susceptible IDH1 mutation. This form of AML is a progressive and deadly cancer of the bone marrow and blood. Approval was based on an open-label, single-arm, multicenter clinical trial where 12 (42.9%) of the 27 patients achieved complete remission with partial hematologic recovery, while 7 (41.2%) of the 17 transfusion-dependent patients achieved transfusion independence lasting at least eight weeks. Ivosidenib is administered orally at a dose of 500 mg daily for a minimum of six months or until disease progression or unacceptable toxicity. Ivosidenib, a seemingly complex molecule, can be synthesized in just two steps, the first step being an Ugi-4CR and the second being a Buchwald reaction (Scheme 9.9). Crystallization gave the desired diastereomer in 99.7% de [54].

9.3.3 Rapid Lead Optimization with Ugi 4CR

A structure–activity relationship (SAR) is the relationship between a chemical structure and its biological properties and is used in different stages of drug discovery to achieve optimal physicochemical properties for a given drug. The drug is designed to better interact and fit its target protein's binding pocket. Typically, one site on the drug is changed at a time until an analog is found with optimal binding capabilities, and this method was used for ivosidenib (Scheme 9.10).

While **8** contained favorable potency (70 nM) against IDH1, it had a rather poor hepatic extraction ratio (E_h) of 0.93. This was thought to be because of the benzylic methyl group and potentially the cyclohexane. Therefore, fluorocycloalkyl groups were used to replace the cyclohexane, and bulky chlorine was replaced with the methyl resulting in a fivefold loss in activity, but raising the metabolic stability by 200%. Additionally, the imidazole motif was switched out for a methyl-ester proline derivative **9**, which further raised activity. It was believed that metabolism occurred at the methyl ester and a-amino position. Therefore, 2-pyrrolidinone **10**

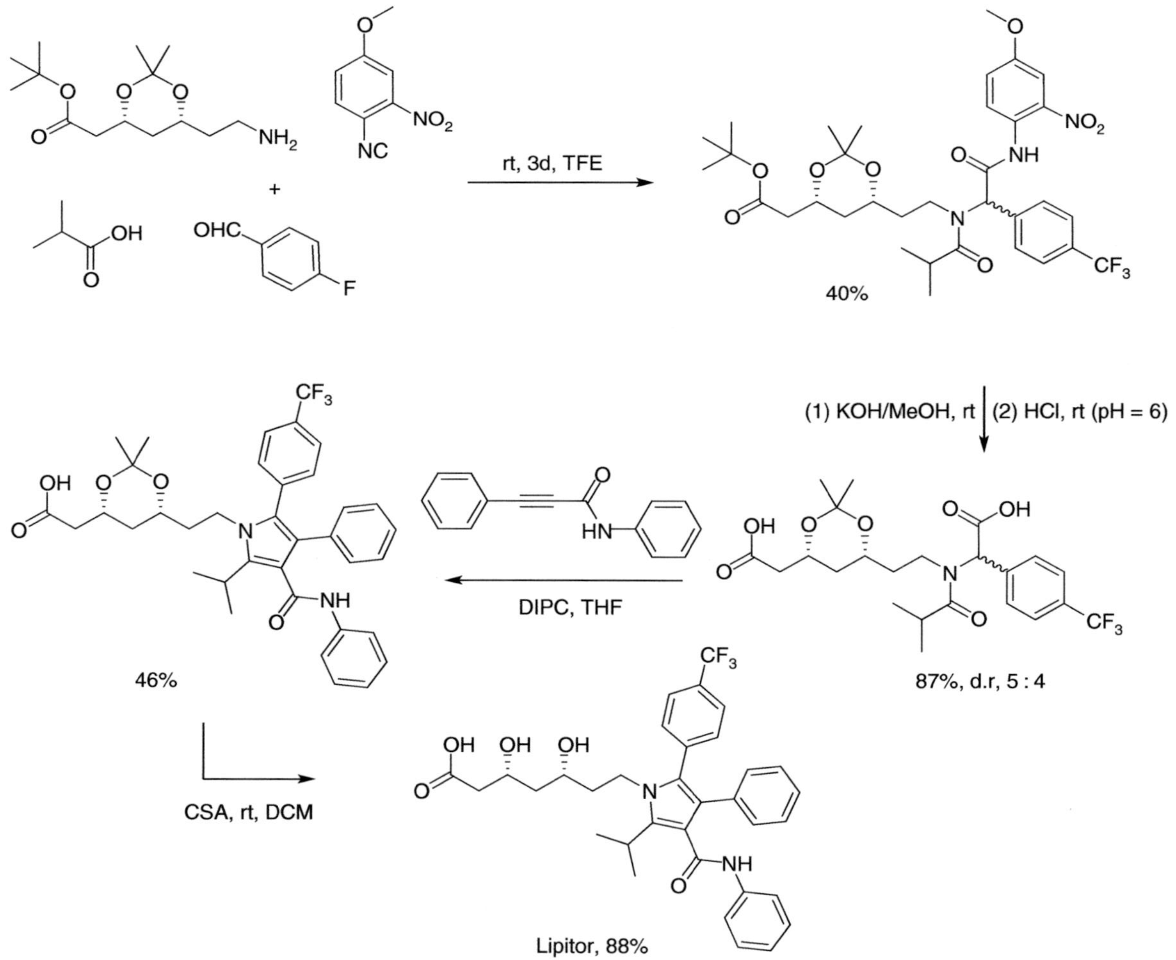

Scheme 9.8 A four-step synthesis of Lipitor utilizing Ugi 4CR methodology. Source: Based on Zarganes-Tzitzikas et al. [43].

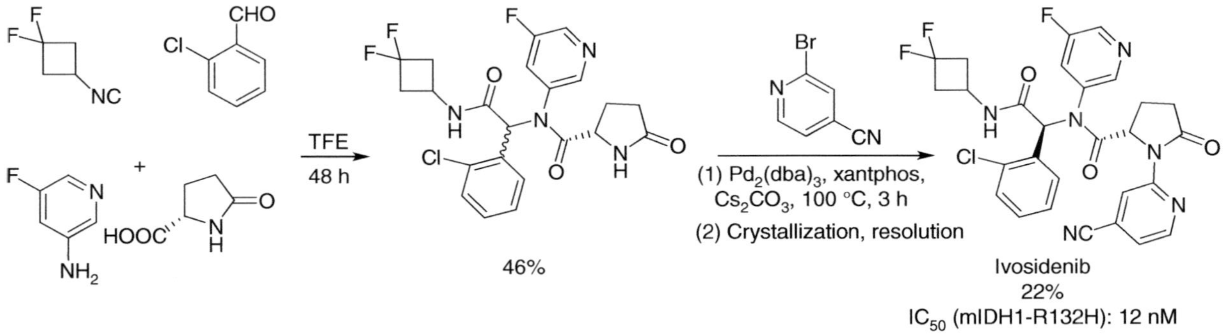

Scheme 9.9 A two-step synthesis of Ivosidenib utilizing Ugi 4CR methodology.

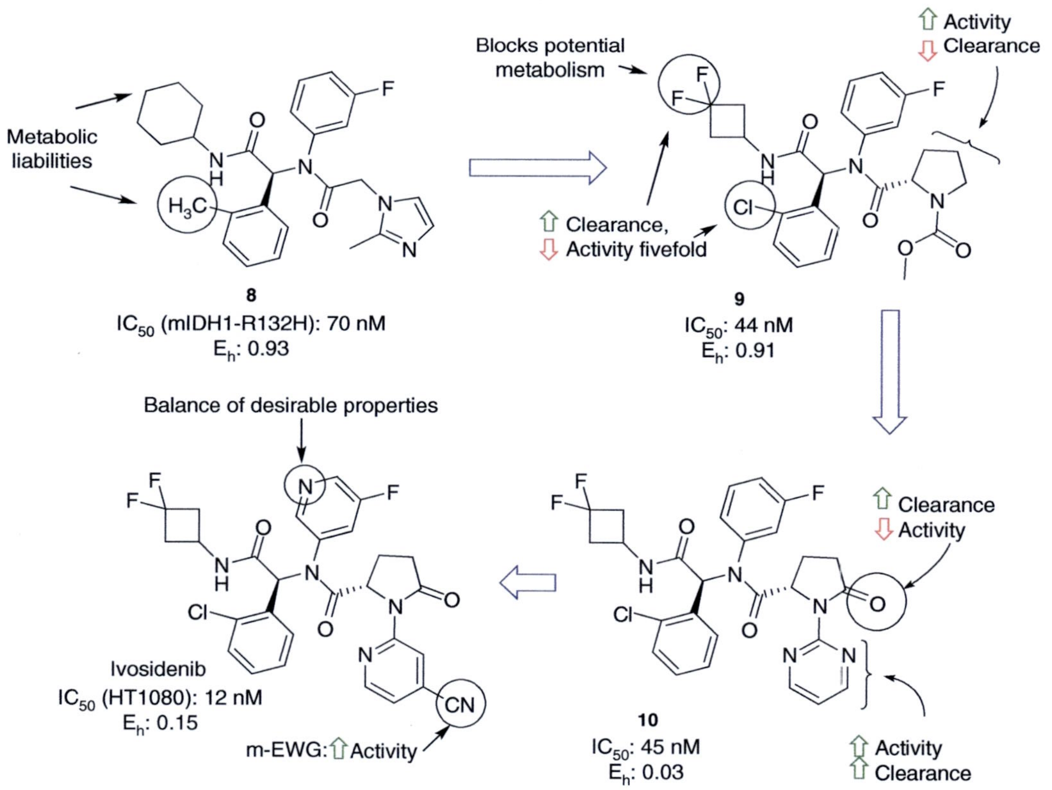

Scheme 9.10 Exploring the SAR of IDH1 inhibitors toward the lead compound, Ivosidenib.

was used to prevent these metabolic issues which contained high metabolic stability while having great activity. It was found that aromatic substitution, specifically pyrimidine, about the amide of the pyrrolidinone motif restored activity and clearance. Last, changing the C—F bond for N and the adjacent C—H bond for C—F raised metabolic stability, while the change of the pyrimidine to a pyridylcyano group raised activity while also decreasing clearance. With this, ivosidenib was obtained which contained a strong balance of desirable properties [54].

9.4 The Passerini Reaction

The Passerini reaction was first discovered by Italian chemist Mario Passerini in 1921 [55] and is the first isocyanide-based MCR developed. This 3CR includes the use of an isocyanide, ketone, or aldehyde, and a carboxylic acid component (Scheme 9.11). The carboxylic acid component can be substituted for water to form α-hydroxycarboxamides or with an azide source to form α-hydroxyalkyltetrazoles.

Like the Ugi reaction, the isocyanide acts bivalently by adding to carbonyl component followed by nucleophilic attack of the carboxylic acid on the carbon of this isonitrile forming an imidoyl intermediate. This then proceeds through a Mumm rearrangement to form the Passerini product (Scheme 9.12). Along with an isocyanide, a ketone or aldehyde, and a carboxylic acid is also used to afford an α-acyloxy amide. It should be noted that there has been a recent discovery through high-level density functional theory calculations that a nitrilium intermediate is stable in solution, its formation is rate-determining, catalyzed by a second carboxylic acid molecule, and therefore can be considered a 4-CR [56].

9.4.1 The Passerini Reaction in Natural Products

A major limitation of the Passerini reaction is the possible generation of a stereocenter that usually leads to a racemic mixture. This can be a problem in medicinal chemistry and natural product synthesis because challenging separations are necessary if a pure stereoisomer is desired [57]. Fortunately, Lewis acids [58] and chiral phosphoric acid catalysts [59] have been recently developed to bypass the tedious separations for enantiopurity. This approach was used in the total synthesis of N^{14}-desacetoxytubulysin H (Scheme 9.13) [60], a member of the Tubulysins that disrupt tubulin binding that leads to cytotoxicity in cancer cells [61]. The authors postulated that the origin of the stereoselectivity is due to the bulkiness of **L1** in conjunction with $ZnBr_2$ forms a Cram-chelated complex that favors isocyanide attack from the easier to access *si* face of the carbonyl [60].

9.5 Groebke–Blackburn–Bienaymé (GBB-3CR) MCR

The Groebke–Blackburn–Bienaymé (GBB-3CR) reaction was originally reported in 1998 by three groups, who demonstrated a simplistic way of synthesizing

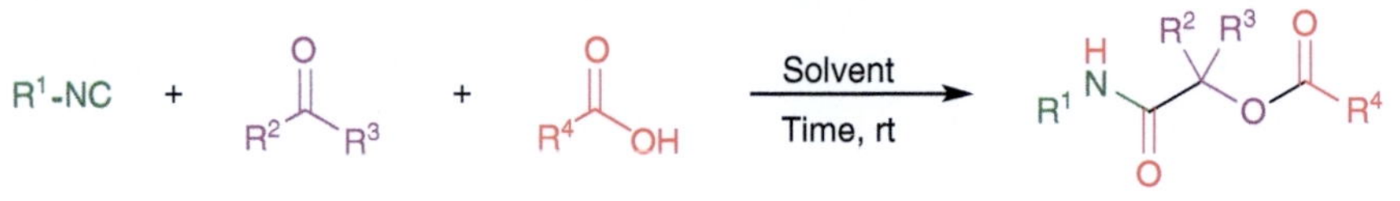

Scheme 9.11 The three-component Passerini reaction.

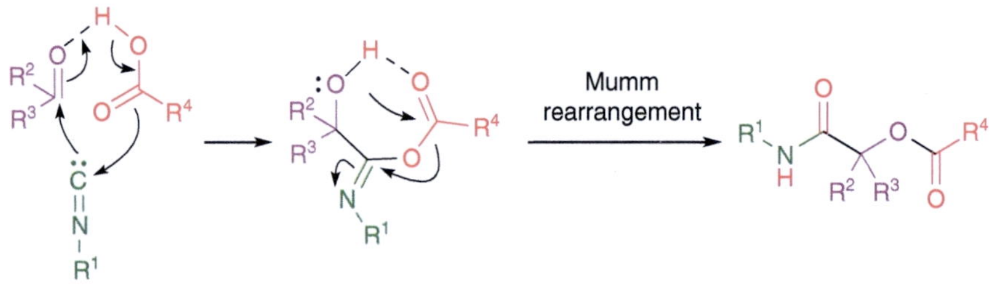

Scheme 9.12 Proposed mechanism of the Passerini reaction.

3-amino imidazo[1,2,x]-heterocyclic compounds (Scheme 9.14) [62–64]. All groups commonly used isocyanides **11**, aldehydes **12**, and amidines **13** but varied in choice of Lewis or Brønstead acid catalyst and solvent. The GBB-3CR undergoes an imine formation followed by a [4 + 1] cycloaddition to produce a useful imidazo-heterocycle with a bridgehead nitrogen **14** pharmacophore.

Medicinally, the bridgehead nitrogen bi- and tricyclic cores within drug-like scaffolds are of interest because of their interactions within biological pathways and therapeutic targets [65]. Imidazo-systems add rigidity to drug-like scaffolds and contain built-in physicochemical properties, which can overcome the entropic barriers present with drug and receptor binding [66]. Also, the basicity imidazo-heterocycles possess can aid in solubility issues, making them sought-after structures for drug design.

One study focused on designing and synthesizing selective inhibitors for urokinase plasminogen activator (uPA) [67]. uPA is a promising oncological target based on its overexpression in metastasizing solid tumors and connection to tumor growth promotion through activation of growth factors and interactions of proteins involved in cell adhesion and signal transduction [68–71]. Assisted by the modified substrate activity screening (MSAS), they uncovered fragments that demonstrated an affinity for the active center of uPA.

Through initial evaluation, it revealed the importance of the guanidine side chain from isocyanide **15** (Scheme 9.15) [67]. This sidechain alone accounts for hydrogen bonding between Gly219, Ser190, and Asp189 in the binding pocket. Further, the amine produced off the imidazo[1,2,α]-pyridine core **16**, allowed for a critical hydrogen bond interaction with Ser195. Optimization of the 4-pyridyl position allowed for additional hydrogen bonding from the amine with Tyr151, leading to a highly selective uPA inhibitor **17** (Scheme 9.15).

The synthesized compound exhibited comparable affinity values to reference compounds UK122 **18**, Amiloride **19**, and Gavexate **20** (Figure 9.3). The GBB-3CR used

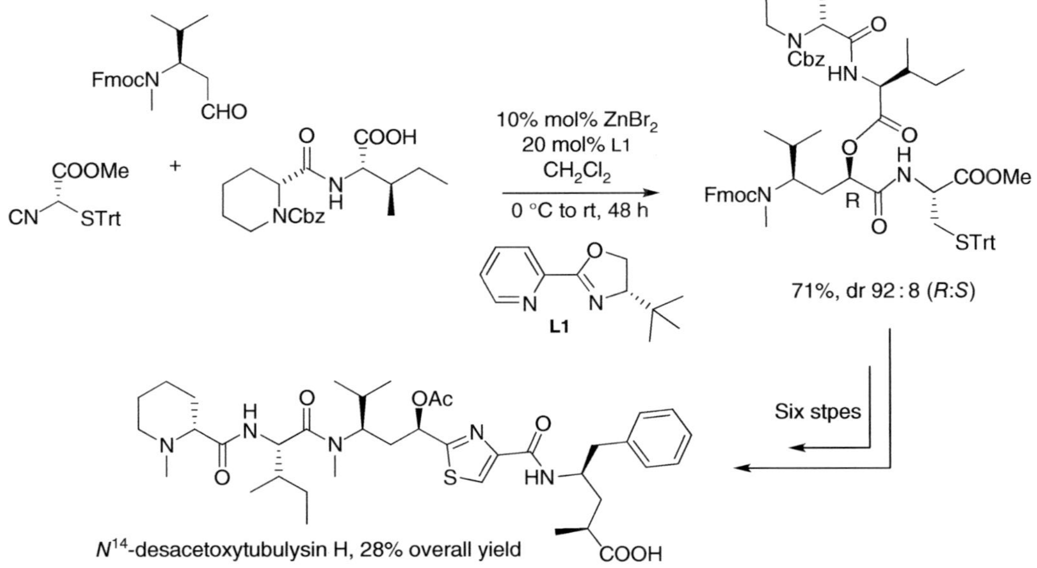

Scheme 9.13 Synthesis of N^{14}-desacetoxytubulysin H via a diastereoselective Passerini-3CR.

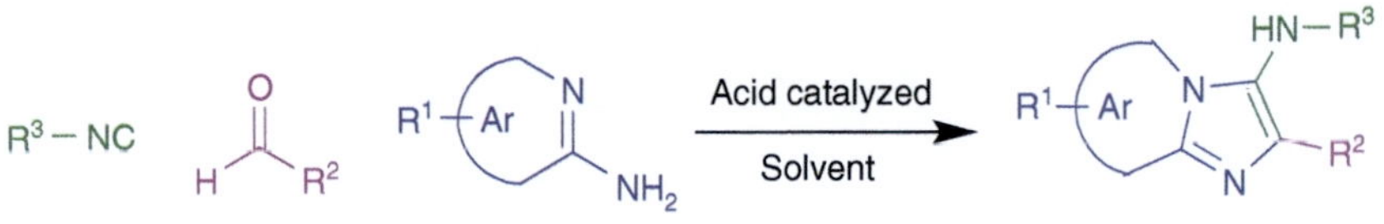

Scheme 9.14 General GBB-3CR reaction.

in this study mirrors the reaction type discovered by Bienaymé et al. (Rhône-Poulenc Technologies (France)), where it was demonstrated that this MCR can proceed with methanol and catalytic amounts of perchloric acid [53].

Designed multiple ligands (DMLs) containing an imidazo[1,2,α]-pyridine core and varied urea substituents have been explored for dual inhibition against 5-lipoxygenase (5-LO) and soluble epoxide hydrolase (sEH) [72, 73]. Both enzymes share a commonality in the metabolism of arachidonic acid and dual inhibition of 5-LO and sEH which could pose an effect on inflammation/immune and hypertension diseases, respectively [74–78]. **21** was synthesized through a GBB-3CR [72] (Scheme 9.16) where the acid catalyst was acetic acid and demonstrated comparable affinity values against known inhibitors EP6 **22**, Zileuton **23**, and AUDA **24** (Figure 9.4). Imidazo[1,2,α]-pyridine cores' ability to act as antimicrobial agents in drug-like scaffolds has also been under exploration [79]. This has drawn research interests because of bacterial resistance to established antimicrobial agents and increased occurrences of fungal infections in immunosuppressed individuals [80–82]. One study discovered that oxadiazole-quinoline scaffolds with imidazo[1,2,α]-heterocycle cores had comparable affinity values with marketed antibiotics ampicillin **25**, cefixime **26**, and fluconazole **27** against Gram-positive/negative bacteria and active fungal strands (Figure 9.5) [79].

The oxadiazole-quinoline derivative 29 was initially synthesized through a GBB-3CR where scandium triflate acts as the Lewis acid catalyst (Scheme 9.17) [82] that was originally discovered by Bienaymé et al. [53]. The quinoline derivative **28** subsequently underwent a coupling with 2-(1H-Benzotriazole-1-yl)-1,1,3,3-tetramethylaminium tetrafluoroborate Novabiochem® (TBTU) and diisopropylethylamine (DIPEA), following a reaction with $POCl_3$ to cyclize the oxadiazole **29**.

The GBB-3CR reaction is a valuable MCR that provides accessibility in creating bridgehead nitrogen bi- and tricyclic cores within drug-like scaffolds. As shown, the medicinal applications for these scaffolds are broad and have shown to be comparable to marketed drugs. Beyond biological and therapeutical treatment, the imidazo[1,2,x]-heterocyclic scaffolds have also been investigated as probes in the use of fluorescent imaging [83]. The methodology for the GBB-3CR is continually evolving, permitting imidazo[1,2,x]-heterocyclic compounds to be synthesized currently under various catalysts and solvent procedures [84], solvent-free conditions [85, 86], catalyst-free conditions [87], and a combined solvent and catalyst-free conditions [88–90].

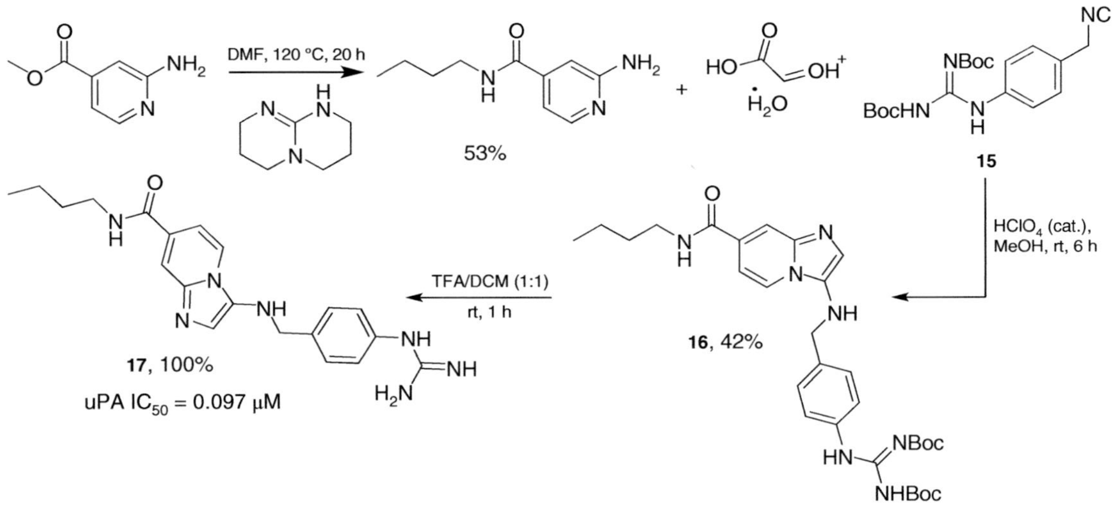

Scheme 9.15 Synthesis of highly selective uPA inhibitor **17**.

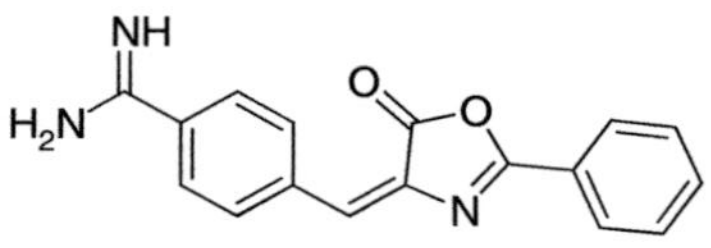

18
UK122
IC_{50} = 8.67 μM

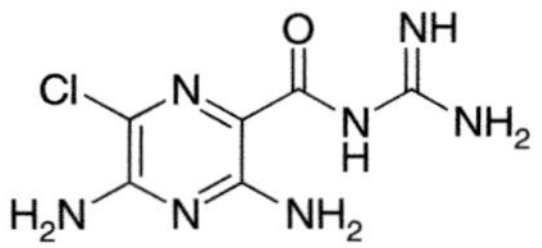

19
Amiloride
IC_{50} = 11.98 μM

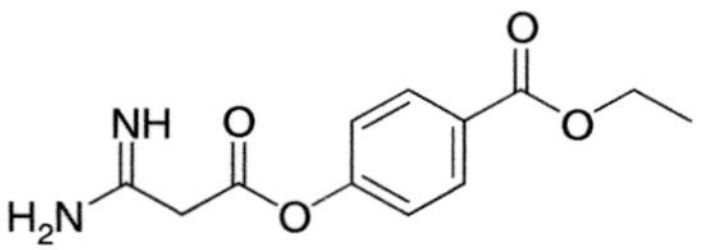

20
Gabexate
IC_{50} = 0.431 μM

Figure 9.3 Reference compounds and their inhibitory effect against uPA.

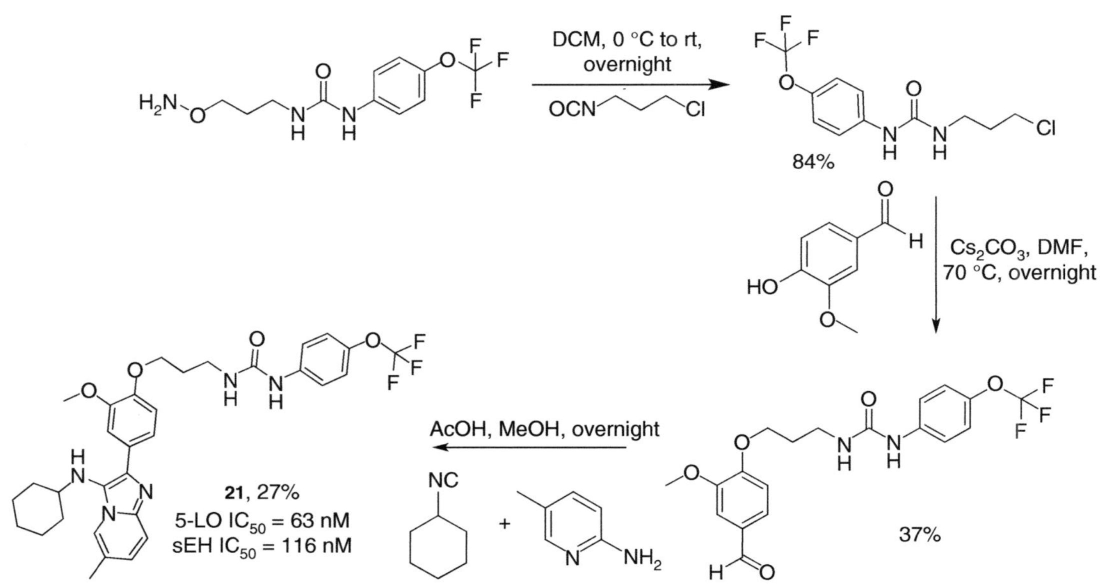

Scheme 9.16 Synthesis of the proposed inhibitor of 5-lipoxygenase.

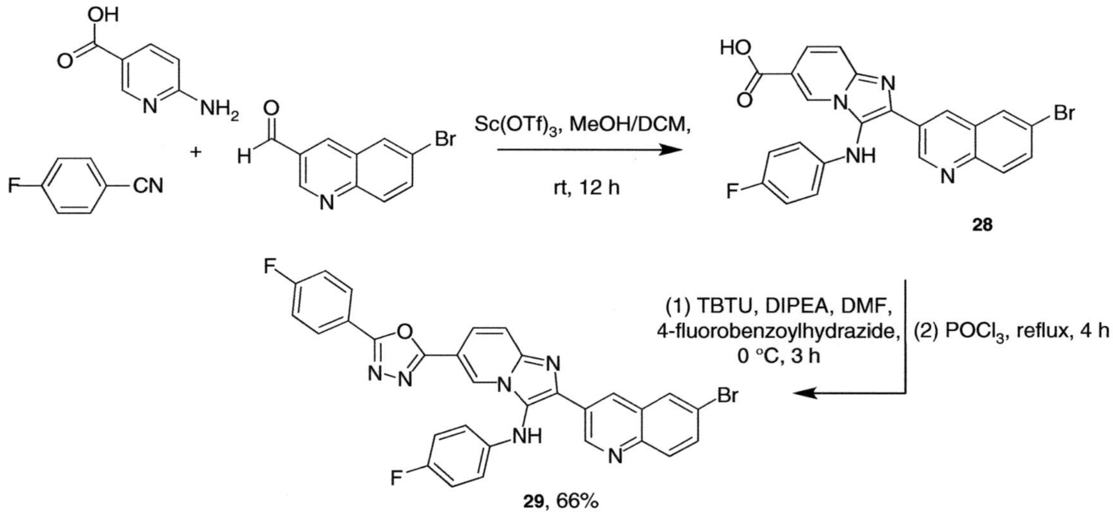

Scheme 9.17 Synthesis of oxadiazole-quinoline derivative **29**.

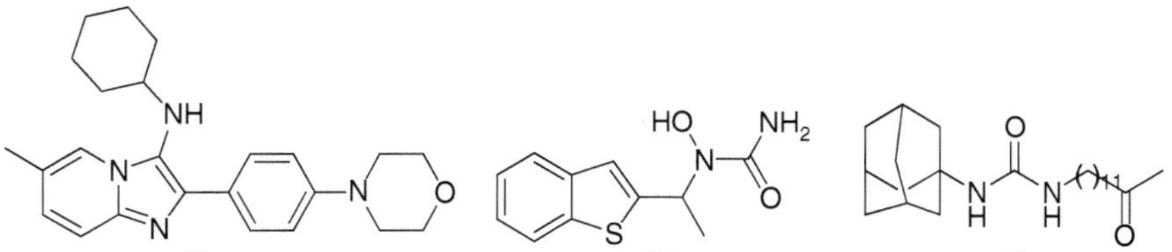

22
EP6
5-LO IC_{50} = 59 nM
sEH IC_{50} = >10 000 nM

23
Zileuton
5-LO IC_{50} = 591 nM
sEH IC_{50} = – nM

24
AUDA
5-LO IC_{50} = >10 000 nM
sEH IC_{50} = 107 nM

Figure 9.4 Reference compounds and their inhibitory effect against 5-LO and sEH.

9.6 Gewald (G-3CR) Reaction

The Gewald reaction (G-3CR) is a simplistic yet advantageous MCR that can produce several 2-aminothiophene **33** analogs with pharmacological properties, utilizing carbonyl compounds **30**, activated nitriles **31**, and elemental sulfur **32** (Scheme 9.18) [91].

The reaction proceeds through a Knoevenagel condensation with the carbonyl compound **30** and activated nitrile **31**, followed by a thiolation, based on an S_NX mechanism, with elemental sulfur **32**. A ring closure proceeds with the sulfurated compound by a nucleophilic mercaptide attack, ending with a prototropic rearrangement to produce the 2-aminothiophene **33** [92, 93]. This scaffold derived from the G-3CR has already displayed pharmacological capabilities through the development of Olanzapine [94] **34** and Tinoridine [95] **35** (Figure 9.6).

Numerous studies have since utilized the 2-aminothiophene scaffold, further supporting its importance in small molecule inhibitors, probes of target proteins, and importance in a SAR producing lead compound design [96]. One study has focused on the versatility of the 2-aminothiophene scaffold, within poly(ADP-ribose) polymerase-1 (PARP-1) inhibitors [97]. This enzyme is a part of numerous cell processes, and overexpression of the PARP-1 enzyme has been linked to various cancers (skin, lung, and breasts) [98–102]. Identification of this overexpression can be considered deadly as it is correlated with a low survival prognosis [103].

A one-pot synthesis of thieno[3,2-*e*]pyrrolo[1,2-*a*]pyrimidine analogs **38** was produced [97] from 2-aminothiophene scaffold **36** and 2-hydroxy-4-oxo-4-phenylbut-2-enoic acid **37** (Scheme 9.19). Fused heterocycles thieno[3,2-*e*]pyrrolo[1,2-*a*] pyrimidine scaffolds have previously displayed PARP-1 inhibition in other studies [104]. The synthesized analogs from this study are of interest, and further studies of their pharmaceutical capabilities are being carried out.

Another study examined the versatility of the 2-aminothiophene scaffold in the development of small molecules for the treatment of cystic fibrosis (CF) patients with F508del mutations [105]. CF can be caused by mutations within the cystic fibrosis transmembrane regulator (CFTR) [106–109]. With over 2000 mutations [110] that can occur within the *CFTR* gene, treatment options have been researched and one marketed drug, Ivacaftor **39**, has been approved for the treatment of CF with G551D mutation (Figure 9.7) [111, 112].

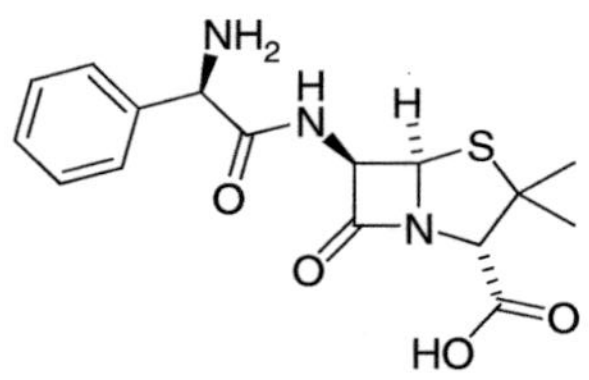

25
Ampicillin

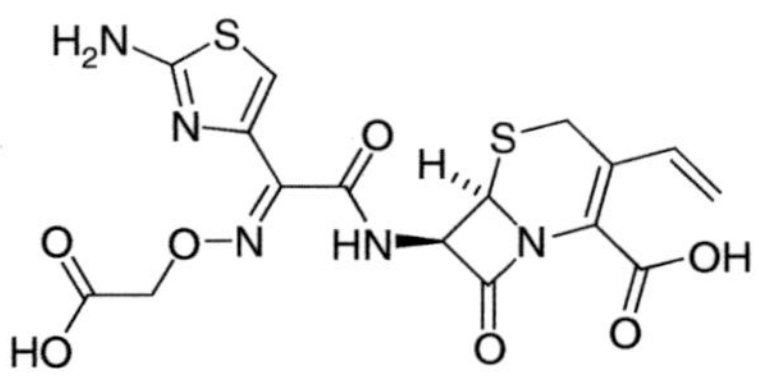

26
Cefixime

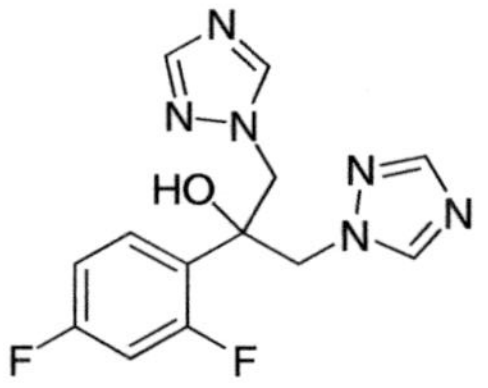

27
Fluconazole

Figure 9.5 Reference antibiotics used in comparison with synthesized compounds.

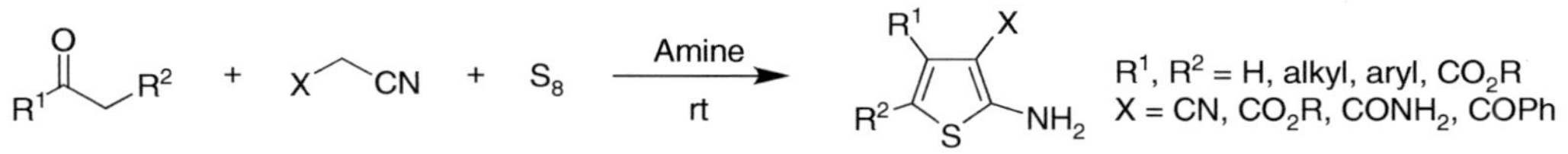

30 **31** **32** **33**

Scheme 9.18 General G-3CR reaction.

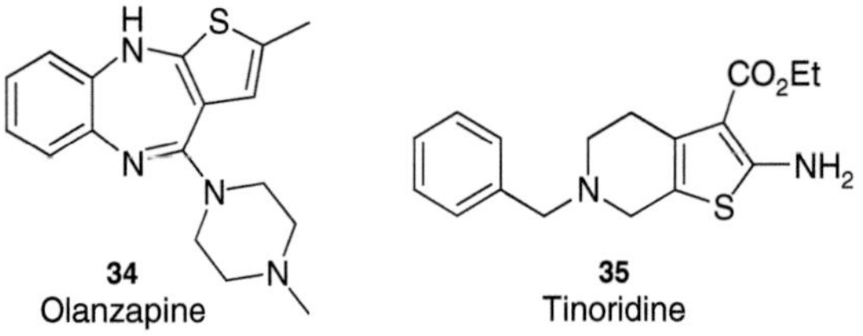

Figure 9.6 Investigated drugs containing 2-aminothiophene scaffold based on G-3CR.

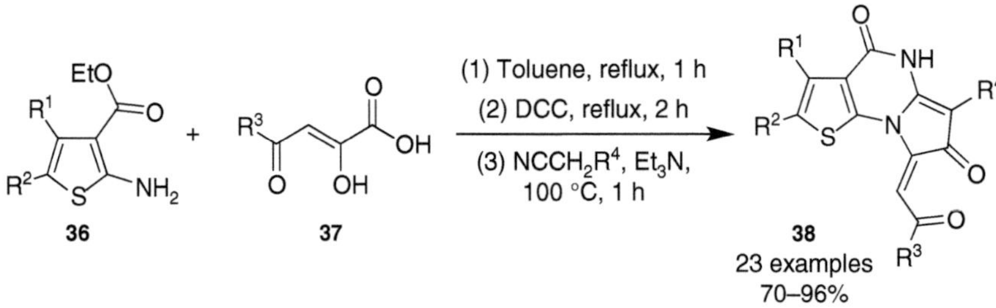

Scheme 9.19 One-pot synthesis of thieno[3,2-*e*]pyrrolo[1,2-*a*]pyrimidine analogs.

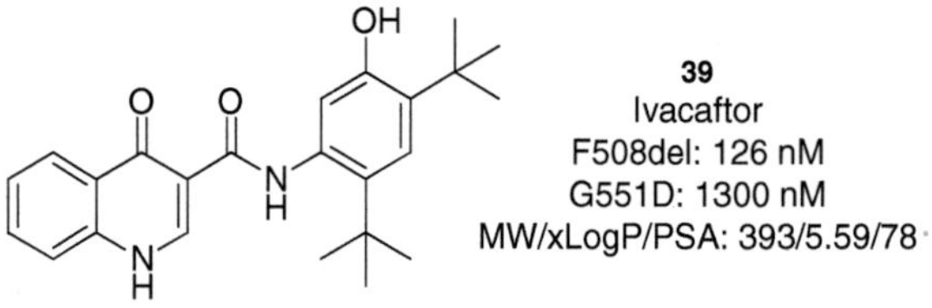

Figure 9.7 Marketed drug for CF.

The explorations into other mutations in the CFTR have led to the investigation of a series of 2-aminothiophene compounds. Through initial investigation and optimization, GLPG1837 **40** was discovered containing a 2-aminothiophene scaffold that was synthesized originally from a Gewald type reaction (Scheme 9.20). Physicochemical and efficacy properties were investigated for GLPG1837 **40**, showing that it had increased potency and efficacy compared to Ivacaftor [105].

9.7 The Hantzsch Dihydropyridine (DHP) Synthesis

The Hantzsch reaction was established in 1882 by Arthur R. Hantzsch when he condensed two equivalents of acetoacetate **41** and one equivalent of ammonia **42** and aldehyde **43** to produce a 1,4 dihydropyridine (DHP) **44** [103]. The general reaction conditions of the Hantzsch reaction include aldehydes, amines, and 1,3-dicarbonyl compounds to produce substituted, symmetrical 1,4-DHPs (Scheme 9.21) and often the 1,4-DHPs freely oxidize to **45**; however, an oxidizing agent must be used for stable DHPs [114]. Cyclic 1,3-diketones can also participate in the Hantzsch reaction, and nonsymmetrical 1,4 DHPs **46** can be synthesized (Scheme 9.22) [115].

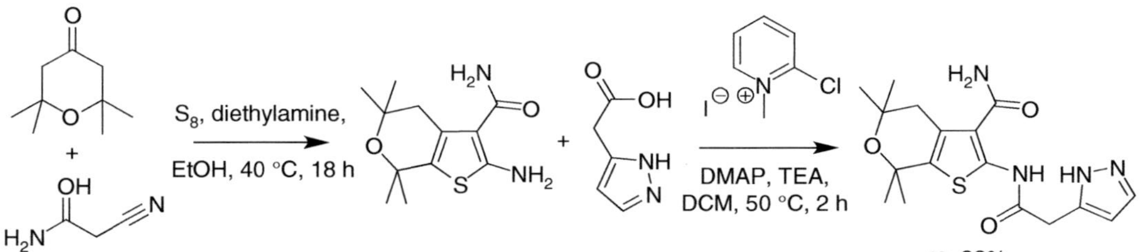

40, 32%
GLPG1837, F508del: 3 nM
G551D: 339 nM, MW/xLogP/PSA: 348/1.8/138

Scheme 9.20 Synthesis of GLPG1837.

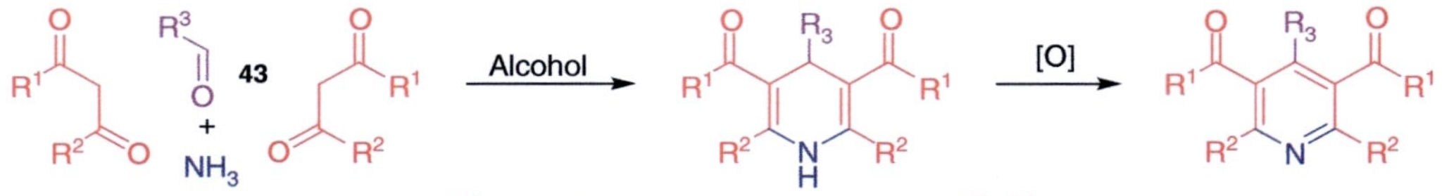

41 42 41 R^1: alkyl, O-alkyl 44 R^2, R^3: alkyl, aryl 45

Scheme 9.21 Hantzsch dihydropyridine synthesis.

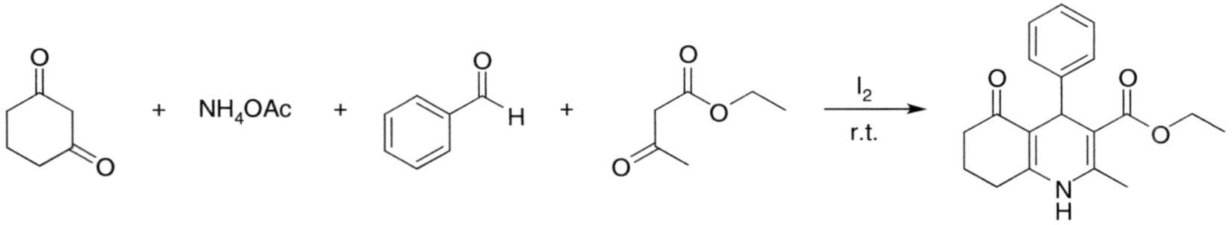

46

Scheme 9.22 Nonsymmetrical Hantzsch dihydropyridine synthesis.

DHP drug, application, year FDA approved

Nifedipine, hypertension, 1981
R^1:R^3:CO_2Me
R^2:2-NO_2-Ph, R^4:Me

Felodipine, hypertension, 1991
R^1:CO_2Me, R^2:2,3-Cl-Ph,
R^3:CO_2Et, R^4:Me

Nicardipine, hypertension, 1988
R^1:CO_2Me, R^2:3-NO_2-Ph,
R^3:$CO_2CH_2CH_2N(CH_3)Bn$, R^4:Me

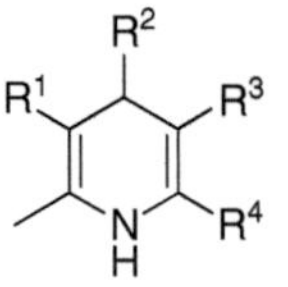

Amlodipine, hypertension, 1992
R^1:CO_2Me, R^2:2-Cl-Ph, R^3:CO_2Et,
R^4:$CH_2OCH_2Ch_2NH_2$

Nimodipine, subarachnoid hemorrhage, 1988
R^1:CO_2i-Pr, R^2:3-NO_2-Ph,
R^3:$CO_2CH_2CH_2OMe$, R^4:Me

Nisoldipine, hypertension, 1995
R^1:CO_2Me, R^2:2-NO_2-Ph,
R^3:$CO_2CH_2CH(CH_3)_2$, R^4:Me

Isradipine, hypertension, 1990
R^1:CO_2Me, R^2:benzo[c][1,2,5]oxadiazole,
R^3:CO_2i-Pr, R^4:Me

Clevidipine, hypertension, 2008
R^1:CO_2Me, R^2:(S)-2,3-Cl-Ph,
R^3:$CO_2CH_2CH(CH_3)_2$, R^4:Me

Figure 9.8 FDA-approved DHP drugs by year.

9.7.1 FDA-approved Hantzsch Dihydropyridines

The information in this section highlights compounds that have been FDA approved as drugs and biological applications for DHPs. The first FDA-approved drug synthesized by the Hantzsch reaction was Nifedipine, which works as a calcium antagonist, in 1981 [115, 116] (Figure 9.8). Slowly new generations of higher selectivity and longer half-life molecules were created almost exclusively for hypertension [117, 118].

9.7.2 Anti-bacterial Hantzsch DHPs

The DHP heterocycles consist of an influential category of bioactive molecules in pharmaceutical chemistry. They play a role as reducing agents [119], valuable intermediates [117], and natural products. Some of their various biological activities include hydrogenase coenzyme analogs [120], calcium channel blockers [121], anti-ulcer [122], anti-inflammatory [123], and anti-ischemic [124]. Recently, a new set of 1,4-DHP derivatives showed antibacterial activity *in vitro* [125] (Figure 9.9). These derivatives were then reacted with thiosemicarbazide by the hydrazinolysis method [121] that increased their anti-bacterial activity. Although not as potent as the control Ciprofloxacin across the board of bacteria, **48a** was more potent against *Staphylococcus aureus*, **48b** slightly less potent against *Klebsiella pneumoniae*, and **48d** slightly less potent against *Pseudomonas aeruginosa*.

The antibacterial effects of DHPs have also been evaluated against tuberculosis (*Mycobacterium tuberculosis*) [126]. The compounds were prepared from 3-oxo-*N*-phenylbutanamide, an imidazole derivative aldehyde, and ammonia in methanol. It was found the DHPs had a lower minimum inhibitory concentration (IC) than the antibiotic isoniazid that is used to treat tuberculosis with the compounds that possessed an electron-withdrawing group in the *para* position **49a–b** being the most potent formed from the Hantzsch DHP route (Figure 9.10).

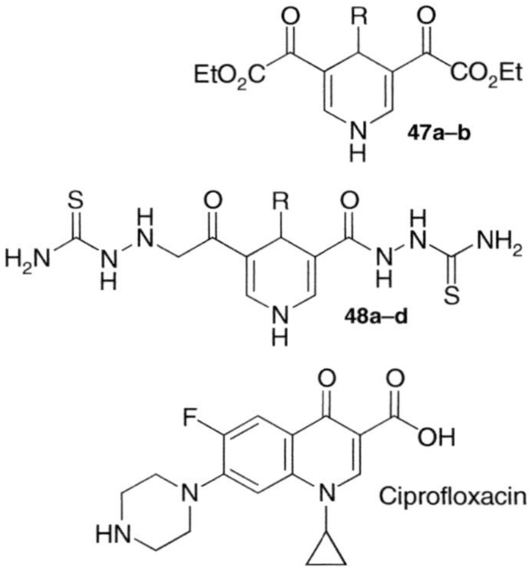

	S. aureus	*K. pneumoniae*	*E. coli*	*P. aeruginosa*
47a–b / 47a–d	**47a** R = 4-Cl-C_6H_4 10 mm ZOI	**47b** R = –Ph 7 mm ZOI	**47c** R = 4-$(CH_3)_2N$-C_6H_4 10 mm ZOI	**47d** R = 2-Fury 14 mm ZOI
48a–d	**48a** R = 4-Cl-C_6H_4 25 mm ZOI	**48b** R = –Ph 18 mm ZOI	**48c** R = 4-$(CH_3)_2N$-C_6H_4 16 mm ZOI	**48d** R = 2-Fury 30 mm ZOI
Ciprofloxacin	22 mm ZOI	19 mm ZOI	27 mm ZOI	32 mm ZOI

Figure 9.9 Zone of inhibition (ZOI) of DHPs at 100 μg/ml DMSO (disk diameter: 7 cm).

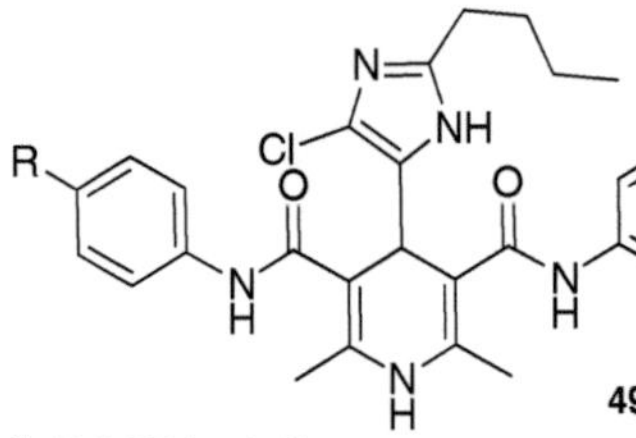

49a–b

Compound	R	% Inhibition[a]	MIC[b]	IC_{50}[c]
49a	F	100	0.02	>10
49b	NO_2	99	0.02	>10

[a] At 6.25 (μg/ml)
[b] Minimum inhibitory concentration (MIC) aganist H37Rv strain of *M. tuberculosis* (μg/ml)
[c] Measurement of cytotoxicity in VERO cells: 50% inhibitory concentrations (μg/ml)

Figure 9.10 Anti-tuberculosis 1,4-dihydropyridines.

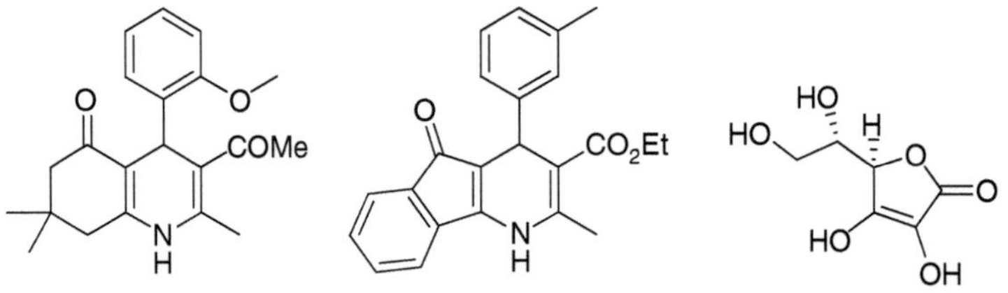

50a 1.65 IC_{50}
17.61 @ 0.2(mg/ml)
53.00 @ 0.6(mg/ml)
81.32 @ 1.0(mg/ml)

50b 0.47 IC_{50}
27.34 @ 0.2(mg/ml)
64.16 @ 0.6(mg/ml)
93.65 @ 1.0(mg/ml)

Ascorbic acid 0.15 IC_{50}
65.74 @ 0.2(mg/ml)
76.85 @ 0.6(mg/ml)
80.24 @ 1.0(mg/ml)

Figure 9.11 Comparison of DPPH radical scavenging activity of DHPs and ascorbic acid.

The antioxidant properties 1,4-DHPs have also been evaluated [127] with compounds **50a** and **50b** showing promise at higher concentrations than ascorbic acid (Figure 9.11). The compounds were prepared in a 1-pot 4-component Hantzsch DHP reaction with several aldehydes, cyclic and acyclic 1,3-dicarbonyls, and ammonium acetate with *N*,*N*,*N*′,*N*′-Tetrachlorobenzene-1,3-disulfonamide (TCBDA) or poly(*N*,*N*′-dichloro-*N*-ethyl-benzene-1,3-disulfonamide) (PCBS) as recoverable catalysts in mild conditions [127].

The Hantzsch DHP synthesis has come a long way since 1882, especially in the last 40 years. Most biological syntheses rely heavily upon different scaffolds for the dicarbonyl and aldehyde substituents of the DHP [115, 117–119, 121, 122, 125–127]. Opportunities exist in study of post-Hantsch modifications, encompassing construction of fused ring systems, and the use of spiro-substituents of the DHP which are under-explored.

9.8 The Biginelli Reaction

The Biginelli reaction was first discovered in 1893 by Pietro Biginelli with the MCR of ethyl acetoacetate, an aromatic aldehyde, and urea with cat. HCl refluxing in ethanol (Scheme 9.23) to form a substituted 3,4-dihydropyrimidin-2(1H)-ones (DHPM) [114]. The reaction sat somewhat unused until the 1990s [114] and has been

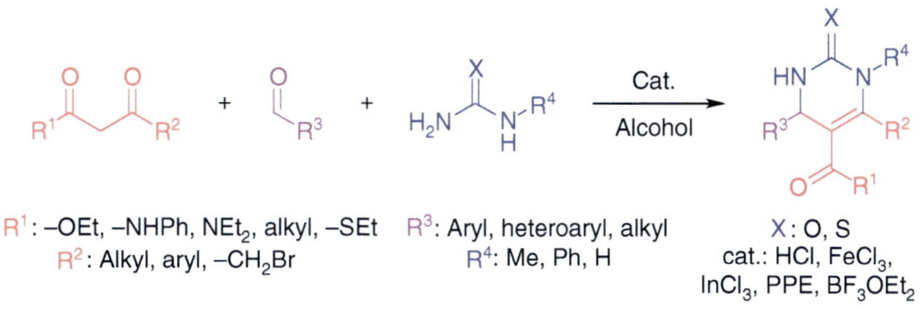

Scheme 9.23 Biginelli reaction.

significantly used, modified, and reviewed [128] due to DHPMs pharmacological activity associated with antihypertension, antibiotic, antimicrobial, anti-HIV, antitumor, calcium channels, and potassium channels [129]. Its application toward pharmaceuticals has also been surveyed [130]. This chapter will highlight some recent applications of Biginelli DHPMs regarding modifications, natural products, and biological activity.

9.8.1 Biginelli Reactions and Natural Products

In 1993, the Overman group successfully prepared an advanced intermediate of the natural product (+)-Ptilomycalin A with a novel modification of the Biginelli reaction with urea tethered to an aldehyde (Scheme 9.24) [122]. This modification led to the total synthesis of crambescidin and batzelladine alkaloids [132].

The tethered Biginelli was recently adapted to the racemic synthesis of Crambescin A in which a tethered guanidine formed a pyrrolinium **51** as a key intermediate [133] (Scheme 9.25).

This spurred the idea that Crambescin A could be synthesized without tethering, which was accomplished by an enantioselective Biginelli reaction (Scheme 9.26) to form (+)-Crambescin A [134].

9.8.2 Biginelli DHPMs as CNS Agents

DHPMs have recently been tested for CNS activity with differing success [135]. This is likely due to their structural similarity to barbiturates, sedatives derived from barbituric acid (Figure 9.12). Monastrol, a cell-permeable small molecule inhibitor of kinesin 5 [136], was compared with 13 other DHPMs for their modulation of $GABA_A$ receptors. In particular, Monastrol and JM-II-43A (Figure 9.12) showed similar current potentiation to other anti-epileptic drugs such as phenobarbital [135].

9.8.3 Biginelli Products Antitumor Capabilities

Biginelli DHPMs have also been implicated in antitumor cell lines (Figures 9.13–9.25) [137–152]. For blood cancer simulating cell lines dihydropyrimidinethione (DHPS), **52** was found to have the lowest concentration for 50% max inhibition of cells (GI_{50}) for MOLT-4 (acute lymphoblastic leukemia), while DHPS **53** had the lowest concentration for half-max effectiveness (EC_{50}) for K562 (chronic myelogenous leukemia) and DHPM **54** had the lowest half-max inhibitory

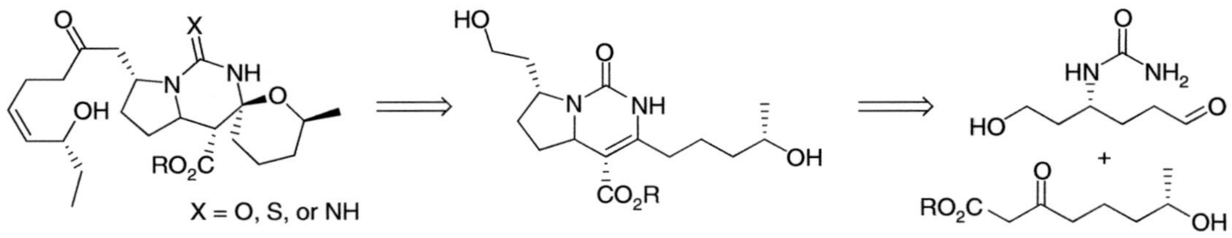

Scheme 9.24 Retrosynthetic synthesis of (+)-Ptilomycalin A. with tethered Biginelli

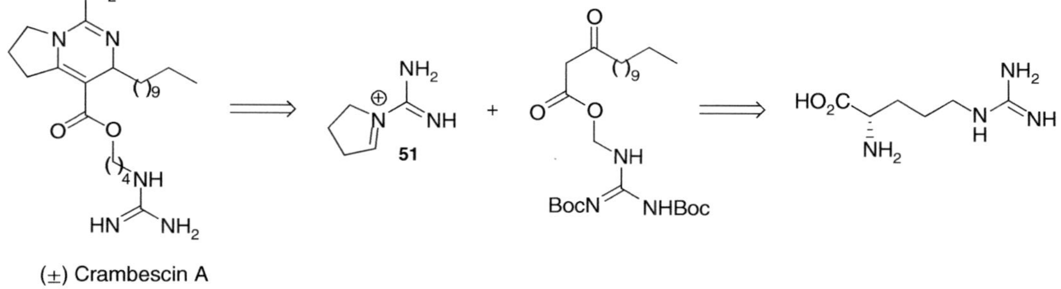

Scheme 9.25 Retrosynthetic Synthesis of (+/−)-Crambescin A via tethered Biginelli.

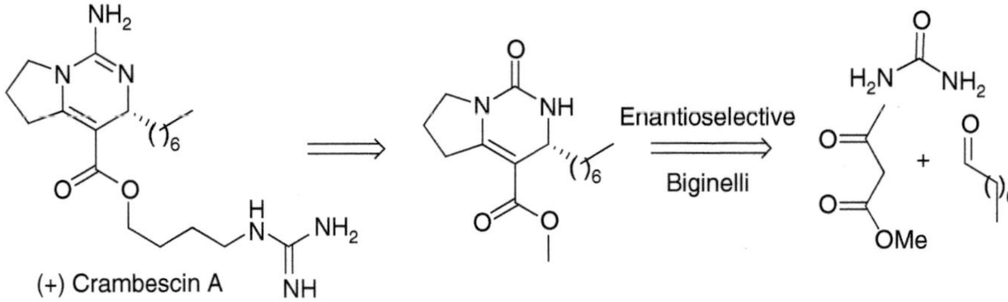

Scheme 9.26 Retrosynthetic synthesis of (+)-Crambescin A via enantioselective Biginelli.

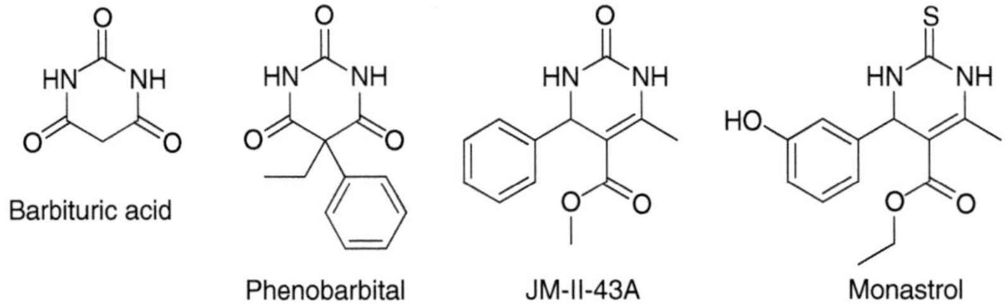

Figure 9.12 Structural similarities between barbituric acid DHPMs.

concentration (IC_{50}) for Jurkat (leukemic T-cell lymphoblast) (Figure 9.13) [137–139].

In breast cancer mimicking cell lines, DHPS **52** was also found to have the lowest GI_{50} for T-47 (ductal carcinoma), while DHPS **53** also had the lowest EC_{50} for hTERT-HME1 (mammary gland) and DHPM **55** had the lowest IC_{50} for BT-549 (ductal carcinoma) (Figure 9.14) [137, 138, 140].

For brain cancer imitating cell lines, DHPM **56** was found to have the lowest % cell viability for U138-MG (glioblastoma), while DHPM **57** had the lowest GI_{50} for U251 (glioblastoma) and DHPS **58** had the lowest IC_{50} for U138-MG (glioblastoma) (Figure 9.15) [139, 141, 142].

For cervical cancer cell lines, DHPM **59** was found to have the lowest GI_{50} for HeLa (human cervix), while DHPM **60** and **61** had less potency (Figure 9.16) [143, 144].

For colorectal cancer cell lines, DHPS **53** was also found to have the lowest EC_{50} for HCT-116 (human colon), while DHPM **62** had the lowest GI_{50} for HT29 (human colon) and DHPM **63** had the lowest IC_{50} for HCT-116 (Figure 9.17) [138, 141, 145].

For kidney cancer cell lines, DHPM **62** was found to have the lowest GI_{50} for 786-O (renal cell adenocarcinoma), while DHPM analog **64** substitution of *para*-chloro for methylthio decreased the GI_{50} fourfold for 786-O and DHPM analog **65** had the lowest GI_{50} for VERO (African green monkey kidney) (Figure 9.18) [141, 146].

For liver cancer cell lines, DHPM **66** was also found to have the lowest GI_{50} for HepG2 (human liver), while DHPS **67** had the lowest IC_{50} for HepaRG (liver tumor of a female patient diagnosed with hepatocarcinoma and hepatitis C) and DHPS **68**'s change from ethyl to methyl ester increased IC_{50} for HepaRG almost ninefold (Figure 9.19) [144, 147].

Blood

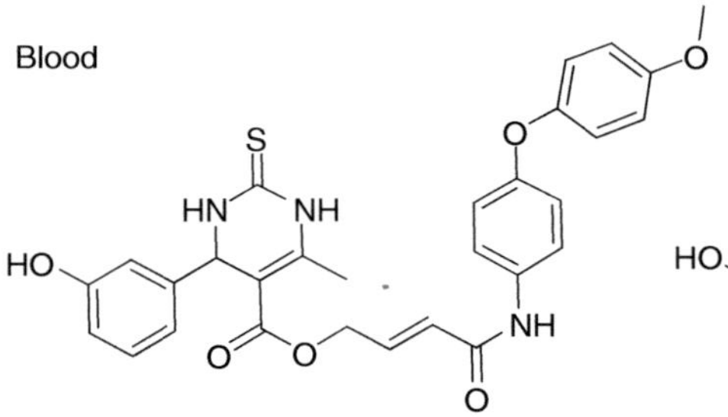

52
10 μM GI_{50} MOLT-4, 48 h [140]

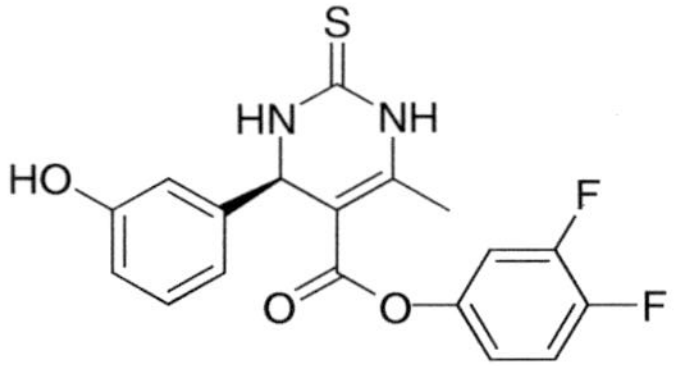

53
820 nM EC_{50} K562 [141]

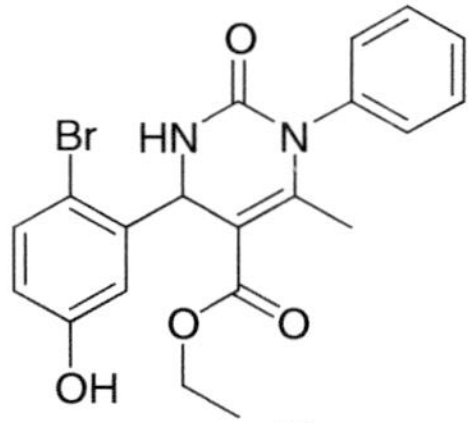

54
15.2 μM IC_{50} Jurkat, 48 h [142]

Figure 9.13 Biginelli products antitumor properties for blood cancer.

Breast

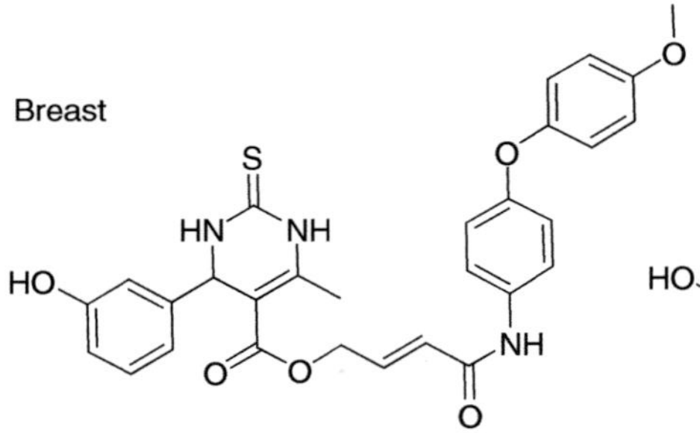

52
<10 μM GI_{50}
T-47D, 48 h [140]

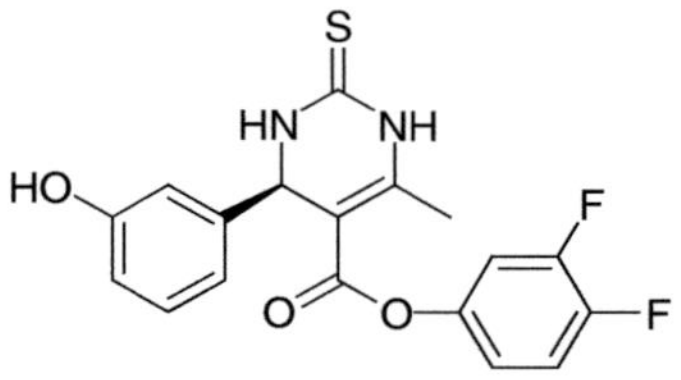

53
223 nM EC_{50}
hTERT-HME1 [141]

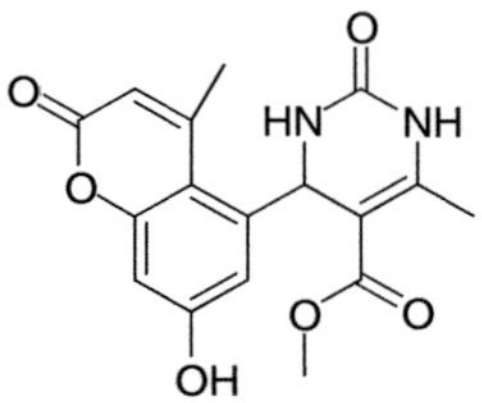

55
20.5 μM IC_{50}
BT 549 [143]

Figure 9.14 Biginelli products antitumor properties for breast cancer.

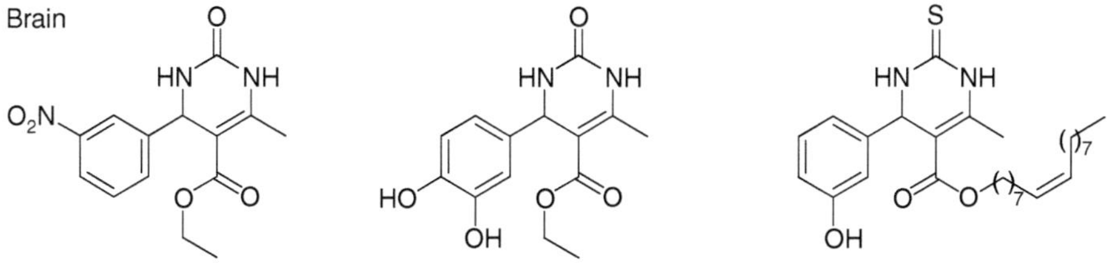

56
27% U138-MG Cell Viability
150 μmol/L 48h [142]

57
0.25 μg/ml GI_{50}
U251, 48 h [144]

58
10 μM IC_{50}
U138-MG, 24 h [145]

Figure 9.15 Biginelli products antitumor properties for brain cancer.

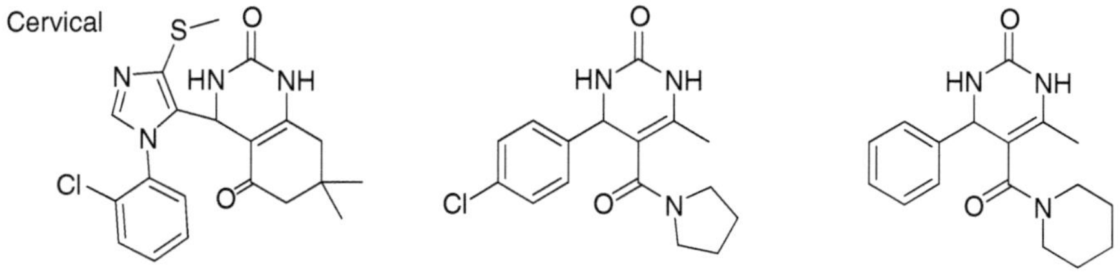

59
98 mg/ml GI_{50}
HeLa, 24 h [146]

60
187 mg/ml GI_{50}
HeLa, 48 h [147]

61
189 mg/ml GI_{50}
HeLa, 48 h [147]

Figure 9.16 Biginelli products antitumor properties for cervical cancer.

In lung cancer modeling cell lines, DHPM **62** was also found to have the lowest GI_{50} for NCI-H460 (lung: pleural effusion), while DHPM **69** had the lowest IC_{50} for A549 (human lung carcinoma) out of its analogs and DHPS dimer **70** had the lowest GI_{50} for A549 out of its analogs (Figure 9.20) [141, 145, 148].

For lymphoma cancer cell lines, DHPM **71** was also found to have the lowest GI_{50} for U937 (human lymphocyte), while DHPM **72**'s change from nitrothiophene to *para*-isopropylphenyl increased the GI_{50} for U937 almost threefold and DHPM **73** had the highest fluorescence activity ratio in I5178-y (*mdr1*-gene transfected mouse lymphoma) (Figure 9.21) [149, 150].

For ovarian cancer-producing cell lines, DHPM **74** was found to have the lowest GI_{50} for NCI-ADR/RES (ovarian tumor), while DHPS **75** had the lowest GI_{50} for OVCAR-03 (human ovary) out of its analogs and DHPS **76**'s change from *para*-methoxyphenyl to benzo-dioxol increased the GI_{50} more than sixfold in OVCAR-03 (Figure 9.22.22) [141, 151].

In pancreas cancer modeling cell lines, DHPS **77** was found to have the lowest EC_{50} for BxPC3 (human pancreas) out of all of its analogs, while DHPS **78**'s removal of 3,4-difluoro greatly increased the EC_{50} for BxPC3 and DHPS **79** change to dimethylhexenone restored almost all of the potency in BxPC3 (Figure 9.22.23) [138].

For prostate cancer cell lines, DHPS **80** was found to have the lowest GI_{50} for PC-3 (Human Caucasian prostate adenocarcinoma) out of its analogs, while DHPM **81**

Colorectal

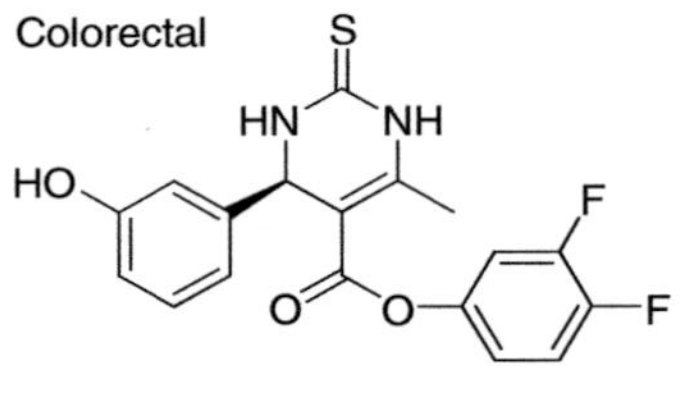

53
327 nM EC_{50}
HCT116 [141]

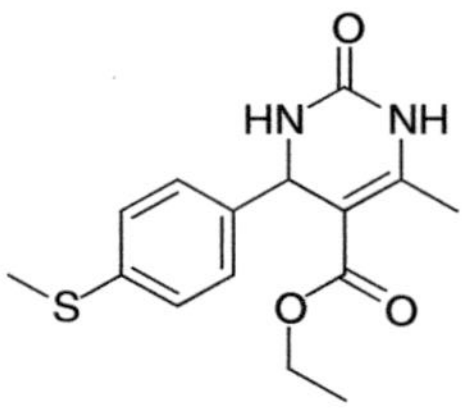

62
10.2 μg/ml GI_{50}
HT29, 48 h [144]

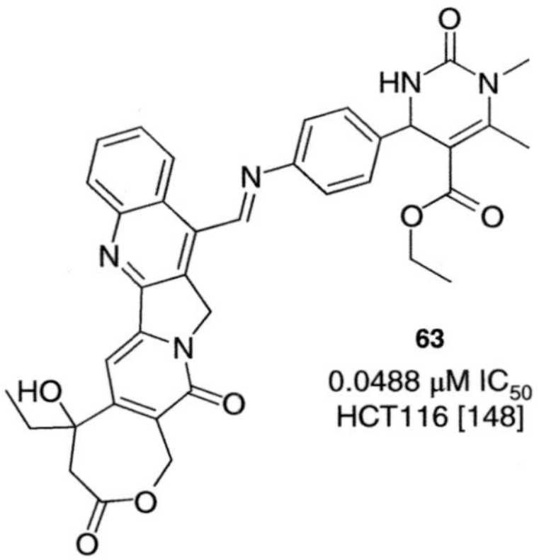

63
0.0488 μM IC_{50}
HCT116 [148]

Figure 9.17 Biginelli products antitumor properties for colorectal cancer.

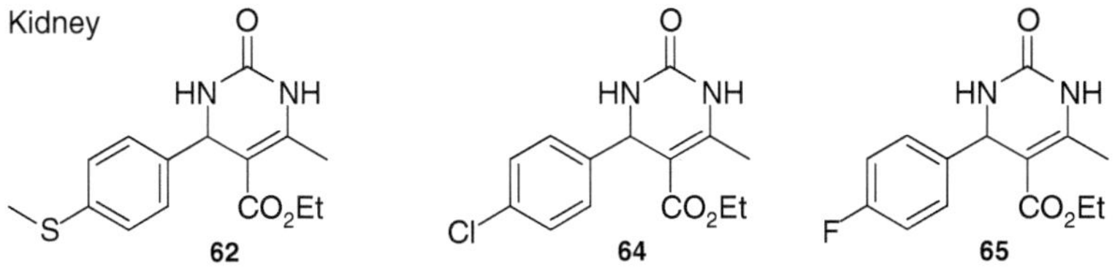

62 — 2.5 mg/ml GI_{50} 786-O, 48 h [144]

64 — 10.16 mg/ml GI_{50} 786-O, 48 h [149]

65 — 14.13 mg/ml GI_{50} VERO, 48 h [149]

Figure 9.18 Biginelli products antitumor properties for kidney cancer.

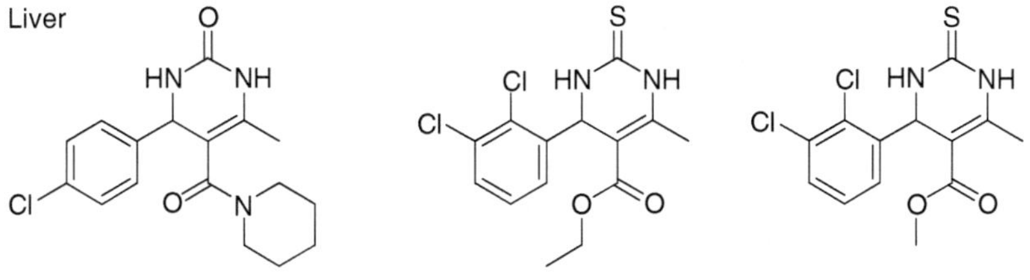

66 — 120 mg/ml GI_{50} HepG2, 48 h [147]

67 — 0.75 mM IC_{50} HepaRG, 72 h [150]

68 — 6.53 mM IC_{50} HepaRG, 72 h [150]

Figure 9.19 Biginelli products antitumor properties for liver cancer.

had the lowest IC_{50} for PC-3 out of its analogs and DHPS **82** had an even lower IC_{50} in PC-3 (Figure 9.24) [130, 131, 142].

For skin cancer-producing cell lines, DHPS **83** was found to have 86.5% growth inhibition at 10 mM in MDA-MB-435 (melanoma), while DHPM **54** also had the lowest IC_{50} for A375 (malignant melanoma) and DHPS dimer **84** had the lowest GI_{50} in A431 (Human squamous carcinoma) (Figure 9.25) [137, 139, 148].

9.9 van Leusen Reaction

9.9.1 Tosmic-mediated Cyclization Toward Nitrogen-containing Heterocycles

Among isocyanide-based MCRs, *p*-tolylsulfonylmehyl isocyanide (TosMIC **85**) is considered as a prototypical isocyanide of all functionalized isocyanides. TosMIC contributes to a variety of reactions due to the following characteristics: the isocyano functionality possesses the classical α-addition reactions, the acidic α-carbon atom ($pK_a = 14$), and the sulfonyl group in the α-position that plays two roles as a sulfinyl leaving group and an electron-withdrawing group for enhancing the acidity of the α-carbon. As shown in Scheme 9.27, nitrogen-containing heterocyclic chemotypes such as oxazoles **86**, imidazoles **87**, 1,2,4-triazoles **88**, thiazoles **89**, pyrroles **90**, and pyrrolo[3,4-*c*]quinolones **91** could be easily obtained via reactions between **85** and the corresponding electrophiles such as aldehydes, imines, aryldiazonium salts, isothiocyanates, and Michael acceptors [153–159].

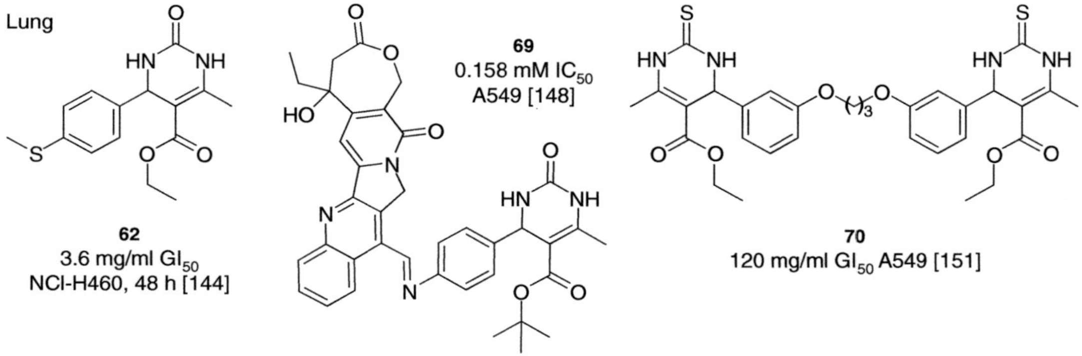

Figure 9.20 Biginelli products antitumor properties for lung cancer.

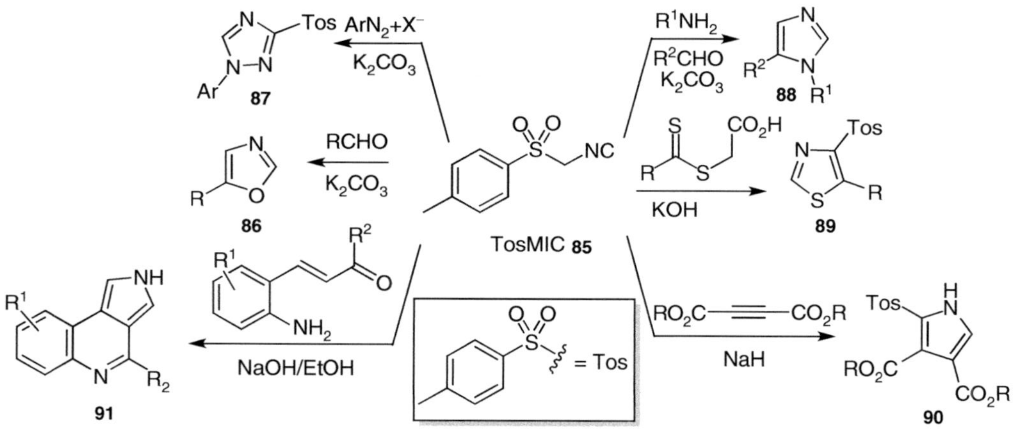

Scheme 9.27 TosMIC-mediated cyclization to construct heterocyclic chemotypes.

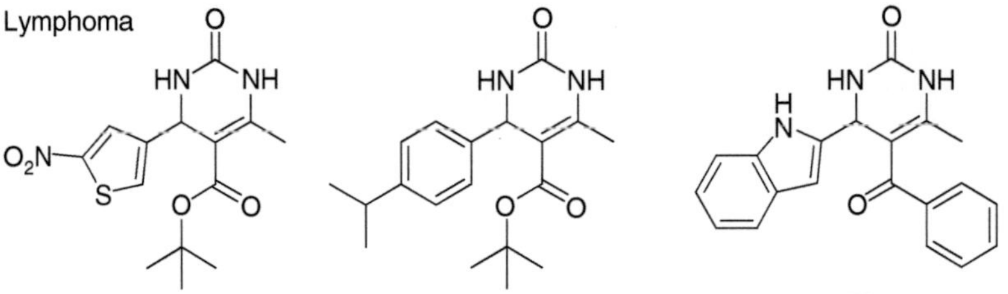

71 10.51 mg/ml GI_{50} U937, 24 h [152]

72 27.55 mg/ml GI_{50} U937, 24 h [152]

73 78.56 fluorescence activity ratio at 4 mM in l5178-y [153]

Figure 9.21 Biginelli products antitumor properties for lymphoma cancer.

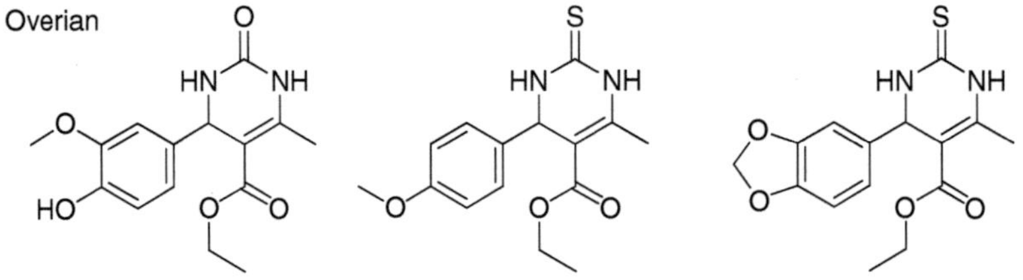

74 0.25 mg/ml GI_{50} NCI-ADR/RES, 48 h [144]

75 0.99 mg/ml GI_{50} OVCAR-03, 48 h [144]

76 6.6 mg/ml GI_{50} OVCAR-03, 48 h [154]

Figure 9.22 Biginelli products antitumor properties for ovarian cancer.

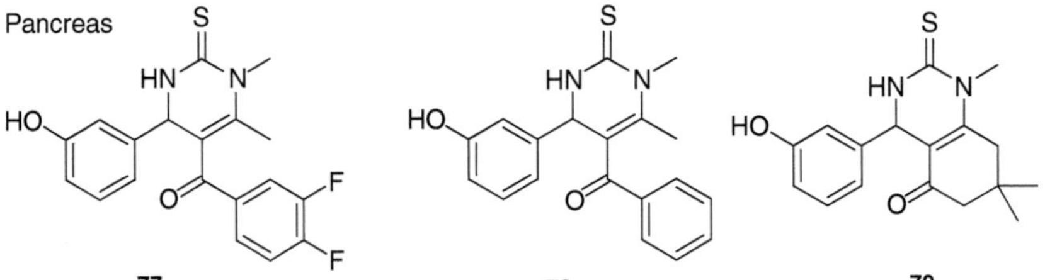

77 601 nM EC_{50} BxPC3 [141]

78 5000 nM EC_{50} BxPC3 [141]

79 743 nM EC_{50} BxPC3 [141]

Figure 9.23 Biginelli products antitumor properties for pancreas cancer.

Moreover, imidazole **88** can be prepared through a one-pot three-component manner called van Leusen reaction (Scheme 9.28), in which an imine **94** is generated *in situ* from the condensation of an amine **92** and an aldehyde **93**, followed by the nucleophilic addition/cyclization/β-elimination process mediated by an anionic TosMIC **95** in the presence of a base. Upon deprotonation at α-carbon in the presence of potassium carbonate, the anionic TosMIC **95** serves as a nucleophile to undergo nucleophilic addition to obtain **96**. The intermediate **96** further undergoes intramolecular cyclization to generate intermediate **97** and proton transfer to construct a dihydroimidazole **98**. Consequently, the intermediate **98** undergoes

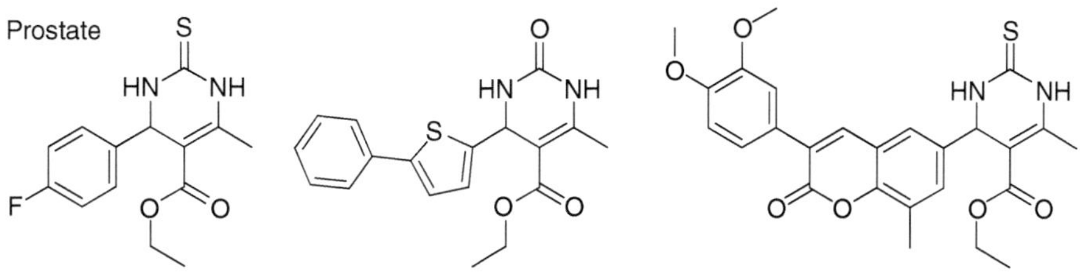

80
1.7 mg/ml GI_{50}
PC-3, 48 h [144]

81
19.3 mM IC_{50}
PC-3 [143]

82
9.82 mM IC_{50}
PC-3 [155]

Figure 9.24 Biginelli products antitumor properties for prostate cancer.

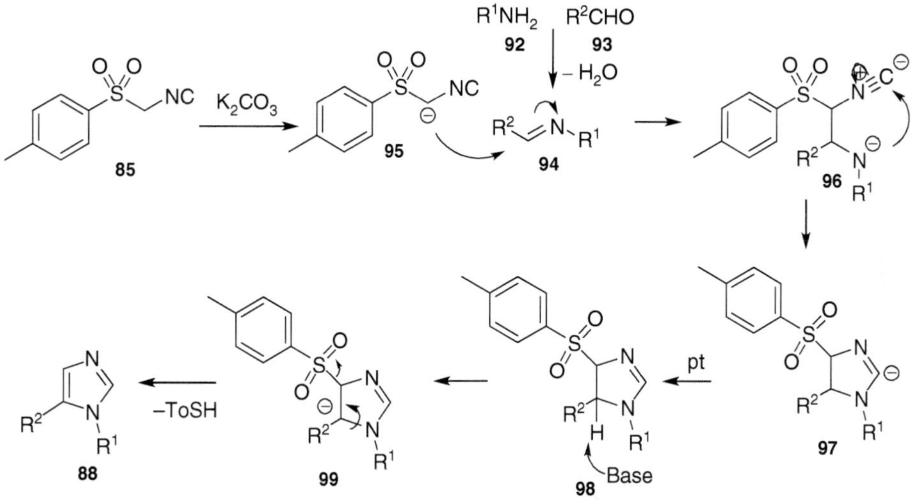

Scheme 9.28 Proposed mechanism of 3C-van Leusen reaction.

deprotonation and β-elimination through anionic **99** to afford the imidazole chemotype **88**.

9.9.2 Applications of the van Leusen Reaction

9.9.2.1 Sequential One-pot Three-step 3C-van Leusen Reaction/Deprotection/Cyclization

Similar to post-condensation in Ugi reactions, the van Leusen products are amenable for further modifications mainly attributed to variable functionalized building blocks. As shown in Scheme 9.29, De Moliner employed *ortho-N*-Boc-phenylenediamine **100**, arylglyoxal **101**, and TosMIC **102** to carry out a sequential one-pot three-step van Leusen reaction/de-Boc/cyclization process under microwave irradiation [160]. As shown, both the carbonyl group in arylglyoxal **101** and the *N*-Boc-amino group in *ortho-N*-Boc-phenylenediamine **100** are embedded

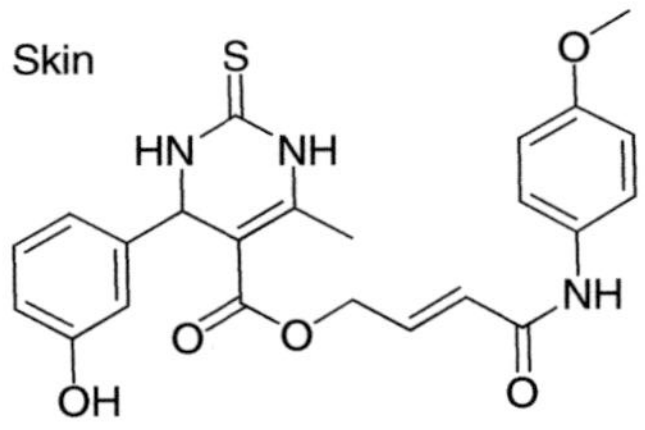

83

86.5% growth inhibition
at 10 mM, MDA-MB-435, 48 h [140]

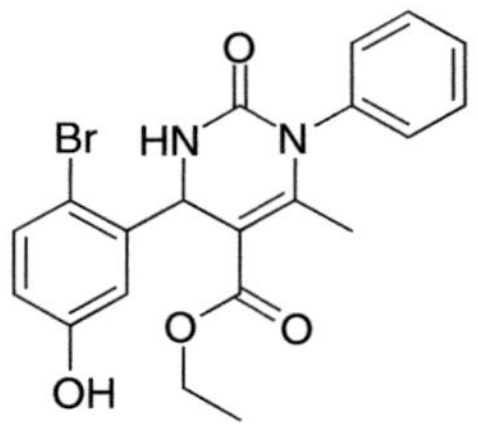

54

70.5 mM IC_{50} A-375, 48 h [142]

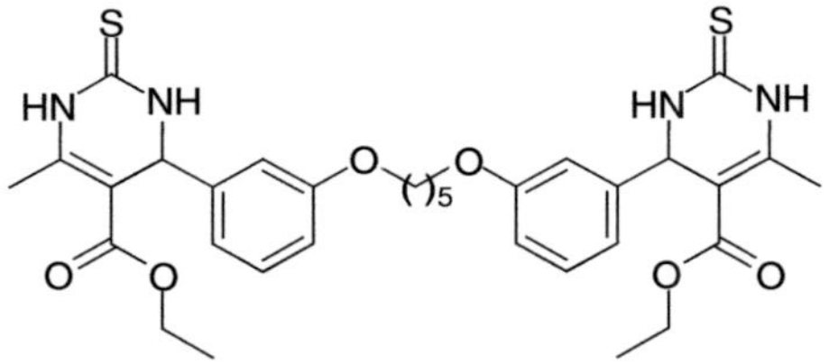

84

74 mg/ml GI_{50} A431 [151]

Figure 9.25 Biginelli products antitumor properties for skin cancer.

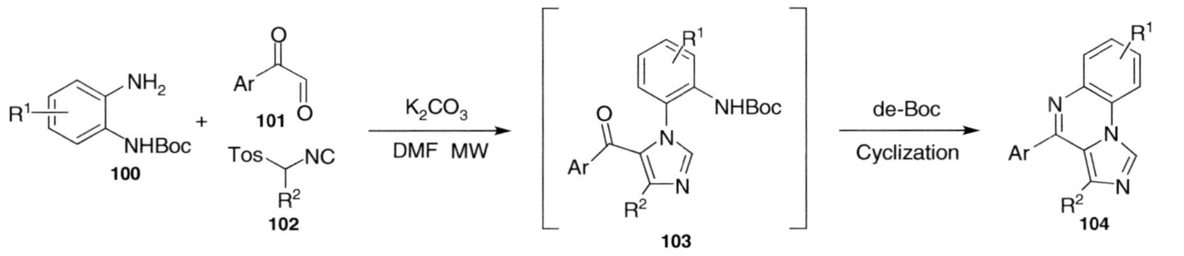

Scheme 9.29 Synthesis of imidazole[1,5-*a*]quinoxaline **104** through van Leusen reaction/de-Boc/cyclization.

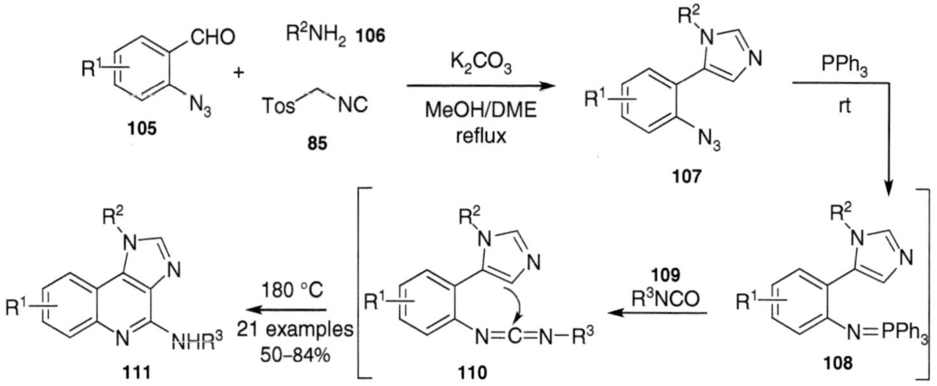

Scheme 9.30 Post-modifications of van Leusen reaction toward imidazoquinolines.

as an electrophile and a nucleophile, respectively, for further cyclization to construct imidazole[1,5-a]quinoxaline **104**.

9.9.2.2 Sequential van Leusen Reaction/Staudinger/aza-Wittig/Cyclization

In 2018, Guan et al. demonstrated a stepwise van Leusen/Staudinger/aza-Wittig/carbodiimide-mediated cyclization operation to prepare a series of imidazoquinolines **111** [151]. As shown in Scheme 9.30, the van Leusen reaction was carried out by the transformation of 2-azidobenzaldehyde **105**, amine **106**, and TosMIC **85** in MeOH/(dimethoxyethane) DME at reflux in the presence of potassium carbonate. The van Leusen product imidazole **107** was further treated with triphenyl phosphine (PPh_3) to produce iminophosphorane **108**. A tandem aza-Wittig reaction of iminophosphorane **108** with isocyanate **109** gave rise to intermediate carbodiimide **110**, accompanied by cyclization at 180 °C to afford 1H-imidazo-[4,5-c]quinolines **111** with a range of 50–86% isolated yield.

9.9.2.3 DNA-conjugated van Leusen Reaction

DNA-encoded library (DEL) is an emerging technology that has drawn a great deal of attention in the pharmaceutical industry. It is a powerful tool required from interdisciplinary knowledge and skills of molecular biology and combinatorial chemistry for early drug discovery [162–164]. In contrast to solid-phase combinatorial synthesis in which building blocks are conjugated to a resin-bound linkage [165, 166], DEL employs a short DNA sequence as a functionalized linker for providing conjugation with building blocks. Furthermore, rather than having a physical barcode attached to individual solid-phase reaction vessels, a DNA codon barcode in DEL synthesis is embedded through DNA ligation (Scheme 9.31). Despite its short history as a high-throughput screening (HTS) platform for hit identification, DEL technology has witnessed progress in developing preclinical and clinical candidates in drug discovery. Nevertheless, it should be noted that the DNA compatibility of organic reaction is still a critical issue for developing DEL-derived drug-like chemical diversity.

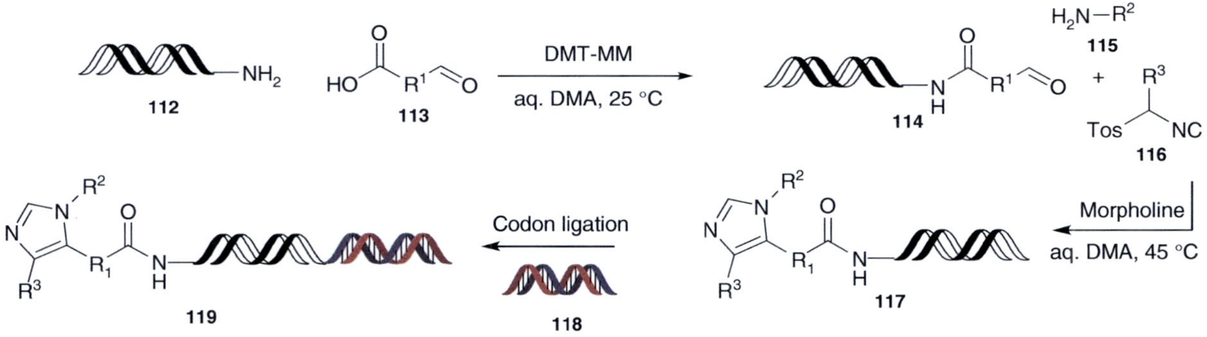

Scheme 9.31 DNA-conjugated one-pot three-component van Leusen reaction.

Consequently, it is of great value to explore DNA-compatible reactions for expanding the scope of chemical diversity. In 2019, Satz et al. reported a DNA-compatible van Leusen reaction to generate highly functionalized DNA-conjugated imidazoles **117** [157]. The 20-mer DNA-conjugated aldehyde **114** was prepared by the amide bond coupling reaction between DNA-conjugated amine **112** and formyl acid **113**. With DNA-conjugated aldehyde **114** in hand, the van Leusen reaction was carried out with optimal conditions to afford a higher conversion of 1,4,5-substituted imidazole **117**. The optimized reaction conditions are shown as follows: a mixture of 20-mer DNA-conjugated aldehyde **114** (1 mM in water), amine **115** (40 eq), and TosMIC **116** (80 eq) in aqueous dimethylacetamide (DMA, 62%) solution in the presence of morpholine (200 eq) under mild heating at 45 °C for 16 hours. Furthermore, the DNA-conjugated imidazole **118** was ligated to a 12-mer base pair DNA codon **118** to construct 45 base-pair DNA ligation product **119**. In summary, the authors claimed that a million new highly functionalized imidazoles **119** may potentially be produced to expand chemical space in DEL production.

9.9.3 Applications of the van Leusen Reaction in Drug Discovery

Undoubtedly, imidazole is one of the privileged chemotypes in medicinal chemistry. It is widely found in natural products and pharmaceutical drugs. Applications of imidazole chemotype in the drug discovery industry have been extensively developed into clinical stages. As such, many imidazole-derived biologically active compounds are implemented in a variety of disease indications, such as antibacterial, antifungal, anti-inflammatory, antiviral, antiparasitic, anticancer, and so on [168]. Some of the selected imidazole-containing clinical drugs are listed (Figure 9.26). The following sections are intentionally recorded from up-to-date publications regarding the van Leusen reaction-inspired early drug discovery campaigns.

9.9.3.1 Purinergic P2X7 Receptor Antagonists

Purinergic P2X7 receptor (P2X7R) is a non-selective cation channel activated by extracellular adenosine 5′-triphosphate (ATP). Meanwhile, it is shown that a higher concentration of ATP in the tumor microenvironment has been found in comparison to healthy tissues. P2X7R seems to be a potential therapeutic target for the treatment of cancer metastasis due to its overexpression and activation by the higher concentration of ATP in various cancer cells for survival, invasiveness, and or metastasis [169, 170]. **KN62** (IC_{50}, 51 nM) was one of the first-generation antagonists of P2X7R discovered [171], which is shown to inhibit the invasiveness of cancer cells *in vitro* and *in vivo* [172]. Park et al. initiated structural modifications of conformationally flexible **KN62** (Scheme 9.32) by rigidification strategy to construct a hydantoin core structure **120** (IC_{50}, 54 nM) with comparable activity [163].

The authors further embarked on a re-scaffolding approach from hydantoin to imidazole core structure through a series of van Leusen reactions [174]. Furthermore, a series of SAR studies were investigated from comparable **121** (IC_{50}, 45 nM) to the optimized **122** (IC_{50}, 0.11 nM) against P2X7R activity. Compound **122** was further

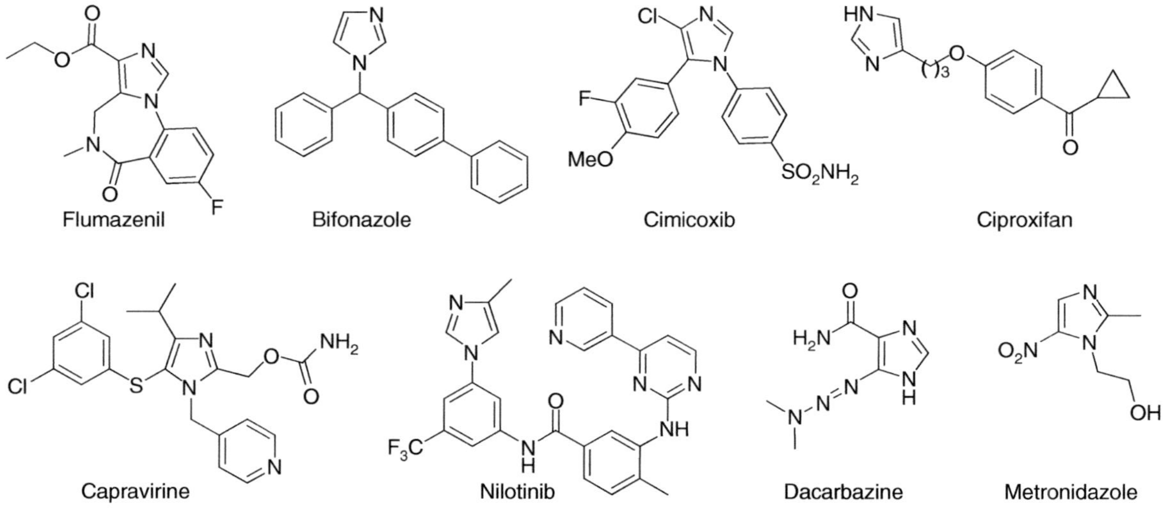

Figure 9.26 Selected imidazole-containing clinical drugs.

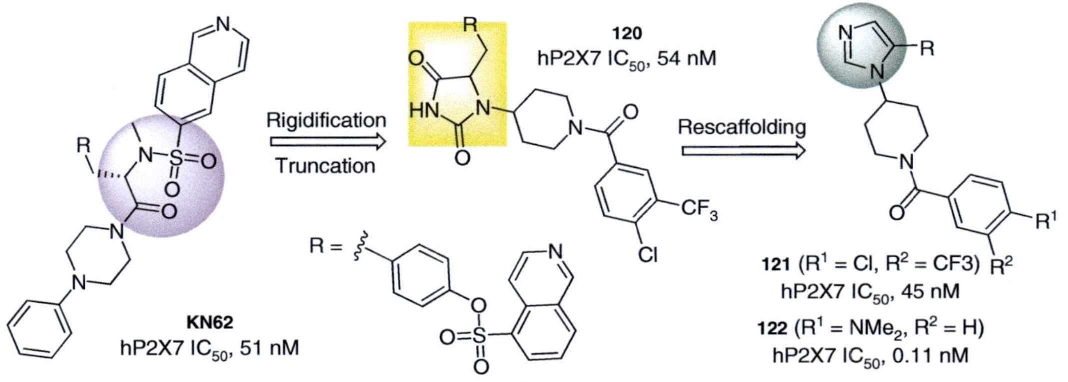

Scheme 9.32 Structural modification of PX27R antagonists via van Leusen reaction.

demonstrated to exhibit anti-invasive activity in MDA-MB-231 triple-negative breast cancer cells and showed an anti-metastatic effect in a zebrafish xenograft model.

9.9.3.2 Indoleamine 2,3-Dioxygenase (IDO1) Inhibitors

Indoleamine 2,3-dioxygenase (IDO1) is a heme-containing tryptophan catabolic enzyme that catalyzes the oxidative cleavage of the C2–C3 indole double bond to produce kynurenine (Kyn). IDO1 is indicated in the regulation of innate and adaptive immunity in the microenvironment of tumors. Dysregulation of IDO1 expression has also been implicated in the progression of inflammatory diseases, autoimmune diseases, and neurological disorders. As such, IDO1 is an attractive immunotherapeutic target for the treatment of a wide range of cancers in addition to other disease indications [175]. Currently, several small-molecule IDO1 inhibitors are under investigations in clinical studies as either standalone therapeutics or adjuvants in combination with traditional chemo- and radiotherapies. Herein, Weaver et al. utilized **Indoximod** (IC_{50}, 34 μM) and **4-PI** (IC_{50}, 48 μM) as molecular templates to implement a structure-based scaffold-hopping approach to develop indole-imidazole hybrid **123** (IC_{50}, 30.5 μM) with comparable activity [179]. The authors further employed a series of the van Leusen reaction to explore imidazole analogs for SAR study. Initial results indicated that compound **124** (IC_{50}, 1.6 μM) exhibits a 20-fold improvement in comparison with compound **123** (Scheme 9.33). Remarkably, compound **125** (IC_{50}, 33.8 nM) is shown to exhibit 47-fold activity compared to that of compound **124** upon the interconversion between indole and phenyl moieties. Compound **125** also demonstrates *in vitro* microsomal stability (Cl_{int}, 26.8 μl/min/mg) potentially for future investigations to suppress cancer metastasis.

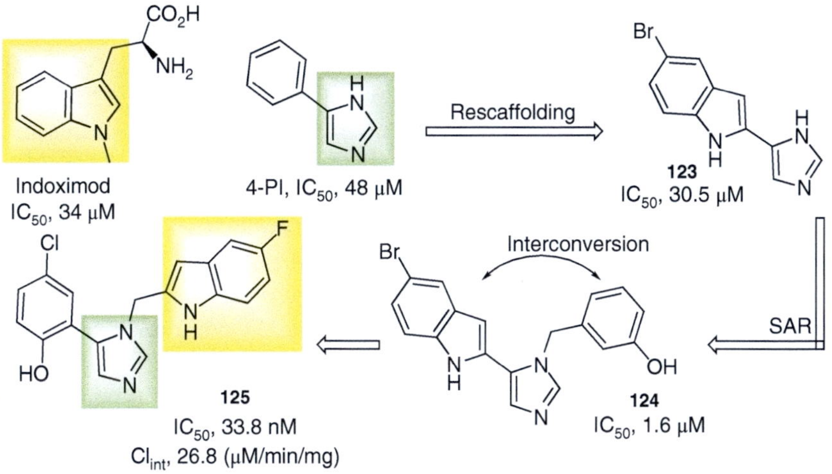

Scheme 9.33 Structure-based scaffold-hopping approach to synthesize IDO1 inhibitors.

9.9.3.3 Disruptors of P53/MDM2 Protein–Protein Interactions

The tumor suppressor p53 is inactivated in most cancers, which is attributed to either mutation in the *TP53* gene or overexpression of p53 negative regulator of MDM2. Reactivation of p53 by disrupting the interaction with MDM2 protein is a promising strategy for the treatment of wild-type p53 cancers [180]. Currently, seven small-molecule disruptors of p53/MDM2 protein–protein interaction have advanced into clinical trials for the treatment of human cancers [181]. Herein, Twarda-Clapa et al. employed van Leusen reaction to generate a series of imidazole analogs for antagonizing p53/MDM2 interaction. In particular, compound **129** (Ki, 58 nM, fluorescence polarization (FP) assay) prepared by TosMIC **126**, amine **127**, and aldehyde **128** in ethanol in the presence of triethylamine (TEA) was further examined (Scheme 9.34) to demonstrate the accumulation of *p21* gene expression for the induction of cell cycle arrest and growth inhibition with an IC_{50} value of 11.08 μM in wild-type p53 ($p53^{wt}$) U-2 osteosarcoma cells. The co-crystal structure of compound **129**/MDM2 confirmed a classical three-finger pharmacophore binding mode. Interestingly, compound **129** is further validated to show the induction of MDM2 dimerization in solution through nuclear magnetic resonance (NMR) spectra analysis [182].

9.9.3.4 Disruptors of PCSK9/LDLR Protein–Protein Interactions

Proprotein convertase subtilisin/kexin type 9 (PCSK9) plays an important role in the regulation of cholesterol homeostasis. On the other hand, hepatic low-density lipoprotein receptors (LDLRs) are the major intracellular carriers of cholesterol in the blood. Upon binding to hepatic LDLR, PCSK9 promotes their lysosomal degradation, subsequently reduces LDL uptake, and results in an increase in LDL cholesterol (LDL-C) concentrations. The accumulation of LDL-C in plasma is known to cause the formation of atherosclerotic plaque, a key risk factor of CVD. As such, PCSK9 is a validated therapeutic target for hypercholesterolemia treatment and CVD [183]. Stucchi et al. reported an imidazole-based peptidomimetic **MeIm** (IC_{50}, 11.2 μM) to show PCSK9 inhibitory activity in the micromolar range as shown in Scheme 9.35 [184]. Furthermore, the authors employed supervised molecular dynamics (SuMD) simulations to predict the binding mechanism of peptidomimetic imidazoles. As a result, target compounds with the highest theoretical affinity were synthesized via multiple van Leusen/formylation synthetic routes to develop tri- and tetra-imidazole scaffolds for examining in PCSK9/LDLR binding assay. As shown, the van Leusen reaction was carried out by aldehyde **130**, methylamine **131**, and TosMIC **85** to synthesize mono-imidazole intermediate **132**. Next, formylation of **132** was generated by the treatment of *n*-BuLi and DMF to obtain formylimidazole **133**. Continuing two or three rounds of the van Leusen/formylation gave rise to the tri- or tetra-imidazole products. As indicated, both **RIm13** (IC_{50}, 1.4 μM) and **RIm14** (IC_{50}, 3.1 μM) were shown to display an improved PCSK9 inhibitory activity in a biochemical assay. Furthermore, upon treatment of **RIm13** at 1 μM on HepG2 cells, the LDLR was

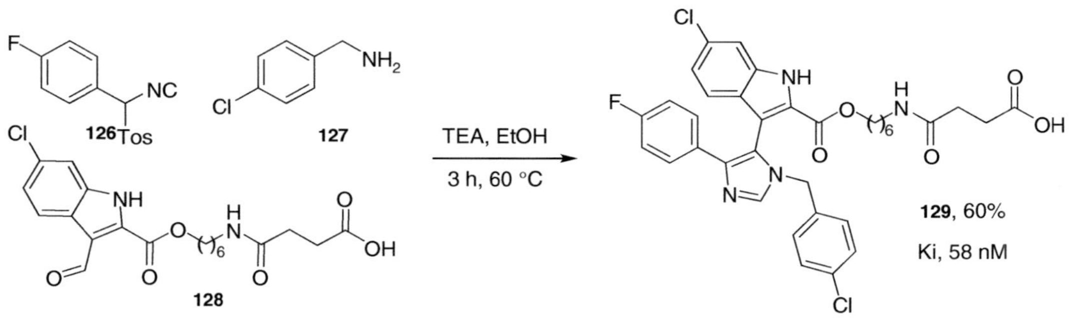

Scheme 9.34 Synthesis of **129** as a disruptor of p53/MDM2 interaction via van Leusen reaction.

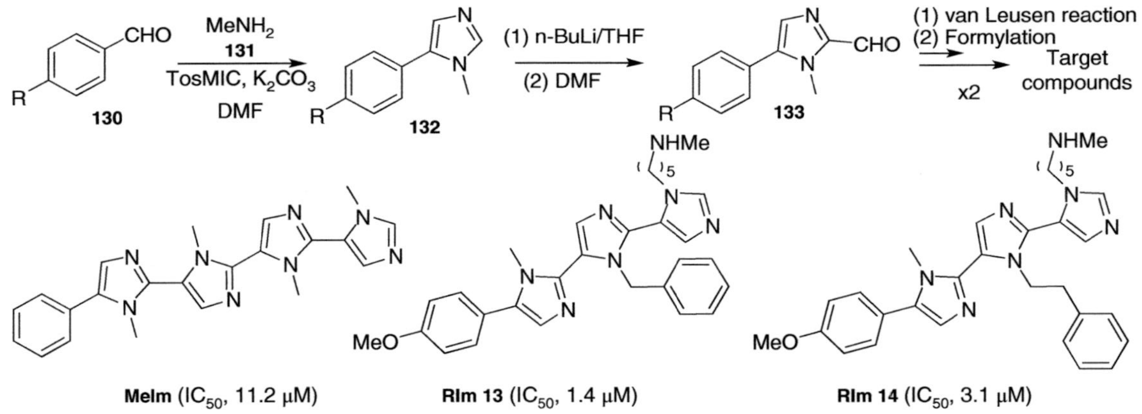

Scheme 9.35 Repetitive 3C-van Leusen reaction/formylation cycle to construct multi-imidazoles.

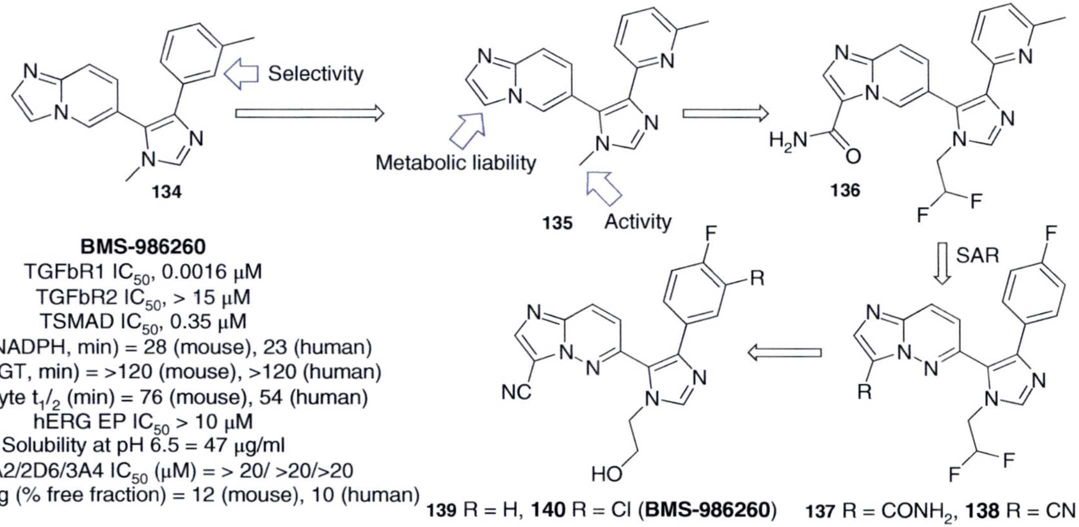

BMS-986260
TGFbR1 IC_{50}, 0.0016 μM
TGFbR2 IC_{50}, > 15 μM
TSMAD IC_{50}, 0.35 μM
LM $t_1/_2$ (NADPH, min) = 28 (mouse), 23 (human)
LM $t_1/_2$ (UGT, min) = >120 (mouse), >120 (human)
Hepatocyte $t_1/_2$ (min) = 76 (mouse), 54 (human)
hERG EP IC_{50} > 10 μM
Solubility at pH 6.5 = 47 μg/ml
CYP1A2/2D6/3A4 IC_{50} (μM) = > 20/ >20/>20
Protein binding (% free fraction) = 12 (mouse), 10 (human)

Scheme 9.36 Hit-to-lead optimization of TGFβR inhibitors.

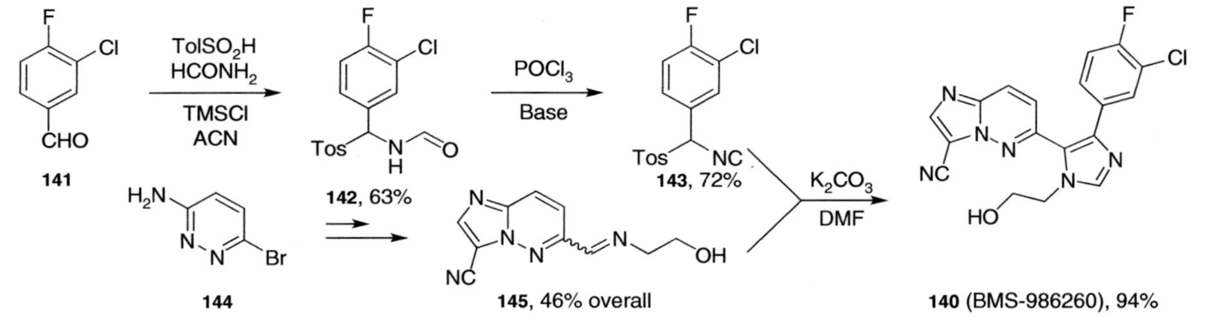

Scheme 9.37 Synthesis of imidazole **140** (**BMS-986260**) through van Leusen reaction.

shown to prevent PCSK9-mediated degradation, accompanied by the increase of the LDL uptake, which is currently representing one of the most potent small molecules targeting the PCSK9/LDLR protein–protein interaction warranted for further study.

9.9.3.5 Inhibitors of TGFβR1 as Immuno-oncology Therapeutics

Transforming growth factor β (TGFβ) is a cytokine that plays a multifunctional role in the biological processes, including cell proliferation, differentiation, migration and adhesion, and extracellular immunosuppression. Overexpression of TGFβ in cancers leads to T cell exclusion and poor response to PD-1/PD-L1 blockade. The TGFβ signaling pathway has therefore become a popular target for different disease indications [185, 186]. Currently, three small-molecule TGFβR1 inhibitors, Galunisertib, Vactoserti, and LY3200882 are being evaluated in clinical trials. Velaparthi et al. from Bristol-Myers Squibb recently reported the hit-to-lead optimization of novel imidazole-based TGFβR1 inhibitors (Scheme 9.36).

As shown in Scheme 9.36, the hit compound **134** was identified to show a good affinity for TGFβR1 inhibition. Replacement of 2-pyridyl moiety to generate compound **135** (IC_{50}, 0.010 μM) increased both selectivity and activity. Incorporation of the polar group at three-position and elongation of the methyl group in compound **135** to obtain compound **136** improved metabolic stability and activity. Further intensive SAR study has been investigated through compounds **137** and **138** to the optimized compound **140** (**BMS-986260**) with improved potency, ADME properties, toxicity, and solubility. **BMS-986260** was further shown to exhibit efficacy through oral administration in combination with anti-PD-1 antibody treatment in the mouse MC38 model for advanced drug development [187].

The preparation of target compound **140** via multi-step sequence including the final step of van Leusen reaction is depicted in Scheme 9.37. As shown, TosMIC **143** was obtained from the dehydration of tosylformamide **142** originally converted from the commercially available aldehyde **141**. The imine **145** was prepared through multi-step synthesis efforts started from the commercially available 6-bromopyridazine-3-amine **144**. Consequently, the van Leusen reaction was carried out to prepare compound **140** (BMS-986260) by TosMIC **143** and imine **145** in the presence of K_2CO_3 in DMF solution.

References

1 Hulme, C., Ayaz, M., Martinez-Ariza, G. et al. (2016). Recent advances in multicomponent reaction chemistry: applications in small molecule drug discovery. In: *Small Molecule Medicinal Chemistry* (eds. W. Czechtizky and P. Hamley), 145–187. Wiley.

2 Akritopoulou-Zanze, I. and Djuric, S. (2010). Applications of MCR-derived heterocycles in drug discovery. In: *Synthesis of Heterocycles via Multicomponent Reactions II* (eds. B. Maes, R. Orru and E. Ruijter), 231–287. Springer.

3 Strecker, A. (1850). Ueber die künstliche Bildung der Milchsäure und einen neuen, dem Glycocoll homologen Körper. *Ann. Chem.* 75: 27–45.

4 Dömling, A., Wang, W., and Wang, K. (2013). Chemistry and biology of multicomponent reactions. *Chem. Rev.* 112: 3083–3135.

5 Dömling, A. and Ugi, I. (2000). Multicomponent reactions with isocyanides. *Angew. Chem. Int. Ed. Eng.* 39: 3168–3210.

6 Müller, T.J.J. (ed.) (2014). General discussion and reactions involving a carbonyl compound as electrophilic component. In: *Multicomponent Reactions; Science of Synthesis*. Thieme.

7 Zhu, J. and Bienaymé, H. (eds.) (2005). *Multicomponent Reactions*. Wiley-VCH.

8 Ugi, I., Meyr, R., Fetzer, U., and Steinbruckner, C. (1959). Studies on isonitriles. *Angew. Chem.* 71: 386.

9 Ugi, I. and Kaufhold, G. (1967). Stereoselektive Synthesen IV. Der Reaktionsmechanismus stereoselektiver Vierkomponenten-Kondensationen. *Justus Liebig Ann. Chem.* 709: 11–28.

10 Ugi, I., Offernmann, K., Herlinger, H., and Marquarding, D. (1967). The reaction of (S)-alpha-phenylethylamine and isobuteraldehyde with benzoic acid and tert-butyl-isocyanide as model reactions for stereoselective peptide-synthesis intermediate. Four component condensations. *Justus Liebigs Ann. Chem.* 709: 1–10.

11 Basso, A., Banfi, L., Guati, G. et al. (2004). Ugi multicomponent reaction with hydroxylamines: an efficient route to hydroxamic acid derivates. *Tetrahedron* 45 (32): 6109–6111. https://doi.org/10.1016/j.tetlet.2004.06.068.

12 Zinner, G., Moderhack, D., and Kliegel, W. (1969). Hydroxylamine derivates. XXXVII. Hydroxylamines in the Ugi four-component condensation. *Chem. Ber.* 102: 2536–2546.

13 Gargano, A.F.G., Buchinger, S., Kohout, M., and Lindner, W. (2013). Single-step Ugi multicomponent reaction for the synthesis of phosphopeptidomimetics. *J. Organomet. Chem.* 78: 10077–10087. https://doi.org/10.1021/jo401372x.

14 Ugi, I. and Bodesheim, F. (1963). Isonitriles. XIV. Reactions of isonitriles with hydrazones and carboxylic acids. *Ann. Chem.* 666: 61–64.

15 Zinner, G. and Kliegel, W. (1966). Ugi reaction with hydrazines. I. *Arch. Pharm.* 299: 746–756.

16 Chandgude, A.L. and Dömling, A. (2017). *N*-hydroxyimide Ugi reaction toward α-hydrazino amides. *Org. Lett.* 19 (5): 1228–1231. https://doi.org/10.1021/acs.orglett.7b00205.

17 Heck, S. and Dömling, A. (2000). A versatile multi-component one-pot thiazole synthesis. *Synlett* 3: 424–426.

18 El Kaim, L., Grimaud, L., and Oble, J. (2005). Phenol Ugi-smiles systems: strategies for the multicomponent *N*-arylation of primary amines with isocyanides, aldehydes, and phenols. *Angew. Chem. Int. Ed.* 44: 7961–7964.

19 Aknin, K., Gauriot, M., Totobenazara, J. et al. (2012). Squaric acid is a suitable building-block in 4C-Ugi reaction: access to original bivalent compounds. *Tetrahedron* 53: 458–461.

20 Pan, S.C. and List, B. (2008). Catalytic three-component Ugi reaction. *Angew. Chem. Int. Ed.* 47: 3622–3625.

21 Zhang, J., Yu, P., Li, S.Y. et al. (2018). Asymmetric phosphoric acid-catalyzed four-component Ugi reaction. *Science* 361: 1087.

22 Ayoup, M., Fouad, M., Abdel-Hamid, H. et al. (2020). Two decades of recent advances of Ugi reactions: synthetic and pharmaceutical applications. *Eur. J. Med. Chem.* 186: 111875.

23 Kreye, O., Türünç, O., Sehlinger, H. et al. (2012). Structurally diverse polyamides obtained from monomers derived via the Ugi multicomponent reaction. *Chem. Eur. J.* 18: 5767–5776.

24 Gracias, V., Moore, J.D., and Djuric, S.W. (2004). Sequential Ugi/Heck cyclization strategies for the facile construction of highly functionalized *N*-heterocyclic scaffolds. *Tetrahedron Lett.* 45: 417–420. https://doi.org/10.1016/j.tetlet.2003.10.147.

25 Xiang, Z., Luo, T., Lu, K. et al. (2004). Concise synthesis of isoquinoline via the Ugi and Heck reactions. *Org. Lett.* 6 (18): 3155–3158. https://doi.org/10.1021/ol048791n.

26 Umkehrer, M., Kalinski, C., Kolb, J., and Burdack, C. (2006). A new and versatile one-pot synthesis of indol-2-ones by a novel Ugi-four-component-Heck reaction. *Tetrahedron Lett.* 47: 2391–2393. https://doi.org/10.1016/j.tetlet.2006.01.149.

27 Kalinski, C., Umkehrer, M., Schmidt, J. et al. (2006). A novel one-pot synthesis of highly diverse indole scaffolds by the Ugi/Heck reaction. *Tetrahedron Lett.* 47: 4683–4686. https://doi.org/10.1016/j.tetlet.2006.04.127.

28 Hulme, C., Peng, J., Morton, G. et al. (1998). Novel safety-catch linker and its application with a Ugi-De-Boc/cyclization (UDC) strategy to access carboxylic acids, 1,4-benzodiazepines, diketopiperazines, ketopiperazines and dihydroquinoxalinones. *Tetrahedron* 39 (40): 7227–7230.

29 Tempest, P., Ma, V., Thomas, S. et al. (2001). Two-step solution-phase synthesis of novel benzimidazoles utilizing a UDC (Ugi/de-Boc/cyclize) strategy. *Tetrahedron* 42 (30): 4959–4962.

30 Nixey, T., Tempest, P., and Hulme, C. (2002). Two-step solution-phase synthesis of novel quinoxalinones utilizing a UDC (Ugi/de-Boc/cyclize) strategy. *Tetrahedron* 43 (9): 1637–1639.

31 Nixey, T., Kelly, M., Semin, D., and Hulme, C. (2002). Short solution phase preparation of fused azepine-tetrazoles via a UDC (Ugi/de-Boc/cyclize) strategy. *Tetrahedron* 41 (45): 8729–8733.

32 Schofield, K., Foley, C., and Hulme, C. (2020). 5-endo trig oxidative radical cyclizations of Ugi-3CR products toward 1,4-imidazolidinones. *Org. Lett.* 23 (1): 107–112. https://doi.org/10.1021/acs.orglett.0c03785.

33 Foley, C., Shaw, A., and Hulme, C. (2018). Aza-riley oxidation of Ui-Azide and Ugi-3CR products toward vicinal tricarbonyl amides: two-step MCR-oxidation methodology accessing functionalized α,β-diketoamides and α,β-diketotetrazoles. *Org. Lett.* 20 (5): 1275–1278. https://doi.org/10.1021/acs.orglett.7b03977.

34 Foley, C., Shaw, A., and Hulme, C. (2017). Oxidative deaminations and deisatinylations of Ugi-Azide and Ugi-3CR products: a two-step MCR-oxidation protocol toward functionalized a-ketoamides and a-ketotetrazoles. *Org. Lett.* 19 (9): 2238–2241.

35 Collet, J.W., Foley, C., Shaw, A.Y. et al. (2017). Copper(I) catalyzed oxidative hydrolysis of Ugi 3-component and Ugi-azide reaction products towards 2 α-ketoamides and α-ketotetrazoles. *Org. Biomol. Chem.* 15: 6131–6135.

36 Zaman, M., Hasan, M., Peshkov, A.A. et al. (2020). Silver(I) triflate-catalyzed protocol for the post-Ugi synthesis of spiroindolines. *Adv. Synth. Catal.* 362: 261–268.

37 Brown, A.L., Churches, Q.I., and Hutton, C.A. (2015). Total synthesis of ustiloxin D utilizing an ammonia-Ugi reaction. *J. Organomet. Chem.* 80: 9831–9837.

38 Li, P., Evans, C.D., and Joullié, M.M. (2005). A convergent total synthesis of ustiloxin D via an unprecedented copper-catalyzed ethynyl aziridine ring-opening by phenol derivatives. *Org. Lett.* 7: 5325.

39 Varadi, A., Palmer, T.C., Haselton, N. et al. (2015). Synthesis of carfentanil amide opioids using the Ugi multicomponent reaction. *ACS Chem. Neurosci.* 6 (9): 1570–1577. https://doi.org/10.1021/acschemneuro.5b00137.

40 Kawato, Y., Iwata, M., Yazaki, R. et al. (2011). A simplified catalytic system for direct catalytic asymmetric aldol reactions of thiamides; application to an enantioselective synthesis of atorvastatin. *Tetrahedron* 67: 6539–6546.

41 Estevez, V., Villacampa, M., and Menendez, J.C. (2014). Concise synthesis of atorvastatin lactone under high-speed vibration milling conditions. *Org. Chem. Front.* 1: 458–463.

42 Lopchuk, J.M. and Gribble, G.W. (2015). Total synthesis of atorvastatin via a late-stage regioselective 1,3-dipolar munchnone cycloaddition. *Tetrahedron Lett.* 56: 3208–3211.

43 Zarganes-Tzitzikas, T., Neochoritis, C., and Dömling, A. (2019). Atorvastatin (Lipitor) by MCR. *ACS Med. Chem. Lett.* 10: 389–392.

44 Lemieux, R.M., Popovici-Muller, J., Travins, J.M., et al. (2013). Therapeutically activity compounds and use thereof. WO2015010626, filed 25 July 2012 and issued 1 January 2015.

45 Popovici-Muller, J., Lemieux, R.M., Artin, E. et al. (2018). Discovery of AG-120 (ivosidenib): a first-in-class mutant IDH1 inhibitor for the treatment of IDH1 mutant cancers. *ACS Med. Chem. Lett.* 9: 300–305.

46 Passerini, M. and Simone, L. (1921). Composto del p-isonitril-azobenzolo con acetone ed acido acetico. *Gazz. Chim. Ital.* 51: 126–129.

47 Hulme, C. and Gore, V. (2003). Multi-component reactions: emerging chemistry in drug discovery from xylocain to crixivan. *Curr. Med. Chem.* 10: 51–58.

48 Moni, L., Banfi, L., Basso, A. et al. (2016). Diastereoselective passerini reaction of biobased chiral aldehydes: divergent synthesis of various polyfunctionalized heterocycles. *Org. Lett.* 18: 1638.

49 Zhang, J., Lin, S., Cheng, D. et al. (2015). Phosphoric acid-catalyzed asymmetric classic passerini reaction. *J. Am. Chem. Soc.* 137: 14039.

50 Vishwanatha, T., Gieomans, B., Goda, S., and Dömling, A. (2020). Tubulysin synthesis featuring stereoselective catalysis and highly convergent multicomponent assembly. *Org. Lett.* 22: 5396–5400.

51 Kaur, G., Hollingshead, M., Holbeck, S. et al. (2006). Biological evaluation of tubulysin A: a potential anticancer and antiangiogenic natural product. *Biochem. J.* 396: 235–242.

52 Groebke, K., Weber, L., and Mehlin, F. (1998). Synthesis of imidazo[1,2-a] annulated pyridines, pyrazines and pyrimidines by a novel three-component condensation. *Synlett* 6: 661–663.

53 Bienaymé, H. and Bouzid, K. (1998). A new heterocyclic multicomponent reaction for the combinatorial synthesis of fused 3-aminoimidazoles. *Angew. Chem. Int. Ed. Eng.* 37: 2234–2237.

54 Blackburn, C. (1998). A three-component solid-phase synthesis of 3-aminoimidazo[1,2-a]azines. *Tetrahedron Lett.* 39: 5469–5472.

55 Boltjes, A. and Domling, A. (2019). The Groebke-Blackburn-Bienaymé reaction. *Eur. J. Org. Chem.* 42: 7007–7049. https://doi.org/10.1002/ejoc.201901124.

56 Hulme, C. and Lee, Y.-S. (2008). Emerging approaches for the synthesis of bicyclic imidazo[1,2-x]-heterocycles. *Mol. Divers.* 12: 1–15. https://doi.org/10.1007/s11030-008-9072-1.

57 Gladysz, R., Adriaenssens, Y., Winter, H.D. et al. (2015). Discovery and SAR of novel and selective inhibitors of urokinase plasminogen activator (uPA) with an imidazo[1,2-a]pyridine scaffold. *J. Med. Chem.* 58: 9238–9257. https://doi.org/10.1021/acs.jmedchem.5b01171.

58 Vuillermet, F., Bourret, J., and Pelletier, G. (2021). Synthesis of imidazo[1,2-a]pyridines: triflic anhydride-mediated annulation of 2H-azirines with 2-chloropyridines. *J. Organomet. Chem.* 86: 388–402. https://doi.org/10.1021/acs.joc.0c02148.

59 He, Q., Liang, Y., Xu, C. et al. (2019). Highly regioselective, acid-catalyzed, three-component cascade reaction for the synthesis of 2-aminopyridine-decorated imidazo[1,2-a]pyridine. *ACS Comb. Sci.* 21: 149–153. https://doi.org/10.1021/acscombsci.8b00149.

60 Fischer, P.M. (2018). Design of small-molecule active-site inhibitors of the S1A family proteases as procoagulant and anticoagulant drugs. *J. Med. Chem.* 61: 3799–3822. https://doi.org/10.1021/acs.jmedchem.7b00772.

61 Lu, S., Zhu, X., Li, K. et al. (2016). Reactivity of *p*-toluenesulfonylmethyl isocyanide: iron-involved C–H tosylmethylation of imidazopyridines in nontoxic media. *J. Organomet. Chem.* 81: 8370–8377. https://doi.org/10.1021/acs.joc.6b01552.

62 Meirer, K., Rödl, C.B., Wisniewska, J.M. et al. (2013). Synthesis and structure-activity relationship studies of novel dual inhibitors of soluble epoxide hydrolase and 5-lipoxygenase. *J. Med. Chem.* 56: 1777–1781. https://doi.org/10.1021/jm301617j.

63 Liu, J., Yang, J., Inceoglu, B. et al. (2010). Inhibition of soluble epoxide hydrolase enhances the anti-inflammatory effects of aspirin and 5-lipoxygenase activation protein inhibitor in a murine model. *Biochem. Pharmacol.* 79: 880–887. https://doi.org/10.1016/j.bcp.2009.10.025.

64 Hyde, C.A. and Missailidis, S. (2009). Inhibition of arachidonic acid metabolism and its implication on cell proliferation and tumour-angiogenesis. *Int. Immunopharmacol.* 9: 701–715. https://doi.org/10.1016/j.intimp.2009.02.003.

65 Radmark, O., Werz, O., Steinhilber, D. et al. (2007). 5-Lipoxygenase: regulation of expression and enzyme activity. *Trends Biochem. Sci.* 7: 332–341. https://doi.org/10.1016/j.tibs.2007.06.002.

66 Imig, J.D. (2012). Epoxides and soluble epoxide hydrolase in cardiovascular physiology. *Physiol. Rev.* 92: 101–130. https://doi.org/10.1152/physrev.00021.2011.

67 Morisseau, C. and Hammock, B.D. (2013). Impact of soluble epoxide hydrolase and epoxyeicosanoids on human health. *Annu. Rev. Pharmacol. Toxicol.* 53: 37–58. https://doi.org/10.1146/annurev-pharmtox-011112-140244.

68 Shen, C.H. (2010). Soluble epoxide hydrolase inhibitors: a patent review. *Expert Opin. Ther. Pate.* 7: 941–956. https://doi.org/10.1517/13543776.2010.484804.

69 Al-Tel, T.H., Al-Qawasmeh, R.A., and Zaarour, R. (2011). Design, synthesis and *in vitro* antimicrobial evaluation of novel Imidazo[1,2-*a*] pyridine and imidazo[2,1-*b*][1.3]benzothiazole motifs. *Eur. J. Med. Chem.* 46: 1874–1881. https://doi.org/10.1016/j.ejmech.2011.02.051.

70 Davies, J. (1996). Bacteria on the rampage. *Nature* 383: 219–220.

71 Chu, D.T.W., Plattner, J.J., and Katz, L. (1996). New directions in antibacterial research. *J. Med. Chem.* 39: 3853–3874. https://doi.org/10.1021/jm960294s.

72 Grare, M., Mourer, M., Fontanay, S. et al. (2007). *In vitro* activity of para-guanidinoethylcalix[4]areene against susceptible and antibiotic-resistant Gram-negative and Gram-positive bacteria. *J. Antimicrob. Chemother.* 60: 575–581. https://doi.org/10.1093/jac/dkm244.

73 Burchak, O.N., Mugherli, L., Ostuni, M. et al. (2011). Combinatorial dicovery of fluorescent pharmacophores by multicomponent reactions in droplet arrays. *J. Am. Chem. Soc.* 133: 10058–10061. https://doi.org/10.1021/ja204016e.

74 Shaaban, S. and Abdel-Wahab, B.F. (2016). Groebke-Blackburn-Bienaymé multicomponent reaction: emerging chemistry for drug discovery. *Mol. Divers.* 20: 233–254. https://doi.org/10.1007/s11030-015-9602-6.

75 Ansari, A.J., Sharma, S., Pathare, R.S. et al. (2016). Solvent-free multicomponent synthesis of biologically-active fused-imidazo heterocycles catalyzed by reusable $Yb(OTf)_3$ under microwave irradiation. *ChemistrySelect* 5: 1016–1021. https://doi.org/10.1002/slct.201600241.

76 Paneri, M. and Khan, S. (2016). A mild and efficient microwave-assisted solvent-free regioselective synthesis of imiazo[2,1-*b*]thiazole scaffold. *Chem. Biol. Interface* 6: 162–170.

77 Adib, M., Mahdavi, M., Noghani, M.A. et al. (2007). Catalyst-free three component reaction betweem 2-aminopyridines (or 2-amino-thiazoles), aldehydes and isocyanides in water. *Tetrahedron Lett.* 48: 7263–7265. https://doi.org/10.1016/j.tetlet.2007.08.049.

78 Vidyacharan, S., Shinde, A.H., Satpathi, B. et al. (2014). A facile protocol for the synthesis of 3-aminoimidazo-fuzed heterocycles via the Grobeke-Blackburn-Bieayme reaction under catalyst-free and solvent free conditions. *Green Chem.* 16: 1168.

79 Claudio-Catalán, M.Á., Pharande, S.G., Quezada-Soto, A. et al. (2018). Solvent- and catalyst-free one-pot green bound-type fused bis-heterocycles synthesis Groebke-Blackburn-Bienaymé reaction/SNAr/ring-chain azido-tautomerization strategy. *ACS Omega* 3: 5177–5186. https://doi.org/10.1021/acsomega.8b00170.

80 Wu, P., Bjørn-Yoshimoto, W.E., Staudt, M. et al. (2019). Indentification and structure-activity relationship study of imidazo[1,2-*a*]pyridine-3-amines as first selective inhibitors of excitatory amino acid transporter subtype 3 (EAAT3). *ACS Chem. Neurosci.* 10: 4414–4429. https://doi.org/10.1021/acschemneuro.9b00447.

81 Gewald, K., Schinke, E., and Böttcher, H. (1966). 2-Amino-thiophene aus methylenaktiven nitrilen carbonylverbindungen und schwefel. *Chem. Ber.* 99: 94–100.

82 Tinsley, J.M. (2005). Gewald aminothiophene synthesis. In: *Named Reactions in Heterocyclic Chemistry* (eds. J.J. Li and E.J. Corey), 193–198. New York: Wiley.

83 Peet, N.P., Sunder, S., Barbuch, R.J. et al. (1986). Mechanistic observations in the Gewald syntheses of 2-aminothiophenes. *J. Heterocyclic Chem.* 23: 129–134. https://doi.org/10.1002/jhet.5570230126.

84 Meltzer, H.Y. and Fibiger, H.C. (1996). Olazapine: a new atypical antipsychotic drug. *Neuropsychopharmacology* 14: 83–85. https://doi.org/10.1016/0893-133X(95)00197-L.

85 Eli Lilly and Company (2008). Annual report. https://investor.lilly.com/annuals.cfm (accessed 26 November 2009).

86 Huang, Y. and Dömling, A. (2011). The Gewald multicomponent reaction. *Mol. Divers.* 15: 3–33. https://doi.org/10.1007/s11030-010-9229-6.

87 Shipilovskikh, S.A. and Rubtsov, A.E. (2019). One-pot synthesis of thieno[3,2-*e*]pyrrolo[1,2,*a*]pyrimidine derivative scaffold: a valuable source of PARP-1 inhibitors. *J. Organomet. Chem.* 84: 15788–15796. https://doi.org/10.1021/acs.joc.9b00711.

88 Rodríguez, M.I., Peralta-Leal, A., O'Valle, F. et al. (2013). PARP-1 regulates metastatic melanoma through modulation of Vientin-induced malignant transformation. *PLoS Genet.* 9 (6): e1003531. https://doi.org/10.1371/journal.pgen.1003531.

89 Nowsheen, S., Cooper, T., Stanley, J.A. et al. (2012). Synthetic lethal interactions between EGFR and PARP inhibition in human triple negative breast cancer cells. *PLoS One* 7 (10): e46614. https://doi.org/10.1371/journal.pone.0046614.

90 Galia, A., Calogero, A.E., and Condorelli, R. (2012). PARP-1 proten expression in glioblastoma multiforme. *Eur. J. Histochem.* 56 (1): e9.

91 Telli, M.L. and Ford, J.M. (2010). Novel treatment approaches for triple-negative breast cancer. *Clin. Breast Cancer* 10 (1): E16–E22. https://doi.org/10.3816/CBC.2010.s.003.

92 Shimizu, S., Nomura, F., Tomonaga, T. et al. (2004). Expression of poly(ADP-ribose) polymerase in human hepatocellular carcinoma and analysis of biopsy specimens obtained under sonographic guidance. *Oncol. Rep.* 12: 821–825. https://doi.org/10.3892/or.12.4.821.

93 Rojo, F., Garcia-Parra, J., Zazo, S. et al. (2012). Nuclear PARP-1 protein overexpression is associated with poor overall survival in early breast cancer. *Ann. Oncol.* 23: 1156–1164. https://doi.org/10.1093/annonc/mdr361.

94 Honda, T.E.H., Kawashima, K., Takaoka, S. et al. (2013). Novel compound having PARP inhibitory activity. WO Patent WO2013008872.

95 Van der Plas, S.E., Kelgtermans, H., De Munck, T. et al. (2018). Discovery of *N*-(3-carbamoyl-5,5,7,7-tetramethyl-5,7-dihydro-4H-thieno[2,3-*c*]pyran-2-yl)-IH-pyrazole-5-carboxamide (GLPG1837), a novel potentiator which can open class III mutant cystic fibrosis transmembrane conductance regulator (CFTR) channels to a high extent. *J. Med. Chem.* 61: 1425–1435. https://doi.org/10.1021/acs.jmedchem.7b01288.

96 Lubamba, B., Dhooghe, B., Noel, S., and Leal, T. (2012). Cystic fibrosis: insight into CFTR pathophysiology and pharmacotherapy. *Clin. Biochem.* 45 (15): 1132–1144.

97 Hanrahan, J.W., Sampson, H.M., and Thomas, D.Y. (2013). Novel pharmacological strategies to treat cystic fibrosis. *Trends Pharmacol. Sci.* 34 (2): 119–125.

98 Merk, D. and Schubert-Zsilavecz, M. (2013). Repairing mutated proteins–development of small molecules targeting defects in the cystic fibrosis transmembrane conductance regulator. *Expert Opin. Drug Discovery* 8 (6): 691–708. https://doi.org/10.1517/17460441.2013.788495.

99 Riordan, J.R., Rommens, J.M., Kerem, R.S. et al. (1989). Identification of the cystic fibrosis gene: cloning and characterization of complementary DNA. *Science* 245 (4922): 1066–1073.

100 Cystic Fibrosis Centre at the Hospital for Sick Children: Toronto 2011. Cystic fibrosis mutation database. http://www.genet.sickkids.on.ca/Home.html (accessed 13 November 2017).

101 De Boeck, K., Munck, A., Walker, S. et al. (2014). Efficacy and safety of Ivacaftor in patients with cystic fibrosis and a non-G551D gating mutation. *J. Cyst. Fibros.* 13 (6): 674–680.

102 Moss, R.B., Flume, P.A., Elborn, J.S. et al. (2015). Efficacy and safety of Ivacaftor in patients with cystic fibrosis who have an Arg117His-CFTR mutation: a double-blind, randomized controlled trial. *Lancet Respir. Med.* 3 (7): 524–533.

103 Hantzsch, A. (1882). Synthesis of pyridine derivatives from acetoacetic ester and aldehydeammoniak. *Liebigs Ann. Chem.* 215: 1–82.

104 Kürti, L. and Czakó, B. (2005). *Strategic Applications of Named Reactions in Organic Synthesis: Background and Detailed Mechanisms*. Amsterdam: Elsevier Academic Press.

105 Wang, A.L. (2017). New generations of dihydropyridines for treatment of hypertension. *J. Geriatr. Cardiol.* 14 (1): 67–72. https://doi.org/10.11909/j.issn.1671-5411.2017.01.006.

106 U.S. Food & Drug Administration (2021). Drugs@FDA: FDA-approved drugs https://www.accessdata.fda.gov/scripts/cder/daf/index.cfm?event=reportsSearch.process (accessed 2 January 2021).

107 Sharma, V.K. and Singh, S.K. (2017). Utility and medicinal importance of 1,2- & 1,4-dihydropyridines. *RSC Adv.* 7: 2682–2732. https://doi.org/10.1039/c6ra24823c.

108 Bossert, V.F., Meyer, H., and Wehinger, E. (1981). 4-Aryldihydropyridine, a new class of highly active calcium antagonists. *Angew. Chem.* 93: 755–763.

109 Huang, Y. (2007). Hantzsch 1,4-dihydropyridine - an effective and convenient reducing agent. *Synlett* 14: 2034–2035. https://doi.org/10.1055/s-2007-985576.

110 Mauzeral, D. (1955). 1-Benzyldihydronicotinamide—a model for reduced DPN. *J. Am. Chem. Soc.* 77: 2261.

111 Swarnalatha, G. (2011). 1,4-Dihydropyridines: a multifunctional molecule- a review. *Int. J. ChemTech Res.* 3: 75–89.

112 Godfraid, T. (1986). Calcium antagonism and calcium entry blockade. *Pharm. Rev.* 38: 321.

113 Khadilkar, B. (1998). Silica gel supported ferric nitrate: a convenient oxidizing reagent. *Synth. Commun.* 28: 207.

114 Kumar, S. (2011). Synthesis and antimicrobial activity of a new series 1,4-dihydropyridine derivatives. *J. Serb. Chem. Soc.* 76: 1.

115 Srivastava, S. (2002). Synthesis of new 1, 2, 4-triazolo-thidiazoles and its 2-oxoazetidines as antimicrobial, anticonvulsant and anti-inflammatory agents. *Indian J. Chem.* 41B: 2357–2363.

116 Desai, N. (2016). Design, synthesis, and biological evaluation of 1,4-dihydropyridine derivatives as potent antitubercular agents. *Chem. Biol. Drug Des.* 86: 370–377.

117 Ghorbani-Vaghei, R. (2016). Synthesis and biological evaluation of new series 1,4-dihydropyridines. *Res. Chem. Intermed.* 42: 4715–4731.

118 Vdovina, S. (2008). New potential of the classical Biginelli reaction. *Russ. Chem. Rev.* 77: 1017–1053.

119 Padmaja, K. (2013). A review on biological activities of dihydro pyrimidinones/thiones. *Indian J. Pharm. Sci. Res.* 3 (2): 69–73.

120 Kappe, C. (2000). Biologically active dihydropyrimidones of the Biginelli-type – a literature survey. *Eur. J. Med. Chem.* 35: 1043–1052.

121 Lewis, R. (2010). Dihydropyrimidinone positive modulation of δ-subunit-containing γ-aminobutyric acid type A receptors, including an epilepsy-linked mutant variant. *Biochemistry* 49: 4841–4851.

122 Overman, L. (1993). Studies toward the total synthesis of (+)-Ptilomycalin A. Use of a tethered Biginelli condensation for the preparation of an advanced tricyclic intermediate. *J. Organomet. Chem.* 58: 3235–3237.

123 Aron, Z. and Overman, L. (2004). The tethered Biginelli condensation in natural product synthesis. *Chem. Commun.* 3: 253–265.

124 Silva, S. (2019). Insights into the biosynthesis of cyclic guanidine alkaloids from crambeidae marine sponges. *Angew. Chem. Int. Ed.* 58: 520–525.

125 Gao, Z. (2020). Eight-step total synthesis of (+)-crambescin A. *RSC Adv.* 10: 39266.

126 Mayer, T. (1999). Small molecule inhibitor of mitotic spindle bipolarity identified in a phenotype-based screen. *Science* 286 (5441): 971–974.

127 Abdou, A. (2015). Useful four-carbon synthons en route to monastrol analogs. *Tetrahedron* 71: 139–146.

128 Kaan, H. (2010). Structural basis for inhibition of Eg5 by dihydropyrimidines: stereoselectivity of antimitotic inhibitors enastron, dimethylenastron and fluorastrol. *J. Med. Chem.* 53: 5676–5683.

129 Terracciano, S. (2016). New dihydropyrimidin-2(1H)-one based Hsp90 C-terminal inhibitors. *RSC Adv.* 6: 82330–82340.

130 Guggilapu, S. (2016). MoO_2Cl_2 catalyzed efficient synthesis of functionalized 3,4-dihydropyrimidin-2(1H)-ones/thiones and polyhydroquinolines: recyclability, fluorescence and biological studies. *New J. Chem.* 40: 838–843.

131 da Silva, D. (2012). Free radical scavenging and antiproliferative properties of Biginelli adducts. *Bioorg. Med. Chem.* 20: 2645–2650.

132 Treptow, T. (2015). Novel hybrid DHPM-fatty acids: synthesis and activity against glioma cell growth *in vitro*. *Eur. J. Med. Chem.* 95: 552–562.

133 Abnous, K. (2013). Synthesis and molecular modeling of six novel monastrol analogues: evaluation of cytotoxicity and kinesin inhibitory activity against HeLa cell line. *Daru* 21: 70.

134 Soumyanarayanan, U. (2012). Monastrol mimic Biginelli dihydropyrimidinone derivatives: synthesis, cytotoxicity screening against HepG2 and HeLa cell lines and molecular modeling study. *Org. Med. Chem. Lett.* 2: 23.

135 Zhu, L. (2011). Synthesis and biological evaluation of novel homocamptothecins conjugating with dihydropyrimidine derivatives as potent topoisomerase I inhibitors. *Arch. Pharm.* 344: 726–734.

136 Malani, K. (2016). Synthesis, characterization and *in silico* designing of diethyl-3-methyl-5-(6-methyl-2-thioxo-4-phenyl-1,2,3,4-tetrahydropyrimidine-5-carboxamido) thiophene-2,4-dicarboxylate derivative as anti-proliferative and anti-microbial agents. *Bioorg. Chem.* 68: 264–274.

137 Matias, M. (2016). Synthesis, *in vitro* evaluation and QSAR modelling of potential antitumoral 3,4-dihydropyrimidin-2-(1H)-thiones. *Arab. J. Chem.* 12: 5086–5102.

138 Kamal, A. (2011). Synthesis and biological evaluation of conformationally flexible as well as restricted dimers of monastrol and related dihydropyrimidones. *J. Med. Chem.* 46: 3274–3281.

139 Reddy, O. (2013). Synthesis and cytotoxic evaluation for some new dihydropyrimidinone derivatives for anticancer activity. *Lett. Drug Des. Discov.* 10: 699–705.

140 Bariwal, J. (2012). Synthesis, characterization and anticancer activity of 3-aza-analogues of DP-7. *Med. Chem. Res.* 21: 4002–4009.

141 Russowsky, D. (2006). Synthesis and differential antiproliferative activity of Biginelli compounds against cancer cell lines: monastrol, oxo-monastrol and oxygenated analogues. *Bioorg. Chem.* 34: 173–182.

142 Sashidhara, K. (2013). Discovery of coumarine-monastrol hybrid as potential antibreast tumor-specific agent. *Eur. J. Med. Chem.* 60: 120–127.

143 Van Leusen, A.M., Boerma, G.J.M., Helmholdt, R.B. et al. (1972). Chemistry of sulfonylmethylisocyanides. Simple synthetic approaches to a new versatile chemical building block. *Tetrahedron Lett.* 13: 2367–2368.

144 Van Leusen, A.M., Wildeman, J., and Oldenziel, O.H. (1977). Base-induced cycloaddition of sulfonylmethyl isocyanides to C, N double bonds. Synthesis of 1,5-disubstituted and 1,4,5-trisubstituted imidazoles from aldimines and imidoyl chlorides. *J. Org. Chem.* 42 (7): 1153–1159.

145 Van Leusen, A.M., Hoogenboom, B.E., and Siderius, H. (1972). Novel and efficient synthesis of oxazoles from tosylmethyl isocyanide and carbonyl compounds. *Tetrahedron Lett.* 13: 2369–2372.

146 Oldenziel, O.H. and Van Leusen, A.M. (1972). Chemistry of sulfonylmethylisocyanides. 4. New synthesis of thiazoles from tosylmethylisocyanide and carboxymethyl dithioates. *Tetrahedron Lett.* 13: 2777–2778.

147 Van Leusen, A.M., Hoogenboom, B.E., and Houwing, H.A. (1976). Chemistry of sulfonylmethyl isocyanides. 11. Synthesis of 1,2,4- triazoles from tosylmethyl isocyanide and aryldiazonium compounds. *J. Org. Chem.* 41: 711–713.

148 Adib, M., Mohammadi, B., Sheikhi, E. et al. (2011). 1-Methylimidazole-catalyzed reaction between tosylmethyl isocyanide and dialkyl acetylenedicarboxylates: an efficient synthesis of functionalized pyrroles. *Chin. Chem. Lett.* 22: 314–317.

149 Hu, Z., Li, Y., Pan, L. et al. (2014). Direct synthesis of pyrrolo[3,4-*c*]quinolines from the domino reaction of tosylmethyl isocyanides and aminochalcones. *Adv. Synth. Catal.* 356: 2974–2978.

150 De Moliner, F. and Hulme, C. (2012). Straightforward assembly of phenylimidazoquinoxalines via a one-pot two-step MCR process. *Org. Lett.* 14 (5): 1354–1357.

151 Guan, Z.R., Liu, Z.M., and Ding, M.W. (2018). New efficient synthesis of 1H-imidazo-[4,5-*c*]quinolines by a sequential van Leusen/Staudinger/aza-Wittig/carbodiimide-mediated cyclization. *Tetrahedron* 74: 7186–7192.

152 Decurtins, W., Wichert, M., Franzini, R.M. et al. (2016). Automated screening for small organic ligands using DNA-encoded chemical libraries. *Nat. Protoc.* 11: 764–780.

153 Song, M. and Hwang, G.T. (2020). DNA-encoded library screening as core platform technology in drug discovery: its synthetic method development and applications in DEL synthesis. *J. Med. Chem.* 63: 6578–6599.

154 Satz, A.L., Cai, J., Chen, Y. et al. (2015). DNA compatible multistep synthesis and applications to DNA encoded libraries. *Bioconjug. Chem.* 26: 1623–1632.

155 Xiao, X.-Y., Li, R., Zhuang, H. et al. (2000). Solid-phase combinatorial synthesis using MicroKan reactors, Rf tagging, and directed sorting. *Biotechnol. Bioeng.* 71 (1): 44–50.

156 Braeckmans, K., De Smedt, S.C., Leblans, M. et al. (2002). Encoding microcarriers: present and future technologies. *Nat. Rev. Drug Discov.* 1: 447–456.

157 Geigle, S.N., Petersen, A.C., and Satz, A.L. (2019). Development of DNA-compatible van Leusen three-component imidazole synthesis. *Org. Lett.* 21 (22): 9001–9004.

158 Zhang, L., Peng, X.M., Damu, G.L. et al. (2014). Comprehensive review in current developments of imidazole-based medicinal chemistry. *Med. Res. Rev.* 34: 340–437.

159 Burnstock, G. (2006). Pathophysiology and therapeutic potential of purinergic signaling. *Pharmacol. Rev.* 58 (1): 58–86.

160 Burnstock, G. and Verkhratsky, A. (2010). Long-term (trophic) purinergic signaling: purinoceptors control cell proliferation, differentiation and death. *Cell Death Dis.* 1 (1): e9.

161 Gargett, C.E. and Wiley, J.S. (1997). The isoquinoline derivative KN-62 a potent antagonist of the P2Z-receptor of human lymphocytes. *Br. J. Pharmacol.* 120: 1483–1490.

162 Jelassi, B., Chantôme, A., Alcaraz-Perez, F. et al. (2011). P2X7 receptor activation enhances SK3 channels- and cystein cathepsin-dependent cancer cells invasiveness. *Oncogene* 30: 2108–2122.

163 Park, J.H., Lee, G.E., Lee, S.D. et al. (2015). Discovery of novel 2,5-dioxoimidazolidine-based P2X7 receptor antagonists as constrained analogues of KN62. *J. Med. Chem.* 58: 2114–2134.

164 Park, J.-H., Williams, D.R., Lee, J.-H. et al. (2016). Potent suppressive effects of 1-piperidinylimidazole based novel P2X7 receptor antagonists on cancer cell migration and invasion. *J. Med. Chem.* 59: 7410–7430.

165 Liu, M., Wang, X., Wang, L. et al. (2018). Targeting the IDO1 pathway in cancer: from bench to bedside. *J. Hematol. Oncol.* 11: 100.

166 Brant, M.G., Goodwin-Tindall, J., and Stover, K.R. (2018). Identification of potent indoleamine 2,3-dioxygenase 1 (Ido1) inhibitors based on a phenylimidazole scaffold. *ACS Med. Chem. Lett.* 9: 131–136.

167 Wade, M., Li, Y.C., and Wahl, G.M. (2013). MDM2, MDMX and p53 in oncogenesis and cancer therapy. *Nat. Rev. Cancer* 13: 83–96.

168 Zhao, Y., Aguilar, A., Bernard, D. et al. (2015). Small-molecule inhibitors of the MDM2-p53 protein-protein interaction (MDM2 Inhibitors) in clinical trials for cancer treatment. *J. Med. Chem.* 58: 1038–1052.

169 Twarda-Clapa, A., Krzanik, S., Kubica, K. et al. (2017). 1,4,5-Trisubstituted imidazole-based p53–MDM2/MDMX antagonists with aliphatic linkers for conjugation with biological carriers. *J. Med. Chem.* 60 (10): 4234–4244.

170 Bergeron, N., Phan, B.P., Ding, Y. et al. (2015). Convertase Subtilisin/Kexin type 9 inhibition: a new therapeutic mechanism for reducing cardiovascular disease risk. *Circulation* 132 (17): 1648–1666.

171 Stucchi, M., Grazioso, G., Lammi, G. et al. (2016). Disrupting the PCSK9/LDLR protein–protein interaction by an imidazole-based minimalist peptidomimetic. *Org. Biomol. Chem.* 14: 9736–9740.

172 Lammi, G., Sgrignani, J., Arnoldi, A. et al. (2019). Computationally driven structure optimization, synthesis, and biological evaluation of imidazole-based

proprotein convertase subtilisin/kexin 9 (PCSK9) inhibitors. *J. Med. Chem.* 62 (13): 6163–6174.
173 Ganesh, K. and Massague, J. (2018). TGF-β inhibition and immunotherapy: checkmate. *Immunity* 48: 626–628.
174 Akhurst, R.J. and Hata, A. (2012). Targeting the TGFβ signaling pathway in disease. *Nat. Rev. Drug Discov.* 11: 790–811.
175 Velaparthi, U., Darne, C.P., Warrier, J. et al. (2020). Discovery of BMS-986260, a potent, selective, and orally bioavailable TGFβR1 inhibitor as an immuno-oncology agent. *ACS Med. Chem. Lett.* 11 (2): 172–178.
176 Ugi, I. and Steinbruckner, C. (1960). Concerning a new condensation principle. *Angew. Chem.* 72: 267–268.
177 Ugi, I. (1962). The α-addition of immonium ions and anions to isonitriles coupled with secondary reactions. *Angew. Chem.* 74: 9–22.
178 Ugi, I. and Bodesheim, F. (1961). Isonitriles. VIII. Reaction of isonitriles with hydrazones and hydrazoic acid. *Chem. Ber.* 94: 2797–2801.
179 Brant, M. G., Goodwin-Tindall, J., and Stover, K. R. (2018). Identification of potent indoleamine 2,3-dioxygenase 1(Ido1) inhibitors based on a phenylimidazole scaffold. *ACS Medicinal Chemistry Letters* 9: 131–136.
180 Wade, M., Li, Y. C., and Wahl, G. M. (2013). MDM2, MDMX and p53 in oncogenesis and cancer therapy. *Nature Reviews Cancer* 13: 83–96.
181 Zhao, Y., Aguilar, A., Bernard, D. et al. (2015). Small-molecule inhibitors of the MDM2-p53 protein-protein interaction (MDM2 Inhibitors) in clinical trials for cancer treatment. *Journal of Medicinal Chemistry* 58: 1038–1052.
182 Twarda-Clapa, A., Krzanik, S., Kubica, K. et al. (2017). 1,4,5-Trisubstituted imidazolebasedp53–MDM2/MDMX antagonists with aliphatic linkers for conjugation with biological carriers. *Journal of Medicinal Chemistry* 60 (10): 4234–4244.
183 Bergeron, N., Phan, B. P., Ding, Y. et al. (2015). Convertase Subtilisin/Kexin type 9 inhibition: A New therapeutic mechanism for reducing cardiovascular disease risk. *Circulation* 132 (17): 1648–1666.
184 Stucchi, M., Grazioso, G., Lammi, C., Manara, S., Zanoni, C., Arnoldi, A., Lesma, G., and Silvani, A. (2016). Disrupting the PCSK9/LDLR Protein-Protein Interaction by an Imidazole-based Minimalist Peptidomimetic. *Org. Biomol. Chem.*, 14. 10.1039/C6OB01642A.
185 Akhurst, R. J. and Hata, A. (2012). Targeting the TGFβ signalling pathway in disease. *Nature reviews Drug discovery*, 11(10), 790–811. https://doi.org/10.1038/nrd3810.
186 Kim, B.G., Malek, E., Choi, S.H. et al. (2021). Novel therapies emerging in oncology to target the TGF-β pathway. *J. Hematol. Oncol.* 14, 55 https://doi.org/10.1186/s13045-021-01053-x
187 Vetrichelvan, M., Rakshit, S., Chandrasekaran, S., et. al. (2020). Development of a Scalable Synthesis of the Small Molecule TGFβR1 Inhibitor BMS-986260. *Org. Process Res. Dev.*, 24, 7, 1310–1320.

10

Multicomponent Syntheses of Heterocycles by Catalytic Generation of Alkynoyl Intermediates

Jonas Niedballa and Thomas J.J. Müller

Institut für Organische Chemie und Makromolekulare Chemie, Heinrich-Heine-Universität Düsseldorf, Universitätsstraße 1, D-40225 Düsseldorf, Germany

10.1 Introduction

Multicomponent reactions (MCRs) have proven to be among the most versatile and powerful concepts for the rapid and elegant construction of compound libraries in a straightforward and highly diverse fashion. MCRs are defined as a one-pot transformation where three or more reactants generate a single product by forming at least two new bonds. In addition, most of the deployed atoms of the starting materials are found in the products [1, 2]. As a consequence, the chemistry of MCRs relies on a reactivity-based concept, where reactive functionalities are created and transformed in each step in a strict one-pot process [3]. The urge for efficient and diversity-oriented syntheses underlines the relevance of MCRs as a superior tool, combining economic and ecological aspects, focusing simultaneously on chemo-, regio-, and stereoselectivity, which lead to high degrees of structural and functional diversity. To date, three different classes of MCRs are distinguished [4]. In the domino one-pot reaction, the process commences with all reagents being present from the beginning and all individual steps occur in a programmed order, rendering it impossible to isolate any occurring intermediary products. In the sequential, MCR components are added in a well-defined order without changing the reaction conditions. Therefore, intermediary products can be isolated in principle. The consecutive one-pot reaction, which is closely related to the sequential reaction, involves the subsequent addition of reagents as well, but with the option to vary or change conditions for each single step. All MCRs should begin with readily accessible and highly diverse starting materials to facilitate an entry to a broad range of physical and structural parameters, thus resulting in a high convergence and immense explorative potential. The broad functionalization of carbon–carbon triple bonds has been confirmed especially in the synthesis of heterocycles [5]. Therefore, the key to MCRs are relative reactivities of functional groups, which can be exploited by the catalytic generation of alkynes and has been more frequently studied in the past two decades [6].

Multicomponent Reactions towards Heterocycles: Concepts and Applications, First Edition.
Edited by Erik V. Van der Eycken and Upendra K. Sharma.

Heterocyclic systems are essential structural motifs in the field of MCR, due to their ubiquitousness in nature and chemical research. Besides myriads of application in life sciences, they increasingly find use in molecular electronics [7] and photonics [8–11], e.g. as organic field effect transistors (OFETs) [12, 13], as dye-sensitized solar cells (DSSCs),[14–16] or as organic light-emitting diodes (OLEDs) [17–20]. Especially multicomponent syntheses of fluorophores via metal catalysis [21] or utilization of alkynoyl intermediates[22] aroused highest interest in recent years. Hence, transition metal-catalyzed MCRs [23], which are based on the *in situ* formation of reactive electrophilic intermediates, like alkynones [24, 25], related alkyl propiolates, and alkyne-1,2-diones, received considerable interest (Scheme 10.1). Therefore, this review focuses on recent developments of catalytic alkynoyl generation and application to MCR heterocycle synthesis in a flashlight fashion, updating our previous overview [26].

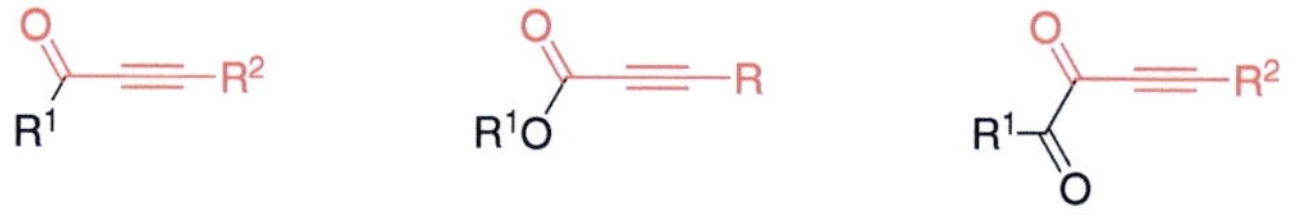

Scheme 10.1 Alkynones, alkyl propiolates, and alkyne-1,2-diones as alkynoyl functionalities under consideration.

10.2 Catalytic Generation of Alkynones

In the past decades, our group has established a variety of complementary methods for the synthesis of alkynones based on Sonogashira alkynylations. These versatile three-carbon building blocks act as crucial intermediates in the synthesis of five- to seven-membered heterocyclic ring systems [27]. Especially in the context of diversity-oriented synthesis, the interplay of six different comprehensive methods completes a powerful toolbox for the synthesis of alkynones (Scheme 10.2).

The modified Sonogashira cross-coupling **I** [28, 29] uses the palladium- and copper-catalyzed transformation of alkynes and acid chlorides [30, 31], enhancing the reaction conditions by reducing the amounts of triethylamine required to one stoichiometric equivalent. The inherent problem of alkyne homocoupling is circumvented by the copper-free variant **VI** [32] in an elegant fashion utilizing an electron-rich palladium complex. It is worth mentioning that the solvent can be varied in order to concatenate singular reactions to one-pot sequences. Toluene, 1,4-dioxane, dichloromethane, and acetonitrile are well tolerated, whereas the latter two proved to be highly effective for subsequent Michael additions. Utilizing oxalyl chloride for activation of carboxylic acids **V** [33] – thus allowing *N*-heterocyclic substituents – as well as electron-rich π-nucleophiles **III** [34] has proven to be an efficient pathway to significantly broaden the substrate spectrum.

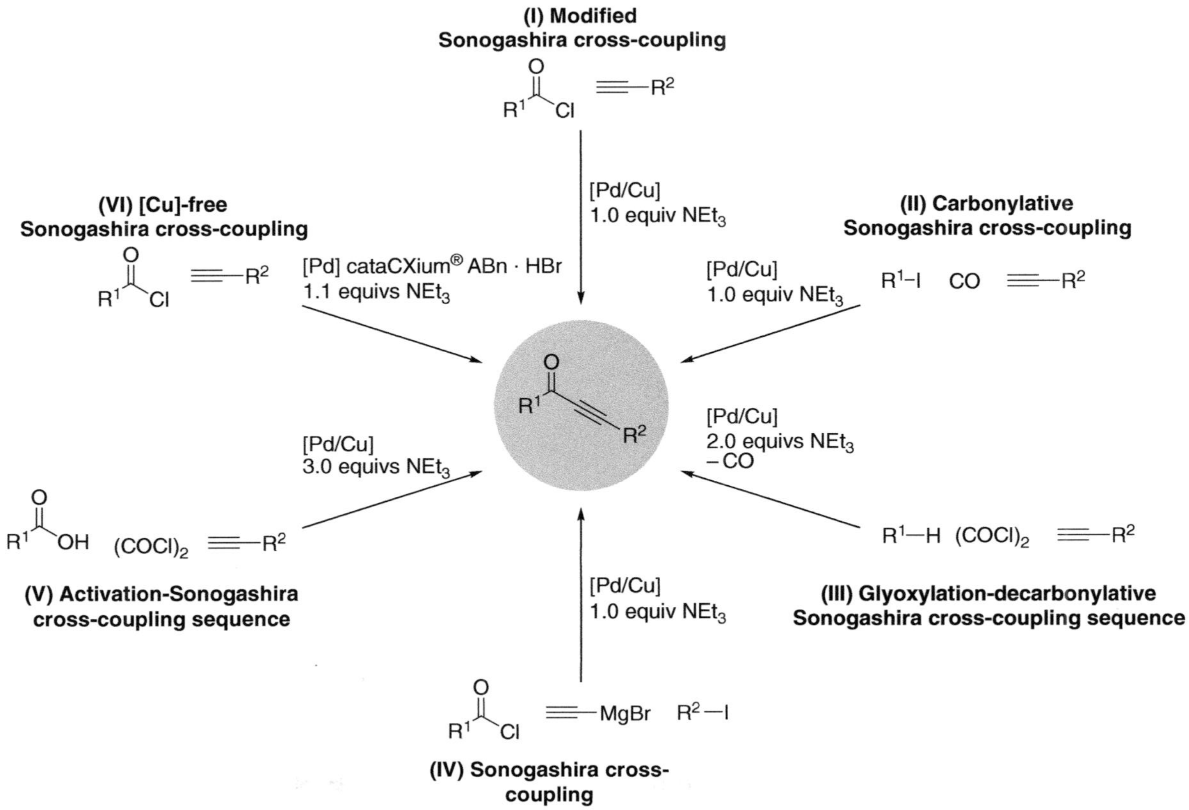

Scheme 10.2 Six Sonogashira cross-coupling-based protocols for the synthesis of alkynones.

While these approaches directly rely on the coupling of terminal alkynes, a rather unusual tool is disclosed by coupling of aryl halides and ethynylmagnesium bromide and subsequent Sonogashira coupling with acid chlorides **IV** [35]. Indeed, the substrate scope could be additionally increased for alkyne substrates replenishing possible pathways to unsymmetrical substituted alkynones. In the carbonylative cross-coupling Sonogashira reaction **II** [36] aryl iodides are as well utilized as starting materials, but the introduced aryl moiety is bound on the carbonyl side of the alkynone, thereby further expanding the synthetic routes to alkynones with a plethora of possible variations.

Another approach to ynones is presented by Zeng and coworkers, executing a palladium-catalyzed Sonogashira coupling of *N*-acylsaccharins, which proceeds via a selective C—N bond cleavage in the presence of triethylamine as a base in THF in good to excellent yield. This protocol retains from copper complexes and high-catalyst loadings resulting in 29 examples with a broad spectrum of substituents, encompassing electron-rich to electron-poor (hetero)aryls, allyl as well as alkyl groups (Scheme 10.3) [37].

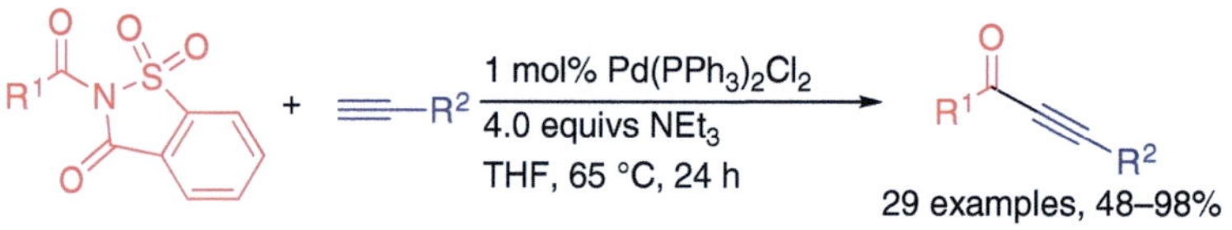

Scheme 10.3 Sonogashira coupling to ynones via C—N bond cleavage of *N*-acylsaccharin substrates. Source: Based on Cui et al. [37].

A general and direct approach from aryl halides and alkyl propiolates in a straightforward Sonogashira coupling reaction leads to 12 examples of aryl propiolates in 68–97% yield (Scheme 10.4). This novel access compensated the reduced alkyne reactivity and diminished stability of ethyl propiolate by applicating a syringe pump, thus circumventing its inherent propensity to undergo oligomerization under basic conditions [38].

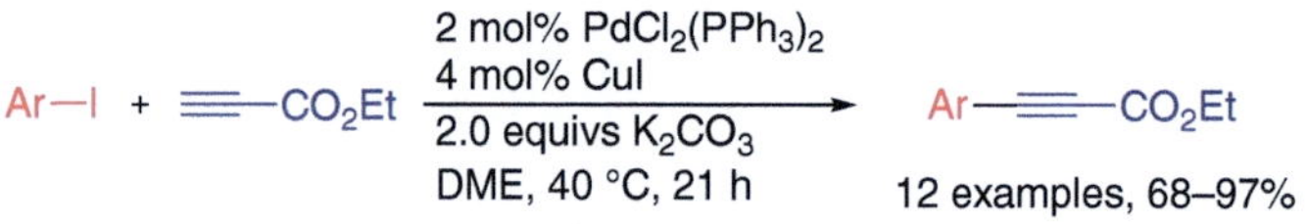

Scheme 10.4 Sonogashira synthesis of 3-aryl propiolates from ethyl propiolate and aryl iodides. Source: Based on Götzinger et al. [38].

A rapidly growing catalytic approach to ynones is the palladium nanoparticle coupling of acyl chlorides and terminal alkynes. This variation implies the advantage of copper-free and aerobic conditions [39] and additionally mild conditions without heating [40, 41]. Besides generally very high yields, the efficiency of these catalysts underlines by 4–6 recycling circles.

Ynediones are structurally related to ynones. Guo and coworkers presented a copper-catalyzed synthesis of ynediones from terminal alkynes and α-carbonyl

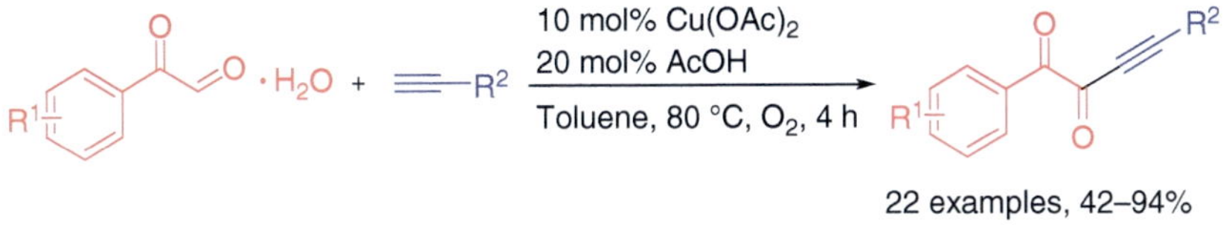

Scheme 10.5 Copper-catalyzed aerobic oxidative coupling to ynediones. Source: Zhou et al. [42].

aldehydes under oxygen atmosphere leading to 22 examples in moderate to excellent yields (Scheme 10.5) [42].

10.3 Multicomponent Syntheses of Five-membered Heterocycles

10.3.1 Pyrazolines

Among the most investigated heterocycles emerging from MCRs via alkynones are five-membered nitrogen containing pyrazoles and pyrazolines. These scaffolds are particularly attractive targets for a wide spectrum of applications, both in life and materials sciences [43–45]. By using oxalyl chloride as activation reagent, alkynediones become accessible by copper-catalyzed alkynylation. In a sequence of activation, alkynylation and cyclization glyoxylic acids, alkynes, and acylhydrazines react to give 1,5-diacyl-5-hydroxypyrazolines in a consecutive three-component fashion [46]. By this method, 17 examples of these pyrazolines were synthesized in 29–78% yield bearing different (hetero)aryl and alkyl groups at three points of diversity (Scheme 10.6).

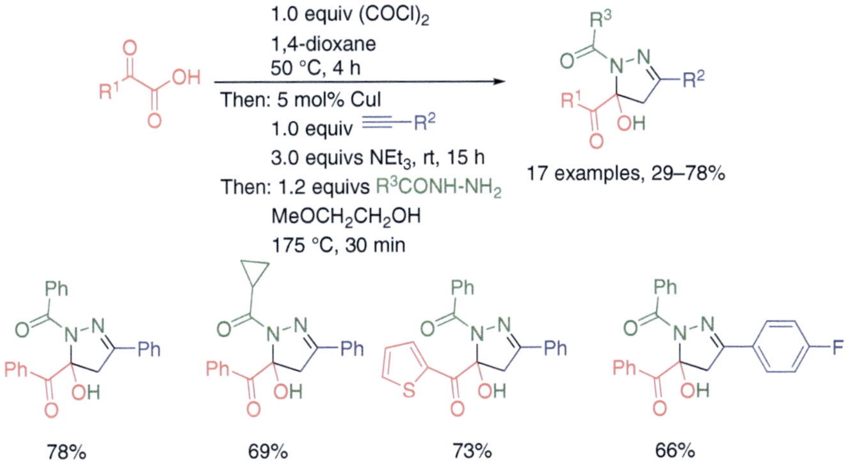

Scheme 10.6 Three-component activation-alkynylation-cyclocondensation synthesis of 1,5-diacyl-5-hydroxypyrazolines. Source: Görgen et al. [46].

10.3.2 Pyrazoles

Götzinger et al. established a novel four-component reaction starting with a sequentially Pd-catalyzed alkynone formation followed by cyclocondensation for accessing functional pyrazoles and pyrimidines in moderate to good yields (Scheme 10.7) [47]. In particular, donor-acceptor-substituted 3,5-diarylpyrazoles show interesting photophysical properties, such as huge emission solvatochromicity. Starting with 1,4-diidobenzene a pseudo-seven-component coupling–coupling-cyclocondensation synthesis furnishes a symmetrical dipyrazole. Benzamidine hydrochloride as a binucleophile precursor leads under conventional heating to six examples of pyrimidines in moderate yields.

Applying the Sonogashira reaction of aryliodides with ethyl propiolate and subsequent cyclocondensation with hydrazines furnishes 17 examples of 3-hydroxy pyrazoles in good to excellent yields (Scheme 10.8) [48]. The substrate scope of the aryl groups ranges from electron-poor to electron-rich (hetero)aryls, and methyl, alkyl, and benzyl substituted hydrazines underline the diversity of *N*-substituents on the 3-hydroxy pyrazoles. While the classic approach to 3-hydroxy pyrazoles relies on cyclocondensation of hydrazines and 1,3-diketones, where often an activation of the carbonyl group is required, 3-aryl propiolates as synthetic equivalents of β-keto esters circumvent this problem due to their higher electrophilicity. As previously mentioned, the crucial step is the slow addition of ethyl propiolate via syringe pump [49], due to its volatility and inherent sensitivity to undergo oligomerization under basic conditions.

Niesobski et al. presented a novel synthetic pathway to 4-pyrazolyl-1,2,3-triazoles utilizing (triisopropylsilyl)butadiyne as a four-carbon building block in a one-pot fashion and moderate yields [50]. The consecutive four-component sequentially Pd–Cu-catalyzed process proceeds via alkynylation, cyclocondensation, desilylation, and alkyne-azide cycloaddition, thereby combining Sonogashira coupling and Medal-Sharpless copper-click reaction with a single-catalyst source in a one-pot fashion. Starting from acid chlorides, 15 examples with a large variety of substituents were obtained (Scheme 10.9).

Cai and coworkers utilized 3-(2-aminoethylamino)propyl-functionalized MCM-41-immobilized palladium(II) complex (MCM-41-2N-Pd$(OAc)_2$) and CuI to couple acid chlorides with terminal alkynes in the presence of triethylamine at 50 °C giving ynones, which were converted into pyrazoles by the cyclocondensation of hydrazines at room temperature with acetonitrile as a cosolvent (Scheme 10.10) [51]. Thereby, 27 examples with a broad range of substituents were generated in 25–88% yield. The immobilized palladium catalyst could be recovered and reused for 10 consecutive trials without a significant depression of activity.

Another promising approach to 1,3,5-substituted pyrazoles was recently described by Thirukovela et al. where *in situ* generated Pd-nanoparticles (PdNPs) catalyze the Cu-free reaction of aroyl chlorides with aryl acetylenes in aqueous PEG-400 solution. The pyrazoles were generated by adding (phenyl)hydrazine to the reaction mixture, thus furnishing 19 examples of diarylated pyrazoles in good to excellent yields (Scheme 10.11). A benefit of this procedure is that homogeneously dispersed PdNPs

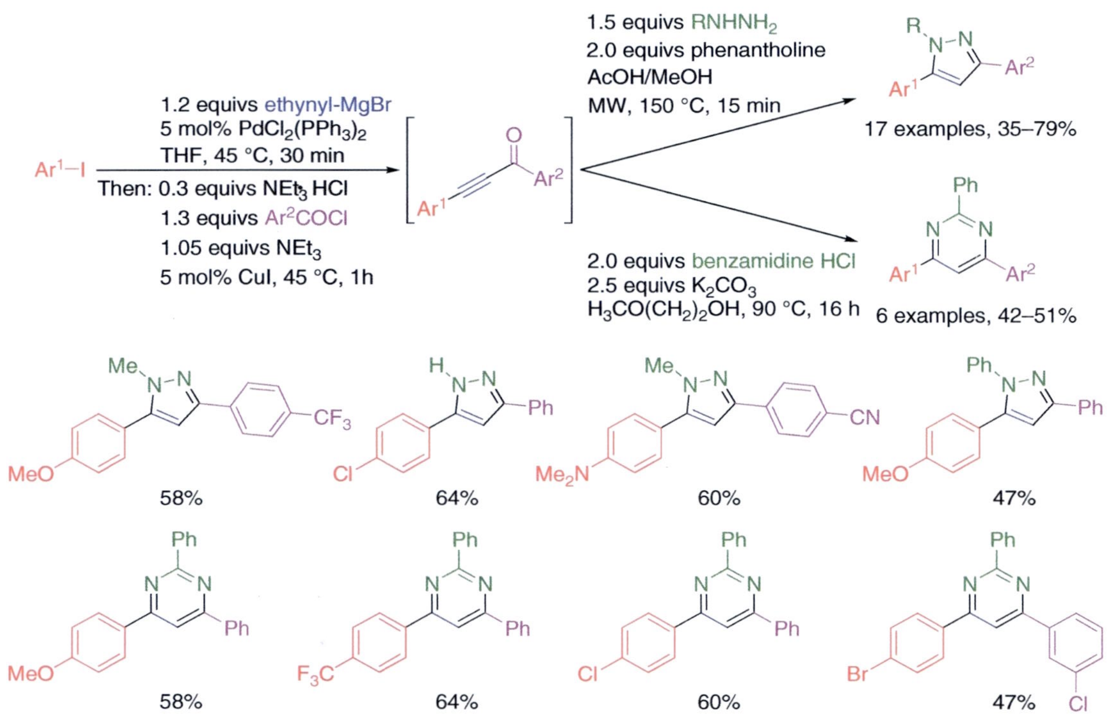

Scheme 10.7 Four-component coupling–coupling-cyclocondensation syntheses of pyrazoles and pyrimidines. Source: Based on Götzinger et al. [47].

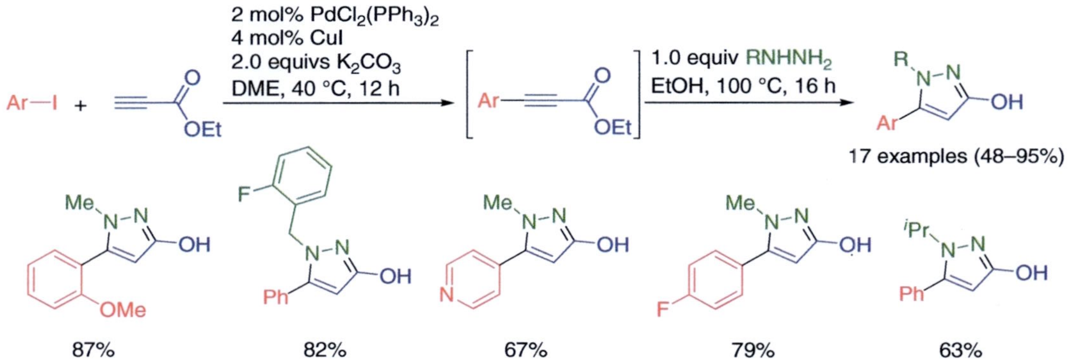

Scheme 10.8 Three-component alkynylation-cyclocondensation synthesis of 3-hydroxy pyrazoles. Source: Based on Niedballa et al. [48].

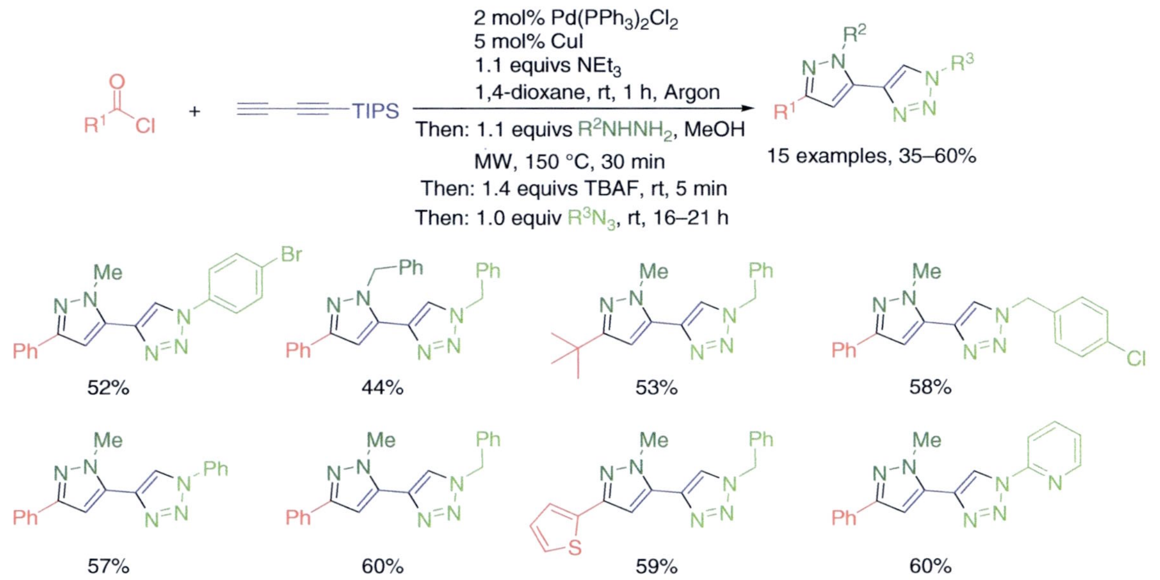

Scheme 10.9 Four-component alkynylation-cyclocondensation-desilylation-CuAAC synthesis of 4-pyrazolyl-1,2,3-triazoles. Source: Niesobski et al. [50].

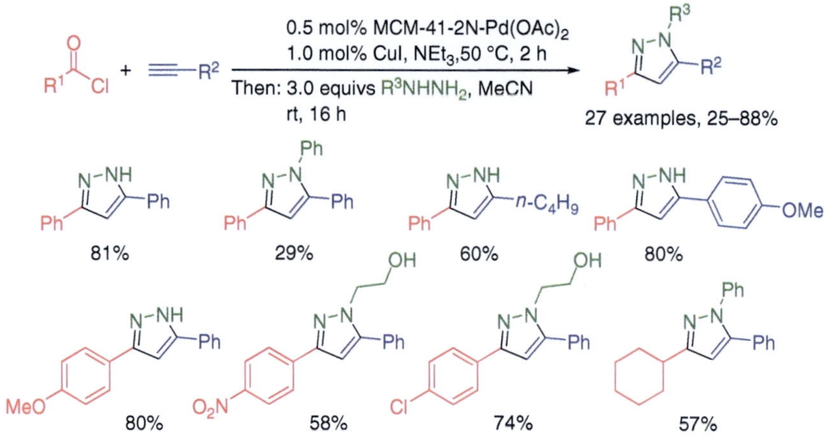

Scheme 10.10 Three-component alkynylation-cyclocondensation synthesis of pyrazoles. Source: Chen et al. [51].

exhibit notable ability to be recycled after the reaction depending on the conditions. The nanoparticles were reused up to five successive runs showing significant catalytic efficiency [52].

The same strategy was also employed in the three-component synthesis of isoxazoles, where hydroxylammonium chloride and sodium acetate were added to the mixture reacting with the *in situ* generated alkynone and isoxazoles were obtained in 76–92% yield (Scheme 10.11).

10.3.3 Isoxazoles

3,5-Di(hetero)aryl-substituted isoxazoles can be synthesized in a consecutive three-component sequence consisting of alkynylation and cyclization starting from aroyl chloride, alkyne, and sodium azide with acetic acid under copper-free palladium catalysis. The novelty of this procedure is the utilization of an azide at room temperature, preserving the oxygen of the aroyl chloride in the final product, as well as the considerably low catalyst loading of the palladium catalyst. It is worth mentioning that in the presence of azide anions, Cu(I) forms a Cu(I)N_3 complex, which loses nitrogen to give a copper nitride, allowing a transformation of the alkynone to give enaminones. As a consequence, the alkynylation must be performed under copper-free conditions giving rise to nine examples in 10–72% yield (Scheme 10.12) [53].

10.3.4 Triazoles

The synthesis of 2-(1,2,3-triazolyl)benzamides was achieved by using a copper(I)-catalyzed MCR of 2-iodobenzamides, sodium azide, and terminal alkynes in the sense of a sequentially copper-catalyzed azidation-alkyne-azide cycloaddition process. This one-step reaction proceeds with short reaction times. It is noteworthy

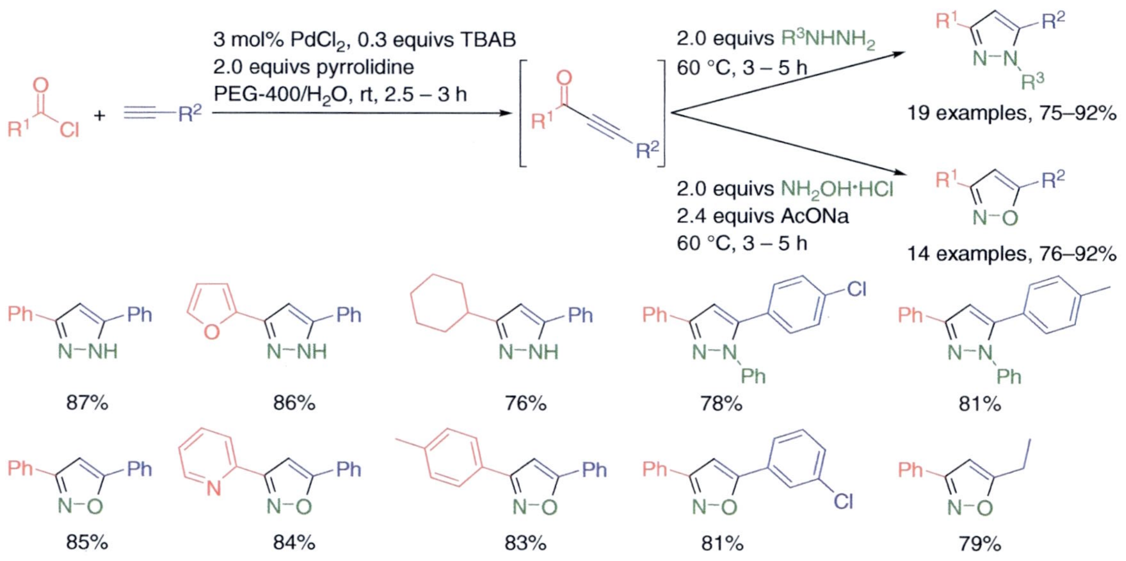

Scheme 10.11 Three-component alkynylation-cyclocondensation syntheses of substituted pyrazoles and isoxazoles. Source: Based on Thirukovela et al. [52].

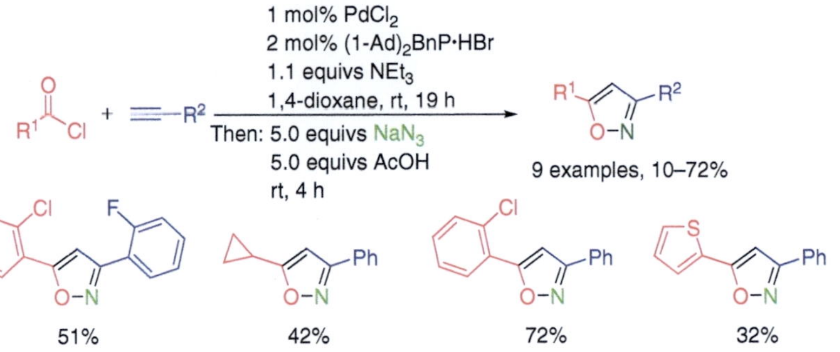

Scheme 10.12 Three-component alkynylation-cyclocondensation synthesis of 3,5-disubstituted isoxazoles. Source: Based on Görgen et al. [53].

that the absence of base in the mixture was crucial for the successful transformation. 2-Iodobenzamides containing electron-donating as well as electron-withdrawing groups and a variety of terminal alkynes with aryl or alkyl groups were applicable in this reaction leading to 21 examples in good to excellent yield (Scheme 10.13) [54].

A different synthetic strategy was applied by Schreiner et al. introduced propargyl bromide as a three-carbon building block and substrate for the terminating Cu-catalyzed alkyne-azide cycloaddition (CuAAC). In this case, the sequentially catalyzed alkyne carboxylation-propagylation-CuAAC process furnished 15 examples of 1,2,3-triazolylmethylarylpropiolates in a four-component synthesis in 28–67% yield. The process could be even further expanded to a five-component synthesis of 1,2,3-triazolylmethyl 3-amino 3-arylacrylates by concatenating a concluding Michael addition (Scheme 10.14). The process takes advantage of mild reaction conditions of 50 °C including two sequentially Cu-catalyzed steps with the same catalyst source, forming four new bonds resulting in a yield per bond-forming step of 73–89% [55].

10.3.5 Thiophenes

Thiophenes are important heterocyclic building blocks in natural products, pharmaceutical active compounds [56] as well as in materials for electronic and optoelectronic devices [57]. The considerable demand for efficient synthetic strategies to these compounds has been recently reviewed with a focus on ring-forming MCRs [58]. A pseudo five-component synthesis of diethyl terthiophene-5,5″-dicarboxylates was previously disclosed by Teiber [59], beginning from 2,5-bis(trimethylsilyl-ethynyl) thiophene as a stable storage form of the sensitive 2,5-diethynyl thiophene. *In situ* desilylation and quick aqueous work up provided the diyne, followed by Sonogashira coupling with acid chlorides gave dialkynones that were transformed via Fiesselmann reaction with ethyl 2-mercaptoacetate in a one-pot fashion to give three terthiophene diesters in moderate to good yield (Scheme 10.15) [60].

The bisethyl dicarboxylates were converted into the corresponding amides, leading to 11 examples of 5,5″-diamide substituted terthiophenes in 64–99% yield. For

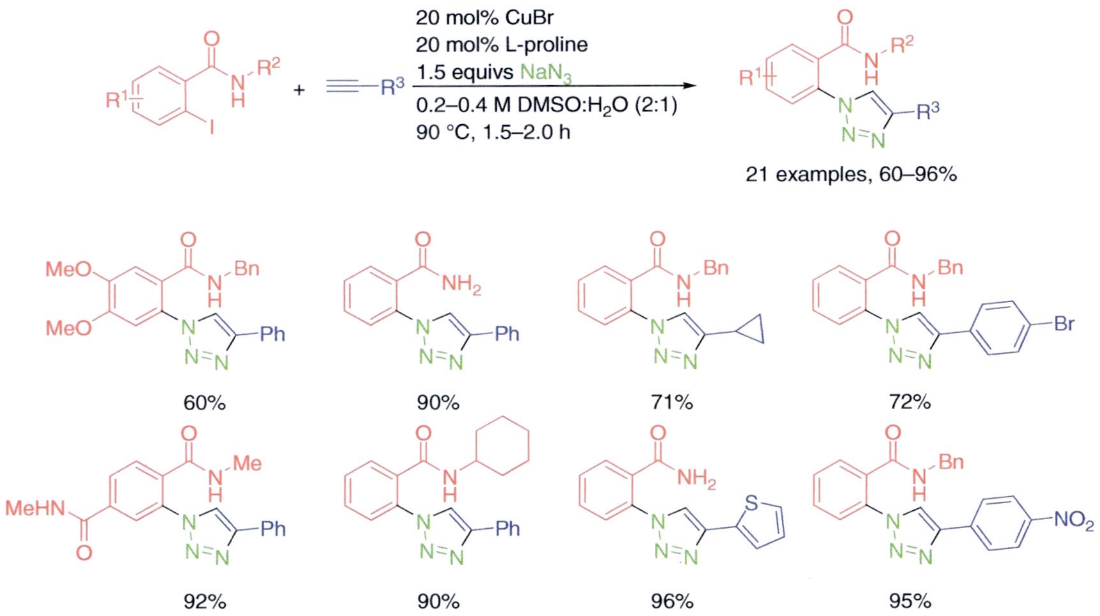

Scheme 10.13 Copper-catalyzed azidation-alkyne-azide cycloaddition synthesis of 2-(1,2,3-triazolyl)benzamide. Source: Based on Hayeebueraheng et al. [54].

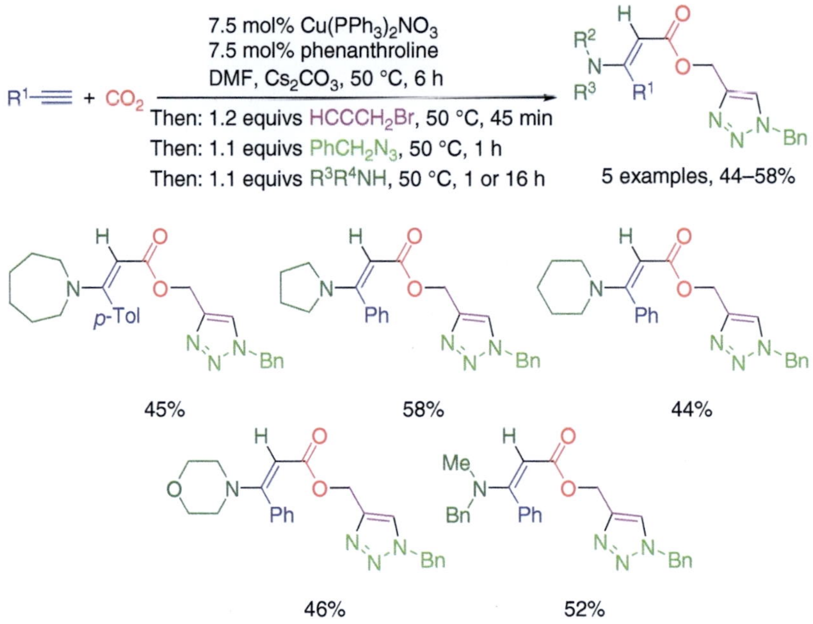

Scheme 10.14 Five-component carboxylation-propargylation-CuAAC-Michael addition synthesis of 1,2,3-triazolylmethyl 3-amino 3-arylacrylates. Source: Based on Schreiner et al. [55].

illustration, one example was synthesized by a pseudo-seven component reaction in a one-pot fashion. Upon increasing the loading of $AlCl_3$ and amine as well as the reaction time of the amidation step, an overall yield of 54% was obtained. The electronic properties of the synthesized 5,5′′-diacceptor substituted terthiophenes reveal reversible oxidation potentials between 300 and 1550 mV qualifying these dyes as redox-switchable systems. In addition, they display intensive blue luminesce in solution, opening further possibilities for employing them in bioanalytical or surface labeling.

10.3.6 Indolones

Conformationally rigidified and often intensively emissive spiro-indolones are generated from intermediary formed propynylidene indolones, which are generated from terminal alkynes and *N*-halophenylalkynylamides by insertion-alkynylation sequence. The domino synthesis for 24 titled examples in 34–99% yield has been performed by Schönhaber et al. applying a broad range of substituents (Scheme 10.16) [61].

The plausible mechanistic rationale starts with an oxidative addition of amide **I** to a Pd(0) complex (Scheme 10.17). The *syn*-carbopalladation of the tethered alkynoyl moiety of the formed aryl-Pd complex **II** furnishes the *E*-configured vinyl-Pd complex **III**. Transmetallation of the Cu-alkynyl complex **VI** to **III** results in the *E*-configured vinyl-alkynyl-Pd complex **VII**, which gives rise to the *E*-configurated

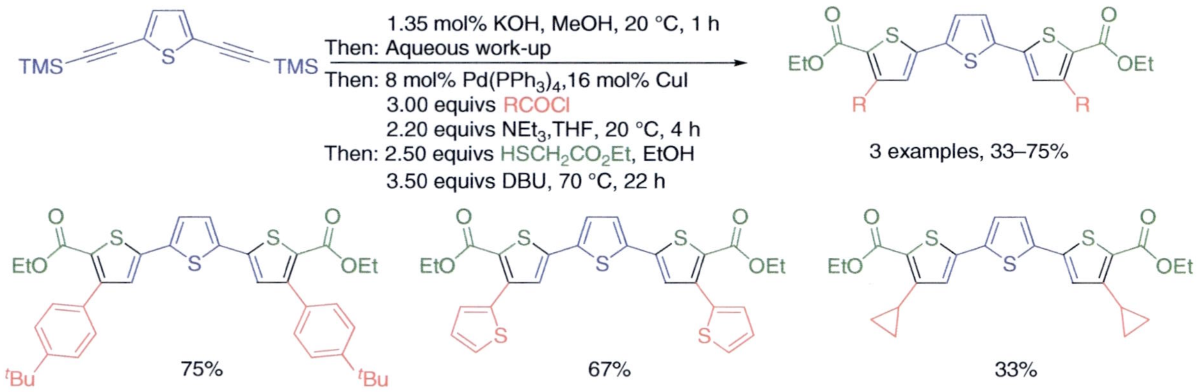

Scheme 10.15 Pseudo five-component coupling-cyclocondensation synthesis of diethyl terthiophene-5,5′′-dicarboxylates. Source: Breuer and Müller [60].

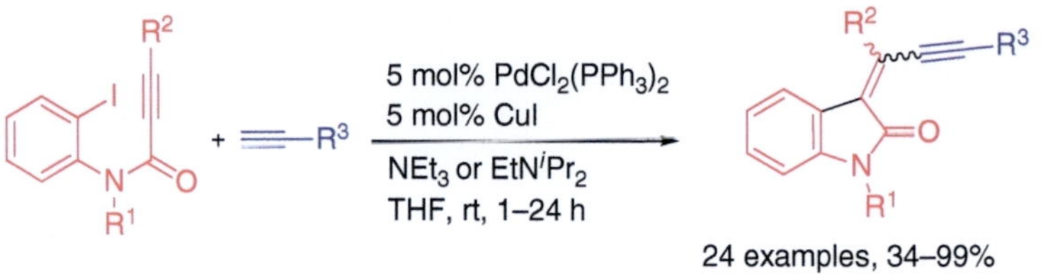

Scheme 10.16 Domino insertion-alkynylation synthesis of propynylidene indolones. Source: Based on Schönhaber et al. [61].

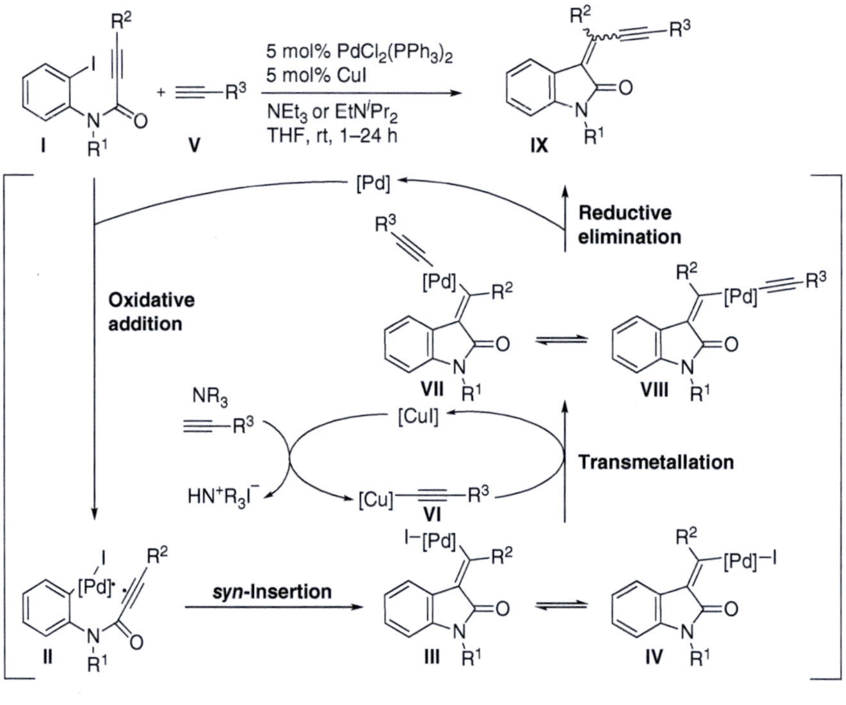

Scheme 10.17 Mechanistic proposal for the domino insertion-alkynylation synthesis of propynylidene indolones. Source: Based on Schönhaber et al. [61].

products **IX** after reductive elimination. The observed *Z* selectivity can be accounted to a stereomutation occurring after the *syn* insertion leading to intermediates **IV** and **VIII** and *Z*-configurated products **IX**. Indeed, propynylidene indolones were isolated with *E*/*Z* ratios varying from 100 : 0 to 0 : 100.

Based on the presented domino synthesis of propynylidene indolones, a consecutive three-component synthesis of 3-piperazinyl propenylidene indolones was established. While the Pd/Cu-catalyzed sequence was originally performed in THF/EtOH as a solvent system, it was later found that dichloromethane and acetonitrile at elevated temperatures lead to comparable yields, however, with considerably shorter reaction times. In this study, 12 examples of propenylidene indolones were obtained in 10–95% yield with a reasonably broad substitution pattern and three points of diversity, ranging from electron-donating to electron-withdrawing substituents on the alkynyl aryl moieties (Scheme 10.18) [62].

This concept for 3-piperazinyl propenylidene indolone merocyanines was expanded to merocyanine-triarylamine bichromophores, superseding boc-protected piperazine with various substituted analogs. Starting from *ortho*-bromo anilides and phenylacetylene, a consecutive insertion-alkynylation-Michael addition sequence was accomplished. In a one-pot three-component synthesis, four examples of 3-piperazinyl propenylidene indolone bichromophores were obtained in 14–69% yield (Scheme 10.19) [63].

An alternative modular strategy employs 1-(4-bromobenzyl)piperazine as amino component, enabling a terminal Suzuki coupling, leading to four 3-piperazinyl propenylidene indolone bichromophores in 30–64% yield in a consecutive four-component reaction (Scheme 10.20). It is worth mentioning that no further catalyst loading is necessary after the catalytic insertion-coupling step. Most interestingly, white light emissive systems based on aggregation-induced dual emission operated by partial energy transfer between both chromophore units are observed and characterized.

10.4 Multicomponent Syntheses of Six-membered Heterocycles

10.4.1 Pyranones

In recent years, one-pot accesses to six-membered heterocycles have received considerable interest due to their promising properties [64, 65]. Heterocyclization to α-pyrone derivates can proceed by starting from (hetero)aroyl chlorides and terminal alkynes to furnish alkynones, which subsequently react with malonates giving these unsaturated lactones. This methodology gave entry to 14 pyranones bearing aryl substituents with electron-donating and electron-withdrawing groups alike as well as heterocyclic groups in yields ranging from 29% to 80% (Scheme 10.21) [66]. It is worth mentoring that the catalyst loading could be considerably lowered to 0.25 mol% $PdCl_2(PPh_3)_2$ and 0.50 mol% CuI deviating from standard alkynone syntheses.

Furthermore, a consecutive four-component alkynylation-Michael addition-cyclocondensation-ammonolysis synthesis of α-pyridones based on the established one-pot protocol for α-pyridones was achieved by addition of an excess of ammonia in the final step and heating at 90 °C for four hours. α-Pyridone products predominantly without ester functionality were obtained because of the concomitant acyl cleavage in moderate yields.

10.4.2 Pyridines

The catalytic ynone formation can equally well applied to consecutive three- and four-component coupling-Bagley–Bohlmann–Rahtz syntheses of tri- and

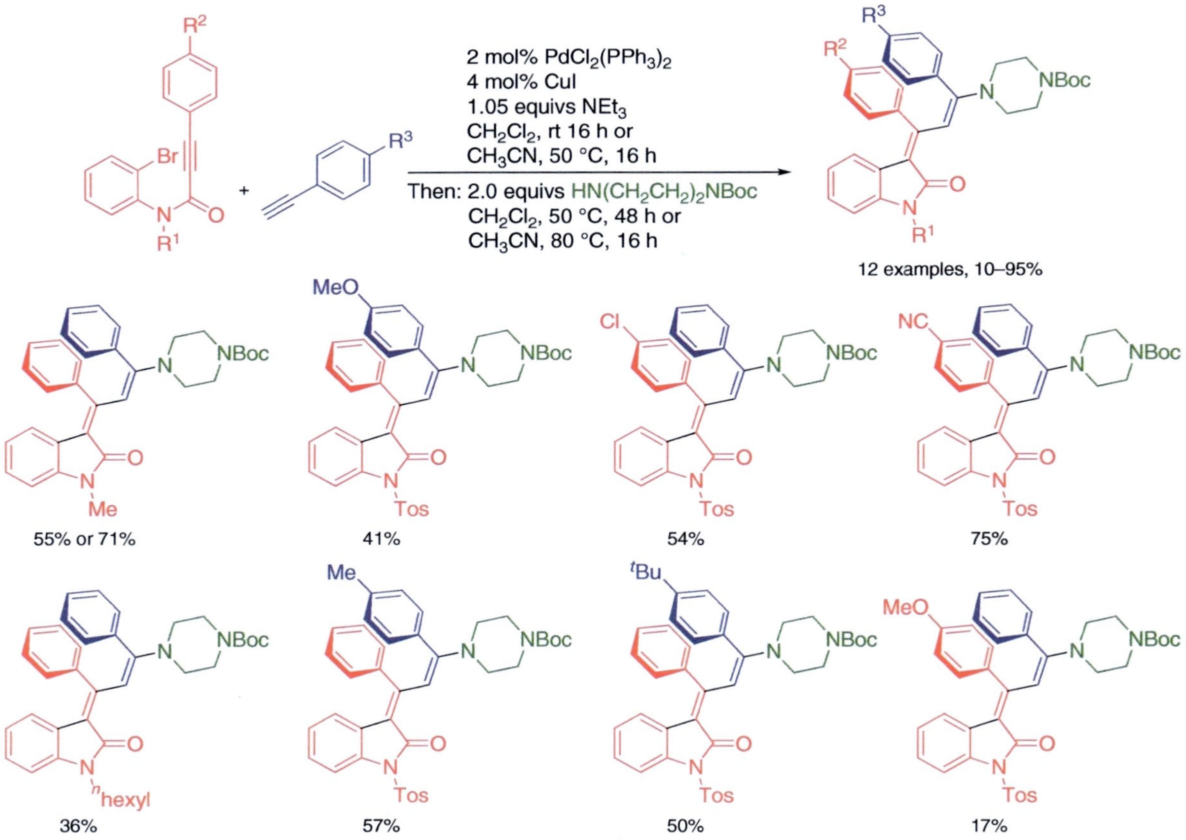

Scheme 10.18 Consecutive three-component insertion-alkynylation-Michael addition synthesis of 3-piperazinyl propenylidene indolone merocyanines. Source: Denißen et al. [62].

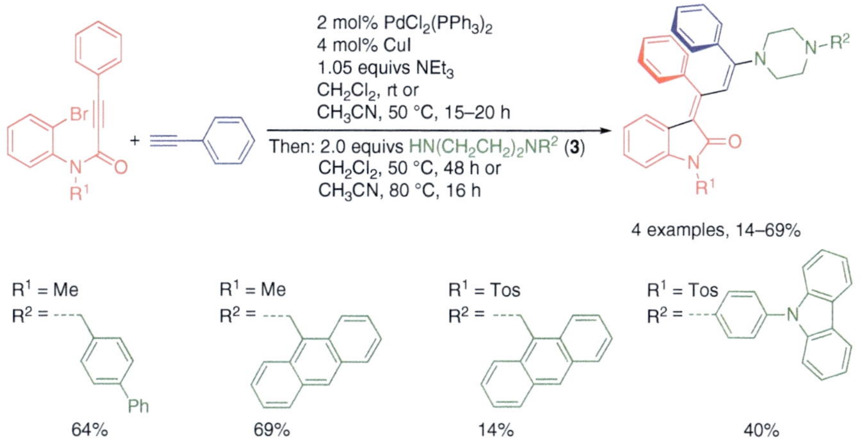

Scheme 10.19 One-pot three-component synthesis of 3-piperazinyl propenylidene indolone bichromophores. Source: Denißen et al. [63].

tetrasubstituted pyridines. The formed ynone was subsequently reacted with ethyl 3-aminocrotonate and acetic acid at 100 °C for 24 hours to give 15 3-ethoxycarbonyl 2-methylpyridines in 19–55% yield, while the yield per bond-forming step accounts to 57–82% (Scheme 10.22) [67].

Employing ethyl acetoacetate and ammonium acetate, respectively, as third and fourth components also leads to tri- and tetrasubstituted pyridines in moderate yields. Considering, that four new bounds are being formed, the yield per bond-forming step accounts to 68–83%.

Another interesting multicomponent approach to pyridines starting with the catalytic alkynone formation is presented by Bakulina et al. The generated ynones are reacted by subsequent addition of 1,8-diazabicyclo[5.4.0]undec-7-ene (DBU) and aluminium chloride to the reaction mixture, to give tricyclic 2-aminopyridinium salts. It was possible to isolate 15 examples in mostly good yield, with a large variety of substituents both for acid chloride and alkyne components (Scheme 10.23). The structures were additionally corroborated by X-ray structure analyses [68].

The structure of the bicyclic amidines was also varied in the sequence employing hexahydro-2*H*-pyrido[1,2-a]pyrimidine, 1,5-diazabicyclo[4.3.0]non-5-ene (DBN), and tetrahydro-3*H*-pyrrolo[1,2-a]imidazole. All products are intensely blue to turquoise luminescent in solution and both hygroscopic and highly soluble in water and organic solvents. With exception of one example (R^1 = *p*-BrC_6H_4, R^2 = biphenyl), all derivatives are potent luminophores that can be applied under physiological conditions.

10.4.3 Pyrimidines

With 5′-ethynyllappaconitine, Cheremnykh et al. chose a quite unusual substrate for the alkynone coupling (Scheme 10.24). Nevertheless, subsequent cyclocondensation with acetamidine hydrochloride or guanidine carbonate under consecutive one-pot

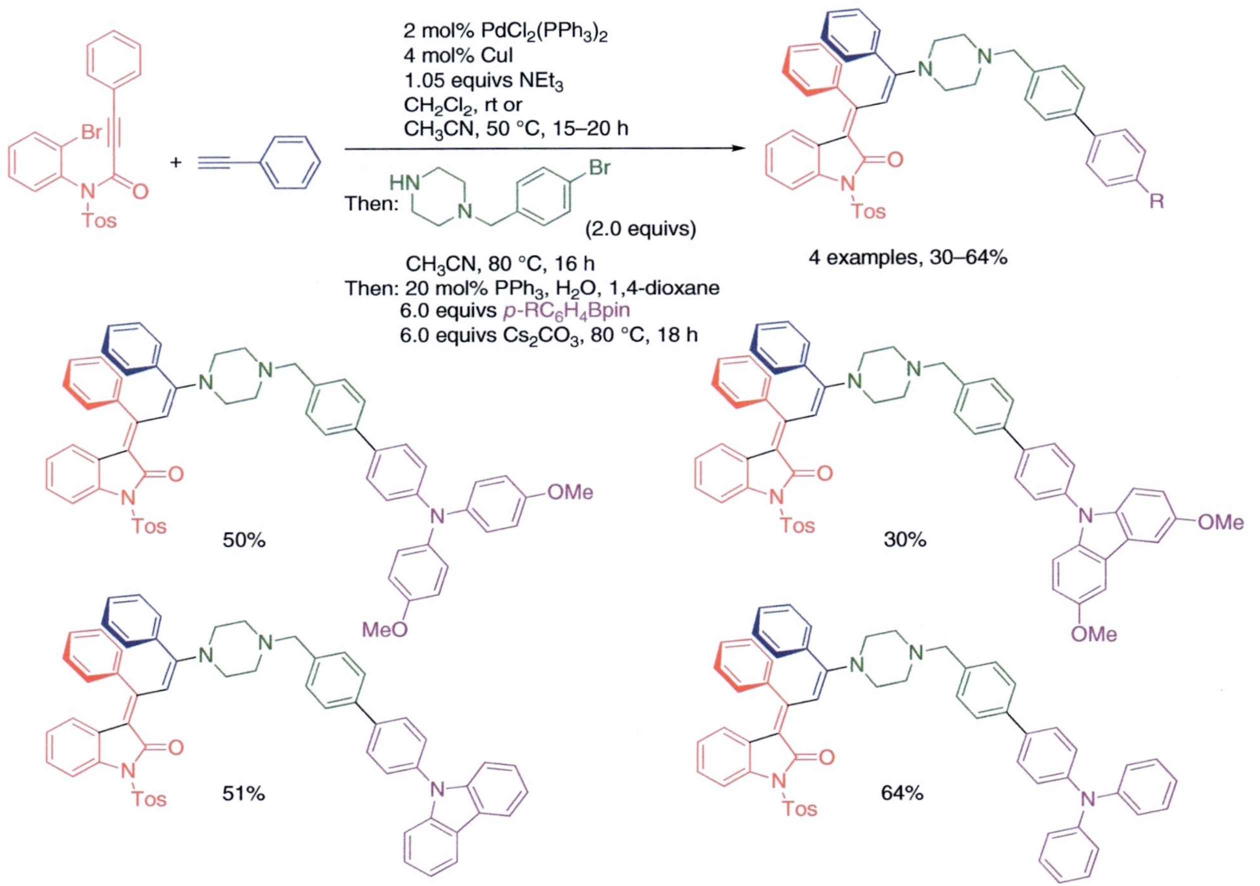

Scheme 10.20 Consecutive four-component insertion-alkynylation-Michael addition-Suzuki synthesis of 3-piperazinyl propenylidene indolone bichromophores. Source: Denißen et al. [63].

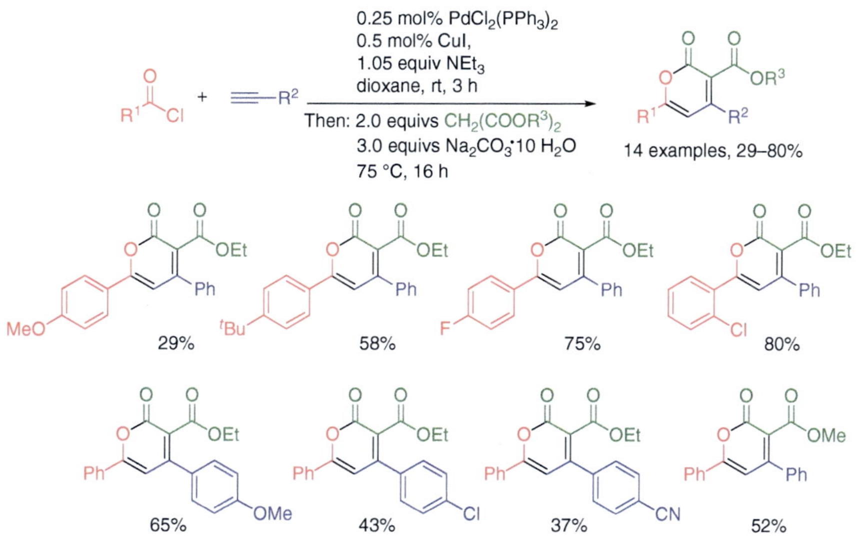

Scheme 10.21 Three-component coupling-cyclocondensation synthesis of α-pyridones. Source: Breuer and Müller [66].

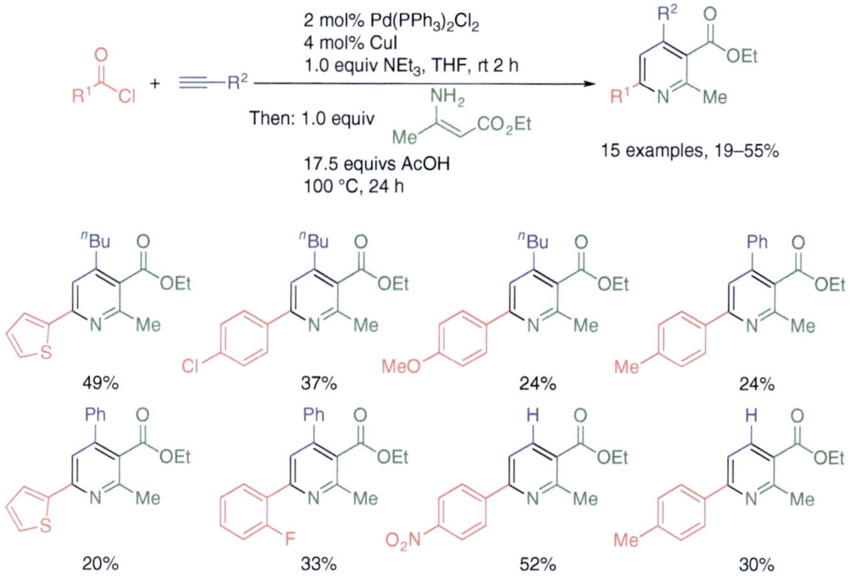

Scheme 10.22 Consecutive three- and four-component coupling-cyclocondensation syntheses of tri- and tetrasubstituted pyridines. Source: Dohe and Müller [67].

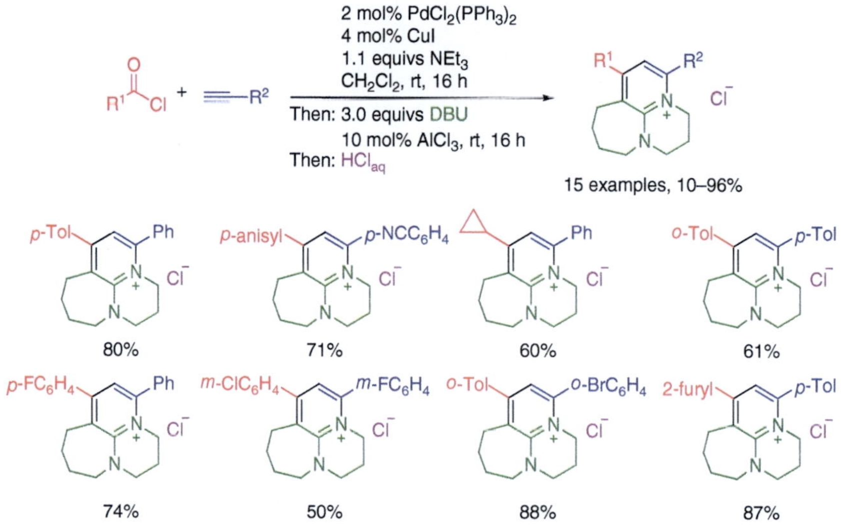

Scheme 10.23 Three-component coupling-(3+3)-anellation synthesis of tricyclic 2-aminopyridinium salts. Source: Based on Bakulina et al. [68].

conditions result in hybrid structures containing diterpene alkaloid and pyrimidine fragments. Two examples were obtained in 70–75% yield, however, in a two-step fashion evaporation of benzene under reduced pressure before adding acetonitrile for the cyclocondensation step [69]. Seven additional examples were obtained by a two-step protocol with intermediate isolation of the alkynones.

10.4.4 Oxazaborinines

Subsequent amine addition to catalytically generated alkynones in the presence of boron trifluoride as an activating Lewis acid results in a diversity-oriented four-component one-pot approach to difluoro oxazaborinines with 29 examples in 5–84% yield (Scheme 10.25). The domino formation with electron-rich amines was hampered but circumvented by performing the enaminone generation prior to the addition of boron trifluoride. Heteronuclear ^{11}B and ^{19}F NMR spectra additionally support the formation of oxazaborinine structures. The title compounds display solid-state emission under ultraviolet (UV) light, while being nonemissive in solution [70].

10.4.5 Coumarines

In a consecutive three-component coupling-addition sequence starting from triflyl coumarins, terminal alkynes and secondary amines, 23 examples of *E*-configured merocyanines were synthesized in 23–90% yield (Scheme 10.26). This process takes advantage of a viniyloguous Michael addition of the amino component. All three reactants and the auxiliary base are employed in an almost strictly equimolar ratio.

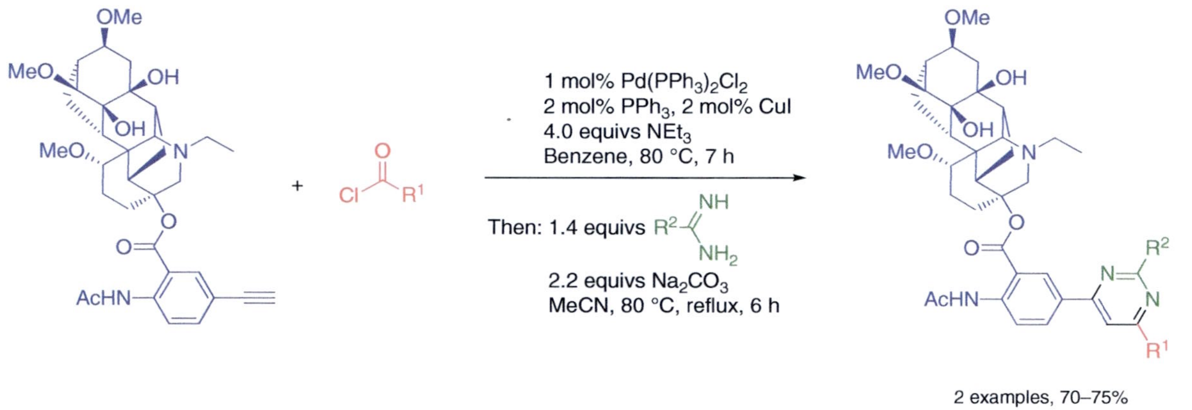

Scheme 10.24 Two-step coupling-cyclocondensation synthesis of alkaloid functionalized pyrimidines. Source: Based on Cheremnykh et al. [69].

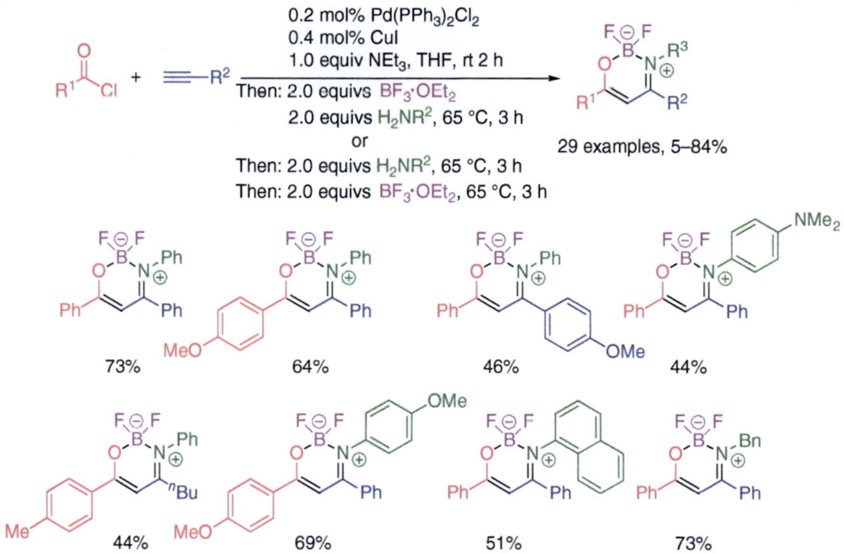

Scheme 10.25 Four-component coupling-addition-borylation synthesis of difluoro oxazaborinines. Source: Dohe et al. [70].

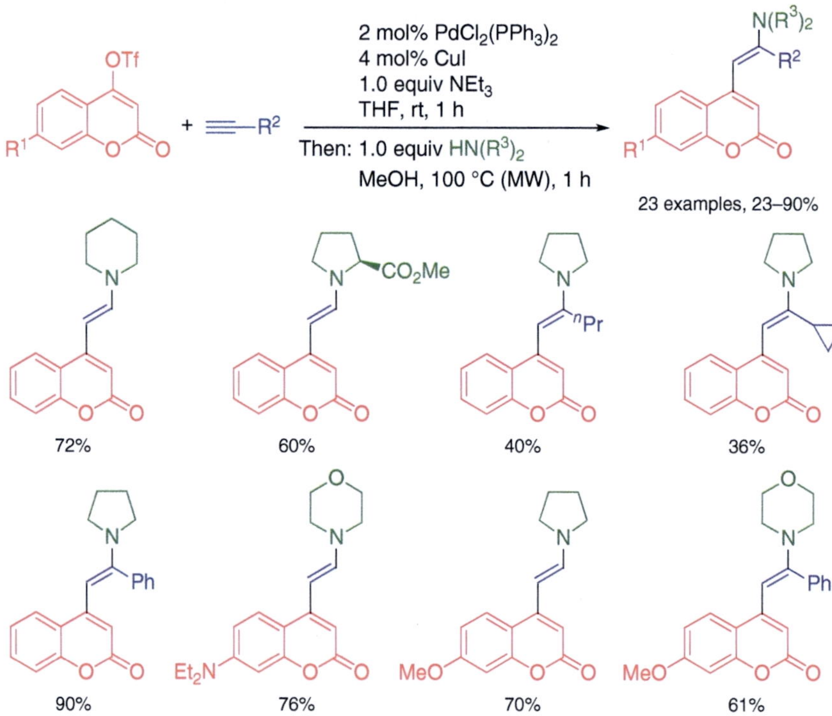

Scheme 10.26 Three-component alkynylation-addition synthesis of coumarin-based merocyanines. Source: Papadopoulos et al. [71].

Neutral to electron-rich coumarins, alkyl, silyl, and aryl acetylenes and acyclic and ali(hetero)cyclic secondary amines are well tolerated as substrates [71].

10.4.6 Quinolines

The Ugi four-component reaction (Ugi-4CR) is one of the most prominent domino MCRs inspiring many fields of research to date [72]. A diversity-oriented approach to highly functionalized 3-hydroxyisoquinolines was reported by the groups of Riva, Müller and coworkers starting with a Ugi-4CR resulting in eight examples of isoquinolinones that were synthesized in a one-pot fashion in moderate to excellent yield (Scheme 10.27). The final aromatization to the desired 3-hydroxyisoquinoline can also proceed in a one-step process. However, under the same optimized conditions, the yields crucially depended on the nature of starting materials. After careful investigation, it was concluded that the first three steps (Ugi/Heck/DMB removal) worked well for all substituents, while the final aromatization requires a fine-tuning of the conditions for each substrate. Therefore, often the two-step protocol proves to be the process of choice [73].

Eight examples of isoquinolinoles were obtained in 52–84% yield showing intensively blue luminescence upon excitation (Scheme 10.27).

Inspired by this process, a five-component synthesis of furoisoquinolines was developed and illustrated by 15 examples in 10–83% yield (Scheme 10.28). The rather unusual formation of the titled product can be explained by starting with an Ugi-4CR of aryl aldehyde, propiolic acid, isonitrile, and 2,4-dimethoxybenzylamine followed by an insertion-alkynylation reaction, debenzylation, and a base-catalyzed terminating 5-*exo-dig* cycloisomerization [74].

10.4.7 Quinoxalines

Two complementary diversity-oriented approaches for the synthesis of 3-ethynylquinoxaline derivatives were disclosed by our group [75]. First, glyoxylic acid chlorides are generated by electrophilic glyoxylation of electron-rich heteroarenes and subsequently converted through copper-catalyzed Stephens–Castro alkynylation into ynediones. The subsequent Hinsberg cyclization with 1,2-diaminoarenes produces 3-ethynylquinoxalines, where the triple bond can be smoothly addressed by a concluding Michael addition with secondary amines furnishing 10 examples of 3-(2-aminovinyl)quinoxalines in 46–72% yield (Scheme 10.29) [76].

The photophysical properties of the title compounds display solvatochromicity and protochromicity underlining the charge transfer character of absorption and emission. In addition, hydrogen bonding and protonation induce fluorescence quenching.

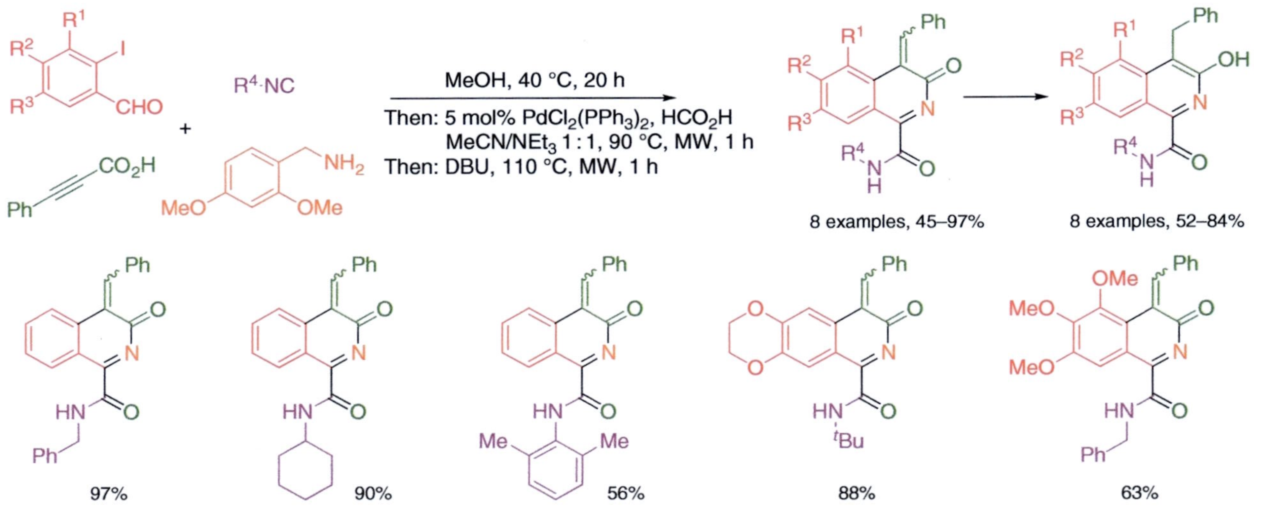

Scheme 10.27 Synthesis of 3-hydroxyisoquinolines by sequential Ugi four-component reaction/reductive Heck cyclization. Source: Based on Moni et al. [73].

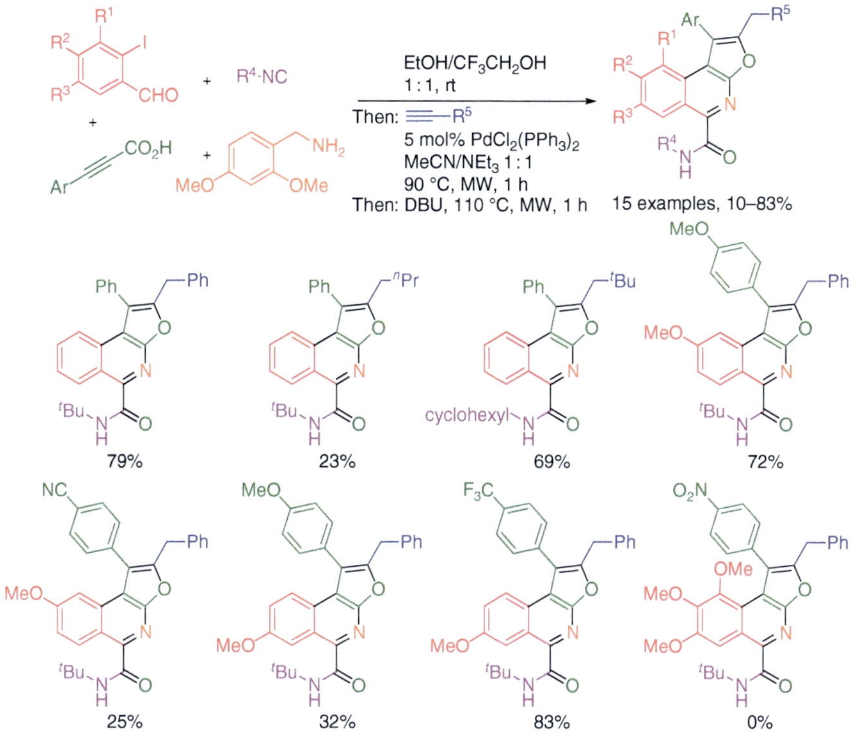

Scheme 10.28 Consecutive five-component Ugi-insertion-alkynylation-deprotection-cycloisomerization synthesis of furoisoquinolines. Source: Moni et al. [74].

For more general application, the consecutive activation-alkynylation-cyclocondensation synthesis 3-ethynylquinoxaline chromophores take advantage of a direct activation of glyoxylic acids furnishing 24 examples in 28–100% yield (Scheme 10.30). The emission solvatochromicity covers a broad spectral range from blue-green to deep-orange by introducing in 2-position *N*,*N*-dimethylaniline as a strong donor [77].

This methodology allows preparing 5-bromothienyl derivatives in a three-component reaction furnishing the corresponding 3-ethynylquinoxalines in 74–91% yield (Scheme 10.31). As pivotal points, these products enable level 2 functionalizations to extended π-system libraries by applying Suzuki and Buchwald–Hartwig coupling [78].

A further expansion of the diversity-oriented synthesis of quinoxalines is possible via CuAAC as terminating step utilizing the *in situ* formed terminal triple bond. This deprotection is easily achieved by the addition of potassium fluoride in the presence of azides and their conversion to give 18 examples of 3-triazolylquinoxalines in 38–82% yield (Scheme 10.32). The average yields per bond-forming step of around 90% underlining the superiority of this one-pot concept over possible stepwise transformations [79].

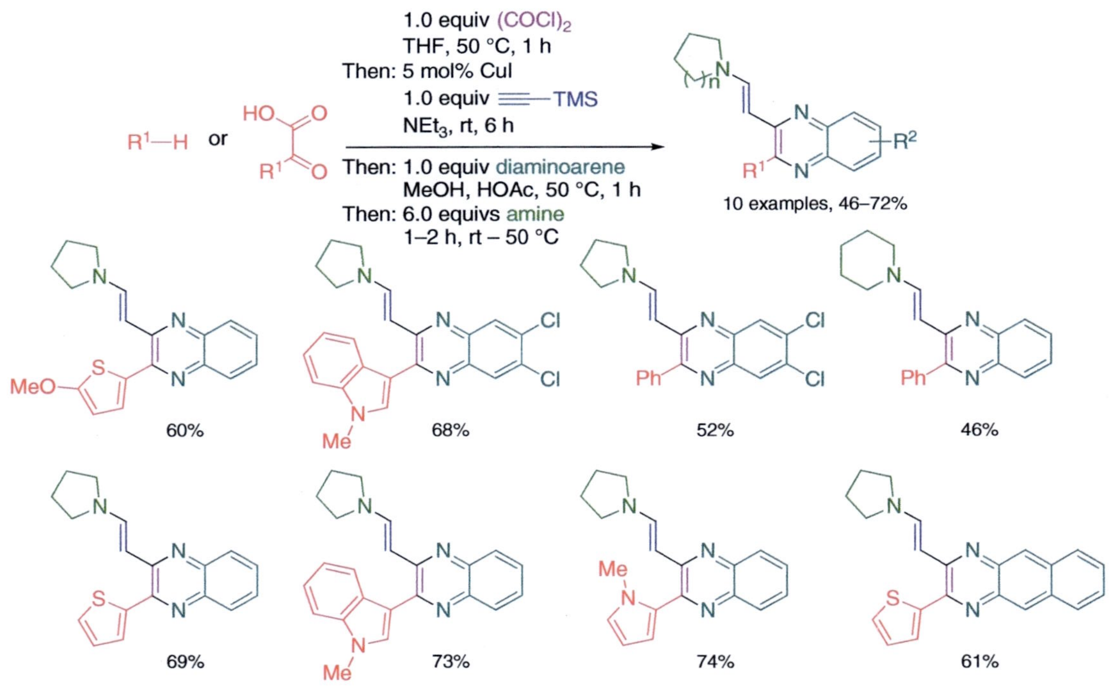

Scheme 10.29 Consecutive four- and five-component syntheses of 3-(2-aminovinyl)quinoxalines. Source: Gers-Panther et al. [76].

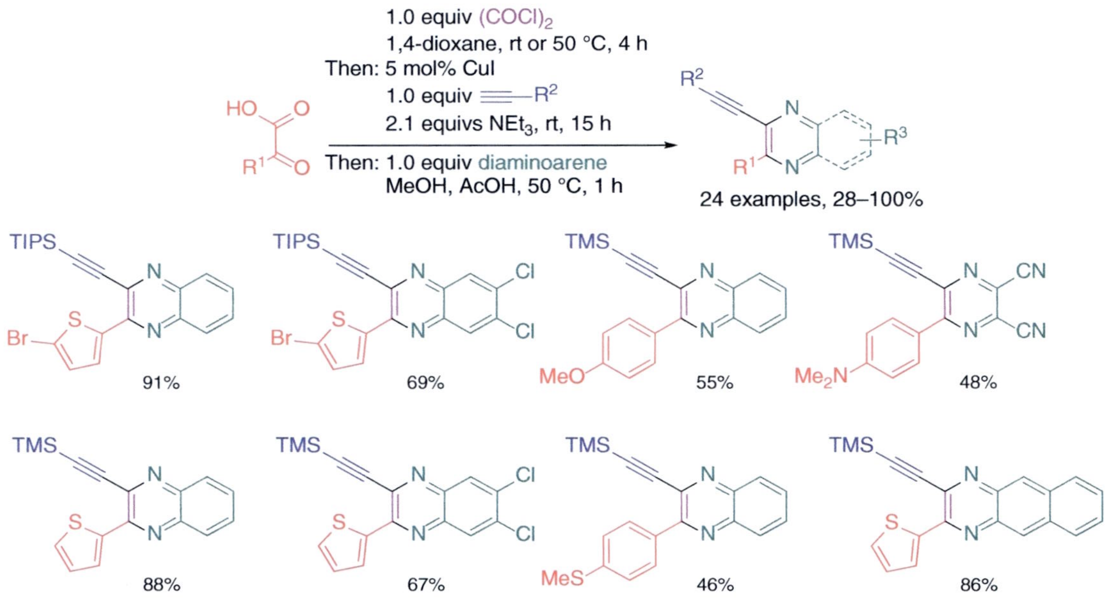

Scheme 10.30 Three-component activation-alkynylation-cyclocondensation synthesis of 3-ethynylquinoxalines. Source: Merkt et al. [77].

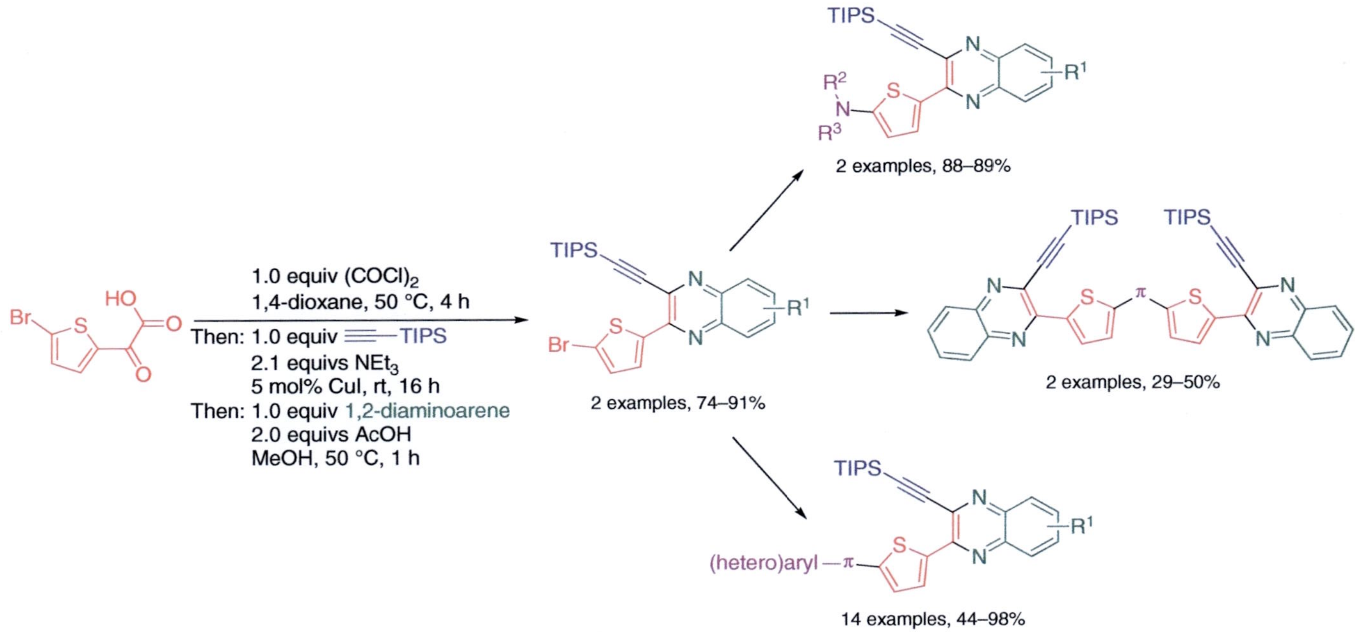

Scheme 10.31 Synthesis of expanded 5-(hetero)aryl-thien-2-yl substituted 3-ethynyl quinoxalines by level 2 functionalization. Source: Merkt et al. [78].

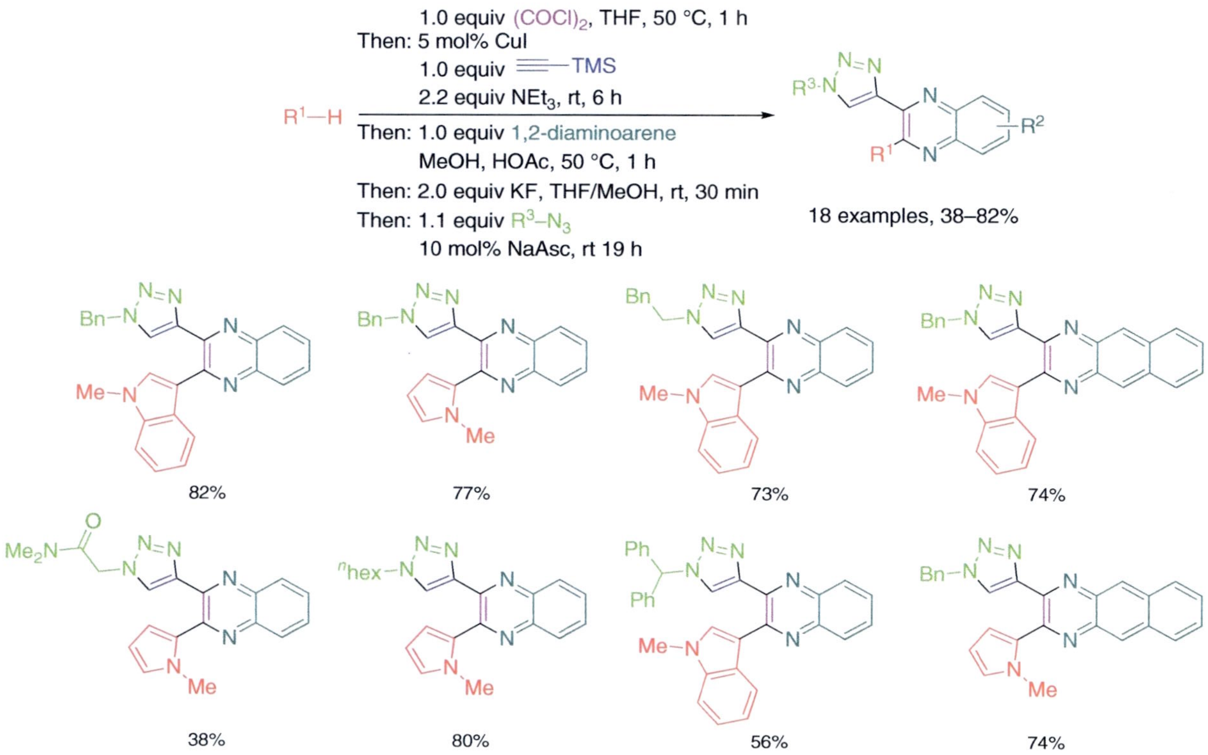

Scheme 10.32 Sequentially Cu-catalyzed five-component glyoxylation-alkynylation-cyclocondensation-CuAAC syntheses of 3-triazolylquinoxalines. Source: Based on Merkt et al. [79].

10.5 Conclusion and Outlook

Transition metal-catalyzed generation of alkynoyl derivatives as an entry to MCRs opens numerous diversity-oriented syntheses of heterocycles. Furthermore, the innocuous reaction conditions of alkynone formations are perfectly compatible with many polar functional groups and, hence, opening new avenues to consecutive multicomponent syntheses of chromophores. Here, we have underlined the methodological advantages over classical approaches, in particular, emphasizing the versatile concept of consecutive one-pot syntheses. Apparently, catalytic generation of alkynoyl functionalities just has opened avenues for heterocycles in many applications.

References

1 Zhu, J. and Bienaymé, H. (eds.) (2006). *Multicomponent Reactions*, 1e. Weinheim: Wiley-VCH Verlag GmbH & Co. KGaA.
2 Levi, L. and Müller, T.J.J. (2016). *Chem. Soc. Rev.* 45: 2825–2846.
3 Müller, T.J.J. (2019). *Beilstein J. Org. Chem.* 15: 1974–1975.
4 Müller, T.J.J. (2014). *Multicomponent Reactions 1. General Discussion and Reactions Involving a Carbonyl Compound as Electrophilic Component*, Science of Synthesis Series (ed. T.J.J. Müller), 5–27. Stuttgart: Georg Thieme Verlag KG.
5 Deilhof, K. and Müller, T.J.J. (2015). *Multicomponent Reactions in Organic Synthesis* (eds. J. Zhu, Q. Wang and M.-X. Wang), 333–378. Weinheim: Wiley-VCH Verlag GmbH & Co. KGaA.
6 Müller, T.J.J. (2018). *Applied Homogeneous Catalysis with Organometallic Compounds* (ed. M. Beller), 1463–1483. Weinheim: Wiley-VCH Verlag GmbH & Co. KGaA.
7 Müllen, K. and Wegner, G. (eds.) (2008). *Electronic Materials: The Oligomer Approach*, 1e. Weinheim: Wiley.
8 Carroll, R.L. and Gorman, C.B. (2002). *Angew. Chem. Int. Ed.* 41: 4378–4400.
9 Shirota, Y. and Kageyama, H. (2007). *Chem. Rev.* 107: 953–1010.
10 Walzer, K., Maennig, B., Pfeiffer, M., and Leo, K. (2007). *Chem. Rev.* 107: 1233–1271.
11 Coropceanu, V., Cornil, J., da Silva Filho, D.A. et al. (2007). *Chem. Rev.* 107: 926–952.
12 Kymissis, I. (2008). *Organic Field Effect Transistors: Theory, Fabrication and Characterization*, 1e. New York: Springer Science and Business Media.
13 Torsi, L., Magliulo, M., Manoli, K., and Palazzo, G. (2013). *Chem. Soc. Rev.* 42: 8612–8628.
14 Grätzel, M. (2001). *Nature* 414: 338–344.
15 Grätzel, M. (2003). *Nature* 421: 586–587.
16 Mishra, A., Fischer, M.K., and Bäuerle, P. (2009). *Angew. Chem. Int. Ed.* 48: 2474–2499.
17 Ma, B., Knowles, D.B., Brown, C.S. et al. (2004). Google Patents.

18 Li, Z., Li, Z.R., and Meng, H. (2006). *Organic Light-Emitting Materials and Devices*, 1e. New York: CRC press.

19 Müllen, K. and Scherf, U. (eds.) (2006). *Organic Light Emitting Devices: Synthesis, Properties and Applications*, 1e. Weinheim: Wiley.

20 Sun, S.-S. and Dalton, L.R. (eds.) (2016). *Introduction to Organic Electronic and Optoelectronic Materials and Devices*, 2e. New York: CRC Press.

21 Levi, L. and Müller, T.J.J. (2016). *Eur. J. Org. Chem.*: 2907–2918.

22 Müller, T.J.J. (2018). *Drug Discov. Today Technol.* 29: 19–26.

23 D'Souza, D.M. and Müller, T.J.J. (2007). *Chem. Soc. Rev.* 36: 1095–1108.

24 Müller, T.J.J. (2010). *Top. Heterocycl. Chem.* 25: 25–94.

25 Whittaker, R.E., Dermenci, A., and Dong, G. (2015). *Synthesis* 48: 161–183.

26 Gers-Panther, C.F. and Müller, T.J.J. (2016). *Advances in Heterocyclic Chemistry: Heterocyclic Chemistry in the 21st Century: A Tribute to Alan Katritzky*, vol. 120 (eds. E.F.V. Scriven and C.A. Ramsden), 67–98, Academic Press.

27 Willy, B. and Müller, T.J.J. (2009). *Curr. Org. Chem.* 13: 1777–1790.

28 D'Souza, D.M. and Müller, T.J.J. (2008). *Nat. Protoc.* 3: 1660–1665.

29 Karpov, A.S. and Müller, T.J.J. (2003). *Org. Lett.* 5: 3451–3454.

30 Sonogashira, K., Tohda, Y., and Hagihara, N. (1975). *Tetrahedron Lett.* 16: 4467–4470.

31 Tohda, Y., Sonogashira, K., Hagihara, N. (1977). *Synthesis*, 777–778.

32 Nordmann, J., Breuer, N., and Müller, T.J.J. (2013). *Eur. J. Org. Chem.* 2013: 4303–4310.

33 Boersch, C., Merkul, E., and Müller, T.J.J. (2011). *Angew. Chem. Int. Ed.* 50: 10448–10452.

34 Merkul, E., Oeser, T., and Müller, T.J.J. (2009). *Chem. Eur. J.* 15: 5006–5011.

35 Götzinger, A.C. and Müller, T.J.J. (2016). *Org. Biomol. Chem.* 14: 3498–3500.

36 Karpov, A.S., Merkul, E., Rominger, F., and Müller, T.J.J. (2005). *Angew. Chem. Int. Ed.* 44: 6951–6956.

37 Cui, M., Wu, H., Jian, J. et al. (2016). *Chem. Commun.* 52: 12076–12079.

38 Götzinger, A.C., Michaelis, C.S., and Müller, T.J.J. (2017). *Dyes Pigments* 143: 308–316.

39 Albano, G., Interlandi, S., Evangelisti, C., and Aronica, L.A. (2020). *Catal. Lett.* 150: 652–659.

40 Sharma, R.K., Yadav, M., Gaur, R. et al. (2016). *ChemPlusChem* 81: 1312–1319.

41 Loganathan, R.K., Ramachandra, S.N., Shekharappa, and Sureshbabu, V.V. (2017). *ChemistrySelect* 2: 8059–8062.

42 Zhou, B., Guo, S., Fang, Z. et al. (2019). *Tetrahedron Lett.* 60: 150914.

43 Götzinger, A.C. and Müller, T.J.J. (2017). *Science of synthesis, Houben-Weyl*. In: *Science of Synthesis Knowledge Updates* ***2017***, vol. 3 (ed. E.M. Carreira), 1–228. Stuttgart, New York: Georg Thieme Verlag.

44 Elguero, J., Silva, A.M.S., and Tomé, A.C. (2011). *Modern Heterocyclic Chemistry* (eds. J. Alvarez-Builla, J.J. Vaquero and J. Barluenga), 635–725. Weinheim: Wiley-VCH Verlag GmbH & Co. KGaA.

45 Marella, A., Rahmat, A.M., Tauquir, A.M. et al. (2013). *Mini-Rev. Med. Chem.* 13: 921–931.

46 Görgen, C., Boden, K., Reiss, G.J. et al. (2019). *Beilstein J. Org. Chem.* 15: 1360–1370.
47 Götzinger, A.C., Theßeling, F.A., Hoppe, C., and Müller, T.J.J. (2016). *J. Organomet. Chem.* 81 (21): 10328–10338.
48 Niedballa, J., Reiss, G.J., and Müller, T.J.J. (2020). *Eur. J. Org. Chem.*: 5019–5024.
49 Sakamoto, T., Shiga, F., Yasuhara, A. et al. (1992). *Synthesis*: 746–748.
50 Niesobski, P., Klukas, F., Berens, H. et al. (2018). *J. Organomet. Chem.* 83: 4851–4858.
51 Chen, Q., Yao, F., Yin, L., and Cai, M. (2016). *J. Organomet. Chem.* 804: 108–113.
52 Thirukovela, N.S., Balabiona, R., Botla, V. et al. (2019). *Catal. Sci. Technol.* 9: 6471–6481.
53 Görgen, C. and Müller, T.J.J. (2017). *Chem. Heterocycl. Compd.* 53: 422–429.
54 Hayeebueraheng, A., Kaewmee, B., Rukachaisirikul, V., and Kaeobamrung, J. (2017). *Eur. J. Org. Chem.* 2017: 6714–6721.
55 Schreiner, E., Wilcke, T., and Müller, T.J.J. (2016). *Synlett* 27: 379–382.
56 Bertaina-Anglade, V., La Rochelle, C.D., and Scheller, D.K.A. (2006). *Eur. J. Pharmacol.* 548: 106–114.
57 Masui, K., Mori, A., Okano, K. et al. (2004). *Org. Lett.* 6: 2011–2014.
58 Schaper, K. and Müller, T.J.J. (2019). *Top. Curr. Chem.* 376: 261–283.
59 Teiber, M., Giebeler, S., Lessing, T., and Müller, T.J.J. (2013). *Org. Biomol. Chem.* 11: 3541–3552.
60 Breuer, N. and Müller, T.J.J. (2018). *Dyes Pigments* 149: 676–685.
61 Schönhaber, J., D'Souza, D.M., Glißmann, T. et al. (2018). *Chem. Eur. J.* 24: 14712–14723.
62 Denißen, M., Nirmalanathan, N., Behnke, T. et al. (2017). *Mater. Chem. Front.* 1: 2013–2026.
63 Denißen, M., Hannen, R., Itskalov, D. et al. (2020). *Chem. Commun.* 56: 7407–7410.
64 Khan, I., Ibrar, A., Abbas, N., and Saeed, A. (2016). *Res. Chem. Intermed.* 42: 5147–5196.
65 Driowya, M., Saber, A., Marzag, H. et al. (2016). *Molecules* 21: 492.
66 Breuer, N. and Müller, T.J.J. (2018). *Synthesis* 50: 2741–2752.
67 Dohe, J. and Müller, T.J.J. (2016). *Z. Naturforsch.* 71b: 705–718.
68 Bakulina, O., Merkt, F.K., Knedel, T.O. et al. (2018). *Angew. Chem. Int. Ed.* 57: 17240–17244.
69 Cheremnykh, K.P., Savel'ev, V., Shkurko, O.P., and Shults, E.E. (2018). *Chem. Heterocycl. Compd.* 54: 1131–1138.
70 Dohe, J., Koßmann, J., and Müller, T.J.J. (2018). *Dyes Pigm.* 157: 198–217.
71 Papadopoulos, J., Merkens, K., and Müller, T.J.J. (2018). *Chem. Eur. J.* 24: 974–983.
72 Ugi, I. and Steinbrückner, C. (1960). *Angew. Chem.* 72: 267–268.
73 Moni, L., Denißen, M., Valentini, G. et al. (2015). *Chem. Eur. J.* 21: 753–762.
74 Moni, L., Gers-Panther, C.F., Anselmo, M. et al. (2016). *Chem. Eur. J.* 22: 2020–2031.

75 Gers, C.F., Nordmann, J., Kumru, C. et al. (2014). *J. Organomet. Chem.* 79: 3296–3310.

76 Gers-Panther, C.F., Fischer, H., Nordmann, J. et al. (2017). *J. Organomet. Chem.* 82: 567–578.

77 Merkt, F.K., Höwedes, S.P., Gers-Panther, C.F. et al. (2018). *Chem. Eur. J.* 24: 8114–8125.

78 Merkt, F.K. and Müller, T.J.J. (2018). *Sci. China Chem.* 61: 909–924.

79 Merkt, F.K., Pieper, K., Klopotowski, M. et al. (2019). *Chem. Eur. J.* 25: 9447–9455.

11

Synthesis of Saturated Heterocycles via Multicomponent Reactions

Carlos K.Z. Andrade, Carlos E.M. Salvador, Thaissa P.F. Rosalba, Lucília Z.A. Correa, Luan A. Martinho, and Yuri R.B. Sousa

Universidade de Brasília, Instituto de Química, Campus Asa Norte, Brasília 70904-900, Brazil

11.1 Introduction

This chapter will deal with many synthetic strategies based on multicomponent reactions (MCRs) to prepare saturated heterocycles. As saturated heterocycles, we have also considered compounds containing carbonyl groups (e.g. lactams and thiazolidinones), exocyclic double bonds and aromatic rings embedded in the cycle. The chapter was divided into sections, each one of which will cover different classes of heterocycles classified by ring size (from 3 to 7, and macrocycles), fused rings, and spirocyclic heterocycles. Recent (2011–2021) representative examples of each class were chosen and in all cases the heterocyclic rings were created during the MCR process.

11.2 Three-membered Ring Heterocycles

Small ring heterocycles, in particular aziridines, are important building blocks in organic reactions [1]. Furthermore, they are present in a range of natural products with biological activity, especially antitumor and antibiotics such as Mitomycin C and Azinomycin B [2, 3]. Bierschenk et al. synthesized complex *N*-substituted aziridines via an Aza-Darzens three-component reaction between anilines, aldehydes, and ethyl diazoacetate (EDA) using the Raymond Tetrahedron, $[K_{12}(Ga_4L_6)]$, as catalyst, where the ligand L is a 1,5-disubstituted naphthalene (Scheme 11.1) [4]. This catalyst has a cavity where the reaction occurs diastereoselectively [5]. This cavity can also stabilize the transition state of some intermediates [6]. Whereas a *cis* non-catalyzed product with an α-diazo ester is reported in the literature [7], with this supramolecular catalyst the major product is the *trans* isomer. The cavity does not support bulky reagents so that only anilines could react to furnish the aziridines. Nevertheless, the anilines can be substituted with both electron-donating and electron-withdrawing groups in *meta* and *para* positions.

Multicomponent Reactions towards Heterocycles: Concepts and Applications, First Edition.
Edited by Erik V. Van der Eycken and Upendra K. Sharma.

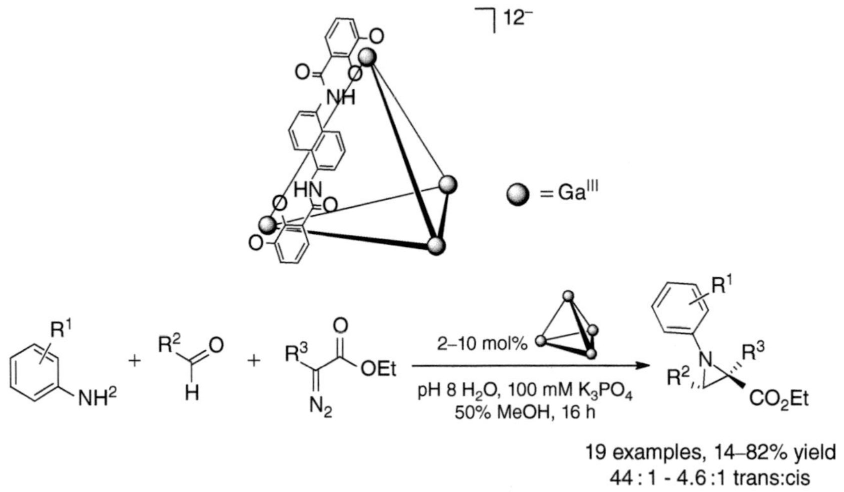

Scheme 11.1 Aza-Darzens three-component synthesis of aziridines.

Jiang et al. [8] reported a three-component reaction to prepare a variety of racemic aziridine-2-carboxylates in good to excellent yields and with excellent diastereomeric ratio (dr) (>20 : 1 dr) (Scheme 11.2). The reagents used in this reaction were an α-ketoester, a protected imine and diethyl phosphite, and the mechanism goes through a [1,2]-phospha-Brook rearrangement followed by a Mannich coupling.

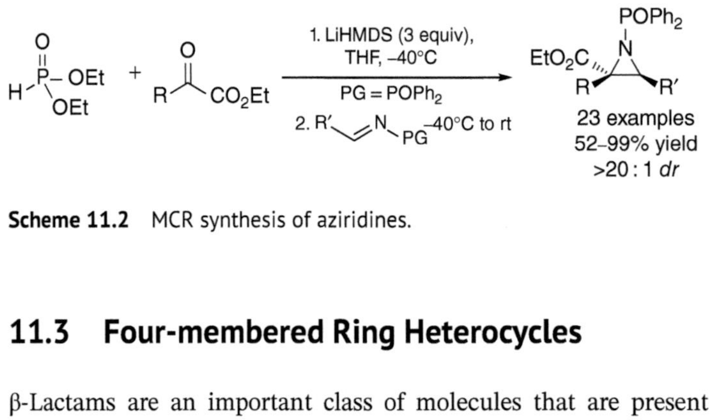

Scheme 11.2 MCR synthesis of aziridines.

11.3 Four-membered Ring Heterocycles

β-Lactams are an important class of molecules that are present in the known antibiotics penicillin and cephalosporine and for this reason are a subject of tremendous study and research [9]. Indeed, many groups have successfully reported new outcomes regarding its reactivity and novel functionalizations [10]. One recent worthmentioned work is the asymmetric interrupted Kinugasa three-component reaction catalyzed by a chiral Cu(I) complex, which is able to furnish α-thiofunctional β-lactams with high diastereo- and enantioselectivity (Scheme 11.3) [11]. The MCR showed a broad scope and involved a terminal alkyne, a nitrone, and a thio-compound, catalysed by a chiral BOX ligand, which, when complexed to copper, is responsible for the observed enantioselectivity. This new

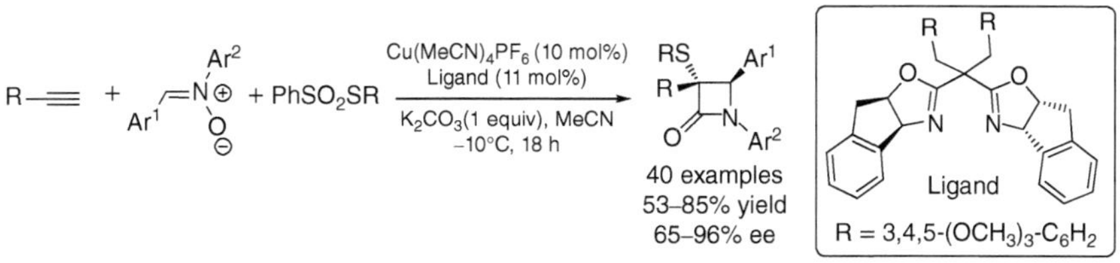

Scheme 11.3 MCR asymmetric synthesis of α-thiofunctional β-lactams.

ligand was developed during this study and was found to be the most efficient among all tested, especially when the reactions are carried out at −10 °C.

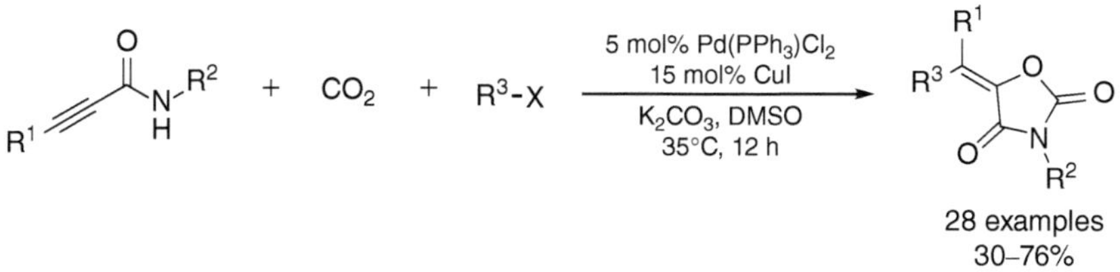

Scheme 11.4 MCR synthesis of oxazolidine-2,4-diones.

11.4 Five-membered Ring Heterocycles

Zhou et al. [12] have reported a palladium/copper-catalyzed oxy-carbonation protocol for the synthesis of oxazolidine-2,4-diones via an MCR of propargylic amides, halohydrocarbons and CO_2 (Scheme 11.4). In general, this protocol provides the target products in moderate to good yields.

In a recent article, Huang et al. [13] have reported the synthesis of 5-((diarylphosphoryl)methyl)oxazolidin-2-ones through a copper-catalyzed three-component reaction of different propargylic amines with CO_2 and phosphine oxides (Scheme 11.5a). As limitations of this method, internal alkynes and aromatic amines resulted in lower product yields. Similarly, an efficient chemical fixation of CO_2 catalyzed by silver has been investigated for the synthesis of oxazolidinones (Scheme 11.5b) [14]. CO_2 was reacted with various propargyl alcohols and primary or secondary amines under neutral and solvent-free conditions to give the products in moderate to high yields. The reaction did not work well using DBU, 2,2′-bipyridine, or TMEDA as additives, internal propargyl alcohols and aromatic amines.

Sun et al. [15] have reported the synthesis of α-alkylidene cyclic carbonates in moderate to excellent yields through a three-component reaction of propargylic alcohols, CO_2 and aryl halides using palladium as catalyst (Scheme 11.6). It was observed that both electron-donating and electron-withdrawing groups led to the formation of the desired product. However, 1,2,4-trichloro-5-iodobenzene did not work well (only 6% yield) and an example of an alkyne bearing an *o*-pyridine moiety did not form the product at all.

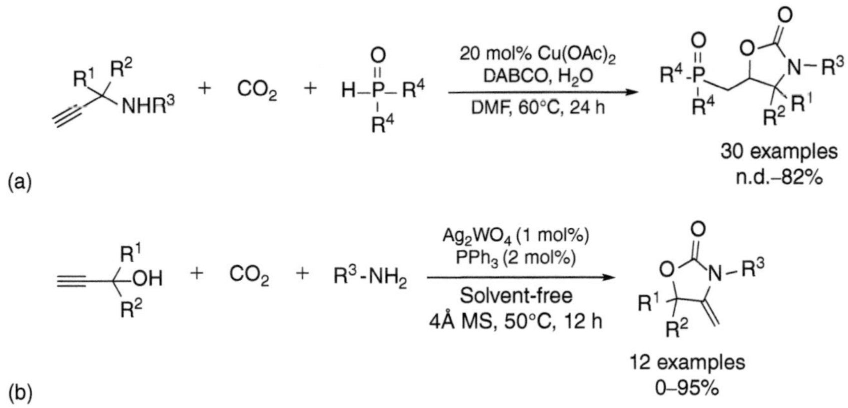

Scheme 11.5 MCR syntheses of oxazolidin-2-ones. (a) Cu-catalyzed three-component synthesis of oxazolidin-2-ones. (b) Ag-catalyzed three-component synthesis of oxazolidin-2-ones.

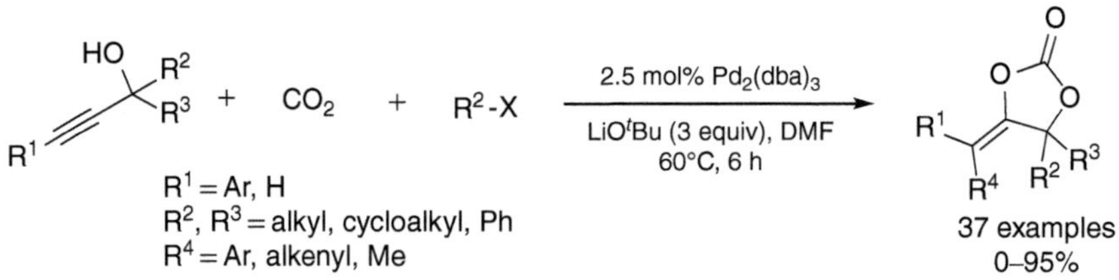

Scheme 11.6 Pd-catalyzed MCR synthesis of cyclic carbonates.

Since alkyl 2-siloxycyclopropanecarboxylates have the potential of being versatile equivalents of functionalized aldehydes or ketones, Gladow et al. [16] performed the Ugi five-center four-component reaction with two amino acids (glycine and L-phenylalanine), 1,1,3,3-tetramethylbutyl isocyanide, methanol and ethyl 1-trifluoromethyl-2-(trimethylsilyloxy) cyclopropanecarboxylate (Scheme 11.7). The reaction with glycine provided directly the two diastereomers of CF_3-substituted γ-lactam derivative in a combined yield of 43% (*cis* = 14%, *trans* = 29%), but the reaction with L-phenylalanine was carried out under milder conditions in the presence of LiCl, making it possible to isolate the linear product instead of the heterocycle in 61% yield as a mixture of the four possible diastereomers (ratio c. 4 : 4 : 1 : 1). The product was then subjected to heating in 1,2-dichlorobenzene to provide the cyclic molecule as a diastereomeric mixture (*dr* c. 55 : 27 : 13 : 5).

In 2015, Vazquez and coworkers [17] reported the synthesis of iso-cotinine analogs via a microwave-assisted Ugi-4CR catalyzed by indium(III) (Scheme 11.8a). The Ugi adduct is not isolated, instead it undergoes a spontaneous cyclization of the S_N2 type without the need of extra base (Scheme 11.8b). According to the authors, this is due to the high acidity of the peptidyl proton in the Ugi adducts and to the presence of basic moieties in the reaction media (the primary amines or the pyridine moiety). Indeed, when and *ortho* or *para* nitrogen are absent in the aromatic ring, the

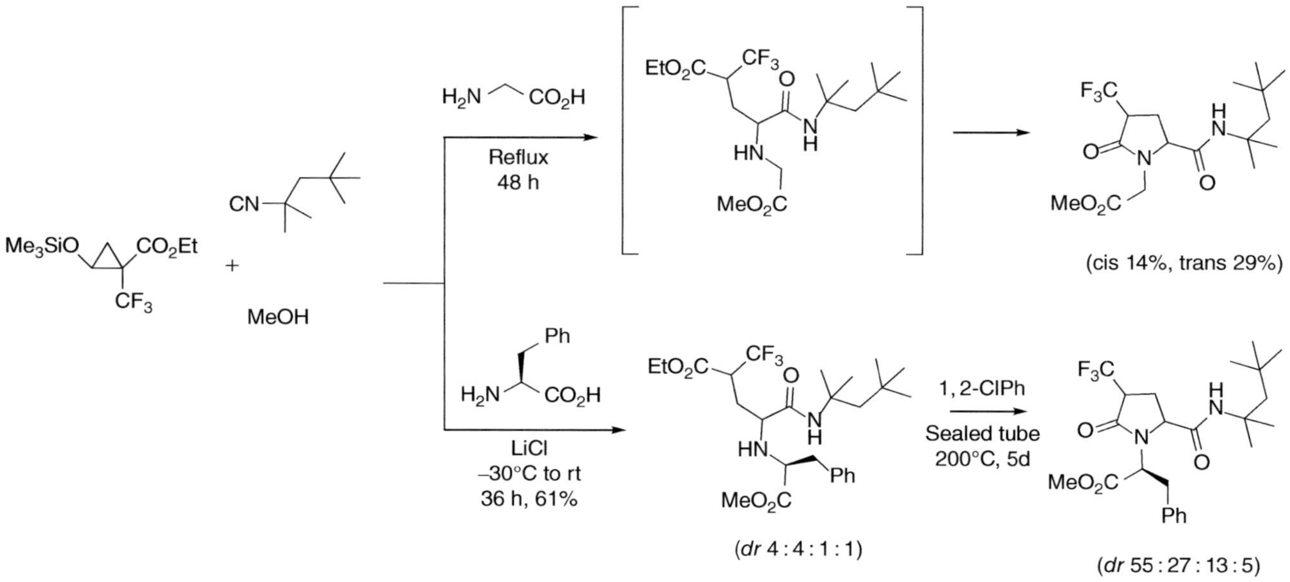

Scheme 11.7 Ugi five-center four-component reaction in the synthesis of γ-lactams.

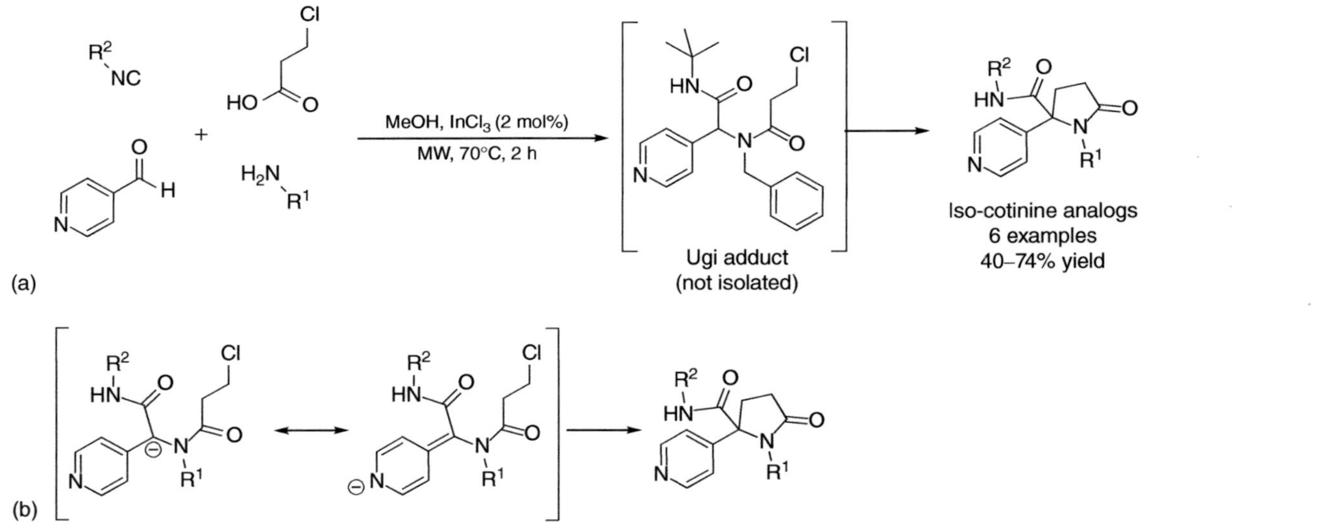

Scheme 11.8 Synthesis of isocotinine analogs via Ugi reactions. (a) Synthesis of isocotinine analogs via Ugi reactions. (b) Spontaneous cyclization of the Ugi adduct.

cyclization only occurs with an excess of an extra base (2 equiv of *t*-BuOK) at 110 °C (Microwave [MW] heating for two hours).

Thiazolidinones are an important pharmacophore in medicinal chemistry, used on the treatment of numerous diseases. For instance, it is present in penicillin antibiotics [18, 19]. Ansari et al. [20] reported a three-component reaction between steroidal ketones, thiosemicarbazide or methyl-thiosemicarbazide and dimethyl acetylenedicarboxylate (DMAD) catalyzed by $AlCl_3$ to prepare steroidal thiazolidinones in very good yields (Scheme 11.9). Spectroscopic and molecular docking methods revealed that some of these molecules were able to bind spontaneously to the human serum albumin (HSA), which has currently been used in the quest of new leading compounds.

Ibrar et al. [21] reported 15 new examples of α-glucosidase inhibitors, i.e. coumarin-iminothiazolidinone hybrids aiming for minimum side effects (Scheme 11.10a). Since α-glucosidase inhibition can lower the exaggerated hypoglycemia increase after meals, it is an important target in diabetes treatment. The first step of the synthesis is the *in situ* formation of intermediate **I** by 2-oxochromene-3-carbohydrizide and an aryl isothiocyanate followed by a three-component reaction between this intermediate, 4-fluorobenzaldehyde, and chloroacetyl chloride, furnishing the coumarin-iminothiazolidinone hybrid in good yields. α-Glucosidase inhibition was tested and one of the compounds (R = 2-OMe) showed an $IC_{50} = 0.09 \pm 0.001$ μM.

The 2-thioxo-1,3-thiazolidin-4-one core, also known as rhodanine, is a five-membered sulfur/nitrogen heterocycle, which is present in molecules with many pharmacological applications [22]. Jacobine and Posner [23] prepared novel thiazolidinone derivatives through a three-step, one-fashion protocol between carbon disulfide, a primary amine and α-chloro-β,γ-alkenoate esters (Scheme 11.10b). Carbon disulfide and the amine generate a dithiocarbamate anion *in situ*, which then reacts with the esters, followed by alkene isomerization and cyclization, furnishing the products in good yields. Some amines have failed in this reaction, such as propargylamine, glycine, and its methyl ester, among others. As an application, an analog of the aldose inhibitory drug epalrestat has been efficiently prepared by this approach.

Nitrones have been frequently used for the synthesis of important biological and synthetic scaffolds such as isoxazolidines [24]. Gupta et al. [25] described a simple and efficient protocol to synthesize fully substituted trifluoromethylated and phosphonylated isoxazolidines via the *in situ* generation of nitrones from the reaction of nitrosobenzene, diazo compounds and allenic esters (Scheme 11.11). In total, 31 compounds were prepared and the yields ranged from moderate to good. The proposed mechanism is quite simple and involves the nitrone formation between nitrosobenzene and the diazo compound with release of N_2 gas, followed by a 1,3-dipolar cycloaddition reaction. Among the several existing methods for the synthesis of isoxazolidines, the method described by this work is one of the most efficient and versatile to selectively obtain *Z* compounds.

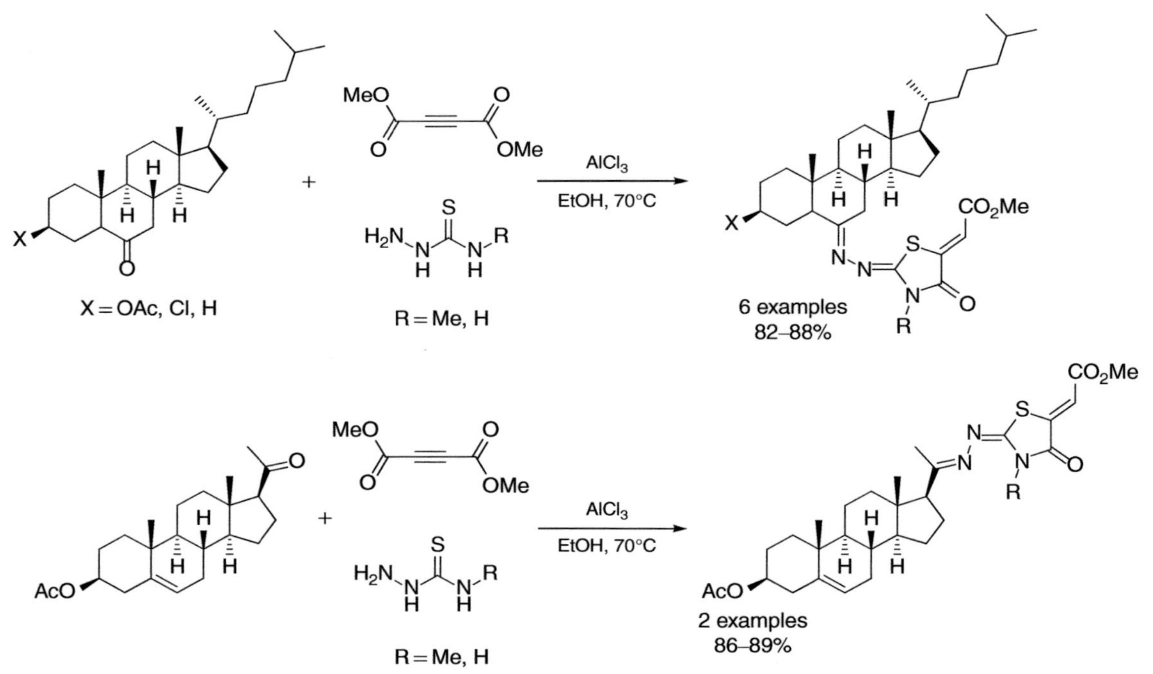

Scheme 11.9 MCR synthesis of steroidal thiazolidinones.

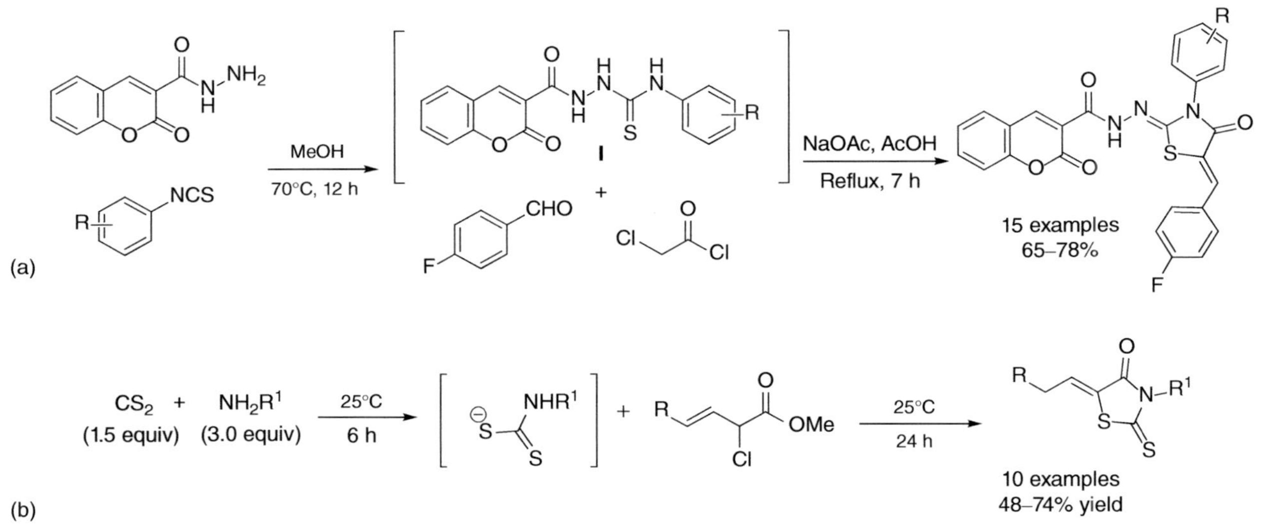

Scheme 11.10 MCR synthesis of novel thiazolidinone derivatives. (a) MCR synthesis of coumarin-iminothiazolidinone hybrids. (b) Thiazolidinone synthesis from carbon disulfide.

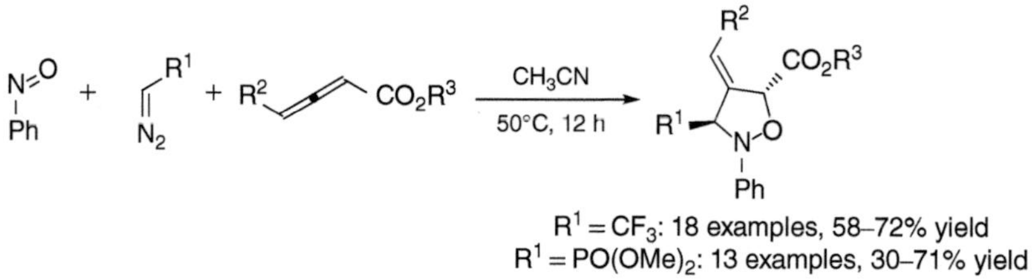

Scheme 11.11 MCR synthesis of isoxazolidines.

Baumann et al. [26] applied continuous flow mode to perform a three-component coupling reaction to synthesize several nitro-pyrrolidines. The reactions were performed in a useful flow setup with two reaction steps (imine formation and cycloaddition) and in-line purification (Scheme 11.12). These approaches allowed the telescope of the reactions on a continuous flow platform using a fixed bed (for imine formation) and a tubular reactor (in which the cycloaddition takes place), respectively. Purification of the residual starting material was possible due to the use of a benzylamine scavenger (QuadraPure benzylamine [QP-BZA]). The target compounds were obtained as a diastereomeric mixture of single regioisomers in shorter reaction times and better yields (above 70%) compared to traditional methods.

The hydantoin nucleus is present in several compounds of high importance for biological and industrial employment and the Bucherer–Bergs MCR is a traditional strategy for synthesizing hydantoins from aldehydes or ketones [27]. Monteiro et al. [28] used the flow technology to obtain hydantoins via this strategy with a high-throughput screening (Scheme 11.13). The reactions were carried out in a pressurized (20 bar) two-phase liquid-liquid system in which one of the phases was aqueous containing KCN and $(NH_4)_2CO_3$ solutions and the other was organic with the carbonyl reactant. The low conversion to the desired product due to the difference of solubility of the starting materials and the loss of the *in situ* generated gaseous component are serious limitations of traditional methods for this reaction. Due to the ability of continuous flow mode to conduct reactions at high temperature and pressure without headspace and promote a high contact surface between two-phase systems, it was possible to overcome these limitations.

11.5 Six-membered Ring Heterocycles

Oxazines and tetrazoles are particularly important organic structures since they have a wide range of applications, being found in many pharmacologically active compounds [29, 30]. Thus, molecules with these two heterocycles linked to each other are remarkably interesting. Chandgude et al. [31] developed a new strategy to synthesize *in situ* oxazine-tetrazole scaffolds by the first use of the Asinger-Ugi-tetrazole union in a five-component reaction (Scheme 11.14). The reactions were carried out with sonication at room temperature, and using NH_4OH as the ammonia source, 3-hydroxypivalaldehyde, $TMSN_3$ and various aldehydes

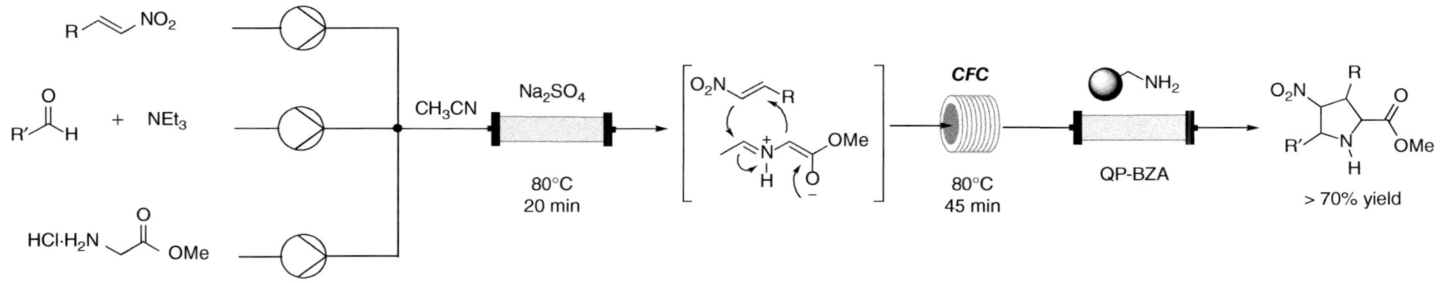

Scheme 11.12 Flow set-up for the synthesis of 3-nitropyrrolidines

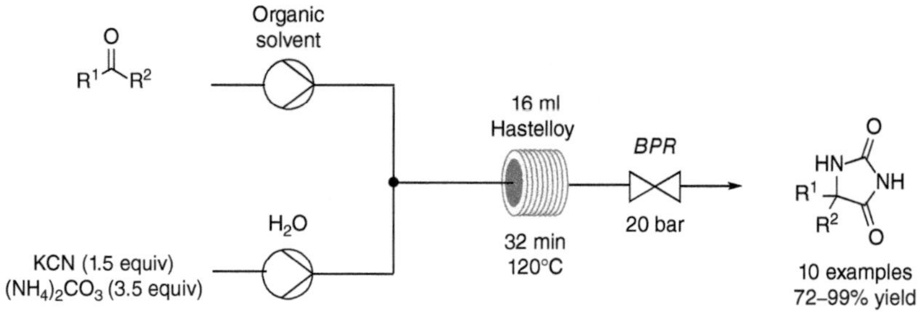

Scheme 11.13 Synthesis of hydantoins via the Bucherer–Bergs reaction in flow mode.

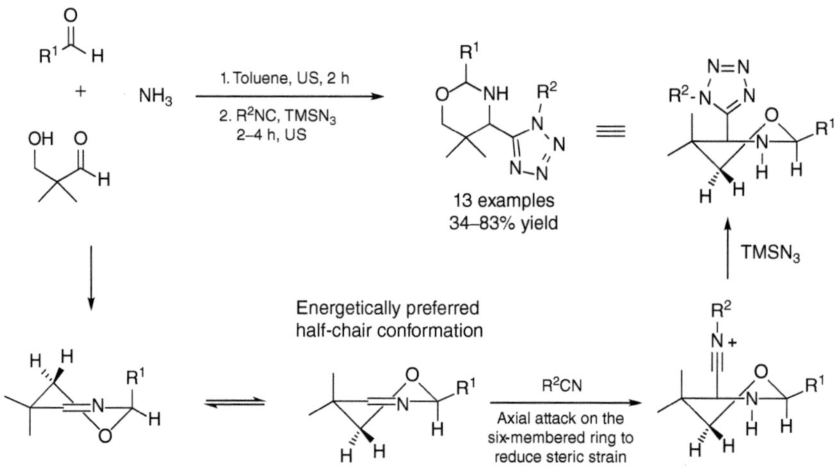

Scheme 11.14 Synthesis of the oxazine-tetrazole scaffold via the Asinger-Ugi-tetrazole union and its proposed mechanism.

and isocyanides in toluene, yielding 13 compounds in modest to high yields (25–83%). This new method for the synthesis of oxazine-tetrazole molecules has many advantages, but the most noteworthy is that an excellent diastereoselectivity was observed (more than 90 : 10) in almost all examples. This observation can be explained by the proposed mechanism: in the first step one of the conformations of 5,6-dihydro-2H-1,3-oxazines is preferable, and the axial attack of isocyanides is favorable for this conformer (Scheme 11.14).

Organophosphorus compounds are essential for life, and drug discovery based on phosphorus is increasing [32–34]. Hossaini and coworkers [35] reported a three-component reaction involving trialkyl phosphites, dialkyl acetylenedicarboxylates and *N*,*N*′-dimethylurea leading to novel organophosphorus heterocycles in high yields (Scheme 11.15). This procedure was carried out under neutral reaction conditions, without any activation or modification of the synthetic precursors.

The several environmental benefits associated with the use of water as solvent and boronic acids as reagents in the formation of new C—C bonds inspired Candeias et al. [36] to develop a green protocol and its application

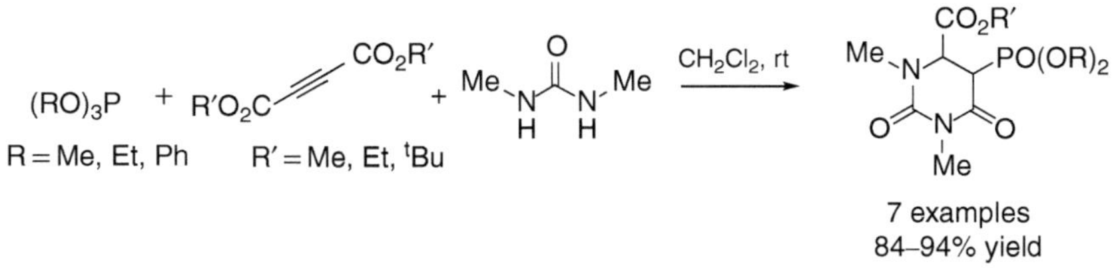

Scheme 11.15 Synthesis of organophosphorus heterocycles via MCR.

in both research and organic teaching laboratories. The six-membered heterocycle 4-benzyl-3-phenylmorpholin-2-ol was synthesized from the Petasis borono–Mannich (PBM) MCR using 2-benzylaminoethanol, glyoxal, phenylboronic acid and water as solvent in an excellent yield of 97% in a research laboratory and 73% average yield (55–90% range) in a teaching laboratory (Scheme 11.16a). This product was obtained in good diastereoselectivity. This was possible because a stereogenic center is formed in the ring-closing step, so the rate determining step of the reaction (the migration of the aryl substituent from the boronic acid to the carbon of the iminium ion intermediate) can be diastereoselective (Scheme 11.16b). This sustainable method is useful for the preparation of libraries of saturated heterocycles such as the morpholines obtained herein.

One of the possible variations in the Biginelli MCR is to replace the traditional ethyl acetoacetate by other 1,3-dicarbonyl compounds, therefore obtaining non-classical products with reinforced pharmacological properties. In 2020, Gein et al. [37] synthesized nine different hexahydropyrimidine-4-carboxylates in good yields (59–78%) via the modified four-component Biginelli reaction, using methyl aroylpyruvate, several aromatic aldehydes, urea (or thiourea) and the solvent as the fourth component in the presence of the nontoxic and cheap $NaHSO_4$ as catalyst (Scheme 11.17). These reactions have a singularity: they stop at the formation of the non-dehydrated hexahydropyrimidine with an alkylated tertiary hydroxyl group, providing new unconventional products of this MCR. The influence of the solvent and the best catalyst for the reaction were investigated and it was observed that $NaHSO_4$ was the only possible catalyst and when MeOH was replaced by EtOH, ethylated tertiary hydroxyl group was isolated, as expected, and partial transesterification of the ester occurred, leading to a mixture of methyl and ethyl esters. When *i*-PrOH was used as solvent, the alkylation of the hydroxyl group at position 4 of the pyrimidine ring did not ocur.

Sathicq et al. [38] have elaborated Preyssler heteropoly acids encapsulated in a silica framework as catalysts under solvent-free conditions for the MCR of aldehydes, trifluoromethylacetoacetate and urea or thiourea to give trifluoromethyl-hexahydropyrimidine derivatives in good to excellent yields (Scheme 11.18a). The solid catalysts were recyclable after the end of the reaction. Similar results using a new vanadium Keggin heteropolyacid (PMo11V) encapsulated in a silica framework as the catalyst have been reported by the same authors [39] to obtain the corresponding hexahydropyrimidines (Scheme 11.18b). This same type of molecules was obtained by Zohdi et al. [40] via a green approach in the

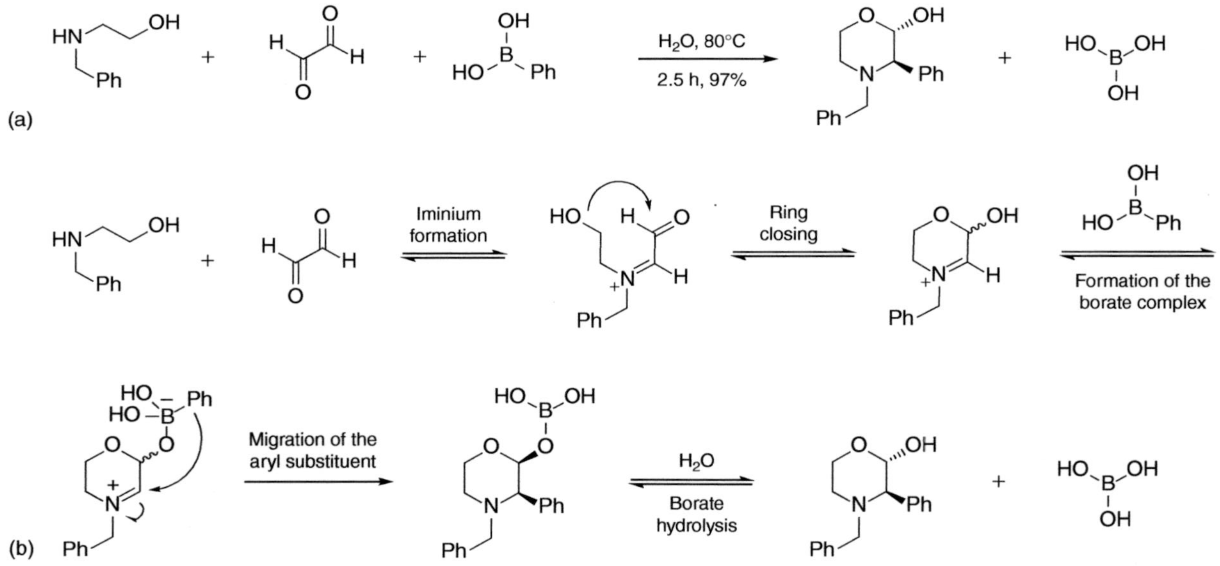

Scheme 11.16 (a) Synthesis of 4-benzyl-3-phenylmorpholin-2-ol from the Petasis borono–Mannich (PBM) MCR. (b) Proposed mechanism.

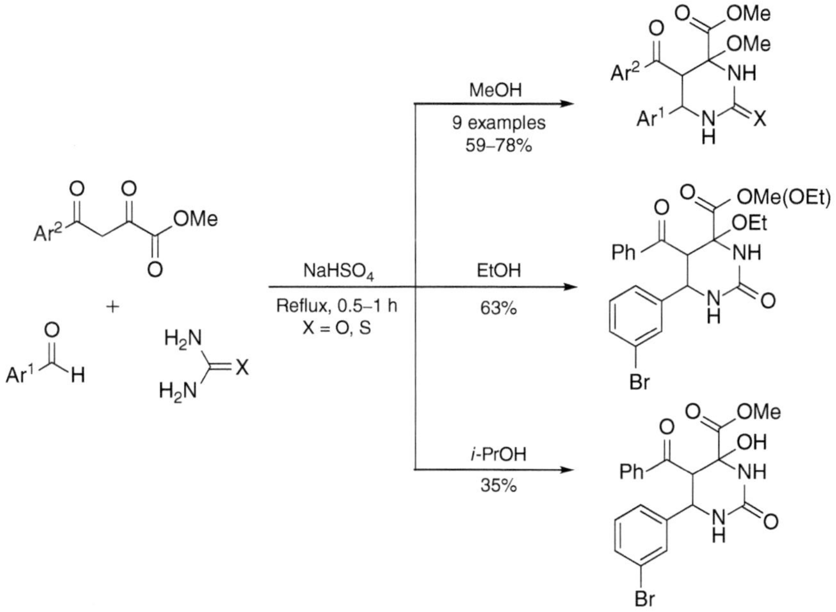

Scheme 11.17 Synthesis of hexahydropyrimidine-4-carboxylates via the modified four-component Biginelli reaction.

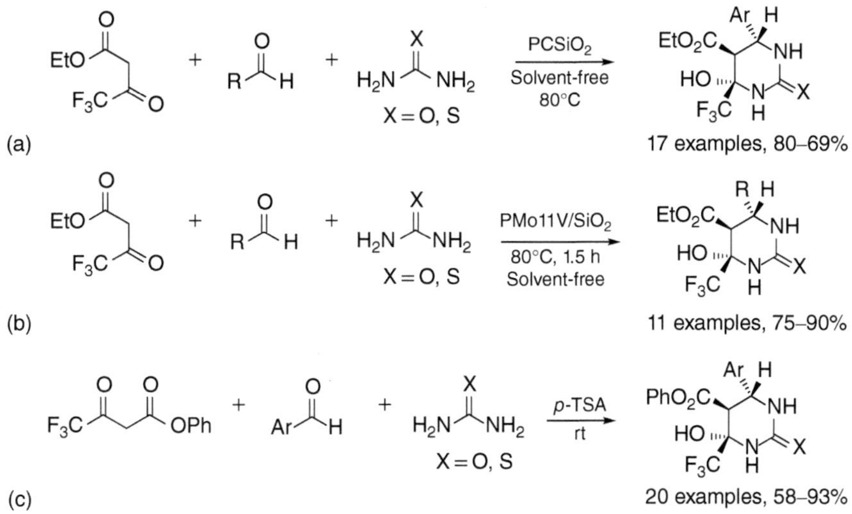

Scheme 11.18 Solvent-free syntheses of trifluoromethyl-hexahydropyrimidines via modified Biginelli reactions using (a) Preyssler heteropolyacids, (b) Keggin heteropolyacid, and (c) *p*-toluenesulfonic acid.

presence of *p*-toluenesulfonic acid under solvent-free conditions at room temperature (Scheme 11.18c). This protocol could be applied to aromatic aldehydes with either electron-withdrawing or electron-donating groups providing excellent yields of the products. These compounds have been demonstrated to have a potential antibacterial activity.

11.6 Seven-membered Ring Heterocycles

Khabibullina et al. [41] reported a one-step green method to synthesize a library of 20 new (1,5,3-dithiazepan-3-yl)alkanoic acids in high yields (62–95%) by the multicomponent cyclocondensation of different amino acids and a preformed mixture of formaldehyde/1,2-ethanedithiol (2 : 1) in water at room temperature (Scheme 11.19). Enantiomerically pure 1,5,3-dithiazepanes were obtained in the reactions using enantiomerically pure amino acids. Other aldehydes were tested but they reacted non-selectively, suggesting that formaldehyde is the best choice. There are several essential features of this method that make it more advantageous than the other methodologies already described in the literature to prepare 1,5,3-dithiazepanes, such as: no catalyst is required; mild reaction conditions are used; the only by-product of the reaction is water; and no hazardous materials are produced. A plausible pathway for the formation of these compounds is depicted in Scheme 11.19b.

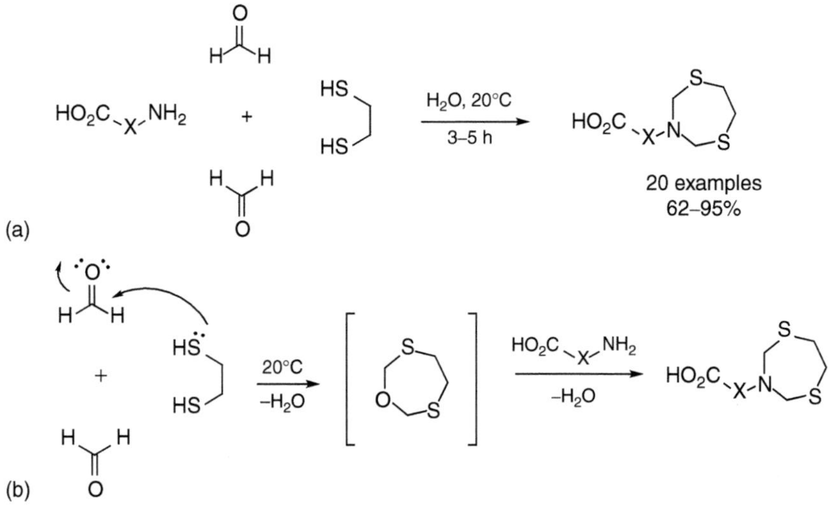

Scheme 11.19 (a) Synthesis of novel (1,5,3-dithiazepan-3-yl)alkanoic acids via the three-component reaction between amino acids, formaldehyde, and 1,2-ethanedithiol. (b) A plausible pathway for the formation of (1,5,3-dithiazepan-3-yl)alkanoic acids.

The azepan-2-one (caprolactam) scaffold is an interesting molecular moiety that can be found in a variety of molecules with pronounced biological activities [42]. Considering that, Rasouli et al. [43] reported a green synthesis of 13 *N*-alkyl-2-(2-oxoazepan-1-yl)-2-arylacetamide derivatives in good to

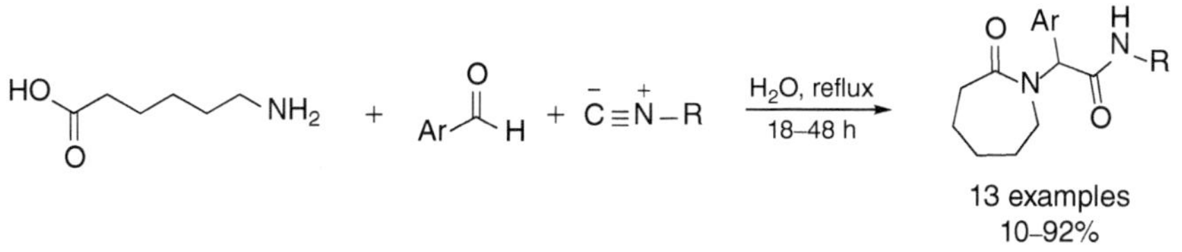

Scheme 11.20 Synthesis of *N*-alkyl-2-(2-oxoazepan-1-yl)-2-arylacetamide derivatives via the Ugi three-component reaction of 6-aminohexanoic acid, aromatic aldehydes and isocyanides.

excellent yields (62–92%) by the Ugi four-center, three-component reaction of 6-aminohexanoic acid and several isocyanides and aromatic aldehydes in water under reflux (Scheme 11.20). In general, the reaction worked well with practically all different aromatic aldehydes, except for 2-methylbenzaldehyde and 2,6-dimethylbenzaldehyde, with which the product was obtained in poor yield (10%) and no product was observed, respectively. This is an advantageous method as it uses water as solvent, it has a one-step synthetic route, does not need complex or expensive catalysts and the starting materials are easy to access.

11.7 Macrocycles

Taking into consideration the applicability of MCRs in the field of hybrid heterocyclic molecules and the variety of pharmacological properties associated with 1,3,5-thiadiazinane-2-thiones (THTTs), a bidirectional multiple multicomponent macrocyclization procedure using two Ugi-4CR was developed by Echemendía et al. [44] in 2018 for the synthesis of four THTT peptide-like macrocycles (Scheme 11.21). This macrocyclization strategy demands the use of two bifunctional building blocks, so in this work the diacid/diamine combination was selected. Initially, the dicarboxylic acid-functionalized THTT was prepared and combined with paraformaldehyde and different diamines and isocyanides in the Ugi-4CR under high dilution conditions, as required for this type of reaction [45], to give the corresponding THTT-linked macrocyclic peptide conjugates in moderate yields.

Peptoides are oligomers of *N*-substituted glycines that mimick the properties and the natural structure of peptides [46, 47]. They lack stereogenic centers, as their lateral chains are conected to the nitrogen atom and not to the α-carbon (as in peptides). The Ugi MCR is one of the most versatile tools to build the peptoids skeleton. Barreto et al. [48] have developed a synthetic strategy towards macrocyclic peptoids based on microwave-mediated consecutive MCRs [49]. For instance, this strategy has been used to prepare six novel cyclic pentadepsipetoids, analogs of the natural product Sansalvamide A (Scheme 11.22). After each MCR (Ugi or Passerini), a basic hydrolysis or an acid treatment unveiled the proper functional groups (amine or carboxylic acid) for the subsequent MCR. The final step involved an intramolecular Ugi reaction, which was carried out on pseudo-high dilution conditions to avoid oligomerization.

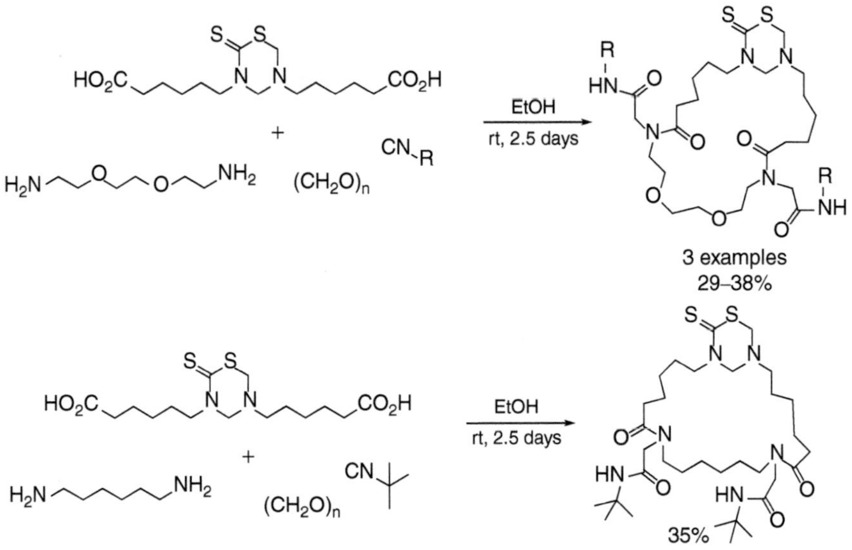

Scheme 11.21 Bidirectional multiple multicomponent macrocyclization procedure using two Ugi-4CR to synthesize THTT peptide-like macrocycles.

A remarkable example of this strategy is the synthesis of a cyclic heptapeptoid in which four consecutive MCRs were used and the whole synthesis (except for the cyclization step) was carried out under microwave heating (Scheme 11.23) [50]. The total time for all MW-mediated steps (disregarding the time spent on isolations and purifications) was only 27 minutes.

11.8 Fused Heterocycles

Taxol, Strychnine, Brevetoxin B, Manzamine A, Azaspiracid-1, Palau'amine, Halistatin 1, and other uncountable complex natural products possess at least one heterocycle fused to other (hetero)cycle in their structures. This fact provides some direction to understand how biological processes work and how nature synthesis works. Therefore, understanding nature, new methodologies, protocols, and greener routes can be developed for augmenting yields and reducing steps with a huge increase in efficiency and/or ideality [51]. In this respect, a bioinspired Ugi/Michael/Aza-Michael cascade reaction (UMAM) protocol was established by Santra and Andreana [52] for the synthesis of fused azaspiro tricycles and azaspiro tetracycles involving a four-component reaction under microwave heating with water as the solvent (Scheme 11.24). The method furnished the products in good to excellent yields and regioselectivities with an appreciable diastereoselectivity in most cases. The authors noted that the formation of the 5,5,6-fused azaspiro tricycle and 5,5,6-fused azaspiro tetracycle systems was influenced by the substituents in R^4. The bulky groups of R^4 assist the Bürgi–Dunitz angle of attack by blocking the 6-exo-trig aza-Michael addition, favoring an unprecedented 5-exo-trig aza-Michael

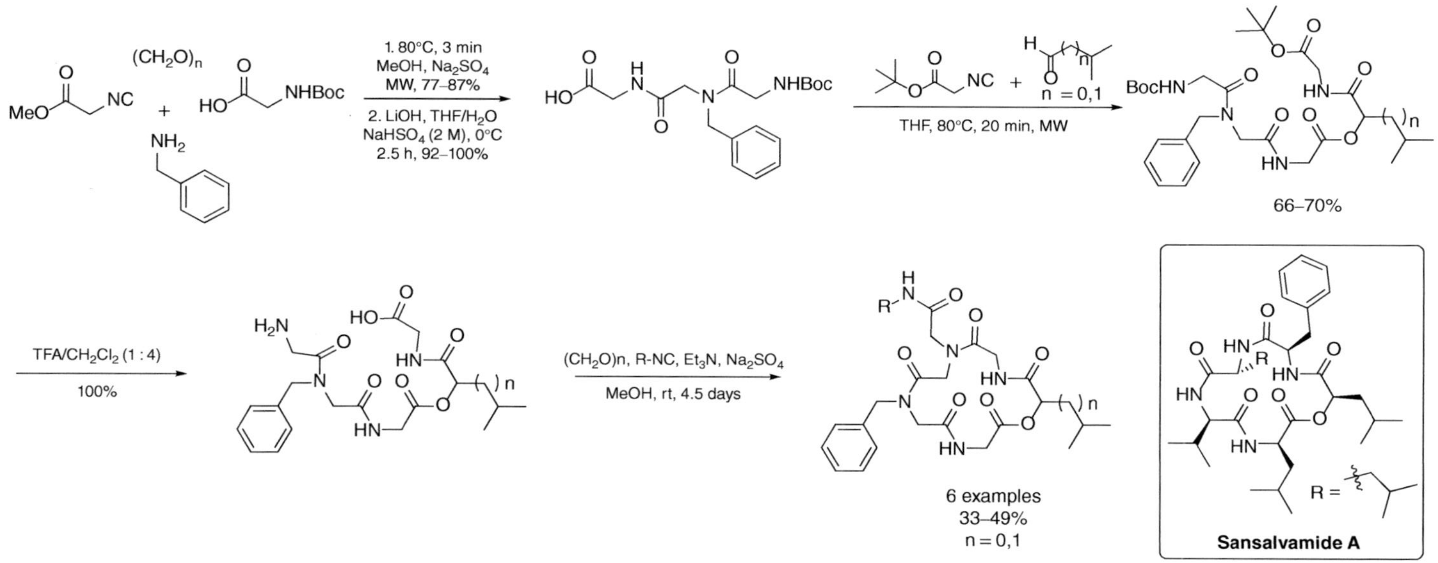

Scheme 11.22 Synthesis of cyclic pentadepsipeptoids through consecutive MCRs.

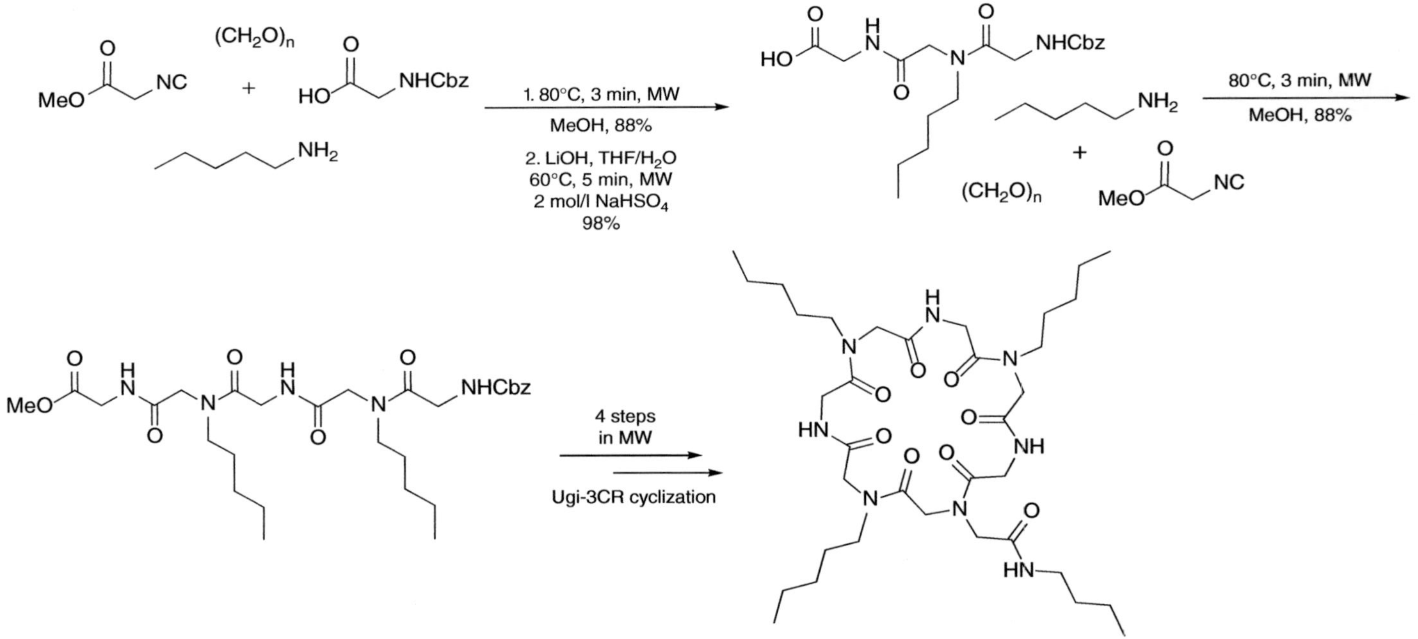

Scheme 11.23 Microwave-mediated synthesis of a cyclic pentadepsipeptoid through consecutive MCRs.

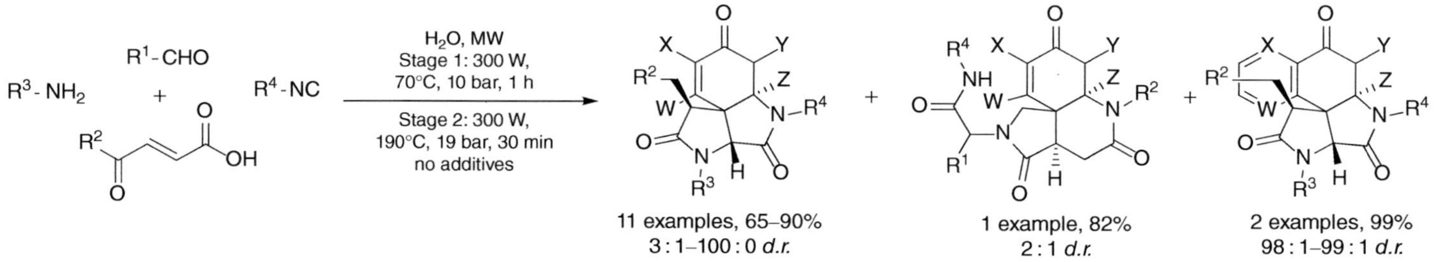

Scheme 11.24 MCR synthesis of fused azaspiro tricycles and azaspiro tetracycles.

addition. However, the 6-exo-trig aza-Michael addition was favored with less bulky R^2, leading to the formation of 5,6,6-fused azaspiro tricycles.

A number of novel fused heterocyclic compounds has been reported by Jin et al. [53]. The three-component reaction of sulfonyl azides, alkynes, and allylamines, followed by two consecutive Ullmann-type C–N coupling reactions via the Cu^I-catalyzed azide/alkyne cycloaddition (CuAAC), resulted in the formation of 2,3-dihydro-1H-imidazo[1,2-a]indoles in moderate to good yields (Scheme 11.25).

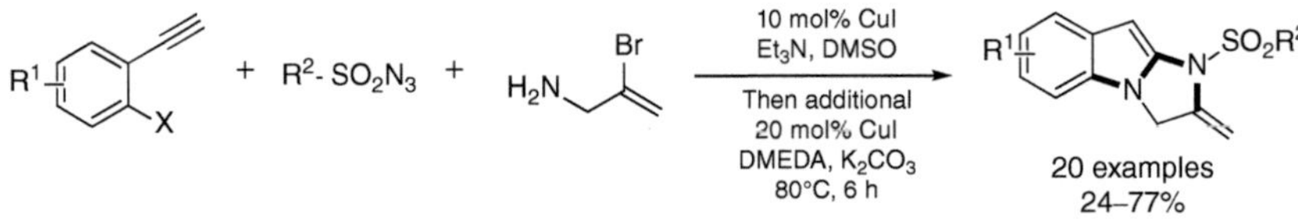

Scheme 11.25 MCR synthesis of imidazoindoles by Ulmann-type reaction.

Yugandhar et al. [54] have elaborated a one-pot synthesis of alkaloid-mimicking tricyclic skeletons through an Ugi four-component condensation (U4CC) of *p*-anisidine, carbonyl compounds, 3-phenylpropiolic acid and *tert*-butyl isocyanide in methanol, followed by an acid mediated *ipso*-cyclization and an aza-Michael addition (Scheme 11.26).

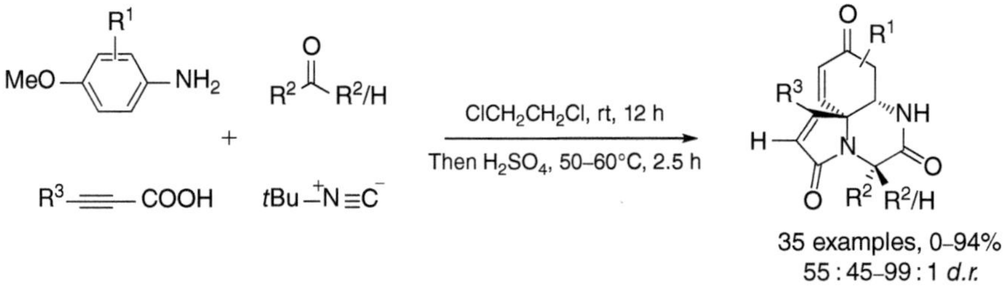

Scheme 11.26 Ugi-mediated synthesis of tricyclic compounds.

Hydrotiazole fused rings are structures of great interest due to their biological activity [55, 56] and one of the main difficulties found in their synthesis is related to the control of the stereochemistry. However, some interesting methodologies have appeared to overcome this problem. For instance, Xiong et al. [57] developed a stereoselective multicomponent approach using a ligand derived from bicyclic proline (L-RaPr$_2$) as catalyst in the reaction of benzothiazoles, isocyanides and alkenes (Scheme 11.27a). The key step is the formation of a zwitterionic intermediate, which is obtained from the nucleophilic attack of the isocyanide to the alkylidene malonates, forming a chiral complex. The same authors [58] described a similar approach towards the stereoselective synthesis of dihydroisoquinoline derivatives from isoquinolines through a [3+2] dearomative annulation cycloaddition fashion, producing a single diastereomer product in good yields with varied enantioselectivities (Scheme 11.27b).

A catalytic cycle was proposed to account for the observed stereoselectivity based on a transition state in which the nuclephilic attack of the isocyanide occurs from the *Si* face of the malonate, once the *Re* face is blocked by the bulky aromatic group of

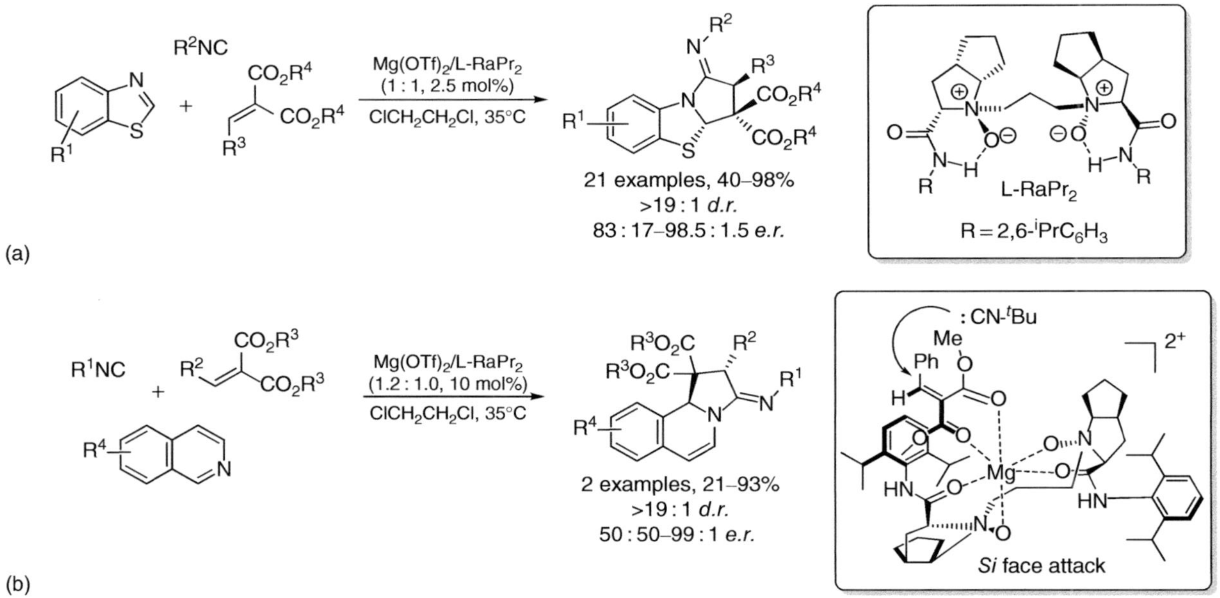

Scheme 11.27 Stereoselective synthesis of hydrotiazole fused rings by a multicomponent [2+1+2] cycloaddition. (a) Bicyclic proline-catalyzed synthesis of hydrothiazole fused rings. (b) Stereoselective synthesis of dihydroisoquinoline derivatives.

L-RaPr$_2$ (Scheme 11.27b). This attack is followed by an intramolecular attack (again on the *Si* face) and ring closing of the heterocycle. This is a nice example of a formal [2+1+2] cycloaddition of benzothiazoles via an MCR in the presence of a chiral catalyst.

The introduction of multicatalysis process in MCRs is much similar to the enzymatic machinery. Complex products are formed from simple structures allowing assymmetric synthesis to occur. Cala et al. [59] developed a unique double catalytic system for the asymmetric synthesis of furo[2,3-b]pyrrole derivatives from simple substrates such as 3-butynamines, glioxylic acid and anilines (Scheme 11.28). The catalytic system is formed by a gold complex and a BrØnsted acid (a chiral phosphoric acid, CPA). Noteworthy is that three contiguous stereogenic centers are generated from achiral starting materials and the products were obtained in excellent yields. A computational study showed the influence of the non-covalent interactions between the CPA and the reagents on the stereoselectivity control. These interactions are as important as the steric effects and include the linear geometry of the anthracenyl, the electron density on aniline, the van der Waals interactions of the aromatic rings and the presence of aryl (R^2) substituents on the 3-butynamine derivatives, justifying the excellent yields and stereoselectivities.

In 2015, Kröger et al. [60] showed that bisamides containing a thiazolidine substructure are suitable precursors in the diastereoselective Cu-mediated rearrangement reaction to obtain annulated and bridged tricyclic systems combining both the imidazolidin-4-one and thiomorpholine cycle (Scheme 11.29). These precursors were easily accessible by a two-step synthetic pathway using the modified Asinger four-component reaction (A-4CR) and the Ugi three-component reaction (U-3CR). In the first step, six known 2,5-dihydro-1,3-thiazoles (3-thiazolines) were prepared in good yields (up to 73%) by the A-4CR using an α-chloro aldehyde, a carbonyl compound (aldehyde or ketone), ammonia and sodium hydrosulfide. Sequentially, the U-3CRs were carried out combining the 3-thiazolines with different isocyanides and *trans*-bromoacrylic acid. In this step, 14 bisamides were isolated in modest to high yields (28–88%). The main feature of this methodology is the fact that the synthetic route of the precursors consists of a sequential combination of two MCRs, making possible the fast synthesis of diversified complex heterocycles. Furthermore, the rearrangement step has tolerated many substituents at position R^5.

Jana and coworkers [61] reported the synthesis of fused heteropolycycles under metal and oxidant free conditions by a novel C–H functionalization via diastereoselective MCRs of *N*-heterocycles (Scheme 11.30). The reactions were carried out with three components: pyrrolidine, aromatic aldehydes with electron withdrawing groups as carbonyl compounds (the reactions did not work with electron donating groups as substituents), and several substituted cinnamic acids as dipolarophiles. 15 pyrrolizidine derivatives were isolated in moderate yields (35–53%) and with excellent stereoselectivity as all products were obtained as a single diastereoisomer with four contiguous stereocenters.

Shi and coworkers [62] synthesized with excellent regio- and diastereoselectivity 26 fused polycyclic bispiroindolines containing two contiguous spiral atoms in moderate to excellent yields (41–90%) by a new catalyst-free one-pot Ugi-type reaction

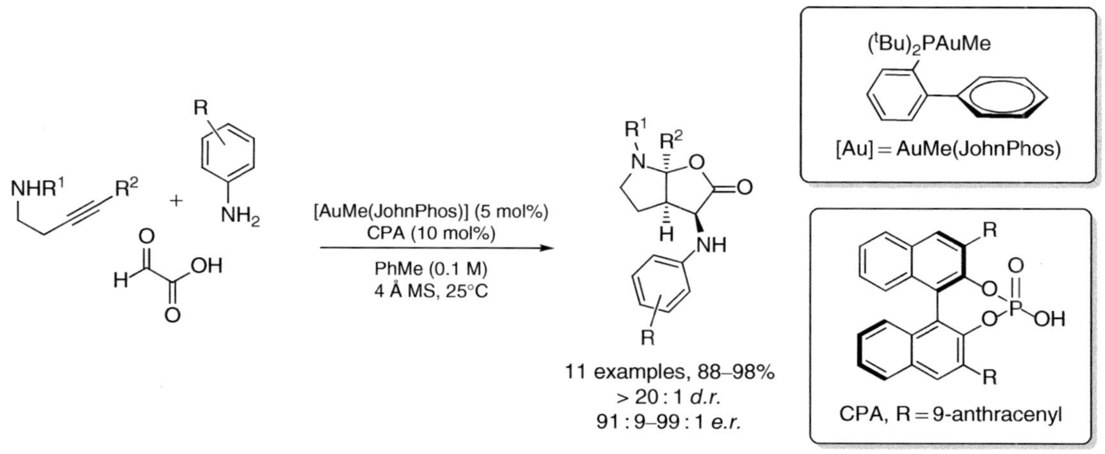

Scheme 11.28 Multicatalytic MCR synthesis of furo[2,3-b]pyrrole derivatives.

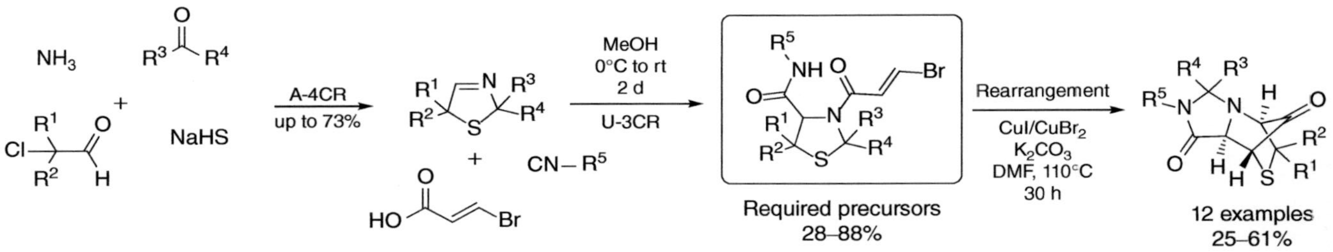

Scheme 11.29 Synthetic pathway using the modified A-4CR and U-3CR to obtain bisamides containing a thiazolidine substructure.

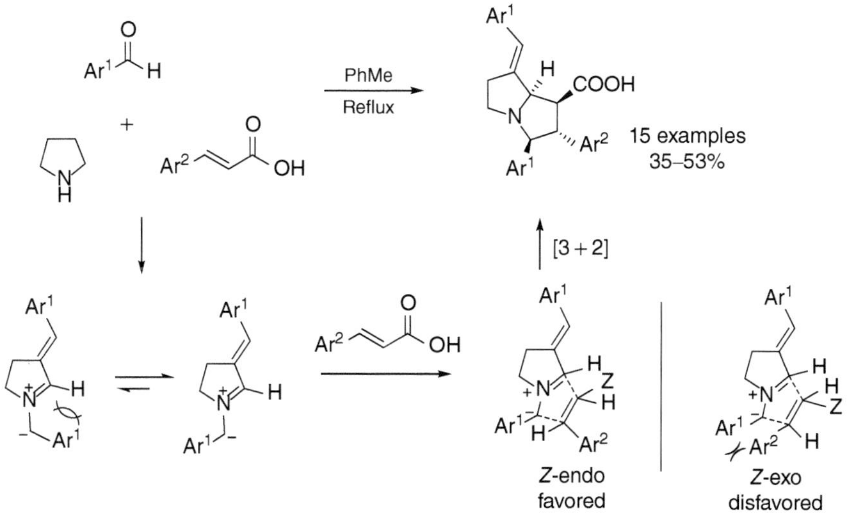

Scheme 11.30 Synthesis of fused heteropolycycles by CH–MCR with substituted cinnamic acids as dipolarophiles and proposal on the origin of diastereoselectivity.

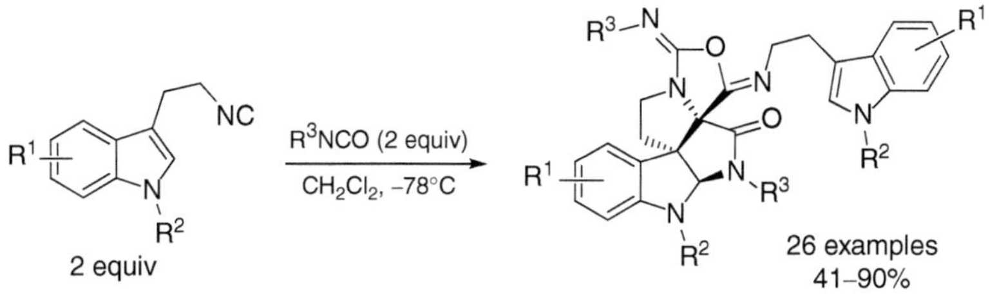

Scheme 11.31 Synthesis of fused polycyclic bispiroindolines containing two contiguous spiral atoms via an Ugi-type reaction.

using two molecules of 2-isocyanoethylindole and isocyanates in dichloromethane (Scheme 11.31). This strategy provided complex products in a simple and easy pathway. Moreover, this method stands out because two spiro atoms are incorporated in the final products, unlike the other similar methodologies to afford spiroindolines described in the literature [63, 64]. Furthermore, the bispirocyclic compounds were obtained as single diastereoisomers.

Arumugam et al. [65] have reported an expedient regio-, stereo- and product-selective synthesis of novel hybrid heterocyclic systems comprising [1,2-c] oxazolidine, pyrrolidine and piperidine units. This was possible through a three-component 1,3-dipolar cycloaddition reaction of the azomethine ylide generated *in situ* with acenapthenequinone and L-phenylalanine, followed by a reaction of annulation with paraformaldehyde (Scheme 11.32). This procedure provides a novel hybrid heptacyclic ring system with seven contiguous stereocenters in good to excellent yields.

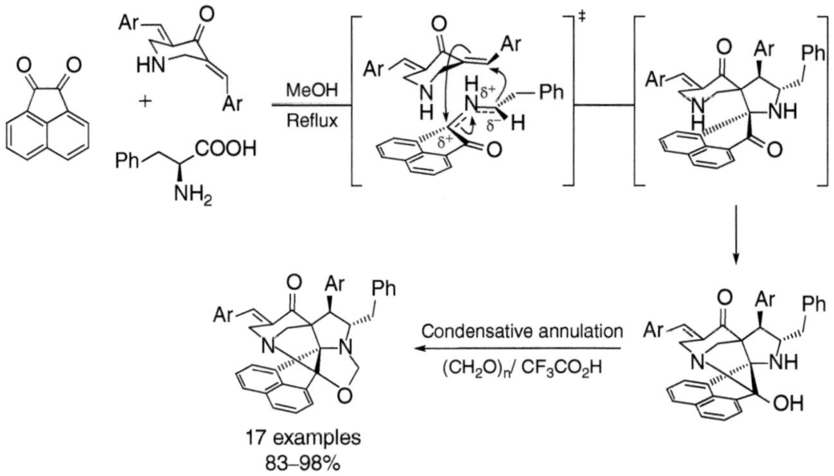

Scheme 11.32 MCR synthesis of a novel heptacyclic ring system.

Coldham and Choi [66] synthesized novel hetero-tetracycles as single racemic diastereomers by a three-component reaction involving an azomethine intermediate, which reacts in a 1,3-dipolar cycloaddition fashion (Scheme 11.33), though a sequential mechanism is not discarded by the authors. This work is relevant since it provides a faster pathway to furnish the piperazinone core and expands the scope for medicinal chemistry. Indeed, there are some drugs in tests with a piperazinone-type core for the treatment of Hepatitis C virus (HCV) and other pathogens [67].

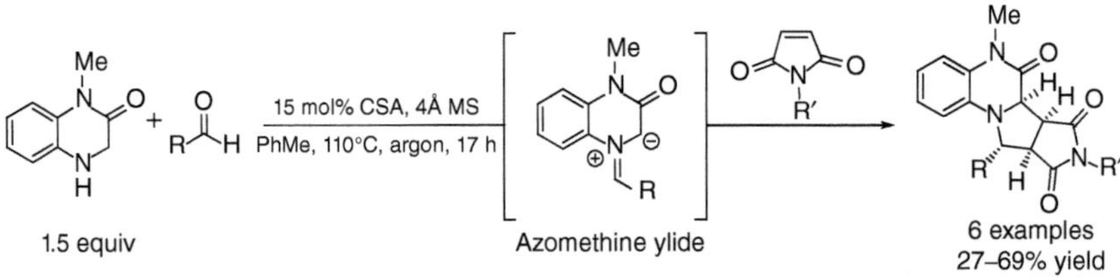

Scheme 11.33 Stereoselective MCR synthesis of hetero-tetracycles.

Another example of azomethine ylides was reported by Samala et al. [68] to achieve polycyclic fused pyrrolizidines by 1,3-dipolar cycloaddition in a three-component reaction fashion, using as reagents amino acids (mostly L-proline), aldehydes and *N*-aromatic zwitterions (Scheme 11.34a). The mechanism driving force is the azomethine ylide intermediate formed *in situ* by a decarboxylation of proline. The reaction is regiospecific and some calculations pointed out that the orbitals involved in the transformation is the highest occupied molecular orbital (HOMO) of azomethine ylide (1,3-dipole) and the lowest unoccupied molecular orbital (LUMO) of pyridinium zwitterion (dipolarophile) (Scheme 11.34b). Then, the expected octacyclic product is not observed, but instead a 1 : 1 ratio of **A** and

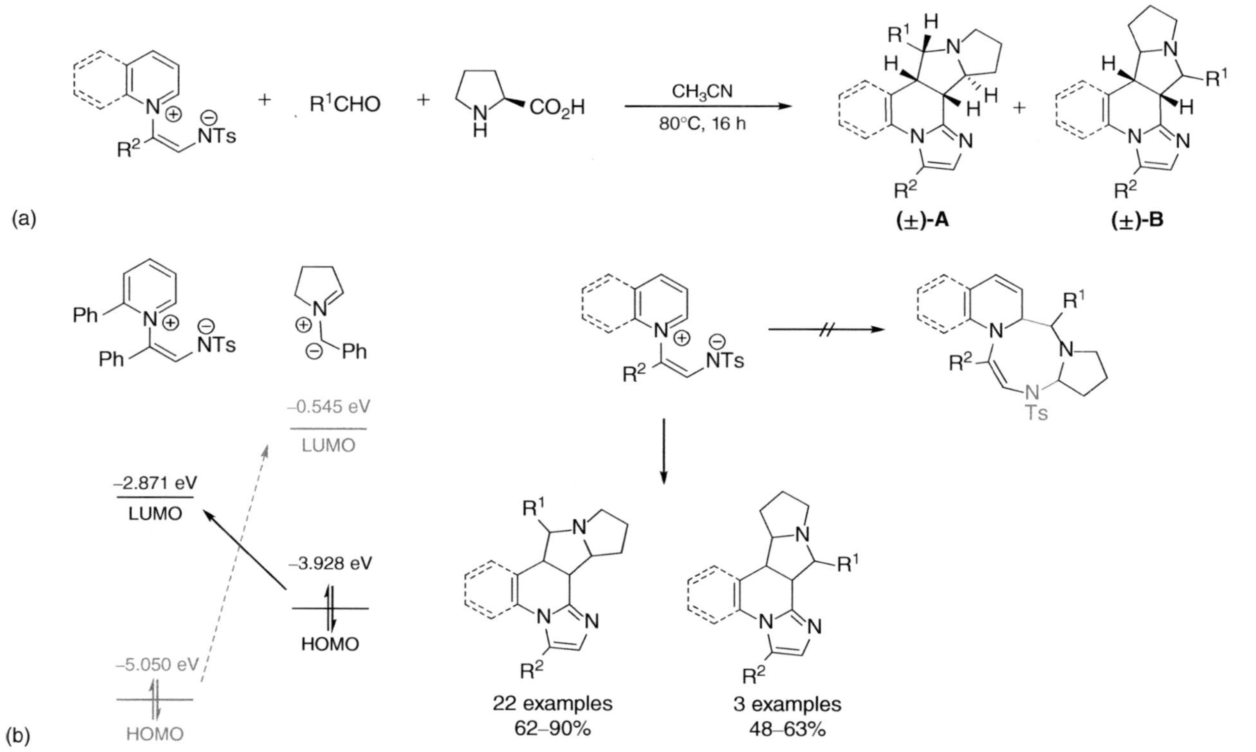

Scheme 11.34 Stereoselective MCR synthesis of hetero-tetracycles. (a) Synthesis of polycyclic fused pyrrolizidines through azomethine ylides. (b) Molecular orbital energies for the transformation.

B are produced as a mixture of 1,3-dipolar cycloadducts. When an aldehyde with an electron-withdrawing group such as cyanide or nitro is present, **A** is obtained exclusively (22 examples).

Xu et al. [69] have developed an efficient method for the synthesis of substituted tropane derivatives via an asymmetric dearomatizing double Mannich reaction of isoquinolines with enals, catalyzed by a chiral *N*-heterocyclic carbene (NHC) (Scheme 11.35a). A variety of aromatic enals and isoquinolinium salts was successfully used to give the desired compounds in moderate to excellent yields with excellent enantio- and diastereoselectivities. Lower yields and enantioselectivities were observed when the dearomatizing double Mannich reaction was performed with aliphatic enals. Similar results were obtained by Kang et al. [70] under mild reaction conditions to afford multisubstituted tetra- and dihydroquinoline derivatives through palladium-catalyzed three-component reactions of ammonium ylides with *N*-alkylquinolinium salts (Scheme 11.35b).

Guo et al. [71] reported an unexpected multicomponent one-pot cascade reaction with "bis-nucleophilic/electrophilic" reagents (Nu–Nu–E), offering a synthetic strategy to the formation of furanobenzodihydropyran-fused polycyclic heterocycles (Scheme 11.36). The protocol had a wide tolerance to *ortho*-hydroxychalcones and the desired products were obtained in all cases, except when steric bulky ketones were used (R = Ph, X = O; R = Ph, X = NTs). In this latter case, only bridged ketals were formed.

Min et al. [72] reported the synthesis of chiral functionalized hydroepoxyisochromenes via one-pot enantioselective multi-component cascade reactions of diazo compounds, ethyl 4-hydroxybutenoate and furfurals, followed by a Diels–Alder reaction (Scheme 11.37). This enantioselective three-component/Diels–Alder process furnished hydroepoxyisochromene derivatives in high diastereo- and enantioselectivities with six contiguous chiral centers, albeit in long reaction times.

The Povarov reaction is a typical example of a method applied to combinatorial chemistry that has several examples in obtaining complex molecules of high added value [73]. Traditionally, it occurs primarily by the formation of the imine component by the condensation reaction between an aniline and an aromatic aldehyde [74]. The resulting imine reacts with the electron-rich dienophiles by an aza-Diels–Alder electrophilic cycloaddition reaction. Cerra et al. [75] developed a remarkable flow chemistry configuration for the synthesis of tricyclic tetrahydroquinolines containing a furan core via the Povarov reaction (Scheme 11.38). This flow mode was designed with an integrated reaction step, a work-up by liquid-liquid membrane separator, purification with automated silica gel flash chromatography and analysis via high precision liquid chromatography (HPLC). Another work from the same group [76] also used an MCR approach for the synthesis of drugs containing fused heterocycles (Scheme 11.39). An integrated flow chemistry mode set up was developed towards the synthesis of chiral tetracyclic tetrahydroquinolines via the multicomponent Povarov reaction.

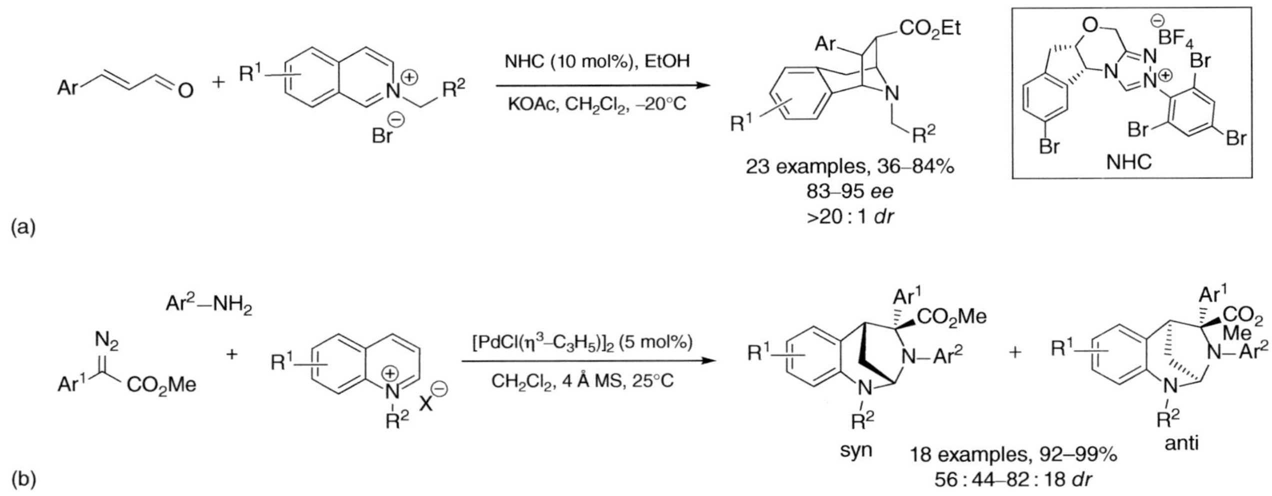

Scheme 11.35 (a) MCR synthesis of tropane derivatives. (b) MCR synthesis of tetra- and dihydroquinoline derivatives.

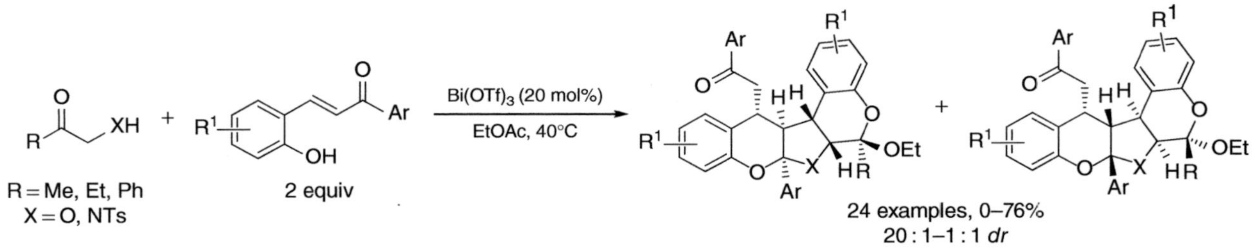

Scheme 11.36 MCR approach to furanobenzodihydropyran-fused polycyclic heterocycles.

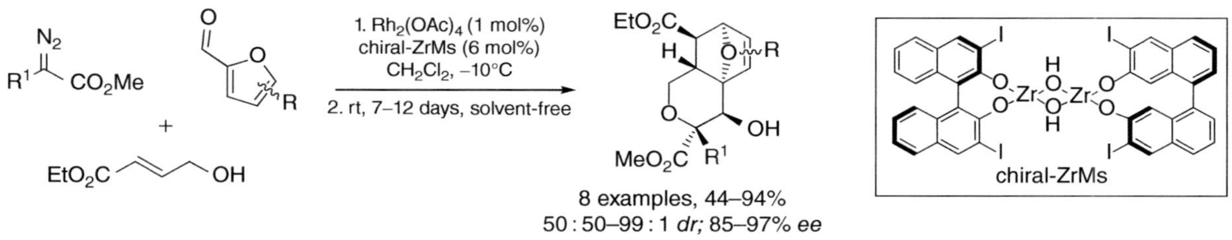

Scheme 11.37 A three-component/Diels–Alder process to obtain hydroepoxyisochromene derivatives.

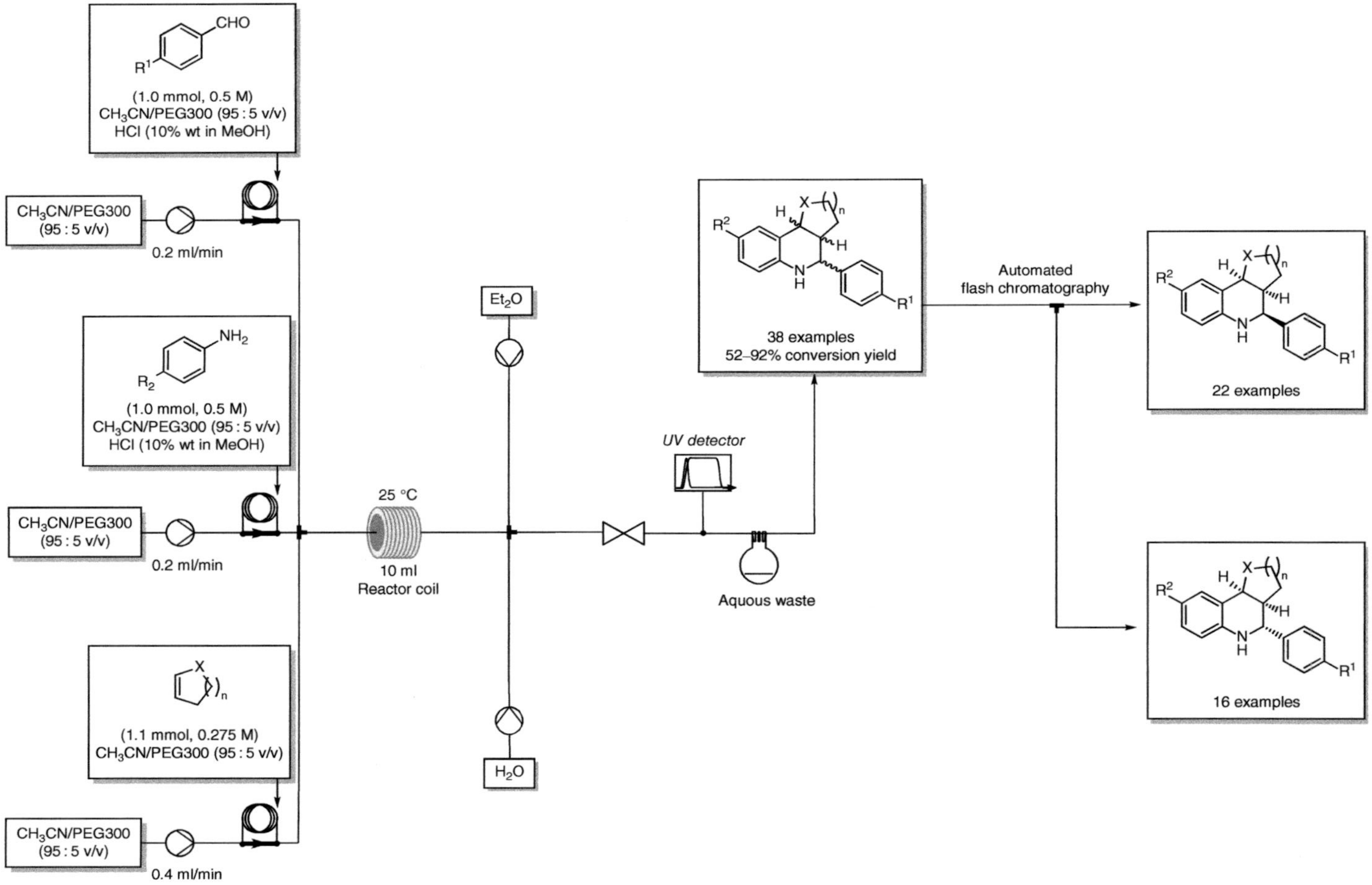

Scheme 11.38 Integrated flow chemistry set up for the synthesis of tricyclic tetrahydroquinolines.

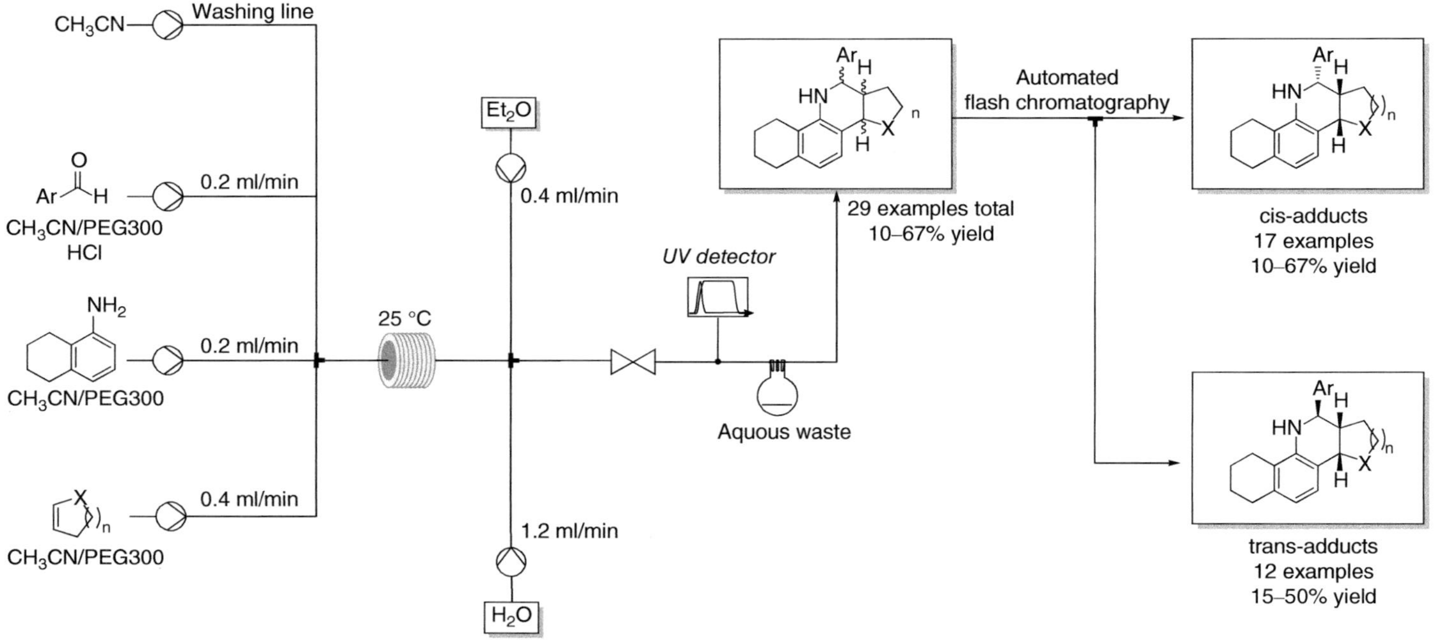

Scheme 11.39 Integrated flow chemistry set up for the synthesis of chiral tetracyclic tetrahydroquinolines.

11.9 Spiro Heterocycles

Isatin has been used in many organic syntheses of different types of spiro-fused heterocyclic compounds, which usually have a significant biological activity for potential pharmaceutical products. The structure of isatin has a reactive α-keto amide (electrophilicity similar to aldehydes) making it an important reagent for chemical transformations such as Knoevenagel-type reactions, dipolarophile formation, ring-opening, etc. Furthermore, the application of isatin on MCRs provides a wide diversity of spiroheterocyclic compounds with different biological activities reported [77–79]. An interesting approach for the synthesis of novel potent inhibitors of advanced glycation end product has been reported by Kaur et al. [80] via reaction of azomethine ylides, generated *in situ* from isatin, and benzyl amine with substituted α,β-unsaturated carbonyl compounds (Scheme 11.40). This process furnished 4-aryl-3-benzoyl-5-phenylspiro[pyrrolidine-2,3′-indolin]-2′-one derivatives in good yields and with a high degree of regio- and stereoselectivity.

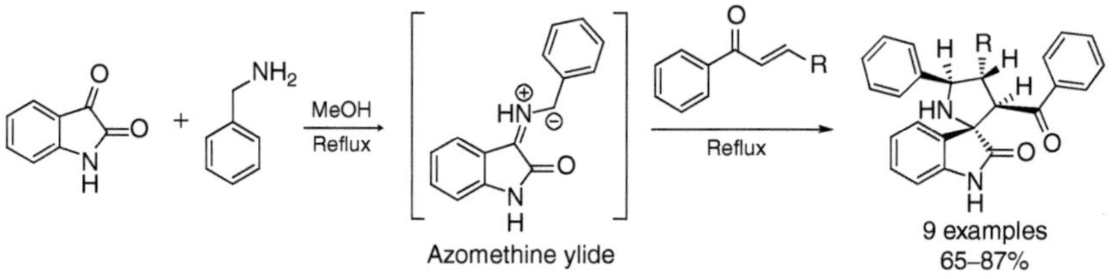

Scheme 11.40 MCR synthesis of spiro-fused heterocycles from isatin.

Kumar et al. [81] have described a novel diastereoselective synthesis of spiropyrrolidineoxindole derivatives through a one-pot MCR via 1,3-dipolar cycloaddition of isatin, 2-phenylthiazolidine-4-carboxylic acid and chalcones under mild reaction conditions with high diastereoselectivity and high yields (Scheme 11.41a). The

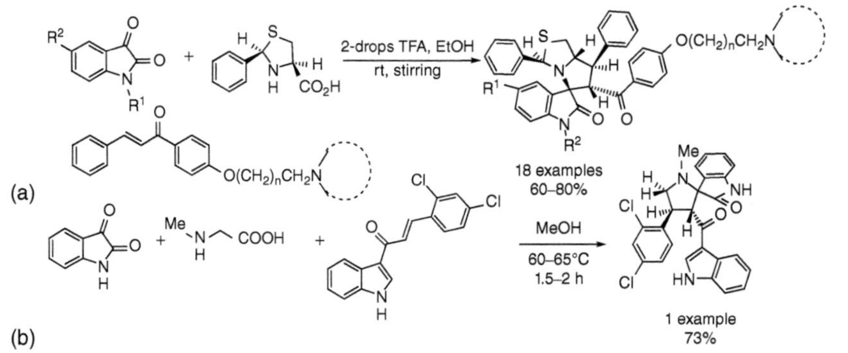

Scheme 11.41 MCRs between isatin, chalcones and aminoacids. (a) Synthesis of spiropyrrolidineoxindole derivatives through MCRs between isatin, chalcones, and aminoacids. (b) Synthesis of spiroindolone analogue through MCRs between isatin, chalcones, and aminoacids.

synthesized compounds showed potential anticancer activity against human breast cancer cell lines. Also from an MCR envolving isatin, chalcone and an aminoacid, Al-Majid et al. [82] have developed a synthesis of a spiroindolone analogue, which showed high selectivity against colon cancer, weak to moderate activity against breast cancer and weak activity against liver cancer (Scheme 11.41b).

Lin et al. [83] have reported a one-pot MCR of different 3-methyl-4-nitro-5-isatylidenyl-isoxazoles with azomethine ylides, generated *in situ* from isatins and proline or thioproline, to produce polycyclic 3,3′-pyrrolidinyl-dispirooxindoles in high yields with good diastereoselectivity (Scheme 11.42). Interestingly, the presence of 3-methyl-4-nitro-5-isatylidenyl-isoxazoles as 1,3-dipolarophiles showed considerable anticancer activity against leukemia cells, prostate cancer cells and human lung cancer cells.

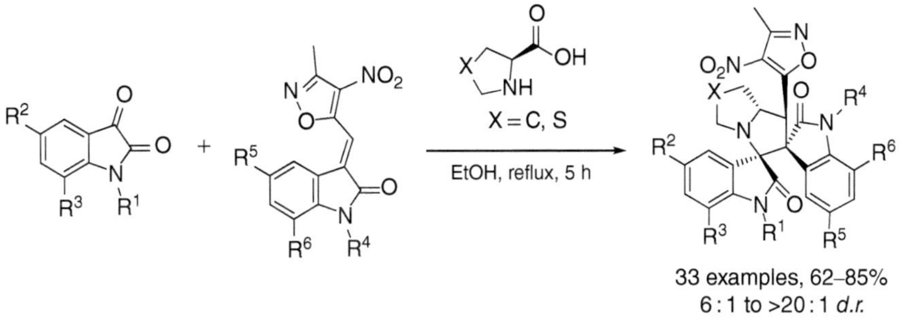

Scheme 11.42 MCR synthesis of polycyclic 3,3′-pyrrolidinyl-dispirooxindoles.

A series of spirooxindolo-pyrrolidines, pyrrolizidines, and pyrrolothiazoles hybrid compounds has been reported by Wu et al. [84] via regioselective three-component

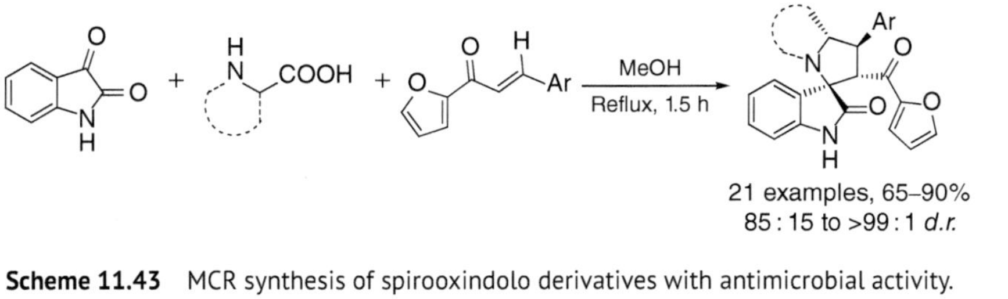

Scheme 11.43 MCR synthesis of spirooxindolo derivatives with antimicrobial activity.

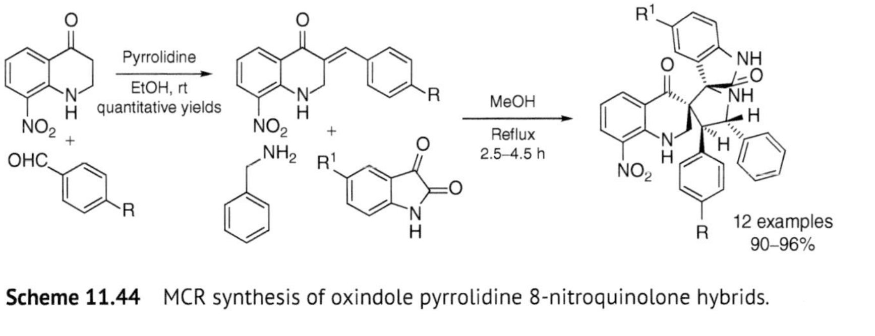

Scheme 11.44 MCR synthesis of oxindole pyrrolidine 8-nitroquinolone hybrids.

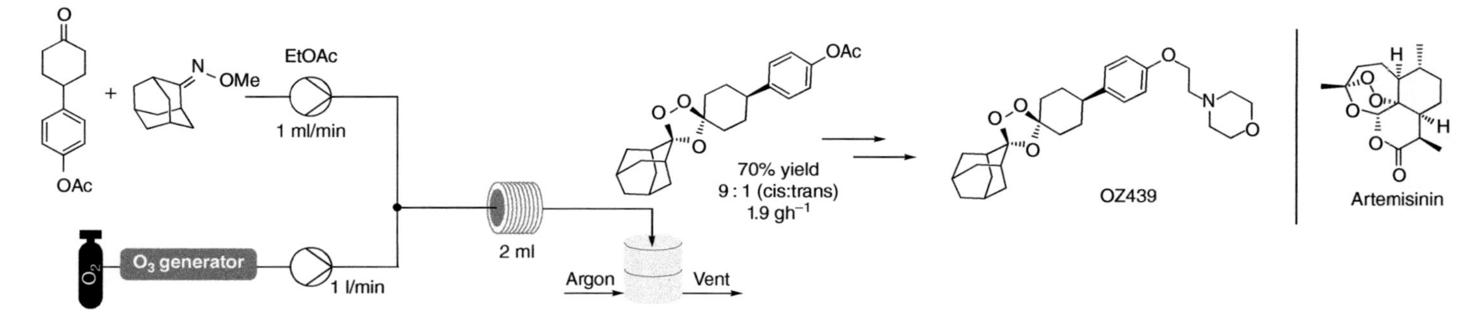

Scheme 11.45 Flow mode for the Griesbaum co-ozonolysis.

reactions between α,β-unsaturated ketones and azomethine ylides, generated *in situ* from amino acids and isatin in methanol (Scheme 11.43). When the reaction was performed in protic solvents, the yields were greater if compared to reactions performed in aprotic solvents, however, when water was used, no product was observed. Some of these compounds have been reported to exhibit antimicrobial activities compared to drugs such as norfloxacin, levofloxacin, and ciprofloxacin.

Recently, Shyamsivappan et al. [85] reported the synthesis of a series of oxindole pyrrolidine 8-nitroquinolone hybrids as single regioisomers in excellent yields via a one-pot three-component reaction of azomethine ylides generated *in situ* from isatins and benzylamine with (*E*)-3-arylidene-2,3-dihydro-8-nitro-4-quinolones (Scheme 11.44). Some of the synthesized compounds have exhibited potent inhibition against human cervical cancer cells.

Lau et al. [86] developed a one-step gas/liquid flow setup for the synthesis of the antimalarial drug candidate OZ439 (artefenomel), which is currently in phase II clinical trials (Scheme 11.45). This compound shows improved pharmacokinetic properties compared with other antimalarial drugs, such as artemisinin. The biggest challenge to synthesize this medicinal compound is the formation of the trioxolane five-membered ring. Traditionally, this nucleus is obtained by the Griesbaum co-ozonolysis reaction [87]. A solution containing the oxime and ketone component in diluted concentrations was combined with a stream of ozone. In this way the desired product was obtained in good yield (70%) and high productivity over three hours as a mixture of isomers favoring the desired *cis* isomer. Compared to the batch reaction, this flow mode is a much safer way of handling ozone due to the exchange of the pentane solvent for ethyl acetate, which is more environmentally friendly. Furthermore, two other steps of the synthetic sequence were carried out under flow conditions to prepare the starting ketone: partial hydrogenation and acetylation. This methodology can be considered a more efficient way to scale up the synthesis of trioxolane as a desired product.

References

1 Weatherly, C.D., Rigoli, J.W., and Schomaker, J.M. (2012). Synthesis of 1,3-diaminated stereotriads via rearrangement of 1,4-diazaspiro[2.2]pentanes. *Org. Lett.* 14: 1704–1707.

2 Paz, M.M., Zhang, X., Lu, J. et al. (2012). A new mechanism of action for the anticancer drug mitomycin C: mechanism-based inhibition of thioredoxin reductase. *Chem. Res. Toxicol.* 25: 1502–1511.

3 Corre, C. and Lowden, P.A.S. (2004). The first biosynthetic studies of the azinomycins: acetate incorporation into azinomycin B. *Chem. Commun.* 8: 990–991.

4 Bierschenk, S.M., Bergman, R.G., Raymond, K.N. et al. (2020). A nanovessel-catalyzed three-component Aza-Darzens reaction. *J. Am. Chem. Soc.* 142: 733–737.

5 Hong, C.M., Bergman, R.G., Raymond, K.N. et al. (2018). Self-assembled tetrahedral hosts as supramolecular catalysts. *Acc. Chem. Res.* 51: 2447–2455.

6 Miyamura, H., Bergman, R.G., Raymond, K.N. et al. (2020). Heterogeneous supramolecular catalysis through immobilization of anionic M4L6 assemblies on cationic polymers. *J. Am. Chem. Soc.* 142: 19327–19338.
7 Hashimoto, T., Nakatsu, H., Yamamoto, K. et al. (2011). Chiral Brønsted acid-catalyzed asymmetric trisubstituted aziridine synthesis using α-diazoacyl oxazolidinones. *J. Am. Chem. Soc.* 133: 9730–9733.
8 Jiang, J., Liu, H., Lu, C.-D. et al. (2016). Diethyl phosphite initiated coupling of α-ketoesters with imines for synthesis of α-phosphonyloxy-β-amino acid derivatives and aziridine-2-carboxylates. *Org. Lett.* 18: 880–883.
9 Pitts, C.R. and Lectka, T. (2014). Chemical synthesis of β-lactams: asymmetric catalysis and other recent advances. *Chem. Rev.* 114: 7930–7953.
10 Ham, J.S., Park, B., Son, M. et al. (2020). C–H/C–C functionalization approach to *N*-fused heterocycles from saturated azacycles. *J. Am. Chem. Soc.* 142: 13041–13050.
11 Qi, J., Wei, F., Huang, S. et al. (2021). Copper(I)-catalyzed asymmetric interrupted Kinugasa reaction: synthesis of α-thiofunctional chiral β-lactams. *Angew. Chem. Int. Ed.* 60: 4561–4565.
12 Zhou, C., Dong, Y., Yu, J.-T. et al. (2019). Palladium/copper-catalyzed multicomponent reactions of propargylic amides, halohydrocarbons and CO_2 toward functionalized oxazolidine-2,4-diones. *Chem. Commun.* 55: 13685–13688.
13 Huang, W.-B., Ren, F.-Y., Wang, M.-W. et al. (2020). Cu(II)-catalyzed phosphonocarboxylative cyclization reaction of propargylic amines and phosphine oxide with CO_2. *J. Org. Chem.* 85: 14109–14120.
14 Song, Q.-W., Yu, B., Li, X.-D. et al. (2014). Efficient chemical fixation of CO_2 promoted by a bifunctional Ag_2WO_4/Ph_3P system. *Green Chem.* 16: 1633–1638.
15 Sun, S., Wang, B., Gu, N. et al. (2017). Palladium-catalyzed arylcarboxylation of propargylic alcohols with CO_2 and aryl halides: access to functionalized α-alkylidene cyclic carbonates. *Org. Lett.* 19: 1088–1091.
16 Gladow, D., Senf, D., Wiecko, J. et al. (2017). New trifluoromethyl-substituted heterocycles by multicomponent reactions of siloxycyclopropanes. *Chem. Heterocycl. Compd.* 53: 416–421.
17 Polinara-García, L.A., Montesinos-Miguel, D., and Vazquez, A. (2015). An efficient microwave-assisted synthesis of cotinine and iso-cotinine analogs from an Ugi-4CR approach. *Org. Biomol. Chem.* 13: 9065–9071.
18 Kaminskyy, D., Khyluk, D., Vasylenko, O. et al. (2012). An efficient method for the transformation of 5-ylidenerhodanines into 2,3,5-trisubstituted-4-thiazolidinones. *Tetrahedron Lett.* 53: 557–559.
19 Li, Y., Giulianotti, M., and Houghten, R.A. (2009). Solid-phase synthesis of 2-alkylidene-6-alkyl-imidazo[2,1-b]thiazole-3,5[2H,6H]-dione derivatives. *Tetrahedron Lett.* 50: 5857–5859.
20 Ansari, A., Ali, A., Asif, M. et al. (2019). Steroidal thiazolidinone derivatives: design, synthesis and their molecular interaction with human serum albumin. *Steroids* 148: 99–113.
21 Ibrar, A., Zaib, S., Khan, I. et al. (2017). New prospects for the development of selective inhibitors of α-glucosidase based on coumarin-iminothiazolidinone

hybrids: synthesis, *in vitro* biological screening and molecular docking analysis. *J. Taiwan Inst. Chem. Eng.* 81: 119–133.

22 Tomasic, T. and Masic, L. (2009). Rhodanine as a privileged scaffold in drug discovery. *Curr. Med. Chem.* 16: 1596–1629.

23 Jacobine, A.M. and Posner, G.H. (2011). Three-component, one-flask synthesis of rhodanines (thiazolidinones). *J. Org. Chem.* 76: 8121–8125.

24 Berthet, M., Cheviet, T., Dujardin, G. et al. (2016). Isoxazolidine: a privileged scaffold for organic and medicinal chemistry. *Chem. Rev.* 116: 15235–15283.

25 Gupta, E., Zaheer, M.K., Kant, R. et al. (2019). Additive-free regio- and diastereoselective construction of fully-substituted isoxazolidines employing diazo compounds. *Org. Chem. Front.* 6: 1109–1113.

26 Baumann, M., Baxendale, I.R., Kirschning, A. et al. (2011). Synthesis of highly substituted nitropyrrolidines, nitropyrrolizines and nitropyrroles via multicomponent-multistep sequences within a flow reactor. *Heterocycles* 82: 1297–1316.

27 Meusel, M. and Gütschow, M. (2004). Recent developments in hydantoin chemistry. A review. *Org. Prep. Proced. Int.* 36: 391–443.

28 Monteiro, J.L., Pieber, B., Corrêa, A.G. et al. (2016). Continuous synthesis of hydantoins: intensifying the Bucherer–Bergs reaction. *Synlett* 27: 83–87.

29 Kim, R., Bihud, V., Mohamad, K. et al. (2012). Cytotoxic and antioxidant compounds from the stem bark of *Goniothalamus tapisoides Mat Salleh. Molecules* 18: 128–139.

30 Chandgude, L. and Dömling, A. (2016, 2016). Convergent three-component tetrazole synthesis. *Eur. J. Org. Chem.*: 2383–2387.

31 Chandgude, A.L., Narducci, D., Kurpiewska, K. et al. (2017). Diastereoselective one pot five-component reaction toward 4-(tetrazole)-1,3-oxazinanes. *RSC Adv.* 7: 49995–49998.

32 Kim, H., Chin, J., Choi, H. et al. (2013). Phosphoiodyns A and B, unique phosphorus-containing iodinated polyacetylenes from a Korean sponge *Placospongia sp. Org. Lett.* 15: 5614.

33 Knouse, K.W., deGruyter, J.N., Schmidt, M.A. et al. (2018). Unlocking P(V): reagents for chiral phosphorothioate synthesis. *Science* 361: 1234–1238.

34 Xu, D., Rivas-Bascón, N., Padial, N.M. et al. (2020). Enantiodivergent formation of C–P bonds: synthesis of P-chiral phosphines and methylphosphonate oligonucleotides. *J. Am. Chem. Soc.* 142: 5785–5792.

35 Yavari, I., Nematpour, M., and Hossaini, Z. (2010). A synthesis of phosphorylated dioxohexahydropyrimidines from *N,N′*-dimethylurea, activated acetylenes, and trialkyl phosphites. *Mol. Divers.* 14: 617–620.

36 Candeias, N.R., Paterna, R., Cal, P.M.S.D. et al. (2012). A sustainable protocol for the aqueous multicomponent Petasis borono–Mannich reaction. *J. Chem. Educ.* 89: 799–802.

37 Gein, V.L., Zamaraeva, T.M., Gorgopina, E.V. et al. (2020). A four-component Biginelli reaction: new opportunities for the synthesis of functionalized pyrimidines. *Chem. Heterocycl. Compd.* 56: 339–346.

38 Sathicq, A.G., Ruiz, D.M., Constantieux, T.T. et al. (2014). Preyssler heteropoly acids encapsulated in a silica framework for an efficient preparation of fluorinated hexahydropyrimidine derivatives under solvent-free conditions. *Synlett* 25: 881–883.

39 Palermo, V., Sathicq, A.G., Constantieux, T. et al. (2015). New vanadium Keggin heteropolyacids encapsulated in a silica framework: recyclable catalysts for the synthesis of highly substituted hexahydropyrimidines under suitable conditions. *Catal. Lett.* 145: 1022–1032.

40 Zohdi, H.F., Rateb, N.M., and Elnagdy, S.M. (2011). Green synthesis and antimicrobial evaluation of some new trifluoromethyl-substituted hexahydropyrimidines by grinding. *Eur. J. Med. Chem.* 46: 5636–5640.

41 Khabibullina, G.R., Fedotova, E.S., Akhmetova, V.R. et al. (2016). A green synthesis in water of novel (1,5,3-dithiazepan-3-yl)alkanoic acids by the multicomponent reaction of amino acids, CH_2O, and 1,2-ethanedithiol. *Mol. Divers.* 20: 557–565.

42 Fox, D.J., Reckless, J., Wilbert, S.M. et al. (2005). Identification of 3-(acylamino)azepan-2-ones as stable broad spectrum chemokine inhibitors resistant to metabolism *in vivo*. *J. Med. Chem.* 48: 867–874.

43 Rasouli, M.A., Mahdavi, M., Ranjbar, P.R. et al. (2012). A green one-pot synthesis of *N*-alkyl-2-(2-oxoazepan-1-yl)-2-arylacetamide derivatives via an Ugi four-center, three-component reaction in water. *Tetrahedron Lett.* 53: 7088–7092.

44 Echemendía, R., Rabêlo, W.F., López, E.R. et al. (2018). A bidirectional access to novel thiadiazine hybrid molecules by double multicomponent reactions. *Tetrahedron Lett.* 59: 4050–4053.

45 Wessjohann, L.A., Rivera, D.G., and Vercillo, O.E. (2009). Multiple multicomponent macrocyclizations (MiBs): a strategic development toward macrocycle diversity. *Chem. Rev.* 109: 796–814.

46 Webster, A.M. and Cobb, S.L. (2018). Recent advances in the synthesis of peptoid macrocycles. *Chem. Eur. J.* 24: 7560–7573.

47 Dohm, M.T., Kapoor, R., and Barron, A.E. (2011). Peptoids: bio-inspired polymers as potential pharmaceuticals. *Curr. Pharm. Des.* 17: 2732–2747.

48 Barreto, A.F.S., Vercillo, O.E., Wessjohann, L.A. et al. (2014). Consecutive isocyanide-based multicomponent reactions: synthesis of cyclic pentadepsipeptoids. *Beilstein J. Org. Chem.* 10: 1017–1022.

49 Barreto, A.F.S. and Andrade, C.K.Z. (2019). Synthesis of (macro)heterocycles by consecutive/repetitive isocyanide-based multicomponent reactions. *Beilstein J. Org. Chem.* 15: 906–930.

50 Barreto, A.F.S. and Andrade, C.K.Z. (2018). Microwave-mediated synthesis of a cyclic heptapeptoid through consecutive Ugi reactions. *Tetrahedron* 75: 6861–6865.

51 Gaich, T. and Baran, P.S. (2010). Aiming for the ideal synthesis. *J. Org. Chem.* 75: 4657–4673.

52 Santra, S. and Andreana, P.R. (2011). A bioinspired Ugi/Michael/aza-Michael cascade reaction in aqueous media: natural-product-like. *Mol. Divers.* 50: 9418–9422.

53 Jin, H., Liu, D., Zhou, B. et al. (2020). One-pot copper-catalyzed three-component reaction of sulfonyl azides, alkynes, and allylamines to access 2,3-dihydro-1H-imidazo[1,2-a]indoles. *Synthesis* 52: 1417–1424.

54 Yugandhar, D., Kuriakose, S., Nanubolu, J.B. et al. (2018). Synthesis of alkaloid-mimicking tricyclic skeletons by diastereo- and regioselective Ugi/ipso-cyclization/aza-Michael cascade reaction in one-pot. *Org. Lett.* 18: 1040–1043.

55 Mertens, A., Zilch, H., König, B. et al. (1993). Selective nonnucleoside HIV-1 reverse transcriptase inhibitors. New 2,3-dihydrothiazolo [2,3-a]isoindol-5(9bH)-ones and related compounds with anti-HIV-1 activity. *J. Med. Chem.* 36: 2526–2535.

56 Trapani, G., Franco, M., Latrofa, A. et al. (1994). Synthesis and anticonvulsant activity of some 1,2,3,3a-tetrahydropyrrolo[2,1-b]benzothiazol-1-ones and pyrrolo[2,1-b]-thiazole analogues. *Eur. J. Med. Chem.* 29: 197–204.

57 Xiong, Q., Li, G., Dong, S. et al. (2019). Enantioselective synthesis of hydrothiazole derivatives via an isocyanide-based multicomponent reaction. *Org. Lett.* 21: 8771–8775.

58 Xiong, Q., Dong, S., Chen, Y. et al. (2019). Asymmetric synthesis of tetrazole and dihydroisoquinoline derivatives by isocyanide-based multicomponent reactions. *Nat. Commun.* 10: 1–10.

59 Cala, L., Villar, P., de Lera, A.R. et al. (2020). Multicomponent and multicatalytic asymmetric synthesis of furo[2,3-b]pyrrole derivatives: further insights into the mode of action of chiral phosphoric acid catalysts. *Chem. Sci.* 11: 9181–9190.

60 Kröger, D., Franz, M., Schmidtmann, M. et al. (2015). Sequential multicomponent reactions and a Cu-mediated rearrangement: diastereoselective synthesis of tricyclic ketones. *Org. Lett.* 17: 5866–5869.

61 Mandal, S., Dwari, S., and Jana, C.K. (2018). Metal free C–H functionalization enabled diastereoselective multicomponent reaction of *N*-heterocycles to fused heteropolycycles. *J. Org. Chem.* 83: 8874–8887.

62 Li, L., Liu, J., and Shi, M. (2018). A highly regio- and diastereoselective four-component reaction to construct polycyclic bispiroindolines from 2-isocyanoethylindoles and isocyanates. *Org. Lett.* 20: 7076–7079.

63 Zhao, X., Liu, X., Mei, H. et al. (2015). Asymmetric dearomatization of indoles through Michael/Friedel-Crafts-type cascade to construct polycyclic spiroindolines. *Angew. Chem. Int. Ed.* 54: 4032–4035.

64 Wang, X., Wang, S.-Y., and Ji, S.-J. (2014). Chemoselective synthesis of polycyclic spiroindolines and polysubstituted pyrroles via the domino reaction of 2-isocyanoethylindoles. *J. Org. Chem.* 79: 8577–8583.

65 Arumugam, N., Almansour, A.I., Kumar, R.S. et al. (2013). A 1,3-dipolar cycloaddition-annulation protocol for the expedient regio-, stereo- and product-selective construction of novel hybrid heterocycles comprising seven rings and seven contiguous stereocentres. *Tetrahedron Lett.* 53: 2515–2519.

66 Choi, A. and Coldham, I. (2019). Three-component couplings for the synthesis of pyrroloquinoxalinones by azomethine ylide 1,3-dipolar cycloaddition chemistry. *Tetrahedron Lett.* 60: 151023.

67 Kakarla, R., Liu, J., Naduthambi, D. et al. (2014). Discovery of a novel class of potent HCV NS4B inhibitors: SAR studies on piperazinone derivatives. *J. Med. Chem.* 57: 2136–2160.

68 Samala, S., Ryu, D.H., Song, C.E. et al. (2019). Multicomponent dipolar cycloadditions: efficient synthesis of polycyclic fused pyrrolizidines via azomethine ylides. *Org. Biomol. Chem.* 17: 1773–1777.

69 Xu, J.-H., Zheng, S.-C., Zhang, J.-W. et al. (2016). Construction of tropane derivatives by the organocatalytic asymmetric dearomatization of isoquinolines. *Angew. Chem. Int. Ed.* 55: 1–7.

70 Kang, Z., Zhang, D., Hu, W. et al. (2017). Regio- and diastereoselective three-component reactions via trapping of ammonium ylides with *N*-alkylquinolinium salts: synthesis of multisubstituted tetra- and dihydroquinoline derivatives. *Org. Lett.* 19: 3783–3786.

71 Guo, J., Miao, H., Zhao, Y. et al. (2019). An unexpected multi-component one-pot cascade reaction to access furanobenzodihydropyran-fused polycyclic heterocycles. *Chem. Commun.* 55: 5207–5210.

72 Min, T., Yong, W., Yuan, L. et al. (2016). One-pot enantioselective multicomponent cascade reactions for synthesis of chiral functionalized hydro-epoxyisochromenes: a rapid access to molecular complexity. *Acta Chim. Sin* 74: 54–60.

73 Glushkov, V.A. and Tolstikov, A.G. (2008). Synthesis of substituted 1,2,3,4-tetrahydroquinones by the Povarov reaction. New potentials of the classical reaction. *Russ. Chem. Rev.* 77: 137–159.

74 Sridharan, V., Suryavanshi, P.A., and Menéndez, J.C. (2011). Advances in the chemistry of tetrahydroquinolines. *Chem. Rev.* 111: 7157–7259.

75 Cerra, B., Mostarda, S., Custodi, C. et al. (2016). Integrating multicomponent flow synthesis and computational approaches for the generation of a tetrahydroquinoline compound based library. *Med. Chem. Commun.* 7: 439–446.

76 Cerra, B., Carotti, A., Passeri, D. et al. (2019). Exploiting chemical toolboxes for the expedited generation of tetracyclic quinolines as a novel class of PXR agonists. *ACS Med. Chem. Lett.* 10: 677–681.

77 Liu, Y., Wang, H., and Wan, J. (2013). Recent advances in diversity oriented synthesis through isatin-based multicomponent reactions. *Asian J. Org. Chem.* 2: 374–386.

78 Ziarani, G.M., Moradi, R., and Lashgari, N. (2016). Synthesis of spiro-fused heterocyclic scaffolds through multicomponent reactions involving isatin. *ARKIVOC* i: 1–81.

79 Brandão, P., Marques, C., Burke, A.J. et al. (2021). The application of isatin-based multicomponent reactions in the quest for new bioactive and druglike molecules. *Eur. J. Med. Chem.* 211: 113102.

80 Kaur, A., Singh, B., Vyas, B. et al. (2014). Synthesis and biological activity of 4-aryl-3-benzoyl-5-phenylspiro[pyrrolidine-2.3′-indolin]-2′-one derivatives as novel potent inhibitors of advanced glycation end product. *Eur. J. Med. Chem.* 79: 282–289.

81 Kumar, A., Gupta, G., Srivastava, S. et al. (2013). Novel diastereoselective synthesis of spiropyrrolidineoxindole derivatives as anti-breast cancer agents. *RSC Adv.* 3: 4731–4735.

82 Al-Majid, A.M., Ghawas, H.M., Islam, M.S. et al. (2020). Synthesis of spiroindolone analogue via three components reaction of olefin with isatin and sarcosine: anti-proliferative activity and computational studies. *J. Mol. Struct.* 1204: 127500.

83 Lin, B., Zhang, W.-H., Wang, D.-D. et al. (2017). 3-Methyl-4-nitro-5-isatylidenyl-isoxazoles as 1,3-dipolarophiles for synthesis of polycyclic 3,3′-pyrrolidinyl-dispirooxindoles and their biological evaluation for anticancer activities. *Tetrahedron* 73: 5176–5188.

84 Wu, G., Ouyang, L., Liu, J. et al. (2013). Synthesis of novel spirooxindolo-pyrrolidines, pyrrolizidines, and pyrrolothiazoles via a regioselective three-component [3+2] cycloaddition and their preliminary antimicrobial evaluation. *Mol. Divers.* 17: 271–283.

85 Shyamsivappan, S., Vivek, R., Saravanan, A. et al. (2019). Synthesis and X-ray study of dispiro 8-nitroquinolone analogues and their cytotoxic properties against human cervical cancer HeLa cells. *Med. Chem. Commun.* 10: 439–449.

86 Lau, S.-H., Galván, A., Merchant, R.R. et al. (2015). Machines vs malaria: a flow-based preparation of the drug candidate OZ439. *Org. Lett.* 17: 3218–3221.

87 Griesbaum, K., Liu, X., Kassiaris, A. et al. (1997). Ozonolyses of O-alkylated ketoximes in the presence of carbonyl groups: a facile access to ozonides. *Liebigs Ann.* 1997: 1381–1390.

12

Multicomponent Reactions and Asymmetric Catalysis

Melody E. Boëtius and Eelco Ruijter

Vrije Universiteit Amsterdam, Amsterdam Institute of Molecular & Life Sciences (AIMMS), Department of Chemistry and Pharmaceutical Sciences, De Boelelaan 1105, Amsterdam 1081 HV, The Netherlands

12.1 Introduction

Multicomponent reactions (MCRs) involve the combination of three or more reactants in a one-pot fashion to form a single product that contains considerable fragments of all reactants [1–4]. MCRs reduce the number of steps required to provide highly diverse and complex structures and thereby avoid time-consuming isolation and purification of intermediates. Other benefits that MCRs provide over conventional multistep reactions include high atom efficiency and the use of inexpensive starting materials [1–3, 5, 6]. Moreover, the chemical structures that MCRs provide often display a wide array of biological activities. Consequently, MCRs have gained considerable research interest and have been applied in the synthesis of complex biologically active compounds, including natural products [1–3, 5].

Owing to the different behavior and/or activities of enantiomers in chiral environments, the need for optically pure compounds in the pharmaceutical and agrochemical industries has prompted intensive research toward efficient asymmetric synthetic routes to access these complex structures [1, 3, 4]. Efforts to develop such asymmetric strategies include the development of chiral auxiliaries and chiral reactants, which give rise to diastereoselective MCRs. In enantioselective MCRs, on the other hand, chiral catalysts are employed to control the formation of stereogenic centers in the reaction of achiral reactants. Chiral catalysts thus have a dual function of both activating reactants and creating a chiral reaction environment [5].

The catalytic approach offers flexibility in the choice of the catalyst, access to either enantiomer by use of the enantiomeric catalyst, and a broader substrate scope compared to the use of chiral reactants [7]. Though both (transition) metal catalysts as well as organocatalysts can be used in asymmetric multicomponent reactions (AMCRs), organocatalysts offer several advantages over metal catalysts as they are less prone to redox processes, cheaper, less toxic, renewable, and readily available [5, 8, 9].

Nevertheless, various reactions still lack satisfactory control over stereochemistry [10]. In this light, contributions of catalytic AMCRs since 2012 will be summarized,

Multicomponent Reactions towards Heterocycles: Concepts and Applications, First Edition.
Edited by Erik V. Van der Eycken and Upendra K. Sharma.

providing insight into new strategies to overcome drawbacks, such as limited scopes, and poor chemo- and enantioselectivities. The types of reactions that will be highlighted are classified by the type of mechanism and are divided into MCRs based on additions to imines, Michael additions, and isocyanide reactivity. The MCRs belonging to these classes are among the most popular and widely used in the synthesis of biologically active compounds [3]. Special attention will be given to the isocyanide-based Passerini- and Ugi(-type) MCRs as their asymmetric versions have recently witnessed a major breakthrough [2].

12.2 Imine-based MCRs

Given the favorable kinetics of the condensation of aldehydes (or ketones) and amines, nucleophilic additions to the resulting imines (or iminium ions) form the basis for several classical MCRs as well as modern variations. The following section will cover recent developments in catalytic asymmetric addition reactions to *in situ*-formed imines.

12.2.1 Strecker Reaction

The Strecker MCR involves the condensation of aldehydes or ketones and primary or secondary amines, followed by addition of a cyanide source to the resulting imine or iminium intermediate to afford α-amino nitriles, which can subsequently be converted into amino acids [1, 5]. Catalysts that have been reported for this reaction include proline-based catalysts [11, 12], chiral phosphoric acids (CPAs) [13], and cinchona alkaloid-based catalysts [14, 15]. A recent example described by Liu et al. utilizes a urea catalyst, affording C^{α}-tetrasubstituted amino nitriles in moderate to good ee's (Scheme 12.1). In this reaction, *p*-TsOH is employed as a Brønsted acid catalyst to mediate ketimine formation from α-fluorinated α-aryl ketones and aromatic amines [15].

In this reaction, water is formed as the byproduct, which converts TMSCN into the more reactive HCN. However, this method was incompatible with aliphatic amines, as only traces of the corresponding products were obtained.

12.2.2 Mannich Reaction

The classical Mannich reaction comprises the condensation of non-enolizable aldehydes, enolizable carbonyl compounds, and amine derivatives to access β-amino carbonyl compounds. These products are also known as Mannich bases and have been used as building blocks in, e.g., anticancer, antimalarial, and antifungal drugs [16].

The first direct enantioselective Mannich MCR was described by Yamasaki et al. who developed an aluminum-lithium bis(binaphthoxide)-complex combined with lanthanum salts giving low to moderate yields (6–64% ee) [4, 17]. Although this reaction was only applied to a limited range of substrates, it paved the way for other catalysts for the asymmetric Mannich MCRs to be developed [4], such as Lewis acids

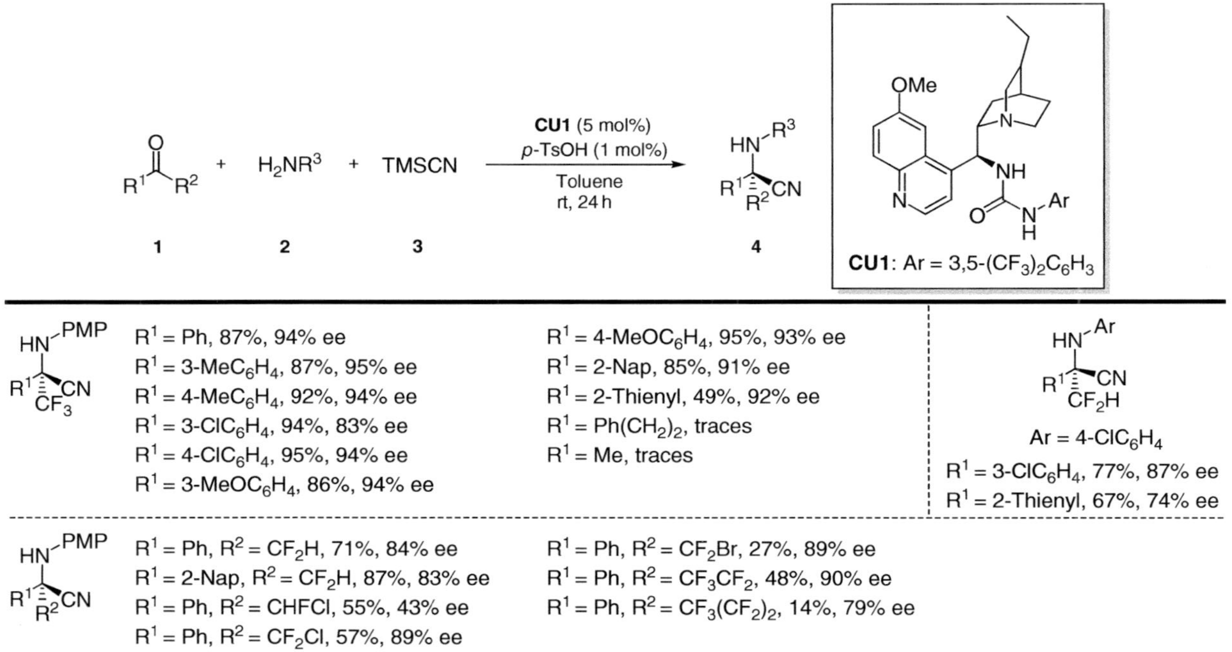

Scheme 12.1 Scope of the Strecker reaction of **4** catalyzed by **CU1**. Source: Based on Liu et al. [15].

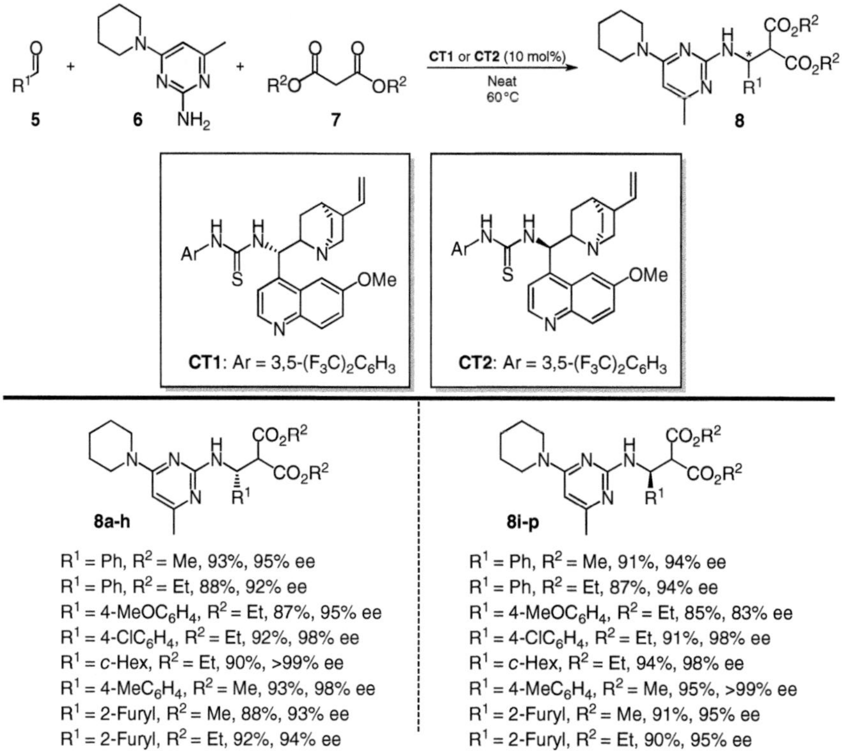

Scheme 12.2 Scope of the Mannich reaction toward **8**. (*S*)-enantiomers (**8a-h**) catalyzed by **CT1**; reaction time: 3d. (*R*)-enantiomers (**8i-p**) catalyzed by **CT2**; reaction time: 36 h. Source: Based on Bai et al. [53].

[18–22], proline derivatives [23–40] and other amine- or amino acid-based catalysts [41–47], proline-based chiral ionic liquids (CILs) [48–50], cinchona-based thioureas [51–54], and CPAs [55, 56]. Several recently developed catalytic systems for Mannich MCRs will be described in the following section.

Bai et al. [53] recently developed two cinchona-based thiourea catalysts (**CT1** and **CT2**) to access chiral β-amino acid ester derivatives, which exhibit antiviral activity against the tobacco mosaic virus (TMV) (Scheme 12.2).

The mode of action of **CT1** and **CT2** involves activation of the aldimine that is formed by the condensation of **5** and **6** by their thiourea moiety, and activation of **7** by the tertiary amine of the cinchona alkaloid moiety. These catalysts were compatible with both aliphatic and aromatic aldehydes and afforded either the (*S*)- or (*R*)-enantiomer with the pseudoenantiomeric catalysts **CT1** or **CT2**, respectively, with excellent ee's.

Another efficient catalyst is L-prolinate, used by Hayashi et al. [57] to access propargylamine derivatives (Scheme 12.3). L-Prolinate was superior to proline in terms of reactivity and enantioselectivity and could be used with ammonium, an alkali metal, or an alkaline earth metal ion as the counterion.

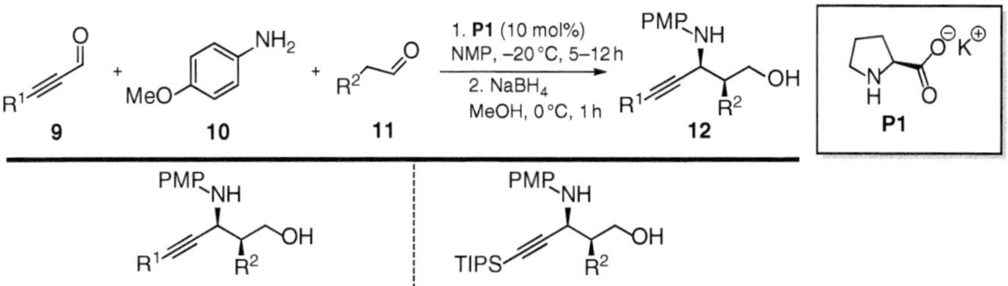

R^1 = Ph, R^2 = Me, 89%, >99 : 1 e.r.
R^1 = Ph, R^2 = Et, 85%, >99 : 1 e.r.
R^1 = *n*-Hex, R^2 = Me, 66%, >99 : 1 e.r.
R^1 = *n*-Hex, R^2 = Et, 62%, >99 : 1 e.r.
R^1 = TBDPSO, R^2 = Me, 82%, >99 : 1 e.r.
R^1 = TBDPSO, R^2 = Et, 78%, >99 : 1 e.r.

R^2 = Me, 84%, >99 : 1 e.r.
R^2 = Et, 80%, >99 : 1 e.r.
R^2 = *n*-Pr, 90%, >99 : 1 e.r.
R^2 = *i*-Pr, 74%, >99 : 1 e.r.
R^2 = PhMe, 83%, >99 : 1 e.r.
R^2 = (*E*)-Et$(CH)_2$Et, 75%, >99 : 1 e.r.

Scheme 12.3 Potassium L-prolinate -catalyzed (**P1**) synthesis of propargyl amine derivatives **12**. Source: Based on Hayashi et al. [57].

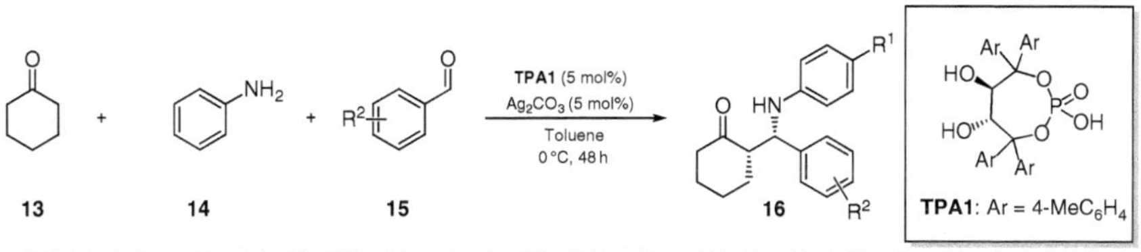

R^1 = OMe, R^2 = 4-NO_2, 94%, 90% ee
R^1 = OMe, R^2 = 4-Cl, 68%, 97% ee
R^1 = H, R^2 = 4-Cl, 87%, 96% ee
R^1 = H, R^2 = 4-F, 47%, 85% ee
R^1 = H, R^2 = 3-F, 87%, 85% ee
R^1 = H, R^2 = 2-F, 75%, 84% ee

R^1 = H, R^2 = 4-Br, 85%, 85% ee
R^1 = H, R^2 = 4-CF_3, 54%, 77% ee
R^1 = H, R^2 = 4-CN, 35%, 73% ee
R^1 = Me, R^2 = 4-Cl, 73%, 74% ee
R^1 = Me, R^2 = 4-Br, 75%, 70% ee

Scheme 12.4 ACDC three-component Mannich reaction toward **16** catalyzed by **TPA1**. Source: Based on Yin et al. [59].

The generality of this reaction was tested with potassium as the counterion. The counterion of the L-prolinate did not affect the enantioselectivity, as excellent enantiomeric ratios (e.r.'s) were obtained in this reaction (Scheme 12.3). This reaction represents a rare example of a *syn*-selective asymmetric Mannich reaction of alkynyl imines and aldehydes.

Another amine-based catalyst was developed by Ayats et al. [58] to access *anti*-β-amino-α-hydroxycarbonyls in 49–95% ee. The authors developed an immobilized primary amino acid catalyst for the reaction of hydroxyacetone and a variety of anilines and aromatic aldehydes.

Yin et al. [59] developed the first asymmetric counterion-directed catalytic (ACDC) system for the asymmetric Mannich three-component reaction. This catalyst consists of a tartaric acid-derived phosphate (**TPA1**) and a silver counterion and afforded β-amino-ketone derivatives **16** with good to high ee's starting from anilines, aromatic aldehydes, and simple ketones (Scheme 12.4). Other ketones that could be used include acetone and acetophenone.

Lewis acids are less commonly used as catalysts in the Mannich reaction, as they tend to coordinate strongly to the Mannich product and cause product inhibition. They must also be water-compatible, since water is generated during imine formation. Nevertheless, several groups have succeeded in developing asymmetric Lewis acid-catalyzed Mannich MCRs [18–22]. Feng et al. [22], for example, developed a chiral-at-metal rhodium catalyst for the reaction of *N*-acylpyrazoles, aldehydes, and primary or secondary amines. This method provided the products with high ee's (91–98%) at low catalytic loadings and is tolerant toward moisture, air, and light.

12.2.2.1 Aza-Henry Reaction

The aza-Henry or nitro-Mannich reaction is a variation of the Mannich reaction in which a nitroalkane substitutes the enolizable aldehyde or ketone to afford β-nitroamine derivatives [5, 60]. These nitroamines can be transformed to vicinal diamines by reduction, or to α-amino acids by oxidation [60, 61].

One such aza-Henry reaction was developed by Cruz-Acosta et al. who used a combination of chiral squaramide **SQ1** and a Brønsted base (*N,N*-dimethylcyclohexyla mine) to afford β-nitroamine derivatives (**18**) with reasonable to high enantioselectivities (Scheme 12.5) [62]. This strategy was compatible with both aromatic and aliphatic aldehydes and could be performed under aqueous conditions.

This reaction could also be catalyzed by two thiourea catalysts (**T1** and **T2**) under the same conditions. Catalyst **T1** was more suitable for aromatic aldehydes and gave e.r.'s of up to 99.5 : 0.5, whereas **T2** was more appropriate for aliphatic aldehydes and gave slightly lower e.r.'s (55 : 45–83 : 17, Scheme 12.5).

12.2.2.2 Petasis Reaction

The borono-Mannich or Petasis three-component reaction is a condensation of amines, carbonyl derivatives, and boronic acids to afford amine or α-amino acid derivatives [63]. In contrast to the conventional Mannich reaction, the Petasis reaction typically requires the presence of a Lewis basic oxygen functionality at the α-position of the aldehyde (or ketone) reactant. This MCR has become a powerful synthetic tool owing to the accessibility of its reactants, its mild and robust reaction conditions, flexible scalability, and the biological activities of its products [64, 65]. The homoallylic amine products resulting from the reaction with allyl boronates have been reported as useful building blocks in the synthesis of a variety of bioactive compounds, including natural products [66, 67].

Nevertheless, the Petasis reaction has two limitations. Firstly, the reaction is limited to activated aldehydes or boron-activating aldehydes. Secondly, the scope of boronic acids is limited to electron-rich (hetero)aryl- and vinylboronic acids. Various attempts to overcome these hurdles have been made through transition metal catalysis, which activates the nucleophilic boronic acid [65]. Several asymmetric Petasis MCRs have been developed using chiral Pd catalysts [65, 68, 69]. The scope of the remaining components can be broadened by Lewis acid catalysis or organocatalysis. Chiral organocatalysts that have been developed for the Petasis three-component reaction include BINOL- [66, 67, 70], VAPOL- [63], and thiourea-based catalysts [66, 71, 72].

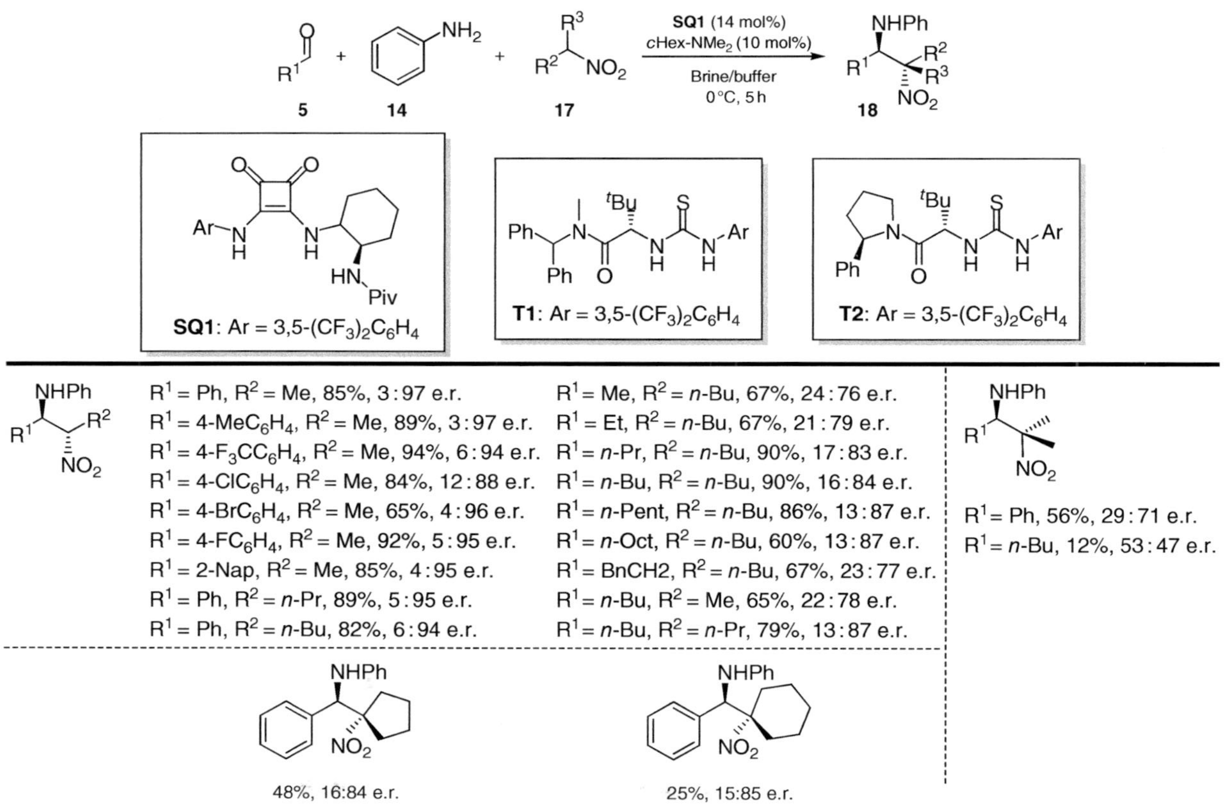

Scheme 12.5 Scope of the aza-Henry reaction toward **18** catalyzed by **SQ1**. Source: Based on Cruz-Acosta et al. [62].

A recent example of an organocatalytic Petasis MCR was reported by Jiang and Schaus [67], who developed a BINOL catalyst (**B1**) to access homoallylic amines **20** with high enantioselectivity (Scheme 12.6). This method was compatible with aromatic amines and both aromatic and aliphatic aldehydes.

12.2.2.3 Aza-Diels–Alder Via Mannich Reaction Pathway

The aza-Diels–Alder reaction has become a valuable tool in the synthesis of nitrogen-containing heterocycles, such as alkaloids [73]. A CPA-catalyzed (**CPA1**) formal aza-Diels–Alder reaction was recently developed by the Schneider group to access 4-alkyl-substituted 2-piperidones (**22**) with good to excellent enantioselectivity from (hetero)aromatic aldehydes, *para*-anisidine, and β-alkyl-substituted vinylketene acetals **21** (Scheme 12.7). These piperidones could be easily converted into highly substituted piperidines with full control over the stereochemistry [73].

This reaction proceeds through a vinylogous Mannich pathway, in which **CPA1** protonates the imine – formed from **5** and **10** – that subsequently undergoes an nucleophilic attack by **21** to form an acyclic Mannich product with a *Z*-configuration. This Mannich intermediate then cyclizes and is desilylated by the phosphate anion of **CPA1** to afford the lactams **22**. Other catalysts that have been described for asymmetric hetero-Diels–Alder MCRs include proline-based catalysts [74–76], CPAs [77, 78], a CIL [79], and several Lewis acids [80–83].

Povarov Reaction The Povarov reaction is a condensation of an aniline and carbonyl compound, and subsequent inverse electron-demand [4 + 2]-cycloaddition of the resulting *N*-arylimine with an electron-rich olefin to afford 1,2,3,4-tetrahydroquinolines (THQs) [84, 85]. THQs are found in numerous natural and synthetic bioactive compounds. As a result, much effort has been spent on finding efficient synthetic routes to access this scaffold. Among the various strategies that have been reported, the Povarov reaction offers the most straightforward and efficient route, moreover allowing high structural diversity [86, 87].

Huang et al. [87] reported a highly enantioselective synthesis of THQs through a Povarov pseudo-three-component reaction between anilines and two equivalents of an enolizable aldehyde catalyzed by the Hayashi–Jørgensen catalyst (*S*)-**P2** (Scheme 12.8). This was the first asymmetric Povarov three-component reaction in which the aldehyde was directly utilized instead of a preformed imine. This reaction is tolerant to a variety of aliphatic aldehydes (**23**) but gave side reactions when two different aldehydes were employed. The only exception was the replacement of one of the enolizable aldehydes by electron-deficient nitrobenzaldehydes, resulting in two examples of a true three-component reaction (Scheme 12.8).

The Povarov reaction is also termed as the inverse electron-demand aza-Diels–Alder reaction and could proceed via this pathway. However, it is now generally accepted that it rather proceeds via a stepwise Mannich and intramolecular Friedel–Crafts pathway [86]. In the protocol developed by Huang et al., this intramolecular Friedel–Crafts reaction proceeded spontaneously with anilines bearing electron-donating groups. However, for anilines bearing electron-withdrawing groups, the reaction tended to stop after the Mannich addition and required

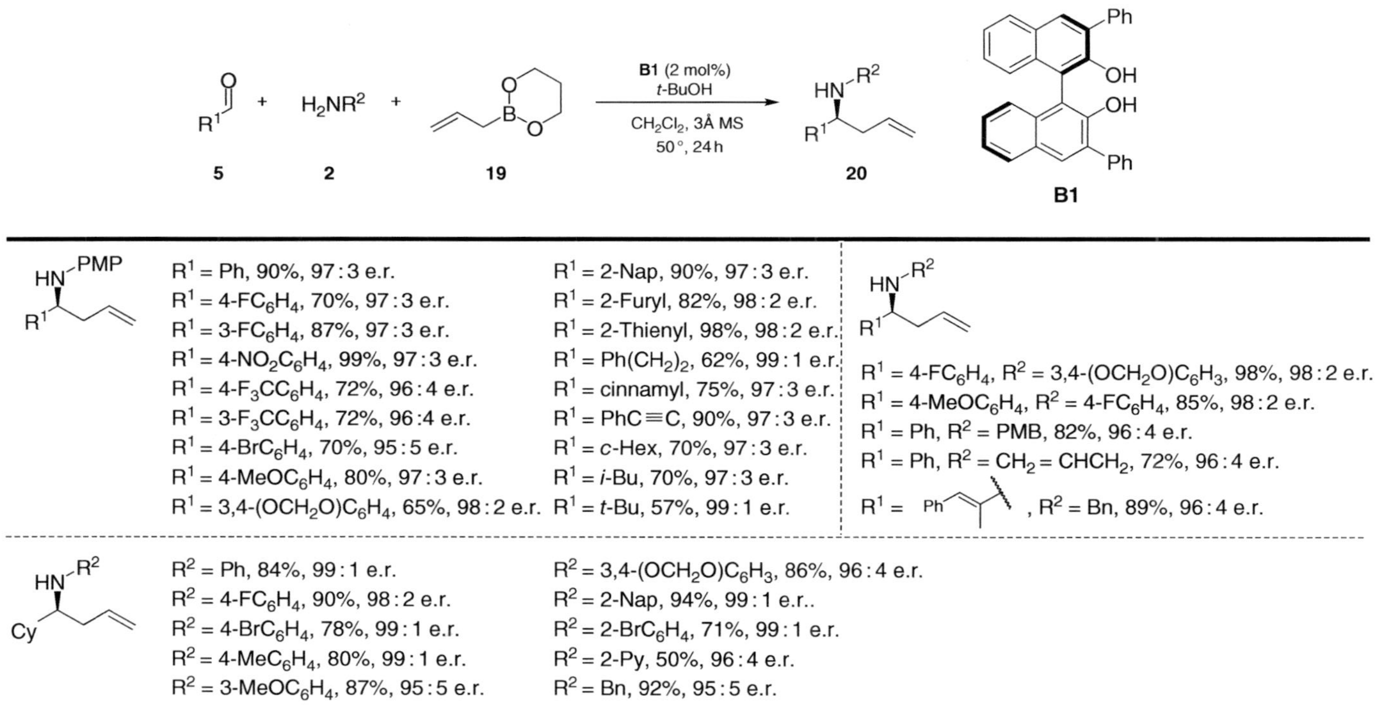

Scheme 12.6 Scope of the Petasis reaction toward **20** catalyzed by **B1**. Source: Based on Jiang et al. [67].

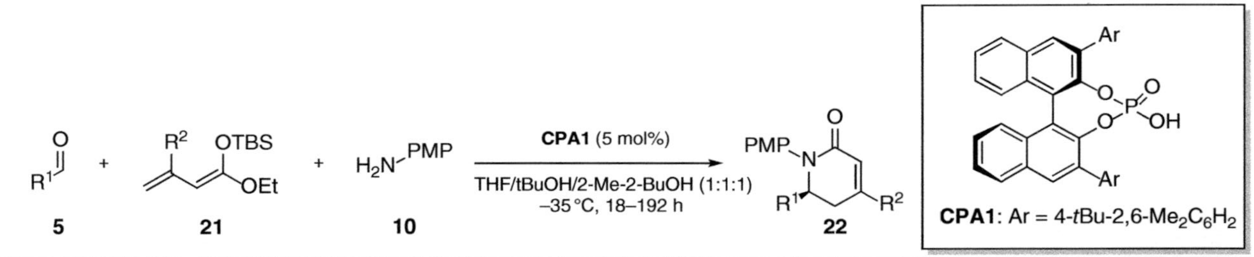

R^1 = Ph, R^2 = Me, 66%, 98 : 2 e.r.
R^1 = 4-MeC_6H_4, R^2 = Me, 75%, 99 : 1 e.r.
R^1 = 3-MeC_6H_4, R^2 = Me, 69%, 98 : 2 e.r.
R^1 = 2-MeC_6H_4, R^2 = Me, 73%, 98 : 2 e.r.
R^1 = 4-EtC_6H_4, R^2 = Me, 82%, >99 : 1 e.r.
R^1 = 4-EtC_6H_4, R^2 = Et, 78%, 99 : 1 e.r.
R^1 = 4-EtC_6H_4, R^2 = *n*-Bu, 73%, 99 : 1 e.r.
R^1 = 4-$O_2NC_6H_4$, R^2 = Me, 68%, 90 : 10 e.r.
R^1 = 4-NCC_6H_4, R^2 = Me, 67%, 97 : 3 e.r.
R^1 = 4-FC_6H_4, R^2 = Me, 67%, 96 : 4 e.r.
R^1 = 3-ClC_6H_4, R^2 = Me, 71%, 99 : 1 e.r.
R^1 = 2-BrC_6H_4, R^2 = Me, 70%, 98 : 2 e.r.
R^1 = 1-Nap, R^2 = Me, 76%, 99 : 1 e.r.
R^1 = 2-Nap, R^2 = Me, 70%, 99 : 1 e.r.
R^1 = 2-Nap, R^2 = Et, 67%, 98 : 2 e.r.
R^1 = 2-Nap, R^2 = *n*-Bu, 71%, 98 : 2 e.r.
R^1 = 2-Furyl, R^2 = Me, 57%, 85 : 15 e.r.
R^1 = 3-Furyl, R^2 = Et, 51%, 89 : 11 e.r.
R^1 = 3-Furyl, R^2 = *n*-Bu, 40%, 89 : 11 e.r.
R^1 = 3-Thienyl, R^2 = Me, 65%, 97 : 3 e.r.
R^1 = 1*H*-indol-2-yl, R^2 = Me, 54%, 99 : 1 e.r.
R^1 = 1*H*-indol-2-yl, R^2 = Et, 55%, 98 : 2 e.r.
R^1 = 1*H*-indol-2-yl, R^2 = *n*-Bu, 78%, >99 : 1 e.r.

Scheme 12.7 Scope of aza-Diels–Alder reaction of **21** catalyzed by **CPA1**. Source: Based on Weilbeer et al. [73].

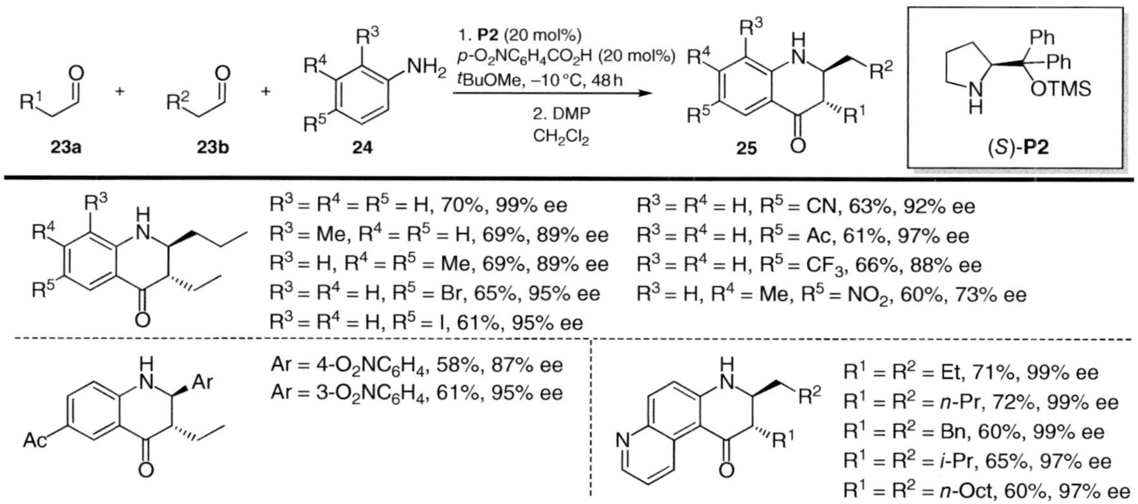

Scheme 12.8 Scope of the Povarov reaction toward **25** catalyzed by (*S*)-**P2**. Source: Based on Huang et al. [87].

additional treatment with the Lewis acid phosphorus oxychloride ($POCl_3$) prior to treatment with Dess–Martin periodinane (DMP). Other chiral catalysts that have been developed for the Povarov MCR include CPAs [78, 88–96] and the Lewis acid $Sc(OTf)_3$ [97, 98].

12.2.2.4 [2 + 2 + 2]-Cycloaddition

The term [2+2+2]-cycloaddition is used to describe a wide range of mechanistically unrelated reactions, which moreover rarely involve a true cycloaddition step. Nevertheless, these formal [2+2+2]-cycloaddition reactions allow for the synthesis of various highly substituted six-membered rings [99, 100]. One recent example developed by Kuang et al. is the Cu(II)/BOX catalyzed reaction of 2,3-dihydropyran (**26**), methylene malonates (**27**), and a variety of substituted indoles (**28**) to access the tetracyclic indolines **29** (Scheme 12.9). Such tetracyclic indoles are common motifs in natural products and pharmaceuticals [101].

The mechanism of this reaction is proposed to proceed through a formal [2+2]-cycloaddition of **26** and **27** to afford **30** (Scheme 12.10). Then, a Lewis acid-promoted ring-opening takes place to form the zwitterionic intermediate **31**. This ring-opening proceeds without loss of stereochemistry at the α-position of the oxonium ion in **31**, whereas the neighboring oxonium carbon atom is (almost) planarized. Finally, **31** undergoes a Friedel–Crafts-type acylation and subsequent intramolecular Mannich reaction to afford the final product **29** [101, 102].

12.2.3 Hantzsch Reaction

The Hantzsch MCR involves the cyclocondensation between an aldehyde, ammonia, and two equivalents of a β-keto ester (or a comparable reactant) to form 1,4-dihydropyridines (DHPs). DHPs, in particular 4-aryl-substituted 1,4-dihydropyridines, have shown to exhibit various pharmacological activities. They have been recognized as Ca(II)-channel modulators which could be used to treat, e.g., cardiovascular diseases, Parkinson's disease, and Alzheimer's disease. These compounds have also been found to exhibit other biological properties, such as antitumor, antimicrobial, anti-inflammatory, and antioxidant activities [103–107]. Despite their therapeutic utility, only few examples of their catalytic enantioselective synthesis have been reported [107].

Two organocatalyzed asymmetric Hantzsch-type reactions have recently been developed by the groups of Rodriguez [108] and Zu [109]. The group of Rodriguez developed a thiourea catalyst that afforded 4-arylpyridines – after oxidation of the initially formed DHPs – with reasonable to excellent atroposelectivity (58–96%). Prior to their findings, the only other catalytic asymmetric method to access atropisomeric 4-arylpyridines was through a Suzuki–Miyaura cross-coupling, which afforded these pyridines in 42% yield [108, 110]. Due to this lack of enantioselective approaches, 4-arylpyridines have received less attention, despite their promising biological activities [108].

The group of Zu also developed an efficient method using the Hayashi–Jørgensen catalyst (*R*)-**P2** that gave access to 2,5-unsubstituted tetrahydropyridines **36** with high ee's (Scheme 12.11) [109].

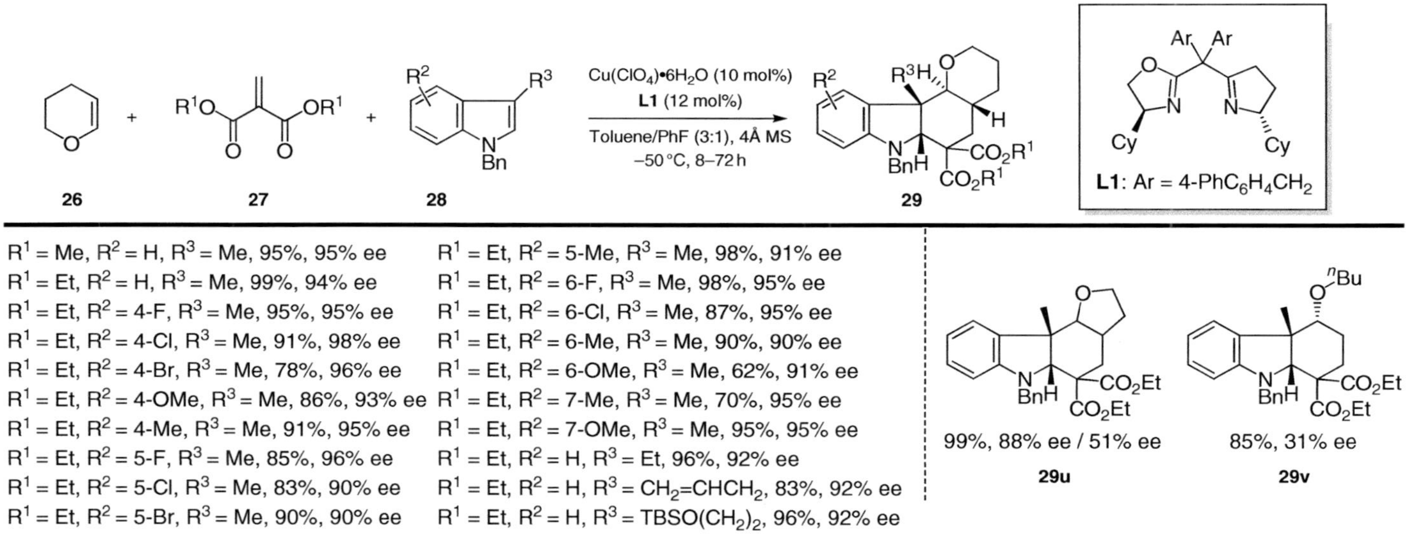

Scheme 12.9 Scope of the [2 + 2 + 2]-cycloaddition toward **29** catalyzed by $Cu(ClO_4)\cdot 6\ H_2O$/**L1**. The relative configuration of **29u** was not determined. Source: Based on Kuang et al. [101].

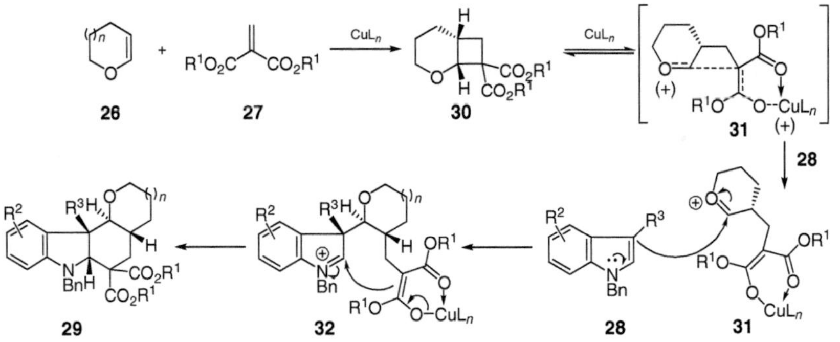

Scheme 12.10 Plausible mechanism for the formation of **29** through a formal [2 + 2]-cycloaddition, Lewis acid-mediated ring-opening, Friedel–Crafts-type acylation, and intramolecular Mannich-type reaction. Source: Based on Kuang et al. [101]; Zhu et al. [102].

In this Hantzsch reaction, a Michael addition of **33** to **34** takes place to form the intermediate **37**, which undergoes another Michael addition with **38** to form **39** (Scheme 12.12). In the final step, a hemiaminal formation produces **36**. These Hantzsch products (**36**) could serve as intermediates in the synthesis of the antidepressant drug paroxetine. Although many asymmetric routes have been developed for the synthesis of paroxetine, none of these include the Hantzsch reaction, despite the straightforward route it offers toward dihydropyridines from a retrosynthetic point of view. This may be due to difficulties associated with the use of enamino esters as substrates, since amines produced by their hydrolysis could compete with the catalyst in the imine formation. The group of Zu addresses this problem by replacing the β-keto ester component with a propiolic ester, allowing enehydrazino esters **37b** to be generated through Michael addition with arylhydrazines. Older Hantzsch reactions were described by Franke et al. [106] and Jiang et al. [104].

12.2.4 Biginelli Reaction

The Biginelli reaction is the condensation of aldehydes, (thio)urea, and β-keto esters to afford dihydropyrimidones (DHPMs) or their 2-thio analogs. DHPMs have also shown to exhibit diverse pharmacological properties, such as antiviral, anti-inflammatory, and antitumor activities.

Huang et al. [111] developed the first asymmetric synthesis of DHPMs through a Biginelli reaction catalyzed by a chiral Yb(III) complex. This reaction afforded DHMPs in high ee's (up to 99%) under mild reaction conditions. Following this report, various organocatalysts have been developed for the Biginelli MCR [112], including proline-based catalysts [113–116], CPAs [117–122], a chiral sulfonic acid [123], a BINOL-based bisphosphorylimide [112], thiourea catalysts [5, 124–126], primary amines [124, 127], and CILs [128]. Since two water molecules are typically generated during the reaction, chiral metal complexes – which tend to be more sensitive to moisture – have been less explored in the Biginelli reaction [120]. Unfortunately, many of the organocatalytic reactions suffer from long reaction times (two to seven days) and employ catalysts that require multistep synthesis [10, 112].

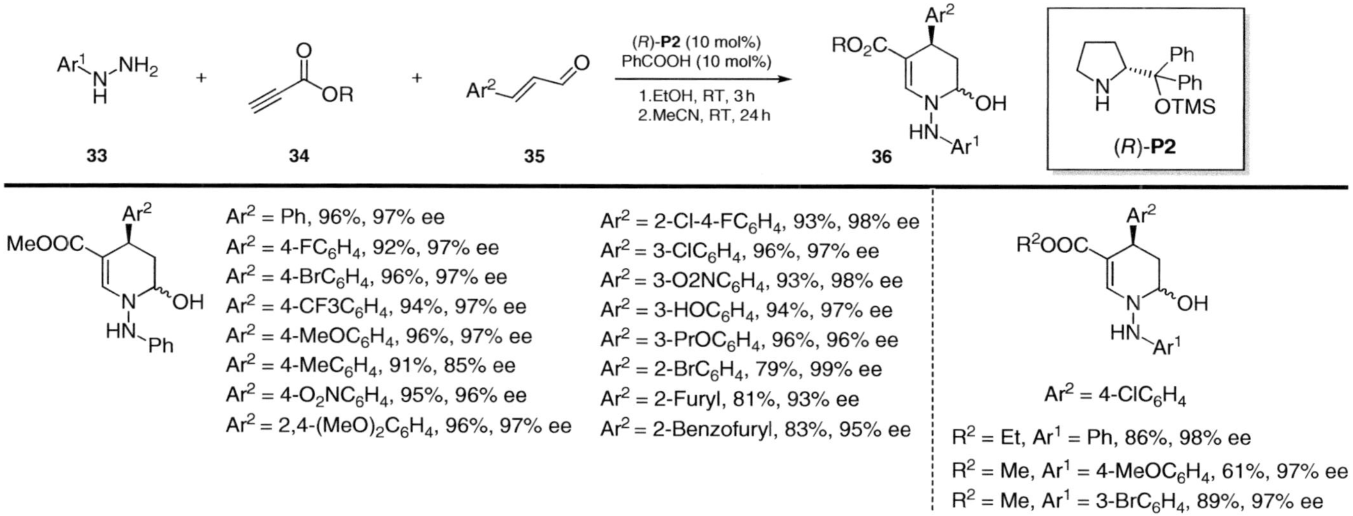

Scheme 12.11 Scope of the Hantzsch reaction toward **36** catalyzed by (*R*)-**P2**. Source: Based on Chen et al. [109].

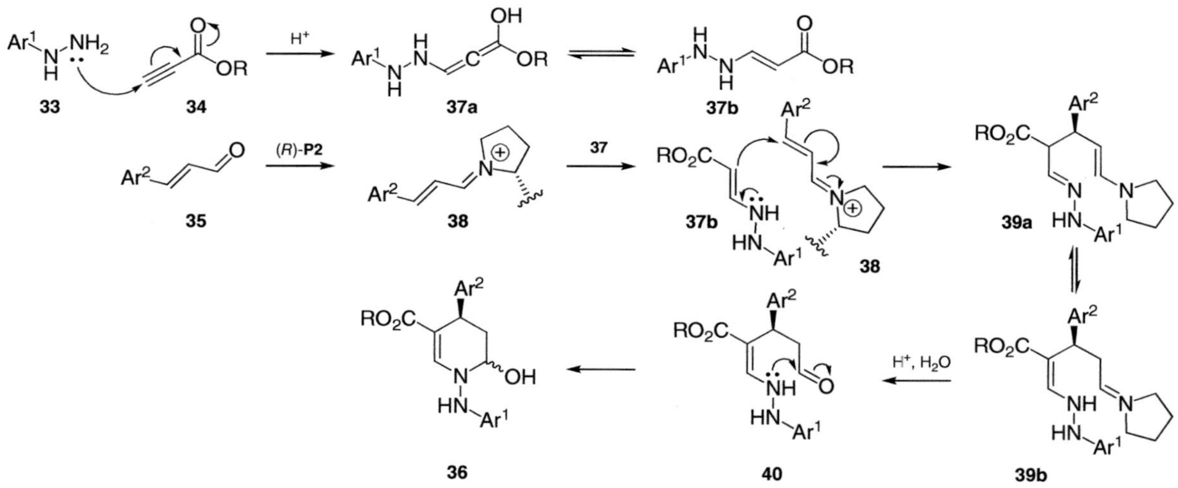

Scheme 12.12 Plausible mechanism of the Hantzsch reaction toward **36**. Source: Based on Chen et al. [109].

Hang et al. [129] reported the use of a supramolecular methanoproline–thiourea catalyst resulting from self-assembly of **P3** and **CT3** to afford 6-isopropyl-3,4-DHPMs **43** with excellent ee's (Scheme 12.13). Compatible aldehydes **5** include *n*-butanal and various (mostly *para*-substituted) benzaldehyde derivatives.

The authors discovered that this self-assembled catalyst was superior to the individual pre-catalysts **P3** and **CT3**. A similar dual catalytic system developed by Yu et al. [130] features a similar substrate scope and enantioselectivity (92–99% ee).

Alvim et al. [10] developed a different method by combining ACDC with the ionic liquid effect (ILE). (Scheme 12.14). The ILE is the ion-pairing ability of CILs and their ability to form larger supramolecular aggregates through hydrogen bond networks [131]. The asymmetric Biginelli reaction typically requires an acid co-catalyst. The catalyst **CPA.IS** (chiral phosphoric acid and imidazolium sulfonic acid), on the other hand, has a dual function of chiral inductor (CPA) and Brønsted acid (sulfonic acid), thereby eliminating the need for a separate co-catalyst. This method affords DHPMs in moderate to high yields from aromatic aldehydes, aliphatic 1,3-dicarbonyl compounds, and simple (thio)ureas (Scheme 12.14).

The Biginelli reaction can, in principle, proceed via three different mechanisms that may occur simultaneously under typical Biginelli reaction conditions, i.e. via the Knoevenagel, iminium, or enamine pathway. Of these pathways, the enantioselective version of the Biginelli reaction proceeds through the iminium pathway. In order to induce iminium ion formation, the aldehyde and urea are typically mixed first before adding the 1,3-dicarbonyl. Alvim et al. found that **CPA.IS** favors the iminium pathway, allowing all three reactants to be added simultaneously in a true MCR setup.

12.3 Michael Addition-based MCRs

One of the current challenges in organic synthesis is efficient access to complex chiral polycyclic structures. Such frameworks can be constructed through organocatalyzed cascade reactions, occasionally in combination with transition metal catalysis [8]. In the following sections, several multicomponent cascade reactions will be highlighted that involve at least one Michael addition.

12.3.1 Oxa-Michael/Michael/Michael/Aldol Condensation Cascade Reactions

Dochain et al. developed series of unprecedented complex tricyclic polyethers through an oxa-Michael/Michael/Michael/aldol condensation/inverse electron-demand hetero-Diels–Alder (IEDHDA) reaction catalyzed by (*R*)-**P2** and the Lewis acid Yb(fod)$_3$ (Scheme 12.15). This highly chemo- and enantioselective protocol is compatible with a variety of aromatic alcohols and nitroalkenes [8].

The first four steps of this domino sequence were catalyzed by (*R*)-**P2**, with benzoic acid as a co-catalyst (Scheme 12.16). The last step (IEDHDA) was catalyzed by Yb(fod)$_3$ [8, 132].

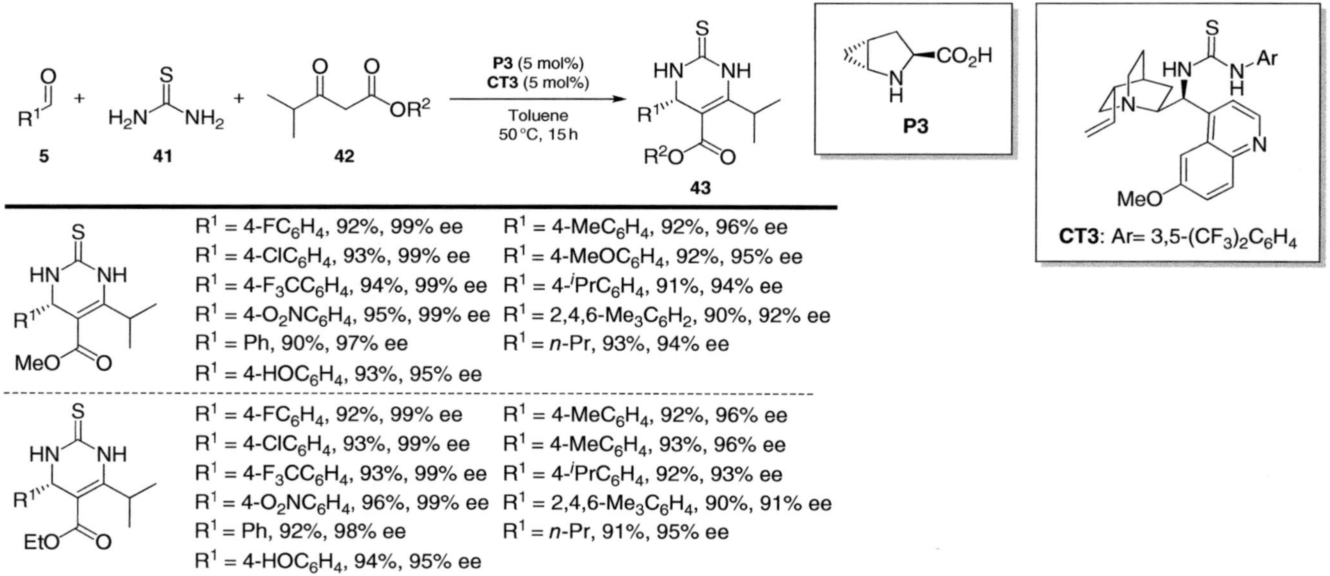

Scheme 12.13 Scope of the Biginelli reaction toward **43** catalyzed by the self-assembled catalyst **P3 CT3**. Source: Based on Hang et al. [129].

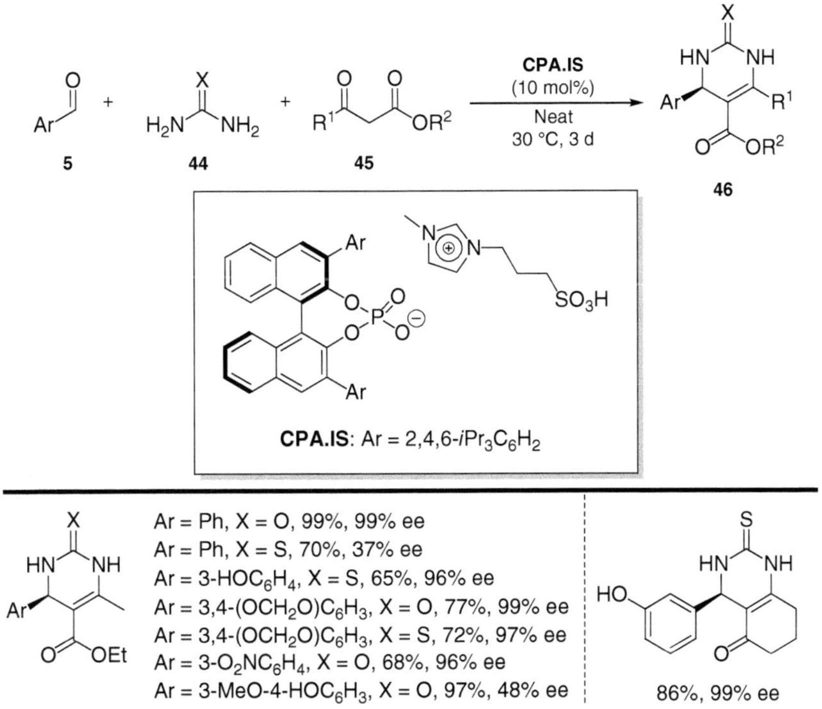

Scheme 12.14 Scope of the Biginelli reaction toward **46** catalyzed by **CPA.IS**. Source: Based on Alvim et al. [10].

Although combining metal catalysts with organocatalysts is challenging, Dochain et al. [8] succeeded in combining both in a one-pot reaction. The Jørgensen group reported a similar approach for the construction of tetrahydroisochromenes, which are valuable motifs of natural products that possess, e.g., antimalarial and anticancer properties [133].

A similar oxa-Michael/Michael/Michael/aldol condensation cascade reaction was developed by the Enders group to afford functionalized tricyclic chromanes that contain three contiguous stereocenters [134]. Tricyclic chromanes are found in a variety of natural and pharmaceutical compounds owing to their broad range of biological properties, such as antimalarial and hepatoprotective activities. This same group also developed a triple Michael/Michael/aldol condensation reaction to afford tricyclic chromanes containing four contiguous stereogenic centers [135]. Several other (*R*)- or (*S*)-**P2**-catalyzed tandem Michael/Michael/aldol condensation reactions have also been reported [132, 136–140].

12.3.2 Knoevenagel–Michael Cascade Reaction

γ-Butyrolactones are found in a diverse array of biologically active natural products displaying a wide range of activities, including anticancer and antibacterial

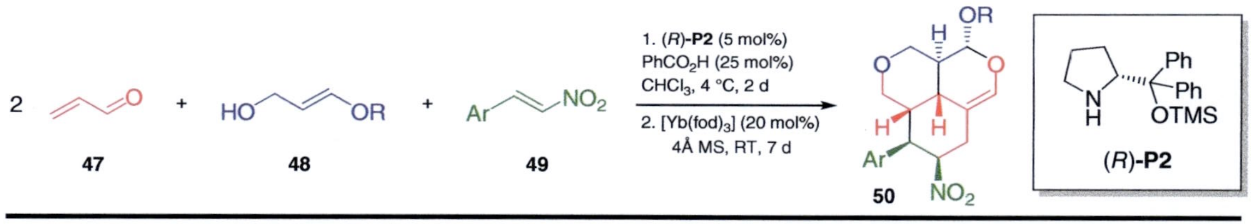

R = Et, Ar = Ph, 21%, >99% ee
R = Et, Ar = 4-FC_6H_4, 16%, >99% ee
R = Et, Ar = 4-MeC_6H_4, 21%, >99% ee
R = *n*-Bu, Ar = Ph, 29%, >99% ee
R = *n*-Bu, Ar = 4-FC_6H_4, 16%, >99% ee
R = *n*-Bu, Ar = 4-MeC_6H_4, 20%, >99% ee
R = *n*-Bu, Ar = 2-ClC_6H_4, 25%, >99% ee
R = *n*-Bu, Ar = 2-Thienyl, 19%, >99% ee
R = Me, Ar = Ph, 27%, >99% ee
R = Bn, Ar = Ph, 14%, >99% ee

Scheme 12.15 Oxa-Michael/Michael/Michael/aldol condensation sequence followed by an inverse electron-demand hetero-Diels–Alder reaction. Source: Based on Dochain et al. [8]; Zhang et al. [132].

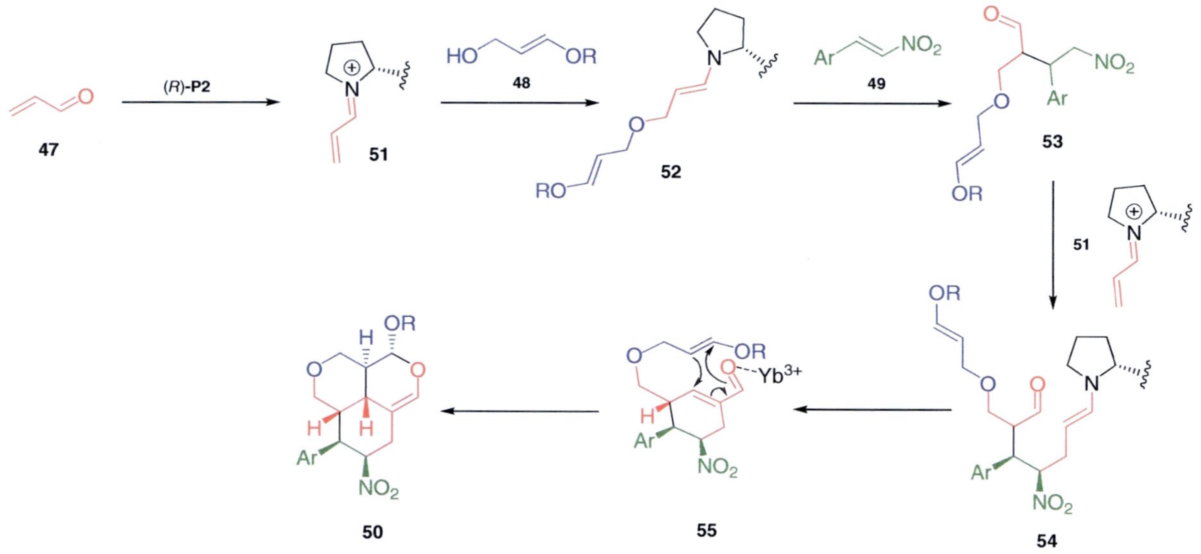

Scheme 12.16 Proposed mechanism of the oxa-Michael/Michael/Michael/aldol/inverse-electron demand hetero-Diels–Alder reaction to give **50**. Additional ligands on Yb were omitted for clarity. Source: Based on Dochain et al. [8]; Zhang et al. [132].

activities. Many of the efforts that have been spent on the enantioselective synthesis of these motifs rely on the use of scarce transition metal catalysts, expensive ligands, and pre-functionalized, noncommercial starting materials. Moreover, the organocatalytic protocols that have been reported are limited to aromatic Michael acceptors or to α-hydroxy- or α,α,α-trifluoroacetophenone. Khopade et al. [141] described the first organocatalytic one-pot asymmetric synthesis of these γ-butyrolactones without the use of expensive transition metals or pre-functionalized substrates (Scheme 12.17). This reaction tolerates a wide variety of aldehydes with diverse steric and electronic properties, although slightly lower ee's were observed for aliphatic aldehydes.

The first step in this reaction is the formation of **60** through a Knoevenagel condensation (Scheme 12.18). **CA1** then catalyzes the formal [4 + 2]-cycloaddition of **60** and **61** through a Michael addition/iminium addition pathway to form **62**. The last two steps comprise a lactonization and subsequent decarboxylation to afford the final product **58**.

The same group also developed a similar enantioselective reaction of aldehydes, ketones, and Meldrum's acid catalyzed by **Q2** to afford δ-ketoacid derivatives in 49–99% ee. Prior to these findings, no asymmetric version of this reaction with these substrates had been reported [142]. Other catalysts that have been reported for Knoevenagel–Michael cascade reactions include proline-based catalysts [143–147], (thiourea-based) cinchona alkaloids [148–153], and Lewis acids [154].

12.3.3 Michael–Henry Cascade Reaction

Chen et al. [155] developed a bifunctional guanidine–amide organocatalyst (**GA1**) in a pseudo-three-component Michael–Michael–Henry cascade reaction to access cyclohexanes with six contiguous stereocenters (Scheme 12.19).

Several trends were observed while evaluating the scope with regard to **49** (Scheme 12.19). Firstly, electron-donating substituents on the aryl of **49** seemed to give higher yields compared to electron-withdrawing substituents. Secondly, *ortho*-substituents seemed to give slightly higher ee's. While exploring the scope with regard to **66,** no significant difference in enantioselectivity was observed between linear/branched alkyl or alkenyl groups.

The guanidine unit of **GA1** acts as a Brønsted base and promotes the deprotonation of **66** and **68**, while the amide and sulfonamide units in **GA1** activate the electrophilic nitroalkenes **49** through hydrogen bond formation (Scheme 12.20). A similar reaction reported by Shi et al. [156] is catalyzed by a Cu complex with a chiral diamine ligand.

12.4 Isocyanide-Based MCRs

Two prominent isocyanide-based MCRs are the Passerini three-component reaction (P-3CR) and the Ugi four-component reaction (U-4CR) [157]. The classical P-3CR involves a reaction between an aldehyde, carboxylic acid, and isocyanide to afford

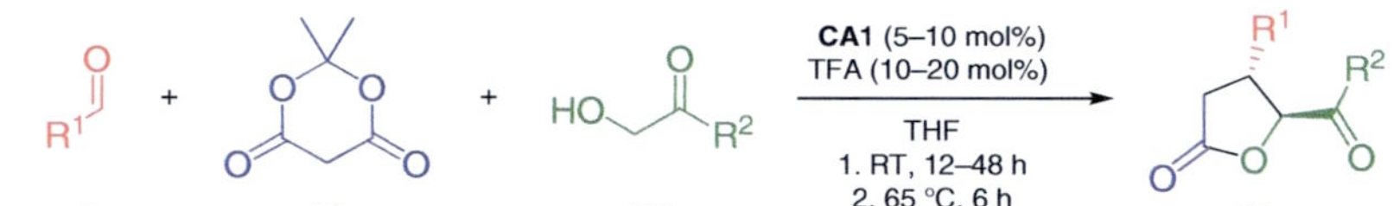

5 **56** **57** **58**

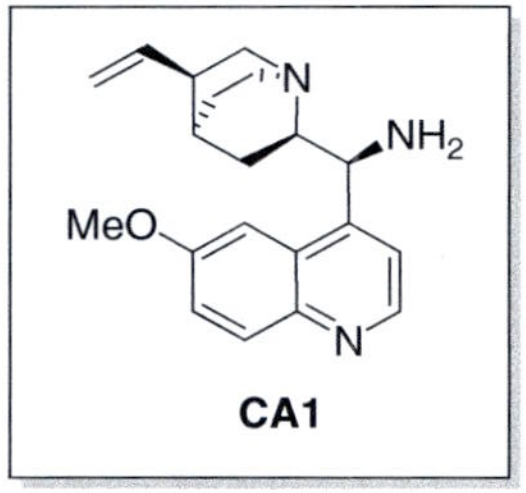

R^1 = Ph, R^2 = Me, 78%, 94% ee	R^1 = 2-Nap, R^2 = Me, 65%, 94% ee
R^1 = 4-MeOC_6H_4, R^2 = Me, 76%, 91% ee	R^1 = 2-Furyl, R^2 = Me, 59%, 87% ee
R^1 = 4-MeC_6H_4, R^2 = Me, 65%, 84% ee	R^1 = 2-Thienyl, R^2 = Me, 64%, 90% ee
R^1 = 4-FC_6H_4, R^2 = Me, 82%, >99% ee	R^1 = 3-Thienyl, R^2 = Me, 59%, 94% ee
R^1 = 4-ClC_6H_4, R^2 = Me, 61%, 91% ee	R^1 = Ph, R^2 = *n*-Pr, 64%, 91% ee
R^1 = 4-NCC_6H_4, R^2 = Me, 86%, 97% ee	R^1 = 4-MeOC_6H_4, R^2 = *n*-Pr, 58%, >99% ee
R^1 = 4-$CF_3C_6H_4$, R^2 = Me, 55%, 95% ee	R^1 = 3,5-$(MeO)_2C_6H_3$, R^2 = Ph, 69%, 80% ee
R^1 = 2-$O_2NC_6H_4$, R^2 = Me, 51%, 86% ee	R^1 = 2-$O_2NC_6H_4$, R^2 = Ph, 59%, 61% ee
R^1 = 2-BrC_6H_4, R^2 = Me, 65%, 91% ee	R^1 = CO_2Et, R^2 = Ph, 64%, 67% ee
R^1 = 3,5-$(MeO)_2C_6H_3$, R^2 = Me, 81%, 94% ee	R^1 = *i*-Bu, R^2 = Ph, 62%, 70% ee

Scheme 12.17 Scope of the Knoevenagel/Michael/lactonization cascade synthesis of **58** catalyzed by **CA1**. Source: Based on Khopade et al. [141].

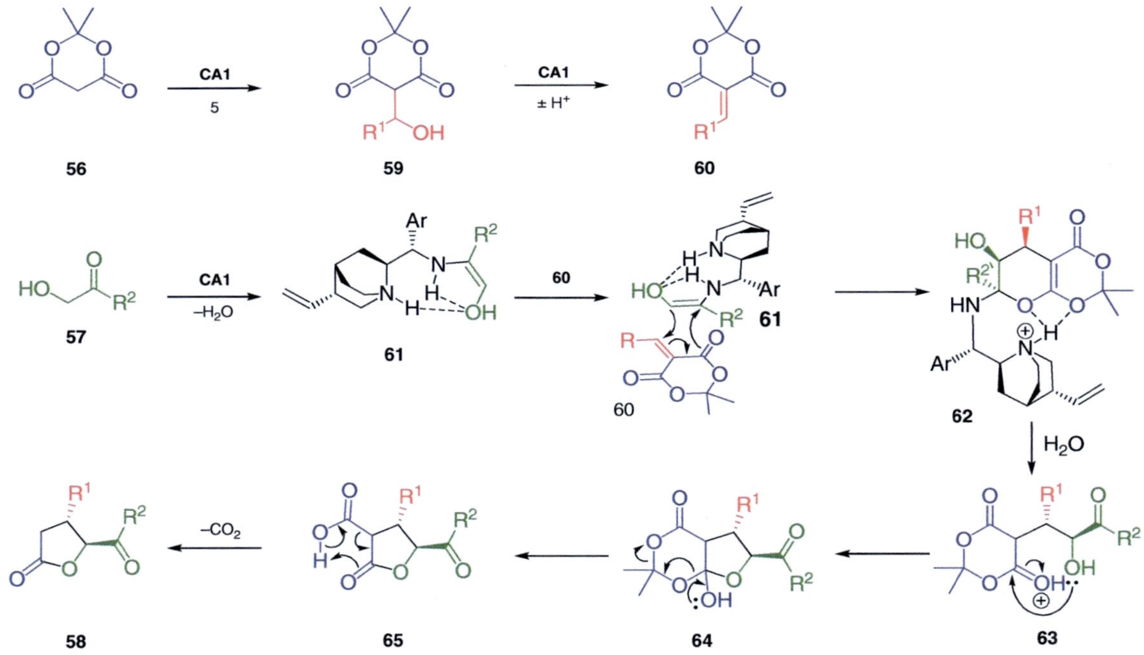

Scheme 12.18 Proposed mechanism of the Knoevenagel/Michael/lactonization cascade reaction. Source: Based on Khopade et al. [141]; Khopade et al. [142].

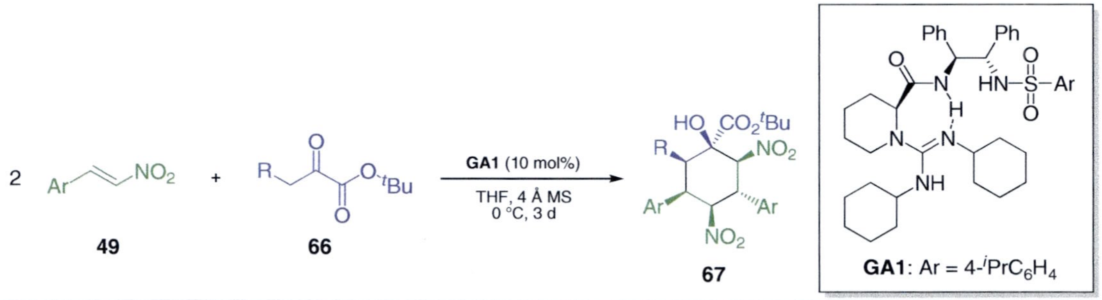

Ar = 4-MeOC_6H_4, R = 4-FC_6H_4, 99%, 90% ee	Ar = Ph, R = Bn, 93%, 90% ee	Ar = 2-MeOC_6H_4, R = Bn, 99%, 91% ee
Ar = 4-MeOC_6H_4, R = Ph$(CH_2)_2$, 99%, 89% ee	Ar = 2-FC_6H_4, R = Bn, 95%, 92% ee	Ar = 3-MeOC_6H_4, R = Bn, 88%, 90% ee
Ar = 4-MeOC_6H_4, R = Ph$(CH_2)_3$, 83%, 88% ee	Ar = 4-FC_6H_4, R = Bn, 88%, 88% ee	Ar = 4-MeOC_6H_4, R = Bn, 96%, 92% ee
Ar = 4-MeOC_6H_4, R = CH_2=CHCH_2, 79%, 90% ee	Ar = 4-ClC_6H_4, R = Bn, 73%, 90% ee	Ar = 3-Furyl, R = Bn, 94%, 85% ee
Ar = 4-MeOC_6H_4, R = CH_2=CH$(CH_2)_2$, 88%, 88% ee	Ar = 4-BrC_6H_4, R = Bn, 77%, 91% ee	Ar = 2-Furyl, R = Bn, 93%, 83% ee
Ar = 4-MeOC_6H_4, R = *n*-Bu, 87%, 90% ee	Ar = 4-F_3CC_6H_4, R = Bn, 84%, 88% ee	Ar = 1-Nap, R = Bn, 99%, 90% ee
Ar = 4-MeOC_6H_4, R = *i*-Bu, 92%, 90% ee	Ar = 3,4-$Cl_2C_6H_3$, R = Bn, 56%, 84% ee	Ar = 2-Nap, R = Bn, 86%, 87% ee
Ar = 4-MeOC_6H_4, R = *n*-Pent, 74%, 88% ee	Ar = 2-MeC_6H_4, R = Bn, 91%, 95% ee	

Scheme 12.19 Scope Michael/Michael/Henry cascade synthesis of **67** catalyzed by **GA1**. Source: Based on Chen et al. [155].

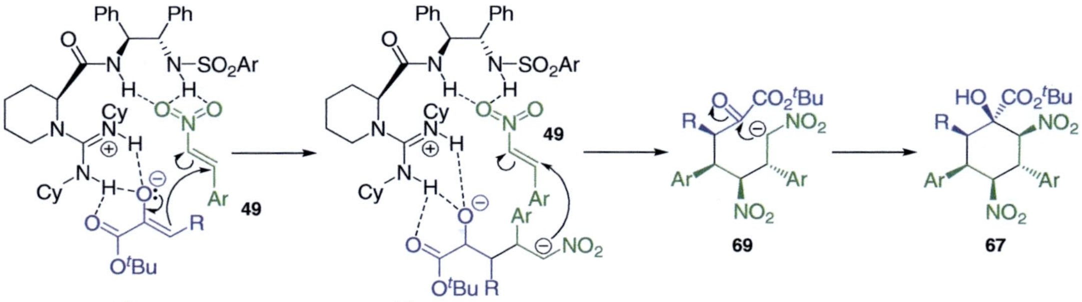

Scheme 12.20 Proposed mechanism of the Michael/Michael/Henry cascade reaction catalyzed by **GA1**. Source: Based on Chen et al. [155]; Shi et al. [156].

α-acyloxy carboxamides, whereas the classical U-4CR involves an additional primary amine to afford α-amido carboxamides [157, 158]. The proposed mechanisms of these isocyanide-based multicomponent reaction (IMCRs) are closely related (Schemes 12.21 and 12.22) [157, 161, 162]. In these reactions, the aldehyde (in P-3CR) or imine (in U-4CR) is activated by the carboxylic acid before undergoing a nucleophilic attack by the isocyanide, in which a stereogenic center is created. In the P-3CR, the attack by the isocyanide and trapping by the carboxylate proceed through a concerted process involving a hydrogen-bonded intermediate, whereas the U-4CR proceeds via the ionic intermediate **76**. A Mumm rearrangement takes place in the final step of both reactions to afford the peptide-like structures **74** or **78** [157, 163]. Consequently, both the P-3CR and U-4CR can be used for the synthesis of chiral (depsi)peptides and natural products [7].

It was previously thought that the Mumm rearrangement was the only irreversible step in these reactions. However, Chéron et al. [164] found that the addition of isocyanide to the carbonyl or imine (creating the stereogenic center) is also irreversible. This mechanistic finding is critical to developing asymmetric strategies, as it indicates that the stereochemical outcome of these reactions can be controlled kinetically [157, 164]. Developing an asymmetric version of these IMCRs should involve activation of the carbonyl or of the imine and could be realized by using either a chiral Lewis acid or Brønsted acid catalyst [165].

However, development of the enantioselective versions of both the P-3CR and U-4CR has proven extremely challenging. One of these challenges is the formal divalency of the nucleophilic isocyanide in Lewis acid-mediated α-additions. The resulting intermediate of this α-addition (**73** or **76**) combines with a carboxylate in the classical Passerini and Ugi reactions. In the absence of this carboxylate, however, these intermediates are susceptible to further addition by the isocyanide, leading to side reactions [166]. Even if the desired α-addition is successful, the dissociation of **73** or **76** that is required for catalyst turnover could be kinetically unfavorable and has thus seldom been successful [157, 166, 167]. Besides controlling the chemoselectivity of isocyanide-based MCRs, it is also difficult to control the stereochemistry of the α-addition [7].

The isocyanide forms yet another complication in metal-catalyzed reactions, as it can coordinate to the metal center and negatively impact the reaction pathway, leading to side reactions [157, 165]. This could be addressed by using a nonchelating Lewis acid or a Brønsted acid. However, the conjugate base of a Brønsted acid may compete with the carboxylate in the addition to the nitrilium intermediate [157, 165, 168]. Moreover, there is competition from uncatalyzed background reactions as well as competition between the Passerini and Ugi reaction, since both can occur spontaneously in an appropriate solvent at ambient temperature [157]. The Lewis basicity of the reactants poses another challenge in the chemoselective activation of the carbonyl or imine by Lewis acids [157, 165, 168].

Development of an asymmetric version of the U-4CR is even more challenging, since the carboxylic acid plays a dual role of both reactant and promotor in catalyzing imine formation and nucleophilic addition of the isocyanide [168]. As a result, this precludes the possibility of using a catalytic amount of chiral carboxylic acid [169].

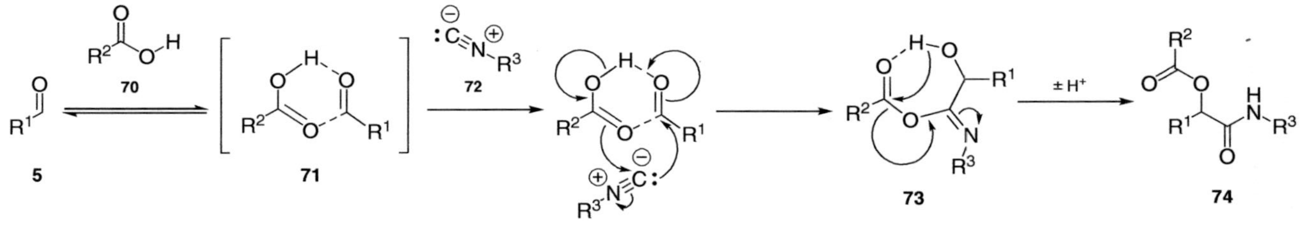

Scheme 12.21 Concerted mechanism of the Passerini reaction. Source: Based on Ramozzi et al. [159]; Banfi et al. [160].

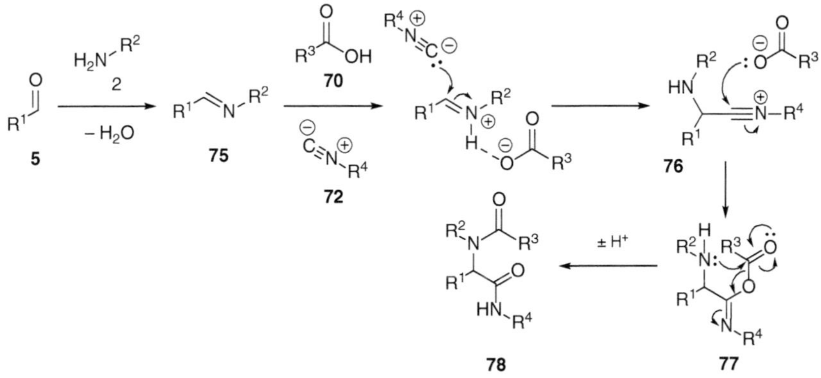

Scheme 12.22 Mechanism of the Ugi reaction through the ionic intermediate **76**. Source: Based on Wang et al. [157].

Although this review focuses on asymmetric MCRs, we will start this section with a two-component variation to provide historic context.

12.4.1 Passerini Reactions

12.4.1.1 Passerini-type Two-component Reactions

Denmark and Fan addressed the issue of the Lewis acid activation by activating the weak Lewis acid $SiCl_4$ with a chiral binaphthyldiamine-derived bisphosphoramide, which acts as a Lewis base. The kinetic species are only formed upon binding of $SiCl_4$ to the chiral catalyst, thereby decreasing the likelihood of $SiCl_4$ leading to achiral pathways [167]. This led to the first enantioselective α-addition of isocyanides to aldehydes in a truncated P-2CR to afford enantioenriched α-hydroxy carboxamides. In order to reduce background reactions caused by the isocyanide – since it also increases the Lewis acidity of $SiCl_4$ – they proposed to use it in a lower concentration. Moderate to excellent enantioselectivities (e.r. 67.4 : 32.6–99.9 : 0.01) were observed for (hetero)aromatic, α,β-unsaturated, and aliphatic aldehydes.

Since this pioneering work, others have also developed asymmetric (truncated) Passerini reactions. Zhu and co-workers, for example, have developed the catalyst (PyBOX)Sn(OTf)$_2$ to afford 2-(1-hydroxyalkyl)-5-aminooxazoles from α-isocyanoacetamides and aldehydes. They discovered that nonchelating aldehydes led to racemic products, since the 5-aminooxazoles products are bidentate ligands that chelate Lewis acids [170]. This led them to develop a chiral (salen)Al(III)Cl complex for this reaction with simple aldehydes instead, which gave moderate to good ee's ranging from 53% to 80% [171]. They further enhanced the selectivity of this reaction by developing a catalyst combination of Et_2AlCl and a CPA, with which ee's of 57–87% were achieved. This Al–CPA complex formed a *trans*-metal complex *in situ*, which seemed to be important for enantioselectivity.

The most efficient synthesis of 5-aminooxazoles was reported by Mihara et al. [172], who developed a heterodinuclear Ga/Yb-Schiff base complex for the addition of α-isocyanoacetamides to various (hetero)aromatic and alkenyl aldehydes with

95–98% ee. Finally, Zeng et al. [173] developed an asymmetric method using a CPA catalyst for the conversion of aliphatic aldehydes with 68–99% ee. Aromatic aldehydes, however, were less reactive under these conditions.

12.4.1.2 Passerini Three-component Reaction

The group of Dömling developed the first Lewis acid-mediated P-3CR with a Ti-TADDOL complex, which was used in stoichiometric amounts. This method afforded α-acyloxy carboxamides in low to moderate moderate enantioselectivity (32–42% ee) [174]. Following this report, Andreana et al. developed a tridentate indane (PyBOX)Cu(II) complex capable of catalyzing the P-3CR of both aliphatic and aromatic aldehydes, with 60–98% ee. Still, this method seemed to be limited to bidentate coordinating aldehydes [175]. Zhu et al. subsequently developed another Lewis acid-mediated P-3CR with a chiral (salen)Al(III)Cl complex [171] for the reaction with various nonchelating aliphatic aldehydes, which gave ee's of 68% up to >99% [176].

Tan and co-workers [163] developed the most efficient P-3CR to date with ee's up to 99%. This CPA-catalyzed (**CPA2**) method was compatible with both aromatic and aliphatic aldehydes and carboxylic acids (Scheme 12.23).

This method is superior to earlier methods developed by Kusebauch [174], Andreana [175], and Wang [176] in terms of enantioselectivity and compatible substrate range. Unlike these previous methods, this method developed by Tan and co-workers did not require an additional chiral Lewis acid for stereochemical control.

Ugi et al. [177] reported that the carboxylic acid in the P-3CR could be replaced by hydrazoic acid to afford tetrazoles. Yue et al. [178] reported the first enantioselective synthesis of tetrazoles catalyzed by a (salen)Al(III)Me complex. The authors substituted HN_3 for $TMSN_3$ in a Passerini-type reaction of a wide range of aliphatic aldehydes, and aromatic and aliphatic isocyanides to give the tetrazole products **80** with 51–97% ee.

Another strategy that could be employed to afford tetrazoles was reported by Xiong et al. [179], who used a chiral Mg(II)-*N*,*N'*-dioxide Lewis acid catalyst for the Passerini-type three-component synthesis of tetrazoles from (hetero)aromatic isocyanides (Scheme 12.24).

This reaction tolerates various β-alkyl or -aryl substituents (R^2 on **79**) and afforded the tetrazole products **80** with e.r.'s 77 : 23–97 : 3 (Scheme 12.24). The steric bulk of β-alkyl substituents seemed to influence the reactivities and stereoselectivity, as both lower yields and enantioselectivities were observed for three- to five-membered rings. When R^3 = aryl, *para*-substituted substrates provided higher yields and enantioselectivities than *ortho*- or *meta*-substituted substrates.

These tetrazoles are generated through an addition of **73** to **79** and subsequent addition of $TMSN_3$, which is converted to HN_3 by water (Scheme 12.25). In the final step, an intramolecular cyclization takes place to form **80**. The same authors also reported a second method to afford tetrazoles through a four-component reaction, in which an additional isocyanide is used. These tetrazoles were obtained with e.r.'s of 76 : 24–95 : 5 for heteroaromatic isocyanides.

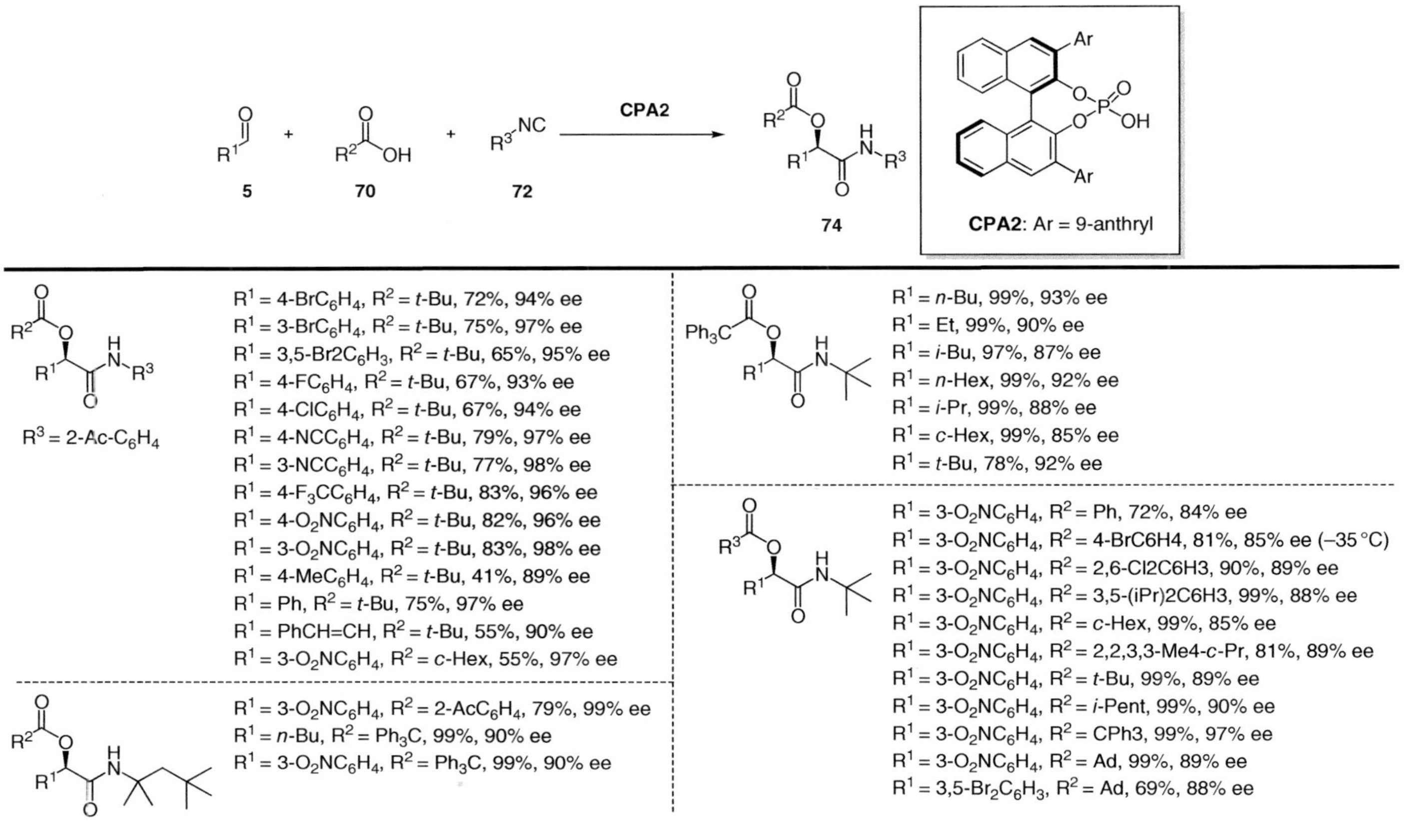

Scheme 12.23 Substrate scope of aliphatic carboxylic acids, aliphatic aldehydes (10 mol% **CPA2**, $CHCl_3$, −20 °C), and aromatic aldehydes (15 mol% **CPA2**, $CHCl_3$, RT) [157, 163] in asymmetric P-3CR.

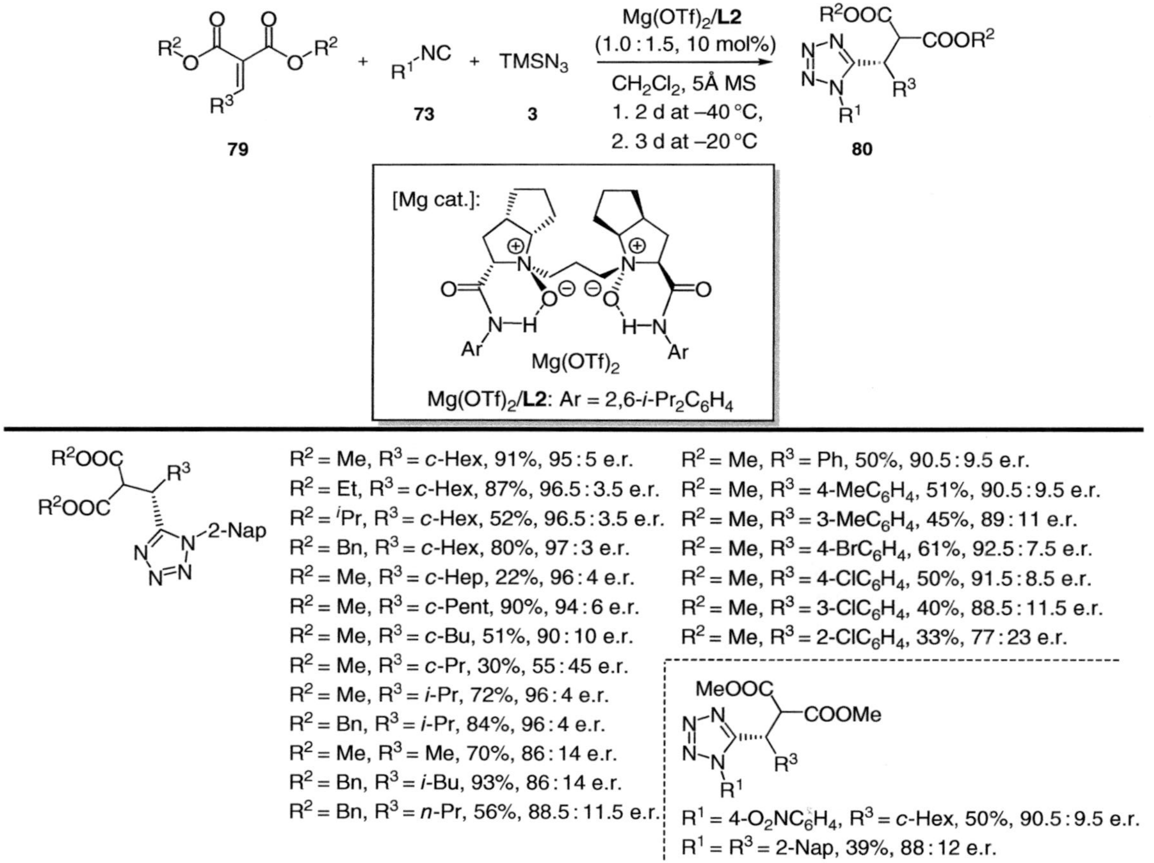

Scheme 12.24 Substrate scope of asymmetric three-component synthesis of tetrazoles **80** catalyzed by $Mg(OTf)_2$/**L2**. Source: Based on Xiong et al. [179].

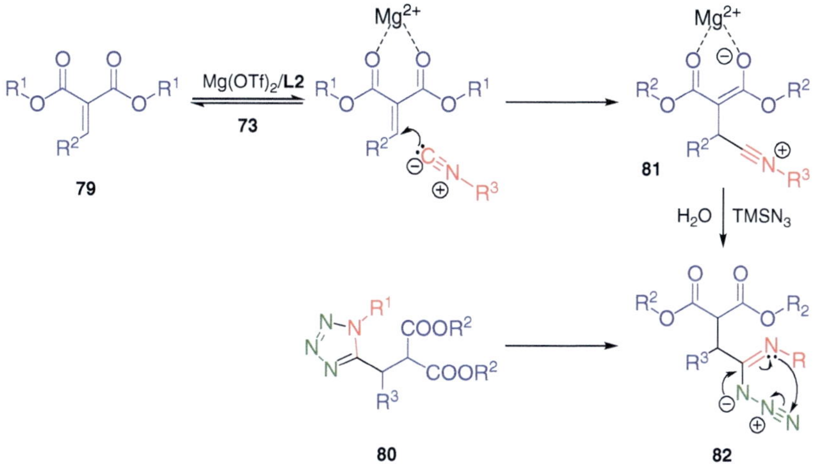

Scheme 12.25 Proposed mechanism of the Passerini-type reaction with $TMSN_3$. Source: Based on Xiong et al. [179].

12.4.2 Isocyanide-Based [3 + 2]-Cycloaddition

Xiong et al. [179] also reported a highly enantioselective [3 + 2]-cycloaddition of isocyanides, alkylidene malonates, and isoquinolines catalyzed by $Mg(OTf)_2$/**L2** to afford 1,2-dihydroisoquinolines (Scheme 12.26).

This reaction proceeds through a similar mechanism as the Passerini-type three-component reaction in Scheme 12.25. In this reaction, however, the common intermediate (**81**) undergoes a dearomative [3 + 2]-annulation with unactivated quinolines to form **84**. Although this reaction was less selective for aliphatic aldehyde-derived alkylidenes (R^3 = alkyl on **79**, e.r. 77 : 23), it was suitable for a number of isoquinolines, regardless of the electronic nature of its substituents (Scheme 12.26). Moreover, bulky isocyanides and aliphatic isocyanides seemed to give higher enantioselectivities than their less bulky or aromatic counterparts. [3 + 2]-Cycloadditions, such as this example, are a powerful tool for the synthesis of five-membered ring systems. Through these cycloadditions biologically important unnatural amino acids, chiral pyrrolidines, and other complex natural or medicinal compounds can be accessed [180–182].

12.4.3 Ugi-type Reactions

The first enantioselective Ugi-type reaction was reported by Zhu and co-workers [183], who used a CPA to afford chiral 2-(1-aminoalkyl)-5-aminooxazoles with 56–90% ee. This strategy was compatible with various aliphatic aldehydes, including linear and α-branched aldehydes, but was less efficient for aromatic aldehydes. Wang et al. also developed a more efficient CPA-catalyzed reaction for the chiral synthesis of 5-alkoxyoxazoles from isocyanides, aliphatic aldehydes, and aromatic amines, with 87–94% ee [184]. Hashimoto et al. [169], on the other hand, developed a chiral dicarboxylic acid catalyst (**CDA1**) to access heterocycles from aromatic

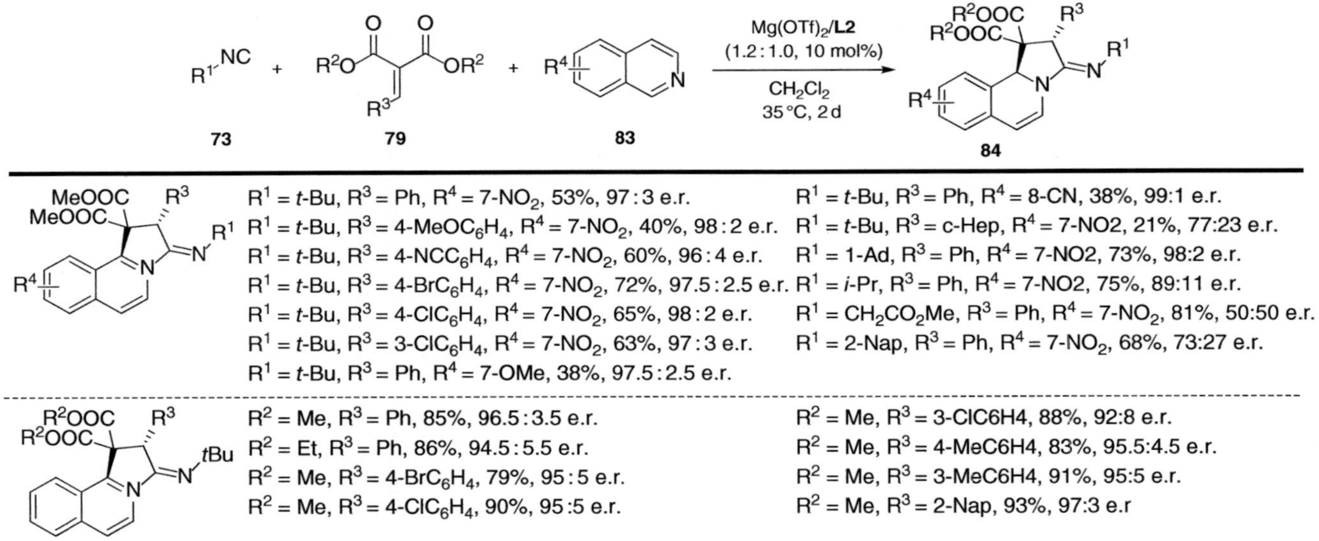

Scheme 12.26 Substrate scope of the asymmetric synthesis of 1,2-dihydroisoquinoline derivatives **84** catalyzed by $Mg(OTf)_2$/**L2**. Source: Based on Xiong et al. [179].

aldehydes (Scheme 12.27) or benzoxazoles from aromatic isocyanides with moderate to good enantioselectivity (Scheme 12.28). In the latter case, the second reaction stage involving transesterification of the benzoate in **87** was required to form the benzoxazole ring. The mild base K_2CO_3 was used to prevent racemization.

A rather different type of catalyst that can be used in Ugi-type reactions is the BOROX catalyst **BX1**, consisting of a chiral boroxinate ion paired with a protonated secondary amine substrate (Scheme 12.29). This catalyst, developed by Wulff and co-workers [185], promotes the Ugi-type three-component reaction of isocyanides, aromatic aldehydes, and secondary amines with high enantioselectivity.

In all of the above-mentioned Ugi reactions, the carboxylic acid is absent. Since this component acts as both a catalyst and reactant in this reaction, the development of an asymmetric U-4CR is very challenging. Zhang et al. [186] attempted to include the carboxylic acid in an intramolecular fashion through the use of 2-formylbenzoic acids using an H_8-BINOL-derived CPA catalyst H_8-BINOL-derived CPA catalyst (Scheme 12.30).

Finally, a true enantioselective U-4CR was developed by Tan and co-workers utilizing SPINOL-derived CPAs, of which **CPA4** was more suitable for aromatic aldehydes, while **CPA5** performed better for aliphatic aldehydes [7, 161]. The mode of action of these CPAs is proposed to involve the formation of a chiral receptor site – with a tunable chiral pocket – in which the phosphoric acid, carboxylic acid, and imine form a stabilized hydrogen-bonding network. This results in increased acidity of the imine and enhanced nucleophilicity of the carboxylic acid. This in turn results in increased chemoselectivity of the nucleophilic attack by the isocyanide on the imine (Scheme 12.31) [7, 161].

Using **CPA4**, aromatic aldehydes could be combined with a range of isocyanides, carboxylic acids, and aliphatic amines (Scheme 12.32). The use of **CPA5** allowed the combination of aliphatic aldehydes with a range of isocyanides, carboxylic acids, and aromatic amines (Scheme 12.33). Besides being more compatible with aliphatic aldehydes, **CPA5** also afforded the opposite enantiomer. Although the highest enantioselectivity observed with either of these catalysts (97% ee) is still lower than what is required for the development of chiral drugs (>98% ee), this method represents a major breakthrough in asymmetric Ugi reactions. Moreover, such high initial ee's can likely be further improved by crystallization.

Finally, the group of Tan also developed an U-3CR with **CPA5** to afford α-amino acids from aliphatic aldehydes [162]. Such three-component reactions, in which the nitrilium ion (**76**, Scheme 12.22) is trapped by water instead of a carboxylate ion, afford structures that are more tunable than the U-4CR would afford. By employing aliphatic aldehydes, α-amino acids could be accessed that are more common in natural products and pharmaceuticals. The developed conditions were tolerated by various aliphatic aldehydes, aromatic amines, and a variety of sterically hindered isocyanides, affording the target α-amino acids with high ee's of 81–>99%.

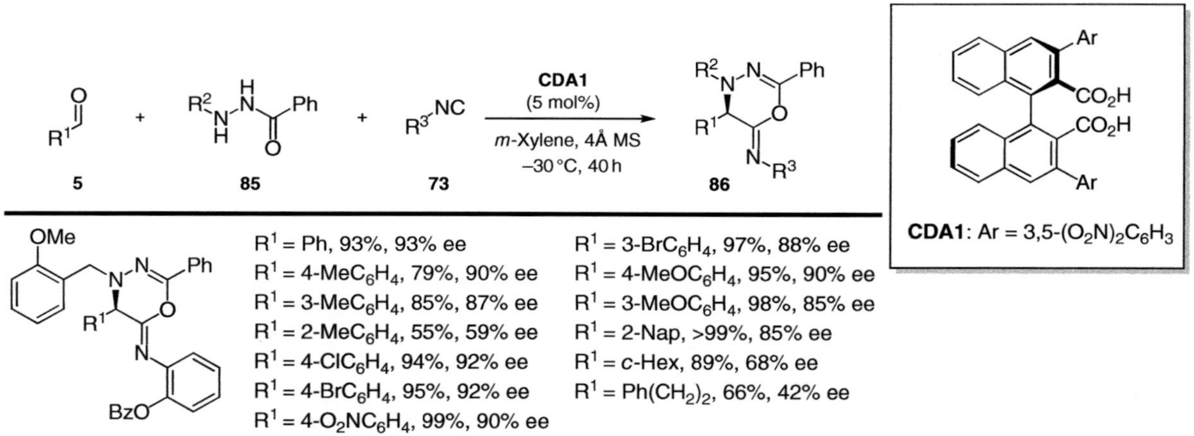

Scheme 12.27 Multicomponent synthesis of oxadiazines catalyzed by **CDA1**. Source: Based on Hashimoto et al. [169].

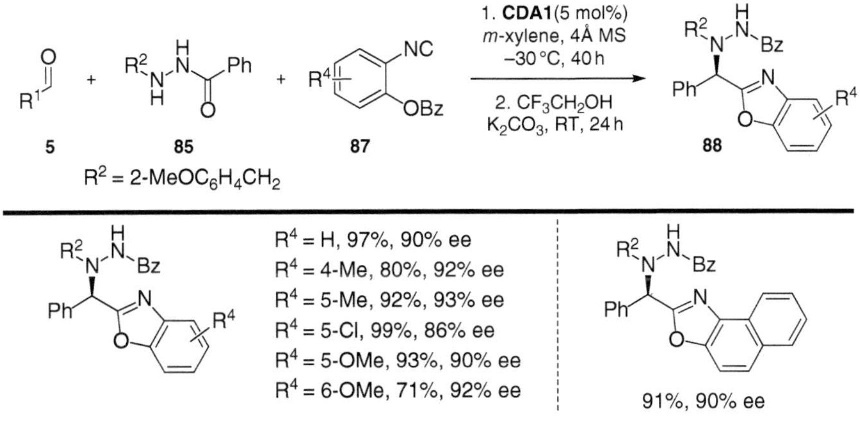

Scheme 12.28 Multicomponent synthesis of benzoxazoles catalyzed by **CDA1**. Source: Based on Hashimoto et al. [169].

12.5 Conclusion

In this review, enantioselective catalytic systems for the most popular MCRs have been covered, of which the most significant contributions since 2012 have been highlighted. These catalysts have led to many improvements in the transformations of complex systems, such as a broadened scope and increased chemo- and stereoselectivity. These developments are crucial to drug discovery, as the frameworks that can be accessed through these AMCRs possess biological activities.

Among the most frequently used catalytic systems in imine and Michael addition-based MCRs are proline- and other amine-based catalysts. Proline derivatives typically act as nucleophilic catalysts through an enamine or iminium ion formation [8, 109, 132, 187–189]. The enamine pathway allows for reactions of the activated aldehyde with electrophiles, whereas the iminium pathway facilitates reaction with nucleophiles [189]. Other amine-based organocatalysts, such as chiral thiourea-based catalysts and guanidine–amides, have been a useful tools in the dual activation of both the nucleophile (through their basic functionality) and electrophile (through double hydrogen bond formation by the thiourea or amide moieties). Through this double functionality, they enhance both reactivity and stereoselectivity [5, 53, 129, 155].

Other commonly used catalytic systems are the Brønsted acidic CPAs, which are useful in Mannich reactions for the activation of the imine. These catalysts have also proven useful in isocyanide-based MCRs as they enhance both the chemo- and stereoselectivity in the nucleophilic attack by the isocyanide. The most frequently used CPAs have a (H_8-)BINOL or SPINOL backbone and contain substituents that affect the pK_a of these Brønsted acids in the electrophilic activation of reactants. These substituents also play a role in chiral induction by appropriate arrangement of reactants through hydrogen-bonding and/or ion-pairing interactions [2, 190].

Despite the many efforts spent on developing AMCRs, developing highly efficient systems is still challenging [2]. A thorough understanding of MCR mechanisms

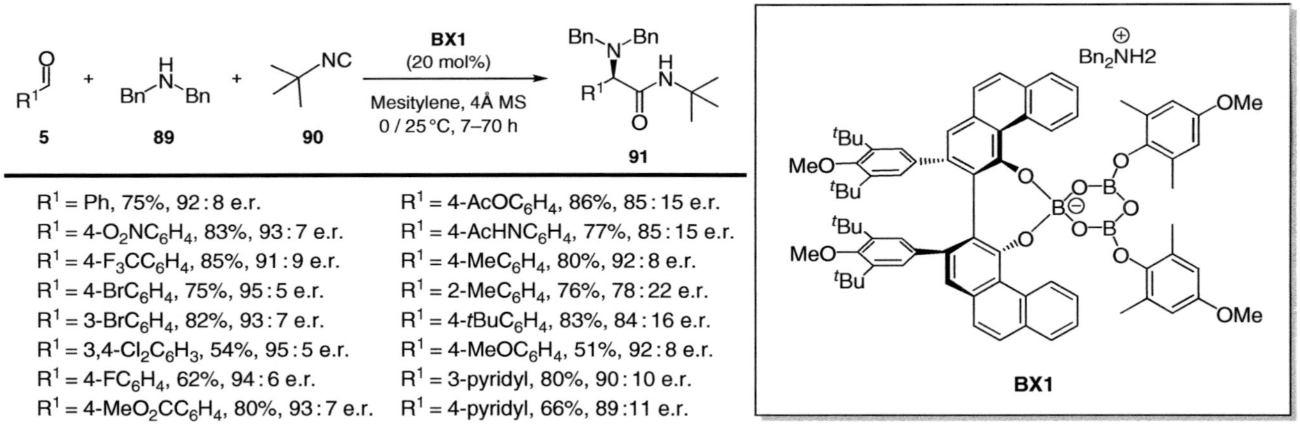

Scheme 12.29 Substrate scope of aromatic aldehydes for the catalytic asymmetric three-component Ugi reaction catalyzed by **BX1**. Source: Based on Zhao et al. [185].

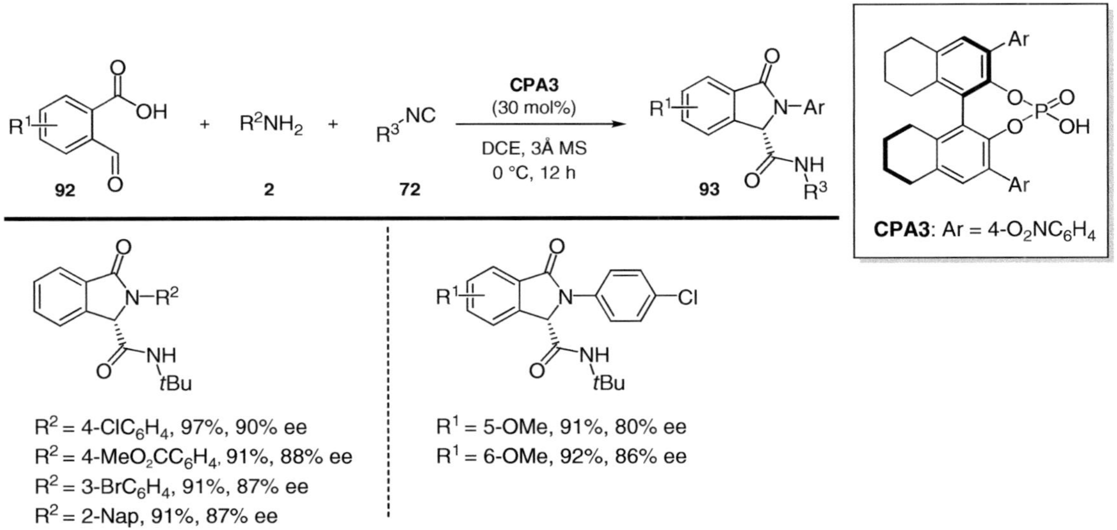

Scheme 12.30 Scope of the Ugi-type three-component reaction catalyzed by **CPA3**. Source: Based on Zhang et al. [186].

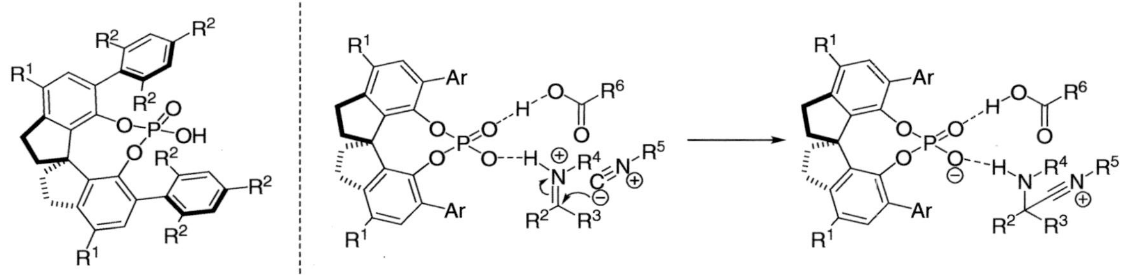

CPA4, R^1 = Br, R^2 = *i*-Pr
CPA5, R^1 = H, R^2 = Cy

Scheme 12.31 Chemical structure of **CPA4** and **CPA5** and the chiral pocket of the CPAs with the carboxylic acid and imine. Source: Based on Zhang et al. [7].

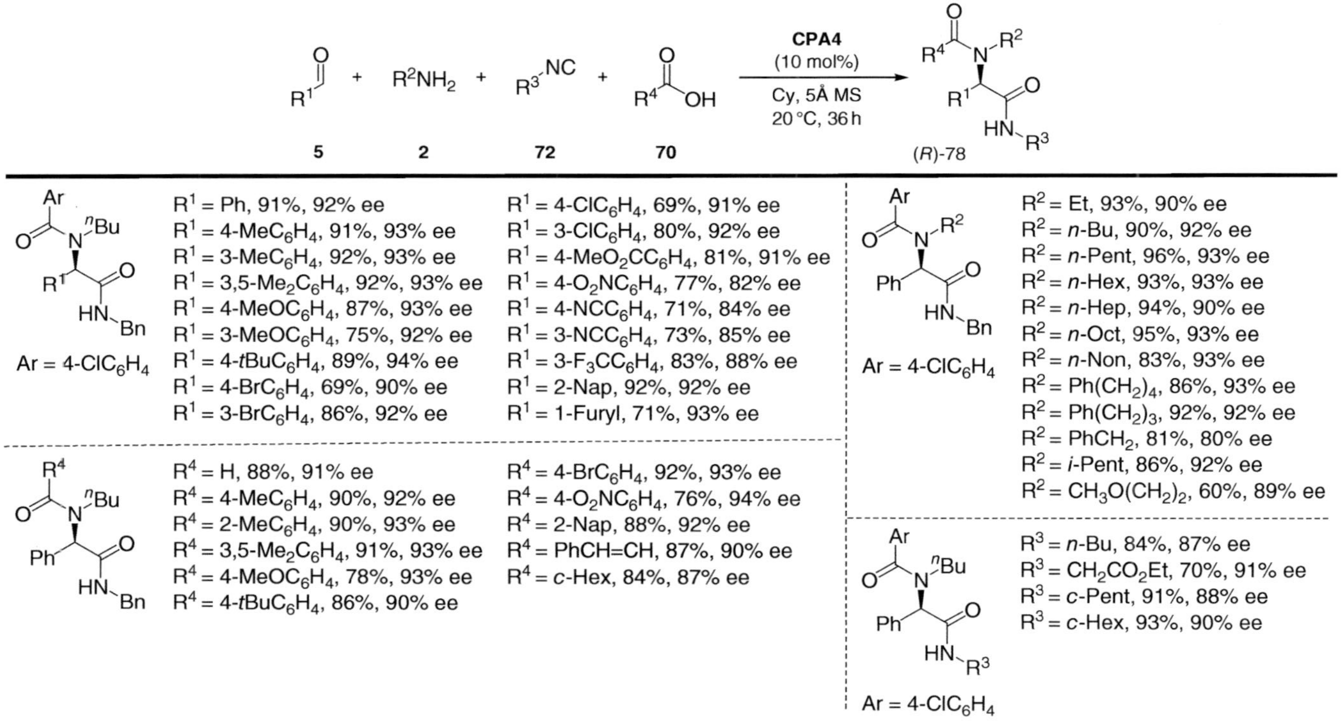

Scheme 12.32 Substrate scope of the Ugi four-component reaction with **CPA4** to afford (*R*)-**78**. Source: Based on Zhang et al. [7].

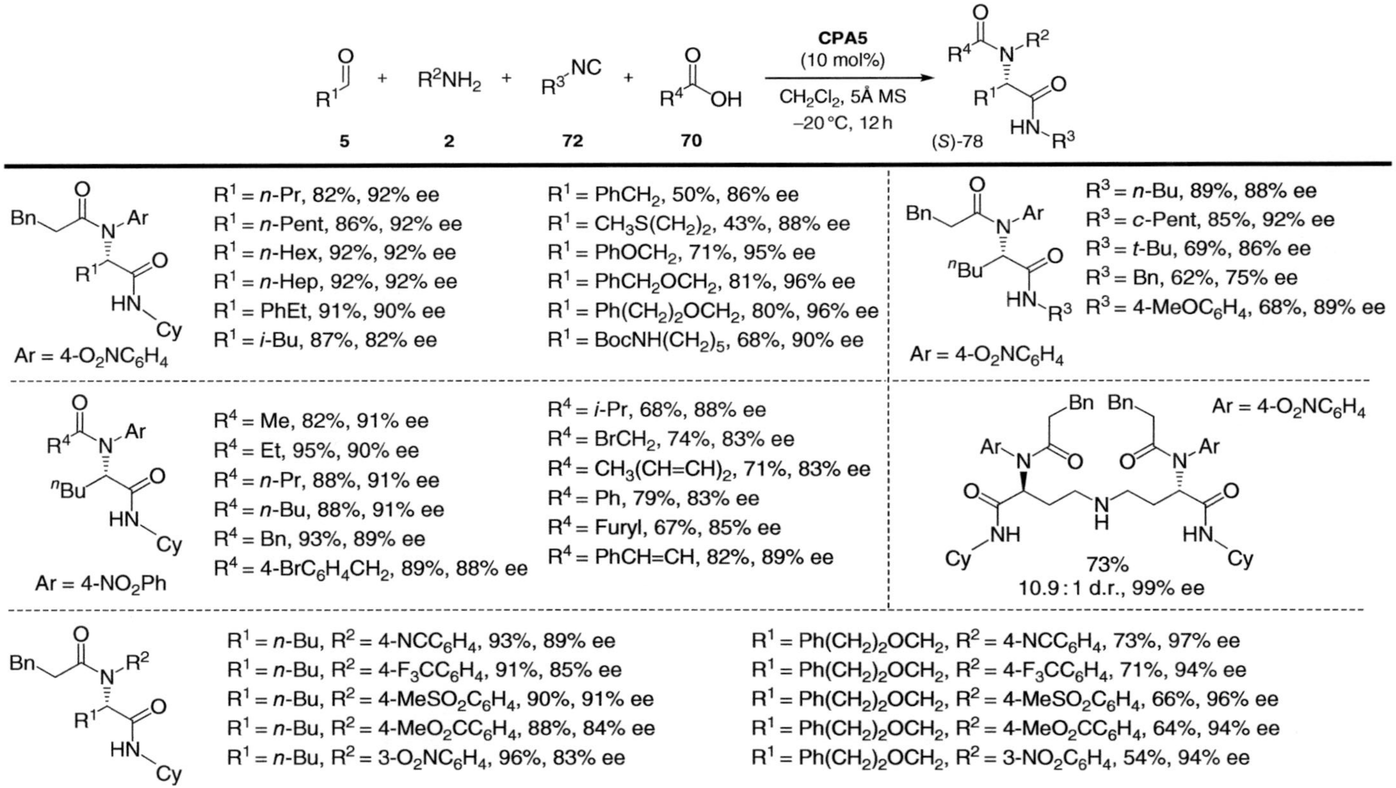

Scheme 12.33 Substrate scope of the Ugi four-component reaction with **CPA5** to afford (*S*)-**78**. Source: Based on Zhang et al. [7].

is of paramount importance for the development of enantioselective versions [3, 7]. MCRs that still require improvement include the Biginelli, Hantzsch, and isocyanide-based reactions. The Biginelli reactions still require improvement in reaction conditions, as high catalyst loadings, and in some cases long reaction times, are still an issue [3, 10]. The Hantzsch reaction not only experiences similar drawbacks as the Biginelli reaction, but can also proceed through several different pathways. This leads to side reactions and difficulties in developing asymmetric versions of this reaction [3].

The Passerini and Ugi reactions have both witnessed major breakthroughs in the past five years. Previously, truncated asymmetric versions of these reactions were developed, the scope of which was also limited [162, 163]. The CPAs developed for these reactions allowed for both the carboxylic acid to be incorporated into the asymmetric reaction and for the scope to be improved. These CPAs operate via a similar mechanism: enhancing the nucleophilicity of the carboxylic acid and activating the isocyanide through hydrogen bonding [162, 163]. Although high enantioselectivities were achieved in both reactions, higher ee's (>98%) are a remaining target for the Ugi 4CR [168].

Abbreviations

Ac	Acetyl
ACDC	Asymmetric counterion-directed catalysis
Ad	1-Adamantyl
AMCR	Asymmetric multicomponent reaction
BINOL	1,1′-Bi-2-naphthol
Bn	Benzyl
BOX	Bis(oxazoline)
Bz	Benzoyl
CA	Cinchona alkaloid
CDA	Chiral dicarboxylic acid
CIL	Chiral ionic liquid
CPA	Chiral phosphoric acid
CPA.IS	Chiral phosphoric acid and imidazolium sulfonic acid
CT	Cinchona-based thiourea
CU	Cinchona-based urea
Cy	Cyclohexane
DCE	1,2-Dichloroethane
DHP	1,4-Dihydropyridines
DHPM	Dihydropyrimidines
DMP	Dess–Martin periodinane
ee	Enantiomeric excess
e.r.	Enantiomeric ratio
fod	2,2-Dimethyl-6,6,7,7,8,8,8-heptafluorooctane-3,5-dionato
GA	Guanidine–amide

IEDHDA	Inverse electron-demand hetero-Diels–Alder reaction
ILC	Ionic liquid catalysis
ILE	Ionic liquid effect
IMCR	Isocyanide-based multicomponent reaction
MCR	Multicomponent reaction
MS	Molecular sieves
Nap	Naphthyl
NMP	*N*-Methylpyrrolidone
OTf	Triflate
P-3CR	Passerini three-component reaction
PMB	*p*-Methoxybenzyl
PMP	*p*-Methoxyphenyl
Py	Pyridine, pyridyl
p-TsOH	*p*-Toluenesulfonic acid
Salen	*N,N'*-bis(salicylidene)ethylenediamine dianion
SQ	Squaramide
TADDOL	α,α,α,α-tetraaryl-1,3-dioxolane-4,5-dimethanol
TBS	*tert*-Butyldimethylsilyl
TBDPS	*tert*-Butyldiphenylsilyl
TFA	Trifluoroacetic acid
THF	Tetrahydrofuran
THQ	1,2,3,4-Tetrahydroquinolines
TMS	Trimethylsilyl
TPA	Tartaric acid-derived phosphate
U-4CR	Ugi four-component reaction

References

1 de Graaff, C., Ruijter, E., and Orru, R.V.A. (2012). Recent developments in asymmetric multicomponent reactions. *Chem. Soc. Rev.* 41 (10): 3969–4009.

2 Nunes, P.S.G., Vidal, H.D.A., and Correâ, A.G. (2020). Recent advances in catalytic enantioselective multicomponent reactions. *Org. Biomol. Chem.* 18 (39): 7751–7773.

3 Alvim, H.G.O., Da Silva Júnior, E.N., and Neto, B.A.D. (2014). What do we know about multicomponent reactions? Mechanisms and trends for the Biginelli, Hantzsch, Mannich, Passerini and Ugi MCRs. *RSC Adv.* 4 (97): 54282–54299.

4 Ramón, D.J. and Yus, M. (2005). Asymmetric multicomponent reactions (AMCRs): the new frontier. *Angew. Chem. Int. Ed.* 44 (11): 1602–1634.

5 Parvin, T., Yadav, R., and Choudhury, L.H. (2020). Recent applications of thiourea-based organocatalysts in asymmetric multicomponent reactions (AMCRs). *Org. Biomol. Chem.* 18 (29): 5513–5532.

6 Ahmadi, T., Mohammadi Ziarani, G., Gholamzadeh, P., and Mollabagher, H. (2017). Recent advances in asymmetric multicomponent reactions (AMCRs). *Tetrahedron Asymm.* 28 (5): 708–724.

7 Zhang, J., Yu, P., Li, S.Y. et al. (2018). Asymmetric phosphoric acid-catalyzed four-component Ugi reaction. *Science* 361 (6407): eaas8707.

8 Dochain, S., Vetica, F., Puttreddy, R. et al. (2016). Combining organocatalysis and lanthanide catalysis: a sequential one-pot quadruple reaction sequence/hetero-diels-alder asymmetric synthesis of functionalized tricycles. *Angew. Chem.* 128 (52): 16387–16389.

9 Gasperi, T., Orsini, M., Vetica, F., and de Figueiredo, R.M. (2015). Organocatalytic asymmetric multicomponent reactions. In: *Multicomponent Reactions* (eds. R.P. Herrera and E. Marqués-López), 16–71. Hoboken, NJ: John Wiley & Sons, Inc.

10 Alvim, H.G.O., Pinheiro, D.L.J., Carvalho-Silva, V.H. et al. (2018). Combined role of the asymmetric counteranion-directed catalysis (ACDC) and ionic liquid effect for the enantioselective Biginelli multicomponent reaction. *J. Org. Chem.* 83 (19): 12143–12153.

11 Wen, Y., Xiong, Y., Chang, L. et al. (2007). Chiral bisformamides as effective organocatalysts for the asymmetric one-pot, three-component strecker reaction. *J. Org. Chem.* 72 (20): 7715–7719.

12 Wen, Y., Gao, B., Fu, Y. et al. (2008). Asymmetric three-component Strecker reactions catalyzed by *trans*-4-hydroxy-L-proline-derived N,N′-dioxides. *Chem. Eur. J.* 14 (22): 6789–6795.

13 Zhang, G.W., Zheng, D.H., Nie, J. et al. (2010). Brønsted acid-catalyzed efficient Strecker reaction of ketones, amines and trimethylsilyl cyanide. *Org. Biomol. Chem.* 8 (6): 1399–1405.

14 Pan, S.C. and List, B. (2007). Catalytic asymmetric three-component acyl-strecker reaction. *Org. Lett.* 9 (6): 1149–1151.

15 Liu, Y.L., Yin, X.P., and Zhou, J. (2018). Internally reuse waste: catalytic asymmetric one-pot strecker reaction of fluoroalkyl ketones, anilines and TMSCN by sequential catalysis. *Chin. J. Chem.* 36 (4): 321–328.

16 Roman, G. (2015). Mannich bases in medicinal chemistry and drug design. *Eur. J. Med. Chem.* 89: 743–816.

17 Yamasaki, S., Iida, T., and Shibasaki, M. (1999). Direct catalytic asymmetric Mannich reaction of unmodified ketones: cooperative catalysis of an AlLibis(binaphthoxide) complex and $La(OTf)_3 \cdot nH_2O$. *Tetrahedron* 55 (29): 8857–8867.

18 Kobayashi, S., Kobayashi, J., Yazaki, R., and Ueno, M. (2007). Toward the total synthesis of onchidin, a cytotoxic cyclic depsipeptide from a mollusc. *Chem. Asian J.* 2 (1): 135–144.

19 Zhang, Q., Hui, Y., Zhou, X. et al. (2010). Highly efficient asymmetric three-component vinylogous mannich reaction catalyzed by a chiral Scandium(III)N,N’-dioxide complex. *Adv. Synth. Catal.* 352 (6): 976–980.

20 Ueno, M., Kitagawa, H., Ishitani, H. et al. (2001). Catalytic enantioselective synthesis of a novel inhibitor of ceramide trafficking, (1R,3R)-N-(3-hydroxy-1-hydroxymethyl-3-phenylpropyl)-dodecanamide (HPA-12). *Tetrahedron Lett.* 42 (44): 7863–7865.

21 Zhao, J., Liu, X., Luo, W. et al. (2013). Asymmetric synthesis of β-amino nitriles through a Sc III-catalyzed three-component mannich reaction of silyl ketene imines. *Angew. Chem. Int. Ed.* 52 (12): 3473–3477.

22 Feng, L., Dai, X., Meggers, E., and Gong, L. (2017). Three-component asymmetric mannich reaction catalyzed by a lewis acid with rhodium-centered chirality. *Chem. Asian J.* 12 (9): 963–967.

23 Ibrahem, I., Casas, J., and Córdova, A. (2004). Direct catalytic enantioselective α-aminomethylation of ketones. *Angew. Chem. Int. Ed.* 43 (47): 6528–6531.

24 Ibrahem, I. and Córdova, A. (2005). Amino acid-catalyzed direct enantioselective synthesis of β-amino-α-oxyaldehydes. *Tetrahedron Lett.* 46 (16): 2839–2843.

25 Chowdari, N.S., Ahmad, M., Albertshofer, K. et al. (2006). Expedient synthesis of chiral 1,2- and 1,4-diamines: protecting group dependent regioselectivity in direct organocatalytic asymmetric Mannich reactions. *Org. Lett.* 8 (13): 2839–2842.

26 Gu, Q., Gong, J.J., Feng, J. et al. (2008). Organocatalytic direct asymmetric Mannich reactions of hydroxyacetone by (2S,5S)-pyrrolidine-2,5-dicarboxylic acid. *Chin. J. Chem.* 26 (10): 1902–1906.

27 Yang, H. and Carter, R.G. (2009). Enantioselective Mannich reactions with the practical proline mimetic *N*-(*p*-dodecylphenyl-sulfonyl)-2-pyrrolidinecarboxamide. *J. Org. Chem.* 74 (5): 2246–2249.

28 Gu, Q., Jiang, L.X., Yuan, K. et al. (2008). Organocatalytic, asymmetric, one-pot, three-component Mannich reaction of hydroxyacetone. *Synth. Commun.* 38 (23): 4198–4206.

29 List, B. (2000). The direct catalytic asymmetric three-component Mannich reaction. *J. Am. Chem. Soc.* 122 (38): 9336–9337.

30 List, B., Pojarliev, P., Biller, W.T., and Martin, H.J. (2002). The proline-catalyzed direct asymmetric three-component Mannich reaction: scope, optimization, and application to the highly enantioselective synthesis of 1,2-amino alcohols. *J. Am. Chem. Soc.* 124 (5): 827–833.

31 Hayashi, Y., Tsuboi, W., Ashimine, I. et al. (2003). The direct and enantioselective, one-pot, three-component, cross-Mannich reaction of aldehydes. *Angew. Chem. Int. Ed.* 42 (31): 3677–3680.

32 Hayashi, Y., Urushima, T., Aratake, S. et al. (2008). Organic solvent-free, enantio- and diastereoselective, direct Mannich reaction in the presence of water. *Org. Lett.* 10 (1): 21–24.

33 Notz, W., Watanabe, S.I., Chowdari, N.S. et al. (2004). The scope of the direct proline-catalyzed asymmetric addition of ketones to imines. *Adv. Synth. Catal.* 346 (9–10): 1131–1140.

34 Ibrahem, I. and Córdova, A. (2005). Amino acid catalyzed direct enantioselective formation of carbohydrates: one-step de novo synthesis of ketoses. *Tetrahedron Lett.* 46 (19): 3363–3367.

35 Ibrahem, I., Zou, W., Xu, Y., and Córdova, A. (2006). Amino acid-catalyzed asymmetric carbohydrate formation: organocatalytic one-step de novo synthesis of keto and amino sugars. *Adv. Synth. Catal.* 348 (1–2): 211–222.

36 Córdova, A. (2004). The direct catalytic asymmetric cross-Mannich reaction: a highly enantioselective route to 3-amino alcohols and α-amino acid derivatives. *Chem. Eur. J.* 10 (8): 1987–1997.

37 Kantam, M.L., Rajasekhar, C.V., Gopikrishna, G. et al. (2006). Proline catalyzed two-component, three-component and self-asymmetric Mannich reactions promoted by ultrasonic conditions. *Tetrahedron Lett.* 47 (33): 5965–5967.

38 Westermann, B. and Neuhaus, C. (2005). Dihydroxyacetone in amino acid catalyzed mannich-type reactions. *Angew. Chem. Int. Ed.* 44 (26): 4077–4079.

39 Enders, D., Grondal, C., Vrettou, M., and Raabe, G. (2005). Asymmetric synthesis of selectively protected amino sugars and derivatives by a direct organocatalytic mannich reaction. *Angew. Chem. Int. Ed.* 44 (26): 4079–4083.

40 Hayashi, Y., Yamaguchi, J., Hibino, K. et al. (2004). A highly active 4-siloxyproline catalyst for asymmetric synthesis. *Adv. Synth. Catal.* 346 (12): 1435–1439.

41 Ibrahem, I., Zou, W., Engqvist, M. et al. (2005). Acyclic chiral amines and amino acids as inexpensive and readily tunable catalysts for the direct asymmetric three-component Mannich reaction. *Chem. Eur. J.* 11 (23): 7024–7029.

42 Cheng, L., Wu, X., and Lu, Y. (2007). Direct asymmetric three-component organocatalytic anti-selective Mannich reactions in a purely aqueous system. *Org. Biomol. Chem.* 5 (7): 1018–1020.

43 Teo, Y.C., Lau, J.J., and Wu, M.C. (2008). Direct asymmetric three-component Mannich reactions catalyzed by a siloxy serine organocatalyst in water. *Tetrahedron Asymm.* 19 (2): 186–190.

44 Ramasastry, S.S.V., Zhang, H., Tanaka, F., and Barbas, C.F. (2007). Direct catalytic asymmetric synthesis of *anti*-1,2-amino alcohols and syn-1,2-diols through organocatalytic anti-Mannich and syn-aldol reactions. *J. Am. Chem. Soc.* 129 (2): 288–289.

45 Dziedzic, P., Ibrahem, I., and Córdova, A. (2008). Direct catalytic asymmetric three-component Mannich reactions with dihydroxyacetone: enantioselective synthesis of amino sugar derivatives. *Tetrahedron Lett.* 49 (5): 803–807.

46 Blümel, M., Chauhan, P., Hahn, R. et al. (2014). Asymmetric synthesis of tetrahydropyridines via an organocatalytic one-pot multicomponent Michael/aza-Henry/cyclization triple domino reaction. *Org. Lett.* 16 (22): 6012–6015.

47 Lin, H., Tan, Y., Liu, W.J. et al. (2013). A highly efficient access to enantiopure tetrahydropyridines: dual-organocatalyst-promoted asymmetric cascade reaction. *Chem. Commun.* 49 (38): 4024–4026.

48 Zheng, X., Qian, Y.B., and Wang, Y. (2010). 2-pyrrolidinecarboxylic acid ionic liquid as a highly efficient organocatalyst for the asymmetric one-pot Mannich reaction. *Eur. J. Org. Chem.* 3: 515–522.

49 Liu, B., Xu, D., Dong, J. et al. (2007). Highly efficient AILs/L-proline synergistic catalyzed three-component asymmetric Mannich reaction. *Synth. Commun.* 37 (17): 3003–3010.

50 Chowdari, N.S., Ramachary, D.B., and Barbas, C.F. (2003). Organocatalysis in ionic liquids: highly efficient L-proline-catalyzed direct asymmetric Mannich reactions involving ketone and aldehyde nucleophiles. *Synlett* 12: 1906–1909.

51 Bai, S., Liang, X., Song, B. et al. (2011). Asymmetric Mannich reactions catalyzed by cinchona alkaloid thiourea: enantioselective one-pot synthesis of novel β-amino ester derivatives. *Tetrahedron Asymm.* 22 (5): 518–523.

52 Guo, Q. and Zhao, J.C.G. (2013). Highly enantioselective three-component direct mannich reactions of unfunctionalized ketones catalyzed by bifunctional organocatalysts. *Org. Lett.* 15 (3): 508–511.

53 Bai, S., Liu, S., Zhu, Y. et al. (2018). Antiviral bioactivity of chiral β-amino acid ester derivatives synthesized through a one-pot, solvent-free asymmetric Mannich reaction. *Synlett* 29 (14): 1921–1925.

54 Hou, W., Wei, Q., Liu, G. et al. (2015). Asymmetric multicomponent sulfa-michael/Mannich cascade reaction: synthetic access to 1,2-diamino-3-organosulfur compounds and 2-nitro allylic amines. *Org. Lett.* 17 (19): 4870–4873.

55 Guo, Q.X., Liu, H., Guo, C. et al. (2007). Chiral Brønsted acid-catalyzed direct asymmetric Mannich reaction. *J. Am. Chem. Soc.* 129 (13): 3790–3791.

56 Jiang, J., Xu, H.D., Xi, J.B. et al. (2011). Diastereoselectively switchable enantioselective trapping of carbamate ammonium ylides with imines. *J. Am. Chem. Soc.* 133 (22): 8428–8431.

57 Hayashi, Y., Yamazaki, T., Kawauchi, G., and Sato, I. (2018). Prolinate salt as a catalyst in the syn -selective, asymmetric Mannich reaction of alkynyl imine. *Org. Lett.* 20 (8): 2391–2394.

58 Ayats, C., Henseler, A.H., Dibello, E., and Pericás, M.A. (2014). Continuous flow enantioselective three-component anti -mannich reactions catalyzed by a polymer-supported threonine derivative. *ACS Catal.* 4 (9): 3027–3033.

59 Yin, Z., Guo, J., Zhang, R. et al. (2020). Direct asymmetric three-component mannich reaction catalyzed by chiral counteranion-assisted silver. *J. Org. Chem.* 85 (16): 10369–10377.

60 Okino, T., Hoashi, Y., Furukawa, T. et al. (2005). Enantio- and diastereoselective michael reaction of 1,3-dicarbonyl compounds to nitroolefins catalyzed by a bifunctional thiourea. *J. Am. Chem. Soc.* 127 (1): 119–125.

61 Palomo, C., Oiarbide, M., Laso, A., and López, R. (2005). Catalytic enantioselective aza-Henry reaction with broad substrate scope. *J. Am. Chem. Soc.* 127 (50): 17622–17623.

62 Cruz-Acosta, F., De Armas, P., and García-Tellado, F. (2013). Water-compatible hydrogen-bond activation: a scalable and organocatalytic model for the stereoselective multicomponent aza-Henry reaction. *Chem. Eur. J.* 19 (49): 16550–16554.

63 Lou, S. and Schaus, S.E. (2008). Asymmetric petasis reactions catalyzed by chiral biphenols. *J. Am. Chem. Soc.* 130 (22): 6922–6923.

64 Wu, P., Givskov, M., and Nielsen, T.E. (2019). Reactivity and synthetic applications of multicomponent petasis reactions. *Chem. Rev.* 119 (20): 11245–11290.

65 Beisel, T. and Manolikakes, G. (2013). A lewis acid palladium(II)-catalyzed three-component synthesis of α-substituted amides. *Org. Lett.* 15 (23): 6046–6049.

66 Han, W.Y., Zuo, J., Zhang, X.M., and Yuan, W.C. (2013). Enantioselective petasis reaction among salicylaldehydes, amines, and organoboronic acids catalyzed by BINOL. *Tetrahedron* 69 (2): 537–541.

67 Jiang, Y. and Schaus, S.E. (2017). Asymmetric petasis borono-mannich allylation reactions catalyzed by chiral biphenols. *Angew. Chem. Int. Ed.* 56 (6): 1544–1548.

68 Beisel, T. and Manolikakes, G. (2015). Palladium-catalyzed enantioselective three-component synthesis of α-substituted amines. *Org. Lett.* 17 (12): 3162–3165.

69 Beisel, T., Diehl, A.M., and Manolikakes, G. (2016). Palladium-catalyzed enantioselective three-component synthesis of α-arylglycines. *Org. Lett.* 18 (16): 4116–4119.

70 Jiang, Y., Thomson, R.J., and Schaus, S.E. (2017). Asymmetric traceless petasis borono-mannich reactions of enals: reductive transposition of allylic diazenes. *Angew. Chem.* 129 (52): 16858–16862.

71 Inokuma, T., Suzuki, Y., Sakaeda, T., and Takemoto, Y. (2011). Synthesis of optically active N-aryl amino acid derivatives through the asymmetric petasis reaction catalyzed by a novel hydroxy-thiourea catalyst. *Chem. Asian J.* 6 (11): 2902–2906.

72 Yamaoka, Y., Miyabe, H., and Takemoto, Y. (2007). Catalytic enantioselective petasis-type reaction of quinolines catalyzed by a newly designed thiourea catalyst. *J. Am. Chem. Soc.* 129 (21): 6686–6687.

73 Weilbeer, C., Sickert, M., Naumov, S., and Schneider, C. (2017). The Brønsted acid-catalyzed, enantioselective aza-diels–alder reaction for the direct synthesis of chiral piperidones. *Chem. Eur. J.* 23 (3): 513–518.

74 Ibrahem, I., Zou, W., Casas, J. et al. (2006). Direct organocatalytic enantioselective α-aminomethylation of ketones. *Tetrahedron* 62 (2–3): 357–364.

75 Sundén, H., Ibrahem, I., Eriksson, L., and Córdova, A. (2005). Direct catalytic enantioselective aza-Diels-Alder reactions. *Angew. Chem. Int. Ed.* 44 (31): 4877–4880.

76 Ramachary, D.B., Reddy, Y.V., and Prakash, B.V. (2008). Double cascade reactions based on the Barbas dienamine platform: highly stereoselective synthesis of functionalized cyclohexanes for cardiovascular agents. *Org. Biomol. Chem.* 6 (4): 719–726.

77 Liu, H., Cun, L.F., Mi, A.Q. et al. (2006). Enantioselective direct aza hetero-Diels-Alder reaction catalyzed by chiral Brønsted acids. *Org. Lett.* 8 (26): 6023–6026.

78 Chen, Z., Wang, B., Wang, Z. et al. (2013). Complex bioactive alkaloid-type polycycles through efficient catalytic asymmetric multicomponent aza-Diels-Alder reaction of indoles with oxetane as directing group. *Angew. Chem. Int. Ed.* 52 (7): 2027–2031.

79 Zheng, X., Qian, Y., and Wang, Y. (2010). Direct asymmetric aza Diels-Alder reaction catalyzed by chiral 2-pyrrolidinecarboxylic acid ionic liquid. *Catal. Commun.* 11 (6): 567–570.
80 Gao, X. and Hall, D.G. (2003). 3-Boronoacrolein as an exceptional heterodiene in the highly enantio- and diastereoselective Cr(III)-catalyzed three-component [4+2]/allylboration. *J. Am. Chem. Soc.* 125 (31): 9308–9309.
81 Gao, X., Hall, D.G., Deligny, M. et al. (2006). Catalytic enantioselective three-component hetero-[4+2] cycloaddition/allylboration approach to α-hydroxyalkyl pyrans: scope, limitations, and mechanistic proposal. *Chem. Eur. J.* 12 (11): 3132–3142.
82 Deng, Y., Kumar, S., Wheeler, K., and Wang, H. (2015). Trio catalysis merging enamine, Brønsted acid, and metal lewis acid catalysis: asymmetric three-component aza-Diels-Alder reaction of substituted cinnamaldehydes, cyclic ketones, and arylamines. *Chem. Eur. J.* 21 (21): 7874–7880.
83 Deng, Y., Karunaratne, C.V., Csatary, E. et al. (2015). Chiral bimetallic catalysts derived from chiral metal phosphates: enantioselective three-component asymmetric Aza-Diels-Alder reactions of cyclic ketones. *J. Org. Chem.* 80 (16): 7984–7993.
84 Bello, D., Ramon, R., and Lavilla, R. (2010). Mechanistic variations of the povarov multicomponent reaction and related processes. *Curr. Org. Chem.* 14 (4): 332–356.
85 Bergonzini, G., Gramigna, L., Mazzanti, A. et al. (2010). Organocatalytic asymmetric povarov reactions with 2- and 3-vinylindoles. *Chem. Commun.* 46 (2): 327–329.
86 Fochi, M., Caruana, L., and Bernardi, L. (2014). Catalytic asymmetric Aza-Diels-Alder reactions: the Povarov cycloaddition reaction. *Synthesis* 46 (2): 135–157.
87 Huang, J.X., Hou, K.Q., Hu, Q.L. et al. (2020). Organocatalytic asymmetric three-component Povarov reactions of anilines and aldehydes. *Org. Lett.* 22 (5): 1858–1862.
88 Liu, H., Dagousset, G., Retailleau, P. et al. (2009). Chiral Brønsted acid-catalyzed enantioselective three-component Povarov reaction. *J. Am. Chem. Soc.*: 4598–4599.
89 Dagousset, G., Retailleau, P., Masson, G., and Zhu, J. (2012). Chiral phosphoric acid-catalyzed enantioselective three-component Povarov reaction using cyclic enethioureas as dienophiles: stereocontrolled access to enantioenriched hexahydropyrroloquinolines. *Chem. Eur. J.* 18 (19): 5869–5873.
90 He, L., Bekkaye, M., Retailleau, P., and Masson, G. (2012). Chiral phosphoric acid catalyzed inverse-electron-demand aza-Diels-Alder reaction of isoeugenol derivatives. *Org. Lett.* 14 (12): 3158–3161.
91 Shi, F., Xing, G.J., Tao, Z.L. et al. (2012). An asymmetric organocatalytic povarov reaction with 2-hydroxystyrenes. *J. Org. Chem.* 77 (16): 6970–6979.
92 Shi, F., Xing, G.J., Zhu, R.Y. et al. (2013). A catalytic asymmetric isatin-involved Povarov reaction: diastereo- and enantioselective construction of spiro[indolin-3,2'-quinoline] scaffold. *Org. Lett.* 15 (1): 128–131.

93 Luo, C. and Huang, Y. (2013). A highly diastereo- and enantioselective synthesis of tetrahydroquinolines: quaternary stereogenic center inversion and functionalization. *J. Am. Chem. Soc.* 135 (22): 8193–8196.
94 Dai, W., Jiang, X.L., Tao, J.Y., and Shi, F. (2016). Application of 3-methyl-2-vinylindoles in catalytic asymmetric Povarov reaction: diastereo- and enantioselective synthesis of indole-derived tetrahydroquinolines. *J. Org. Chem.* 81 (1): 185–192.
95 Wang, C., Han, Z.Y., Luo, H.W., and Gong, L.Z. (2010). Highly enantioselective relay catalysis in the three-component reaction for direct construction of structurally complex heterocycles. *Org. Lett.* 12 (10): 2266–2269.
96 Dagousset, G., Zhu, J., and Masson, G. (2011). Chiral phosphoric acid-catalyzed enantioselective three-component Povarov reaction using enecarbamates as dienophiles: highly diastereo- and enantioselective synthesis of substituted 4-aminotetrahydroquinolines. *J. Am. Chem. Soc.* 133 (37): 14804–14813.
97 Xie, M., Chen, X., Zhu, Y. et al. (2010). Asymmetric three-component inverse electron-demand Aza-Diels-Alder reaction: efficient synthesis of ring-fused tetrahydroquinolines. *Angew. Chem.* 122 (22): 3887–3890.
98 Vicente-García, E., Catti, F., Ramón, R., and Lavilla, R. (2010). Unsaturated lactams: new inputs for Povarov-type multicomponent reactions. *Org. Lett.* 12 (4): 860–863.
99 Domínguez, G. and Pérez-Castells, J. (2011). Recent advances in [2+2+2] cycloaddition reactions. *Chem. Soc. Rev.* 40 (7): 3430–3444.
100 Shibata, T. and Tsuchikama, K. (2008). Recent advances in enantioselective [2 + 2 + 2] cycloaddition. *Org. Biomol. Chem.* 6 (8): 1317–1323.
101 Kuang, X.K., Zhu, J., Zhou, L. et al. (2018). Synergetic tandem enantiomeric enrichment in catalytic asymmetric multi-component reactions (AMCRs): highly enantioselective construction of tetracyclic indolines with four continuous stereocenters. *ACS Catal.* 8 (6): 4991–4995.
102 Zhu, J., Cheng, Y., Kuang, X. et al. (2016). Highly efficient formal [2+2+2] strategy for the rapid construction of polycyclic spiroindolines: a concise synthesis of 11-demethoxy-16-epi-myrtoidine. *Angew. Chem.* 128 (32): 9370–9374.
103 da Costa Cabrera, D., Santa-Helena, E., Leal, H.P. et al. (2019). Synthesis and antioxidant activity of new lipophilic dihydropyridines. *Bioorg. Chem.* 84 (2018): 1–16.
104 Jiang, J., Yu, J., Sun, X.-X. et al. (2008). Organocatalytic asymmetric three-component cyclization of cinnamaldehydes and primary amines with 1,3-dicarbonyl compounds: straightforward access to enantiomerically enriched dihydropyridines. *Angew. Chem.* 120 (13): 2492–2496.
105 Molleti, N., Allu, S., Ray, S.K., and Singh, V.K. (2013). Bifunctional chiral urea catalyzed highly enantioselective Michael addition of cyclic 1,3-dicarbonyl compounds to 2-enoylpyridines. *Tetrahedron Lett.* 54 (25): 3241–3244.
106 Franke, P.T., Johansen, R.L., Bertelsen, S., and Jørgensen, K.A. (2008). Organocatalytic enantioselective one-pot synthesis and application of substituted 1,4-dihydropyridines - Hantzsch ester analogues. *Chem. Asian J.* 3 (2): 216–224.

107 An, D., Zhu, Z., Zhang, G. et al. (2015). H_8-BINOL chiral imidodiphosphoric acids catalyzed cyclization reactions of β,γ-unsaturated α-ketoesters, arylamines and 1,3-dicarbonyl compounds: enantioselective synthesis of 1,4-dihydropyridines. *Tetrahedron Asymm.* 26 (17): 897–906.

108 Quinonero, O., Jean, M., Vanthuyne, N. et al. (2016). Combining organocatalysis with central-to-axial chirality conversion: atroposelective hantzsch-type synthesis of 4-arylpyridines. *Angew. Chem. Int. Ed.* 55 (4): 1401–1405.

109 Chen, L., Zhang, Z., and Zu, L. (2020). Organocatalytic hantzsch type reaction using aryl hydrazines, propiolic acid esters and enals: enantioselective synthesis of paroxetine. *Adv. Synth. Catal.* 362 (23): 5385–5390.

110 Donohoe, T.J., Jones, C.R., Kornahrens, A.F. et al. (2013). Total synthesis of the antitumor antibiotic (±)-streptonigrin: first- and second-generation routes for de novo pyridine formation using ring-closing metathesis. *J. Org. Chem.* 78 (24): 12338–12350.

111 Huang, Y., Yang, F., and Zhu, C. (2005). Highly enantioseletive biginelli reaction using a new chiral ytterbium catalyst: asymmetric synthesis of dihydropyrimidines. *J. Am. Chem. Soc.* 127 (47): 16386–16387.

112 An, D., Sen, F.Y., Gao, Y. et al. (2014). Highly enantioselective Biginelli reaction catalyzed by double axially chiral bisphosphorylimides. *Eur. J. Org. Chem.* 2014 (2): 301–306.

113 Xu, D.Z., Li, H., and Wang, Y. (2012). Highly enantioselective Biginelli reaction catalyzed by a simple chiral primary amine catalyst: asymmetric synthesis of dihydropyrimidines. *Tetrahedron* 68 (38): 7867–7872.

114 Saha, S. and Moorthy, J.N. (2011). Enantioselective organocatalytic Biginelli reaction: dependence of the catalyst on sterics, hydrogen bonding, and reinforced chirality. *J. Org. Chem.* 76 (2): 396–402.

115 Xin, J., Chang, L., Hou, Z. et al. (2008). An enantioselective Biginelli reaction catalyzed by a simple chiral secondary amine and achiral Brønsted acid by a dual-activation route. *Chem. Eur. J.* 14 (10): 3177–3181.

116 Wu, Y.Y., Chai, Z., Liu, X.Y. et al. (2009). Synthesis of substituted 5-(pyrrolidin-2-yl)tetrazoles and their application in the asymmetric Biginelli reaction. *Eur. J. Org. Chem.* 6: 904–911.

117 Chen, X.H., Xu, X.Y., Liu, H. et al. (2006). Highly enantioselective organocatalytic Biginelli reaction. *J. Am. Chem. Soc.* 128 (46): 14802–14803.

118 Huang, D., Xu, F., Lin, X., and Wang, Y. (2012). Highly enantioselective pictet-spengler reaction catalyzed by SPINOL-phosphoric acids. *Chem. Eur. J.* 18 (11): 3148–3152.

119 Guo, Y., Gao, Z., Meng, X. et al. (2017). Highly enantioselective Biginelli reaction of aliphatic aldehydes catalyzed by chiral phosphoric acids. *Synlett* 28 (15): 2041–2045.

120 Gong, L.Z., Chen, X.H., and Xu, X.Y. (2007). Asymmetric organocatalytic biginelli reactions: a new approach to quickly access optically active 3,4-dihydropyrimidin-2-(1H)-ones. *Chem. Eur. J.* 13 (32): 8920–8926.

121 Qu, H., Li, X., Mo, F., and Lin, X. (2013). Efficient synthesis of dihydropyrimidinones via a three-component Biginelli-type reaction of urea, alkylaldehyde and arylaldehyde. *Beilstein J. Org. Chem.* 9: 2846–2851.

122 Stucchi, M., Lesma, G., Meneghetti, F. et al. (2016). Organocatalytic asymmetric Biginelli-like reaction involving isatin. *J. Org. Chem.* 81 (5): 1877–1884.

123 Barbero, M., Cadamuro, S., and Dughera, S. (2017). A Brønsted acid catalysed enantioselective Biginelli reaction. *Green Chem.* 19 (6): 1529–1535.

124 Wang, Y., Yang, H., Yu, J. et al. (2009). Highly enantioselective Biginelli reaction promoted by chiral bifunctional primary amine-thiourea catalysts: asymmetric synthesis of dihydropyrimidines. *Adv. Synth. Catal.* 351 (18): 3057–3062.

125 Wang, Y., Yu, J., Miao, Z., and Chen, R. (2011). Bifunctional primary amine-thiourea-TfOH (BPAT·TfOH) as a chiral phase-transfer catalyst: the asymmetric synthesis of dihydropyrimidines. *Org. Biomol. Chem.* 9 (8): 3050–3054.

126 Frings, M., Thomé, I., and Bolm, C. (2012). Synthesis of chiral sulfoximine-based thioureas and their application in asymmetric organocatalysis. *Beilstein J. Org. Chem.* 8: 1443–1451.

127 Ding, D. and Zhao, C.-G. (2010). Primary amine catalyzed Biginelli reaction for the enantioselective synthesis of 3,4-dihydropyrimidin-2(1H)-ones. *Eur. J. Org. Chem.* 2010 (20): 3802–3005.

128 Yadav, L.D.S., Rai, A., Rai, V.K., and Awasthi, C. (2008). Chiral ionic liquid-catalyzed Biginelli reaction: stereoselective synthesis of polyfunctionalized perhydropyrimidines. *Tetrahedron* 64 (7): 1420–1429.

129 Hang, Z., Zhu, J., Lian, X. et al. (2016). A highly enantioselective Biginelli reaction using self-assembled methanoproline-thiourea organocatalysts: asymmetric synthesis of 6-isopropyl-3,4-dihydropyrimidines. *Chem. Commun.* 52 (1): 80–83.

130 Yu, H., Xu, P., He, H. et al. (2017). Highly enantioselective Biginelli reactions using methanopyroline/thiourea – based dual organocatalyst systems: asymmetric synthesis of 4-substituted unsaturated aryl dihydropyrimidines. *Tetrahedron Asymm.* 28 (2): 257–265.

131 Dupont, J. (2011). From molten salts to ionic liquids: a "nano" journey. *Acc. Chem. Res.* 44 (11): 1223–1231.

132 Zhang, F.L., Xu, A.W., Gong, Y.F. et al. (2009). Asymmetrie organocatalytic four-component quadruple domino reaction initiated by oxa-Michael addition of alcohols to acrolein. *Chem. Eur. J.* 15 (28): 6815–6818.

133 Hammer, N., Leth, L.A., Stiller, J. et al. (2016). Oxadendralenes in asymmetric organocatalysis for the construction of tetrahydroisochromenes. *Chem. Sci.* 7 (6): 3649–3657.

134 Kumar, M., Chauhan, P., Bailey, S.J. et al. (2018). Organocatalytic oxa-michael/michael/michael/aldol condensation quadruple domino sequence: asymmetric synthesis of tricyclic chromanes. *Org. Lett.* 20 (4): 1232–1235.

135 Kumar, M., Chauhan, P., Valkonen, A. et al. (2017). Asymmetric synthesis of functionalized tricyclic chromanes via an organocatalytic triple domino reaction. *Org. Lett.* 19 (11): 3025–3028.

136 Van, K.N. and Romo, D. (2018). Multicomponent, enantioselective michael-michael-aldol-β-lactonizations delivering complex β-lactones. *J. Org. Chem.* 83 (2): 632–643.

137 Enders, D., Hüttl, M.R.M., Grondal, C., and Raabe, G. (2006). Control of four stereocentres in a triple cascade organocatalytic reaction. *Nature* 441 (7095): 861–863.

138 Enders, D., Hüttl, M.R.M., Runsink, J. et al. (2007). Organocatalytic one-pot asymmetric synthesis of functionalized tricyclic carbon frameworks from a triple-cascade/Diels-Alder sequence. *Angew. Chem. Int. Ed.* 46 (3): 467–469.

139 Carlone, A., Cabrera, S., Marigo, M., and Jørgensen, K.A. (2007). A new approach for an organocatalytic multicomponent domino asymmetric reaction. *Angew. Chem. Int. Ed.* 46 (7): 1101–1104.

140 Kotame, P., Hong, B.C., and Liao, J.H. (2009). Enantioselective synthesis of the tetrahydro-6*H*-benzo[*c*]chromenes via Domino Michael-aldol condensation: control of five stereocenters in a quadruple-cascade organocatalytic multi-component reaction. *Tetrahedron Lett.* 50 (6): 704–707.

141 Khopade, T.M., Sonawane, A.D., Arora, J.S., and Bhat, R.G. (2017). Direct organocatalytic multicomponent synthesis of enantiopure γ-butyrolactones via tandem knoevenagel-michael-lactonization sequence. *Adv. Synth. Catal.* 359 (22): 3905–3910.

142 Khopade, T.M., Mete, T.B., Arora, J.S., and Bhat, R.G. (2018). An adverse effect of higher catalyst loading and longer reaction time on enantioselectivity in an organocatalytic multicomponent reaction. *Chem. Eur. J.* 24 (23): 6036–6040.

143 Ramachary, D.B. and Barbas, C.F. (2004). Towards organo-click chemistry: development of organocatalytic multicomponent reactions through combinations of aldol, Wittig, Knoevenagel, Michael, Diels-Alder and Huisgen cycloaddition reactions. *Chem. Eur. J.* 10 (21): 5323–5331.

144 Ramachary, D.B., Anebouselvy, K., Chowdari, N.S., and Barbas, C.F. (2004). Direct organocatalytic asymmetric heterodomino reactions: the Knoevenagel/Diels-Alder/epimerization sequence for the highly diastereoselective synthesis of symmetrical and nonsymmetrical synthons of benzoannelated centropolyquinanes. *J. Org. Chem.* 69 (18): 5838–5849.

145 Ramachary, D.B., Chowdari, N.S., and Barbas, C.F. (2003). Organocatalytic asymmetric domino Knoevenagel/Diels-Alder reactions: a bioorganic approach to the diastereospecific and enantioselective construction of highly substituted spiro [5,5]undecane-1,5,9-triones. *Angew. Chem. Int. Ed.* 42 (35): 4233–4237.

146 Ramachary, D.B., Chowdari, N.S., and Barbas, C.F. (2003). The first organocatalytic hetero-domino Knoevenagel-Diels-Alder-epimerization reactions: diastereoselective synthesis of highly substituted spiro[cyclohexane-1,2′-indan]-1′,3′,4-triones. *Synlett* 2003 (12): 1910–1914.

147 Yu, D.F., Wang, Y., and Xu, P.F. (2011). Organocatalytic enantioselective multicomponent cascade reaction: facile access to tetrahydropyridines with C3 all-carbon quaternary stereocenters. *Tetrahedron* 67 (18): 3273–3277.

148 Chen, W.B., Wu, Z.J., Pei, Q.L. et al. (2010). Highly enantioselective construction of spiro[4*H*-pyran-3,3′- oxindoles] through a domino Knoevenagel/

Michael/cyclization sequence catalyzed by cupreine. *Org. Lett.* 12 (14): 3132–3135.

149 Zhao, H.W., Li, B., Tian, T. et al. (2015). Highly enantioselective synthesis of chiral pyranonaphthoquinone-fused spirooxindoles through organocatalytic three-component cascade reactions. *Eur. J. Org. Chem.* 2015 (15): 3320–3326.

150 Shi, J., Liu, Y., Wang, M. et al. (2011). Asymmetric synthesis of multi-substituted spiro[5,5]undecane-1,5,9-triones via organocatalytic three-component reaction. *Tetrahedron* 67 (10): 1781–1787.

151 Ding, R., Zheng, B., Wang, Y., and Peng, Y. (2015). A cation-directed enantioselective sulfur-mediated michael/mannich three-component domino reaction involving chalcones as michael acceptors. *Org. Lett.* 17 (17): 4128–4131.

152 Zhu, Q.N., Zhang, Y.C., Xu, M.M. et al. (2016). Enantioselective construction of tetrahydroquinolin-5-one-based spirooxindole scaffold via an organocatalytic asymmetric multicomponent [3 + 3] cyclization. *J. Org. Chem.* 81 (17): 7898–7907.

153 Xiao, W., Zhou, Z., Yang, Q.Q. et al. (2018). Organocatalytic asymmetric four-component [5+1+1+1] cycloadditions via a quintuple cascade process. *Adv. Synth. Catal.* 360 (18): 3526–3533.

154 Miao, Y.H., Hua, Y.Z., and Wang, M.C. (2019). Dinuclear zinc cooperative catalytic three-component reactions for highly enantioselective synthesis of 3,3′-dihydrofuran spirooxindoles. *Org. Biomol. Chem.* 17 (30): 7172–7181.

155 Chen, Y., Liu, X., Luo, W. et al. (2017). Asymmetric organocatalytic michael/michael/henry sequence to construct cyclohexanes with six vicinal stereogenic centers. *Synlett* 28 (8): 966–969.

156 Shi, D., Xie, Y., Zhou, H. et al. (2012). A highly diastereo- and enantioselective reaction for constructing functionalized cyclohexanes: six contiguous stereocenters in one step. *Angew. Chem. Int. Ed.* 51 (5): 1248–1251.

157 Wang, Q., Wang, D.X., Wang, M.X., and Zhu, J. (2018). Still unconquered: enantioselective Passerini and ugi multicomponent reactions. *Acc. Chem. Res.* 51 (5): 1290–1300.

158 Saya, J.M., Berabez, R., Broersen, P. et al. (2018). Hexafluoroisopropanol as the acid component in the Passerini reaction: one-pot access to β-amino alcohols. *Org. Lett.* 20 (13): 3988–3991.

159 Ramozzi, R. and Morokuma, K. (2015). Revisiting the Passerini reaction mechanism: existence of the nitrilium, organocatalysis of its formation, and solvent effect. *J. Org. Chem.* 80 (11): 5652–5657.

160 Banfi, L. and Riva, R. (2005). The Passerini reaction. In: *Organic Reactions* (ed. A.B. Charette), 1–140. Hoboken, NJ, USA: John Wiley & Sons, Inc.

161 Shaabani, S. and Dömling, A. (2018). The catalytic enantioselective ugi four-component reactions. *Angew. Chem. Int. Ed.* 57 (50): 16266–16268.

162 Zhang, J., Wang, Y.Y., Sun, H. et al. (2020). Enantioselective three-component Ugi reaction catalyzed by chiral phosphoric acid. *Sci. China Chem.* 63 (1): 47–54.

163 Zhang, J., Lin, S.X., Cheng, D.J. et al. (2015). Phosphoric acid-catalyzed asymmetric classic Passerini reaction. *J. Am. Chem. Soc.* 137 (44): 14039–14042.

164 Chéron, N., Ramozzi, R., El, K.L. et al. (2012). Challenging 50 years of established views on Ugi reaction: a theoretical approach. *J. Org. Chem.* 77 (3): 1361–1366.

165 Riva, R. (2018). Enantioselective four-component Ugi reactions. *Science* 361 (6407): 1072–1073.

166 Denmark, S.E. and Fan, Y. (2003). The first catalytic, asymmetric α-additions of isocyanides. Lewis-base-catalyzed, enantioselective Passerini-type reactions. *J. Am. Chem. Soc.* 125 (26): 7825–7827.

167 Denmark, S.E. and Fan, Y. (2005). Catalytic, enantioselective α-additions of isocyanides: lewis base catalyzed Passerini-type reactions. *J. Org. Chem.* 70 (24): 9667–9676.

168 Feng, Q.Y., Zhu, J., Wang, M.X., and Tong, S. (2020). Organocatalytic double Ugi reaction with statistical amplification of product enantiopurity: a linker cleavage approach to access highly enantiopure Ugi products. *Org. Lett.* 22 (2): 483–487.

169 Hashimoto, T., Kimura, H., Kawamata, Y., and Maruoka, K. (2012). A catalytic asymmetric Ugi-type reaction with acyclic azomethine imines. *Angew. Chem. Int. Ed.* 51 (29): 7279–7281.

170 Wang, S., Wang, M.X., Wang, D.X., and Zhu, J. (2007). Asymmetric Lewis acid catalyzed addition of isocyanides to aldehydes - synthesis of 5-amino-2-(1-hydroxyalkyl)oxazoles. *Eur. J. Org. Chem.* 2 (24): 4076–4080.

171 Wang, S.X., Wang, M.X., Wang, D.X., and Zhu, J. (2007). Chiral salen - aluminum complex as a catalyst for enantioselective α-addition of isocyanides to aldehydes: asymmetric synthesis of 2-(1-hydroxyalkyl)-5-aminooxazoles. *Org. Lett.* 9 (18): 3615–3618.

172 Mihara, H., Xu, Y., Shepherd, N.E. et al. (2009). A heterobimetallic Ga/Yb-Schiff base complex for catalytic asymmetric α-addition of isocyanides to aldehydes. *J. Am. Chem. Soc.* 131 (24): 8384–8385.

173 Zeng, X., Ye, K., Lu, M. et al. (2010). Chiral brønsted acid catalyzed enantioselective addition of α-isocyanoacetamides to aldehydes. *Org. Lett.* 12 (10): 2414–2417.

174 Kusebauch, U., Beck, B., Messer, K. et al. (2003). Massive parallel catalyst screening: toward asymmetric MCRs. *Org. Lett.* 5 (22): 4021–4024.

175 Andreana, P.R., Liu, C.C., and Schreiber, S.L. (2004). Stereochemical control of the Passerini reaction. *Org. Lett.* 6 (23): 4231–4233.

176 Wang, S.-X., Wang, M.-X., Wang, D.-X., and Zhu, J. (2008). Catalytic enantioselective Passerini three-component reaction. *Angew. Chem.* 120 (2): 394–397.

177 Ugi, I., Meyr, R., and Isonitrile, V. (1961). Erweiterter Anwendungsbereich der Passerini-reaktion. *Chem. Ber.* 94 (8): 2229–2233.

178 Yue, T., Wang, M.-X., Wang, D.-X., and Zhu, J. (2008). Asymmetric synthesis of 5-(1-hydroxyalkyl)tetrazoles by catalytic enantioselective Passerini-type reactions. *Angew. Chem.* 120 (49): 9596–9599.

179 Xiong, Q., Dong, S., Chen, Y. et al. (2019). Asymmetric synthesis of tetrazole and dihydroisoquinoline derivatives by isocyanide-based multicomponent reactions. *Nat. Commun.* 10 (1): 1–10.

180 Liu, Y.K., Liu, H., Du, W. et al. (2008). Reaction control in the organocatalytic asymmetric one-pot, three-component reaction of aldehydes, diethyl α-aminomalonate and nitroalkenes: toward diversity-oriented synthesis. *Chem. Eur. J.* 14 (32): 9873–9877.

181 Požgan, F., Al Mamari, H., Grošelj, U. et al. (2018). Synthesis of non-racemic pyrazolines and pyrazolidines by [3+2] cycloadditions of azomethine imines. *Molecules* 23 (1): 3.

182 Wang, N. and Xing, Y. (2020). Progress in the synthesis of chiral nitrogen heterocycles mainly by asymmetric [3 + 2] cycloadditions. In: *Synthetic Approaches to Nonaromatic Nitrogen Heterocycles* (ed. A.M.F. Phillips), 89–118. John Wiley & Sons Ltd.

183 Yue, T., Wang, M.X., Wang, D.X. et al. (2009). Brønsted acid catalyzed enantioselective three-component reaction involving the α addition of isocyanides to imines. *Angew. Chem. Int. Ed.* 48 (36): 6717–6721.

184 Su, Y., Bouma, M.J., Alcaraz, L. et al. (2012). Organocatalytic enantioselective one-pot four-component Ugi-type multicomponent reaction for the synthesis of epoxy-tetrahydropyrrolo[3,4-*b*]pyridin-5-ones. *Chem. Eur. J.* 18 (40): 12624–12627.

185 Zhao, W., Huang, L., Guan, Y., and Wulff, W.D. (2014). Three-component asymmetric catalytic Ugi reaction – concinnity from diversity by substrate-mediated catalyst assembly. *Angew. Chem. Int. Ed.* 53 (13): 3436–3441.

186 Zhang, Y., Ao, Y.F., Huang, Z.T. et al. (2016). Chiral phosphoric acid catalyzed asymmetric Ugi reaction by dynamic kinetic resolution of the primary multicomponent adduct. *Angew. Chem. Int. Ed.* 55 (17): 5282–5285.

187 Beadham, I., Gurbisz, M., and Gathergood, N. (2013). ChemInform abstract: designing safer organocatalysts – what lessons can be learned when the rebirth of an old research area coincides with the advent of green chemistry? *ChemInform* 44 (16): 159–226.

188 Chowdari, N.S., Ramachary, D.B., and Barbas, C.F. (2003). Organocatalytic asymmetric assembly reactions: one-pot synthesis of functionalized β-amino alcohols from aldehydes, ketones, and azodicarboxylates. *Org. Lett.* 5 (10): 1685–1688.

189 Vachan, B.S., Karuppasamy, M., Vinoth, P. et al. (2020). Proline and its derivatives as organocatalysts for multi- component reactions in aqueous media: synergic pathways to the green synthesis of heterocycles. *Adv. Synth. Catal.* 362 (1): 87–110.

190 Shao, Y.D. and Cheng, D.J. (2021). Chiral phosphoric acid: a powerful organocatalyst for the asymmetric synthesis of heterocycles with chiral atropisomerism. *ChemCatChem* 13 (5): 1271–1289.

13

Recent Trends in Metal-catalyzed MCRs Toward Heterocycles

Lilia Fuentes-Morales and Luis D. Miranda

Instituto de Química Universidad Nacional Autónoma de México, Circuito Exterior S/N Ciudad Universitaria, Coyoacán, Ciudad de México 04510, México

13.1 Introduction

The development of practical methodologies for constructing heterocyclic systems is an area that has had, and continues to have, extraordinary strength. A high percentage of commercial drugs have at least one heterocyclic system in their main core. No less important is that life itself, at its molecular level, is based on a wide range of heterocyclic molecules. Nitrogenous bases in nucleic acids, heterocyclic amino acids, along with a wide range of heterocyclic secondary metabolites, allow living organisms to function.

From the synthetic point of view, multicomponent reactions play an important role in constructing a broad range of heterocyclic systems. These highly convergent processes provide the synthetic chemist access to complex structures from simple starting materials. Indeed, modern synthetic chemistry is based on the production of molecules in a few reaction steps, from easily accessible raw materials, with a minimum consumption of energy and reagents quantities (preferably catalytic).

Metal-catalyzed multicomponent reactions (MC-MCRs) are among the most promising methods for the synthesis of heterocyclic systems in a more sustainable fashion. Metal-catalyzed reactions offer several important features. Among others, by using only catalytic amounts of the metal complex, several functional groups that cannot be activated by traditional methodologies react to generate different types of C—C and C—X bonds. The development of the new metal-catalyzed MCRs field has expanded in recent years and continues to grow. Indeed, this area of research has progressed significantly by the development of new and more efficient metallic catalysts for several new applications.

In this chapter, we have tried to summarize methodologies representative of metal-catalyzed MCRs published recently (2015–2020). The methods included are only those in which the heterocyclic system is assembled directly via a multicomponent process. Protocols in which the heterocyclic system is constructed in a post-condensation process after the MCR are beyond the scope of this chapter. We have not included processes where the metal salt is used as a Lewis acid to

Multicomponent Reactions towards Heterocycles: Concepts and Applications, First Edition.
Edited by Erik V. Van der Eycken and Upendra K. Sharma.

accelerate reactions, or when two or more classical metal-catalyzed coupling processes are carried out in one-pot by adding the reagents consecutively.

Several excellent book chapters and reviews have been published previously and were organized in several different ways (by metal used, by MCR type, etc.) [1–10]. We have chosen to categorize this chapter by the type of heterocyclic system. This organization gives the reader direct access to the information for the heterocyclic scaffold of interest, rather than by the kind of reaction or catalyst used.

13.2 Five-membered Heterocycles with One Heteroatom

Synthetic or natural pyrrole derivatives have found a wide range of applications, from drugs produced in tons to applications in organic electrical conductors and fluorescent materials. Although there are several methodologies available in the synthetic toolbox, the demand for more practical synthetic methods for the pyrrole system is still a very active field. Recently, several contributions to the metal-catalyzed MCRs for the construction of pyrrole-based systems have been reported.

Classic methods (Paal–Knorr, van Leusen, Knorr and Hantzsch) for the construction of the pyrrole system rely on the condensation of carbonyl compounds (aldehydes or ketones) with an amine or another nitrogen source. However, in most cases, the synthesis of the starting carbonyl compound is the most time- and resource-consuming process. Furthermore, the acidic alpha-hydrogens and the high reactivity of the carbonyl group itself generally trigger a number of side products during the construction of the pyrrole system. A promising strategy to alleviate these problems relies on the catalytic dehydrogenation of stable alcohols, a process that delivers *in situ* the carbonyl component. To this end, several catalytic systems ("Ir–"[11], "Ru–"[12], and Pt/C-based [13, 14]) have been developed for the direct conversion of specific alcohols into different pyrrole systems in the presence of another carbonyl compound and an amine source, in a three-component process. Along this line, in 2017, Seayad and coworkers reported an efficient, less expensive Cu-catalyzed synthesis of pyrroles via a three-component process (Scheme 13.1).

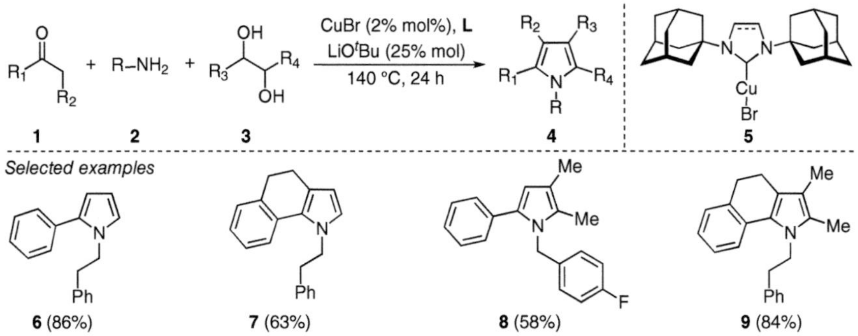

Scheme 13.1 Multicomponent synthesis of pyrroles catalyzed by a [Cu(NHC)].

Two, three, four, and fully substituted pyrroles (**6–9**) were obtained by a suitable choice of the ketone (**1**), the primary amine (**2**), and the diol (**3**) [15]. The process is catalyzed by an *in situ* generated NHC=*N*-heterocyclic carbene (**5**).

Alternatively, Münchnone (**17**, 1,3-oxazolium-5-one), a mesoionic five-membered heterocycle, undergoes spontaneous 1,3-dipolar cycloaddition to alkanes or alkynes bearing one or two electron-withdrawing groups to deliver pyrroles after CO_2 loss (Scheme 13.2). Since the first report by Huisgen et al. [16], different approaches have been described for the *in situ* generation of the Münchnone [17, 18], including a practical use of an Ugi 4-CR [19]. Recently, the group of Arndtsen described a Pd-mediated multicomponent approach for the synthesis of Münchnones (see Scheme 13.11 for its formation mechanism) from aryl iodides (**10**), imines (**11**), and CO (10 atm) [20]. The cycloaddition of the *in situ* generated Münchnones with an appropriate dipolarophile (alkynes (**12**) or olefins (**13**) bearing one or two electron-withdrawing groups) converted this protocol into a novel four-component, one-pot process for the synthesis of highly substituted pyrroles (**14–16**). This highly modular approach has a remarkable synthetic potential because it allows the production of pyrroles with various degrees of substitution from readily accessible starting materials (Scheme 13.2).

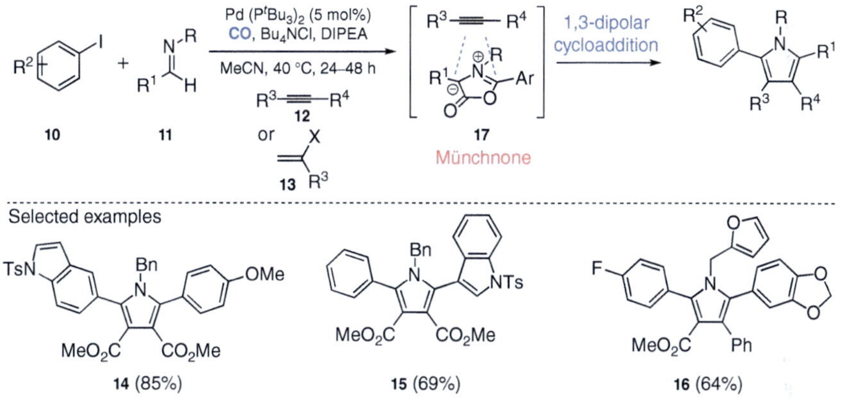

Scheme 13.2 Palladium-catalyzed multicomponent synthesis of multisubstituted pyrroles.

When the dipolarophile is incorporated in the imine (**18a**) in the components input set, the corresponding fused pyrrole (**20a** and **20b**) is generated in good yields [21]. Furthermore, this protocol can be coupled with an *in situ* initial Sonogashira coupling (**18b**) before the reaction with the carbon monoxide. This change in the manifold results in a remarkable five-component reaction which gives direct access to differently substituted tricyclic pyran-fused pyrroles (Scheme 13.3).

In 2015, Tonks and colleagues described a formal [2 + 2 + 1] three-component process involving two equivalents of an alkyne (**12**) with a phenyldiazene (**21**) for the synthesis of tri- and penta-substituted pyrroles (**23–27**, Scheme 13.4). The process is catalyzed by the imido precatalyst, $py_3Cl_2Ti(NPh)$ (**22**) and can be seen as an nitrenoid analogue of the classical Pauson–Khand-type reaction of cyclopentenones

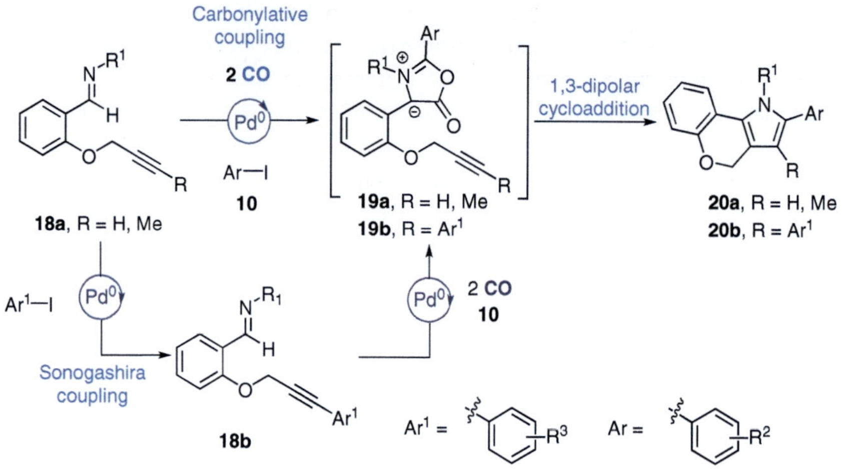

Scheme 13.3 Pd-Catalyzed multicomponent synthesis of fused pyrroles.

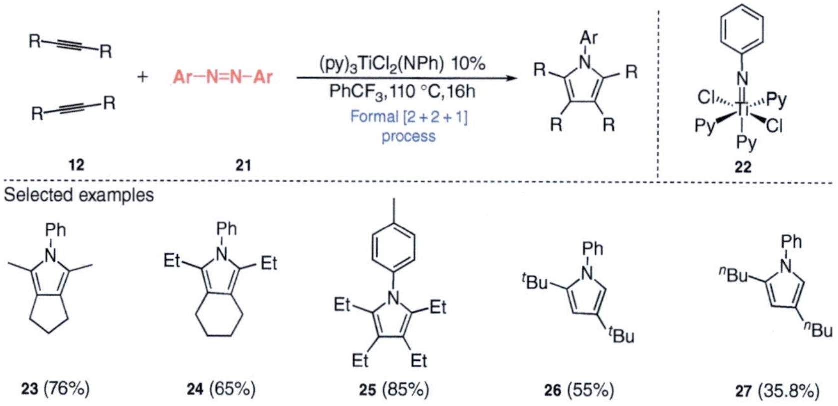

Scheme 13.4 Synthesis of pyrroles via catalytic formal [2 + 2 + 1] reaction catalyzed by titanium.

(Scheme 13.4). This process requires the use of electron-rich alkynes. Those with electron-withdrawing substituents, such as diethyl acetylenedicarboxylate, fail to give rise to the expected pyrrole system. When asymmetric alkynes were utilized, a mixture of regioisomers was produced [22].

A series of substituted furopyrrolamines (specifically 6-imino-4,6-dihydro1*H*-furo[3,4-b]pyrrol-2-amines) have been obtained by a palladium-catalyzed three-component reaction of propargylic carbonates (**28**) with isocyanides (**29**, Scheme 13.5) [23]. The process is based on the successive addition of two isocyanide molecules to an allenylpalladium intermediate to generate **30**. The desired products (**34** and **35**) can be obtained in good yields when $Pd(PPh_3)_2Cl_2$ (5 mol%), PPh_3 (10 mol%), CsF (2 equiv), and H_2O (3 equiv) are utilized in DMSO at room temperature. When the reaction is carried out at 110 °C, and 1,1′-ferrocenediyl-bis(diphenylphosphine) (DPPF) is used as the ligand, 5-iminopyrrolones (**36** and **37**) are isolated as the major products. The different

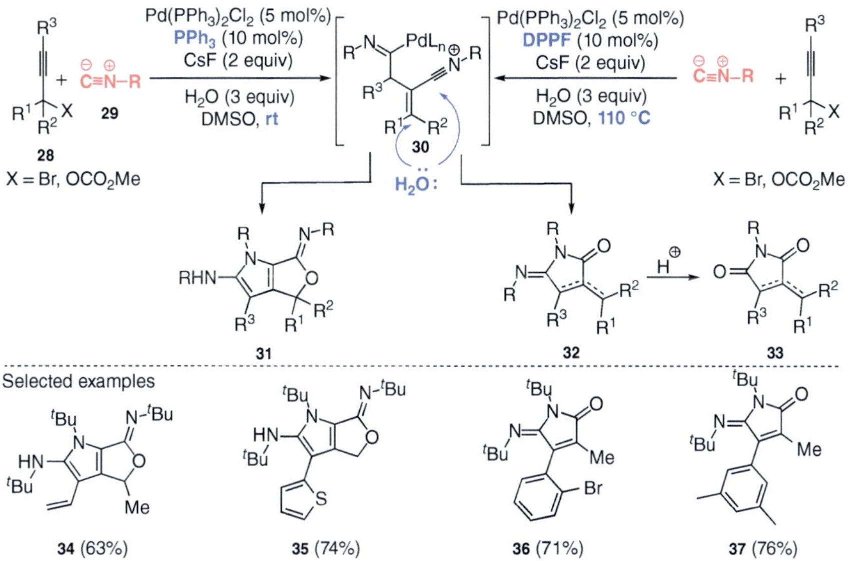

Scheme 13.5 Synthesis of pyrroles by means of MCR between propargylic carbonates and isocyanides.

outcome of the process is an obvious consequence of the divergent regioselective addition of a water molecule to intermediate **30** (Scheme 13.5). Further hydrolysis of the 5-iminopyrrolones (**32**) affords the corresponding maleimides (**33**), an important building block in organic synthesis.

5-Iminopyrrolones have also been obtained through a conceptually different approach. Raghuvanshi et al. described a novel protocol for the synthesis of various 5-iminopyrrolones, via a [2+2+1] UV-photo-induced (125 W mercury lamp) cycloaddition of a terminal alkyne (**38**), with a carbodiimide (**39**) (diisopropyl- and dicyclohexyl-carbodiimide) and CO, in the presence of $Fe(CO)_5$ (25 mol%) (Scheme 13.6). Under these conditions, 5-iminopyrrolones (**41–44**) were obtained in poor to good yields, from several *t*-butyl and methyl-substituted phenylacetylenes.

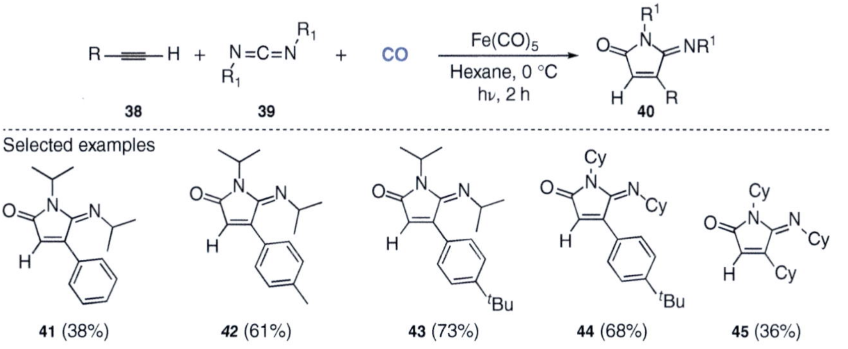

Scheme 13.6 Synthesis of 5-iminopyrrolones via [2 + 2 + 1] cycloaddition catalyzed by $Fe(CO)_5$.

Cyclohexylacetylene also gave the expected 5-iminopyrrolone (**45**) [24]. In principle, hydrolysis of the 5-iminopyrrolones might yield the corresponding maleimides as described in Scheme 13.5 (**32**→**33**).

On the other hand, a four-component palladium-catalyzed carbonylative amidation has been used for the direct construction of the maleimide ring system **47**. Generally, the reaction involves 1,2-dihalo alkenes or 1,2-dihaloarenes (**46**) and uses CO (2 equiv) as the carbonylative reagent and an amine as the nitrogen source [25–31]. In 2015, Fuse et al. extended this methodology to the synthesis of a series of thieno[3,4-c]pyrrole-4,6-diones (**48**, **49**) (TPDs) from 3,4-dibromothiophenes (Scheme 13.7) [32]. The optimal conditions use 10 mol% of $Pd(OAc)_2$, 30 mol% of $[(t\text{-}Bu_3P)H]BF_4$, CO (1 atm), toluene, 80 °C in a 0.250 M solution. It is worth pointing out that TPDs are the most frequently used monomers for the synthesis of functional polymers with applications in organic solar cells [33]. The methodology was successfully applied to the synthesis of the pyridinopyrroledione **50** and also to the preparation of a pyrroloisoindoletetraone **51**, through an impressive quadruple carbonylative amidation from 1,2,4,5-tetrabromobenzene.

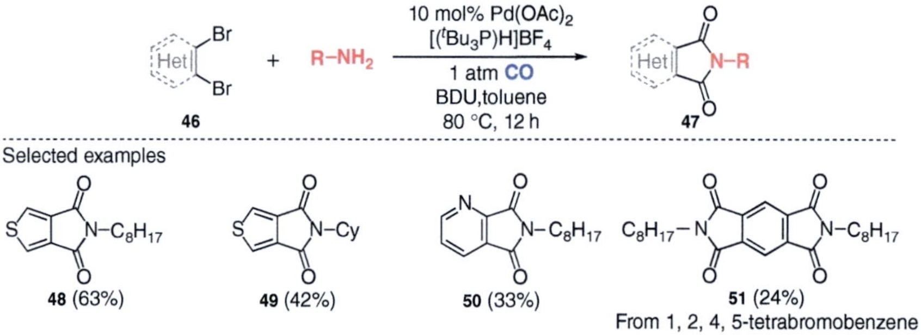

Scheme 13.7 Palladium-catalyzed synthesis of *N*-substituted phthalimides.

In 2009, Abarbri, Parrain et al. described an efficient process to prepare α,β-substituted γ-alkylidenebutenolides (**54**) via a Sonogashira-type coupling between the (*Z*)-3-iodobutenoic acid (**53**), and a terminal alkyne, followed by a cycloisomerization process, the whole process mediated by CuI [34, 35]. More recently, the same group adapted this methodology to the direct construction of the corresponding 5-hydroxy-1*H*-pyrrol-2(5*H*)-ones (**55**–**60**), in a one-pot, consecutive, three-component process (Scheme 13.8) [36]. The protocol involved heating a solution of an amine, an alkyne (**52**), and a 3-iodobutanoic acid (**53**), with CuI (20 mol%) in isopropanol at 50 °C. The methodology was then utilized for the construction of several indole polycyclic derivatives (**61**, **62**) [37]. In these experiments, the acyliminium species, generated under acidic conditions from the hemiaminals, is intramolecularly trapped by the indole system in a Pictet–Spengler type process. Thus, the structure of the polycyclic derivatives depends on the position of the indole nucleus in the initial starting materials of the multicomponent process.

Furan and its derivatives are important structural motifs ubiquitous in nature with a wide range of biological activities. Due to their flexible reactivity, these heterocycles

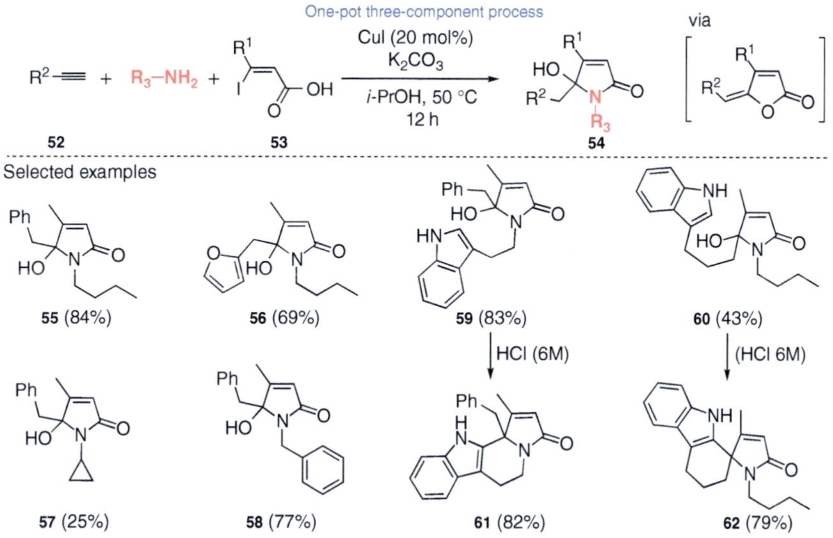

Scheme 13.8 Cu-catalyzed synthesis of γ-alkylidenebutenolides via a Sonogashira-type coupling followed by a cycloisomerization.

are versatile building blocks that serve as templates for the construction of more complex molecules. Consequently, the development of more general and simplified procedures that give access to these heterocyclic systems is still an active field of investigation. In 2015, the group of Zhan and Fan described the synthesis of diversely substituted 2-(furan-3-yl)acetates (**69–72**) via a palladium-catalyzed condensation of allenols (**64**), aryl iodides (**63**), alcohols, and carbon monoxide (two molecules) as an low-cost and atom-economical C1 source (Scheme 13.9). [38] This five-component process involves the carbonylation of the aryl iodide, followed by the alcohoxyl carbonylation of the *in situ* formed allyl palladium complex **67**, and the cyclization of the α-hydroxyl enone intermediate **68**. More recently, He and coworkers managed to adapt the methodology to generate the corresponding 2-(furan-3-yl)acetamides (**73–76**) under similar conditions [39]. To this end, ethanol was replaced by a secondary amine such as *N,N*-diethylamine and *N,N*-diisopropylamine, among others. In this subsequent study, various aryl and heteroaryl allenols and iodides were also used, whereby it was possible to prepare heterotrimers such as **76** in good yields.

In 2015, Wu and Yoshicai reported a modular synthesis of tri- and tetra-substituted furans (**81–84**). Their approach relied on a palladium-catalyzed three-component condensation of alkynylbenziodoxoles (**77**), carboxylic acids (**78**), and imines (**79**) [40]. The process involved cleavage of the triple bond, and only one of its carbons remained in the furan system (C-5) (Scheme 13.10). Earlier, the same group had described a similar condensation reaction to obtain substituted furans with a 2-iodobenzoate moiety at C-5 [41]. With the incorporation of a carboxylic acid in the system, the substituent at C-5 could be varied and a number of substituted furans were prepared, including heptaaryl compound **84**, which was obtained from the corresponding bis-alkynylbenziodoxole.

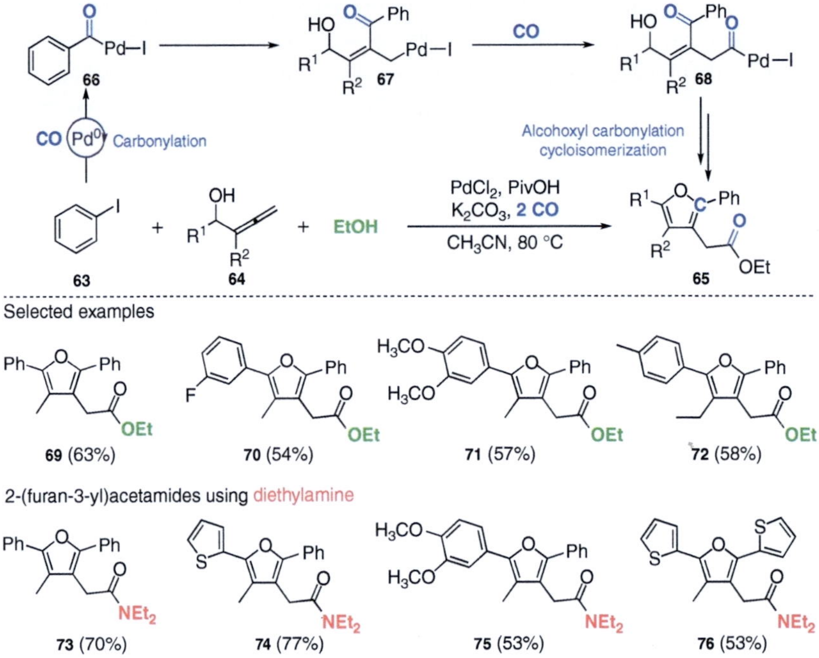

Scheme 13.9 Pd-catalyzed five-component synthesis of substituted 2-(furan-3-yl)acetates.

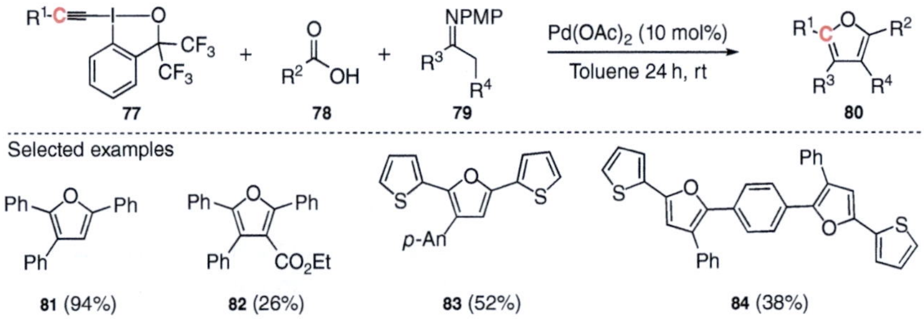

Scheme 13.10 Pd-catalyzed three-component reaction to prepare substituted furans.

13.3 Five-membered Systems with Two Heteroatoms

The imidazole nucleus is present in several important drugs having a broad range of pharmaceutical activities, including the amino acid histidine and the hormone histamine. Furthermore, imidazole derivatives have found applications in electronic materials and ionic liquids and are also the basis of the *N*-heterocyclic carbenes for transition metal ligands. In general, the preparation of imidazole derivatives relies on the use pre-synthetized 1,2-diamines (or their synthetic equivalents) with different electrophiles. In the context of the metal-catalyzed MCRs, the group of Arndtsen described a palladium-catalyzed coupling of imines **86** and **87**, with acyl chlorides (**89**) to generate imidazoles (**91–94**) [42]. Mechanistically, a 1,3-dipolar cycloaddition of the *N*-tosyl imine with an *in situ* generated Münchnone dipole (**90**) has been proposed. In principle, loss of a CO_2 molecule and *p*-TsOH favored the aromatization of the system (Scheme 13.11). More recently, the same group managed to develop reaction conditions in which the aroyl chloride could be generated *in situ,* via a carbonylative process of aryl halides (**85**) [43]. This five-component reaction represents a highly modular approach to a series of fully substituted imidazoles.

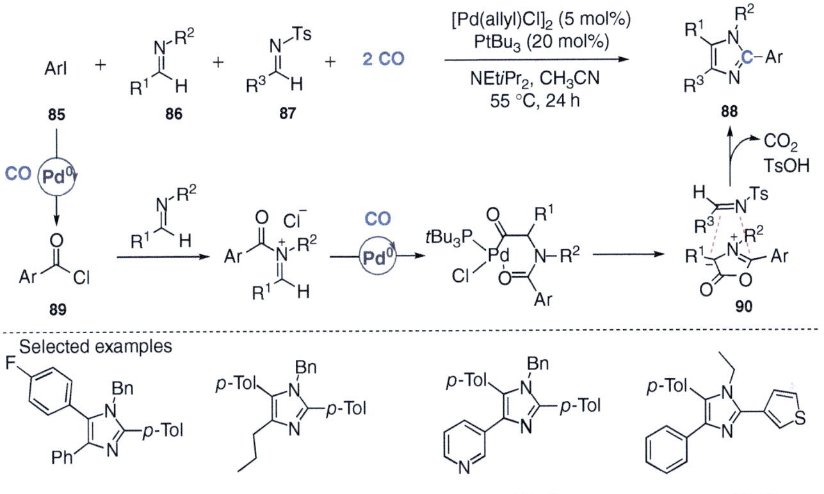

Scheme 13.11 Pd-catalyzed five-component coupling for the synthesis of imidazoles.

Imidazolidine, a saturated congener of the imidazole, is a motif frequently found in natural products but also found to have particular applications as chiral catalysts and ligands for several transformations. Accordingly, several methods for their preparation have been reported throughout the years. Recently, Zhao and Gong's group described a Pd-catalyzed three-component reaction of 2,3-allenylamines (**95**) with aryl iodides (**96**) and imines (**97**) for the synthesis of a series of polysubstituted imidazolidines (**99–102**) [44]. This double addition/cyclization cascade reaction represents a useful route to derivatives of type **98** with high to moderate *trans*-selectivity depending on the aryl substituents (Scheme 13.12).

Another nonaromatic member of the family of five-membered heterocycles with two nitrogen atoms is the imidazolone system (e.g. **106**), which is found in several commercially available drugs. Traditional synthetic methods to construct this heterocyclic system rely on the condensation of α-amino acid derivatives (α-amidoamides, α-isocyanoacetamides, azlactones, among others). Along this line, in 2018, Zhu and coworkers described a novel three-component process for the synthesis of tetra-substituted imidazolones (**106**) from methyl α–isocyanoacetates (**103**), primary amines (**104**), and aryl(vinyl) iodides(bromides) (**105**) (Scheme 13.13)

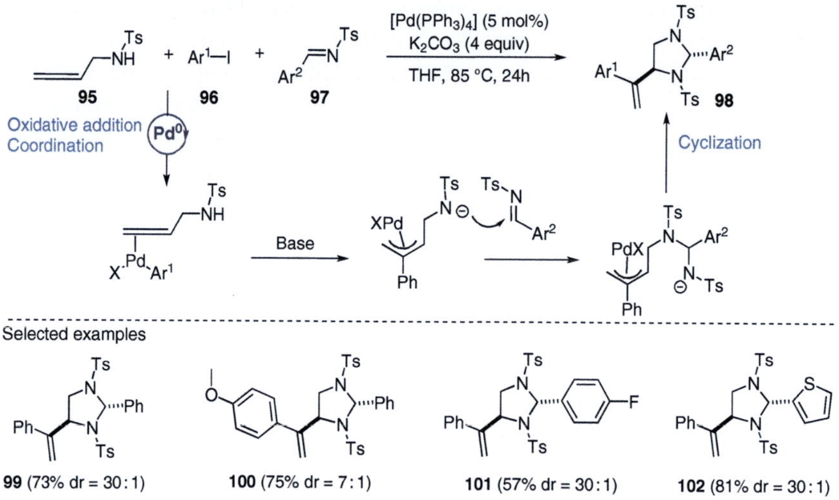

Scheme 13.12 Synthesis of imidazolidines via a Pd-catalyzed three-component reaction.

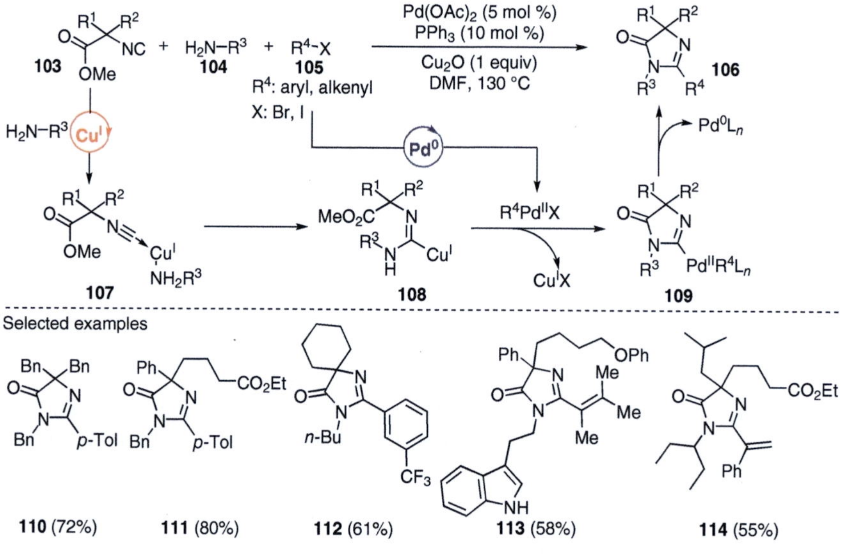

Scheme 13.13 Synthesis of imidazolones by dual-catalyzed, three-component process.

[45]. The reaction was broadly applicable in terms of the aryl iodides and the alkyl amines. Anilines failed to give the expected products. Formation of a copper complex **107** between the isocyano and the amino group, followed by a migratory insertion (**108**), has been proposed as the mechanism of formation of **109**. Transmetalation with ArPd(II)X, generated by oxidative addition of the aryl iodide to Pd(0), might be followed by reductive elimination to produce the observed products (**107–110**).

13.4 Five-membered Systems with Three Heteroatoms

The most common strategy for the synthesis of 1,2,3-triazoles relies on copper-catalyzed 1,3-dipolar cycloaddition involving azides and terminal alkynes (CuAAC), although recent reports of azide alkyne cycloaddition (AAC) reactions catalyzed by Ru [46], Rh [47], Ir [48], and Ag [49] have been disclosed. Click chemistry coupled with other synthetic transformations represents a formidable tool for the design and synthesis of functionalized 1,2,3-triazoles, an important scaffold present in biologically active compounds that exhibit anti-cancer, anti-tubercular, antimicrobial, and antibacterial activity, among other properties. 1,2,3-Triazole derivatives have also found extensive application in fluorescent imaging and in materials science.

Nagesh et al. reported a three-component cascade process for the synthesis of 1,2,3-triazole-tethered benzimidazol[1,2-*a*]quinolones (**118**) starting from 2-(azidomethyl)-1*H*-benzo[d]imidazoles (**115**), acetylenes (**116**), and 2-bromo/chloro aryl aldehydes (**117**) [50]. The three-component reaction is performed via a copper-catalyzed azide alkyne cycloaddition (CuAAC) reaction/N—C bond formation/Knoevenagel condensation sequence in which a copper salt performs two distinct catalytic processes. On the one hand, copper promotes a 1,3-dipolar cycloaddition to produce the corresponding 1,2,3-triazole derivative, a structural motif that enhances the reactivity of the methylene bridge group toward subsequent *N*-arylation/Knoevenagel condensation. On the other hand, a copper-catalyzed Ullmann-type reaction between 2-bromoaryl aldehydes and 1,2,3-triazole derivatives takes place leading to the formation of 1,2,3-triazole-fused

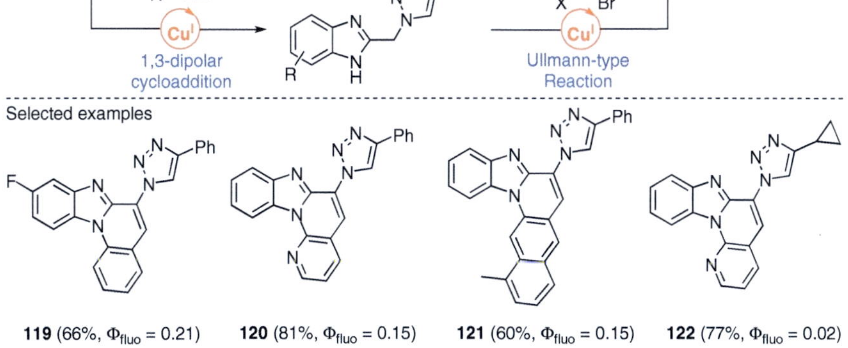

Scheme 13.14 Synthesis of 1,2,3-triazole tethered benzimidazo[1,2-a]quinoline.

heterocyclic molecules (**119–122**) (Scheme 13.14). The authors examined the UV and fluorescent properties of these molecules and found that they could be considered for biological and fluorescent applications given their high extinction coefficients.

The most popular strategy for the synthesis of fully substituted 1,2,3-triazoles is based on the trapping of *in situ* generated 5-cuprated-1,2,3-triazole intermediates with a suitable electrophile. In this sense, Wang et al. disclosed an interrupted click reaction using electrophilic benzenethiosulfonate (**124**) to prepare the 5-thiotriazole **123** in a three-component, one-pot operation. Using *O*-benzoylhydroxylamine (**126**) as an electrophilic amination reagent, an interrupted click reaction afforded 5-amino triazoles (**125**) [51]. The Xu group also examined bromoalkynes (**128**) as suitable electrophilic reagent in the three-component click/alkynylation sequence for the synthesis of 5-alkynyl-1,2,3-triazoles (**127**) [52]. Propargyl carbonates (**130**) have also been considered as appropriate electrophiles for the synthesis of 5-subsituted triazoles as the Cheng group has reported for the synthesis of 5-allenyl-1,2,3-triazoles(**129**) [53]. Zhang et al. reported that a vinyl copper intermediate may be intercepted by a carbene species, generated *in situ* from *N*-tosylhydrazones, to obtain trisubstituted triazoles (**131**). Starting from terminal alkynes, organic azides, and *N*-tosylhydrazones, it is possible to obtain 5-benzyl-1,2,3-triazoles. The key process in this approach involves a copper carbene migratory insertion, enabling the C—C bond formation in the final product [54]. Scheme 13.15 summarizes the aforementioned reports.

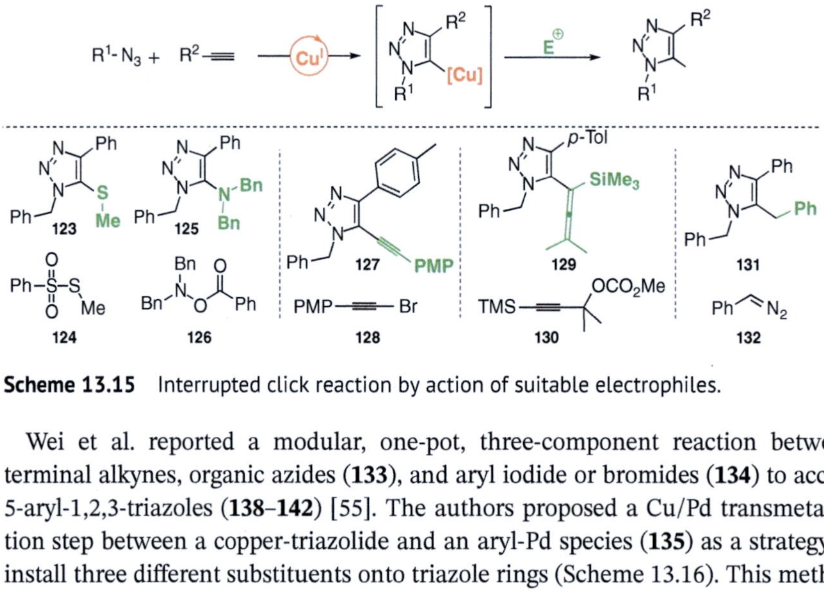

Scheme 13.15 Interrupted click reaction by action of suitable electrophiles.

Wei et al. reported a modular, one-pot, three-component reaction between terminal alkynes, organic azides (**133**), and aryl iodide or bromides (**134**) to access 5-aryl-1,2,3-triazoles (**138–142**) [55]. The authors proposed a Cu/Pd transmetallation step between a copper-triazolide and an aryl-Pd species (**135**) as a strategy to install three different substituents onto triazole rings (Scheme 13.16). This method is a general modular synthesis. However, stoichiometric copper is required in order to favor the designed click reaction pathway over a Sonogashira cross-coupling between copper-(I)-acetylide and aryl palladium intermediates.

5-Iodotriazoles (**145**) can efficiently be transformed into fully substituted triazoles (**144**) by means of a Pd-catalyzed Suzuki coupling, Sonogashira reaction, or Heck

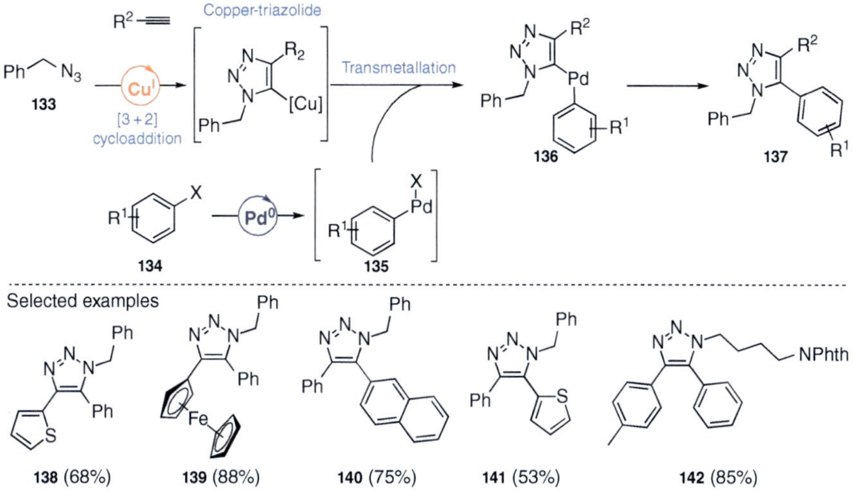

Scheme 13.16 Three-component reaction for the synthesis of trisubstituted triazoles.

reaction. By employing this strategy, a multicomponent-multicatalyst reaction $(MC)^2R$ for the construction of fully substituted 1,2,3-triazoles (**146–150**) was disclosed by the Lautens group in 2016. The CuAAC/Cu–Pd Sonogashira cross-coupling sequence of an organic azide, iodoalkynes, and terminal alkynes proceeded with good yields and displayed a broad scope of reactants. Since the reactivity of iodoalkynes (**143**) toward the CuAAC reaction exceeded that of terminal alkynes, the CuAAC reaction takes place in a chemoselective manner (Scheme 13.17). Furthermore, the latent catalyst activation approach confers control over the sequence of catalytic reactions, preventing possible side reactions from occurring, as well as and enforcing the preferred reaction pathway [56].

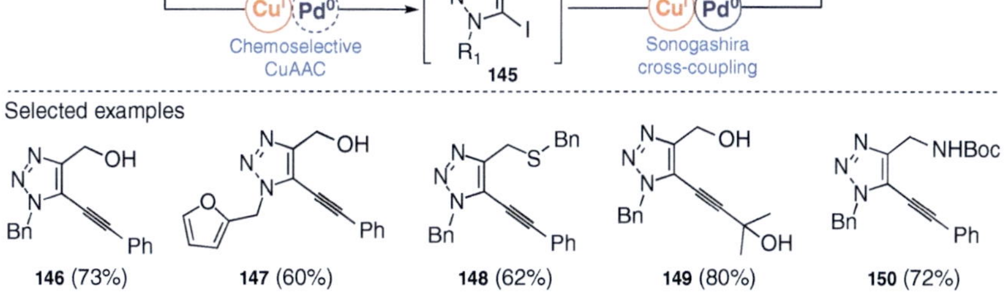

Scheme 13.17 Multicomponent, multicatalyst reaction for the synthesis of substituted 1,2,3-triazoles.

In 2017, Nallagangula and Namitharan reported the synthesis of triazolopyrimidines (**152, 153**) via a copper-catalyzed three-component cascade reaction of sulfonyl azides, alkynes, and 2*H*-azirines (**151**). In this 3-CR, both the copper–triazole **154** and ketenimine **155** intermediates participate in the formation of the targeted bicycle [57]. The chemical sequence proceeded via the nucleophilic addition of the 2*H*-aziridine to the *sp*-hybridized carbon atom in the ketenimine species **155** to furnish the zwitterionic intermediate **156**. The azirine ring opening by copper-triazolide eventually led to [1,2,3]triazolo[1,5-c]pyrimidine (**158** and **159**). It is worth noting that reaction with aliphatic alkynes yields more functionalized triazoles within sulfonamide incorporated (**160** and **161**), likely due to their inability to generate phenonium ion **157** (Scheme 13.18).

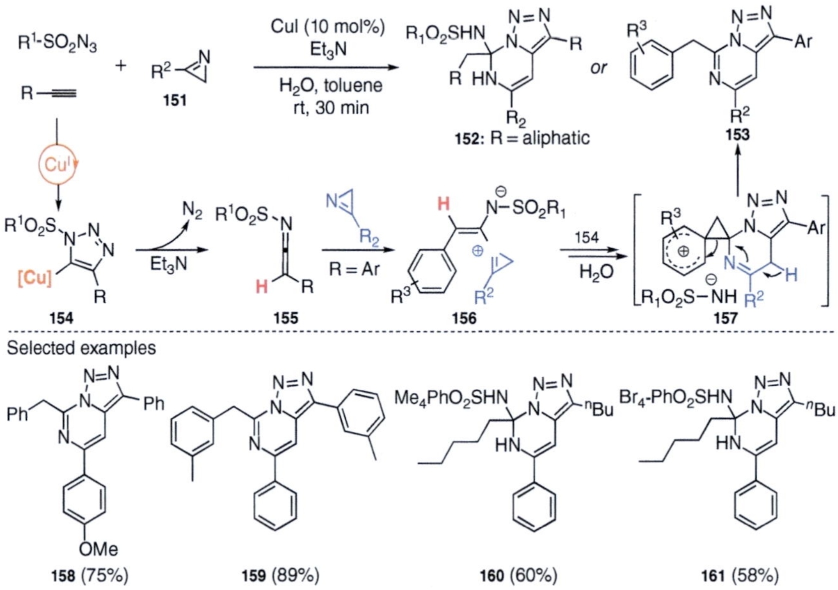

Scheme 13.18 Synthesis of triazolopyrimidines catalyzed by copper through copper-triazolide and ketemine intermediates.

The CuAAC/Ullman-type condensation approach has proven to be efficient for the combinatorial synthesis of fused 1,2,3-triazolo-1,3,6-triazonines (**165** and **166**) in one pot, as Peringer et al. have reported [58]. The products were obtained in moderate to good yields starting from 2-azidobenzaldehydes (**162**), substituted diaminobenzenes (**163**), and terminal alkynes in the presence of a catalytic amount of copper iodide, Et_3N as base, in DMSO at 100 °C for 24 hours (Scheme 13.19). The authors expanded the scope of this methodology to the synthesis of (arylselenyl)- and (arylsulfenyl)-alkyl-1,2,3-triazolo-1,3,6-triazonines (**167**, **168**, respectively) via the reaction of *o*-phenylenediamine, 2-azidobenzaldehyde, and different arylchalcogenyl alkynes [59].

The Lautens Group envisioned a Pd-catalyzed carbopalladation/C–H functionalization for the selective arylation of 1,2,3-triazoles, which would be prepared

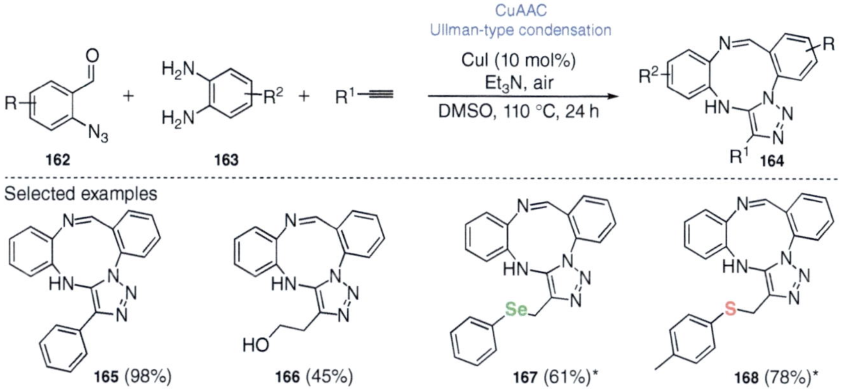

Scheme 13.19 One-pot combinatorial synthesis of fused 1,2,3-triazolo-1,3,6-triazonines. *Conditions: CuI (10 mol%), Et_3N, N_2, 1,4-dioxane, 100 °C, 24 h.

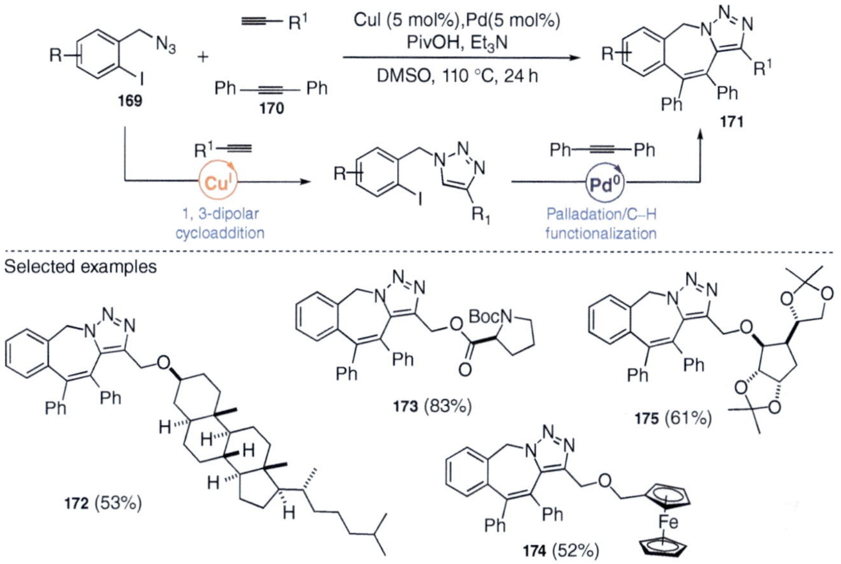

Scheme 13.20 Three-component dual-metal-catalyzed reaction for the synthesis of polycyclic triazoles.

through a CuAAC. In 2017, this research group introduced a three-component reaction that combined organic azides (**169**), terminal alkynes, and internal alkynes (**170**) featuring a Cu/Pd system that led to an efficient, one-pot synthesis of polycyclic triazoles in one pot (Scheme 13.20). ^{1}H NMR studies showed that the three-component reaction begins with a chemoselective copper-catalyzed 1,3-dipolar cycloaddition between the azide and the terminal alkyne, followed by a palladium-catalyzed incorporation of an internal alkyne [60]. The last step of this chemical sequence is the arylation of the triazole motif, which likely proceeds via an electrophilic palladation mechanism, according to deuterium labeling experiments.

This robust, simple, one-pot procedure affords good yields for a diverse collection of substrates, including biologically active motifs such as cholesterol (**172**), proline (**173**), and glucose derivatives (**175**). Additionally, the central seven-membered ring is susceptible to derivatization using *n*-BuLi.

In contrast, the development of related oxidative coupling reactions with nucleophilic reagents to trap copper-triazolides remains less explored, likely due to the risk of the oxidation of Cu(I) to Cu(II), thereby inhibiting the alkyne–azide cycloaddition. In 2019, Zhu described a copper-catalyzed multicomponent synthesis of 5-pentafluoroalkyl-1,2,3-triazoles (**176**) using air as the sole oxidant. These authors synthesized a new class of glycinamide type ligands (**177**), which promote aerobic oxidative coupling of copper triazolides with nucleophilic fluoroalkyl reagents (Scheme 13.21).

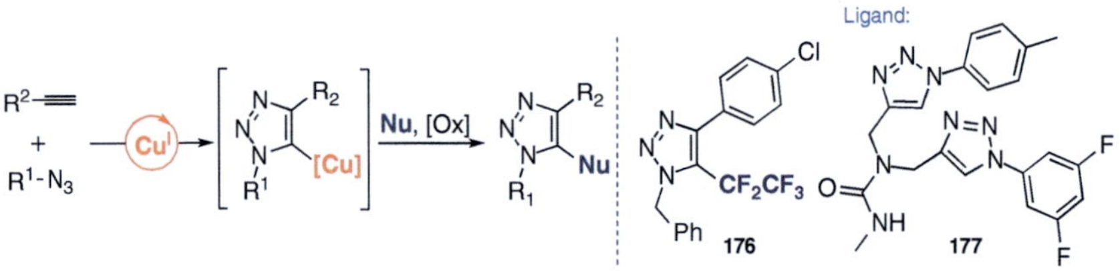

Scheme 13.21 Cu-catalyzed synthesis of fully substituted triazoles by oxidative coupling.

13.5 Six-membered Heterocycles with One Heteroatom and Their Benzo-fused Derivatives

The pyridine system is perhaps one of the most studied aromatic heterocyclic systems. Its derivatives have found application in wide ranging areas of study. They are broadly distributed in nature as the core of several natural products with diverse biological functions. Mono-, di-, and tri-pyridines are versatile donor ligands in coordination chemistry. Furthermore, several derivatives are also the main core in commercially active pharmaceutical ingredients for the treatment of a variety of disorders. As a consequence, a number of synthetic methods have been developed for the construction of the pyridine nucleus, most of them based on the condensation of carbonyl compounds with a nitrogen source (Hantzsch, Chichibabin, and Kröhnke reactions) [61]. More recently, the metal-catalyzed [2+2+2] cycloaddition of two alkynes and nitriles has emerged as an excellent option for the construction of the pyridine system. Catalytic systems based on metals such as Rh, Ru, Co, Au, among others have been reported to promote the cyclocondensations [62, 63]. Along this line, in 2017, Chen and coworkers described a novel copper-catalyzed [2+2+2] alkenylation of nitriles with vinyliodonium salts for the synthesis of multisubstituted pyridines [64]. Mechanistically, this protocol proposed that the process begins with the alkenylation of the nitrile by the vinyliodonium salt **178**, to generate an aza-butadienylium intermediate **180** which collapses to the aza-butadienyl cation **181**. A concerted, or step-wise cycloaddition

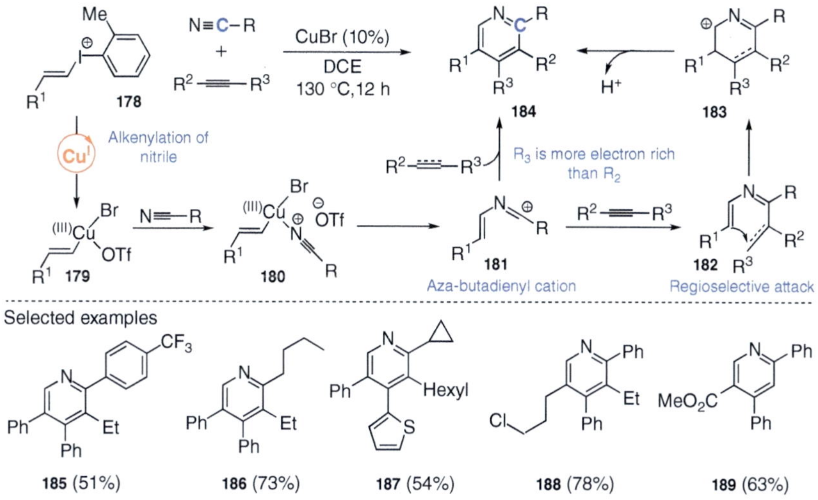

Scheme 13.22 Cu-catalyzed [2 + 2 + 2] alkenylation of nitriles with vinyliodonium salts.

with the alkyne or alkene finally generates the pyridine system (Scheme 13.22). The reaction is broad in scope in terms of the nitrile, alkyne, and vinyliodonium salt species, leading to the preparation of a series of substituted pyridines (e.g. **185–189**). The reaction also worked well when the alkyne was replaced by an alkene, greatly expanding the scope of this synthesis.

On the other hand, the group of Deng has prepared a series of 2,4-diarylsubstituted pyridines through a ruthenium-catalyzed four-component reaction [65]. In this process, two molecules of acetophenone are condensed with ammonium acetate and a molecule of DMF. In the proposed mechanism, oxidation of DMF affords imine **192**, which is trapped by the enamine **191**, formed from autocondensation of acetophenone, followed by imine formation and tautomerization. Then, oxidation of **193** might afford the azatriene **194** which, upon a 6π electron cyclization and methyformamide elimination, provides the expected pyridine. Thus, one of the DMF methyl groups is incorporated as the carbon atom at C-6 of the pyridine system (Scheme 13.23). Under the optimized conditions, eighteen 2,4-diarylsubstituted pyridines, containing electron-donating and electron-attracting groups (**195–198**), were synthesized in moderate to good yields.

Metal-catalyzed cyclocondensation of *O*-acyl ketoximes (**200**, R^4 = OC(O)R) with alkenylboronic acids [66], alkynes [67], or alkenes [68] has emerged as a direct method to assemble the pyridine system. In all those reactions, the N—O bond cleavage of the ketoxime facilitates the aromatization pathway. Thus, in principle, the result of the process could be the pyridinium salt in the event that an imine (**200**) participates in a similar reaction. In this regard, the group of Cheng recently reported a Rh-catalyzed vinylic C–H activation-annulation process for the synthesis of multisubstituted pyridinium salts (**199**) [69]. In the process, a vinyl(aryl) ketone or aldehyde, an amine, and an alkyne react in a three-component reaction. In the proposed mechanism, the imine **200** is trapped by the catalyst to

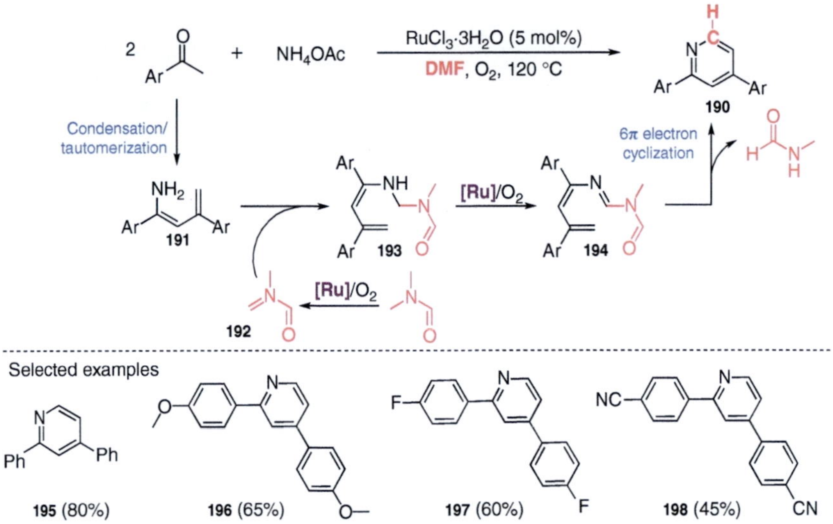

Scheme 13.23 Synthesis of 2,4-diarylsubstituted-pyridines through a Ru-catalyzed four-component reaction.

form a five-membered rhodacycle **201**. Coordination of the alkyne followed by a regioselective insertion into a carbon–rhodium bond gives the intermediate **202**, which upon reductive elimination generates the pyridinium salt (Scheme 13.24). Under these conditions, *N*-alkyl- and *N*-aryl-substituted pyridinium salts (**203–207**) can be synthesized. Furthermore, hetero-fused systems can be prepared when heteroaryl ketones (e.g. 1-(thiophen-2-yl)ethan-1-one) are utilized (**207**).

A series of tetrahydroquinolines have been prepared via a four-component cascade C–H functionalization/cyclization/nucleophilic substitution process. An aniline, a carboxylic anhydride, propenol, and an alcohol participate in the reaction. C–H activation of the aniline was facilitated by *in situ* generation of the corresponding anilide (with the anhydride), which functions as the directing group for the formation of complex **209**. Rh-catalyzed C–C coupling of the allylic alcohol affords the aldehyde **211**, via β-elimination of complex **210**. Then, spontaneous cyclization in the presence of the alcohol affords **208**. Anilines with electron-attracting and electron-donating proved to be efficient components in the reaction. Also, different alkyl anhydrides and a variety of alcohols were efficiently introduced in the process (**212–215**) (Scheme 13.25).

Recently, several reports on the synthesis of vinyl benzoxazine-2-ones (i.e. protonated **219**, R = H) through metal-catalyzed cyclizations of 2-(phenylethynyl)anilines in the presence of CO_2, have been published [70]. It is important to note that when propargyl amines are used in the same process, the corresponding oxazolidinones can be obtained [71]. In this context, Cheng and coworkers recently described a Pd-catalyzed three-component reaction between *o*-alkynylanlines, aryl iodides, under atmospheric pressure of CO_2 [72]. The strategy is based on a sequential carboxylation, *trans*-oxopalladation of the triple bond by ArPdX species to generate intermediate **217**. Under reported conditions, the benzoxazine-2-one **219**

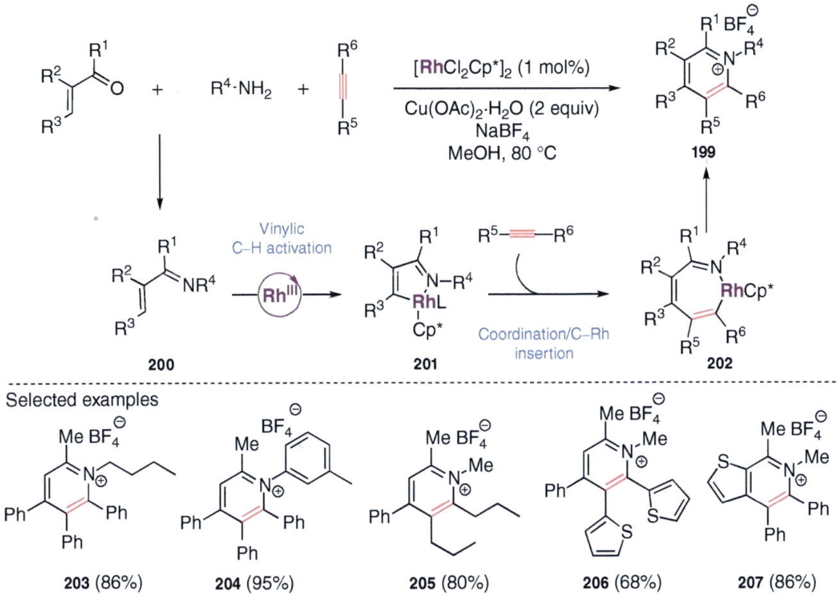

Scheme 13.24 Synthesis of pyridinium salts through a Rh(III)-catalyzed vinylic C–H activation.

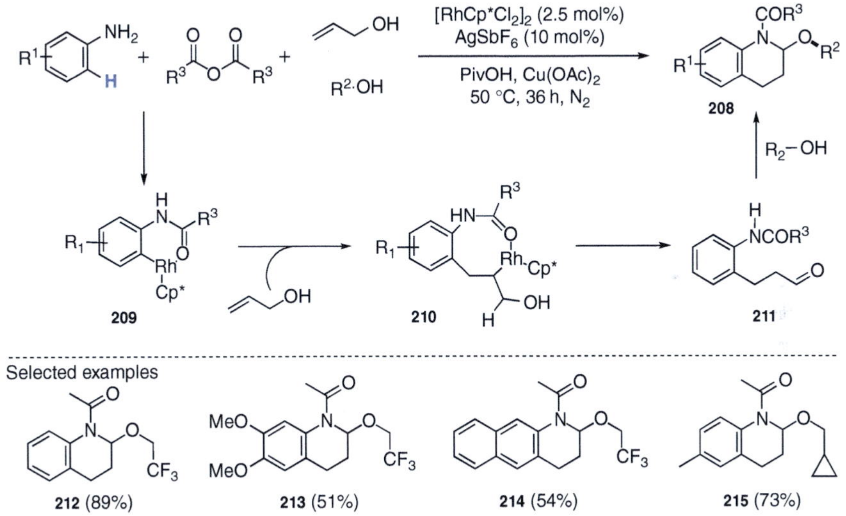

Scheme 13.25 Four-component cascade C–H functionalization/cyclization/nucleophilic substitution process.

rearranges to generate 3,3-diaryl 2,4-quinolinediones bearing a quaternary carbon center (**221–224**).

Similar to the pyridine synthesis strategy (Scheme 13.26), the isoquinoline system can be assembled from the corresponding benzaldoximes through a metal-catalyzed C–H-activation/cyclo condensation with alkynes [73, 74]. This concept was recently

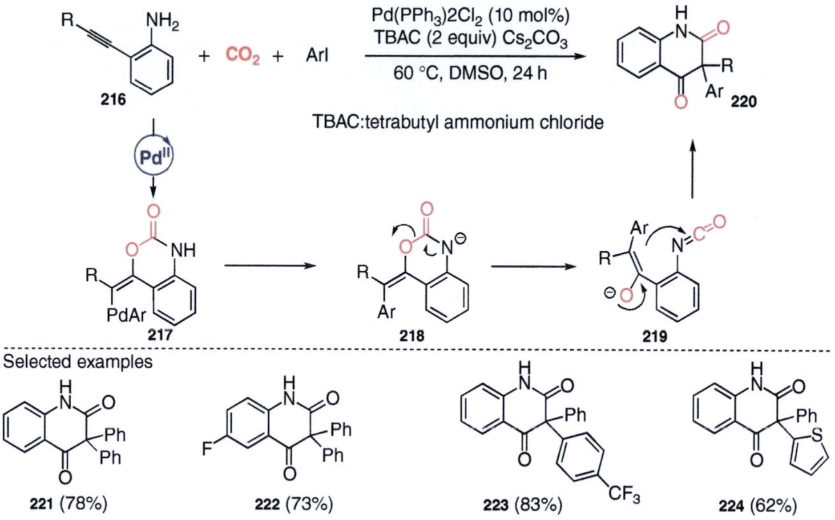

Scheme 13.26 Synthesis of 3-diaryl 2,4-quinolinediones via a Pd-catalyzed multicomponent process of *o*-alkynylanilines, aryl Iodides, and CO_2.

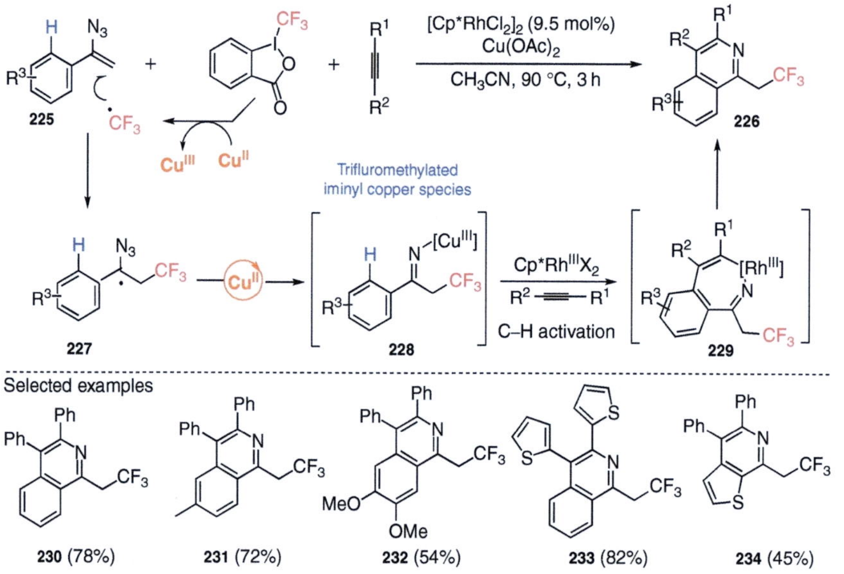

Scheme 13.27 Multicomponent synthesis of trifluoroethyl isoquinolines from alkynes and vinyl azides.

extended to the use of vinyl azides **225** as the nitrogen source to access trifluoroethyl isoquinolines **230–234** (Scheme 13.27) [75]. The reaction was achieved using a Rh(III)–Cu(II) bimetallic system and Togni's reagent as the source of trifluoromethyl radicals. In principle, it supposes that the process is triggered by the addition of the trifluoromethyl radical to vinyl azide to afford radical **227**, which is oxidized to produce a trifluoromethylated iminyl copper species **228**. This three-component reaction tolerates various vinyl azides and internal alkynes.

In an ancillary work, Wang and coworkers described a rhodium(III)-catalyzed three-component reaction of *N*-methoxybenzamides, α-diazoesters, and alkynes to produce isoquinoline derivatives. The reaction proceeds through successive *O*-alkylation and C–H activation processes, both promoted by the same catalyst [76].

13.6 Six-membered *O*-heterocycles and their Benzofused Derivatives

Unlike the CuAAC reaction of aryl and alkyl azides, sulfonyl azides react with terminal alkynes to furnish ketenimines (**236**), reactive intermediates whose *sp*-hybridized carbon atom confers its observed reactivity toward nucleophiles and free radicals. First disclosed by Chang and coworkers [77], the Cu-catalyzed three-component reaction of sulfonyl azides and terminal alkynes with appropriate nucleophiles has been a recurrent strategy for the synthesis of diverse structural motifs. In 2006, Wang and coworkers successfully implemented this strategy for the first time in the Cu-catalyzed synthesis of coumarin derivatives featuring salicylaldehyde a suitable coupling partner in a three-component reaction [78]. This approach relies on the nucleophilic addition to the *sp*-hybridized carbon atom of an *in situ* generated ketenimine followed by intramolecular ring closure. Following this strategy, Murugavel and Punniyamurthy reported a selective microwave-assisted copper(I)-catalyzed synthesis of 3-*N*-sulfonylamidine coumarines (**235**) employing sulfonyl azides, propiolates, secondary amines, and salicylaldehydes (Scheme 13.28) [79]. This four-component process afforded desired *O*-heterocycles (**238**–**241**) in a one-pot operation with high selectivity and good yields.

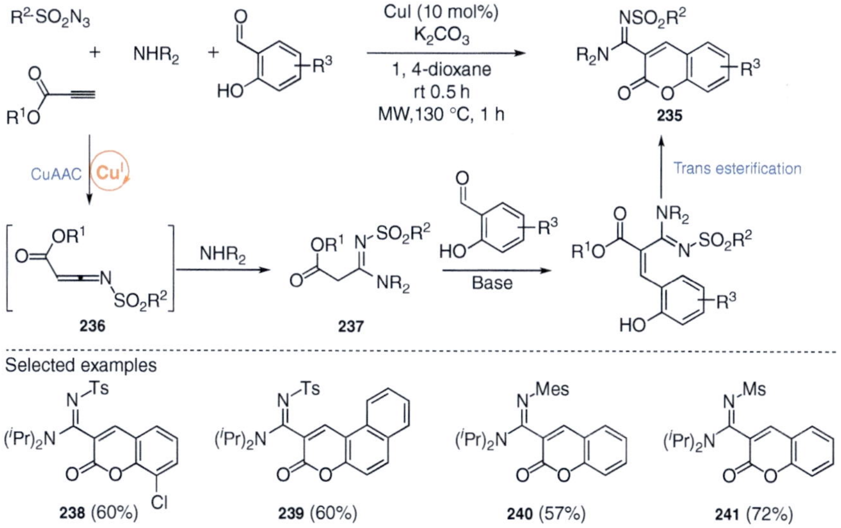

Scheme 13.28 Microwave assisted four-component synthesis of 3-*N*-sulfonylamidine coumarines.

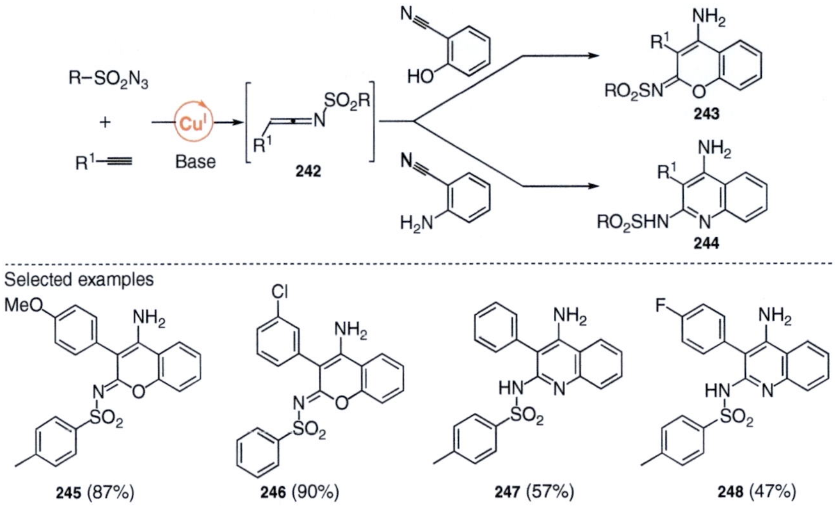

Scheme 13.29 Synthesis of 4-amino iminocoumarines and 4-aminoquinolines via a three-component reaction catalyzed by copper.

Later on, the Yi group developed a copper-catalyzed three-component reaction to synthesize polysubstituted 4-amino iminocoumarines, starting from 2-hydroxybenzonitriles, terminal alkynes, and sulfonyl azides (Scheme 13.29) [80]. The process comprised the CuAAC reaction/nucleophilic addition/isomerization sequence to afford target 4-amino iminocoumarines (**245** and **246**) in good yields. When 2-aminobenzonitrile is used as nucleophile, the corresponding 4-aminoquinolines are obtained (**247** and **248**).

Chen et al. reported the diversity-oriented synthesis of 2-aminochromenone imines via a domino click reaction/Kemp elimination/[4 + 2] cycloaddition (Scheme 13.30). The authors envisioned that a cyanophenol species (**251**), generated *in situ* from benzo[*d*]isoxaozole (**249**) through a Kemp elimination, could participate in a [4 + 2] cycloaddition process with ketenimines (**250**), yielding 4-amino iminocoumarines (**253–257**) in a one-pot procedure. The methodology is operationally simple, has a broad substrate scope, and has proved to be scalable (gram-scale) [81].

Electrophilic ketenimines (**236**, **242**) have been extensively exploited in nucleophilic addition and cycloaddition reactions. In contrast, reports regarding nucleophilic metalated ketenimine (**260**) are less abundant. On this matter, Sun *et al.* explored the reactivity of *in situ* generated *N*-vinyl-copper ketenimines with enaminones to afford iminolactones (**259**); hence, a Cu-catalyzed three-component reaction of aryl acetylene, sulfonyl azide, and enaminone (**258**) was developed (Scheme 13.31) [82]. The overall process involves a CuAAC/Michael addition/elimination/6π electron cyclization sequence. The authors proposed that the cascade process was triggered by the copper coordination of the ketenimine with the enaminone to give intermediate **261** that eventually afforded targeted *O*-heterocycles (**263–267**) via a 6π electron cyclization. It is worth mentioning that exocyclic enaminones are suitable substrates for this 3-CR, furnishing analogs of iminocoumarin **267**.

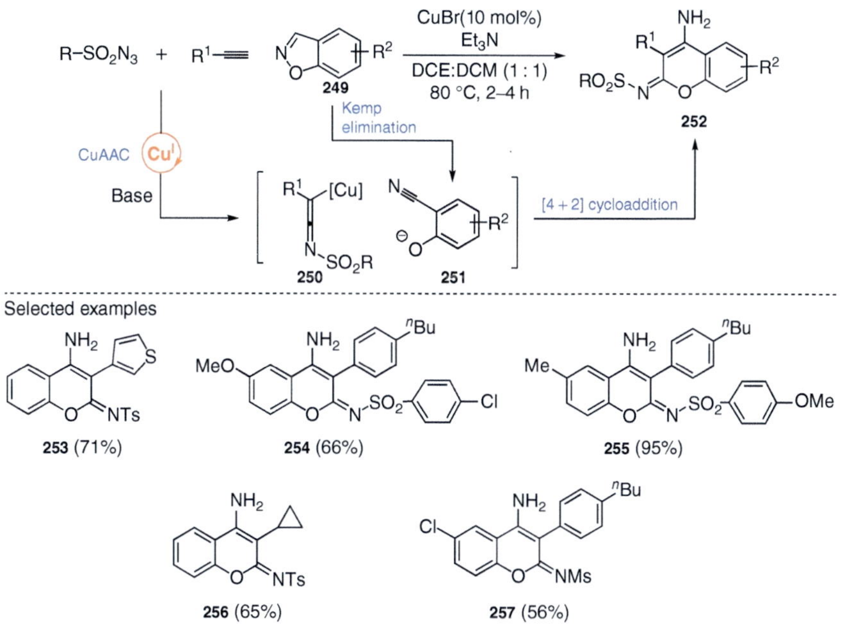

Scheme 13.30 Kemp elimination in the Cu-catalyzed multicomponent synthesis of *O*-heterocycles.

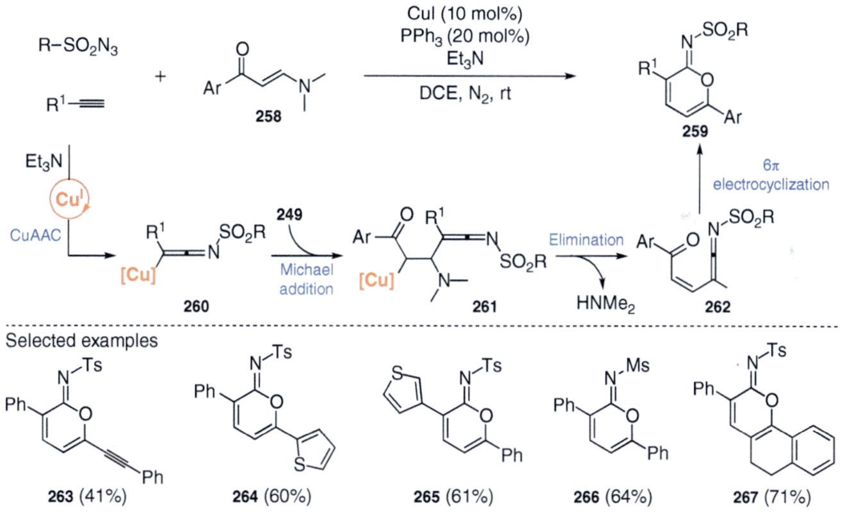

Scheme 13.31 Cu-catalyzed three-component synthesis of iminolactones via 6π-electrocyclization.

13.7 Four-membered *N*-heterocycles and Seven-membered Benzofused *N*-heterocycles

The most straightforward method for the synthesis of β-lactams (**268**) and their derivatives is the [2 + 2] cycloaddition of imines and ketenes (**271**). Alternatively, the group of Arndtsen demonstrated that mesoionic Münchnones (**270**) could be used for the synthesis of these structural motifs by means of palladium-catalyzed coupling between an imine, carbon monoxide, and aryl halides (Scheme 13.32) [83]. The sequence involves two tandem catalytic carbonylation reactions, the first one to generate corresponding acid chloride and the second one to produce Münchnone (**270**) from the corresponding *N*-acyl iminium salt (**269**). The method is amenable to structural diversification allowing the synthesis of diversely substituted β-lactams (**272–275**), including spirocyclic β-lactam **275**.

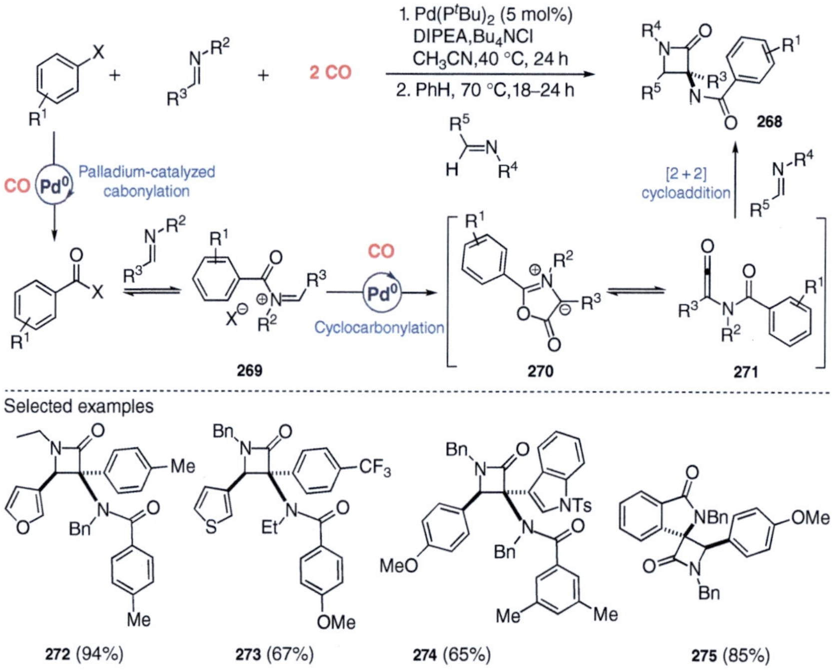

Scheme 13.32 Pd-catalyzed multicomponent synthesis of β-lactams through Münchnones intermediates.

Synthetic strategies for the construction of seven-membered heterocycles based on [5 + 2] cycloadditions have been less explored than their counterparts for the construction of five- and six-membered rings. In 2014, Yoo group found that Rh-catalyzed reaction of 1-sulfonyl-1,2,3-triazoles and 2-substituted pyridines gives isolable azomethine ylides, which act as 1,5-dipoles and can participate in [5 + 2]

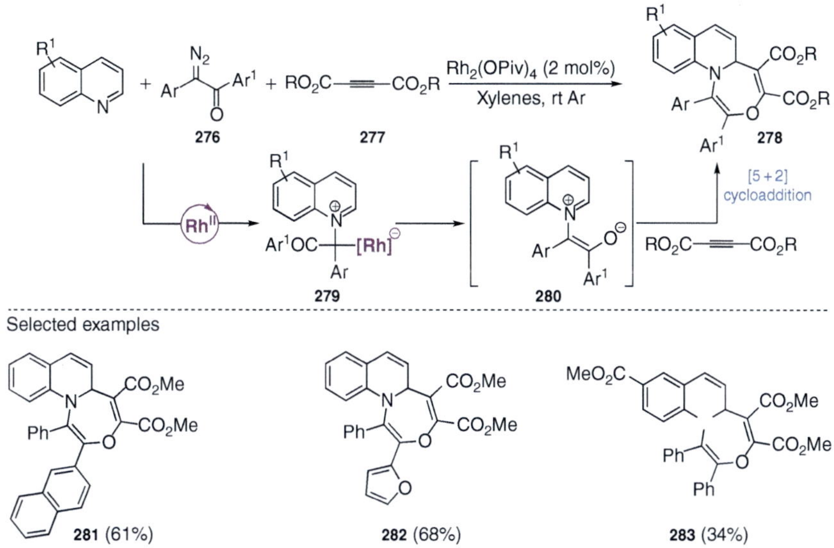

Scheme 13.33 Rh-catalyzed [5 + 2] cycloadditions of quinolinium ylides with alkynes for the construction of 1,4-oxazepines.

cycloaddition reactions with suitable 2π-dipolarophiles to afford 1,4-diazepines [84]. Inspired by the above-mentioned approach and other preceding reports, Peng and co-workers envisioned that pyridinium ylides (**280**) derived from α-diazoketones (**276**) could act as 1,5-dipole in [5 + 2] cycloadditions with electron-deficient alkynes (**277**) and developed a three-component reaction catalyzed by rhodium to prepare 1,4-oxazepine heterocycles (**281**–**283**) (Scheme 13.33) [85]. While quinolinium ylides derived from α-diazoketones lead to seven-membered heterocycles through a [5 + 2] cycloaddition, quinolinium ylides derived from aryl diazoacetates lead to indolizine derivatives via a [3 + 2] cycloaddition/1,3-ester migration (not shown in Scheme 13.33).

In 2015, Li and co-workers reported a novel Rh-catalyzed [3 + 2]/[5 + 2] annulation of 4-aryl 1-tosyl-1,2,3-triazoles (**284**) with internal alkynes (**285**). Inspired by the progress made in the Rh-catalyzed oxidative C–H activation field, particularly regarding Rh-catalyzed annulation with π components, the Li group envisioned that the Rh^{II} azavinyl carbenes (**287**) and the aryl C(sp^2)—H bond could be trapped in an annulation reaction by addition of an appropriate 2π component [86]. This three-component reaction enables the selective synthesis of [1,7-*cd*]azepin-1-ols via dual aryl C(sp^2)–H functionalization with two alkyne molecules (Scheme 13.34). Yang et al. proposed that the cascade reaction proceeds through the addition of rhodium-carbenoid intermediate (**287**) to an alkyne **285**, and subsequent electrophilic cyclization involving one C(sp^2)—H bond to produce intermediate **288**, which in turn undergoes a [5 + 2] cycloaddition with the second molecule of alkyne **285** to eventually furnish the targeted compounds (**290**–**293**).

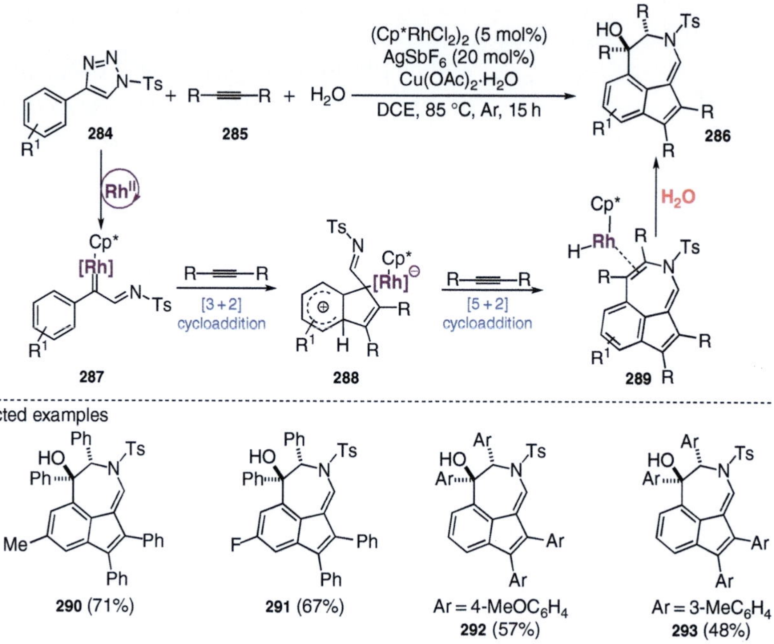

Scheme 13.34 Rh-catalyzed three-component [3 + 2]/[5 + 2] annulation for the synthesis of [1,7-cd]azepin-1-ols.

13.8 Conclusion

In this chapter, we have reviewed a collection of the latest reports of transition MC-MCRs for the construction of heterocycles. This group of reactions continues to grow and has spread to different heterocyclic systems. With the advent of new catalytic systems based on non-precious metals, and processes such as C–H activations, or the fixation of small molecules such as CO and CO_2, these reactions have become increasingly practical and efficient. There is no doubt that many of these processes are or will be the protocols of choice for synthesizing a variety of heterocyclic systems, in most cases, from less elaborated raw materials, and under milder reaction conditions. We hope that with the methodologies, we have selected that the state of the art of these important processes will be put into perspective. We are aware that inadvertently we have probably missed some essential reports, for which we apologize.

References

1 Müller, T.J.J. (2017). Synthesis of heterocycles by Pd-catalyzed and Pd-catalysis initiated multi-component reactions. In: *Applied Homogeneous Catalysis with Organometallic Compounds: A Comprehensive Handbook in Four Volumes*, 3rd edn (eds. B. Cornils, W.A. Herrmann, M. Beller and R. Paciello), 1463–1483. Weinheim, Germany: Wiley-VCH.

2 Sarkar, S., Banerjee, A., and Patel, B.K. (2016). Transition metal catalyzed synthesis of heterocycles via multicomponent reactions. In: *Multicomponent Reactions: Synthesis of Bioactive Heterocycles* (eds. K.L. Ameta and A. Dandia), 139–182. New York: Taylor & Francis.

3 Lorenzini, F., Tjutrins, J., Quesnel, J.S. et al. (2014). Metal-catalyzed multicomponent synthesis of heterocycles. In: *Multicomponent Reactions in Organic Synthesis* (eds. J. Zhu, Q. Wang and M.-X. Wang), 207–230. Weinheim, Germany: Wiley-VCH.

4 Müller, T.J.J. (2010). Palladium-copper catalyzed alkyne activation as an entry to multicomponent syntheses of heterocycles. In: *Synthesis of Heterocycles via Multicomponent Reactions II* (eds. R.V.A. Orru and E. Ruijter), 25–94. Berlin: Springer.

5 Neetha, M., Rohit, K.R., Saranya, S. et al. (2020). Zinc-catalysed multi-component reactions: an overview. *ChemistrySelect* 5 (3): 1054–1070.

6 Saranya, S., Aneeja, T., Neetha, M. et al. (2020). Recent advances in the iron-catalysed multicomponent reactions. *Appl. Organomet. Chem.* 34 (12): e5991.

7 Shilpa, T., Dhanya, R., Saranya, S. et al. (2020). An overview of rhodium-catalysed multi-component reactions. *ChemistrySelect* 5 (2): 898–915.

8 Saranya, S., Rohit, K.R., Radhika, S. et al. (2019). Palladium-catalyzed multicomponent reactions: an overview. *Org. Biomol. Chem.* 17 (35): 8048–8061.

9 Cherian, R.M., Harry, N.A., Saranya, S. et al. (2018). Copper-catalysed multicomponent syntheses of heterocycles. *Asian J. Org. Chem.* 8 (2): 197–233.

10 Odom, A.L., and McDaniel, T.J. (2015). Titanium-catalyzed multicomponent couplings: efficient one-pot syntheses of nitrogen heterocycles. *Acc. Chem. Res.* 48 (11): 2822–2833.

11 Michlik, S., and Kempe, R. (2013). A sustainable catalytic pyrrole synthesis. *Nat. Chem.* 5: 140–144.

12 Srimani, D., Ben-David, Y., and Milstein, D. (2013). Direct synthesis of pyrroles by dehydrogenative coupling of β-aminoalcohols with secondary alcohols catalyzed by ruthenium pincer complexes. *Angew. Chem. Int. Ed.* 52 (14): 4012–4015.

13 Iida, K., Miura, T., Ando, J. et al. (2013). The dual role of ruthenium and alkali base catalysts in enabling a conceptually new shortcut to *N*-unsubstituted pyrroles through unmasked α-amino aldehydes. *Org. Lett.* 15 (7): 1436–1439.

14 Siddiki, S.M.A.H., Touchy, A.S., Chaudhari, C. et al. (2016). Synthesis of 2,5-disubstituted pyrroles *via* dehydrogenative condensation of secondary alcohols and 1,2-amino alcohols by supported platinum catalysts. *Org. Chem. Front.* 3 (7): 846–851.

15 Dang, T.T., and Seayad, A.M. (2017). Efficient [Cu(NHC)]-catalyzed multicomponent synthesis of pyrroles. *Chem. Asian J.* 12 (18): 2383–2387.

16 Huisgen, R., Gotthardt, H., and Bayer, H.O. (1964). Azlactones as 1,3-dipoles; a new pyrrole synthesis. *Angew. Chem. Int. Ed.* 3 (2): 135–136.

17 Huisgen, R., and Maeder, H. (1971). Azomethine ylide from dimethyl 1-(p-methoxyphenyl)aziridine-2,3(cis)-dicarboxylate. *J. Am. Chem. Soc.* 93 (7): 1777–1779.

18 Boyd, G.V. (1968). 5-Oxazolonium perchlorates from α-acylamino-acids. *Chem. Commun. (Camb)* (22): 1410–1412.

19 Keating, T.A. and Armstrong, R.W. (1996). Postcondensation modifications of Ugi four-component condensation products: 1-isocyanocyclohexene as a convertible isocyanide. Mechanism of conversion, synthesis of diverse structures, and demonstration of resin capture. *J. Am. Chem. Soc.* 118 (11): 2574–2583.

20 Torres, G.M., Quesnel, J.S., Bijou, D. et al. (2016). From aryl Iodides to 1,3-dipoles: design and mechanism of a palladium catalyzed multicomponent synthesis of pyrroles. *J. Am. Chem. Soc.* 138 (23): 7315–7324.

21 Firoozi, N., Torres, G.M., and Arndtsen, B.A. (2016). Palladium catalyzed, multicomponent synthesis of fused-ring pyrroles from aryl iodides, carbon monoxide, and alkyne-tethered Imines. *J. Org. Chem.* 81 (22): 11145–11152.

22 Gilbert, Z.W., Hue, R.J., and Tonks, I.A. (2015). Catalytic formal [2+2+1] synthesis of pyrroles from alkynes and diazenes via Ti^{II}/Ti^{IV} redox catalysis. *Nat. Chem.* 8 (1): 63–68.

23 Peng, J., Gao, Y., Hu, W. et al. (2016). Palladium-catalyzed multicomponent reaction (MCR) of propargylic carbonates with isocyanides. *Org. Lett.* 18 (22): 5924–5927.

24 Raghuvanshi, A., Singh, A.K., Mobin, S.M. et al. (2017). $Fe(CO)_5$ catalyzed [2+2+1] cycloaddition of alkyne, carbodiimide and CO for the synthesis of 5-iminopyrrolones. *ChemistrySelect* 2 (29): 9245–9248.

25 Perry, R.J. and Turner, S.R. (1991). Preparation of N-substituted phthalimides by the palladium-catalyzed carbonylation and coupling of *o*-dihalo aromatics and primary amines. *J. Org. Chem.* 56 (23): 6573–6579.

26 Cao, H., and Alper, H. (2010). Palladium-catalyzed double carbonylation reactions of *o*-dihaloarenes with amines in phosphonium salt ionic liquids. *Org. Lett.* 12 (18): 4126–4129.

27 Khedkar, M.V., Khan, S.R., Sawant, D.N. et al. (2011). Palladium on carbon: an efficient, heterogeneous and reusable catalytic system for carbonylative synthesis of *N*-substituted phthalimides. *Adv. Synth. Catal.* 353 (18): 3415–3422.

28 Marosvoelgyi-Hasko, D., Petz, A., Takács, A. et al. (2011). Synthesis of tetrahydrophthalazine and phthalamide (phthalimide) derivatives via palladium-catalysed carbonylation of iodoarenes. *Tetrahedron* 67 (47): 9122–9128.

29 Wu, X.-F., Oschatz, S., Sharif, M. et al. (2013). Palladium-catalyzed carbonylative synthesis of phthalimides from 1,2-dibromoarenes with molybdenum hexacarbonyl as carbon monoxide source. *Adv. Synth. Catal.* 355 (18): 3581–3585.

30 Khedkar, M.V., Shinde, A.R., Sasaki, T. et al. (2014). Immobilized palladium metal containing ionic liquid catalyzed one step synthesis of isoindole-1,3-diones by carbonylative cyclization reaction. *J. Mol. Catal. A Chem.* 385: 91–97.

31 Chen, J., Natte, K., Spannenberg, A. et al. (2014). Efficient palladium-catalyzed double carbonylation of *o*-dibromobenzenes: synthesis of thalidomide. *Org. Biomol. Chem.* 12 (30): 5578–5581.

32 Fuse, S., Takahashi, R., and Takahashi, T. (2015). Facile, one-step synthesis of 5-substituted thieno[3,4-c]pyrrole-4,6-dione by palladium-catalyzed carbonylative amidation. *Eur. J. Org. Chem.* 2015 (16): 3430–3434.

33 Warnan, J., El Labban, A., Cabanetos, C. et al. (2014). Ring substituents mediate the morphology of PBDTTPD-PCBM bulk-heterojunction solar cells. *Chem. Mater.* 26 (7): 2299–2306.

34 Inack-Ngi, S., Rahmani, R., Commeiras, L. et al. (2009). Copper-catalyzed preparation of γ-alkylidenebutenolides and isocoumarins under mild palladium-free conditions. *Adv. Synth. Catal.* 351 (5): 779–788.

35 Inack-Ngi, S., Cherry, K., Héran, V. et al. (2011). Carboxylate-directed tandem functionalisations of α,β-dihaloalkenoic acids with 1-alkynes: a straightforward access to (Z)-configured, α,β-substituted γ-alkylidenebutenolides. *Chem. Eur. J.* 17 (49): 13692–13696.

36 Mardjan, M.I.D., Parrain, J.-L., and Commeiras, L. (2016). Copper(I)-catalysed multicomponent reaction: straightforward access to 5-hydroxy-1H-pyrrol-2(5H)-ones. *Adv. Synth. Catal.* 358 (4): 543–548.

37 Mardjan, M.I.D., Perie, S., Parrain, J.-L. et al. (2017). A tunable copper-catalyzed multicomponent reaction towards alkaloid-inspired indole/lactam polycycles. *Org. Biomol. Chem.* 15 (15): 3304–3309.

38 He, Y., Zhang, X., and Fan, X. (2015). Synthesis of diversely substituted 2-(furan-3-yl)acetates from allenols through cascade carbonylations. *Chem. Commun.* 51 (90): 16263–16266.

39 He, Y., Zheng, Z., Liu, Q. et al. (2018). Tunable synthesis of 2-ene-1,4-diones, 4-hydroxycyclopent-2-en-1-ones, and 2-(furan-3-yl)acetamides via palladium-catalyzed cascade reactions of allenols. *J. Org. Chem.* 83 (20): 12514–12526.

40 Wu, J. and Yoshikai, N. (2015). Modular synthesis of multisubstituted furans through palladium-catalyzed three-component condensation of alkynylbenziodoxoles, carboxylic acids, and imines. *Angew. Chem. Int. Ed.* 54 (38): 11107–11111.

41 Lu, B., Wu, J., and Yoshikai, N. (2014). Palladium-catalyzed condensation of N-aryl imines and alkynylbenziodoxolones to form multisubstituted furans. *J. Am. Chem. Soc.* 136 (33): 11598–11601.

42 Siamaki, A.R., and Arndtsen, B.A. (2006). A direct, one step synthesis of imidazoles from imines and acid chlorides: a palladium catalyzed multicomponent coupling approach. *J. Am. Chem. Soc.* 128 (18): 6050–6051.

43 Tjutrins, J., and Arndtsen, B.A. (2017). A palladium-catalyzed synthesis of (hetero)arylsubstituted imidazoles from aryl halides, imines and carbon monoxide. *Chem. Sci.* 8 (2): 1002–1007.

44 Hu, J., Kong, B., Liu, Y. et al. (2017). Highly Stereoselective synthesis of imidazolidines through the palladium(0)-catalyzed three-component reaction of 2,3-allenylamines, organic halides, and imines. *ChemCatChem* 9 (3): 403–406.

45 Clemenceau, A., Wang, Q., and Zhu, J. (2018). Cooperative Pd/Cu catalysis: multicomponent synthesis of tetrasubstituted imidazolones from methyl α-isocyanoacetates, primary amines, and aryl(vinyl) iodides. *Org. Lett.* 20 (1): 126–129.

46 Johansson, J.R., Beke-Somfai, T., Stålsmeden, A.S. et al. (2016). Ruthenium-catalyzed azide alkyne cycloaddition reaction: scope, mechanism, and applications. *Chem. Rev.* 116 (23): 14726–14768.

47 Song, W., Zheng, N., Li, M. et al. (2018). Rhodium(I)-catalyzed azide-alkyne cycloaddition (RhAAC) of internal alkynylphosphonates with high regioselectivities under mild conditions. *Adv. Synth. Catal.* 360 (13): 2429–2434.

48 Song, W., and Zheng, N. (2017). Iridium-catalyzed highly regioselective azide–ynamide cycloaddition to access 5-amido fully substituted 1,2,3-triazoles under mild, air, aqueous, and bioorthogonal conditions. *Org. Lett.* 19 (22): 6200–6203.

49 McNulty, J., Keskar, K., and Vemula, R. (2011). The first well-defined silver(I)-complex-catalyzed cycloaddition of azides onto terminal alkynes at room temperature. *Chem. Eur. J.* 17 (52): 14727–14730.

50 Nagesh, H.N., Suresh, A., Reddy, M.N. et al. (2016). Multicomponent cascade reaction: dual role of copper in the synthesis of 1,2,3-triazole tethered benzimidazo[1,2-a]quinoline and their photophysical studies. *RSC Adv.* 6 (19): 15884–15894.

51 Wang, W., Peng, X., Wei, F. et al. (2016). Copper(I)-catalyzed interrupted click reaction: synthesis of diverse 5-hetero-functionalized triazoles. *Angew. Chem. Int. Ed.* 55 (2): 649–653.

52 Wang, W., Wei, F., Ma, Y. et al. (2016). Copper(I)-catalyzed three-component click/alkynylation: one-pot synthesis of 5-alkynyl-1,2,3-triazoles. *Org. Lett.* 18 (17): 4158–4161.

53 Wu, F., Zhou, W., Chen, K. et al. (2018). Copper mediated three-component reactions of alkynes, azides, and propargylic carbonates: synthesis of 5-allenyl-1,2,3-triazoles. *Syn. Catal.* 360 (13): 2435–2439.

54 Zhang, Z., Zhou, Q., Ye, F. et al. (2015). Copper(I)-catalyzed three-component coupling of N-tosylhydrazones, alkynes and azides:synthesis of trisubstituted 1,2,3-triazoles. *Adv. Synth. Catal.* 357 (10): 2277–2286.

55 Wei, F., Li, H., Song, C. et al. (2015). Cu/Pd-catalyzed, three-component click reaction of azide, alkyne, and aryl halide: one-pot strategy toward trisubstituted triazoles. *Org. Lett.* 17 (11): 2860–2863.

56 Yamamoto, K., Bruun, T., Kim, J.Y. et al. (2016). A new multicomponent multicatalyst reaction $(MC)^2R$: chemoselective cycloaddition and latent catalyst activation for the synthesis of fully substituted 1,2,3-triazoles. *Org. Lett.* 18 (11): 2644–2647.

57 Nallagangula, M., and Namitharan, K. (2017). Copper-catalyzed sulfonyl azide–alkyne cycloaddition reactions: simultaneous generation and trapping of copper–triazoles and –ketenimines for the synthesis of triazolopyrimidines. *Org. Lett.* 19 (13): 3536–3539.

58 Peringer, F., do Nascimento, J.E.R., Abib, P.B. et al. (2017). Copper-catalyzed multicomponent reactions: synthesis of fused 1,2,3-triazolo-1,3,6-triazonines. *Eur. J. Org. Chem.* 2017 (18): 2579–2586.

59 Aquino, T.B., do Nascimento, J.E.R., Dias, I.F.C. et al. (2018). Synthesis of (arylselanyl)- and (arylsulfenyl)-alkyl-1,2,3-triazolo-1,3,6-triazonines via a copper-catalyzed multicomponent reaction. *Tetrahedron Lett.* 59 (12): 1080–1083.

60 Qureshi, Z., Kim, J.K., Bruun, T. et al. (2016). Cu/Pd-catalyzed synthesis of fully decorated polycyclic triazoles: introducing C–H functionalization to multicomponent multicatalytic reactions ($(MC)^2R$). *ACS Catal.* 6 (8): 4946–4952.

61 Allais, C., Grassot, J.-M., Rodriguez, J. et al. (2014). Metal-free multicomponent syntheses of pyridines. *Chem. Rev.* 114 (21): 10829–10868.

62 For reviews: Okamoto, S. (2012). Synthesis of 2,2'-bipyridines by transition metal-catalyzed alkyne/nitrile [2 + 2 + 2] cycloaddition reactions. *Heterocycles* 85 (7): 1579–1602.

63 Chopade, P.R., and Louie, J. (2006). [2+2+2] cycloaddition reactions catalyzed by transition metal complexes. *Adv. Synth. Catal.* 348 (16-17): 2307–2327.

64 Sheng, J., Wang, Y., Su, X. et al. (2017). Copper-catalyzed [2+2+2] modular synthesis of multisubstituted pyridines: alkenylation of nitriles with vinyliodonium salts. *Angew. Chem. Int. Ed.* 56 (17): 4824–4828.

65 Bai, Y., Tang, L., Huanga, H. et al. (2015). Synthesis of 2,4-diarylsubstituted-pyridines through a Ru-catalyzed four component reaction. *Org. Biomol. Chem.* 13 (15): 4404–4407.

66 Liu, S., and Liebeskind, L.S. (2008). A simple, modular synthesis of substituted pyridines. *J. Am. Chem. Soc.* 130 (22): 6918–6919.

67 Parthasarathy, K., Jeganmohan, M., and Cheng, C.-H. (2008). Rhodium-catalyzed one-pot synthesis of substituted pyridine derivatives from α,β-unsaturated ketoximes and alkynes. *Org. Lett.* 10 (2): 325–328.

68 Neely, J.M., and Rovis, T. (2013). Rh(III)-catalyzed regioselective synthesis of pyridines from alkenes and α,β-unsaturated oxime esters. *J. Am. Chem. Soc.* 135 (1): 66–69.

69 Luo, C.-Z., Jayakumar, J., Gandeepan, P. et al. (2015). Rhodium(III)-catalyzed vinylic C–H activation: a direct route toward pyridinium salts. *Org. Lett.* 17 (4): 924–927.

70 Song, Q.-W., Zhou, Z.-H., and He, L.-N. (2017). Efficient, selective and sustainable catalysis of carbon dioxide. *Green Chem.* 19 (16): 3707–3728.

71 García-Domínguez, P., Fehr, L., Rusconi, G. et al. (2016). Palladium-catalyzed incorporation of atmospheric CO_2: efficient synthesis of functionalized oxazolidinones. *Chem. Sci.* 7 (6): 3914–3918.

72 Wang, B., Sun, S., Yu, J.-T. et al. (2017). Palladium-catalyzed multicomponent reactions of *o*-alkynylanilines, aryl iodides, and CO_2 toward 3,3-diaryl 2,4-quinolinediones. *Org. Lett.* 19 (16): 4319–4322.

73 Too, P.C., Wang, Y.-F., and Chiba, S. (2010). Rhodium(III)-catalyzed synthesis of isoquinolines from aryl ketone *O*-acyloxime derivatives and internal alkynes. *Org. Lett.* 12 (24): 5688–5691.

74 Gerfaud, T., Neuville, L., and Zhu, J. (2009). Palladium-catalyzed annulation of acyloximes with arynes (or alkynes): synthesis of phenanthridines and isoquinolines. *Angew. Chem. Int. Ed.* 48 (3): 572–577.

75 Liu, K., Chen, S., Li, X.G. et al. (2015). Multicomponent cascade synthesis of trifluoroethyl isoquinolines from alkynes and vinyl azides. *J. Org. Chem.* 81 (1): 265–270.

76 Zhou, C., Jiang, J., and Wang, J. (2019). Three-component synthesis of isoquinoline derivatives by a relay catalysis with a single rhodium(III) catalyst. *Org. Lett.* 21 (13): 4971–4975.

77 Bae, I., Han, H., and Chang, S. (2005). Highly efficient one-pot synthesis of N-sulfonylamidines by Cu-catalyzed three-component coupling of sulfonyl azide, alkyne, and amine. *J. Am. Chem. Soc.* 127 (7): 2038–2039.

78 Cui, S.-L., Lin, X.-F., and Wang, Y.-G. (2006). Novel and efficient synthesis of iminocoumarins via copper-catalyzed multicomponent reaction. *Org. Lett.* 8 (20): 4517–4520.

79 Murugavel, G., and Punniyamurthy, T. (2015). Microwave-assisted copper-catalyzed four-component tandem synthesis of 3-N-sulfonylamidine coumarins. *J. Org. Chem.* 80 (12): 6291–6299.

80 Yi, F., Zhang, S., Huang, Y. et al. (2017). An efficient one-pot protocol for the synthesis of polysubstituted 4-amino-iminocoumarins and 4-aminoquinolines by a copper-catalyzed three-component reaction. *Eur. J. Org. Chem.* 2017 (1): 102–110.

81 Chen, Z., Han, C., Fan, C. et al. (2018). Copper-catalyzed diversity-oriented synthesis (DOS) of 4-amino-2H-chromen-2-imines: application of Kemp elimination toward *O*-heterocycles. *ACS Omega* 3 (7): 8160–8168.

82 Sun, J., Cheng, X., Mansaray, J.K. et al. (2018). A copper-catalyzed three component reaction of aryl acetylene, sulfonyl azide and enaminone to form iminolactone *via* 6π electrocyclization. *Chem. Commun.* 54 (99): 13953–13956.

83 Torres, G.M., De la Higuera, M., Quesnel, J.S. et al. (2016). Palladium-catalyzed, multicomponent approach to β-lactams via aryl halide carbonylation. *J. Org. Chem.* 81 (24): 12106–12115.

84 Lee, D.J., Han, H.S., Shin, J. et al. (2014). Multicomponent [5 + 2] cycloaddition reaction for the synthesis of 1,4-diazepines: isolation and reactivity of azomethine ylides. *J. Am. Chem. Soc.* 136 (33): 11606–11609.

85 He, M., Chen, N., Wang, J. et al. (2019). Rhodium-catalyzed regiodivergent [3 + 2] and [5 + 2] cycloadditions of quinolinium ilides with alkynes. *Org. Lett.* 21 (13): 5167–5171.

86 Yang, Y., Zhou, M.-B., Ouyang, X.-H. et al. (2015). Rhodium(III)-catalyzed [3+2]/[5+2] annulation of 4-aryl 1,2,3-triazoles with internal alkynes through dual C(sp^2)-H functionalization. *Angew. Chem. Int. Ed.* 54 (22): 6595–6599.

Index

a

c

e

j

k

n

q

r

s